The term "holomorphic spaces" is short for "spaces of holomorphic functions." It refers not so much to a branch of mathematics as to a common thread running through much of modern analysis—through functional analysis, operator theory, harmonic analysis, and, of course, complex analysis.

This is a collection of expository articles arising from MSRI's Fall 1995 program on holomorphic spaces. The opening article, by Donald Sarason, gives an overview of several aspects of the subject. The remaining articles, while more specialized, are nevertheless designed in varying degrees to be accessible to the nonexpert; some are minicourses in themselves. A range of topics is addressed:

- Bergman spaces (Hedenmalm, Stroethoff)
- Hankel operators in various guises (Gorkin, Peller, Rochberg, Saccone)
- the Dirichlet space (Wu)
- subnormal operators (Conway, Yang)
- operator models, interpolation problems, systems theory (Alpay, Dijksma, Dym, Kheifets, Nikolski, Rovnyak, Sadosky, de Snoo, Vasyunin, Young)

The concluding article, by Victor Vinnikov, describes an approach to certain commuting families of nonselfadjoint operators in which operator theory is linked with algebraic geometry.

Mathematical Sciences Research Institute
Publications

33

Holomorphic Spaces

Mathematical Sciences Research Institute
Publications

Volumes 1 through 27 are available from Springer-Verlag

Holomorphic Spaces

Edited by

Sheldon Axler

San Francisco State University

John E. McCarthy

Washington University in St. Louis

Donald Sarason

University of California, Berkeley

CAMBRIDGE
UNIVERSITY PRESS

Sheldon Axler
Mathematics Department
San Francisco State University
San Francisco, CA 94132
USA

John E. McCarthy
Department of Mathematics
Campus Box 1146
One Brookings Drive
Washington University
St. Louis MO 63130
USA

Donald Sarason
Department of Mathematics
University of California
Berkeley, CA 94720-3840
USA

Mathematical Sciences Research
    Institute
1000 Centennial Drive
Berkeley, CA 94720

The Mathematical Sciences Research Institute wishes to acknowledge
support by the National Science Foundation.

Published by the Press Syndicate of the University of Cambridge
The Pitt Building, Trumpington Street, Cambridge CB2 1RP
40 West 20th Street, New York, NY 10011–4211, USA
10 Stamford Road, Oakleigh, Melbourne 3166, Australia

© Mathematical Sciences Research Institute 1998

First Published 1998

Printed in the United States of America

Library of Congress cataloging-in-publication data available

A catalogue record for this book is available from the British Library

ISBN 0–521–63193–9 Hardback

Holomorphic Spaces
MSRI Publications
Volume **33**, 1998

Contents

Holomorphic Spaces
MSRI Publications
Volume **33**, 1998

Preface

The term "Holomorphic Spaces" is short for "Spaces of Holomorphic Functions." It refers not so much to a branch of mathematics as to a common thread running through much of modern analysis—through functional analysis, operator theory, harmonic analysis, and, of course, complex analysis.

In the fall of 1995 the Mathematical Sciences Research Institute in Berkeley sponsored the program Holomorphic Spaces. Over forty participants came for periods of two weeks to a full semester; an additional forty or so attended a week-long workshop in October. Spaces of holomorphic functions arise in many contexts. The MSRI program focused predominantly on operator-theoretic aspects of the subject. A series of minicourses formed the program's centerpiece.

This volume consists of expository articles by participants in the program (plus collaborators, in two cases), including several articles based on minicourses. The opening article, by Donald Sarason, gives an overview of several aspects of the subject. The remaining articles, while more specialized, are nevertheless designed in varying degrees to be accessible to the nonexpert. A range of topics is addressed: Bergman spaces (Hakan Hedenmalm, Karl Stroethoff); Hankel operators in various guises (Vladimir Peller, Pamela Gorkin, Scott Saccone, Richard Rochberg); the Dirichlet space (Zhijian Wu); subnormal operators (John B. Conway and Liming Yang); operator models and related areas, especially interpolation problems and systems theory (Nikolai Nikolski and Vasily Vasyunin, Cora Sadosky, Nicholas Young, Alexander Kheifets, Harry Dym, James Rovnyak and coauthors). The concluding article, by Victor Vinnikov, describes an approach to certain commuting families of nonself-adjoint operators in which operator theory is linked with algebraic geometry.

The program committee, in addition to the editors of this volume, consisted of Joseph Ball, Nikolai Nikolski, Mihai Putinar, and Cora Sadosky. On behalf of all participants, the program committee wishes to thank the staff of MSRI, especially Director William Thurston, Associate Director Tsit-Yuen Lam, Alisa Colloms, and Kim Garrett, for their many efforts in our behalf.

This volume benefited greatly from the expertise of Silvio Levy, the series editor.

The editors

Holomorphic Spaces
MSRI Publications
Volume **33**, 1998

Holomorphic Spaces:
A Brief and Selective Survey

DONALD SARASON

ABSTRACT. This article traces several prominent trends in the development
of the subject of holomorphic spaces, with emphasis on operator-theoretic
aspects.

The term "Holomorphic Spaces," the title of a program held at the Mathematical Sciences Research Institute in the fall semester of 1995, is short for "Spaces of Holomorphic Functions." It refers not so much to a branch of mathematics as to a common thread running through much of modern analysis—through functional analysis, operator theory, harmonic analysis, and, of course, complex analysis. This article will briefly outline the development of the subject from its origins in the early 1900's to the present, with a bias toward operator-theoretic aspects, in keeping with the main emphasis of the MSRI program. I hope that the article will be accessible not only to workers in the field but to analysts in general.

Origins

The subject began with the thesis of P. Fatou [1906], a student of H. Lebesgue. The thesis is a study of the boundary behavior of certain harmonic functions in the unit disk (those representable as Poisson integrals). It contains a proof, for example, that a bounded holomorphic function in the disk has a nontangential limit at almost every point of the unit circle. This initial link between function theory on the circle (real analysis) and function theory in the disk (complex analysis) recurred continually in the ensuing years. Some of the highlights are the paper of F. Riesz and M. Riesz [1916] on the absolute continuity of analytic measures; F. Riesz's paper [1923] in which he christened the Hardy spaces, H^p, and introduced the technique of dividing out zeros (i.e., factoring by a Blaschke product); G. Szegö's investigations [1920; 1921] of Toeplitz forms; M. Riesz's proof [1924] of the L^p boundedness of the conjugation operator ($1 < p < \infty$);

A. N. Kolmogorov's proof [1925] of the weak-L^1 boundedness of the conjugation operator; G. H. Hardy and J. E. Littlewood's introduction of their maximal
function [1930]; and R. Nevanlinna's development [1936] of his theory of functions of bounded characteristic. By the late 1930's the theory had expanded to
the point where it could become the subject of a monograph. The well-known
book of I. I. Privalov appeared in 1941, the year of the author's death, and was
republished nine years later [Privalov 1950], followed by a German translation
[Privalov 1956].

Beurling's Paper

Simultaneously, the general theory of Banach spaces and their operators had
been developing. By mid-century, when A. Beurling published a seminal paper
[1949], the time was ripe for mutual infusion. After posing two general questions
about Hilbert space operators with complete sets of eigenvectors, Beurling's
paper focuses on the (closed) invariant subspaces of the unilateral shift operator
on the Hardy space H^2 of the unit disk (an operator whose adjoint is of the kind
just mentioned).

For the benefit of readers who do not work in the field, here are a few of
the basic definitions. For $p > 0$ the Hardy space H^p consists of the holomorphic
functions f in the unit disk, $\mathbb{D}$, satisfying the growth condition $\sup_{0<r<1} \|f_r\|_p <
\infty$, where f_r is the function on the unit circle defined by $f_r\left(e^{i\theta}\right) = f\left(re^{i\theta}\right)$, and
$\|f_r\|_p$ denotes the norm of f_r in the L^p space of normalized Lebesgue measure
on the circle, hereafter denoted simply by L^p. As noted earlier, the spaces H^p
were introduced by F. Riesz [1923]; they were named by him in honor of G. H.
Hardy, who had proved [1915] that $\|f_r\|_p$ increases with r (unless f is constant).
From the work of Fatou and his successors one knows that each function in H^p
has an associated boundary function, defined almost everywhere on $\partial\mathbb{D}$ in terms
of nontangential limits. Because of this, one can identify H^p with a subspace
of L^p; in case $p \geq 1$, the subspace in question consists of the functions in L^p
whose Fourier coefficients with negative indices vanish (i.e., the functions whose
Fourier series are of power series type). A function in H^p, for $p \geq 1$, can be
reconstructed from its boundary function by means of the Poisson integral, or
the Cauchy integral. The space H^2 can be alternatively described as the space
of holomorphic functions in $\mathbb{D}$ whose Taylor coefficients at the origin are square
summable. In the obvious way it acquires a Hilbert space structure in which the
functions z^n, for $n = 0, 1, 2, \ldots$, form an orthonormal basis.

The unilateral shift is the operator S on H^2 of multiplication by z, the identity
function. It is an isometry, sending the n-th basis vector, z^n, to the $(n + 1)$-st,
z^{n+1}. It is, in fact, the simplest pure isometry. (A Hilbert space isometry is
called pure if it has no unitary direct summand. Every pure isometry is a direct
sum of copies of S.) Beurling showed that the invariant subspace structure of S
mirrors the factorization theory of H^2 functions.

From the work of F. Riesz and Nevanlinna it is known that every nonzero function in H^p can be written as the product of what Beurling called an outer function and an inner function. The factors are unique to within unimodular multiplicative constants. An outer function is a nowhere vanishing holomorphic function f in $\mathbb{D}$ such that $\log|f|$ is the Poisson integral of its boundary function. Beurling showed that the outer functions in H^2 are the cyclic vectors of the operator S (i.e., the functions contained in no proper S-invariant subspaces). An inner function is a function in H^∞ whose boundary function has unit modulus almost everywhere. Beurling showed that if the H^2 function f has the factorization $f = uf_0$, with u an inner function and f_0 an outer function, then the S-invariant subspace generated by f is the same as that generated by u, and it equals uH^2. Finally, Beurling showed that every invariant subspace of S is generated by a single function, and hence by an inner function. Thus, understanding the invariant subspace structure of S is tantamount to understanding the structure of inner functions.

There are two basic kinds of inner functions: Blaschke products and singular functions. Only the constant inner functions are of both kinds, and every inner function is the product of a Blaschke product and a singular function, the factors being unique to within unimodular multiplicative constants. Blaschke products (products of Blaschke factors) are associated with zero sequences. The zero sequence of a function in H^2 (in fact, of a function in any H^p) is a so-called Blaschke sequence, a finite sequence in $\mathbb{D}$ or an infinite sequence $(z_n)_{n=1}^\infty$ satisfying $\sum(1 - |z_n|) < \infty$ (the Blaschke condition). The Blaschke factor corresponding to a point w of $\mathbb{D}$ is, in case $w \neq 0$, the linear-fractional map of $\mathbb{D}$ onto $\mathbb{D}$ that sends w to 0 and 0 to the positive real axis, and in case $w = 0$ it is the identity function. A Blaschke product is the product of the Blaschke factors corresponding to the terms of a Blaschke sequence, or a unimodular constant times such a function. In the case of a finite sequence it is obviously an inner function, and in the case of an infinite sequence, the Blaschke condition is exactly what one needs to prove that the corresponding infinite product of Blaschke factors converges locally uniformly in $\mathbb{D}$ to an inner function. If the inner function associated with an S-invariant subspace is a Blaschke product, then the subspace is just the subspace of functions in H^2 that vanish at the points of the corresponding Blaschke sequence (with the appropriate multiplicities at repeated points).

A singular function is an inner function without zeroes in $\mathbb{D}$. The logarithm of the modulus of such a function, if the function is nonconstant, is a negative harmonic function in $\mathbb{D}$ having the nontangential limit 0 at almost every point of $\partial\mathbb{D}$. One can conclude on the basis of the theory of Poisson integrals that the logarithm of the modulus of a nonconstant singular function is the Poisson integral of a negative singular measure on $\partial\mathbb{D}$. The most general nonconstant singular function can thus be represented as

$$\lambda \, \exp\left(-\int_{\partial \mathbb{D}} \frac{e^{i\theta} + z}{e^{i\theta} - z} \, d\mu\left(e^{i\theta}\right)\right) \qquad \text{for } z \in \mathbb{D},$$

where λ is a unimodular constant and μ is a positive singular measure on $\partial \mathbb{D}$. The simplest such function is

$$\exp\left(\frac{z+1}{z-1}\right),$$

corresponding to the case where $\lambda = 1$ and μ consists of a unit point mass at the point 1. If the inner function associated with an S-invariant subspace is a singular function, then the functions in the subspace have no common zeroes in $\mathbb{D}$, but the common singular inner factor they share forces them all to have the nontangential limit 0 almost everywhere on $\partial \mathbb{D}$ with respect to the associated singular measure.

Because of Beurling's theorem, the preceding description of inner functions translates into a description of the invariant subspaces of the operator S. The theorem is a splendid early example of how a natural question in operator theory can lead deeply into analysis.

Multiple Shifts and Operator Models

Beurling's work was extended to multiple shifts by P. D. Lax [1959] and P. R. Halmos [1961]. Here one naturally encounters vector-valued function theory.

For $1 \leq n \leq \aleph_0$, the unilateral shift of multiplicity n (that is, the direct sum of n copies of S) can be modeled as the operator of multiplication by z on a vector-valued version of H^2; the functions in this space have values belonging to an auxiliary Hilbert space $\mathcal{E}$ of dimension n. The space, usually denoted by $H^2(\mathcal{E})$, can be defined, analogously to the scalar case, as the space of holomorphic $\mathcal{E}$-valued functions in $\mathbb{D}$ whose Taylor coefficients at the origin are square summable. The shift-invariant subspaces of $H^2(\mathcal{E})$ have a description analogous to that in Beurling's theorem, the inner functions in that theorem being replaced by operator-valued analogues. Something is lost in the generalization, because the latter functions are not generally susceptible to a precise structural description like the one discussed above for scalar inner functions. (An exception is afforded by what are usually called matrix inner functions, bounded holomorphic matrix-valued functions in $\mathbb{D}$ having unitary nontangential limits almost everywhere on $\partial \mathbb{D}$. For this class of functions, and a symplectic analogue, V. P. Potapov [1955] has developed a beautiful structure theory.)

Multiple shifts play a prominent role in model theories for Hilbert space contractions. The prototypical theory of operator models is, of course, the classical spectral theorem, which in its various incarnations provides canonical models for self-adjoint operators, normal operators, one-parameter unitary groups, and commutative C^*-algebras. Model theories that go beyond the confines of the classical spectral theorem developed on several parallel fronts beginning in the 1950's. The theory originated by M. S. Livshitz [1952] and M. S. Brodskii [1956] focuses on operators that are "nearly" self-adjoint. The theories of B. Sz.-Nagy and C. Foiaş [1967] and L. de Branges and J. Rovnyak [1966, Appendix, pp. 347–392] apply to general contractions but are most effective for "nearly" unitary ones.

The spirit of these "nonclassical" model theories can be illustrated with the Volterra operator, the operator V on $L^2[0,1]$ of indefinite integration:

$$(Vf)(x) = \int_0^x f(t)\, dt \qquad \text{for } 0 \le x \le 1.$$

The adjoint of V is given by

$$(V^*f)(x) = \int_x^1 f(t)\, dt \qquad \text{for } 0 \le x \le 1,$$

from which one sees that $V + V^*$ is a positive operator of rank one, and hence that the operator $(I - V)(I + V)^{-1}$ is a contraction and a rank-one perturbation of a unitary operator.

The invariant subspaces of V were determined, by different methods, by Brodskii [1957] and W. F. Donoghue [1957]. The result is also a corollary of earlier work of S. Agmon [1949]; it says that the only invariant subspaces of V are the obvious ones, the subspaces $L^2[a,1]$ for $0 \le a \le 1$. (Here, $L^2[a,1]$ is identified with the subspace of functions in $L^2[0,1]$ that vanish off $[a,1]$.) It was eventually recognized that the Agmon–Brodskii–Donoghue result is "contained" in Beurling's theorem [Sarason 1965].

To explain the last remark we consider, for $a > 0$, the singular inner function

$$u_a(z) = \exp\left(a\left(\frac{z+1}{z-1}\right)\right),$$

and the orthogonal complement of its corresponding invariant subspace, which we denote by K_a:

$$K_a = H^2 \ominus u_a H^2.$$

We look in particular at K_1, and on K_1 we consider the operator S_1 one obtains by compressing the shift S. Thus, to apply S_1 to a function in K_1, one first multiplies the function by z and then projects the result onto K_1. (The adjoint of S_1 is the restriction of S^* to K_1.)

There is a natural isometry, involving the Cayley transform, that maps L^2 (of $\partial\mathbb{D}$) onto $L^2(\mathbb{R})$. If one follows that isometry by the Fourier–Plancherel transformation, one obtains again an isometry of L^2 onto $L^2(\mathbb{R})$. The latter isometry maps H^2 onto $L^2[0,\infty)$ and maps K_a onto $L^2[0,a]$. And it transforms the operator S_1 on K_1 to the operator $(I-V)(I+V)^{-1}$ on $L^2(0,1)$. The operator S_1 is thus a "model" of the operator $(I - V)(I + V)^{-1}$.

By Beurling's theorem, the invariant subspaces of S_1 are exactly the subspaces $uH^2 \ominus u_1H^2$ with u an inner function that divides u_1 (divides, that is, in the algebra H^∞ of bounded holomorphic functions in $\mathbb{D}$). From the structure theory for inner functions described earlier one can see that the only inner functions that divide u_1 are the functions u_a with $0 \le a \le 1$ (and their multiples by unimodular constants). On the basis of the transformation described above, one concludes that the invariant subspaces of $(I-V)(I+V)^{-1}$ are the subspaces $L^2[a,1]$ with $0 \le a \le 1$. Finally, each of the operators V and $(I-V)(I+V)^{-1}$ is easily seen to be approximable in norm by polynomials in the other, implying that these two operators have the same invariant subspaces. The Agmon–Brodskii–Donoghue result follows.

One can sum up the preceding remarks by saying that the Volterra operator, V, is "contained in" the shift operator, S. A simple and elegant observation of G. C. Rota [1960], which provides a hint of the Sz.-Nagy–Foiaş and de Branges–Rovnyak model theories, shows, startlingly, that all Hilbert space operators are "contained in" multiple shifts. Consider an operator T of spectral radius less than 1 on a Hilbert space $\mathcal{E}$. With each vector x in $\mathcal{E}$ we associate the $\mathcal{E}$-valued holomorphic function f_x given by the power series $\sum_{n=0}^\infty z^n T^{*n}x$. Because of the spectral condition imposed on T, the function f_x is holomorphic on $\overline{\mathbb{D}}$, so in particular it belongs to $H^2(\mathcal{E})$. The space of all such functions f_x is a subspace K of $H^2(\mathcal{E})$, invariant under the adjoint of the shift operator on $H^2(\mathcal{E})$. The map $x \to f_x$ from $\mathcal{E}$ onto K is a bounded, invertible operator that intertwines T^* with the adjoint of the shift operator. It follows that the operator T is similar to the compression of the shift operator to K. In a sense, then, multiple shifts provide replicas of all operators.

To be a bit more precise, Rota's observation provides a similarity model for every Hilbert space operator whose spectral radius is less than 1. The model space is the orthogonal complement of a shift-invariant subspace of a vector-valued H^2 space, and the model operator is the compression of the shift to the model space. The more powerful Sz.-Nagy–Foiaş and de Branges–Rovnyak theories provide unitarily equivalent models, not merely similarity models, for Hilbert space contractions. The theory of Sz.-Nagy and Foiaş springs from the subject of unitary dilations. Their model spaces include, among a wider class, the orthogonal complements of all shift-invariant subspaces of vector-valued H^2 spaces, the corresponding model operators being compressions of shifts. The model spaces of de Branges–Rovnyak are certain Hilbert spaces that live inside vector-valued H^2 spaces, not necessarily as subspaces but as contractively contained spaces, that is, spaces whose norms dominate the norms of the containing spaces.

The connection between the model theories of Sz.-Nagy–Foiaş and de Branges–Rovnyak has been explained by J. A. Ball and T. L. Kriete [1987]. Further insight was provided by N. K. Nikolskii and V. I. Vasyunin [1989], who developed what they term a coordinate-free model theory that contains, as particular cases, the Sz.-Nagy–Foiaş and de Branges–Rovnyak theories.

Interpolation

The operators S and S^* are themselves model operators in the Sz.-Nagy–Foiaş theory, and rather transparent ones. They model irreducible pure isometries and their adjoints, respectively. Next in simplicity are the compressions of S to proper S^*-invariant subspaces of H^2. For u a nonconstant inner function, let K_u denote the orthogonal complement in H^2 of the Beurling subspace uH^2, and let S_u denote the compression of S to K_u. (The action of S_u is thus multiplication by z followed by projection onto K_u.) J. W. Moeller [1962] showed that the spectrum of S_u consists of the zeros of u in $\mathbb{D}$ plus the points on $\partial\mathbb{D}$ where u has 0 as a cluster value. Moeller's paper and other considerations led the author to suspect that every operator commuting with S_u should be obtainable as the compression of an operator commuting with S. The operators of the latter kind are just the multiplication operators on H^2 induced by H^∞ functions. The result was eventually proved in a more precise form: an operator T that commutes with S_u is the compression of an operator that commutes with S and has the same norm as T [Sarason 1967]. There is a close link with two classical interpolation problems, the problems of Carathéodory–Fejér and Nevanlinna–Pick.

In the first of these problems [Carathéodory and Fejér 1911], one is given as data a finite sequence $c_0, c_1, \ldots, c_{N-1}$ of complex numbers, and one wants to find a function in the unit ball of H^∞ having those numbers as its first N Taylor coefficients at the origin. To recast this as a problem about operators, let u be the inner function z^N. The functions $1, z, \ldots, z^{N-1}$ form an orthonormal basis for the corresponding space K_u, and the matrix in this basis for the operator S_u has the entry 1 in each position immediately below the main diagonal and 0 elsewhere. Let $T = \sum_{j=0}^{N-1} c_j S_u^j$, so the matrix for T is lower triangular with the entry c_j in each position j steps below the main diagonal. Then T commutes with S_u, and the question of whether the Carathéodory–Fejér problem has a solution for the data $c_0, c_1, \ldots, c_{N-1}$ is the same as the question of whether T is the compression of an operator commuting with S and having norm at most 1. According to the result from [Sarason 1967], T has such a compression if and only if its norm is at most 1. One recaptures in this way a solvability criterion for the Carathéodory–Fejér problem attributed by those authors to O. Toeplitz.

In the second classical interpolation problem [Nevanlinna 1919; Pick 1916], one is given as data a finite sequence $z_1, \ldots, z_N$ of distinct points in $\mathbb{D}$ and a finite sequence $w_1, \ldots, w_N$ of complex numbers. One wants to find a function in the unit ball of H^∞ taking the value w_j at z_j, for $j = 1, \ldots, N$. For an operator reinterpretation, let u be the finite Blaschke product with zero sequence $z_1, \ldots, z_N$. The space K_u is spanned by the kernel functions for the points $z_1, \ldots, z_N$, the functions $k_j(z) = (1 - \bar{z}_j z)^{-1}$, where $j = 1, \ldots, N$. The distinctive property of k_j is that the linear functional it induces on H^2 is the functional of evaluation at z_j. From this one sees that $S^* k_j = \bar{z}_j k_j$, so the functions $k_1, \ldots, k_N$ form an eigenbasis for S_u^*. Let the operator T on K_u be defined by $T^* k_j = \bar{w}_j k_j$.

Then T commutes with S_u, and the question of whether the Nevanlinna–Pick problem is solvable for the given data is the same as the question of whether T is the compression of an operator that commutes with S and has norm at most 1. By the result from [Sarason 1967], the latter happens if and only if the norm of T is at most 1, which is easily seen to coincide with Pick's solvability criterion, namely, the positive semidefiniteness of the matrix

$$\left(\frac{1 - w_j \bar{w}_k}{1 - z_j \bar{z}_k}\right)_{j,k=1}^N.$$

Subsequently, Sz.-Nagy and Foiaş established their famous commutant-lifting theorem [1968; 1967], according to which the result from [Sarason 1967] is a special case of a general property of unitary dilations. The commutant-lifting theorem provides an operator approach to a variety of interpolation problems. The books [Rosenblum and Rovnyak 1985; Foiaş and Frazho 1990], are good sources for this material. H. Dym's review of the latter book [Dym 1994] is also recommended.

The commutant-lifting approach is just one of several operator approaches to interpolation problems. In the same year that the commutant-lifting theorem appeared, V. M. Adamyan, D. Z. Arov and M. G. Krein published the first two of a remarkable series of papers on the Nehari interpolation problem [Adamyan et al. 1968b; 1968a]. In the Nehari problem one is given as data a sequence $(c_n)_{n=1}^\infty$ of complex numbers, and one wants to find a function f in the unit ball of L^∞ (on the unit circle) having these numbers as its negatively indexed Fourier coefficients (i.e., $\hat{f}(-n) = c_n$ for $n = 1, 2, \dots$). Z. Nehari [1957] proved that the problem is solvable if and only if the Hankel matrix $(c_{j+k+1})_{j,k=0}^\infty$ has norm at most 1 as an operator on l^2. Using a method akin to the operator approach to the Hamburger moment problem, Adamyan, Arov and Krein proved that finding a solution f, in case one exists, is tantamount to finding a unitary extension of a certain isometric operator constructed from the data. What is more, the family of all such solutions is in one-to-one correspondence with the family of all such unitary extensions (satisfying a minimality requirement), a connection that enabled them to derive a linear-fractional parameterization of the set of all solutions in case the problem is indeterminate. For the indeterminate Nevanlinna–Pick problem, a linear-fractional parameterization of the solution set was found by Nevanlinna [1919] on the basis of the Schur algorithm, a technique invented by I. Schur [1917] in connection with the Carathéodory–Fejér problem. Nevanlinna's parameterization, and the corresponding one implicit in Schur's paper, can be deduced from the one of Adamyan, Arov and Krein, because one can show that the Nehari problem embraces the Carathéodory–Fejér and Nevanlinna–Pick problems.

There is a close connection between the commutant-lifting theorem and the work of Nehari and Adamyan, Arov and Krein. Nehari's theorem is in fact a corollary of the commutant-lifting theorem; the details can be found, for example, in [Sarason 1991]. In the other direction, it was recognized by D. N. Clark (unpublished notes) and N. K. Nikolskii [1986] that the theorem from [Sarason 1967] can be deduced very simply from Nehari's theorem, and more recently R. Arocena [1989] has shown how to give a proof of the commutant-lifting theorem using the methods of Adamyan, Arov and Krein. Further discussion can be found in [Sarason 1987; 1991].

There follows a brief description of some other approaches to interpolation problems.

• The Abstract Interpolation Problem of V. E. Katsnelson, A. Ya. Kheifets and P. M. Yuditskii [Katsnelson et al. 1987; Kheifets and Yuditskii 1994] is based on the approach of V. P. Potapov and coworkers to problems of Nevanlinna–Pick type [Kovalishina and Potapov 1974; Kovalishina 1974; 1983]. It abstracts the key elements of Potapov's theory to an operator setting and yields, upon specialization, a wide variety of classical problems. The model spaces of de Branges–Rovnyak play an important role in this approach. As was the case with the Adamyan–Arov–Krein treatment of the Nehari problem, the solutions of the Abstract Interpolation Problem correspond to the unitary extensions of a certain isometric operator. There is a unified derivation of the linear-fractional parameterizations of the solution sets of indeterminate problems.

• The approach favored by H. Dym [1989] emphasizes reproducing kernel Hilbert spaces, especially certain de Branges–Rovnyak spaces, and J-inner matrix functions. (The J here refers to a signature matrix, a square, self-adjoint, unitary matrix. A meromorphic matrix function in $\mathbb{D}$, with values of the same size as J, is called J-inner if it is J-contractive at each point of $\mathbb{D}$ where it is holomorphic, and its boundary function is J-unitary almost everywhere on $\partial\mathbb{D}$. These are the symplectic analogues of inner functions that, as mentioned earlier, have been analyzed by Potapov [1955].)

• J. A. Ball and J. W. Helton [1983] have developed a Krein space approach to interpolation problems. In their approach an interpolation problem, rather than being reinterpreted as an operator extension problem, is reinterpreted as a subspace extension problem in a suitable Krein space. Shift-invariant subspaces of vector H^2 spaces that are endowed with a Krein space structure arise. One of the key results is a Beurling-type theorem for such subspaces. A treatment of the Nehari problem using this method can be found in [Sarason 1987].

• J. Agler [1989] has, in a sense, axiomatized the Nevanlinna–Pick problem and obtained the analogue of Pick's criterion in two new contexts, interpolation by multipliers of the Dirichlet space (the space of holomorphic functions in $\mathbb{D}$ whose derivatives are square integrable with respect to area), and interpolation by bounded holomorphic functions in the bidisk.

• M. Cotlar and C. Sadosky [1994] have used their theory of Hankel forms to attack problems of Nevanlinna–Pick and Nehari type in the polydisk.

• B. Cole, K. Lewis and J. Wermer have attacked problems of Nevanlinna–Pick type from the perspective of uniform algebras [Cole et al. 1992].

The foregoing list is but a partial sample of the enormous activity surrounding interpolation problems.

Systems Theory

In the early 1970's J. W. Helton became aware that there is a large overlap between the mathematics of linear systems theory and the operator theory that had grown around dilation theory and interpolation problems. In an April 17, 1972 letter to the author, he wrote: "I've spent the year learning engineering systems which at some levels is almost straight operator theory. Some of the best functional analysis (Krein, Livsic) has come from engineering institutes and I'm beginning to see why. Such collaboration does not exist in this country.... The [Sz.-]Nagy–Foiaş canonical model theory is precisely a study of infinite dimensional discrete time systems which lose and gain no energy."

Helton embarked upon a program to bridge the chasm between operator theorists and engineers in the United States. The result has been an enrichment of both mathematics and engineering. The systems theory viewpoint now permeates a large part of operator theory. On the engineering side, a new subject, H^∞ control, has sprung up [Francis 1987].

Bergman Spaces and the Bergman Shift

The mathematics discussed above flows, in large part, from Beurling's theorem via its generalization to vector H^2 spaces, in other words, to multiple shifts. Another natural direction for exploration unfolds when one replaces the shift, not by a multiple version of itself, but by its analogue (multiplication by z) on a holomorphic space of scalar functions other than H^2. There are countless possibilities for this other space; one that has turned out to be especially interesting is the Bergman space.

What was just referred to as "the" Bergman space is really just the most immediate member of a large family of spaces. Given a bounded domain in the complex plane and a positive number p, the Bergman space with exponent p for the domain consists of the holomorphic functions in the domain that are p-th power integrable with respect to area. These spaces are named for S. Bergman because the ones with exponent 2, which are Hilbert spaces, played a fundamental role in much of his work [Bergman 1970]. In the unit disk, the Bergman space with exponent p is denoted by A^p (or B^p, or L^p_a), and it is given the norm (or "norm," if $p < 1$) inherited from L^p of normalized area measure on the disk.

The powers of z form an orthogonal basis for the Hilbert space A^2, the norm of z^n being $1/\sqrt{n+1}$. Thus, a holomorphic function in $\mathbb{D}$ belongs to A^2 if and only if its Taylor coefficients at the origin are square summable when weighted against the sequence $\left(1/(n+1)\right)_{n=0}^{\infty}$. From this one can see that functions in A^2 need not possess boundary values in the usual sense of nontangential limits. For example, the function $\sum_{n=0}^{\infty}(n+1)^{-1/2}z^n$ is in A^2, but its coefficients are not square summable. A probabilistic argument shows that by randomly changing the signs of the coefficients of such a function, one almost surely obtains a function, obviously also in A^2, failing at almost every point of $\partial\mathbb{D}$ to have a radial limit (details are in [Duren 1970, p. 228]). With more work one can make a similar argument for any of the spaces A^p. Thus, in the study of Bergman spaces, one of the main techniques used to study Hardy spaces, the reliance on boundary functions, is lacking.

The Bergman shift is the operator on A^2 of multiplication by z. It acts on the orthonormal basis $\left(\sqrt{n+1}\,z^n\right)_{n=0}^{\infty}$ by sending the n-th basis vector to $\sqrt{(n+1)/(n+2)}$ times the $(n+1)$-st basis vector. It thus belongs to the class of weighted shifts, a seemingly restricted family of operators that exhibit surprisingly diverse behavior [Shields 1974]. A natural problem, in view of Beurling's theorem, is that of classifying the shift-invariant subspaces of A^2 (hereafter called just invariant subspaces of A^2). Part of that problem, and a natural starting place, is the problem of describing the zero sequences for A^2 functions, because associated with each zero sequence is the invariant subspace of functions in A^2 that vanish on it.

Significant progress in understanding A^p zero sequences was made by C. Horowitz [1974]. Among his results: (1) For $p < q$, there are A^p zero sequences that are not A^q zero sequences. This contrasts with the situation for the Hardy spaces: for every p, the H^p zero sequences are just the Blaschke sequences. (2) There exist two A^p zero sequences whose union is not an A^p zero sequence. Taking $p = 2$, one obtains an example of a pair of nontrivial invariant subspaces of A^2 whose intersection is trivial, a phenomenon that does not occur in the space H^2. (3) Every subsequence of an A^p zero sequence is an A^p zero sequence. To elaborate, suppose $(z_k)_{k=1}^{\infty}$ is a subsequence of the zero sequence of the A^p function f. For each k, let b_k be the Blaschke factor for the point z_k. Horowitz showed that the infinite product $h = \prod b_k\,(2 - b_k)$ converges, and that f/h is again in A^p. This furnishes an analogue of F. Riesz's Hardy space technique of dividing out zeros, but with a divisor h that need not itself be in A^p.

The Bergman shift belongs to a class, called $\mathbf{A}_{\aleph_0}$, studied by C. Apostol, H. Bercovici, C. Foiaş, C. Pearcy, and others [Apostol et al. 1985; Bercovici et al. 1985]. On the basis of the theory of what the preceding authors call dual algebras, they established certain properties of the lattice of invariant subspaces of A^2 that raise pessimism over the prospects of ever understanding that lattice well. Apostol, Bercovici, Foiaş and Pearcy showed that for each n between 1 and $\aleph_0$ there is an invariant subspace M of A^2 such that zM has codimension n in M; that the lattice of invariant subspaces of A^2 has a sublattice isomorphic to the lattice of all the subspaces of a Hilbert space of dimension $\aleph_0$; and, to strengthen one of Horowitz's results, that there is a family of invariant subspaces of A^2 with the cardinality of the continuum any two of which have a trivial intersection. The lattice of invariant subspaces of A^2 thus differs in striking ways from Beurling's lattice. It appears to be considerably "wilder."

Despite the preceding results, a substantially clearer picture of the invariant subspaces of A^2 has emerged in the past few years. The initial breakthrough came in a paper of H. Hedenmalm [1991]. In H^2, the normalized inner function (normalized in the sense that its first nonvanishing Taylor coefficient at 0 is positive) associated with a nontrivial invariant subspace solves a certain extremal problem: if N is the smallest integer such that some function in the subspace has a nonvanishing N-th derivative at the origin, then it maximizes $\mathrm{Re}\, f^{(n)}(0)$ among all functions f in the subspace having unit norm. Hedenmalm examined the corresponding functions associated with invariant subspaces of A^2. In case the invariant subspace consists of the functions that vanish along a given A^2 zero sequence, he showed that every function in the subspace is divisible in A^2 by the extremal function, the norm of the quotient being no larger than the norm of the original function. One can thus divide out the zeros of an A^2 function in such a way that both terms of the factored function are in A^2. For a general invariant subspace, Hedenmalm showed that the contractive divisibility property holds at least for functions in the invariant subspace generated by the extremal function. Hedenmalm's extremal functions appear to be natural A^2 analogues of inner functions.

Hedenmalm's results were quickly extended to general A^p spaces. This required new techniques and insights, provided by P. Duren, D. Khavinson, H. S. Shapiro, and C. Sundberg [Duren et al. 1993; 1994]; see also [Hedenmalm 1994; Khavinson and Shapiro 1994]. An interesting aspect of this work is the role played by the biharmonic Green function, the Green function in $\mathbb{D}$ for the square of the Laplacian. The positivity of that function turns out to be the key to contractive divisibility. A. Aleman, S. Richter and Sundberg [Aleman et al. 1996], using related techniques, have shown that at least one vestige of Beurling's theorem carries over to A^2: if M is an invariant subspace of A^2, then M is generated as an invariant subspace by $M \ominus zM$.

By the index of an invariant subspace M of A^2 one means the dimension of $M \ominus zM$. As mentioned above, Apostol, Bercovici, Foiaş and Pearcy showed using the theory of dual algebras that this number can take any value between 1 and $\aleph_0$. Their argument, since it applies to a general class of operators, does not give insight into the mechanism behind the phenomenon for the Bergman shift. Hedenmalm [1993] responded by giving an explicit example of an invariant subspace of A^2 with index 2; his argument is based on K. Seip's characterization of sampling and interpolating sequences in A^2 [Seip 1993]. Subsequently, Hedenmalm, Richter and Seip [Hedenmalm et al. 1996] gave explicit examples of invariant subspaces of all indices in A^2. In the other direction, Aleman and Richter [1997] showed that an invariant subspace of A^2 will have index 1 if it contains a reasonably well-behaved function, for example, a function in the Nevanlinna class.

There has been much additional recent progress on invariant subspaces of A^2, and on other aspects of Bergman spaces, and many important contributions and contributors (B. I. Korenblum and K. Zhu, to name just two) are not mentioned in the discussion above, which is meant merely to present a sample of the work from this intensely active field.

Dirichlet Spaces and the Dirichlet Shift

The Dirichlet integral of a holomorphic function f in the unit disk is defined by

$$D(f) = \int_{\mathbb{D}} |f'|^2 \, dA \, ,$$

where A is normalized area measure on $\mathbb{D}$. The Dirichlet space, D, consists of those functions f for which $D(f)$ is finite. It is contained in H^2 and is a Hilbert space under the norm defined by

$$\|f\|_D^2 = \|f\|_2^2 + D(f) \, ,$$

(where $\| \cdot \|_2$ denotes the norm in H^2). The powers of z form an orthogonal basis for D, the norm of z^n being $\sqrt{n+1}$.

The Dirichlet shift is the operator on D of multiplication by z. It is another weighted shift, sending the n-th vector in the orthonormal basis $\left(z^n/\sqrt{n+1} \right)_{n=1}^{\infty}$ to $\sqrt{(n+2)/(n+1)}$ times the $(n+1)$-st vector. Because D is contained in the well-understood space H^2, one would expect the Dirichlet shift to be more manageable than the Bergman shift. Although that turns out to be the case, we are still a long way from a thorough understanding of the Dirichlet space and its invariant subspaces. For example, L. Carleson [1952] and later H. S. Shapiro and A. L. Shields [1962] long ago obtained information about the zero sequences of functions in D, but we still lack a characterization. More recently, L. Brown and Shields [1984] obtained information about the cyclic vectors of the Dirichlet shift, but, again, we still lack a characterization.

The Dirichlet shift belongs to a class of operators, called two-isometries, that arose in the work of J. Agler [1990]. A Hilbert space operator T is called a two-isometry if it satisfies $T^{*2}T^2 + I = 2T^*T$. S. Richter, in trying to understand the invariant subspaces of D, naturally began exploring general properties of two-isometries. From [Richter and Shields 1988] he knew that every nontrivial invariant subspace of D has index 1. In [Richter 1988] he proved a general result about two-isometries, which, together with the result just mentioned, implies that every nontrivial invariant subspace of D is cyclic; more precisely, if M is a nontrivial invariant subspace of D then M is generated as an invariant subspace by any nonzero vector in $M \ominus zM$. Thus, the restriction of the Dirichlet shift to any of its nontrivial invariant subspaces is a cyclic two-isometry. It is also what Richter calls analytic, meaning that the intersection of the ranges of its powers is trivial. A model theory for cyclic analytic two-isometries is developed in [Richter 1991].

Richter's model spaces are variants of the space D. Given a positive measure μ on $\partial\mathbb{D}$, let $P\mu$ denote its Poisson integral. For f a holomorphic function in $\mathbb{D}$, the Dirichlet integral of f with respect to μ is defined by

$$D_\mu(f) = \int_\mathbb{D} |f'|^2 \, P\mu \, dA \ .$$

The space $D(\mu)$ consists of all f such that $D_\mu(f)$ is finite. It is contained in H^2 and is a Hilbert space under the norm defined by

$$\|f\|_\mu^2 = \|f\|_2^2 + D_\mu(f) \ .$$

One obtains the space D by taking μ to be normalized Lebesgue measure on $\partial\mathbb{D}$. The space $D(0)$, corresponding to the zero measure, is defined to be just H^2. Richter's structure theorem states that S_μ, the operator of multiplication by z on $D(\mu)$, is a cyclic analytic two-isometry, and that any cyclic analytic two-isometry is unitarily equivalent to S_μ for a unique μ. The invariant spaces of D are thus modeled by certain of the spaces $D(\mu)$.

In collaboration with C. Sundberg, Richter continued the study of the spaces $D(\mu)$ [Richter and Sundberg 1991; 1992]. Among many other interesting results in those papers is a structure theorem for the invariant subspaces of $D(\mu)$: If M is a nontrivial invariant subspace of $D(\mu)$, then M has index 1, and if φ is a function of unit norm in $M \ominus zM$, then M is the isometric image of $D(|\varphi|^2\mu)$ under the operator of multiplication by φ. The functions φ in the preceding statement are the analogues in $D(\mu)$ of inner functions in H^2 and of Hedenmalm's extremal functions in A^2. The result of Richter and Sundberg reduces the problem of understanding the invariant subspace structure of $D(\mu)$ to the problem of understanding the structure of these extremal functions. This awaits further study, although interesting progress for the space D was made in [Richter and Sundberg 1994]. In particular, the authors showed that the extremal functions in D are contractive multipliers of D. This contrasts with Hedenmalm's extremal functions in A^2, which are expansive multipliers in the sense that multiplication by one of them does not decrease the A^2 norm of any polynomial. A study of S. M. Shimorin [1995] sheds further light on this phenomenon.

Hankel Operators

A Hankel matrix is a square matrix (finite or infinite) with constant cross diagonals, in other words, a matrix whose (j, k)-th entry depends only on the sum $j + k$. A famous example is the Hilbert matrix, $\left(1/(j + k + 1)\right)_{j,k=0}^\infty$. According to [Hardy et al. 1952, p. 226], D. Hilbert proved in his lectures that this matrix induces a bounded operator on l^2. I. Schur [1911] showed that the sharp bound is π and W. Magnus [1950] that the spectrum is $[0, \pi]$ and purely continuous.

It was noted earlier that semi-infinite Hankel matrices arise in the Nehari interpolation problem. The theorem of Nehari (page) characterizes their boundedness as operators on l^2. Nehari's result will now be restated in slightly different language. By a Hankel operator we shall mean an operator on H^2, possibly unbounded, whose domain contains the vectors in the standard orthonormal basis, and whose matrix in that basis is a Hankel matrix. With each function φ in L^2 of the circle we associate such an operator, which we denote by Γ_φ; it is the one whose matrix has as its (j, k)-th entry the Fourier coefficient $\hat{\varphi}(-j - k - 1)$, where $j, k = 0, 1, 2, \ldots$. Thus, Γ_φ depends only on $P_-\varphi$, the projection of φ onto $L^2 \ominus H^2$. One calls φ a symbol of Γ_φ. Each Hankel operator Γ is Γ_φ for some φ; one can take for φ the image of $\Gamma 1$ under the "reflection" operator R, the operator on L^2 defined by $(Rf)\left(e^{i\theta}\right) = e^{-i\theta} f\left(e^{-i\theta}\right)$ (which sends H^2 onto $L^2 \ominus H^2$, and vice versa). But as a symbol for Γ one can also take any function differing from $R\Gamma 1$ by a function in H^2. For example, the function $\varphi\left(e^{i\theta}\right) = i(\theta - \pi)$ $(0 < \theta < 2\pi)$ is a symbol for the operator whose matrix is the Hilbert matrix.

One easily sees that the action of Γ_φ, on polynomials, say, consists of multiplication by φ, followed by projection onto $L^2 \ominus H^2$ (i.e., application of P_-), followed by reflection (i.e., application of R). In particular, if φ is bounded then so is Γ_φ, with norm at most $\|\varphi\|_\infty$. (Thus, because the operator corresponding to the Hilbert matrix has a symbol of supremum norm π, the norm of the Hilbert matrix is at most π, which is part of Schur's result.) Nehari's theorem, expressed qualitatively, states that a Hankel operator is bounded if and only if it has a bounded symbol. (The quantitative version adds that there exists a symbol whose supremum norm is the norm of the operator.) A companion theorem of P. Hartman [1958] states that a Hankel operator is compact if and only if it has a continuous symbol (of supremum norm arbitrarily close to the norm of the operator). Together with C. Fefferman's characterization of functions of bounded mean oscillation [Fefferman 1971; Garnett 1981] and a related characterization of functions of vanishing mean oscillation [Sarason 1975], these two theorems give the following boundedness and compactness criteria for Hankel operators: Γ_φ is bounded if and only if $P_-\varphi$ is in BMO, the space of functions of bounded mean oscillation on $\partial \mathbb{D}$, and Γ_φ is compact if and only if $P_-\varphi$ is in VMO, the space of functions of vanishing mean oscillation on $\partial \mathbb{D}$. The condition for a Hankel operator to have finite rank was found by L. Kronecker long ago [1881]: Γ_φ has finite rank if and only if $P_-\varphi$ is a rational function.

By the conjugate-analytic symbol of Γ_φ one means the function $P_-\varphi$, the unique symbol in $L^2 \ominus H^2$. The results of Kronecker–Nehari–Hartman thus relate certain basic properties of a Hankel operator to the structure of its conjugate-analytic symbol. This illustrates a recurrent theme in concrete operator theory. Typically in this subject, one is given a natural class of operators induced in some way or other by certain functions, often referred to as the symbols of the operators they induce. One wants to understand how the properties of the operators are encoded by their inducing functions. The Kronecker–Nehari–Hartman results suggest the problem of characterizing the Hankel operators that belong to the Schatten class S_p, the class of compact operators whose singular values are p-th-power summable, where $0 < p < \infty$.

The preceding question for S_2, the Hilbert–Schmidt class, is easy, because an operator is in S_2 if and only if its matrix entries (in any orthonormal basis) are square summable. It follows that a Hankel operator is in S_2 if and only if its conjugate-analytic symbol is the image under the reflection operator R of (the boundary function of) a function in D, the Dirichlet space. The question for general S_p remained a mystery for a long time despite some suggestive work of M. Rosenblum and J. Howland [Howland 1971] pertaining to S_1, the trace class. The breakthrough came in 1979 when V. V. Peller found the condition for a Hankel operator to be in S_1. A short time later he handled the case of S_p for $p \geq 1$ [Peller 1980]. His result says that Γ_φ belongs to S_p if and only if $P_-\varphi$ belongs to a certain Besov space (the space $B_p^{1/p}$). This was extended to $0 < p < 1$ independently by Peller [1983] and S. Semmes [1984]. The results have interesting applications to prediction theory and to rational approximation, which can be found in [Peller and Khrushchëv 1982; Khrushchëv and Peller 1986].

Since Peller's work it has been recognized that results like his and those of his predecessors hold in many other settings. For Hankel operators on the Bergman space, A^2, for instance, boundedness and compactness criteria have been established by S. Axler [1986], K. Zhu [1987], and K. Stroethoff [1990], and Schatten class criteria by J. Arazy, S. Fisher and J. Peetre [Arazy et al. 1988]. More information on this and related matters can be found in [Zhu 1990].

The spectral theory of Hankel operators has been developed to an extent, notably by S. C. Power, whose book [1982] can be consulted for information and references. There have been some interesting developments since that book appeared. Power [1984] showed that there are no nontrivial nilpotent Hankel operators and raised the question whether there are any quasinilpotent ones. A. V. Megretskii [1990] used a clever construction to answer that question in the affirmative. In a very nice paper [Megretskii et al. 1995], Megretskii, Peller and S. R. Treil have given a spectral characterization of self-adjoint Hankel operators, that is, a set of necessary and sufficient conditions, expressed solely in terms of spectral data, for a self-adjoint operator to be unitarily equivalent to a Hankel operator. Ideas from systems theory, especially the notion of a balanced realization, play a prominent role in their analysis.

Toeplitz Operators

This vast subject cannot be adequately addressed in an article such as this one. Only a few highlights will be touched on. The excellent book of A. Böttcher and B. Silbermann [1990] can be consulted for further information.

A Toeplitz matrix is a square matrix with constant diagonals, in other words, a matrix whose (j, k)-th entry depends only on the difference $j - k$. Toeplitz operators can be introduced in many settings, but here the focus will be on the classical Toeplitz operators, which are the bounded operators on H^2 whose matrices with respect to the standard orthonormal basis are Toeplitz matrices. With each such operator is associated a unique symbol, namely, the function φ in L^2 such that the (j, k)-th entry of the corresponding Toeplitz matrix equals the Fourier coefficient $\hat{\varphi}(j - k)$. It is a result of P. Hartman and A. Wintner [1950] that φ is actually in L^∞, with supremum norm equal to the norm of the corresponding operator. The Toeplitz operator with symbol φ is denoted by T_φ. It is the compression to H^2 of the operator on L^2 of multiplication by φ; in other words, it acts on a function in H^2 as multiplication by φ followed by projection onto H^2.

Toeplitz operators are discrete versions of Wiener–Hopf operators, that is, integral operators on $L^2(0, \infty)$ whose kernels, which are functions on $(0, \infty) \times (0, \infty)$, depend only on the difference of the arguments (and so have the form $(x, y) \mapsto K(x - y)$, where K is a function on $\mathbb{R}$). There is in fact more than an analogy here, because the unitary map from H^2 to $L^2(0, \infty)$ mentioned earlier (in the discussion of Beurling's theorem) transforms Toeplitz operators to Wiener–Hopf operators (whose kernels, in general, can be distributions). The preceding observation was first made by M. Rosenblum [1965] and A. Devinatz [1967].

G. Szegő [1920; 1921] has already been mentioned as one of the pioneers of our subject. Among other things, he studied the asymptotic behavior of finite sections of Toeplitz matrices, a line of investigation that continues to the present [Böttcher and Silbermann 1990; Basor and Gohberg 1994].

Self-adjoint Toeplitz operators are well understood. The operator T_φ is self-adjoint if and only if φ is real valued. In that case, according to a theorem of Hartman and Wintner [1954], the spectrum of T_φ is the closed interval whose lower endpoint is the essential infimum of φ and whose upper endpoint is the essential supremum of φ. A concrete spectral representation of T_φ, for φ real, has been given in [Rosenblum 1965; Rosenblum and Rovnyak 1985]. If the essential range of φ is contained in a line, then T_φ is a linear function of a self-adjoint operator and so is described by Rosenblum's theorem, but that is the only time T_φ can be a normal operator, according to a result of A. Brown and P. R. Halmos [1963]. In investigating Toeplitz operators, therefore, one is largely beyond the scope of classical spectral theory.

There is no description of the spectrum of T_φ for general φ. Two general facts are known. One is the spectral inclusion theorem of Hartman and Wintner [1950], which states that the spectrum of T_φ always contains the essential range of φ. The other is a deep result of H. Widom [1964]: the spectrum of T_φ is always connected.

L. A. Coburn [1966] has observed that a nonzero Toeplitz operator cannot have both a trivial kernel and a trivial cokernel. Thus, a Toeplitz operator will be invertible if it is a Fredholm operator of index 0, a fact that has been useful in attempts to determine the spectrum of T_φ. A great many theorems have been obtained saying, roughly, that if φ belongs to such-and-such a class, then T_φ is a Fredholm operator if and only if the origin is not in the "range" of φ, in which case the index of T_φ is the negative of the "winding number" of φ about the origin. Here, "range" is usually interpreted in some generalized sense, and "winding number" is interpreted accordingly. The original and simplest version is where φ is continuous, in which case range and winding number have their usual meanings. That result goes back to I. C. Gohberg [1952], M. G. Krein [1958], Widom [1960], and Devinatz [1964]. The book of Böttcher and Silbermann [1990] contains many other versions.

Toeplitz operators have interacted strongly with the theory of operator algebras. The works of R. G. Douglas [1972; 1973] are early examples of the interaction. There has been much subsequent work, a discussion of which would be beyond this author's competence.

In another direction, a theory of similarity models for Toeplitz operators with rational symbols has been developed by D. N. Clark [1981; 1982] (see also earlier papers referenced there). The following theorem from [Clark and Morrel 1978] illustrates the kind of results obtained: Let φ be a rational function that is univalent in some annulus $\rho \leq |z| \leq 1$ and whose restriction to $\partial\mathbb{D}$ has winding number 1 about each point of the interior of $\varphi(\partial\mathbb{D})$. Then T_φ is similar to T_ψ, where ψ is a Riemann map of the unit disk onto the interior of $\varphi(\partial\mathbb{D})$. The theory has been extended by D. M. Wang [1984] and D. V. Yakubovich [1989; 1991] to encompass nonrational symbols satisfying certain smoothness and topological requirements.

It is unknown whether every Toeplitz operator has a nontrivial invariant subspace. The strongest result to date, due to V. V. Peller [1993], gives an affirmative answer for Toeplitz operators with piecewise continuous symbols satisfying certain extra conditions.

I hope that the scattered remarks above give at least an inkling of the richness of the subject of Toeplitz operators. There is much more that could be said. Toeplitz operators on vector H^p spaces have been studied extensively and are discussed in the book of Böttcher and Silbermann. Besides that, Toeplitz operators arise naturally in many other settings, and one is likely to find several papers on them reviewed in Section 47 of any recent issue of *Mathematical Reviews*.

Holomorphic Composition Operators

Associated with each holomorphic self-map φ of $\mathbb{D}$ is the corresponding composition operator, C_φ; it acts on any function f defined in $\mathbb{D}$ according to the formula $(C_\varphi f)(z) = f(\varphi(z))$. A classical result, the surbordination principle of J. E. Littlewood [1925], guarantees that C_φ acts boundedly on the Hardy spaces H^p and on the Bergman spaces A^p.

The study of holomorphic composition operators, from an operator-theoretic viewpoint, does not date back as far as the study of Toeplitz operators or the study of Hankel operators. It is now thriving, thanks in large part to the influence of J. H. Shapiro, especially to a beautiful theorem of his characterizing when C_φ acts compactly on the spaces H^p. Shapiro began studying compact composition operators in [Shapiro and Taylor 1973/74], where, among other results, the authors proved that if C_φ is compact on one of the spaces H^p then it is compact on every H^p. They also obtained a necessary condition for C_φ to be compact on H^p, namely, that φ not possess an angular derivative in the sense of Carathéodory (an ADC, for short) at any point of $\partial\mathbb{D}$. (One says φ has an ADC at the point λ of $\partial\mathbb{D}$ if φ has a nontangential limit of unit modulus at λ, and the difference quotient $(\varphi(z) - \varphi(\lambda))/(z - \lambda)$ has a nontangential limit at λ.) B. D. MacCluer and Shapiro [1986] showed that the angular derivative condition is not sufficient for the compactness of C_φ on the spaces H^p, although it is both necessary and sufficient in the spaces A^p. They also showed that if φ is univalent then the angular derivative condition does imply C_φ acts compactly on the spaces H^p.

Shapiro's characterization came a year later [1987]. It is quite easy to see that C_φ is compact on the spaces H^p if φ assumes no values near $\mathbb{D}$, in other words, if $\|\varphi\|_\infty < 1$. Shapiro's necessary and sufficient condition is, roughly, that φ not take too many values near $\mathbb{D}$ too often. This is quantitized by means of the Nevanlinna counting function, which is a device from Nevanlinna theory that gives a biased measure of the number of times φ assumes a given value w. The Nevanlinna counting function of φ is defined by

$$N_\varphi(w) = \sum_{z \in \varphi^{-1}(w)} \log \frac{1}{|z|} \qquad \text{for } w \in \mathbb{D}.$$

In the sum on the right side, the points z in $\varphi^{-1}(w)$ are counted with multiplicities, and the sum is interpreted to be 0 if w is not in $\varphi(\mathbb{D})$. The convergence of the series, in case $\varphi^{-1}(w)$ is infinite, follows by the Blaschke condition (assuming φ is not the constant function w). Shapiro's theorem: C_φ *is compact on the spaces H^p if and only if $N_\varphi(w)/\log \frac{1}{|w|} \to 0$ as $|w| \to 1$*. This remarkable theorem has inspired much additional interesting work. To mention just two results, D. H. Luecking and K. Zhu [1992] have characterized the functions φ for which C_φ belongs to one of the Schatten classes as an operator on H^2 or on A^2, and P. Poggi–Corradini [1997] has characterized those univalent φ for which C_φ is a Riesz operator (an operator with essential spectral radius 0) on H^2.

Shapiro's book [1993] is a delightful elementary account of his theorem and related issues. C. C. Cowen and MacCluer [1995] have written a comprehensive treatment of holomorphic composition operators.

Composition operators played a key role in L. de Branges's renowned proof of the Bieberbach–Robertson–Milin conjectures [de Branges 1985; 1987]. If the function φ is univalent and vanishes at the origin, the operator C_φ acts contractively in D, the Dirichlet space, and in certain closely related spaces (with indefinite metrics). De Branges recognized that the Robertson and Milin conjectures can be interpreted as norm inequalities involving composition operators, an observation that underlies his approach. Although the operator methodology that led de Branges to his proof has been discarded in many accounts, its appearance is really quite natural, at least to someone reared as a functional analyst. De Branges's ideas have been further developed in [Vasyunin and Nikolskii 1990; 1991].

Subnormal Operators

A Hilbert space operator is said to be subnormal if it has a normal extension, in other words, if there is a normal operator acting on a space containing the given one as a subspace and coinciding with the given operator in that subspace. (The given Hilbert space is thus an invariant subspace of the normal operator.) The notion was introduced by P. R. Halmos [1950].

Among subnormal operators are of course all normal ones, but also all isometries, and all analytic Toeplitz operators, i.e., Toeplitz operators whose symbols are in H^∞, on both H^2 and A^2. In particular, the unilateral shift and the Bergman shift are subnormal. These examples are all fairly evident, but some subnormal operators appear in disguised form. For instance, there are Toeplitz operators on H^2 that are subnormal yet neither normal nor analytic [Cowen and Long 1984; Cowen 1986]. There are composition operators on H^2 whose adjoints are subnormal, for nonobvious reasons [Cowen and Kriete 1988]. An unexpected and particularly striking example of a subnormal operator is the Cesàro operator, the operator on l^2 that sends the sequence $(x_n)_{n=0}^\infty$ to the sequence $(y_n)_{n=0}^\infty$ defined by $y_n = \frac{1}{n+1} \sum_{k=0}^n x_k$ [Kriete and Trutt 1971; Cowen 1984].

With each positive compactly supported measure μ in the complex plane, there is naturally associated a subnormal operator, the operator of multiplication by z on $P^2(\mu)$, the closure of the polynomials in $L^2(\mu)$. One easily sees on the basis of the spectral theorem that such operators model all cyclic subnormal operators. Thus, the study of subnormal operators quickly leads to questions about polynomial approximation and thence to questions about approximation by rational functions. A large portion of J. B. Conway's comprehensive account of subnormal operators [1991] is devoted to the subject of rational approximation.

The invariant subspace problem for subnormal operators, the question whether every subnormal operator has a nontrivial invariant subspace, because it has an obvious positive answer for noncyclic operators, is equivalent to the question whether, for every measure μ as above, the space $P^2(\mu)$ has a nontrivial invariant subspace (under multiplication by z). In case $P^2(\mu) = L^2(\mu)$ the answer is clear. In the contrary case, one would like to understand the mechanism behind the inequality $P^2(\mu) \neq L^2(\mu)$. For example, does it force the functions in $P^2(\mu)$ to behave, in some sense or other, like holomorphic functions, as happens when μ is Lebesgue measure on $\partial \mathbb{D}$ (in which case $P^2(\mu) = H^2$) and when μ is area measure on $\mathbb{D}$ (in which case $P^2(\mu) = A^2$)? One is thus led to the question: If $P^2(\mu) \neq L^2(\mu)$, does $P^2(\mu)$ carry some kind of analytic structure?

To settle the invariant subspace question, though, one could settle for much less than analytic structure. A point w in the plane is called a bounded point evaluation for $P^2(\mu)$ if the linear functional $p \mapsto p(w)$ on polynomials is bounded in the norm of $L^2(\mu)$. In that case the functional extends boundedly to $P^2(\mu)$; in other words, it is meaningful to evaluate the functions in $P^2(\mu)$ at w. Moreover, the subspace of functions in $P^2(\mu)$ that vanish at w is a nontrivial invariant subspace. (We are ignoring here the irrelevant case where μ is a point mass.) The question thus arises: If $P^2(\mu) \neq L^2(\mu)$, does $P^2(\mu)$ possess bounded point evaluations?

Until recently the preceding question was open, the strongest partial results being due to J. E. Brennan [1979a; 1979b] (and earlier papers). It is now known that the answer is positive—more on that presently. The invariant subspace problem for subnormal operators was eventually settled using a different tack in S. W. Brown's dissertation [1978a], surely one of the most influential dissertations in operator theory ever written. (The published version is [Brown 1978b].) Brown sidestepped $P^2(\mu)$ by working instead with $P^\infty(\mu)$, the weak-star closure of the polynomials in $L^\infty(\mu)$. The structure of $P^\infty(\mu)$ was well understood at the time of Brown's work, and in particular it was known that $P^\infty(\mu)$ possesses weak-star continuous point evaluations whenever it is not all of $L^\infty(\mu)$. By making various reductions, Brown was able to narrow the invariant subspace question for $P^2(\mu)$ to the case where $P^\infty(\mu)$ is just H^∞ of the unit disk, and $P^2(\mu)$ admits no bounded point evaluations. He showed in that case that the evaluation functionals on H^∞ at the points of $\mathbb{D}$ have spatial representations of a certain simple kind in $P^2(\mu)$, from which the existence of nontrivial invariant subspaces follows immediately.

The underlying concept guiding Brown's work was that of an algebra of operators on a Hilbert space H that is closed in the weak-star topology that the algebra of all operators on H acquires as the dual of $\mathcal{S}_1$, the space of trace-class operators on H. Such an algebra is then the dual of a certain quotient space of $\mathcal{S}_1$. Of particular interest is the unital weak-star-closed algebra $\mathcal{A}_T$ generated by a single operator T on H. For example, if T is the unilateral shift on H^2 then $\mathcal{A}_T$ consists of the algebra of analytic Toeplitz operators and so is a replica of

H^∞. Brown's philosophy was that enough information about the structure of $\mathcal{A}_T$ should enable one to find a nontrivial invariant subspace of T.

It was quickly realized that Brown's basic ideas, including his method for constructing spatial representations, apply far beyond the realm of subnormal operators. The outcome has been the theory of dual algebras [Bercovici et al. 1985], mentioned earlier in connection with the Bergman shift. Many existence theorems for invariant subspaces have resulted from this program. One striking example: A Hilbert space contraction whose spectrum contains the unit circle has a nontrivial invariant subspace, as shown by Brown, B. Chevreau and C. Pearcy [Brown et al. 1988]; see also H. Bercovici [1990].

After Brown's breakthrough, the existence question for bounded point evaluations on $P^2(\mu)$ remained open for over ten years. It was settled, with a vengeance, by J. E. Thomson [1991] (see also the last chapter of [Conway 1991]). A point in the plane is called an analytic bounded point evaluation of $P^2(\mu)$ if it belongs to an open set of bounded point evaluations on which the functions in $P^2(\mu)$ are holomorphic. Thomson showed that, if $P^2(\mu) \neq L^2(\mu)$, then $P^2(\mu)$ not only has bounded point evaluations, it has an abundance of analytic bounded point evaluations. He obtained a structure theorem for $P^2(\mu)$ saying, very roughly, that $P^2(\mu)$ can be decomposed into the direct sum of an L^2 space and a space of holomorphic functions. His proof is a tour de force involving powerful and delicate techniques from the theory of rational approximation plus a variant of Brown's basic construction.

Thomson's result shows, paradoxically, that the situation in which Brown originally applied his technique ($P^\infty(\mu) = H^\infty$, yet $P^2(\mu)$ has no bounded point evaluations) is in fact void. Even theorems about the empty set, it seems, can contain interesting ideas.

Another long-standing question about subnormal operators was recently settled. It concerns the relation between subnormality and a related concept, hyponormality, also introduced by Halmos [1950] (although the current terminology was fixed later). A Hilbert space operator T is called hyponormal if the self-adjoint operator $T^*T - TT^*$ is positive semidefinite. A simple argument shows that every subnormal operator is hyponormal.

Although the inequality $T^*T - TT^* \geq 0$ might seem at first glance a rather weak condition to impose on an operator, it has unexpectedly strong implications. A substantial and very interesting theory of hyponormal operators has grown over the years, which, however, will not be discussed here. See [Putnam 1967; Clancey 1979; Vol'berg et al. 1990; Martin and Putinar 1989].

If an operator is subnormal then so are all of its powers. Halmos [1950] gave an example of a hyponormal operator whose square is not hyponormal, thus showing that hyponormality does not imply subnormality. S. K. Berberian raised the question of whether an operator is subnormal if all of its powers are hyponormal. This was answered in the negative by J. G. Stampfli [1965]; his counterexample is a bilateral weighted shift. At about that time the question arose whether an operator T is subnormal if it is polynomially hyponormal, in other words, if $p(T)$ is hyponormal for every polynomial T. This question resisted attack for over 25 years, until R. Curto and M. Putinar [1993] obtained a strong negative answer. Their analysis shows a close relation between the question and classical moment problems.

Several Variables

The discussion so far has dealt almost exclusively with the one-dimensional theory. The theory in several variables, while less well developed, is being energetically pursued and is maturing. Because this author's knowledge is limited, the remarks to follow are brief and incomplete.

The basic theory of Hardy spaces in the polydisk and the ball of $\mathbb{C}^N$ can be found in two books of W. Rudin [1969; 1980]. Some properties from one dimension, such as existence of boundary values, extend nicely. Lacking, however, is a version of the inner-outer factorization. In fact, inner functions in several variables are hard to deal with. In the polydisk it is easy to produce examples, but the general inner function is not well understood; some information is in [Rudin 1969]. In the ball it was an open question for a long time whether there are any nonconstant inner functions. That there are was eventually proved by A. B. Aleksandrov [1982] and E. Løw [1982]. Although inner functions do not play the same central role in the ball that they do in the disk, the ideas needed to prove their existence have had interesting repercussions; further information can be found in [Rudin 1986].

By an invariant subspace of H^2 of the polydisk one means a subspace that is invariant under multiplication by all of the coordinate functions. Many studies of these invariant subspaces have been made, with still very incomplete results. This is part of the emerging theory of multivariable spectral theory, the study of commuting N-tuples of operators. An interesting approach, the theory of Hilbert modules, was initiated by R. G. Douglas and is developed in [Douglas and Paulsen 1989]. The recent book of J. Eschmeier and M. Putinar [1996] emphasizes sheaf-theoretic methods. See also the articles in [Curto et al. 1995].

Hankel, Toeplitz, and composition operators in several variables have received a great deal of attention. A recent study of Hankel operators is [Arazy 1996]. The book [Upmeier 1996] concerns Toeplitz operators. Information on holomorphic composition operators in several variables can be found in [Cowen and MacCluer 1995].

The corona problem, solved for the unit disk by L. Carleson [1962], is one of the basic open problems in several complex variables. The corona problem for a domain in the plane or in $\mathbb{C}^N$ is the problem of deciding whether the points of the domain (more accurately, the evaluation functionals at these points) are dense in the Gelfand space of the Banach algebra of bounded holomorphic functions in the domain. In more concrete terms, it asks whether a finite set of

bounded holomorphic functions in the domain must generate the whole algebra
of bounded holomorphic functions as an ideal, if the given functions satisfy the
obvious necessary condition, namely, that they do not tend to 0 simultaneously
on any sequence in the domain. Carleson's positive solution for the unit disk,
despite not having been previously mentioned in this narrative, is a landmark of
twentieth century function theory. It would be hard to overestimate the amount
of mathematics that has flowed from Carleson's proof.

Carleson's theorem was quickly extended to finitely connected domains in the
plane and to finite bordered Riemann surfaces by various people. The problem
for general planar domains is still open, although a positive solution is known
for some infinitely connected domains, notably domains whose complements lie
on the real axis [Garnett and Jones 1985]. B. Cole has constructed a Riemann
surface for which the solution is negative; his example can be found in [Gamelin
1978, Chapter IV].

A few years after Carleson's proof, L. Hörmander [1967] pointed out the connection between the corona problem and the $\bar{\partial}$-equation. A positive solution to
a corona problem can be reduced to the existence of bounded solutions of certain
$\bar{\partial}$-equations. Despite many advances in $\bar{\partial}$-technology, the corona problems for
the polydisk and the ball in several variables remain open.

References

[Adamyan et al. 1968a] V. M. Adamjan, D. Z. Arov, and M. G. Kreĭn, "Бесконечные
ганкелевы матрицы и обобщённые задачи Каратеодори–Фейера и И. Шура",
Funkcional. Anal. i Priložen. **2**:4 (1968), 1–17. Translated as "Infinite Hankel
matrices and generalized Carathéodory–Fejér and I. Schur problems", *Functional
Anal. Appl.* **2** (1968), 269–281.

[Adamyan et al. 1968b] V. M. Adamjan, D. Z. Arov, and M. G. Kreĭn, "О бесконечных
ганкелевых матрицах и обобщённых задачах Каратеодори–Фейера и Ф.
Рисса", *Funkcional. Anal. i Priložen.* **2**:1 (1968), 1–19. Translated as "Infinite
Hankel matrices and generalized problems of Carathéodory–Fejér and F. Riesz",
Functional Anal. Appl. **2** (1968), 1-18.

[Agler 1989]   J. Agler, "Some interpolation theorems of Nevanlinna–Pick type",
preprint, 1989.

[Agler 1990] J. Agler, "A disconjugacy theorem for Toeplitz operators", *Amer. J.
Math.* **112**:1 (1990), 1–14.

[Agmon 1949] S. Agmon, "Sur un problème de translations", *C. R. Acad. Sci. Paris*
229 (1949), 540–542.

[Aleksandrov 1982] A. B. Aleksandrov, "The existence of inner functions in the ball",
Mat. Sb. **118(160)** (1982), 147–163. In Russian.

[Aleman and Richter 1997] A. Aleman and S. Richter, "Some sufficient conditions for
the division property of invariant subspaces in weighted Bergman spaces", *J. Funct.
Anal.* **144**:2 (1997), 542–556.

[Aleman et al. 1996] A. Aleman, S. Richter, and C. Sundberg, "Beurling's theorem for the Bergman space", *Acta Math.* **177**:2 (1996), 275–310.

[Apostol et al. 1985] C. Apostol, H. Bercovici, C. Foiaş, and C. Pearcy, "Invariant subspaces, dilation theory, and the structure of the predual of a dual algebra, I", *J. Funct. Anal.* **63**:3 (1985), 369–404.

[Arazy 1996] J. Arazy, "Boundedness and compactness of generalized Hankel operators on bounded symmetric domains", *J. Funct. Anal.* **137**:1 (1996), 97–151.

[Arazy et al. 1988] J. Arazy, S. D. Fisher, and J. Peetre, "Hankel operators on weighted Bergman spaces", *Amer. J. Math.* **110**:6 (1988), 989–1053.

[Arocena 1989] R. Arocena, "Unitary extensions of isometries and contractive intertwining dilations", pp. 13–23 in *The Gohberg anniversary collection* (Calgary, AB, 1988), vol. 2, edited by H. Dym et al., Oper. Theory Adv. Appl. **41**, Birkhäuser, Basel, 1989.

[Axler 1986] S. Axler, "The Bergman space, the Bloch space, and commutators of multiplication operators", *Duke Math. J.* **53**:2 (1986), 315–332.

[Ball and Helton 1983] J. A. Ball and J. W. Helton, "A Beurling-Lax theorem for the Lie group $U(m, n)$ which contains most classical interpolation theory", *J. Operator Theory* **9** (1983), 107–142.

[Ball and Kriete 1987] J. A. Ball and T. L. Kriete, III, "Operator-valued Nevanlinna–Pick kernels and the functional models for contraction operators", *Integral Equations Operator Theory* **10**:1 (1987), 17–61.

[Basor and Gohberg 1994] E. L. Basor and I. Gohberg (editors), *Toeplitz operators and related topics* (Santa Cruz, CA, 1992), Oper. Theory Adv. Appl. **71**, Birkhäuser, Basel, 1994.

[Bercovici 1990] H. Bercovici, "Notes on invariant subspaces", *Bull. Amer. Math. Soc. (N.S.)* **23**:1 (1990), 1–36.

[Bercovici et al. 1985] H. Bercovici, C. Foiaş, and C. Pearcy, *Dual algebras with applications to invariant subspaces and dilation theory*, CBMS Regional Conference Series in Mathematics **56**, Amer. Math. Soc., Providence, 1985.

[Bergman 1970] S. Bergman, *The kernel function and conformal mapping*, revised ed., Mathematical Surveys **5**, American Mathematical Society, Providence, 1970.

[Beurling 1949] A. Beurling, "On two problems concerning linear transformations in Hilbert space", *Acta Math.* **81** (1949), 239–255.

[Böttcher and Silbermann 1990] A. Böttcher and B. Silbermann, *Analysis of Toeplitz operators*, Springer, Berlin, 1990.

[de Branges 1985] L. de Branges, "A proof of the Bieberbach conjecture", *Acta Math.* **154**:1-2 (1985), 137–152.

[de Branges 1987] L. de Branges, "Underlying concepts in the proof of the Bieberbach conjecture", pp. 25–42 in *Proceedings of the International Congress of Mathematicians* (Berkeley, 1986), Amer. Math. Soc., Providence, 1987.

[de Branges and Rovnyak 1966] L. de Branges and J. Rovnyak, "Canonical models in quantum scattering theory", pp. 295–392 in *Perturbation theory and its application in quantum mechanics* (Madison, 1965), edited by C. H. Wilcox, Wiley, New York, 1966.

[Brennan 1979a] J. E. Brennan, "Invariant subspaces and subnormal operators", pp. 303–309 in *Harmonic analysis in Euclidean spaces* (Williamstown, 1978), Part 1, edited by G. Weiss and S. Wainger, Proc. Sympos. Pure Math. **35**, Amer. Math. Soc., Providence, 1979.

[Brennan 1979b] J. E. Brennan, "Point evaluations, invariant subspaces and approximation in the mean by polynomials", *J. Functional Anal.* **34** (1979), 407–420.

[Brodskii 1956] M. S. Brodskiĭ, "Characteristic matrix functions of linear operators", *Mat. Sb. N.S.* **39(81)** (1956), 179–200. In Russian.

[Brodskii 1957] M. S. Brodskiĭ, "On a problem of I. M. Gel'fand", *Uspehi Mat. Nauk* (*N.S.*) **12**:2 (=74) (1957), 129–132. In Russian.

[Brown 1978a] S. W. Brown, *Banach algebras that are dual spaces*, Doctoral dissertation, University of California at Santa Barbara, 1978.

[Brown 1978b] S. W. Brown, "Some invariant subspaces for subnormal operators", *Integral Equations Operator Theory* **1**:3 (1978), 310–333.

[Brown and Halmos 1963] A. Brown and P. R. Halmos, "Algebraic properties of Toeplitz operators", *J. Reine Angew. Math.* **213**:1-2 (1963), 89–102.

[Brown and Shields 1984] L. Brown and A. L. Shields, "Cyclic vectors in the Dirichlet space", *Trans. Amer. Math. Soc.* **285**:1 (1984), 269–303.

[Brown et al. 1988] S. W. Brown, B. Chevreau, and C. Pearcy, "On the structure of contraction operators, II", *J. Funct. Anal.* **76**:1 (1988), 30–55.

[Carathéodory and Fejér 1911] C. Carathéodory and L. Fejér, "Über den Zusammenhang der extremen von harmonischen Funktionen mit ihren Koeffizienten und über den Picard-Landau'schen Satz", *Rend. Circ. Mat. Palermo* **32** (1911), 218–239.

[Carleson 1952] L. Carleson, "On the zeros of functions with bounded Dirichlet integrals", *Math. Z.* **56** (1952), 289–295.

[Carleson 1962] L. Carleson, "Interpolations by bounded analytic functions and the corona problem", *Ann. of Math.* (2) **76**:3 (1962), 547–559.

[Clancey 1979] K. Clancey, *Seminormal operators*, Lecture Notes in Math. **742**, Springer, Berlin, 1979.

[Clark 1981] D. N. Clark, "On the structure of rational Toeplitz operators", pp. 63–72 in *Contributions to analysis and geometry* (Baltimore, 1980), edited by D. N. Clark et al., Johns Hopkins Univ. Press, Baltimore, 1981.

[Clark 1982] D. N. Clark, "On Toeplitz operators with loops, II", *J. Operator Theory* **7**:1 (1982), 109–123.

[Clark and Morrel 1978] D. N. Clark and J. H. Morrel, "On Toeplitz operators and similarity", *Amer. J. Math.* **100**:5 (1978), 973–986.

[Coburn 1966] L. A. Coburn, "Weyl's theorem for nonnormal operators", *Michigan Math. J.* **13** (1966), 285–288.

[Cole et al. 1992] B. Cole, K. Lewis, and J. Wermer, "Pick conditions on a uniform algebra and von Neumann inequalities", *J. Funct. Anal.* **107**:2 (1992), 235–254.

[Conway 1991] J. B. Conway, *The theory of subnormal operators*, Mathematical Surveys and Monographs **36**, Amer. Math. Soc., Providence, 1991.

[Cotlar and Sadosky 1994] M. Cotlar and C. Sadosky, "Nehari and Nevanlinna–Pick problems and holomorphic extensions in the polydisk in terms of restricted BMO", *J. Funct. Anal.* **124**:1 (1994), 205–210.

[Cowen 1984] C. C. Cowen, "Subnormality of the Cesàro operator and a semigroup of composition operators", *Indiana Univ. Math. J.* **33**:2 (1984), 305–318.

[Cowen 1986] C. C. Cowen, "More subnormal Toeplitz operators", *J. Reine Angew. Math.* **367** (1986), 215–219.

[Cowen and Kriete 1988] C. C. Cowen and T. L. Kriete, III, "Subnormality and composition operators on H^2", *J. Funct. Anal.* **81**:2 (1988), 298–319.

[Cowen and Long 1984] C. C. Cowen and J. J. Long, "Some subnormal Toeplitz operators", *J. Reine Angew. Math.* **351** (1984), 216–220.

[Cowen and MacCluer 1995] C. C. Cowen and B. D. MacCluer, *Composition operators on spaces of analytic functions*, Studies in Advanced Mathematics, CRC Press, Boca Raton, FL, 1995.

[Curto and Putinar 1993] R. E. Curto and M. Putinar, "Nearly subnormal operators and moment problems", *J. Funct. Anal.* **115**:2 (1993), 480–497.

[Curto et al. 1995] R. E. Curto, R. G. Douglas, J. D. Pincus, and N. Salinas (editors), *Multivariable operator theory* (Seattle, 1993), Contemporary Mathematics **185**, American Mathematical Society, Providence, 1995.

[Devinatz 1964] A. Devinatz, "Toeplitz operators on H^2 spaces", *Trans. Amer. Math. Soc.* **112** (1964), 304–317.

[Devinatz 1967] A. Devinatz, "On Wiener-Hopf operators", pp. 81–118 in *Functional Analysis* (Irvine, CA, 1966), edited by B. R. Gelbaum, Academic Press, London, and Thompson Book Co., Washington, DC, 1967.

[Donoghue 1957] J. Donoghue, W. F., "The lattice of invariant subspaces of a completely continuous quasi-nilpotent transformation", *Pacific J. Math.* **7** (1957), 1031–1035.

[Douglas 1972] R. G. Douglas, *Banach algebra techniques in operator theory*, Pure and Applied Mathematics **49**, Academic Press, New York, 1972.

[Douglas 1973] R. G. Douglas, *Banach algebra techniques in the theory of Toeplitz operators* (Athens, GA, 1972), CBMS Regional Conference Series in Mathematics **15**, American Mathematical Society, Providence, 1973.

[Douglas and Paulsen 1989] R. G. Douglas and V. I. Paulsen, *Hilbert modules over function algebras*, Pitman Research Notes in Mathematics Series **217**, Longman, Harlow, 1989.

[Duren 1970] P. L. Duren, *Theory of H^p spaces*, Pure and Applied Mathematics **38**, Academic Press, New York, 1970.

[Duren et al. 1993] P. Duren, D. Khavinson, H. S. Shapiro, and C. Sundberg, "Contractive zero-divisors in Bergman spaces", *Pacific J. Math.* **157**:1 (1993), 37–56.

[Duren et al. 1994] P. Duren, D. Khavinson, H. S. Shapiro, and C. Sundberg, "Invariant subspaces in Bergman spaces and the biharmonic equation", *Michigan Math. J.* **41**:2 (1994), 247–259.

[Dym 1989] H. Dym, *J contractive matrix functions, reproducing kernel Hilbert spaces and interpolation*, CBMS Regional Conference Series in Mathematics **71**, Amer. Math. Soc., Providence, 1989.

[Dym 1994] H. Dym, "Review of 'The commutant lifting approach to interpolation problems' by C. Foias and A. E. Frazho", *Bull. Amer. Math. Soc. (N.S.)* **31** (1994), 125–140.

[Eschmeier and Putinar 1996] J. Eschmeier and M. Putinar, *Spectral decompositions and analytic sheaves*, London Mathematical Society Monographs (N.S.) **10**, Oxford University Press, New York, 1996.

[Fatou 1906] P. Fatou, "Séries trigonometriques et séries de Taylor", *Acta Math.* **30** (1906), 335–400.

[Fefferman 1971] C. Fefferman, "Characterizations of bounded mean oscillation", *Bull. Amer. Math. Soc.* **77** (1971), 587–588.

[Foiaş and Frazho 1990] C. Foiaş and A. E. Frazho, *The commutant lifting approach to interpolation problems*, Oper. Theory Adv. Appl. **44**, Birkhäuser, Basel, 1990.

[Francis 1987] B. A. Francis, *A course in H_∞ control theory*, Lecture Notes in Control and Information Sciences **88**, Springer, Berlin, 1987.

[Gamelin 1978] T. W. Gamelin, *Uniform algebras and Jensen measures*, London Mathematical Society Lecture Note Series **32**, Cambridge Univ. Press, Cambridge, 1978.

[Garnett 1981] J. B. Garnett, *Bounded analytic functions*, Pure and Applied Mathematics **96**, Academic Press, New York, 1981.

[Garnett and Jones 1985] J. B. Garnett and P. W. Jones, "The corona theorem for Denjoy domains", *Acta Math.* **155**:1-2 (1985), 27–40.

[Gohberg 1952] I. C. Gohberg, "On an application of the theory of normed rings to singular integral equations", *Uspehi Matem. Nauk (N.S.)* **7**:2 (=48) (1952), 149–156. In Russian.

[Halmos 1950] P. R. Halmos, "Normal dilations and extensions of operators", *Summa Brasil. Math.* **2** (1950), 125–134.

[Halmos 1961] P. R. Halmos, "Shifts on Hilbert spaces", *J. Reine Angew. Math.* **208** (1961), 102–112.

[Hardy 1915] G. H. Hardy, "On the mean value of the modulus of an analytic function", *Proc. London Math. Soc.* **14** (1915), 269–277.

[Hardy and Littlewood 1930] G. H. Hardy and J. E. Littlewood, "A maximal theorem with function theoretic implications", *Acta Math.* **54** (1930), 81–116.

[Hardy et al. 1952] G. H. Hardy, J. E. Littlewood, and G. Pólya, *Inequalities*, 2nd ed., Cambridge Univ. Press, 1952.

[Hartman 1958] P. Hartman, "On completely continuous Hankel matrices", *Proc. Amer. Math. Soc.* **9** (1958), 862–866.

[Hartman and Wintner 1950] P. Hartman and A. Wintner, "On the spectra of Toeplitz's matrices", *Amer. J. Math.* **72** (1950), 359–366.

[Hartman and Wintner 1954] P. Hartman and A. Wintner, "The spectra of Toeplitz's matrices", *Amer. J. Math.* **76** (1954), 867–882.

[Hedenmalm 1991] H. Hedenmalm, "A factorization theorem for square area-integrable analytic functions", *J. Reine Angew. Math.* **422** (1991), 45–68.

[Hedenmalm 1993] P. J. H. Hedenmalm, "An invariant subspace of the Bergman space having the codimension two property", *J. Reine Angew. Math.* **443** (1993), 1–9.

[Hedenmalm 1994] H. Hedenmalm, "A factoring theorem for the Bergman space", *Bull. London Math. Soc.* **26**:2 (1994), 113–126.

[Hedenmalm et al. 1996] H. Hedenmalm, S. Richter, and K. Seip, "Interpolating sequences and invariant subspaces of given index in the Bergman spaces", *J. Reine Angew. Math.* **477** (1996), 13–30.

[Hörmander 1967] L. Hörmander, "Generators for some rings of analytic functions", *Bull. Amer. Math. Soc.* **73** (1967), 943–949.

[Horowitz 1974] C. Horowitz, "Zeros of functions in the Bergman spaces", *Duke Math. J.* **41** (1974), 693–710.

[Howland 1971] J. S. Howland, "Trace class Hankel operators", *Quart. J. Math. Oxford Ser.* (2) **22** (1971), 147–159.

[Katsnelson et al. 1987] V. È. Katsnel'son, A. Y. Kheĭfets, and P. M. Yuditskiĭ, "An abstract interpolation problem and the theory of extensions of isometric operators", pp. 83–96, 146 in *Operators in function spaces and problems in function theory*, edited by V. A. Marchenko, Naukova Dumka, Kiev, 1987. In Russian; translation in *Topics in interpolation theory*, edited by Harry Dym et al., Oper. Theory Adv. Appl. **95**, Birkhäuser, Basel, 1997, pp. 283–298.

[Khavinson and Shapiro 1994] D. Khavinson and H. S. Shapiro, "Invariant subspaces in Bergman spaces and Hedenmalm's boundary value problem", *Ark. Mat.* **32**:2 (1994), 309–321.

[Kheifets and Yuditskii 1994] A. Y. Kheĭfets and P. M. Yuditskiĭ, "An analysis and extension of V. P. Potapov's approach to interpolation problems with applications to the generalized bi-tangential Schur–Nevanlinna–Pick problem and *J*-inner-outer factorization", pp. 133–161 in *Matrix and operator valued functions*, edited by I. Gohberg and L. A. Sakhnovich, Oper. Theory Adv. Appl. **72**, Birkhäuser, Basel, 1994.

[Khrushchëv and Peller 1986] S. V. Hruščev and V. V. Peller, *Hankel operators of Schatten–von Neumann class and their application to stationary processes and best approximations*, Grundlehren der mathematischen Wissenschaften **273**, Springer, Berlin, 1986.

[Kolmogorov 1925] A. N. Kolmogorov, "Sur les fonctions harmoniques et les séries de Fourier", *Fund. Math.* **7** (1925), 24–29.

[Kovalishina 1974] I. V. Kovalishina, "*J*-expansive matrix functions in the problem of Carathéodory", *Akad. Nauk Armyan. SSR Dokl.* **59** (1974), 129–135. In Russian.

[Kovalishina 1983] I. V. Kovalishina, "Analytic theory of a class of interpolation problems", *Izv. Akad. Nauk SSSR Ser. Mat.* **47**:3 (1983), 455–497. In Russian.

[Kovalishina and Potapov 1974] I. V. Kovalishina and V. P. Potapov, "An indefinite metric in the Nevanlinna–Pick problem", *Akad. Nauk Armyan. SSR Dokl.* **59** (1974), 17–22. In Russian.

[Kreĭn 1958] M. G. Kreĭn, "Integral equations on the half-line with a kernel depending on the difference of the arguments", *Uspehi Mat. Nauk (N.S.)* **13**:5 (=83) (1958), 3–120.

[Kriete and Trutt 1971] T. L. Kriete, III and D. Trutt, "The Cesàro operator in l^2 is subnormal", *Amer. J. Math.* **93** (1971), 215–225.

[Kronecker 1881] L. Kronecker, "Zur Theorie der Elimination einer Variablen aus zwei algebraischen Gleichungen", *Monatsber. Königl. Preussischen Akad. Wies. (Berlin)* (1881), 535–600. Reprinted as pp. 113-192 in *Mathematische Werke*, vol. 2, B. G. Teubner, Leipzig, 1897 or Chelsea, New York, 1968.

[Lax 1959] P. D. Lax, "Translation invariant spaces", *Acta Math.* **101** (1959), 163–178.

[Littlewood 1925] J. E. Littlewood, "On inequalities in the theory of functions", *Proc. London Math. Soc.* **23** (1925), 481–519.

[Livshitz 1952] M. S. Livšic, "On the reduction of a linear non-Hermitian operator to 'triangular' form", *Dokl. Akad. Nauk SSSR* **84** (1952), 873–876.

[Løw 1982] E. Løw, "A construction of inner functions on the unit ball in $\mathbb{C}^p$", *Invent. Math.* **67**:2 (1982), 223–229.

[Luecking and Zhu 1992] D. H. Luecking and K. H. Zhu, "Composition operators belonging to the Schatten ideals", *Amer. J. Math.* **114**:5 (1992), 1127–1145.

[MacCluer and Shapiro 1986] B. D. MacCluer and J. H. Shapiro, "Angular derivatives and compact composition operators on the Hardy and Bergman spaces", *Canad. J. Math.* **38**:4 (1986), 878–906.

[Magnus 1950] W. Magnus, "On the spectrum of Hilbert's matrix", *Amer. J. Math.* **72** (1950), 699–704.

[Martin and Putinar 1989] M. Martin and M. Putinar, *Lectures on hyponormal operators*, Oper. Theory Adv. Appl. **39**, Birkhäuser, Basel, 1989.

[Megretskii 1990] A. V. Megretskiĭ, "A quasinilpotent Hankel operator", *Algebra i Analiz* **2**:4 (1990), 201–212.

[Megretskii et al. 1995] A. V. Megretskiĭ, V. V. Peller, and S. R. Treil', "The inverse spectral problem for self-adjoint Hankel operators", *Acta Math.* **174**:2 (1995), 241–309.

[Moeller 1962] J. W. Moeller, "On the spectra of some translation invariant spaces", *J. Math. Anal. Appl.* **4** (1962), 276–296.

[Nehari 1957] Z. Nehari, "On bounded bilinear forms", *Ann. of Math.* (2) **65** (1957), 153–162.

[Nevanlinna 1919] R. Nevanlinna, "Über beschränkte Funktionen die in gegebene Punkten vorgeschriebene Werte annehmen", *Ann. Acad. Sci. Fenn. Ser. A* **13**:1 (1919).

[Nevanlinna 1936] R. Nevanlinna, *Eindeutige analytische Funktionen*, Grundlehren der mathematischen Wissenschaften **46**, Springer, Berlin, 1936.

[Nikolskii 1986] N. K. Nikol'skiĭ, *Treatise on the shift operator*, Grundlehren der mathematischen Wissenschaften **273**, Springer, Berlin, 1986.

[Nikolskii and Vasyunin 1989] N. K. Nikolskiĭ and V. I. Vasyunin, "A unified approach to function models, and the transcription problem", pp. 405–434 in *The Gohberg anniversary collection* (Calgary, AB, 1988), vol. 2, edited by H. Dym et al., Oper. Theory Adv. Appl. **41**, Birkhäuser, Basel, 1989.

[Peller 1980] V. V. Peller, "Операторы Ганкеля класса $\mathfrak{S}_p$ и их приложения (рациональная аппроксимация, гауссовские процессы, проблема мажорации операторов)", *Mat. Sb. (N.S.)* **113(155)**:4 (1980), 538–581. Translated as "Hankel operators of class $\mathfrak{S}_p$ and their applications (rational approximation, Gaussian processes, the problem of majorization of operators)", *Math. USSR Sb.* **41** (1982), 443–479.

[Peller 1983] V. V. Peller, "Описание операторов Ганкеля класса $\mathfrak{S}_p$ при $p > 0$, исследование скорости рациональной аппроксимации и другие приложения", *Mat. Sb. (N.S.)* **122(164)**:4 (1983), 481–510. Translated as "Description of Hankel operators of the class $\mathfrak{S}_p$ for $p > 0$, investigation of the rate of rational approximation and other applications", *Math. USSR Sb.* **50** (1985), 465-494.

[Peller 1993] V. V. Peller, "Invariant subspaces of Toeplitz operators with piecewise continuous symbols", *Proc. Amer. Math. Soc.* **119**:1 (1993), 171–178.

[Peller and Khrushchëv 1982] V. V. Peller and S. V. Khrushchëv, "Операторы Ганкеля, наилучшие приближения и стационарные гауссовские процессы", *Uspehi Mat. Nauk (N.S.)* **37**:1 (1982), 53–124. Translated as "Hankel operators, best approximations and stationary Gaussian processes", *Russian Math. Surveys* **37**:1 (1982), 61-144.

[Pick 1916] G. Pick, "Über die Beschränkungen analytischen Funktionen, welche durch vorgegebene Functionswerte bewirkt werden", *Math. Ann.* **77** (1916), 7–23.

[Poggi-Corradini 1997] P. Poggi-Corradini, "The Hardy class of geometric models and the essential spectral radius of composition operators", *J. Funct. Anal.* **143**:1 (1997), 129–156.

[Potapov 1955] V. P. Potapov, "The multiplicative structure of J-contractive matrix functions", *Trudy Moskov. Mat. Obšč.* **4** (1955), 125–236. In Russian; translation in *Amer. Math. Soc. Transl.* (2) **15** (1960), 131–243.

[Power 1982] S. C. Power, *Hankel operators on Hilbert space*, Research Notes in Mathematics **64**, Pitman, Boston, 1982.

[Power 1984] S. C. Power, "Quasinilpotent Hankel operators", pp. 259–261 in *Linear and complex analysis problem book: 199 research problems*, edited by V. P. Havin et al., Lecture Notes in Math. **1043**, Springer, Berlin, 1984.

[Privalov 1950] I. I. Privalov, Граничные свойства аналитических функций, 2nd ed., Gosudarstv. Izdat. Tehn.-Teor. Lit., Moscow-Leningrad, 1950.

[Privalov 1956] I. I. Priwalow, *Randeigenschaften analytischer Funktionen*, VEB Deutscher Verlag, Berlin, 1956.

[Putnam 1967] C. R. Putnam, *Commutation properties of Hilbert space operators and related topics*, Ergebnisse der Mathematik und ihrer Grenzgebiete **36**, Springer, New York, 1967.

[Richter 1988] S. Richter, "Invariant subspaces of the Dirichlet shift", *J. Reine Angew. Math.* **386** (1988), 205–220.

[Richter 1991] S. Richter, "A representation theorem for cyclic analytic two-isometries", *Trans. Amer. Math. Soc.* **328**:1 (1991), 325–349.

[Richter and Shields 1988] S. Richter and A. Shields, "Bounded analytic functions in the Dirichlet space", *Math. Z.* **198**:2 (1988), 151–159.

[Richter and Sundberg 1991] S. Richter and C. Sundberg, "A formula for the local Dirichlet integral", *Michigan Math. J.* **38**:3 (1991), 355–379.

[Richter and Sundberg 1992] S. Richter and C. Sundberg, "Multipliers and invariant subspaces in the Dirichlet space", *J. Operator Theory* **28**:1 (1992), 167–186.

[Richter and Sundberg 1994] S. Richter and C. Sundberg, "Invariant subspaces of the Dirichlet shift and pseudocontinuations", *Trans. Amer. Math. Soc.* **341**:2 (1994), 863–879.

[Riesz 1923] F. Riesz, "Über die Randwerte einer analytischen Funktion", *Math. Z.* **18** (1923), 87–95. Reprinted in F. Riesz, *Oeuvres complètes*, Gauthier-Villars, Paris, 1960.

[Riesz 1924] M. Riesz, "Les fonctions conjugueés et les séries de Fourier", *C. R. Acad. Sci. Paris Ser. A-B* **178** (1924), 1464–1467.

[Riesz and Riesz 1916] F. Riesz and M. Riesz, "Über die Randwerte einer analytischen Funktion", pp. 27–44 in *Comptes Rendus du 4. Congr. des Math. Scand.* (Stockholm, 1916), 1916. Reprinted in F. Riesz, *Oeuvres complètes*, Gauthier-Villars, Paris, 1960.

[Rosenblum 1965] M. Rosenblum, "A concrete spectral theory for self-adjoint Toeplitz operators", *Amer. J. Math.* **87** (1965), 709–718.

[Rosenblum and Rovnyak 1985] M. Rosenblum and J. Rovnyak, *Hardy classes and operator theory*, Oxford Mathematical Monographs, Oxford University Press, New York, 1985. Reprinted with corrections by Dover, Mineola (NY), 1997.

[Rota 1960] G.-C. Rota, "On models for linear operators", *Comm. Pure Appl. Math.* **13** (1960), 469–472.

[Rudin 1969] W. Rudin, *Function theory in polydiscs*, W. A. Benjamin, New York, 1969.

[Rudin 1980] W. Rudin, *Function theory in the unit ball of* $\mathbf{C}^n$, Grundlehren der Mathematischen Wissenschaften **241**, Springer, New York, 1980.

[Rudin 1986] W. Rudin, *New constructions of functions holomorphic in the unit ball of* $\mathbf{C}^n$, CBMS Regional Conference Series in Mathematics **63**, Amer. Math. Soc., Providence, 1986.

[Sarason 1965] D. Sarason, "A remark on the Volterra operator", *J. Math. Anal. Appl.* **12** (1965), 244–246.

[Sarason 1967] D. Sarason, "Generalized interpolation in H^∞", *Trans. Amer. Math. Soc.* **127**:2 (1967), 179–203.

[Sarason 1975] D. Sarason, "Functions of vanishing mean oscillation", *Trans. Amer. Math. Soc.* **207** (1975), 391–405.

[Sarason 1987] D. Sarason, "Moment problems and operators in Hilbert space", pp. 54–70 in *Moments in mathematics* (San Antonio, 1987), edited by H. J. Landau, Proc. Sympos. Appl. Math. **37**, Amer. Math. Soc., Providence, 1987.

[Sarason 1991] D. Sarason, "New Hilbert spaces from old", pp. 195–204 in *Paul Halmos*, edited by J. H. Ewing and F. W. Gehring, Springer, New York, 1991.

[Schur 1911] I. Schur, "Bemerkungen zur Theorie der beschränkten Bilinearformen mit unendlich vielen Veränderlicher", *J. Reine Angew. Math.* **140** (1911), 1–28.

[Schur 1917] I. Schur, "Über Potenzreihen die im Innern des Einheitskreises beschränkt sind", *J. Reine Angew. Math.* **147** (1917), 205–232.

[Seip 1993] K. Seip, "Beurling type density theorems in the unit disk", *Invent. Math.* **113**:1 (1993), 21–39.

[Semmes 1984] S. Semmes, "Trace ideal criteria for Hankel operators, and applications to Besov spaces", *Integral Equations Operator Theory* **7**:2 (1984), 241–281.

[Shapiro 1987] J. H. Shapiro, "The essential norm of a composition operator", *Ann. of Math.* (2) **125**:2 (1987), 375–404.

[Shapiro 1993] J. H. Shapiro, *Composition operators and classical function theory*, Universitext: Tracts in Mathematics, Springer, New York, 1993.

[Shapiro and Shields 1962] H. S. Shapiro and A. L. Shields, "On the zeros of functions with finite Dirichlet integral and some related function spaces", *Math. Z.* **80** (1962), 217–229.

[Shapiro and Taylor 1973/74] J. H. Shapiro and P. D. Taylor, "Compact, nuclear, and Hilbert-Schmidt composition operators on H^2", *Indiana Univ. Math. J.* **23** (1973/74), 471–496.

[Shields 1974] A. L. Shields, "Weighted shift operators and analytic function theory", pp. 49–128 in *Topics in operator theory*, edited by C. Pearcy, Math. Surveys **13**, Amer. Math. Soc., Providence, 1974.

[Shimorin 1995] S. M. Shimorin, "On a family of conformally invariant operators", *Algebra i Analiz* **7**:2 (1995), 133–158. In Russian; translation in *St. Petersburg Math. J.* **7**:2 (1996), 287–306.

[Stampfli 1965] J. G. Stampfli, "Hyponormal operators and spectral density", *Trans. Amer. Math. Soc.* **117** (1965), 469–476. Errata in **115** (1965), 550.

[Stroethoff 1990] K. Stroethoff, "Compact Hankel operators on the Bergman space", *Illinois J. Math.* **34**:1 (1990), 159–174.

[Sz.-Nagy and Foiaş 1967] B. Sz.-Nagy and C. Foiaş, *Analyse harmonique des opérateurs de l'espace de Hilbert*, Masson, Paris, and Akadémiai Kiadó, Budapest, 1967. Translated as *Harmonic analysis of operators on Hilbert space*, North-Holland, Amsterdam, and Akadémiai Kiadó, Budapest, 1970.

[Sz.-Nagy and Foiaş 1968] B. Sz.-Nagy and C. Foiaş, "Dilatation des commutants d'opérateurs", *C. R. Acad. Sci. Paris Sér. A-B* **266** (1968), A493–A495.

[Szegö 1920] G. Szegö, "Beiträge zur Theorie der Toeplitschen Formen I", *Math. Z.* **6** (1920), 167–202.

[Szegö 1921] G. Szegö, "Beiträge zur Theorie der Toeplitschen Formen II", *Math. Z.* **9** (1921), 167–190.

[Thomson 1991] J. E. Thomson, "Approximation in the mean by polynomials", *Ann. of Math.* (2) **133**:3 (1991), 477–507.

[Upmeier 1996] H. Upmeier, *Toeplitz operators and index theory in several complex variables*, Operator Theory: Advances and Applications **81**, Birkhäuser, Basel, 1996.

[Vasyunin and Nikolskii 1990] V. I. Vasyunin and N. K. Nikol'skiĭ, "Quasi-orthogonal decompositions with respect to complementary metrics, and estimates for univalent functions", *Algebra i Analiz* **2**:4 (1990), 1–81. In Russian; translation in *Leningrad Math. J.* **2**:4 (1991), 691–764.

[Vasyunin and Nikolskii 1991] V. I. Vasyunin and N. K. Nikol'skiĭ, "Operator-valued measures and coefficients of univalent functions", *Algebra i Analiz* **3**:6 (1991), 1–75 (1992). In Russian; translation in *St. Petersburg Math. J.* **3**:6 (1992), 1199–1270.

[Vol'berg et al. 1990] A. L. Vol'berg, V. V. Peller, and D. V. Yakubovich, "A brief excursion into the theory of hyponormal operators", *Algebra i Analiz* **2**:2 (1990), 1–38. In Russian; translation in *Leningrad Math. J.* **2**:2 (1991), 211–243.

[Wang 1984] D. M. Wang, "Similarity of smooth Toeplitz operators", *J. Operator Theory* **12**:2 (1984), 319–329.

[Widom 1960] H. Widom, "Inversion of Toeplitz matrices, II", *Illinois J. Math.* **4** (1960), 88–99.

[Widom 1964] H. Widom, "On the spectrum of a Toeplitz operator", *Pacific J. Math.* **14** (1964), 365–375.

[Yakubovich 1989] D. V. Yakubovich, "Riemann surface models of Toeplitz operators", pp. 305–415 in *Toeplitz operators and spectral function theory*, edited by N. K. Nikol'skiĭ, Oper. Theory: Adv. Appl. **42**, Birkhäuser, Basel, 1989.

[Yakubovich 1991] D. V. Yakubovich, "On the spectral theory of Toeplitz operators with a smooth symbol", *Algebra i Analiz* **3**:4 (1991), 208–226. In Russian.

[Zhu 1987] K. Zhu, "VMO, ESV, and Toeplitz operators on the Bergman space", *Trans. Amer. Math. Soc.* **302**:2 (1987), 617–646.

[Zhu 1990] K. H. Zhu, *Operator theory in function spaces*, Monographs and Textbooks in Pure and Applied Mathematics **139**, Marcel Dekker Inc., New York, 1990.

DONALD SARASON
DEPARTMENT OF MATHEMATICS
UNIVERSITY OF CALIFORNIA
BERKELEY, CA 94720
UNITED STATES
sarason@math.berkeley.edu

Holomorphic Spaces
MSRI Publications
Volume **33**, 1998

Recent Progress in the Function Theory
of the Bergman Space

HÅKAN HEDENMALM

ABSTRACT. The recent developments in the function theory of the Bergman space are reviewed. Key ingredients are: factorization based on extremal divisors, an analog of Beurling's invariant subspace theorem, concrete examples of invariant subspaces of index higher than one, a partial description of zero sequences, characterizations of interpolating and sampling sequences, and some remarks on weighted Bergman spaces.

1. The Hardy and Bergman Spaces: A Comparison

The Hardy space H^2 consists of all holomorphic functions on the open unit disk $\mathbb{D}$ such that

$$\|f\|_{H^2} = \sup_{0<r<1} \left(\int_{\mathbb{T}} |f(rz)|^2 \, ds(z) \right)^{\frac{1}{2}} < +\infty, \qquad (1\text{--}1)$$

where $\mathbb{T}$ stands for the unit circle and ds is arc length measure, normalized so that the mass of $\mathbb{T}$ equals 1. In terms of Taylor coefficients, the norm takes a more appealing form: if $f(z) = \sum_n a_n z^n$, then

$$\|f\|_{H^2} = \left(\sum_n |a_n|^2 \right)^{\frac{1}{2}}.$$

The Bergman space L_a^2, on the other hand, consists of all holomorphic functions on $\mathbb{D}$ such that

$$\|f\|_{L_a^2} = \left(\int_{\mathbb{D}} |f(z)|^2 \, dS(z) \right)^{\frac{1}{2}} < +\infty,$$

where dS is area measure, normalized so that the mass of $\mathbb{D}$ equals 1. Though the integral expression of the norm is more straightforward than for the Hardy

Key words and phrases. Bergman spaces, canonical divisors.

This research was supported in part by the Swedish Natural Science Research Council.

space, it is more complicated in terms of Taylor coefficients: if $f(z) = \sum_n a_n z^n$, then

$$\|f\|_{L_a^2} = \left(\sum_n \frac{|a_n|^2}{n+1} \right)^{\frac{1}{2}}.$$

The Bergman space L_a^2 contains H^2 as a dense subspace. It is intuitively clear from the definition of the norm in H^2 that functions in it have well-defined boundary values in $L^2(\mathbb{T})$. This is however not the case for L_a^2. In fact, there is a function in it which fails to have radial limits at every point of $\mathbb{T}$. This is a consequence of a more general statement due to MacLane [1962]; see also [Luzin and Privalov 1925; Cantor 1964]. Apparently the two spaces H^2 and L_a^2 are very different from a function-theoretic perspective.

The Hardy space theory. The classical factorization theory for the Hardy spaces (these are the spaces H^p, with $0 < p \leq +\infty$, which are defined by property (1–1), with 2 replaced by p), which relies on work due to Blaschke, Riesz, and Szegö, requires some familiarity with the concepts of a Blaschke product, a singular inner function, an inner function, and an outer function. Let H^∞ stand for the space of bounded analytic functions in $\mathbb{D}$, supplied with the supremum norm. Given a (finite or infinite) sequence $A = \{a_j\}_j$ of points in $\mathbb{D}$, one considers the product

$$B_A(z) = \prod_j \frac{\bar{a}_j}{|a_j|} \frac{a_j - z}{1 - \bar{a}_j z} \qquad \text{for } z \in \mathbb{D},$$

which converges to a function in H^∞ of norm 1 if and only if the Blaschke condition $\sum_j 1 - |a_j| < +\infty$ is fulfilled, in which case A is said to be a Blaschke sequence, and B_A is said to be a Blaschke product. We note that for Blaschke sequences A, B_A vanishes precisely on the A in $\mathbb{D}$, with appropriate multiplicities, depending on how many times a point is repeated in the sequence. Moreover, the function B_A has boundary values of modulus 1 almost everywhere, provided that the limits are taken in nontangential approach regions. We also note that if the sequence A fails to be Blaschke, the product B_A collapses to 0. Given a finite positive Borel measure μ on the unit circle $\mathbb{T}$, which is singular to arc length Lebesgue measure, one associates a singular inner function

$$S_\mu(z) = \exp\left(-\int_{\mathbb{T}} \frac{\zeta + z}{\zeta - z} \, d\mu(\zeta) \right) \qquad \text{for } z \in \mathbb{D},$$

which is in H^∞, and has norm 1 there. Also, S_μ has no zeros in $\mathbb{D}$, and its nontangential boundary values are almost all 1 in modulus. This is the general criterion for a function in H^∞ to be inner: to have boundary values of modulus 1 almost everywhere. A product of a unimodular constant, a Blaschke product, and a singular inner function, is still inner, and all inner functions are obtained this way. If h is a real-valued L^1 function on $\mathbb{T}$, the associated outer function is

$$O_h(z) = \exp\left(\int_{\mathbb{T}} \frac{\zeta + z}{\zeta - z} \, h(\zeta) \, ds(\zeta) \right) \qquad \text{for } z \in \mathbb{D},$$

which is an analytic function in $\mathbb{D}$ with $|O_h(z)| = \exp\big(h(z)\big)$ almost everywhere on the circle, the boundary values of O_h being thought of in the non-tangential sense. The function O_h is in H^2 if and only if $\exp(h)$ is in $L^2(\mathbb{T})$. The factorization theorem in H^2 then states that every nonidentically vanishing f in H^2 has the form

$$f(z) = \gamma B_A(z) S_\mu(z) O_h(z) \qquad \text{for } z \in \mathbb{D},$$

where γ is a unimodular constant, and $\exp(h) \in L^2(\mathbb{T})$. The natural setting for the factorization theory is a larger class of functions, known as the Nevanlinna class. To make a long story short it consists of all functions of the above type, where no additional requirement is made on h, and where the singular measure μ is allowed to take negative values as well. We denote the Nevanlinna class by N. It is well known that $f \in N$ if and only if the function f is holomorphic in $\mathbb{D}$, and

$$\sup_{0<r<1} \int_{\mathbb{T}} \log^+ |f(rz)|\, ds(z) < +\infty.$$

The Bergman space case: inner functions. The Bergman space L_a^2 contains H^2. How then does it relate to N? It turns out that there are functions in N that are not in L_a^2, and that there are functions in L_a^2 which are not in N. The latter statement follows from the fact alluded to above that there is a function in L_a^2 lacking nontangential boundary values altogether. The functions in N, on the other hand, all do have finite nontangential boundary values almost everywhere. The former statement follows from a much simpler example: take μ equal to a point mass at, say 1, and consider the function $1/S_\mu$. It is in the Nevanlinna class, but it is much too big near 1 to be in L_a^2.

The classical Nevanlinna factorization theory is ill-suited for the Bergman space. This is particularly apparent from the fact that there are zero sequences for L_a^2 that are not Blaschke. The question is which functions can replace the Blaschke products or more general inner functions in the Bergman space setting. There may be several ways to do this, but only one is canonical from the point of view of operator theory.

A subspace M of H^2 is invariant if it is closed and $zM \subset M$. It is well known that inner functions in H^2 are characterized as elements of unit norm in some $M \ominus zM$, where M is a nonzero invariant subspace. Following Halmos, we call $M \ominus zM$ the wandering subspace for M. For a collection L of functions in H^2, we let $[L]$ stand for the smallest invariant subspace containing L. We note that $u \in H^2$ is an inner function if and only if

$$h(0) = \int_{\mathbb{T}} h(z) |u(z)|^2\, ds(z) \qquad \text{for } h \in L_h^\infty(\mathbb{D}), \tag{1--2}$$

$L_h^\infty(\mathbb{D})$ being the Banach space of bounded harmonic functions on $\mathbb{D}$.

We take (1--2) as the starting point in our search for analogues of inner functions for the Bergman space L_a^2. We say that a function $G \in L_a^2$ is L_a^2-inner provided that

$$h(0) = \int_{\mathbb{D}} h(z) \, |G(z)|^2 \, dS(z) \qquad \text{for } h \in L_h^\infty(\mathbb{D}). \tag{1-3}$$

A function G of unit norm in L_a^2 is L_a^2-inner if and only if it is in a wandering subspace $M \ominus zM$ for some nonzero invariant subspace M of L_a^2. In contrast with the H^2 case, where $M \ominus zM$ always has dimension 1 (unless M is the zero subspace), this time the dimension may take any value in the range $1, 2, 3, \ldots, +\infty$. This follows from the dilation theory developed by Apostol, Bercovici, Foiaş, and Pearcy [Apostol et al. 1985]. The dimension of $M \ominus zM$ will be referred to as the index of the invariant subspace M.

For the space H^2, Beurling's invariant subspace theorem yields a complete description:

THEOREM 1.1 [Beurling 1949]. *Let M be an invariant subspace of H^2, and let $M \ominus zM$ be the associated wandering subspace. Then $M = [M \ominus zM]$. If M is not the zero subspace, then $M \ominus zM$ is one-dimensional and spanned by an inner function, call it φ. It follows that $M = [\varphi] = \varphi H^2$.*

A natural question is whether the analogous statement $M = [M \ominus zM]$ (with the brackets referrring to the invariant subspace lattice of L_a^2 this time) holds for general invariant subspaces M of L_a^2. Pleasantly, and perhaps surprisingly, this turns out to be true [Aleman et al. 1996]. We shall return to this matter in Section 3.

2. Factorization of Zeros in the Bergman Space

The treatment of the subject matter of this section is taken from [Hedenmalm 1991; 1994a; Duren et al. 1993; 1994]. It should be mentioned that the first results on factorization in Bergman spaces were obtained by Horowitz [1974], and slightly later, but independently, by Korenblum [1975].

An example. We begin with a simple but illuminating example: a multiple zero at the origin of multiplicity n. Let M_n be the subspace of L_a^2 of all such functions, which is clearly invariant. The associated wandering subspace $M_n \ominus zM_n$ is one-dimensional, and spanned by the unit vector $G_n(z) = \sqrt{n+1} \, z^n$. According to the terminology introduced in the previous section, the function G_n is an L_a^2-inner function. Since it comes from a zero-based invariant subspace, it is a Bergman space analog of a (finite) Blaschke product. Let f be an arbitrary element of M_n, which then has a Taylor expansion $f(z) = \sum_{j=n}^\infty a_j z^j$. It can be factored $f = G_n g$, where $g(z) = (n+1)^{-1/2} \sum_{j=n}^\infty a_j z^{j-n}$. Let us compare the norms of f and g,

$$\|g\|_{L_a^2}^2 = \frac{1}{n+1} \sum_{j=n}^\infty \frac{|a_j|^2}{j-n+1} \leq \sum_{j=n}^\infty \frac{|a_j|^2}{j+1} = \|f\|_{L_a^2}^2;$$

here, we used that

$$j + 1 \leq (n+1)(j - n + 1) \qquad \text{for } j = n,\, n+1,\, \ldots.$$

Since $g = f/G_n$, we see that division by the unit element G_n of the wandering subspace $M_n \ominus zM_n$ is contractive on M_n; in other words, multiplication by G_n is norm expansive on L_a^2.

General zero sets. Now let A be a zero sequence for the space L_a^2, counting multiplicities, and let M_A be the subspace of all functions in L_a^2 that vanish on A, with at least the given multiplicity at each point. It is an invariant subspace, and its wandering subspace $M_A \ominus zM_A$ is one-dimensional. Let G_A denote a unit element of the wandering subspace. Let j be the multiplicity of the origin in the sequence A (which is 0 if the origin is not in A). By multiplying G_A by an appropriate unimodular constant, we may suppose that it solves the extremal problem

$$\sup \left\{ \operatorname{Re} G^{(j)}(0) : \ G \in M_A,\ \|G\|_{L_a^2} = 1 \right\}. \tag{2--1}$$

The above example with a multiple zero at the origin suggests that the function G_A may be a contractive divisor on M_A. This turns out to be the case. To begin with, we must rule out the possibility that the function G_A may have extraneous zeros.

More general invariant subspaces. We consider a more general situation with an invariant subspace M of index 1, and let G_M be a unit element of the one-dimensional wandering subspace $M \ominus zM$. We shall show that for $f \in H^2$, $G_M f$ is in L_a^2, and that

$$\|f\|_{L_a^2} \leq \|G_M f\|_{L_a^2} \leq \|f\|_{H^2}. \tag{2--2}$$

One of the inequalities states that multiplication by G_M expands the L_a^2 norm of functions in H^2; this is what entails, after some work, that division by G_M is well-defined and norm contractive $M \to L_a^2$. As in the above case $M = M_A$, we may assume, by multiplying G_M by an appropriate unimodular constant, that it solves the extremal problem

$$\sup \left\{ \operatorname{Re} G^{(j)}(0) : \ G \in M,\ \|G\|_{L_a^2} = 1 \right\}, \tag{2--3}$$

where j is the multiplicity of the common zero at the origin of all the functions in M. For this reason, we shall refer to G_M as the extremal function for M. Since G_M is an L_a^2-inner function,

$$h(0) = \int_{\mathbb{D}} h(z)\, |G_M(z)|^2 \, dS(z) \qquad \text{for } h \in L_h^\infty(\mathbb{D}),$$

and so

$$\int_{\mathbb{D}} h(z)\big(|G_M(z)|^2 - 1\big)\, dS(z) = 0 \qquad \text{for } h \in L_h^\infty(\mathbb{D}). \tag{2--4}$$

Equation (2–4) has the interpretation that $|G_M|^2 - 1$ annihilates the bounded harmonic functions on $\mathbb{D}$.

Potential theory. Consider the function Φ_M that solves the boundary value problem

$$\begin{cases} \Delta\Phi_M = |G_M|^2 - 1 & \text{on } \mathbb{D}, \\ \Phi_M = 0 & \text{on } \mathbb{T}, \end{cases} \qquad (2\text{–}5)$$

where Δ is one quarter of the usual Laplacian (this is not so important, really, one can use the standard Laplacian, only later would we then have to use slightly different Green functions). If we play around with Green's formula, in the form

$$\int_{\mathbb{D}} (v\Delta u - u\Delta v) \, dS = \int_{\mathbb{T}} \left(v\frac{\partial u}{\partial n} - u\frac{\partial v}{\partial n} \right) \tfrac{1}{2} \, ds, \qquad (2\text{–}6)$$

where the normal derivatives are taken in the outward direction, and forget about regularity requirements, then (2–4) can be reformulated as saying that the normal derivative of Φ_M vanishes on $\mathbb{T}$. If we add this condition to (2–5), this system becomes overdetermined. Elliptic equations of order $2m$ are determined by m boundary data, so we may raise the order of the partial differential equation to 4 and keep a unique solution. This is accomplished by applying a Laplacian to both sides, and we get, in view of $\Delta(|G_M|^2 - 1) = |G'_M|^2$, that

$$\begin{cases} \Delta^2\Phi_M = |G'_M|^2 & \text{on } \mathbb{D}, \\ \Phi_M = 0 & \text{on } \mathbb{T}, \\ \dfrac{\partial}{\partial n}\Phi_M = 0 & \text{on } \mathbb{T}. \end{cases} \qquad (2\text{–}7)$$

The Green function for Δ^2 is the function $U(z, \zeta)$ on $\mathbb{D} \times \mathbb{D}$ that solves for fixed $\zeta \in \mathbb{D}$

$$\begin{cases} \Delta^2 U(\,\cdot\,, \zeta) = \delta_\zeta & \text{on } \mathbb{D}, \\ U(\,\cdot\,, \zeta) = 0 & \text{on } \mathbb{T}, \\ \frac{\partial}{\partial n} U(\,\cdot\,, \zeta) = 0 & \text{on } \mathbb{T}, \end{cases}$$

and it is given explicitly as

$$U(z, \zeta) = |z - \zeta|^2 \Gamma(z, \zeta) + (1 - |z|^2)(1 - |\zeta|^2),$$

where

$$\Gamma(z, \zeta) = 2\log\left|\frac{z - \zeta}{1 - \bar{\zeta}z}\right|$$

is the Green function for Δ. These are the expressions obtained when it is agreed to identify locally integrable functions φ with the corresponding measures $\varphi \, dS$ (recall that dS involved some normalization) to interpret the functions as distributions.

By now it should not require too much of a leap of faith to believe that

$$\Phi_M(z) = \int_{\mathbb{D}} U(z, \zeta) \, |G'_M(\zeta)|^2 \, dS(\zeta),$$

so that in view of the fact that $0 < U(z, \zeta)$ on $\mathbb{D} \times \mathbb{D}$, $0 < \Phi_M$ on $\mathbb{D}$, unless G_M is constant, in which case $\Phi_M = 0$. If we apply Green's formula (2–6) and recall the definition of Φ_M, it follows that

$$\int_{\mathbb{D}} \left(|G_M(z)|^2 - 1 \right) |f(z)|^2 \, dS(z) = \int_{\mathbb{D}} \Phi_M(z) \, |f'(z)|^2 \, dS(z),$$

for, say, polynomials f. We rewrite this as

$$\int_{\mathbb{D}} |G_M(z)f(z)|^2 \, dS(z)$$
$$= \int_D |f(z)|^2 \, dS(z) + \int_{\mathbb{D}} \Phi_M(z) \, |f'(z)|^2 \, dS(z)$$
$$= \int_D |f(z)|^2 \, dS(z) + \int_{\mathbb{D}} \int_{\mathbb{D}} U(z, \zeta) \, |f'(z)|^2 |G'_M(\zeta)|^2 \, dS(z) \, dS(\zeta). \quad (2\text{–}8)$$

Let Ψ solve

$$\begin{cases} \Delta \Psi = -1 & \text{on } \mathbb{D}, \\ \Psi = 0 & \text{on } \mathbb{T}; \end{cases}$$

the solution comes out to be $\Psi(z) = 1 - |z|^2$. The function $\Phi_M - \Psi$ is subharmonic, and has 0 boundary values, so inside $\mathbb{D}$ it must be ≤ 0. In view of what we have already shown, it follows that $0 \leq \Phi_M(z) \leq \Psi(z) = 1 - |z|^2$. It is well known that

$$\|f\|_{H^2}^2 = \|f\|_{L_a^2}^2 + \int_{\mathbb{D}} (1 - |z|^2) \, |f'(z)|^2 \, dS(z) \qquad \text{for } f \in H^2,$$

so that by continuity, identity (2–8) extends to all $f \in H^2$, and (2–2) holds. Let $\mathcal{A}(G_M)$ be the space of all functions $f \in L_a^2$ with

$$\|f\|_{\mathcal{A}(G_M)}^2 = \|f\|_{L_a^2}^2 + \int_{\mathbb{D}} \int_{\mathbb{D}} U(z, \zeta) \, |f'(z)|^2 |G'_M(\zeta)|^2 \, dS(z) \, dS(\zeta) < +\infty,$$

and let $\mathcal{A}_0(G_M)$ be the closure of the polynomials in $\mathcal{A}(G_M)$. Then multiplication by G_M is an isometry $\mathcal{A}_0(G_M) \to M \subset L_a^2$, and $H^2 \subset \mathcal{A}_0(G_M) \subset \mathcal{A}(G_M) \subset L_a^2$. Moreover, the injection mappings $H^2 \to \mathcal{A}_0(G_M)$ and $\mathcal{A}(G_M) \to L_a^2$ are contractions. It follows that the invariant subspace generated by G_M, $[G_M]$, equals $G_M \mathcal{A}_0(G_M)$. A number of questions appear:

- Is $\mathcal{A}_0(G_M) = \mathcal{A}(G_M)$?
- Is $[G_M] = M$?
- Is $\mathcal{A}(G_M) = \{f \in L_a^2 : G_M f \in L_a^2\}$?

The answer to the first two questions is yes [Aleman et al. 1996] (see Section 3). The answer to the third question is no [Borichev and Hedenmalm 1995; 1997].

Extraneous zeros. Let λ be a point of $\mathbb{D}$, and let G_λ be the extremal function associated with M_λ, the invariant subspace of all functions vanishing at λ. In terms of the Bergman kernel function $k(z, \zeta) = (1 - \bar{\zeta}z)^{-2}$, it has the form

$$G_\lambda(z) = \left(1 - \frac{1}{k(\lambda, \lambda)}\right)^{-\frac{1}{2}} \left(1 - \frac{k(z, \lambda)}{k(\lambda, \lambda)}\right),$$

and one quickly verifies that on $\mathbb{T}$, it has modulus bigger than 1, and in $\mathbb{D}$, it has a simple zero at λ, and nowhere else. This means that if $f \in L_a^2$ vanishes at λ, then $f/G_\lambda \in L_a^2$, and since multiplication by G_λ is norm expansive on L_a^2 (see (2–2) and (2–8) for $M = M_\lambda$), $\|f/G_\lambda\|_{L_a^2} \leq \|f\|_{L_a^2}$. Now suppose that G_M has an extraneous zero at λ by which we mean that G_M vanishes at λ with a multiplicity higher than that of some element of M. By inspection of the extremal problem (2–3), to which G_M is the unique solution, we see that λ cannot be 0. If we divide G_M by G_λ, we get an element of L_a^2. If G_M/G_λ is in fact in M, then the function $\widetilde{G} = \gamma G_M/G_\lambda$, with $\gamma = \|G_M/G_\lambda\|_{L_a^2}^{-1}$, is a competing function with G_M in the extremal problem (2–3). It has norm 1, belongs to M, and the j-th derivative at 0 is $\widetilde{G}^{(j)}(0) = \gamma G_M^{(j)}(0)/G_\lambda(0)$, which is bigger than $G_M^{(j)}(0)$, as $1 < \gamma$ and $G_\lambda(0) < 1$. Hence $\widetilde{G}$ is more extremal than G_M itself, which leads to a contradiction. So, the assumption that G_M had an extraneous zero must be false. The weak point thus far is that we have not explained why G_M/G_λ was in M in the first place. Recall that $M \ominus zM$ was one-dimensional: from a perturbation argument it follows that $M \ominus (z - \lambda)M$ is one-dimensional for each $\lambda \in \mathbb{D}$. The subspace $(z - \lambda)M$ having codimension one in M means that it consists of all functions in M having an extra zero (or a zero of multiplicity one higher) at λ, so that G_M, having an extraneous zero at λ, must be in $(z - \lambda)M$. The conclusion that G_M/G_λ is in M follows.

Factorization of zeros. Let us see what kinds of conclusions we can draw from the above. For a finite zero sequence A, G_A has no extraneous zeros in $\mathbb{D}$, extends analytically across $\mathbb{T}$ to a rational function with poles at the reflected points in $\mathbb{T}$ of A, and the expansive multiplier property (2–2) implies that $1 \leq |G_A|$ on $\mathbb{T}$. One shows that $\mathcal{A}_0(G_A) = L_a^2$ (the norms are different, though equivalent), so that G_A is an expansive multiplier on all of L_a^2. It follows that

$$\|f\|_{L_a^2}^2 = \|f/G_A\|_{L_a^2}^2 + \int_{\mathbb{D}} \int_D U(z, \zeta) |(f/G_A)'(z)|^2 |G_A'(\zeta)|^2 \, dS(z) \, dS(\zeta), \quad f \in M_A,$$

and if we go to the limit as A approaches an infinite zero sequence, Fatou's lemma yields a $\geq$ in place of the $=$ sign. In particular, $f/G_A \in \mathcal{A}(G_A) \subset L_a^2$, and division by G_A is norm contractive $M_A \to L_a^2$.

3. General Invariant Subspaces

Most of the results mentioned here are from [Aleman et al. 1996].

The following version of (2–8) will prove useful:

$$\|f\|_{L_a^2}^2 = \|f/G_M\|_{L_a^2}^2 + \int_{\mathbb{D}} \int_{\mathbb{D}} U(z,\zeta) |(f/G_M)'(z)|^2 |G_M'(\zeta)|^2 \, dS(z) \, dS(\zeta)$$

$$\text{for } f \in [G_M]. \quad (3\text{–}1)$$

A skewed projection operator. Let M be an invariant subspace in L^2, and suppose for the moment that it has index 1; G_M denotes the associated extremal function. For $\lambda \in \mathbb{D}$,

$$f = \frac{f(\lambda)}{G_M(\lambda)} G_M + \left(f - \frac{f(\lambda)}{G_M(\lambda)} G_M \right) \qquad \text{for } f \in M$$

offers a unique decomposition of M as a sum, $M = (M \ominus zM) + (z - \lambda)M$, and as the two summands are got by bounded (skewed) projection operators, the subspaces $M \ominus zM$ and $(z - \lambda)M$ are at a positive angle. Note that the first projection,

$$Q_\lambda f = \frac{f(\lambda)}{G_M(\lambda)} G_M \qquad \text{for } f \in M,$$

is well-defined for all $\lambda \in \mathbb{D}$ as we know that G_M has no extraneous zeros. In terms of Q_λ, identity (3–1) can be written as

$$\|f\|_{L_a^2}^2 = \int_{\mathbb{D}} \|Q_\lambda f\|_{L_a^2}^2 \, dS(\lambda) + \int_{\mathbb{D}} \int_{\mathbb{D}} U(z,\zeta) \Delta_z \Delta_\zeta |Q_z f(\zeta)|^2 \, dS(z) \, dS(\zeta)$$

$$\text{for } f \in [G_M]. \quad (3\text{–}2)$$

This form lends itself to generalization to general invariant subspaces M, not necessarily of index 1. Namely, one shows that a skewed decomposition of the type $M = (M \ominus zM) + (z - \lambda)M$ holds in general, so that a corresponding projection $Q_\lambda : M \to M \ominus zM$ can be defined, and it depends analytically on $\lambda \in \mathbb{D}$. Moreover, (3–2) carries over, almost letter by letter:

$$\|f\|_{L_a^2}^2 = \int_{\mathbb{D}} \|Q_\lambda f\|_{L_a^2}^2 \, dS(\lambda) + \int_{\mathbb{D}} \int_{\mathbb{D}} U(z,\zeta) \Delta_z \Delta_\zeta |Q_z f(\zeta)|^2 \, dS(z) \, dS(\zeta)$$

$$\text{for } f \in [M \ominus zM]. \quad (3\text{–}3)$$

Abel summation. Define the bounded linear operator $L : M \to M$ by declaring that $Lf = f/z$ for $f \in zM$, and $Lf = 0$ for $f \in M \ominus zM$. Also, let P be the orthogonal projection $M \to M \ominus zM$. If f is in M, we can decompose f as a sum of an element of $M \ominus zM$ and a "remainder term" by the formula $f = Pf + zLf$. Repeating this for Lf, we obtain $f = Pf + zPLf + z^2 L^2 f$. Continuing this process, we get the formal series

$$f = Pf + zPLf + z^2 PL^2 f + z^3 PL^3 f + \cdots,$$

each term of which is in $[M \ominus zM]$. If the series were to converge to f for each given $f \in M$, the assertion $M = [M \ominus zM]$ would be immediate. However, this is probably not the case in general, so we choose the second best thing: we form the Abel series

$$R_t f = Pf + tzPLf + t^2 z^2 PL^2 f + t^3 z^3 PL^3 f + \cdots \qquad \text{for } 0 \leq t < 1,$$

which does converge to an element of $[M \ominus zM]$, in the hope that $R_t f \to f$ as $t \to 1$. The skewed projection Q_λ has a similar series expansion,

$$Q_\lambda f = Pf + \lambda PLf + \lambda^2 PL^2 f + \lambda^3 PL^3 f + \cdots \qquad \text{for } \lambda \in \mathbb{D},$$

which one can use to show that $Q_\lambda R_t = Q_{t\lambda}$. The operator R_t may also be thought of as given by $R_t f(z) = Q_{tz} f(z)$. As $t \to 1$, $Q_{tz} f(z) \to Q_z f(z)$, and $Q_z f(z) = f(z)$, because by the definition of Q_λ, $f(z) - Q_\lambda f(z)$ is zero when $z = \lambda$. If follows that $R_t f(z) \to f(z)$ as $t \to 1$ pointwise in $\mathbb{D}$. It remains to show that the convergence holds in norm, too.

Controlling the norm of the Abel sum. General functional analysis arguments show that we do not really need to show that $R_t f$ tends to f in norm, weak convergence would suffice. Moreover, weak convergence would follow if we only had a uniform bound of the norm of $R_t f$ as $t \to 1$. This is the crux of the problem. By (3–3) and the identity $Q_\lambda R_t = Q_{t\lambda}$,

$$\|R_t f\|_{L_a^2}^2 = \int_{\mathbb{D}} \|Q_{t\lambda} f\|_{L_a^2}^2 \, dS(\lambda) + \int_{\mathbb{D}} \int_{\mathbb{D}} U(z,\zeta) \Delta_z \Delta_\zeta |Q_{tz} f(\zeta)|^2 \, dS(z) \, dS(\zeta)$$

$$\text{for } f \in M. \quad (3\text{--}4)$$

Certain regularity properties of the biharmonic Green function $U(z,\zeta)$ can be used to show that, as $t \to 1$, the right hand side of (3–4) tends to

$$\int_{\mathbb{D}} \|Q_\lambda f\|_{L_a^2}^2 \, dS(\lambda) + \int_{\mathbb{D}} \int_{\mathbb{D}} U(z,\zeta) \Delta_z \Delta_\zeta |Q_{tz} f(\zeta)|^2 \, dS(z) \, dS(\zeta), \qquad (3\text{--}5)$$

so if we can only bound this expression, we are done. A bound that works is $\|f\|_{L_a^2}^2$. The approach in [Aleman et al. 1996] is based on the identity

$$\|f\|_{L_a^2}^2 = \int_{\mathbb{T}} \|Q_{r\lambda} f\|_{L_a^2}^2 \, ds(\lambda) + \int_{\mathbb{D}} \int_{\mathbb{T}} (|z|^2 - r^2) \left| \frac{f(z) - Q_{r\lambda} f(z)}{z - r\lambda} \right|^2 \, ds(\lambda) \, dS(z),$$

for $0 < r < 1$. By cleverly applying Green's theorem to the above right hand side expression and obtaining estimates of "remainder terms" as $r \to 1$, Aleman, Richter, and Sundberg were able to show that the expression (3–5) is no bigger than $\|f\|_{L_a^2}^2$, whence the following analogue of Beurling's theorem follows.

THEOREM 3.1. *Let M be an invariant subspace of L_a^2. Then $M = [M \ominus zM]$.*

Invertibility and cyclicity. A function $f \in L_a^2$ is said to be cyclic if $[f] = L_a^2$. It has been a long standing problem whether there are noncyclic functions $f \in L_a^2$ that are invertible, that is, $1/f \in L_a^2$. A complicated construction of such functions was recently found by Borichev and Hedenmalm [1995; 1997]. The idea is that if the given function f grows maximally fast on a "big" set, then an H^∞ function must be small there and hence cannot lift the small values of f as required by cyclicity. This example has certain consequences for the uniqueness of inner-outer factorization in L_a^2 [Aleman et al. 1996]. It is not clear whether the invariant subspace associated to the constructed invertible noncyclic function is in the closure of the collection of zero-based invariant subspaces, with respect to any of the topologies suggested by Korenblum [1993].

4. Zero Sequences

The first results on zero sequences for Bergman space functions were obtained by Horowitz [1974; 1977]. For instance, he showed that the union of two zero sequences need not be a zero sequence. If we consider the corresponding zero-based invariant subspaces, call them M_A and M_B, then $M_A \cap M_B = \{0\}$. Actually, this behavior of index one invariant subspaces is the raison d'être for the invariant subspaces of higher index [Richter 1987]. Also, look at the explicit constructions in Section 6, and [Hedenmalm 1993; Hedenmalm et al. 1996b].

The sharpest results so far were obtained by Seip [1994; 1995]. The tools were borrowed from the fundamental work of Korenblum [1975; 1977] on the topological algebra $A^{-\infty}$, which consists of all functions f holomorphic in $\mathbb{D}$ that meet the growth condition $|f(z)| \leq C(1 - |z|)^{-N}$ for some positive constants C and N.

For a finite subset F of $\mathbb{T}$, let $\mathbb{T} \setminus F = \cup_k I_k$ be the complementary arcs, and consider the Beurling-Carleson entropy of F,

$$\widehat{\varkappa}(F) = \sum_k \frac{|I_k|}{2\pi} \left(\log \frac{2\pi}{|I_k|} + 1 \right).$$

Let $d(\,\cdot\,,\,\cdot\,)$ be the curvilinear metric $d(e^{it}, e^{is}) = \pi^{-1}|t - s|$ on $\mathbb{T}$, where it is assumed that $|t - s| \leq \pi$. The Korenblum star associated with the finite set F is

$$G(F) = \left\{ z \in \bar{\mathbb{D}} \setminus \{0\} : d(z/|z|, F) \leq 1 - |z| \right\} \cup \{0\},$$

and for a sequence A of points in $\mathbb{D}$, let

$$\sigma(A, F) = \sum_{z \in A \cap G(F)} \log \frac{1}{|z|}$$

be the local "Blaschke sum". The Korenblum density $\delta(A)$ of the sequence A is the infimum over all β such that

$$\sup_F \left(\sigma(A, F) - \beta \widehat{\varkappa}(F) \right) < +\infty,$$

the supremum being taken over all finite subsets of $\mathbb{T}$.

THEOREM 4.1 ([SEIP 1994]). *Let A be a sequence of points in $\mathbb{D} \setminus \{0\}$. If A is the zero sequence of some function in L_a^2, then $\delta(A) \leq \frac{1}{2}$. On the other hand, if $\delta(A) < \frac{1}{2}$, then A is the zero sequence of some L_a^2 function.*

5. Interpolating and Sampling Sequences

A sequence $A = \{a_j\}_j$ of distinct points in $\mathbb{D}$ is said to be a sampling sequence for L_a^2 if

$$\sum_j (1 - |a_j|^2)^2 |f(a_j)|^2 \asymp \|f\|_{L^2}^2 \qquad \text{for } f \in L_a^2,$$

where the $\asymp$ sign means that the left hand side is bounded from above and below by positive constant multiples of the right hand side. The reason why the factor $(1 - |a_j|^2)^2$ is needed is that in a more general setting, one should use the reciprocal of the reproducing kernel, $k(a_j, a_j)^{-1}$. Similarly, A is interpolating for L_a^2 provided that to every l^2 sequence $\{w_j\}_j$, there exists a function $f \in L_a^2$ such that

$$(1 - |a_j|^2)f(a_j) = w_j \quad \text{for all } j.$$

Generally, sampling sequences are fat, and interpolating sequences are thin. Clearly, a sampling sequence cannot be a zero sequence for L_a^2. However, every interpolating sequence for L_a^2 is also a zero sequence, for the following reason. Take an interpolant for the sequence $w_1 = 1$ and $w_j = 0$ for all other j, and multiply this function by $z - a_1$ to get a nonidentically vanishing function that vanishes on the sequence A. This actually only shows that A must be a subsequence of an L_a^2 zero sequence, but it is well known that every subsequence of a zero sequence for L_a^2 is itself a zero sequence for L_a^2 [Horowitz 1974; Hedenmalm 1991]. Seip [1993] has obtained a complete description of the sampling and interpolating sequences for L_a^2. We shall try to describe the result, but in order to do so, we need some notation.

A sequence $A = \{a_j\}_j$ of points in $\mathbb{D}$ is said to be uniformly discrete if for some $\varepsilon > 0$,

$$\varepsilon \leq \left| \frac{a_j - a_k}{1 - \bar{a}_k a_j} \right| \qquad \text{for } j \neq k.$$

For $\lambda \in \mathbb{D}$, let A_λ be the image of A under the conformal automorphism of the unit disk

$$\varphi_\lambda(z) = \frac{z - \lambda}{1 - \bar{\lambda}z} \qquad \text{for } z \in \mathbb{D}.$$

Associate with A_λ the function $n(r, A_\lambda)$, which counts the number of points of A_λ contained within the disk $r\mathbb{D}$ $(0 < r < 1)$. We shall need the definite integral

$$N(r, A_\lambda) = \int_0^r n(t, A_\lambda)\, dt \qquad \text{for } 0 < r < 1.$$

If $A(r)$ stands for the function

$$A(r) = \log \frac{1+r}{1-r} \qquad \text{for } 0 < r < 1,$$

Seip defines the upper density of A as

$$D^+(A) = \limsup_{r \to 1} \sup_{\lambda \in \mathbb{D}} \frac{N(r, A_\lambda)}{A(r)},$$

and the lower density as

$$D^-(A) = \liminf_{r \to 1} \inf_{\lambda \in \mathbb{D}} \frac{N(r, A_\lambda)}{A(r)}.$$

For the standard Bergman space, his result is as follows.

THEOREM 5.1. *A sequence A of distinct points in $\mathbb{D}$ is sampling for L_a^2 if and only if it can be expressed as a finite union of uniformly discrete sets and it contains a uniformly discrete subsequence A' for which $D^-(A') > \frac{1}{2}$.*

THEOREM 5.2. *A sequence A of distinct points in $\mathbb{D}$ is interpolating for L_a^2 if and only if it is uniformly discrete and $D^+(A') < \frac{1}{2}$.*

6. Invariant Subspaces of Index Two or Higher

Back in Section 1 it was mentioned that an invariant subspace M of the Bergman space L_a^2 may have index $1, 2, 3, \ldots$, whereas the most obvious examples have index 1. For instance, every zero based invariant subspace M_A has index 1, and so does every singly generated one. Invariant subspaces based on singular masses at the boundary have index 1, too [Hedenmalm et al. 1996a]. It is therefore a natural question to ask what these invariant subspaces of higher index look like. A simple constuction was found in [Hedenmalm 1993].

Let A and B be two disjoint zero sequences for L_a^2, and let $C = A \cup B$. Let $M = M_A \vee M_B$, the smallest invariant subspace containing both M_A and M_B; it is obtained as the closure of the sum $M_A + M_B$. It turns out that in this situation, either $M = L_a^2$ or M has index 2. Moreover, what determines which of these alternatives occurs is the fatness of the sequence C. If one of the sequences A and B fails to accumulate on an arc of $\mathbb{T}$, then C is not fat enough, and so $M = L_a^2$. On the other hand, if C is sampling, then M has index 2, because M_A and M_B are at a positive angle. To see this, let $C = \{c_j\}_j$, and use the sampling property

$$\|f\|_{L_a^2}^2 \asymp \sum_j (1 - |c_j|^2)^2 |f(c_j)|^2 \qquad \text{for } f \in L_a^2.$$

Let $f \in M_A$ and $g \in M_B$ be arbitrary. Then for every point c_j of C, we have

$$|f(c_j) + g(c_j)|^2 = |f(c_j)|^2 + |g(c_j)|^2,$$

so that

$$\|f + g\|_{L_a^2}^2 \asymp \sum_j (1 - |c_j|^2)^2 |f(c_j) + g(c_j)|^2$$

$$= \sum_j (1 - |c_j|^2)^2 (|f(c_j)|^2 + |g(c_j)|^2) \asymp \|f\|_{L_a^2}^2 + \|g\|_{L_a^2}^2.$$

Consequently, the sum $M_A + M_B$ is closed, and it is easy to show that M has index two: $zM = zM_A + zM_B$ [Hedenmalm 1993].

7. Green Functions for Weights and Factorization

The results discussed in this section are mostly from [Hedenmalm 1996].

It is clear from Section 2 that Green functions for certain elliptic operators of order 4 play an important role in the study of factorization in Bergman spaces. Let ω be a nonnegative sufficiently smooth function in $\mathbb{D}$, and let $L_a^2(\omega)$ be the corresponding weighted Bergman space of all holomorphic functions f in $\mathbb{D}$ with

$$\|f\|_{L_a^2(\omega)}^2 = \int_{\mathbb{D}} |f(z)|^2 \omega(z) \, dS(z) < +\infty.$$

It is a Hilbert space if ω is not equal to 0 too frequently near $\mathbb{T}$; this is not so precise, but it is enough for here. For the extremal functions to have a chance to be good divisors, we need to ask of ω that

$$h(0) = \int_{\mathbb{D}} h(z)\omega(z) \, dS(z) \qquad \text{for } h \in L_h^\infty(\mathbb{D}).$$

The relevant Green function is that of the operator $\Delta \omega^{-1} \Delta$ on $\mathbb{D}$, and issue is whether it is positive (or at least nonnegative). It was shown in [Hedenmalm 1994b] that it is positive for the weights $\omega_\alpha(z) = (\alpha + 1)|z|^{2\alpha}$, with $\alpha > -1$, and the issue at hand is whether this information can be used to tell us anything about weights that are convex combinations of these. For instance, is it true in general that if ω and μ are two weights, with associated Green functions U_ω and U_μ, that we have, with $\omega[t] = (1 - t)\omega + t\mu$,

$$(1-t)U_\omega(z,\zeta) + tU_\mu(z,\zeta) \le U_{\omega[t]}(z,\zeta) \quad \text{for } (z,\zeta) \in \mathbb{D} \times \mathbb{D} \text{ and } 0 < t < 1? \quad (7\text{--}1)$$

This is probably not so, although I cannot supply an immediate counterexample. However, with some additional information given in terms of the Green functions, it is true. If we apply a Laplacian to $U_\omega(z,\zeta)$, we get

$$\Delta_z U_\omega(z,\zeta) = \omega(z) \big(\Gamma(z,\zeta) + H_\omega(z,\zeta) \big),$$

where $H_\omega(z,\zeta)$ is harmonic in z. We call H_ω the harmonic compensator. If $H_\mu(z,\zeta) \le H_\omega(z,\zeta)$ holds pointwise, then (7–1) holds, and $H_\mu(z,\zeta) \le H_{\omega_t}(z,\zeta)$ also holds pointwise. A consequence of this result is that if

$$\omega(z) = \int_{]-1,+\infty[} (\alpha + 1)|z|^{2\alpha} d\rho(\alpha),$$

where ρ is a probability measure, then

$$0 \leq \int_{]-1,+\infty[} U_{\omega_\alpha}(z,\zeta)\,d\rho(\alpha) \leq U_\omega(z,\zeta) \qquad \text{for } (z,\zeta) \in \mathbb{D} \times \mathbb{D}.$$

For related work see, for instance, [Shimorin 1993; 1995].

References

[Aleman et al. 1996] A. Aleman, S. Richter, and C. Sundberg, "Beurling's theorem for the Bergman space", *Acta Math.* **177**:2 (1996), 275–310.

[Apostol et al. 1985] C. Apostol, H. Bercovici, C. Foiaş, and C. Pearcy, "Invariant subspaces, dilation theory, and the structure of the predual of a dual algebra, I", *J. Funct. Anal.* **63**:3 (1985), 369–404.

[Beurling 1949] A. Beurling, "On two problems concerning linear transformations in Hilbert space", *Acta Math.* **81** (1949), 239–255.

[Borichev and Hedenmalm 1995] A. A. Borichev and P. J. H. Hedenmalm, "Cyclicity in Bergman-type spaces", *Internat. Math. Res. Notices* **5** (1995), 253–262.

[Borichev and Hedenmalm 1997] A. Borichev and H. Hedenmalm, "Harmonic functions of maximal growth: invertibility and cyclicity in Bergman spaces", *J. Amer. Math. Soc.* **10**:4 (1997), 761–796.

[Cantor 1964] D. G. Cantor, "A simple construction of analytic functions without radial limits", *Proc. Amer. Math. Soc.* **15** (1964), 335–336.

[Duren et al. 1993] P. Duren, D. Khavinson, H. S. Shapiro, and C. Sundberg, "Contractive zero-divisors in Bergman spaces", *Pacific J. Math.* **157**:1 (1993), 37–56.

[Duren et al. 1994] P. Duren, D. Khavinson, H. S. Shapiro, and C. Sundberg, "Invariant subspaces in Bergman spaces and the biharmonic equation", *Michigan Math. J.* **41**:2 (1994), 247–259.

[Hedenmalm 1991] H. Hedenmalm, "A factorization theorem for square area-integrable analytic functions", *J. Reine Angew. Math.* **422** (1991), 45–68.

[Hedenmalm 1993] P. J. H. Hedenmalm, "An invariant subspace of the Bergman space having the codimension two property", *J. Reine Angew. Math.* **443** (1993), 1–9.

[Hedenmalm 1994a] H. Hedenmalm, "A factoring theorem for the Bergman space", *Bull. London Math. Soc.* **26**:2 (1994), 113–126.

[Hedenmalm 1994b] P. J. H. Hedenmalm, "A computation of Green functions for the weighted biharmonic operators $\Delta |z|^{-2\alpha}\Delta$, with $\alpha > -1$", *Duke Math. J.* **75**:1 (1994), 51–78.

[Hedenmalm 1996] P. J. H. Hedenmalm, "Boundary value problems for weighted biharmonic operators", *Algebra i Analiz* **8**:4 (1996), 173–192.

[Hedenmalm et al. 1996a] H. Hedenmalm, B. Korenblum, and K. Zhu, "Beurling type invariant subspaces of the Bergman spaces", *J. London Math. Soc.* (2) **53**:3 (1996), 601–614.

[Hedenmalm et al. 1996b] H. Hedenmalm, S. Richter, and K. Seip, "Interpolating sequences and invariant subspaces of given index in the Bergman spaces", *J. Reine Angew. Math.* **477** (1996), 13–30.

[Horowitz 1974] C. Horowitz, "Zeros of functions in the Bergman spaces", *Duke Math. J.* **41** (1974), 693–710.

[Horowitz 1977] C. Horowitz, "Factorization theorems for functions in the Bergman spaces", *Duke Math. J.* **44**:1 (1977), 201–213.

[Korenblum 1975] B. Korenblum, "An extension of the Nevanlinna theory", *Acta Math.* **135**:3-4 (1975), 187–219.

[Korenblum 1977] B. Korenblum, "A Beurling-type theorem", *Acta Math.* **138**:3-4 (1977), 265–293.

[Korenblum 1993] B. Korenblum, "Outer functions and cyclic elements in Bergman spaces", *J. Funct. Anal.* **115**:1 (1993), 104–118.

[Luzin and Privalov 1925] N. N. Luzin and I. I. Privalov, "Sur l'unicité et la multiplicité des fonctions analytiques", *Ann. Sci. Ecole Norm. Sup.* **42** (1925), 143–191.

[MacLane 1962] G. R. MacLane, "Holomorphic functions, of arbitrarily slow growth, without radial limits", *Michigan Math. J.* **9** (1962), 21–24.

[Richter 1987] S. Richter, "Invariant subspaces in Banach spaces of analytic functions", *Trans. Amer. Math. Soc.* **304**:2 (1987), 585–616.

[Seip 1993] K. Seip, "Beurling type density theorems in the unit disk", *Invent. Math.* **113**:1 (1993), 21–39.

[Seip 1994] K. Seip, "On a theorem of Korenblum", *Ark. Mat.* **32**:1 (1994), 237–243.

[Seip 1995] K. Seip, "On Korenblum's density condition for the zero sequences of $A^{-\alpha}$", *J. Anal. Math.* **67** (1995), 307–322.

[Shimorin 1993] S. M. Shimorin, "Factorization of analytic functions in weighted Bergman spaces", *Algebra i Analiz* **5**:5 (1993), 155–177. In Russian; translation in *St. Petersburg Math. J.* **5**:5 (1994), 1005–1022.

[Shimorin 1995] S. M. Shimorin, "On a family of conformally invariant operators", *Algebra i Analiz* **7**:2 (1995), 133–158. In Russian; translation in *St. Petersburg Math. J.* **7**:2 (1996), 287–306.

HÅKAN HEDENMALM
DEPARTMENT OF MATHEMATICS
LUND UNIVERSITY
BOX 118
S-221 00 LUND, SWEDEN
haakan@math.lu.se

Holomorphic Spaces
MSRI Publications
Volume **33**, 1998

Harmonic Bergman Spaces

KAREL STROETHOFF

ABSTRACT. We present a simple derivation of the explicit formula for the harmonic Bergman reproducing kernel on the ball in euclidean space and give an elementary proof that the harmonic Bergman projection is L^p-bounded, for $1 < p < \infty$. We furthermore discuss duality results.

1. Introduction

Throughout the paper n is a positive integer greater than 1. We will be working with functions defined on all or part of $\mathbb{R}^n$. Let D_j denote the partial derivative with respect to the j-th coordinate variable. Recall that $\nabla u(x) = (D_1 u(x), \cdots, D_n u(x))$. The Laplacian of u is $\Delta u(x) = D_1^2 u(x) + \cdots + D_n^2 u(x)$. A real- or complex-valued function u is harmonic on an open subset Ω of $\mathbb{R}^n$ if $\Delta u \equiv 0$ on Ω. The purpose of this article is to present an elementary treatment of some known results for the harmonic Bergman spaces consisting of all harmonic functions on the unit ball in $\mathbb{R}^n$ that are p-integrable with respect to volume measure. Several properties of these spaces are analogous to those of the Bergman spaces of analytic functions on the unit ball in $\mathbb{C}^n$. As in the analytic case, there is a reproducing kernel and associated projection. Duality results follow once we know that the projection is L^p-bounded. Coifman and Rochberg [1980] used deep results from harmonic analysis to establish L^p-boundedness of the harmonic Bergman projection.

An explicit formula for the harmonic Bergman reproducing kernel has only been determined recently; see [Axler et al. 1992]. We give a simple derivation for such a formula in Section 2. In Section 3 we give an elementary proof that the harmonic Bergman projection is L^p-bounded for $1 < p < \infty$. Our proof is similar to Forelli and Rudin's proof of L^p-boundedness of the analytic Bergman projection [Forelli and Rudin 1974/75; Rudin 1980], but as in Axler's argument [1988]

1991 *Mathematics Subject Classification.* 47B35, 47B38.

The author was partially supported by grants from the Montana University System and the University of Montana.

in the context of the analytic Bergman spaces on the unit disk, we will avoid the use of the binomial theorem, the gamma function and Stirling's formula.

In Section 4 we discuss the dual and the predual of the Bergman space of integrable harmonic functions on the unit ball in $\mathbb{R}^n$. Analogous to the analytic case, the dual and predual are identified with the harmonic Bloch and little Bloch space, which are spaces of harmonic functions determined by growth conditions on the gradient. As in the analytic case, these duality results follow from the fact that the harmonic Bergman projection maps L^∞ onto the harmonic Bloch space.

The remainder of this section establishes some of the notation and contains the prerequisites from harmonic function theory needed in the paper.

We will repeatedly make use of Green's identity, which states that if Ω is a bounded open subset of $\mathbb{R}^n$ with smooth boundary $\partial\Omega$ and u and v are continuously twice-differentiable functions in an open set containing $\bar{\Omega}$, the closure of Ω, then

$$\int_\Omega (u\,\Delta v - v\,\Delta u)\,dV = \int_{\partial\Omega} (u\,D_\mathbf{n}v - v\,D_\mathbf{n}u)\,ds, \qquad (1\text{--}1)$$

where V denotes volume measure on $\mathbb{R}^n$, s denotes surface area measure on $\partial\Omega$, and the symbol $D_\mathbf{n}$ denotes differentiation with respect to the outward unit normal vector $\mathbf{n}$: $D_\mathbf{n}w = \nabla w \cdot \mathbf{n}$. A special case of (1–1) is obtained by taking $v \equiv 1$: if u is harmonic in an open neighborhood of $\bar{\Omega}$, then

$$\int_{\partial\Omega} D_\mathbf{n}u\,ds = 0. \qquad (1\text{--}2)$$

For $y \in \mathbb{R}^n$ and $r > 0$ we write $B(y,r) = \{x \in \mathbb{R}^n : |x - y| < r\}$ and $\bar{B}(y,r)$ for its closure. We use B to denote the unit ball $B(0,1)$ and write S for its boundary, the unit sphere $S = \{x \in \mathbb{R}^n : |x| = 1\}$. The area of S is easily determined: if we take $\Omega = B$, $u \equiv 1$ and $v(x) = |x|^2$ in (1–1), then $\Delta v \equiv 2n$ and $D_\mathbf{n}v \equiv 2$, and we obtain $2nV(B) = 2s(S)$; thus $s(S) = nV(B)$. It will be convenient to work with normalized surface area on S, which we denote by σ; thus $ds = nV(B)\,d\sigma$ on S and $\sigma(S) = 1$.

If u is harmonic on an open neighborhood of $\bar{B}(y,r)$, the chain rule gives $(d/dr)u(y + r\zeta) = \nabla u(y + r\zeta) \cdot \zeta = D_\mathbf{n}u(y + r\zeta)$, where the normal derviative $D_\mathbf{n}$ is taken with respect to $B(y,r)$, so that

$$\frac{d}{dr}\int_S u(y + r\zeta)\,d\sigma(\zeta) = \int_S D_\mathbf{n}u(y + r\zeta)\,d\sigma(\zeta) = 0,$$

by (1–2). Hence $\int_S u(y + r\zeta)\,d\sigma(\zeta)$ does *not* depend on r, so that the function u satisfies the following so-called *mean value property*

$$\int_S u(y + r\zeta)\,d\sigma(\zeta) = u(y). \qquad (1\text{--}3)$$

In particular, if u is harmonic on an open set containing $\bar{B}$, we have

$$u(0) = \int_S u(\zeta)\, d\sigma(\zeta).$$

In fact, for any $y \in B$, $u(y)$ is a weighted average of u over S, namely $u(y) = \int_S u(\zeta)\, P(\zeta, y)\, d\sigma(\zeta)$, where $P(\zeta, y)$ is the so called *Poisson kernel* for the ball B. Since this Poisson kernel will in the sequel play a fundamental role, we will derive an explicit formula for P.

Fix $y \in B \setminus \{0\}$, choose any $0 < r < 1 - |y|$, and set

$$\Omega = \{x \in \mathbb{R}^n : r < |x - y| < 1\}.$$

We will only consider here the case $n > 2$. Put $v(x) = |x - y|^{2-n}$. It is easy to verify that v is harmonic on $\mathbb{R}^n \setminus \{y\}$ and $\nabla v(x) = (2 - n)|x - y|^{-n}(x - y)$; consequently, $D_\mathbf{n} v = (2 - n)r^{1-n}$ on $\partial B(y, r)$. For $\zeta \in S$ it is easy to verify that $|\zeta - y|^2 = |y|^2 \left|\zeta - y/|y|^2\right|^2$; thus, on S the function v coincides with the function $w(x) = |y|^{2-n}\left|x - y/|y|^2\right|^{2-n}$. Using that w is harmonic on $\bar{B}$, an application of (1–1) yields $\int_S u D_\mathbf{n} w\, ds = \int_S w D_\mathbf{n} u\, ds = \int_S v D_\mathbf{n} u\, ds$. It follows that

$$\int_S (u D_\mathbf{n} v - u D_\mathbf{n} w)\, ds = \int_S (u D_\mathbf{n} v - v D_\mathbf{n} u)\, ds.$$

By (1–1)

$$\int_S (u D_\mathbf{n} v - v D_\mathbf{n} u)\, ds = \int_{\partial B(y,r)} (u D_\mathbf{n} v - v D_\mathbf{n} u)\, ds$$

$$= (2 - n)r^{1-n} \int_{\partial B(y,r)} u\, ds = (2 - n)n V(B) u(y).$$

We conclude that $u(y) = \int_S u P_y\, d\sigma$, where $P_y = (2 - n)^{-1}(D_\mathbf{n} v - D_\mathbf{n} w)$. A simple calculation shows that $P_y(\zeta) = P(\zeta, y)$ is given by the formula

$$P(\zeta, y) = \frac{1 - |y|^2}{|\zeta - y|^n}.$$

In particular we have

$$\int_S \frac{1 - |y|^2}{|y - \zeta|^n}\, d\sigma(\zeta) = 1 \quad \text{for } y \in B. \tag{1–4}$$

We extend the Poisson kernel P to a function on $B \times B$ as follows: if $x, y \in B$, then we set $P(x, y) = P(x/|x|, |x|y)$. This is called the *extended Poisson kernel*. Thus

$$P(x, y) = \frac{1 - |x|^2|y|^2}{(1 - 2x \cdot y + |x|^2|y|^2)^{n/2}} \quad \text{for } x, y \in B.$$

For each fixed $y \in B$ the function $x \mapsto P(x, y)$ is harmonic, and by symmetry, for each fixed $x \in B$ the function $y \mapsto P(x, y)$ is harmonic.

It will be convenient to use polar coordinates to integrate functions over balls. Heuristically, we have

$$\frac{d}{dr}\int_{B(0,r)} f(x)\,dV(x) = \int_{rS} f\,ds,$$

and because the sphere rS has area $nV(B)r^{n-1}$ this is equal to

$$nV(B)r^{n-1}\int_S f(r\zeta)\,d\sigma(\zeta).$$

Integrating with respect to r we obtain the following formula

$$\int_{B(0,\rho)} f(x)\,dV(x) = nV(B)\int_0^\rho r^{n-1}\int_S f(r\zeta)\,d\sigma(\zeta)\,dr. \tag{1-5}$$

If u is harmonic on B, then we define its *radial derivative* $\mathcal{R}u$ by $\mathcal{R}u(x) = \nabla u(x)\cdot x$. It is easy to verify that $\mathcal{R}u$ is also harmonic on B. In fact, this follows at once from the formula $\Delta(\mathcal{R}u) = 2\Delta u + \mathcal{R}(\Delta u)$, which is easy to verify.

The reader interested in learning more harmonic function theory should consult [Axler et al. 1992].

2. The Harmonic Bergman Spaces

For $1 \le p < \infty$ we denote by $b^p(B)$ the set of all harmonic functions u on B for which

$$\|u\|_p = \left(\int_B |u(x)|^p\,dV(x)\right)^{1/p} < \infty.$$

The spaces $b^p(B)$ are called *harmonic Bergman spaces*. The space $b^2(B)$ is a linear subspace of $L^2(B)$ with inner product given by

$$\langle f,g\rangle = \int_B f(x)\overline{g(x)}\,dV(x) \quad \text{for } f,g \in L^2(B).$$

If u is harmonic on B and $y \in B$ is fixed, it follows from (1–3) and Cauchy-Schwarz's inequality that $|u(y)|^2 \le \int_S |u(y+r\zeta)|^2\,d\sigma(\zeta)$ for $0 < r < 1 - |y|$. Applying (1–5) to the function $f(x) = |u(y+x)|^2$ with $\rho = 1 - |y|$ we see that

$$nV(B)\int_0^\rho r^{n-1}\int_S |u(y+r\zeta)|^2\,d\sigma(\zeta)\,dr = \int_{B(y,\rho)} |u|^2\,dV \le \|u\|_2^2,$$

and conclude that

$$|u(y)| \le \frac{1}{V(B)^{1/2}(1-|y|)^{n/2}}\|u\|_2 \quad \text{for } u \in b^2(B). \tag{2-1}$$

It follows from (2–1) that $b^2(B)$ is a closed subspace of $L^2(B)$, and thus it is a Hilbert space. Inequality (2–1) implies that for each fixed $y \in B$ the linear functional $u \mapsto u(y)$ is bounded. By the Riesz representation theorem there is a unique function $R_y \in b^2(B)$, called *the reproducing kernel at y*, such that $u(y) = \langle u, R_y\rangle$, for all $u \in b^2(B)$. By considering real and imaginary parts, it is easily seen that the function reproducing the value at y for the real Bergman space, also reproduces the value at y for complex-valued functions of $b^2(B)$, and thus (by uniqueness) we conclude that R_y is real-valued. We write $R(x, y) = R_y(x)$, and call this the *Bergman reproducing kernel* of $b^2(B)$. Because the Bergman reproducing kernel is real–valued we have $R(y, x) = \langle R_x, R_y \rangle = \langle R_y, R_x \rangle = R_y(x) = R(x, y)$, for all $x, y \in B$. Thus R is a symmetric function on $B \times B$.

The Bergman kernel for the ball. In this subsection we will derive an explicit formula for the Bergman kernel R of B. We will not make use of so-called zonal and spherical harmonics used in [Axler et al. 1992] to calculate the Bergman kernel R, but instead use Green's identity to relate R to the extended Poisson kernel P.

Suppose u is harmonic on $\bar{B}$ and fix $y \in B$. Let v also be harmonic on $\bar{B}$, and define the function w by $w(x) = (|x|^2 - 1)v(x) = |x|^2 v(x) - v(x)$. Observe that the function Δw is harmonic on $\bar{B}$, for

$$\Delta w(x) = \Delta(|x|^2)\, v(x) + 2\nabla |x|^2 \cdot \nabla v(x) + |x|^2 \Delta v(x)$$

$$= 2nv(x) + 4x \cdot \nabla v(x) = 2nv(x) + 4\mathcal{R}v(x). \qquad (2\text{--}2)$$

Note that $\nabla w(x) = 2v(x)\, x + (|x|^2 - 1)\nabla v(x)$, so that $D_{\mathbf{n}}w(x) = \nabla w(x) \cdot x/|x| = 2v(x)\,|x| + (|x|^2 - 1)D_{\mathbf{n}}v(x)$. In particular, $D_{\mathbf{n}}w \equiv 2v$ on S. Since $\Delta u \equiv 0$ on B and $w \equiv 0$ on S, it follows from (1–1) (applied to u and w) that

$$\int_B u\Delta w\, dV = \int_S u D_{\mathbf{n}}w\, ds = 2\int_S uv\, ds = 2nV(B)\int_S uv\, d\sigma. \qquad (2\text{--}3)$$

It is clear that our choice for v should be the extended Poisson kernel: $v(x) = P(x, y)$ for $x \in B$. Then

$$\int_B u\Delta w\, dV = 2nV(B)\int_S u(\zeta)\, P(\zeta, y)\, d\sigma(\zeta) = 2nV(B)u(y).$$

We conclude that the harmonic function $\Delta w/(2nV(B))$ is the reproducing kernel at y. Using (2–2) we obtain the following formula for the Bergman kernel

$$R(x, y) = \frac{1}{nV(B)}\left(nP(x, y) + 2x \cdot \nabla_x P(x, y)\right). \qquad (2\text{--}4)$$

By elementary calculus, we get the formula

$$R(x, y) = \frac{1}{nV(B)(1 - 2x \cdot y + |x|^2|y|^2)^{n/2}}\left(\frac{n(1 - |x|^2|y|^2)^2}{1 - 2x \cdot y + |x|^2|y|^2} - 4|x|^2|y|^2\right). \qquad (2\text{--}5)$$

In the next section we will need an estimate on $|R(x, y)|$. By Cauchy-Schwarz $x \cdot y \leq |x||y|$. It follows that $(1 - |x||y|)^2 = 1 + |x|^2|y|^2 - 2|x||y| \leq 1 - 2x \cdot y + |x|^2|y|^2$, and thus $(1 - |x|^2|y|^2)^2 = (1 + |x||y|)^2(1 - |x||y|)^2 \leq 4(1 - 2x \cdot y + |x|^2|y|^2)$. Therefore we have

$$|R(x, y)| \leq \frac{4}{V(B)(1 - 2x \cdot y + |x|^2|y|^2)^{n/2}}. \qquad (2\text{--}6)$$

3. L^p-Boundedness of the Bergman Projection and Duality

Let Q denote the orthogonal projection of $L^2(B)$ onto $b^2(B)$. If $f \in L^2(B)$ and $y \in B$, then $Q[f](y) = \langle Qf, R_y \rangle = \langle f, R_y \rangle$ and we have the following formula:

$$Q[f](y) = \int_B R(x,y)\, f(x)\, dV(x). \qquad (3\text{--}1)$$

For fixed $y \in B$ the function $R(\cdot, y)$ is bounded, so that we can use the formula above for $Q[f]$ to extend the domain of Q to $L^p(B)$, where $1 \leq p < \infty$. Our goal is to prove the following theorem.

THEOREM 3.1. *Let $1 < p < \infty$. Then Q maps $L^p(B)$ boundedly onto $b^p(B)$.*

PROOF. We will use the so-called Schur Test. Specifically, we will show the existence of a positive function h and a constant C such that

$$\int_B h(x)^q\, |R(x,y)|\, dV(x) \leq Ch(y)^q \quad \text{for all } y \in B, \quad \text{and} \qquad (3\text{--}2)$$

$$\int_B h(y)^p\, |R(x,y)|\, dV(y) \leq Ch(x)^p \quad \text{for all } x \in B, \qquad (3\text{--}3)$$

where q denotes the conjugate index of p, that is, $q = p/(p-1)$. That this will imply the result is then proved as follows. Given $f \in L^p(B, dV)$, applying Hölder's inequality and (3–2) we have

$$|Q[f](y)| \leq \int_B \frac{|f(x)|}{h(x)}\, h(x)\, |R(x,y)|\, dV(x)$$

$$\leq \left(\int_B \frac{|f(x)|^p}{h(x)^p}\, |R(x,y)|\, dV(x) \right)^{1/p} \left(\int_B h(x)^q\, |R(x,y)|\, dV(x) \right)^{1/q}$$

$$\leq C^{1/q} h(y) \left(\int_B \frac{|f(x)|^p}{h(x)^p}\, |R(x,y)|\, dV(x) \right)^{1/p}.$$

Thus, applying Fubini's theorem to reverse the order of integration, and using (3–3) we obtain

$$\int_B |Q[f](y)|^p\, dV(y) \leq C^{p/q} \int_B h(y)^p \left(\int_B \frac{|f(x)|^p}{h(x)^p}\, |R(x,y)|\, dV(x) \right) dV(y)$$

$$= C^{p/q} \int_B \frac{|f(x)|^p}{h(x)^p} \left(\int_B h(y)^p\, |R(x,y)|\, dV(y) \right) dV(x)$$

$$\leq C^{p/q} \int_B \frac{|f(x)|^p}{h(x)^p}\, (Ch(x)^p)\, dV(x) = C^p \int_B |f(x)|^p\, dV(x),$$

proving the theorem.

We claim that the function $h(x) = (1 - |x|^2)^{-1/(pq)}$ works, that is, satisfies (3–2) and (3–3). By symmetry in p and q, it will suffice to find a constant C_p for which

$$\int_B (1 - |x|^2)^{-1/p} |R(x,y)|\, dV(y) \leq C_p (1 - |y|^2)^{-1/p} \quad \text{for all } y \in B. \qquad (3\text{--}4)$$

Fix $y \in B \backslash \{0\}$. For $0 < r < 1$ and $\zeta \in S$ it follows from (2–6) that

$$|R(r\zeta, y)| \leq \frac{4}{V(B)} \frac{1}{(1 - ry \cdot \zeta + r^2|y|^2)^{n/2}} = \frac{4}{V(B)} \frac{1}{|\zeta - ry|^n}.$$

Using (1–5) and (1–4) we have

$$\int_B \frac{|R(x, y)|}{(1 - |x|^2)^{1/p}} \, dV(x) = nV(B) \int_0^1 r^{n-1}(1 - r^2)^{-1/p} \int_S |R(r\zeta, y)| \, d\sigma(\zeta) \, dr$$

$$\leq 4n \int_0^1 r^{n-1}(1 - r^2)^{-1/p} \int_S \frac{1}{|\zeta - ry|^n} \, d\sigma(\zeta) \, dr$$

$$\leq 2n \int_0^1 2r(1 - r^2)^{-1/p} \frac{1}{1 - r^2|y|^2} \, dr$$

$$= 2n \int_0^1 (1 - t)^{-1/p}(1 - t|y|^2)^{-1} \, dt.$$

Now

$$\int_0^{|y|^2} (1 - t)^{-1/p}(1 - t|y|^2)^{-1} \, dt \leq \int_0^{|y|^2} (1 - t)^{-1-1/p} \, dt \leq p(1 - |y|^2)^{-1/p}.$$

Also (recall that $1 - 1/p = 1/q$)

$$\int_{|y|^2}^1 (1 - t)^{-1/p}(1 - t|y|^2)^{-1} \, dt \leq (1 - |y|^2)^{-1} \int_{|y|^2}^1 (1 - t)^{-1/p} \, dt$$

$$= (1 - |y|^2)^{-1} q(1 - |y|^2)^{1-1/p} = q(1 - |y|^2)^{-1/p}.$$

Addition yields

$$\int_0^1 (1 - t)^{-1/p}(1 - t|y|^2)^{-1} \, dt \leq (p + q)(1 - |y|^2)^{-1/p}.$$

Thus (3–4) is proved with $C_p = 2n(p + q)$. This concludes the proof of the L^p-boundedness of Q. In fact, from the Schur Test, the estimates above and the observation that $C_q = C_p$, we obtain the following bound on the norm of Q as an operator from $L^p(B)$ onto $b^p(B)$: $\|Q\| \leq 2np^2/(p - 1)$. $\qquad \square$

REMARK 1. The norm estimate given above is far from being sharp; it can be improved by estimating the integrals using binomial series, as in [Forelli and Rudin 1974/75; Rudin 1980].

REMARK 2. As we will see in the next section, Theorem 3.1 does *not* hold for $p = 1$ or $p = \infty$.

Duality. It is a consequence of the L^p-boundedness of the Bergman projection that the spaces $b^p(B)$ and $b^q(B)$ are dual to each other: if $v \in b^q(B)$, then the function φ defined by $\varphi(u) = \langle u, v \rangle$ ($u \in b^p(B)$) defines a bounded linear functional on $b^p(B)$, and every bounded linear functional on $b^p(B)$ is of the form above. To prove the latter statement, suppose that $\varphi \in b^p(B)^*$. By the Hahn-Banach theorem, φ extends to a bounded linear functional ψ on $L^p(B)$. There exists a $g \in L^q(B)$ such that $\psi(f) = \langle f, g \rangle$ for all $f \in L^p(B)$. In particular, if $u \in b^p(B)$, then $\varphi(u) = \langle u, g \rangle$. Note that $v = Q[g] \in Q(L^q(B)) = b^q(B)$. Using Fubini's theorem to reverse the order of integration it can then be shown that $\langle u, v \rangle = \langle u, g \rangle$, and we obtain that $\varphi(u) = \langle u, v \rangle$, for all $u \in b^p(B)$.

4. The Bloch Space and the Dual of $b^1(B)$

A harmonic function u on B is said to be a *Bloch function* if

$$\|u\|_{\mathcal{B}} = \sup_{x \in B}(1 - |x|^2)|\nabla u(x)| < \infty.$$

The *harmonic Bloch space* $\mathcal{B}$ is the set of all harmonic Bloch functions on B. We will show that $\mathcal{B}$ is the dual of the Bergman space $b^1(B)$. We will first prove that $\mathcal{B} = Q[L^\infty(B)]$. The hard part of the proof will be to show that each function $u \in \mathcal{B}$ is of the form $u = Q[g]$, where g is a bounded function on B. If $u \in \mathcal{B}$, then it follows from the inequality $|\mathcal{R}u(x)| \le |\nabla u(x)|$ that the function $(1 - |x|^2)\mathcal{R}u$ is bounded on B. Thus we will try to relate $Q[(1 - |x|^2)\mathcal{R}u]$ to u. We will first find an expression for $Q[(1 - |x|^2)u]$. Combining (2–4) and (3–1) we have

$$nV(B)Q[(1 - |x|^2)u](y) = 2\langle(1 - |x|^2)u, \mathcal{R}P_y\rangle + n\langle(1 - |x|^2)u, P_y\rangle.$$

To rewrite $\langle(1 - |x|^2)u, \mathcal{R}P_y\rangle$ we use the same idea as the derivation of formula (2–4): assuming u to be harmonic on an open set containing $\bar{B}$, apply identity (1–1) with u and $v(x) = (1 - |x|^2)^2 P(x, y)$. The outward normal derivative $D_{\mathbf{n}}v$ is 0 on S; thus (1–1) gives us

$$\int_B u(x)\Delta_x\big((1 - |x|^2)^2 P(x, y)\big)\, dV(x) = 0. \tag{4–1}$$

It is easily verified that

$$\Delta_x\big((1 - |x|^2)^2 P(x, y)\big) = 8|x|^2 P(x, y) + 4n(|x|^2 - 1)P(x, y) + 8(|x|^2 - 1)\mathcal{R}_x P(x, y).$$

Thus (4–1) shows that

$$\int_B |x|^2 u(x) P(x, y)\, dV(x)$$

$$= \tfrac{1}{2}n\int_B u(x)(1 - |x|^2)P(x, y)\, dV(x) + \int_B u(x)(1 - |x|^2)\,\mathcal{R}_x P(x, y)\, dV(x),$$

and using (3–1) and (2–4) we can write the equation above as

$$\int_B |x|^2 u(x) P(x,y)\, dV(x) = \tfrac{1}{2} n V(B) Q[(1-|x|^2)u](y). \qquad (4\text{--}2)$$

It follows from (1–1) that $\int_S u(r\zeta) D_{\mathbf{n}} v(r\zeta)\, d\sigma(\zeta) = \int_S D_{\mathbf{n}} u(r\zeta) v(r\zeta)\, d\sigma(\zeta)$ for harmonic functions u and v on B and $0 < r < 1$. Multiplication by $nV(B)r^{n+1}$ and integration over r yields the formula

$$\int_B |x|^2 u(x) \mathcal{R}v(x)\, dV(x) = \int_B |x|^2 \mathcal{R}u(x) v(x)\, dV(x). \qquad (4\text{--}3)$$

Applying (4–2) to the function $\mathcal{R}u$, and making use of (4–3), we also have

$$\int_B |x|^2 u(x) \mathcal{R}_x P(x,y)\, dV(x) = \tfrac{1}{2} n V(B) Q[(1-|x|^2)\mathcal{R}u](y). \qquad (4\text{--}4)$$

Combining (4–2), (4–4) and (2–4) we arrive at

$$Q[|x|^2 u](y) = \tfrac{1}{2} n Q[(1-|x|^2)u](y) + Q[(1-|x|^2)\mathcal{R}u](y),$$

and, writing $u = Q[(1-|x|^2)u] + Q[|x|^2 u]$ we obtain the formula in the following theorem.

THEOREM 4.1. *If $u \in \mathcal{B}$, then*

$$u = Q[(1-|x|^2)\mathcal{R}u + (\tfrac{1}{2}n+1)(1-|x|^2)u]. \qquad (4\text{--}5)$$

PROOF. We have shown the stated result for u harmonic on an open set containing $\bar{B}$. To get the result for general $u \in \mathcal{B}$, let $0 < r < 1$ and consider the dilate u_r of u, defined by $u_r(x) = u(rx)$, $x \in B$. Since u_r is harmonic on the set $\{x \in \mathbb{R}^n : |x| < 1/r\}$, equation (4–5) holds for u_r. It is easily seen that $(\mathcal{R}u)_r = \mathcal{R}u_r$. We leave it as an exercise to show that $(1-|x|^2)\mathcal{R}u_r \to (1-|x|^2)\mathcal{R}u$ and $(1-|x|^2)u_r \to (1-|x|^2)u$ in $L^2(B)$ as $r \to 1^-$. The boundedness of Q and the continuity of point evaluation at y imply that $Q[(1-|x|^2)\mathcal{R}u_r](y) \to Q[(1-|x|^2)\mathcal{R}u](y)$ and $Q[(1-|x|^2)u_r](y) \to Q[(1-|x|^2)u](y)$ as $r \to 1^-$, and since $u_r(y) \to u(y)$ formula (4–5) follows. $\qquad \square$

COROLLARY 4.2. $\mathcal{B} = Q[L^\infty(B)]$.

PROOF. If $g \in L^\infty(B)$, and $u = Q[g]$, then we claim that $u \in \mathcal{B}$. Differentiating $u(x) = \int_B g(y) R(x,y)\, dV(y)$ we obtain

$$D_j u(x) = \int_B g(y) \frac{\partial}{\partial x_j} R(x,y)\, dV(y),$$

and consequently

$$|\nabla u(x)| \le \|g\|_\infty \int_B |\nabla_x R(x,y)|\, dV(y). \qquad (4\text{--}6)$$

Using (2–5) we have

$$\nabla_x R(x,y) = \frac{n(y - |y|^2 x)}{(1 - 2x \cdot y + |x|^2 |y|^2)^{1+n/2}} \left(\frac{n(1 - |x|^2 |y|^2)^2}{(1 - 2x \cdot y + |x|^2 |y|^2)} - 4|x|^2 |y|^2 \right)$$

$$+ \frac{1}{(1 - 2x \cdot y + |x|^2 |y|^2)^{n/2}} \left(\frac{-4n(1 - |x|^2 |y|^2)|y|^2 x}{(1 - 2x \cdot y + |x|^2 |y|^2)} + \right.$$

$$\left. + \frac{2n(1 - |x|^2 |y|^2)^2 (y - x|y|^2)}{(1 - 2x \cdot y + |x|^2 |y|^2)^2} - 8|y|^2 x \right).$$

Noting that $|y - |y|^2 x|^2 = |y|^2 - 2|y|^2 x \cdot y + |y|^4 |x|^2 = |y|^2 (1 - 2x \cdot y + |y|^2 |x|^2)$, we see that $|y - |y|^2 x| \leq (1 - 2x \cdot y + |y|^2 |x|^2)^{1/2}$. Recalling that also $1 - |x|^2 |y|^2 \leq (1 - 2x \cdot y + |y|^2 |x|^2)^{1/2}$ we obtain

$$|\nabla_x R(x,y)| \leq \frac{C}{\left(1 - 2x \cdot y + |x|^2 |y|^2\right)^{(n+1)/2}}. \tag{4–7}$$

Thus

$$\int_B |\nabla_x R(x,y)| \, dV(x) \leq CnV(B) \int_0^1 r^{n-1} \int_S \frac{1}{|\zeta - ry|^{n+1}} \, d\sigma(\zeta) \, dr.$$

Now,

$$\int_S \frac{1}{|\zeta - ry|^{n+1}} \, d\sigma(\zeta) = \frac{1}{1 - r^2 |y|^2} \int_S \frac{1}{|\zeta - ry|} P(ry, \zeta) \, d\sigma(\zeta)$$

$$\leq \frac{1}{1 - r^2 |y|^2} \frac{1}{(1 - r|y|)} \int_S P(ry, \zeta) \, d\sigma(\zeta) \leq \frac{1}{(1 - r|y|)^2}.$$

Multiply by $nV(B)r^{n-1}$ and integrate with respect to r to obtain

$$\int_B \frac{1}{(1 - 2x \cdot y + |x|^2 |y|^2)^{(1+n)/2}} \, dV(x) \leq nV(B) \frac{1}{(1 - |y|)}. \tag{4–8}$$

Using that $1/(1 - |y|) = (1 + |y|)/(1 - |y|^2) \leq 2/(1 - |y|^2)$, we conclude from (4–6), (4–7) and (4–8) that $(1 - |y|^2)|\nabla u(y)| \leq 2nV(B)C\|g\|_\infty$ for all $y \in B$, establishing our claim that $u \in \mathcal{B}$. This proves the inclusion $Q[L^\infty(B)] \subset \mathcal{B}$.

The other inclusion $\mathcal{B} \subset Q[L^\infty(B)]$ follows from Theorem 4.1, for if $u \in \mathcal{B}$, the function $g = (1 - |x|^2)\mathcal{R}u + (\frac{1}{2}n + 1)(1 - |x|^2)u$ is bounded on B.   □

This corollary can be used to show that Theorem 3.1 does *not* hold for $p = 1$ or $p = \infty$. It is not difficult to construct unbounded harmonic Bloch functions (the function $u(x) = \log((1 - x_1)^2 + x_2{}^2)$ provides an example), so Corollary 4.2 shows that the operator Q is not L^∞-bounded. By duality it follows that Q is not L^1-bounded either.

The dual of $b^1(B)$. We will show how Corollary 4.2 can be used to prove that $\mathcal{B}$ is the dual of $b^1(B)$. Our first concern is to define a "pairing": given a function $v \in \mathcal{B}$ we could try to define φ by $\varphi(u) = \langle u, v \rangle$ ($u \in \mathcal{B}$), but we have to be careful here: the function v need not be bounded, so there is no guarantee that $u\bar{v}$ is integrable for all $u \in b^1(B)$ (in fact, if v is unbounded, some standard functional analysis can be used to prove that there must exist a function $u \in b^1(B)$ for which $u\bar{v}$ is not integrable). This problem is easily overcome by using dilates. If u is a harmonic function on B and $0 < r < 1$, the dilate u_r, defined by $u_r(x) = u(rx)$, $x \in B$, is bounded on B. If $u \in C(\bar{B})$, then u is uniformly continuous on B, and it follows that $u_r \to u$ uniformly on B as $r \to 1^-$. Using the fact that $C(\bar{B})$ is dense in $L^1(B)$ it is easy to verify that if $u \in b^1(B)$, then $u_r \to u$ in $b^1(B)$ as $r \to 1^-$.

We are now ready to extend the usual pairing. Given $u \in b^1(B)$ and $v \in \mathcal{B}$ we claim that $\lim_{r \to 1^-} \langle u_r, v \rangle$ exists. To prove this, write $v = Q[g]$, where $g \in L^\infty(B)$. Since $u_r \in b^2(B)$ we have $\langle u_r, v \rangle = \langle u_r, Q[g] \rangle = \langle u_r, g \rangle$. The inequality $|\langle u_r, v \rangle - \langle u_s, v \rangle| = |\langle u_r - u_s, g \rangle| \leq \|u_r - u_s\|_1 \|g\|_\infty$ together with the fact that $\|u_r - u_s\|_1 \to 0$ as $r, s \to 1^-$ (because $\|u_r - u_s\|_1 \leq \|u_r - u\|_1 + \|u - u_s\|_1$) implies that $\lim_{r \to 1^-} \langle u_r, v \rangle$ exists. It is clear that the function φ defined by

$$\varphi(u) = \lim_{r \to 1^-} \langle u_r, v \rangle \quad \text{for } u \in b^1(B)$$

is a linear functional on $b^1(B)$. Using $|\langle u_r, v \rangle| \leq \|u_r\|_1 \|g\|_\infty$ we see that $|\varphi(u)| \leq \|u\|_1 \|g\|_\infty$, so $\varphi \in b^1(B)^*$.

We claim that every element of $b^1(B)^*$ is of the form above. For let $\varphi \in b^1(B)^*$. By the Hahn-Banach theorem, φ extends to a bounded linear functional ψ on $L^1(B, dV)$. There exists a $g \in L^\infty(B)$ such that $\psi(f) = \langle f, g \rangle$ for all $f \in L^1(B)$. In particular, $\varphi(u) = \langle u, g \rangle$ for all $u \in b^1(B)$. Now, if $u \in b^1(B)$, then $u_r \to u$ in $b^1(B)$ as $r \to 1^-$, and thus $\varphi(u) = \lim_{r \to 1^-} \varphi(u_r)$. Since $u_r \in b^2(B)$ we have $\langle u_r, g \rangle = \langle Q[u_r], g \rangle = \langle u_r, Q[g] \rangle$. Thus, if we set $v = Q[g]$, then $\varphi(u) = \lim_{r \to 1^-} \langle u_r, v \rangle$, as was to be proved. $\square$

The predual of $b^1(B)$. In this subsection we will identify the predual of $b^1(B)$ under the above pairing. We define the space $\mathcal{B}_0$, called the *harmonic little Bloch space*, to be the set of all harmonic functions u on B for which $(1-|x|^2)|\nabla u(x)| \to 0$ as $|x| \to 1^-$. Clearly, $\mathcal{B}_0 \subset \mathcal{B}$. We leave it as an exercise for the reader to show that if $u \in \mathcal{B}_0$, then $\|u_r - u\|_\mathcal{B} \to 0$ as $r \to 1^-$ (the converse is also true). Now suppose that $u \in \mathcal{B}_0$ and $v \in b^1(B)$. We claim that $\lim_{r \to 1^-} \langle u_r, v \rangle$ exists. To show this, we use Corollary 4.2 (or rather its proof) to conclude that

$$|\langle u_r, v \rangle - \langle u, v_r \rangle| \leq |\langle u_r - u, v \rangle| + |\langle u, v - v_r \rangle|$$
$$\leq C\|u_r - u\|_\mathcal{B} \|v\|_1 + C\|u\|_\mathcal{B} \|v - v_r\|_1,$$

and the statement follows. So if $v \in b^1(B)$, then $\varphi(u) = \lim_{r \to 1^-} \langle u_r, v \rangle$ defines a bounded linear functional on $\mathcal{B}_0$.

Our claim is that all bounded linear functionals on $\mathcal{B}_0$ arise this way. Suppose φ is a bounded linear functional on $\mathcal{B}_0$. Then $\psi(g) = \varphi(Q[g])$ defines a bounded linear functional on $C(\bar{B})$: $|\psi(g)| = |\varphi(Q[g])| \leq \|\varphi\| \, \|g\|_2 \leq \|\varphi\| \, \|g\|_\infty$. By the Riesz representation theorem $\psi(g) = \int_{\bar{B}} g \, d\mu$, for all $g \in C(\bar{B})$, where μ is a finite complex Borel measure on $\bar{B}$. If $u \in \mathcal{B}_0$, then $g(x) = (1 - |x|^2)\{\mathcal{R}u(x) + (\frac{1}{2}n + 1)u(x)\}$ is in $C(\bar{B})$, and, using Theorem 4.1, we have $\varphi(u) = \psi(g) = \int_{\bar{B}} g(y) \, d\mu(y)$. Define

$$v(x) = \int_{\bar{B}} (1 - |y|^2)\{\mathcal{R}_x R(x, y) + (\tfrac{1}{2}n + 1)R(x, y)\} \, d\bar{\mu}(y) \quad \text{for } x \in B.$$

The function v is clearly harmonic in B. We claim that in fact $v \in b^1(B)$. To show this, use Fubini's theorem to get

$$\int_B |v(x)| \, dV(x) \leq \int_{\bar{B}} (1 - |y|^2) \int_B \{|\mathcal{R}_x R(x, y)| + (\tfrac{1}{2}n + 1)|R(x, y)|\} \, dV(x) \, d|\mu|(y)$$

In the proof of Corollary 4.2 we have seen that there exists a positive constant C such that $\int_B |\mathcal{R}_x R(x, y)| \, dV(x) \leq C/(1 - |y|^2)$. Also, using (2–6) and (1–5), we have

$$\int_B |R(x, y)| \, dV(x) \leq 4n \int_0^1 r^{n-1} \int_S \frac{1}{|\zeta - ry|^n} \, d\sigma(\zeta) \, dr$$

$$= 4n \int_0^1 r^{n-1} \frac{1}{1 - r^2|y|^2} \, dr$$

$$\leq 4n \int_0^1 r^{n-1} \frac{1}{1 - |y|^2} = \frac{4}{1 - |y|^2}.$$

Thus,

$$\int_B |v(x)| \, dV(x) \leq \int_{\bar{B}} (1 - |y|^2) \frac{C'}{1 - |y|^2} \, d|\mu|(y) = C' \int_{\bar{B}} d|\mu|(y) = C' \|\mu\| < \infty,$$

and our claim that $v \in b^1(B)$ is proved. Now, assuming u to be harmonic on $\bar{B}$, applying Fubini's theorem we have

$$\langle u, v \rangle = \int_{\bar{B}} (1 - |y|^2) \int_B u(x)\{\mathcal{R}_x R(x, y) + (\tfrac{1}{2}n + 1)R(x, y)\} \, dV(x) \, d\mu(y).$$

Similarly to (4–3) we have $\int_B u(x)\mathcal{R}_x R(x, y) \, dV(x) = \int_B \mathcal{R}u(x)R(x, y) \, dV(x) = \mathcal{R}u(y)$, thus

$$\int_B u(x)\{\mathcal{R}_x R(x, y) + (\tfrac{1}{2}n + 1)R(x, y)\} \, dV(x) = \mathcal{R}u(y) + (\tfrac{1}{2}n + 1)u(y)$$

and we obtain

$$\langle u, v \rangle = \int_{\bar{B}} (1 - |y|^2) \left\{\mathcal{R}u(y) + (\tfrac{1}{2}n + 1)u(y)\right\} d\mu(y) = \int_{\bar{B}} g(y) \, d\mu(y) = \varphi(u).$$

Hence $\varphi(u_r) = \langle u_r, v \rangle$, and since $u_r \to u$ in $\mathcal{B}_0$ and φ is continuous on $\mathcal{B}_0$ we have $\varphi(u) = \lim_{r \to 1^-} \langle u_r, v \rangle$. Note that this pairing coincides with the one we saw earlier: if $u \in \mathcal{B}_0$ and $v \in b^1(B)$, then $\lim_{r \to 1^-} \langle u_r, v \rangle = \lim_{r \to 1^-} \langle u, v_r \rangle$. $\square$

Acknowledgments

This article was written while visiting the Free University, Amsterdam, The Netherlands, on sabbatical leave; I thank the University of Montana for awarding me a sabbatical and the Mathematics Department of the Free University for its hospitality and support.

References

[Axler 1988] S. Axler, "Bergman spaces and their operators", pp. 1–50 in *Surveys of some recent results in operator theory*, vol. 1, edited by J. B. Conway and B. B. Morrel, Pitman Res. Notes Math. Ser. **171**, Longman Sci. Tech., Harlow, 1988.

[Axler et al. 1992] S. Axler, P. Bourdon, and W. Ramey, *Harmonic function theory*, Graduate Texts in Mathematics **137**, Springer, New York, 1992.

[Coifman and Rochberg 1980] R. R. Coifman and R. Rochberg, "Representation theorems for holomorphic and harmonic functions in L^p", pp. 11–66 in *Representation theorems for Hardy spaces*, Astérisque **77**, Soc. Math. France, Paris, 1980.

[Forelli and Rudin 1974/75] F. Forelli and W. Rudin, "Projections on spaces of holomorphic functions in balls", *Indiana Univ. Math. J.* **24** (1974/75), 593–602.

[Rudin 1980] W. Rudin, *Function theory in the unit ball of* $\mathbf{C}^n$, Grundlehren der Mathematischen Wissenschaften **241**, Springer, New York, 1980.

KAREL STROETHOFF
DEPARTMENT OF MATHEMATICAL SCIENCES
UNIVERSITY OF MONTANA
MISSOULA, MT 59812–1032
UNITED STATES
ma_kms@selway.umt.edu

Holomorphic Spaces
MSRI Publications
Volume **33**, 1998

An Excursion into the Theory
of Hankel Operators

VLADIMIR V. PELLER

ABSTRACT. This survey is an introduction to the theory of Hankel operators, a beautiful area of mathematical analysis that is also very important in
applications. We start with classical results: Kronecker's theorem, Nehari's
theorem, Hartman's theorem, Adamyan–Arov–Krein theorems. Then we
describe the Hankel operators in the Schatten–von Neumann class S_p and
consider numerous applications: Sarason's commutant lifting theorem, rational approximation, stationary processes, best approximation by analytic
functions. We also present recent results on spectral properties of Hankel
operators with lacunary symbols. Finally, we discuss briefly the most recent results involving Hankel operators: Pisier's solution of the problem
of similarity to a contraction, self-adjoint operators unitarily equivalent to
Hankel operators, and approximation by analytic matrix-valued functions.

CONTENTS

The author was partially supported by NSF grant DMS-9304011.

1. Introduction

I would like to invite the reader on an excursion into the theory of Hankel operators, a beautiful and rapidly developing domain of analysis that is important in numerous applications.

It was Hankel [1861] who began the study of finite matrices whose entries depend only on the sum of the coordinates, and therefore such objects are called *Hankel matrices*. One of the first theorems about infinite Hankel matrices was obtained by Kronecker [1881]; it characterizes Hankel matrices of finite rank. Hankel matrices played an important role in many classical problems of analysis, and in particular in the so-called moment problems; for example, Hamburger's moment problem is solvable if and only if the corresponding infinite Hankel matrix is positive semi-definite [Hamburger 1920; 1921].

Since the work of Nehari [1957] and Hartman [1958] it has become clear that Hankel operators are an important tool in function theory on the unit circle. Together with Toeplitz operators they form two of the most important classes of operators on Hardy spaces.

For the last three decades the theory of Hankel operators has been developing rapidly. A lot of applications in different domains of mathematics have been found: interpolation problems [Adamyan et al. 1968b; 1968a; 1971]; rational approximation [Peller 1980; 1983]; stationary processes [Peller and Khrushchëv 1982]; perturbation theory [Peller 1985]; Sz.-Nagy–Foiaş function model [Nikol'-skiĭ 1986]. In the 1980s the theory of Hankel operators was fueled by the rapid development of H^∞ control theory and systems theory (see [Fuhrmann 1981; Glover 1984; Francis 1987]). It has become clear that it is especially important to develop the theory of Hankel operators with matrix-valued (and even operator-valued) symbols. I certainly cannot mention here all applications of Hankel operators. The latest application I would like to touch on here is Pisier's solution of the famous problem of similarity to a contraction; see Section 12 for more detail.

The development of the theory of Hankel operators led to different generalizations of the original concept. A lot of progress has taken place in the study of Hankel operators on Bergman spaces on the disk, Dirichlet type spaces, Bergman and Hardy spaces on the unit ball in $\mathbb{C}^n$, on symmetric domains; commutators, paracommutators, etc. This survey will not discuss such generalizations, but will concentrate on the classical Hankel operators on the Hardy class H^2—or, in other words, operators having Hankel matrices. Even under this constraint it is impossible in a survey to cover all important results and describe all applications. I have chosen several aspects of the theory and several applications to demonstrate the beauty of the theory and importance in applications.

So, if you accept my invitation, fasten seat belts and we shall be off!

In Section 2 we obtain the boundedness criterion and discuss symbols of Hankel operators of minimal L^∞ norm. As an application, we give in Section 3 a proof of Sarason's commutant lifting theorem, based on Nehari's theorem. Section 4 is devoted to the proof of Kronecker's theorem characterizing the Hankel operators of finite rank. In Section 5 we describe the compact Hankel operators. In Section 6 we prove the profound Adamyan–Arov–Krein theorem on finite-rank approximation of Hankel operators. Section 7 is devoted to membership of Hankel operators in Schatten–von Neumann classes S_p. In Section 8 we consider applications of Hankel operators in the theory of rational approximation. Section 9 concerns hereditary properties of the operator of best uniform approximation by functions analytic in the unit disk. Section 10 deals with applications of Hankel operators in prediction theory. In Section 11 we study spectral properties of Hankel operators with lacunary symbols. We conclude the survey with Section 12 which briefly reviews some recent developments of Hankel operators and their applications; namely, we touch on the problem of unitary equivalent description of the self-adjoint Hankel operators, we discuss the problem of approximating a matrix function on the unit circle by bounded analytic functions, and conclude the section with Pisier's solution of the problem of similarity to a contraction.

Preliminaries. An infinite matrix A is called a *Hankel matrix* if it has the form

$$A = \begin{pmatrix} \alpha_0 & \alpha_1 & \alpha_2 & \alpha_3 & \cdots \\ \alpha_1 & \alpha_2 & \alpha_3 & \alpha_4 & \cdots \\ \alpha_2 & \alpha_3 & \alpha_4 & \alpha_5 & \cdots \\ \alpha_3 & \alpha_4 & \alpha_5 & \alpha_6 & \cdots \\ \vdots & \vdots & \vdots & \vdots & \ddots \end{pmatrix},$$

where $\alpha = \{\alpha_j\}_{j\geq 0}$ is a sequence of complex numbers. In other words, a Hankel matrix is one whose entries depend only on the sum of the coordinates.

If $\alpha \in \ell^2$, we can consider the *Hankel operator* $\Gamma_\alpha : \ell^2 \to \ell^2$ with matrix A in the standard basis of ℓ^2 that is defined on the dense subset of finitely supported sequences.

Hankel operators admit the following important realizations as operators from the Hardy class H^2 of functions on the unit circle $\mathbb{T}$ to the space

$$H^2_- \stackrel{\text{def}}{=} L^2 \ominus H^2.$$

Let $\varphi \in L^2$. We define the Hankel operator H_φ on the dense subset of polynomials by

$$H_\varphi f \stackrel{\text{def}}{=} \mathbb{P}_- \varphi f,$$

where $\mathbb{P}_-$ is the orthogonal projection onto H^2_-. The function φ is called a *symbol* of the Hankel operator H_φ; there are infinitely many different symbols that produce the same Hankel operator. It is easy to see that H_φ has Hankel matrix $\{\hat{\varphi}(-j-k)\}_{j\geq 1, k\geq 0}$ in the bases $\{z^k\}_{k\geq 0}$ of H^2 and $\{\bar{z}^j\}_{j\geq 1}$ of H^2_-; here $\hat{\varphi}(m)$ is the m-th Fourier coefficient of φ.

We also need the notion of a Toeplitz operator on H^2. For $\varphi \in L^\infty$ we define the *Toeplitz operator* $T_\varphi : H^2 \to H^2$ by

$$T_\varphi f = \mathbb{P}_+ \varphi f \quad \text{for } f \in H^2,$$

where $\mathbb{P}_+$ is the orthogonal projection of L^2 onto H^2. It is easy to see that $\|T_\varphi\| \leq \|\varphi\|_{L^\infty}$. In fact, $\|T_\varphi\| = \|\varphi\|_{L^\infty}$; see, for example, [Douglas 1972; Sarason 1978].

Notation. The following notation is used throughout the survey:

- z stands for the identity function on a subset of $\mathbb{C}$.
- m is normalized Lebesgue measure on the unit circle $\mathbb{T}$.
- m_2 is planar Lebesgue measure.
- $S : H^2 \to H^2$ is the shift operator; that is, $Sf \stackrel{\text{def}}{=} zf$ for $f \in H^2$.
- $\mathcal{S} : L^2 \to L^2$ is the bilateral shift operator; that is, $\mathcal{S}f \stackrel{\text{def}}{=} zf$ for $f \in L^2$.
- For a function f in $L^1(\mathbb{T})$ we denote by $\tilde{f}$ the harmonic conjugate of f.
- BMO is the space of functions φ on $\mathbb{T}$ of bounded mean oscillation:

$$\sup_{|I|} \frac{1}{|I|} \int_I |\varphi - \varphi_I| \, dm < \infty,$$

  where the supremum is taken over all intervals I of $\mathbb{T}$ and $|I| \stackrel{\text{def}}{=} m(I)$.
- VMO is the closed subspace of BMO consisting of functions φ satisfying

$$\lim_{|I| \to 0} \frac{1}{|I|} \int_I |\varphi - \varphi_I| \, dm = 0.$$

2. Boundedness

The following theorem of Nehari [1957] characterizes the bounded Hankel operators Γ_α on ℓ^2.

THEOREM 2.1. *The Hankel operator Γ_α is bounded on ℓ^2 if and only if there exists a function $\psi \in L^\infty$ on the unit circle $\mathbb{T}$ such that*

$$\alpha_m = \hat{\psi}(m) \quad \text{for } m \geq 0. \tag{2-1}$$

In this case

$$\|\Gamma_\alpha\| = \inf\{\|\psi\|_\infty : \hat{\psi}(n) = \alpha_n \text{ for } n \geq 0\}.$$

PROOF. Let $\psi \in L^\infty$ and set $\alpha_m = \hat{\psi}(m)$ for $m \geq 0$. Given two finitely supported sequences $a = \{a_n\}_{n \geq 0}$ and $b = \{b_k\}_{k \geq 0}$ in ℓ^2, we have

$$(\Gamma_\alpha a, b) = \sum_{j,k \geq 0} \alpha_{j+k} a_j \bar{b}_k. \tag{2-2}$$

Let

$$f = \sum_{j \geq 0} a_j z^j, \quad g = \sum_{k \geq 0} \bar{b}_k z^k.$$

Then f and g are polynomials in the Hardy class H^2. Put $q = fg$. It follows from (2–2) that

$$(\Gamma_\alpha a, b) = \sum_{j,k \geq 0} \hat{\psi}(j+k) a_j \bar{b}_k = \sum_{m \geq 0} \hat{\psi}(m) \sum_{j=0}^{m} a_j \bar{b}_{m-j}$$

$$= \sum_{m \geq 0} \hat{\psi}(m) \hat{q}(m) = \int_{\mathbb{T}} \psi(\zeta) q(\bar{\zeta}) \, d\boldsymbol{m}(\zeta).$$

Therefore

$$|(\Gamma_\alpha a, b)| \leq \|\psi\|_\infty \|q\|_{H^1} \leq \|\psi\|_\infty \|f\|_{H^2} \|g\|_{H^2} = \|\psi\|_{L^\infty} \|a\|_{\ell^2} \|b\|_{\ell^2}.$$

Conversely, suppose that Γ is bounded on ℓ^2. Let L_α be the linear functional defined on the set of polynomials in H^1 by

$$L_\alpha q = \sum_{n \geq 0} \alpha_n \hat{q}(n). \tag{2–3}$$

We show that L_α extends by continuity to a continuous functional on H^1 and its norm $\|L_\alpha\|$ on H^1 satisfies

$$\|L_\alpha\| \leq \|\Gamma_\alpha\|. \tag{2–4}$$

By the Hahn-Banach theorem this will imply the existence of some ψ in L^∞ that satisfies (2–1) and

$$\|\psi\|_\infty \leq \|\Gamma_\alpha\|.$$

Assume first that $\alpha \in \ell^1$. In this case the functional L_α defined by (2–3) is obviously continuous on H^1. We prove (2–4). Let $q \in H^1$ and $\|q\|_1 \leq 1$. Then q admits a representation $q = fg$, where $f, g \in H^2$ and $\|f\|_2 \leq 1$, $\|g\|_2 \leq 1$. We have

$$L_\alpha q = \sum_{m \geq 0} \alpha_m \hat{q}(m) = \sum_{m \geq 0} \alpha_m \sum_{j=0}^{m} \hat{f}(j) \hat{g}(m-j) = \sum_{j,k \geq 0} \alpha_{j+k} \hat{f}(j) \hat{g}(k) = (\Gamma_\alpha a, b),$$

where $a = \{a_j\}_{j \geq 0}$ with $a_j = \hat{f}(j)$ and $b = \{b_k\}_{k \geq 0}$ with $b_k = \overline{\hat{g}(k)}$. Therefore

$$|L_\alpha q| \leq \|\Gamma_\alpha\| \|f\|_2 \|g\|_2 \leq \|\Gamma_\alpha\|,$$

which proves (2–4) for $\alpha \in \ell^1$.

Now assume that α is an arbitrary sequence for which Γ_α is bounded. Let $0 < r < 1$. Consider the sequence $\alpha^{(r)}$ defined by

$$\alpha_j^{(r)} = r^j \alpha_j \quad \text{for } j \geq 0.$$

It is easy to see that $\Gamma_{\alpha^{(r)}} = D_r \Gamma_\alpha D_r$, where D_r is multiplication by $\{r^j\}_{j \geq 0}$ on ℓ^2. Since obviously $\|D_r\| \leq 1$, it follows that the operators $\Gamma_{\alpha^{(r)}}$ are bounded and

$$\|\Gamma_{\alpha^{(r)}}\| \leq \|\Gamma_\alpha\| \quad \text{for } 0 < r < 1.$$

Clearly $\alpha^{(r)} \in \ell^1$, so we have already proved that

$$\|L_{\alpha^{(r)}}\|_{H^1 \to \mathbb{C}} \leq \|\Gamma_{\alpha^{(r)}}\| \leq \|\Gamma_\alpha\|.$$

It is easy to see now that the functionals $L_{\alpha^{(r)}}$ being uniformly bounded converge strongly to L_α; that is, $L_{\alpha^{(r)}}\psi \to L_\alpha \psi$ for any $\psi \in H^1$. This proves that L_α is continuous and satisfies (2–4). $\qquad\qquad\qquad\qquad\qquad\qquad\qquad\qquad\qquad\qquad\square$

Theorem 2.1 reduces the problem of whether a sequence α determines a bounded operator on ℓ^2 to the question of the existence of an extension of α to the sequence of Fourier coefficients of a bounded function. However, after the work of C. Fefferman on the space BMO of functions of bounded mean oscillation it has become possible to determine whether Γ_α is bounded in terms of the sequence α itself.

By C. Fefferman's theorem (see [Garnett 1981], for example), a function φ on the unit circle belongs to the space BMO if and only if it admits a representation

$$\varphi = \xi + \mathbb{P}_+ \eta \quad \text{with } \xi, \eta \in L^\infty.$$

The space BMOA is by definition the space of BMO functions analytic in the unit disc $\mathbb{D}$:

$$\text{BMOA} = \text{BMO} \cap H^1.$$

It is easy to see that Nehari's and Fefferman's theorems imply the following result.

THEOREM 2.2. *The operator Γ_α is bounded on ℓ^2 if and only if the function*

$$\varphi = \sum_{m \geq 0} \alpha_m z^m \tag{2–5}$$

belongs to BMOA.

Clearly Γ_α is a bounded operator if the function φ defined by (2–5) is bounded. However, the operator Γ_α can be bounded even with an unbounded φ. We consider an important example of such a Hankel matrix:

EXAMPLE (THE HILBERT MATRIX). Let $\alpha_n = 1/(n+1)$ for $n \geq 0$. The corresponding Hankel matrix Γ_α is called the *Hilbert matrix*. Clearly the function

$$\sum_{n \geq 0} \frac{1}{n+1} z^n$$

is unbounded in $\mathbb{D}$. However, Γ_α is bounded. Indeed, consider the function ψ on $\mathbb{T}$ defined by

$$\psi(e^{it}) = ie^{-it}(\pi - t) \quad \text{for } t \in [0, 2\pi).$$

It is easy to see that

$$\hat{\psi}(n) = \frac{1}{n+1} \quad \text{for } m \geq 0,$$

and that $\|\psi\|_{L^\infty} = \pi$. It follows from Theorem 2.1 that Γ_α is bounded and $\|\Gamma_\alpha\| \le \pi$. In fact, $\|\Gamma_\alpha\| = \pi$; see [Nikol'skiĭ 1986, Appendix 4, 165.21], for example.

Clearly, Theorem 2.1 admits the following reformulation.

THEOREM 2.3. *Let* $\varphi \in L^2$. *The following statements are equivalent*:

(a) H_φ *is bounded on* H^2.

(b) *There exists a function* ψ *in* L^∞ *such that*

$$\hat{\psi}(m) = \hat{\varphi}(m) \quad \text{for } m < 0. \tag{2-6}$$

(c) $\mathbb{P}_-\varphi \in \text{BMO}$.

If one of the conditions (a)–(c) *is satisfied, then*

$$\|H_\varphi\| = \inf\{\|\psi\|_{L^\infty} : \hat{\psi}(m) = \hat{\varphi}(m) \text{ for } m < 0\}. \tag{2-7}$$

Equality (2–6) is equivalent to the fact that $H_\varphi = H_\psi$. Thus (b) means that H_φ is bounded if and only if it has a bounded symbol. So the operators H_φ with $\varphi \in L^\infty$ exhaust the class of bounded Hankel operators. If $\varphi \in L^\infty$, (2–7) can be rewritten in the following way:

$$\|H_\varphi\| = \inf\{\|\varphi - f\|_\infty : f \in H^\infty\}. \tag{2-8}$$

Let $\varphi \in L^\infty$. It follows easily from a compactness argument that the infimum on the right-hand side of (2–8) is attained for any $\varphi \in L^\infty$. A function f that realizes the minimum on the right-hand side of (2–8) is called *a best approximation of* φ *by analytic functions in the* L^∞-*norm*. The problem of finding, for a given $\varphi \in L^\infty$, a best approximation by analytic functions is called *Nehari's problem*. It plays a significant role in applications, particularly in control theory. If f realizes the minimum on the right-hand side of (2–8), then clearly, $\varphi - f$ is a symbol of H_φ of minimal L^∞-norm. A natural question arises of whether such a symbol of minimal norm is unique.

The first results in this direction were apparently obtained by Khavinson [1951] (see also [Rogosinski and Shapiro 1953]), where it was shown that for a continuous function φ on $\mathbb{T}$ there exists a unique best uniform approximation by analytic functions and that uniqueness fails in general; see also [Garnett 1981, Section IV.1]. However, in the case when the Hankel operator attains its norm on the unit ball of H^2, that is, when $\|H_\varphi g\|_2 = \|H_\varphi\|\|g\|_2$ for some nonzero $g \in H^2$, we do have uniqueness, as the following result shows [Adamyan et al. 1968b].

THEOREM 2.4. *Let* φ *be a function in* L^∞ *such that* H_φ *attains its norm on the unit ball of* H^2. *Then there exists a unique function* f *in* H^∞ *such that*

$$\|\varphi - f\|_\infty = \text{dist}_{L^\infty}(\varphi, H^\infty).$$

Moreover, $\varphi - f$ has constant modulus almost everywhere on $\mathbb{T}$ and admits a representation

$$\varphi - f = \|H_\varphi\| \, \bar{z}\vartheta \, \frac{\bar{h}}{h}, \tag{2--9}$$

where h is an outer function in H^2 and ϑ is an inner function.

PROOF. Without loss of generality we can assume that $\|H_\varphi\| = 1$. Let g be a function in H^2 such that $1 = \|g\|_2 = \|H_\varphi g\|_2$. Let $f \in H^\infty$ be a best approximation of φ, so that $\|\varphi - f\|_\infty = 1$. We have

$$1 = \|H_\varphi g\|_2 = \|\mathbb{P}_-(\varphi - f)g\|_2 \leq \|(\varphi - f)g\|_2 \leq \|g\|_2 = 1.$$

Therefore the inequalities in this chain are in fact equalities. The fact that

$$\|\mathbb{P}_-(\varphi - f)g\|_2 = \|(\varphi - f)\|_2$$

means that $(\varphi - f)g \in H_-^2$, so

$$H_\varphi g = H_{\varphi - f} g = (\varphi - f)g. \tag{2--10}$$

The function g, being in H^2, is nonzero almost everywhere on $\mathbb{T}$, so

$$f = \varphi - \frac{H_\varphi g}{g}.$$

Hence f is determined uniquely by H_φ: the ratio $(H_\varphi g)/g$ does not depend on the choice of g.

Since $\|\varphi - f\|_\infty = 1$, the equality

$$\|(\varphi - f)g\|_2 = \|g\|_2$$

means that $|\varphi(\zeta) - f(\zeta)| = 1$ a.e. on the set $\{\zeta : g(\zeta) \neq 0\}$, which is of full measure since $g \in H^2$. Thus $\varphi - f$ has modulus one almost everywhere on $\mathbb{T}$.

Consider the functions g and $\bar{z}\overline{H_\varphi g}$ in H^2. It follows from (2--10) that they have the same moduli. Therefore they admit factorizations

$$g = \vartheta_1 h, \quad \bar{z}\overline{H_\varphi g} = \vartheta_2 h,$$

where h is an outer function in H^2, and ϑ_1 and ϑ_2 are inner functions. Consequently,

$$\varphi - f = \frac{H_\varphi g}{g} = \frac{\bar{z}\vartheta_2 \bar{h}}{\vartheta_1 h} = \bar{z}\bar{\vartheta}_1\vartheta_2\frac{\bar{h}}{h},$$

which proves (2--9) with $\vartheta = \vartheta_1\vartheta_2$. $\qquad\qquad\square$

COROLLARY 2.5. *If H_φ is a compact Hankel operator, the conclusion of Theorem 2.4 holds.*

PROOF. Any compact operator attains its norm. $\qquad\qquad\square$

We shall see in Section 5 that Hankel operators with continuous symbols are compact, so Corollary 2.5 implies Khavinson's theorem [1951] mentioned above.

Adamyan, Arov, and Krein proved in [Adamyan et al. 1968a] that, if there are at least two best approximations to φ, there exists a best approximation g such that $\varphi - g$ has constant modulus on $\mathbb{T}$. They found a formula that parametrizes all best approximations.

We now show that Hankel operators can be characterized as the operators that satisfy a certain commutation relation. Recall the S and $\mathcal{S}$ are the shift and bilateral shift operators, respectively.

THEOREM 2.6. *Let R be a bounded operator from H^2 to H^2_-. Then R is a Hankel operator if and only if it satisfies the commutation relation*

$$\mathbb{P}_- \mathcal{S} R = RS. \qquad (2\text{–}11)$$

PROOF. Let $R = H_\varphi$, with $\varphi \in L^\infty$. Then

$$\mathbb{P}_- \mathcal{S} R f = \mathbb{P}_- z H_\varphi f = \mathbb{P}_- z \mathbb{P} \varphi f = \mathbb{P}_- z \varphi f = H_\varphi z f.$$

Conversely, suppose that R satisfies (2–11). Let $n \geq 1$, $k \geq 1$. We have

$$(Rz^n, \bar{z}^k) = (RSz^{n-1}, \bar{z}^k) = (\mathbb{P}_- \mathcal{S} R z^{n-1}, \bar{z}^k) = (\mathcal{S} R z^{n-1}, \bar{z}^k) = (Rz^{n-1}, \bar{z}^{k+1}).$$

Therefore R has Hankel matrix in the bases $\{z^n\}_{n \geq 0}$ of H^2 and $\{\bar{z}^k\}_{k \geq 1}$ of H^2_-. It follows from Theorems 2.1 and 2.3 that $R = H_\varphi$ for some φ in L^∞. $\square$

3. Sarason's Theorem

In this section we study the commutant of compressions of the shift operator on H^2 to its coinvariant subspaces, and we prove Sarason's commutant lifting theorem. We use an approach given in [Nikol'skiĭ 1986, Section VIII.1], based on Hankel operators and Nehari's theorem. Then we establish an important formula that relates functions of such a compression with Hankel operators.

Let ϑ be an inner function. Put

$$K_\vartheta = H^2 \ominus \vartheta H^2.$$

By Beurling's theorem (see [Hoffman 1962, Chapter 7] or [Nikol'skiĭ 1986, Section I.1], for example), any nontrivial invariant subspace of the backward shift operator S^* on H^2 coincides with K_ϑ for some inner function ϑ. Denote by S_ϑ the compression of the shift operator S to K_ϑ, defined by

$$S_\vartheta f = P_\vartheta z f \quad \text{for } f \in K_\vartheta, \qquad (3\text{–}1)$$

where P_ϑ is the orthogonal projection from H^2 onto K_ϑ. Clearly, $S_\vartheta^* = S^*|K_\vartheta$.

It can easily be shown that

$$P_\vartheta f = f - \vartheta \mathbb{P}_+ \bar{\vartheta} f = \vartheta \mathbb{P}_- \bar{\vartheta} f \quad \text{for } f \in H^2. \qquad (3\text{–}2)$$

S_ϑ is the *model operator* in the Sz.-Nagy–Foiaş function model. Any *contraction* T (that is, a linear operator such that $\|T\| \leq 1$) for which $\lim_{n\to\infty} T^{*n} = 0$ in the strong operator topology and $\mathrm{rank}(I - T^*T) = \mathrm{rank}(I - TT^*) = 1$ is unitarily equivalent to S_ϑ for some inner function ϑ; see [Sz.-Nagy and Foiaş 1967, Chapter 6; Nikol'skiĭ 1986, Lecture I].

The operator S_ϑ admits an H^∞ functional calculus. Indeed, given $\varphi \in H^\infty$, we define the operator $\varphi(S_\vartheta)$ by

$$\varphi(S_\vartheta)f = P_\vartheta \varphi f \quad \text{for } f \in K_\vartheta. \tag{3–3}$$

Clearly, this functional calculus is linear. It is also easy to verify that it is multiplicative. Hence, for any $\varphi \in H^\infty$, the operator $\varphi(S_\vartheta)$ commutes with S_ϑ, and it follows from (3–3) that

$$\|\varphi(S_\vartheta)\| \leq \|\varphi\|_{H^\infty}.$$

This is known as *von Neumann's inequality*.

The following theorem of Sarason [1967] describes the commutant of S_ϑ. It is a partial case of the commutant lifting theorem of Sz.-Nagy and Foiaş [1967].

THEOREM 3.1. *Let T be an operator that commutes with S_ϑ. Then there exists a function φ in H^∞ such that $T = \varphi(S_\vartheta)$ and $\|T\| = \|\varphi\|_{H^\infty}$.*

LEMMA 3.2. *Let T be an operator on K_ϑ. Consider the operator $\tilde{T} : H^2 \to H^2_-$ defined by*

$$\tilde{T}f = \bar{\vartheta}\, T P_\vartheta\, f. \tag{3–4}$$

Then T commutes with S_ϑ if and only if $\tilde{T}$ is a Hankel operator.

PROOF. $\tilde{T}$ is a Hankel operator if and only if

$$\mathbb{P}_- z\tilde{T}f = \tilde{T}zf \quad \text{for } f \in H^2 \tag{3–5}$$

(see (2–11)), which means that

$$\mathbb{P}_- z\bar{\vartheta} T P_\vartheta f = \bar{\vartheta} T P_\vartheta z f \quad \text{for } f \in H^2,$$

which in turn is equivalent to

$$\vartheta \mathbb{P}_- \bar{\vartheta} z T P_\vartheta f = T P_\vartheta z f \quad \text{for } f \in H^2. \tag{3–6}$$

We have by (3–2)

$$\vartheta \mathbb{P}_- \bar{\vartheta} z T P_\vartheta f = P_\vartheta z T P_\vartheta f = S_\vartheta T P_\vartheta f.$$

Since obviously the left-hand side and the right-hand side of (3–6) are zero for $f \in \vartheta H^2$, it follows from (3–1) that (3–5) is equivalent to the equality

$$S_\vartheta T f = T S_\vartheta f \quad \text{for } f \in K_\vartheta. \qquad \square$$

PROOF OF THEOREM 3.1. By Lemma 3.2 the operator $\tilde{T}$ defined by (3–4) is a Hankel operator. By Nehari's theorem there exists a function ψ in L^∞ such that $\|\psi\|_\infty = \|\tilde{T}\|$ and $H_\psi = \tilde{T}$; that is,

$$\mathbb{P}_-\psi f = \bar{\vartheta} T P_\vartheta f \quad \text{for } f \in H^2.$$

It follows that $\mathbb{P}_-\psi f = 0$ for any $f \in \vartheta H^2$. That means that $H_{\psi\vartheta} = 0$. Put $\varphi = \psi\vartheta$. Clearly $\varphi \in H^\infty$ and $\psi = \bar{\vartheta}\varphi$. We have

$$\bar{\vartheta} T f = \mathbb{P}_-\bar{\vartheta}\varphi f \quad \text{for } f \in K_\vartheta,$$

so

$$T f = \vartheta \mathbb{P}_-\bar{\vartheta}\varphi f = P_\vartheta \varphi f = \varphi(S_\vartheta)f \quad \text{for } f \in K_\vartheta. \tag{3–7}$$

Obviously

$$\|T\| = \|\tilde{T}\| = \|\psi\|_\infty = \|\varphi\|_\infty,$$

which completes the proof.   □

REMARK. Formula (3–7) implies a remarkable relation, due to Nikol'skiĭ [1986], between Hankel operators and functions of model operators: Let ϑ be an inner function, and let $\varphi \in H^\infty$. Then

$$\varphi(S_\vartheta) = \Theta H_{\bar{\vartheta}\varphi}|K_\vartheta, \tag{3–8}$$

where Θ is multiplication by ϑ. This formula shows that $\varphi(S_\vartheta)$ has the same metric properties as $H_{\bar{\vartheta}\varphi}$; compactness, nuclearity, etc.

Formula (3–8) relates the Hankel operators of the form $H_{\bar{\vartheta}\varphi}$ with functions of model operators. It can easily be shown that such Hankel operators are exactly the Hankel operators from H^2 to H^2_- that have a nontrivial kernel. It is worth mentioning that the set of functions of the form $\bar{\vartheta}\varphi$, where ϑ is inner and $\varphi \in H^\infty$, forms a dense subset of L^∞ [Douglas 1972, 6.32].

4. Finite Rank

One of the first results about Hankel matrices was a theorem of Kronecker [1881] that describes the Hankel matrices of finite rank.

Let $r = p/q$ be a rational function where p and q are polynomials. If p/q is in its lowest terms, the degree of r is, by definition,

$$\deg r = \max(\deg p, \deg q),$$

where $\deg p$ and $\deg q$ are the degrees of the polynomials p and q. It is easy to see that $\deg r$ is the sum of the multiplicities of the poles of r (including a possible pole at infinity).

We are going to describe the Hankel matrices of finite rank without any assumption on the boundedness of the matrix.

We identify sequences of complex numbers with the corresponding formal power series. If $a = \{a_j\}_{j \geq 0}$ is a sequence of complex numbers, we associate with it the formal power series

$$a(z) = \sum_{j \geq 0} a_j z^j.$$

The space of formal power series forms an algebra with respect to the multiplication

$$(ab)(z) = \sum_{m \geq 0} \left(\sum_{j=0}^{m} a_j b_{m-j} \right) z^m, \quad \text{with } a = \sum_{j \geq 0} a_j z^j \text{ and } b = \sum_{j \geq 0} b_j z^j.$$

Consider the shift operator S and the backward shift operator S^* defined on the space of formal power series in the following way:

$$(Sa)(z) = za(z), \qquad S^* \sum_{j \geq 0} a_j z^j = \sum_{j \geq 0} a_{j+1} z^j.$$

Let $\alpha = \{\alpha_j\}_{j \geq 0}$ be a sequence of complex numbers, which we identify with the corresponding formal power series

$$\alpha(z) = \sum_{j \geq 0} \alpha_j z^j. \tag{4–1}$$

Denote by Γ_α the Hankel matrix $\{\alpha_{j+k}\}_{j,k \geq 0}$.

THEOREM 4.1. *Γ_α has finite rank if and only if the power series* (4–1) *determines a rational function. In this case*

$$\operatorname{rank} \Gamma_\alpha = \deg z\alpha(z).$$

PROOF. Suppose that $\operatorname{rank} \Gamma_\alpha = n$. Then the first $n+1$ rows are linearly dependent. That means that there exists a nontrivial family $\{c_j\}_{0 \leq j \leq n}$ of complex numbers (*nontrivial* means that not all the c_j are equal to zero) such that

$$c_0 \alpha + c_1 S^* \alpha + \cdots + c_n S^{*n} \alpha = 0. \tag{4–2}$$

It is easy to see that

$$S^n S^{*k} \alpha = S^{n-k} \alpha - S^{n-k} \sum_{j=0}^{k-1} \alpha_j z^j \quad \text{for } k \leq n. \tag{4–3}$$

It follows easily from (4–2) and (4–3) that

$$0 = S^n \sum_{k=0}^{n} c_k S^{*k} \alpha = \sum_{k=0}^{n} c_k S^n S^{*k} \alpha = \sum_{k=0}^{n} c_k S^{n-k} \alpha - p, \tag{4–4}$$

where p has the form

$$p(z) = \sum_{j=0}^{n-1} p_j z^j.$$

Put

$$q(z) = \sum_{j=0}^{n} c_{n-j} z^{j}. \tag{4–5}$$

Then p and q are polynomials and it follows from (4–4) that $q\alpha = p$, so $\alpha(z) = (p/q)(z)$ is a rational function. Clearly,

$$\deg z\alpha(z) \leq \max (\deg zp(z), \deg q(z)) = n.$$

Conversely, suppose that $\alpha(z) = (p/q)(z)$ where p and q are polynomials such that $\deg p \leq n - 1$ and $\deg q \leq n$. Consider the complex numbers c_j defined by (4–5). We have

$$\sum_{j=0}^{n} c_j S^{n-j} \alpha = p.$$

Therefore

$$S^{*n} \sum_{j=0}^{n} c_j S^{n-j} \alpha = \sum_{j=0}^{n} c_j S^{*j} \alpha = 0,$$

which means that the first $n + 1$ rows of Γ_α are linearly dependent. Let $m \leq n$ be the largest number for which $c_m \neq 0$. Then $S^{*m}\alpha$ is a linear combination of the $S^{*j}\alpha$ with $j \leq m - 1$:

$$S^{*m} \alpha = \sum_{j=0}^{m-1} d_j S^{*j} \alpha \quad \text{with } d_j \in \mathbb{C}.$$

We will show by induction that any row of Γ_α is a linear combination of the first m rows. Let $k > m$. We have

$$S^{*k} \alpha = (S^*)^{k-m} S^{*m} \alpha = \sum_{j=0}^{m-1} d_j (S^*)^{k-m+j} \alpha. \tag{4–6}$$

Since $k - m + j < k$ for $0 \leq j \leq m - 1$, by the induction hypothesis each of the terms of the right-hand side of (4–6) is a linear combination of the first m rows. Therefore $\operatorname{rank} \Gamma_\alpha \leq m$, which completes the proof. $\square$

It is easy to see that Kronecker's theorem for Hankel operators H_φ on H^2 admits the following reformulation.

COROLLARY 4.2. *Let $\varphi \in L^\infty$. The Hankel operator H_φ has finite rank if and only if $\mathbb{P}_- \varphi$ is a rational function. In this case*

$$\operatorname{rank} H_\varphi = \deg \mathbb{P}_- \varphi.$$

COROLLARY 4.3. *H_φ has finite rank if and only if there exists a finite Blaschke product B such that $B\varphi \in H^\infty$.*

5. Compactness

In this section we establish Hartman's compactness criterion for Hankel operators. We also compute the essential norm of a Hankel operator and study the problem of approximation of a Hankel operator by compact Hankel operators.

Recall that the essential norm $\|T\|_e$ of an operator T from a Hilbert space $\mathcal{H}_1$ to a Hilbert space $\mathcal{H}_2$ is, by definition,

$$\|T\|_e = \inf\{\|T - K\| : K \text{ is compact}\}. \qquad (5\text{--}1)$$

To compute the essential norm of a Hankel operator, we have to introduce the space $H^\infty + C$.

DEFINITION. The space $H^\infty + C$ is the set of functions φ in L^∞ such that φ admits a representation $\varphi = f + g$, where $f \in H^\infty$ and $g \in C(\mathbb{T})$.

THEOREM 5.1 [Sarason 1978]. *The set $H^\infty + C$ is a closed subalgebra of L^∞.*

To prove the theorem, we need the following elementary lemma, where

$$C_A \overset{\text{def}}{=} H^\infty \cap C(\mathbb{T}).$$

LEMMA 5.2. *Let $\varphi \in C(\mathbb{T})$. Then*

$$\operatorname{dist}_{L^\infty}(\varphi, H^\infty) = \operatorname{dist}_{L^\infty}(\varphi, C_A). \qquad (5\text{--}2)$$

PROOF. The inequality $\operatorname{dist}(\varphi, H^\infty) \le \operatorname{dist}_{L^\infty}(\varphi, C_A)$ is trivial; we prove the opposite one. For $f \in L^\infty$ we consider its harmonic extension to the unit disc and keep the same notation for it. Put $f_r(\zeta) = f(r\zeta)$ for $\zeta \in \mathbb{D}$. Let $\varphi \in C(\mathbb{T})$, $h \in H^\infty$. We have

$$\|\varphi - h\|_\infty \ge \lim_{r \to 1} \|(\varphi - h)_r\|_\infty \ge \lim_{r \to 1}(\|\varphi - h_r\|_\infty - \|\varphi - \varphi_r\|_\infty)$$
$$= \lim_{r \to 1} \|\varphi - h_r\|_\infty \ge \operatorname{dist}_{L^\infty}(\varphi, C_A),$$

since $\|\varphi - \varphi_r\|_\infty \to 0$ for continuous φ. $\qquad \square$

PROOF OF THEOREM 5.1. Equality (5–2) means exactly that the natural imbedding of $C(\mathbb{T})/C_A$ in L^∞/H^∞ is isometric, so $C(\mathbb{T})/C_A$ can be considered as a closed subspace of L^∞/H^∞. Let $\rho : L^\infty \to L^\infty/H^\infty$ be the natural quotient map. It follows that $H^\infty + C = \rho^{-1}(C(\mathbb{T})/C_A)$ is closed in L^∞.

This implies that

$$H^\infty + C = \operatorname{clos}_{L^\infty}(\cup_{n \ge 0} \bar{z}^n H^\infty). \qquad (5\text{--}3)$$

It is easy to see that if f and g belong to the right-hand side of (5–3), then so does fg. Hence $H^\infty + C$ is an algebra. $\qquad \square$

Now we are going to compute the essential norm of a Hankel operator. The following result was apparently discovered by Adamyan, Arov, and Krein [Adamyan et al. 1968b].

THEOREM 5.3. *Let* $\varphi \in L^\infty$. *Then*

$$\|H_\varphi\|_e = \text{dist}_{L^\infty}(\varphi, H^\infty + C).$$

LEMMA 5.4. *Let K be a compact operator from H^2 to H^2_-. Then*

$$\lim_{n\to\infty} \|KS^n\| = 0.$$

PROOF. Since any compact operator can be approximated by finite-rank operators, it is sufficient to prove the assertion for rank-one operators K. Let $Kf = (f, \xi)\eta$, $\xi \in H^2$, $\eta \in H^2_-$. We have $KS^n f = (f, S^{*n}\xi)\eta$, so

$$\|KS^n\| = \|S^{*n}\xi\|_2 \|\eta\|_2 \to 0. \qquad \square$$

PROOF OF THEOREM 5.3. By Corollary 4.2, H_f is compact for any trigonometric polynomial f. Therefore H_f is compact for any f in $C(\mathbb{T})$. Consequently,

$$\text{dist}_{L^\infty}(\varphi, H^\infty + C) = \inf_{f \in C(\mathbb{T})} \|H_\varphi - H_f\| \geq \|H_\varphi\|_e.$$

On the other hand, for any compact operator K from H^2 to H^2_-,

$$\|H_\varphi - K\| \geq \|(H_\varphi - K)S^n\| \geq \|H_\varphi S^n\| - \|KS^n\| = \|H_{z^n\varphi}\| - \|KS^n\|$$

$$= \text{dist}_{L^\infty}(\varphi, \bar{z}^n H^\infty) - \|KS^n\| \geq \text{dist}_{L^\infty}(\varphi, H^\infty + C) - \|KS^n\|.$$

Therefore, in view of Lemma 5.4,

$$\|H_\varphi\|_e \geq \text{dist}_{L^\infty}(\varphi, H^\infty + C). \qquad \square$$

REMARK. In Section 2 we studied the question of existence and uniqueness of a best H^∞ approximant in the L^∞-norm. The same question can be asked about approximation by $H^\infty + C$ functions; it was explicitly posed by Adamyan, Arov, and Krein in [Adamyan et al. 1984]. However, the situation here is quite different. It was shown in [Axler et al. 1979] that, for any $\varphi \in L^\infty \setminus H^\infty + C$, there are infinitely many best approximants in $H^\infty + C$. See also [Luecking 1980] for another proof.

We now obtain Hartman's compactness criterion.

THEOREM 5.5. *Let* $\varphi \in L^\infty$. *The following statements are equivalent.*

(a) H_φ *is compact.*
(b) $\varphi \in H^\infty + C$.
(c) *There exists a function ψ in $C(\mathbb{T})$ such that $H_\varphi = H_\psi$.*

PROOF. Obviously (b) and (c) are equivalent.

Suppose that $\varphi \in H^\infty + C$. Then $\|H_\varphi\|_e = 0$ by Theorem 5.3, which means that H_φ is compact. Thus (b) implies (a).

To show that (a) implies (b), assume H_φ is compact. Then Theorem 5.3 gives $\text{dist}_{L^\infty}(\varphi, H^\infty + C) = 0$, which, in combination with Theorem 5.5, yields $\varphi \in H^\infty + C$. $\qquad \square$

COROLLARY 5.6. *Let $\varphi \in L^\infty$. Then*

$$\|H_\varphi\|_e = \inf\{\|H_\varphi - H_\psi\| : H_\psi \text{ is compact}\}. \tag{5-4}$$

In other words, to compute the essential norm of a Hankel operator we can consider on the right-hand side of (5–1) only compact Hankel operators.

COROLLARY 5.7. *Let $\varphi \in H^\infty + C$. Then for any $\varepsilon > 0$ there exists a function ψ in $C(\mathbb{T})$ such that $H_\psi = H_\varphi$ and $\|\psi\|_\infty \le \|H_\varphi\| + \varepsilon$.*

PROOF. Without loss of generality we can assume that $\varphi \in C(\mathbb{T})$. By Theorem 5.3, $\|H_\varphi\| = \operatorname{dist}_{L^\infty}(\varphi, H^\infty)$. On the other hand, by Lemma 5.2,

$$\operatorname{dist}_{L^\infty}(\varphi, H^\infty) = \operatorname{dist}_{L^\infty}(\varphi, C_A).$$

This means that for any $\varepsilon > 0$ there exists a function $h \in C_A$ such that $\|\varphi - h\| \le \|H_\varphi\| + \varepsilon$. Thus $\psi = \varphi - h$ does the job.   □

EXAMPLE. For a compact Hankel operator, it is not always possible to find a continuous symbol whose L^∞-norm is equal to the norm of the operator. Indeed, let α be a real-valued function in $C(\mathbb{T})$ such that $\tilde{\alpha} \notin C(\mathbb{T})$, where $\tilde{\alpha}$ is the harmonic conjugate of α. Put $\varphi = \bar{z}e^{i\tilde{\alpha}}$. Then $\varphi = \bar{z}e^{\alpha + i\tilde{\alpha}}e^{-\alpha}$. Clearly $e^{\alpha + i\tilde{\alpha}} \in H^\infty$ and $e^{-\alpha} \in C(\mathbb{T})$. It follows from Theorem 5.1 that $\varphi \in H^\infty + C$ and so H_φ is compact. Let us show that $\|H_\varphi\| = 1$. Put

$$h = \exp \tfrac{1}{2}(\tilde{\alpha} - i\alpha).$$

Clearly, h is an outer function. To prove that $h \in H^2$ we need the following theorem of Zygmund [Zygmund 1968, Chapter 7, Theorem 2.11]: If ξ is a bounded real function such that $\|\xi\|_{L^\infty} < \pi/(2p)$, then $\exp \tilde{\xi} \in L^p$. Indeed, approximating α by trigonometric polynomials, we can easily deduce from Zygmund's theorem that $h \in H^p$ for any $p < \infty$. Clearly $\|H_\varphi h\|_2 = \|\bar{z}\bar{h}\|_2 = \|h\|_2$. Hence $\|H_\varphi\| = \|\varphi\|_\infty = 1$. By Corollary 2.4, $\|\varphi + f\|_\infty > 1$ for any nonzero f in H^∞. It is also clear that $\varphi \notin C(\mathbb{T})$. This proves the result.

In Section 2 we gave a boundedness criterion for a Hankel operator H_φ in terms of $\mathbb{P}_- \varphi$. That criterion involves the condition $\mathbb{P}_- \varphi \in \mathrm{BMO}$. We can give a similar compactness criterion if we replace BMO by the space VMO of functions of vanishing mean oscillation.

THEOREM 5.8. *Let $\varphi \in L^2$. Then H_φ is compact if and only if $\mathbb{P}_- \varphi \in \mathrm{VMO}$.*

This can be derived from Theorem 5.5 in the same way as it has been done in Section 2 if we use the following description of VMO due to Sarason:

$$\mathrm{VMO} = \{\xi + \mathbb{P}_+ \eta : \xi, \eta \in C(\mathbb{T})\}.$$

See [Garnett 1981], for example.

6. Approximation by Finite-Rank Operators

DEFINITION. For a bounded linear operator T from a Hilbert space $\mathcal{H}_1$ to a Hilbert space $\mathcal{H}_2$, the *singular values* $s_m(T)$, for $m \in \mathbb{Z}_+$, are defined by

$$s_m(T) = \inf\{\|T - R\| : \operatorname{rank} R \leq m\}. \tag{6-1}$$

Clearly, $s_0(T) = \|T\|$ and $s_{m+1}(T) \leq s_m(T)$.

Adamyan, Arov, and Krein [Adamyan et al. 1971] proved that in order to find $s_m(T)$ for a Hankel operator T we can consider the infimum in (6-1) over only the Hankel operators of rank at most m. This is a deep and important result.

THEOREM 6.1. *Let Γ be a Hankel operator from H^2 to H^2_-, and let $m \geq 0$. Then there exists a Hankel operator Γ_m of rank at most m such that*

$$\|\Gamma - \Gamma_m\| = s_m(\Gamma). \tag{6-2}$$

By Kronecker's theorem, $\operatorname{rank}\Gamma_m$ is at most m if and only if Γ_m has a rational symbol of degree at most m, so Theorem 6.1 admits the following reformulation. Let $\tilde{\mathcal{R}}_m$ be the set of functions f in L^∞ such that $\mathbb{P}_- f$ is a rational function of degree at most m. Clearly, $\tilde{\mathcal{R}}_m$ can be identified with the set of meromorphic functions in $\mathbb{D}$ bounded near $\mathbb{T}$ and having at most m poles in $\mathbb{D}$ counted with multiplicities.

THEOREM 6.2. *Let $\varphi \in L^\infty$, $m \in \mathbb{Z}_+$. There exists a function ψ in $\tilde{\mathcal{R}}_m$ such that*

$$\|\varphi - \psi\|_\infty = s_m(H_\varphi). \tag{6-3}$$

We will prove Theorems 6.1 and 6.2 only for compact Hankel operators. For the general case see [Adamyan et al. 1971] or [Treil' 1985a], where an alternative proof is given. Another fact that we state without proof is that for a compact Hankel operator there exists a *unique* Hankel operator Γ_m of rank at most m that satisfies (6-2); see [Adamyan et al. 1971].

DEFINITION. Let T be a compact linear operator from a Hilbert space $\mathcal{H}_1$ to a Hilbert space $\mathcal{H}_2$. If s is a singular value of T, consider the subspaces

$$E_s^{(+)} = \{x \in \mathcal{H}_1 : T^*Tx = s^2 x\}, \qquad E_s^{(-)} = \{y \in \mathcal{H}_2 : TT^*y = s^2 y\}.$$

Vectors in $E_s^{(+)}$ are called *Schmidt vectors* of T (or, more precisely, s-Schmidt vectors of T). Vectors in $E_s^{(-)}$ are called *Schmidt vectors* of T^* (or s-Schmidt vectors of T^*). Clearly, $x \in E_s^{(+)}$ if and only if $Tx \in E_s^{(-)}$. A pair $\{x, y\}$, with $x \in \mathcal{H}_1$ and $y \in \mathcal{H}_2$, is called a *Schmidt pair* of T (or s-Schmidt pair) if $Tx = sy$ and $T^*y = sx$.

PROOF OF THEOREM 6.1 IN THE COMPACT CASE. Put $s = s_m(\Gamma)$. If $s = \|\Gamma\|$, the result is trivial. Assume that $s < \|\Gamma\|$. Then there exist positive integers k and μ such that $k \leq m \leq k + \mu - 1$ and

$$s_{k-1}(\Gamma) > s_k(\Gamma) = \cdots = s_{k+\mu-1}(\Gamma) > s_{k+\mu}(\Gamma). \qquad (6\text{--}4)$$

Clearly, it suffices to consider the case $m = k$.

LEMMA 6.3. Let $\{\xi_1, \eta_1\}$ and $\{\xi_2, \eta_2\}$ be s-Schmidt pairs of Γ. Then $\xi_1 \bar{\xi}_2 = \eta_1 \bar{\eta}_2$.

To prove the lemma we need the following identity, which is a consequence of (2–11):

$$\mathbb{P}_- (z^n \Gamma f) = \Gamma z^n f \quad \text{for } n \in \mathbb{Z}_+. \qquad (6\text{--}5)$$

PROOF OF LEMMA 6.3. Let $n \in \mathbb{Z}_+$. We have

$$\widehat{\xi_1 \bar{\xi}_2}(-n) = (z^n \xi_1, \xi_2) = s^{-1}(z^n \xi_1, \Gamma^* \eta_2) = s^{-1}(\Gamma z^n \xi_1, \eta_2)$$
$$= s^{-1}(\mathbb{P}_- z^n \Gamma \xi_1, \eta_2) = (z^n \eta_1, \eta_2) = \widehat{\eta_1 \bar{\eta}_2}(-n),$$

by (6–5). Similarly, $\widehat{\xi_1 \bar{\xi}_2}(n) = \widehat{\eta_1 \bar{\eta}_2}(n)$, $n \in \mathbb{Z}_+$, which implies $\xi_1 \bar{\xi}_2 = \eta_1 \bar{\eta}_2$. □

COROLLARY 6.4. Let $\{\xi, \eta\}$ be an s-Schmidt pair of Γ. Then the function

$$\varphi_s = \frac{\eta}{\xi} \qquad (6\text{--}6)$$

is unimodular and does not depend on the choice of $\{\xi, \eta\}$.

PROOF. Let $\xi_1 = \xi_2 = \xi$ and $\eta_1 = \eta_2 = \eta$ in Lemma 6.3. It follows that $|\xi|^2 = |\eta|^2$ and so η/ξ is unimodular for any Schmidt pair $\{\xi, \eta\}$.

Let $\{\xi_1, \eta_1\}$ and $\{\xi_2, \eta_2\}$ be s-Schmidt pairs of Γ. By Lemma 6.3, $\eta_1/\xi_1 = \bar{\xi}_2/\bar{\eta}_2$. Since η_2/ξ_2 is unimodular, $\eta_1/\xi_1 = \eta_2/\xi_2$. □

We resume the proof of Theorem 6.1. Put

$$\Gamma_s = H_{s\varphi_s},$$

where φ_s is defined by (6–6). Clearly $\|\Gamma_s\| \leq s$. The result will be established if we show that $\mathrm{rank}(\Gamma - \Gamma_s) \leq k$.

Let $\{\xi, \eta\}$ be an s-Schmidt pair of Γ. We show that it is also an s-Schmidt pair of Γ_s. Indeed,

$$\Gamma_s \xi = s\mathbb{P}_- \frac{\eta}{\xi} \xi = s\eta, \quad \Gamma_s^* \eta = s\mathbb{P}_+ \frac{\xi}{\eta} \eta = s\xi.$$

Set

$$E_+ = \{\xi \in H^2 : \Gamma^* \Gamma \xi = s^2 \xi\} \quad \text{for } E_- = \{\eta \in H_-^2 : \Gamma \Gamma^* \eta = s^2 \eta\}$$

be the spaces of Schmidt vectors of Γ and Γ^*. Clearly, $\dim E_+ = \dim E_- = \mu$.

It follows easily from (6–5) that if $\Gamma \xi = \Gamma_s \xi$, then $\Gamma z^n \xi = \Gamma_s z^n \xi$ for any $n \in \mathbb{Z}_+$. Since $\Gamma_s | E_+ = \Gamma | E_+$, it follows that Γ and Γ_s coincide on the S-invariant subspace spanned by E_+, where S is multiplication by z on H^2. By Beurling's theorem this subspace has the form ϑH^2, where ϑ is an inner function (see [Nikol'skiĭ 1986, Lecture I, 1], for example). Denote by Θ multiplication by ϑ. We have $\Gamma\Theta = \Gamma_s\Theta$. The proof will be complete if we show that $\dim(H^2 \ominus \vartheta H^2) \leq k$. Put $d = \dim(H^2 \ominus \vartheta H^2)$.

LEMMA 6.5. *The singular value s of the operator $\Gamma\Theta$ has multiplicity at least $d + \mu$.*

Note that $\Gamma\Theta$ is compact, so it will follow from Lemma 6.5 that $d < \infty$.

PROOF OF LEMMA 6.5. Let τ be an inner divisor of ϑ, which means that $\vartheta\tau^{-1} \in H^\infty$. We show that, for any $\xi \in E_+$,

$$(\Gamma_s\Theta)^*(\Gamma_s\Theta)\bar{\tau}\xi = s^2\bar{\tau}\xi \in E_+. \tag{6-7}$$

Indeed it is easy to see that $\Gamma^*\bar{z}\bar{f} = \bar{z}\overline{\Gamma f}$ for any $f \in H^2$. Let J be the transformation on L^2 defined by $Jf = \bar{z}\bar{f}$. It follows that J maps E_+ onto E_-. Since $E_+ \subset \vartheta H^2$, we have $E_- \subset \bar{\vartheta}\bar{H}^2_-$.

Let $\xi \in E_+$ and set $\eta = s^{-1}\Gamma\xi \in E_-$. We can represent η as $\eta = \bar{\vartheta}\eta_*$, where $\eta_* \in H^2_-$. We have

$$(\Gamma_s\Theta)^*(\Gamma_s\Theta)\bar{\tau}\xi = (\Gamma_s\Theta)^* s\mathbb{P}_- \frac{\eta}{\xi}\vartheta\bar{\tau}\xi = s(\Gamma_s\Theta)^*\mathbb{P}_-\eta_*\bar{\tau}$$

$$= s(\Gamma_s\Theta)^*\eta_*\bar{\tau} = s^2\mathbb{P}_+ \frac{\xi}{\eta}\bar{\vartheta}\eta_*\bar{\tau} = s^2\bar{\tau}\xi,$$

which proves (6–7).

Since $d = \dim(H^2 \ominus \vartheta H^2)$, we can find for any $n < d$ inner divisors $\{\vartheta_j\}_{1 \leq j \leq n+1}$ of ϑ such that $\vartheta_{n+1} = \vartheta$, $\vartheta_{j+1}\vartheta_j^{-1} \in H^\infty$, and ϑ_1 and the $\vartheta_{j+1}\vartheta_j^{-1}$ are not constants; see [Nikol'skiĭ 1986, Lecture II, 2], for example. Then it follows from (6–7) that the subspace

$$E_j = \text{span}\{E_+, \bar{\vartheta}_1 E_+, \ldots, \bar{\vartheta}_j E_+\}, \quad \text{for } 1 \leq j \leq n+1,$$

consists of eigenvectors of $(\Gamma\Theta)^*(\Gamma\Theta)$ corresponding to the eigenvalue s^2. Clearly, $E_1 \setminus E_+ \neq \varnothing$ and $E_{j+1} \setminus E_j \neq \varnothing$ for $1 \leq j \leq n$. Therefore

$$\dim \text{Ker}\big((\Gamma\Theta)^*\Gamma\Theta - s^2I\big) \geq \dim E_{n+1} \geq \mu + n + 1.$$

The left-hand side is equal to ∞ if $d = \infty$ and is at least $\mu + d$ if $d < \infty$. $\square$

We can complete now the proof of Theorem 6.1. We have already observed that $s_j(\Gamma\Theta) \leq s_j(\Gamma)$, so by Lemma 6.5 we have

$$s_{k+\mu}(\Gamma) < s_{k+\mu-1}(\Gamma) = \cdots = s_k(\Gamma) = s_{d+\mu-1}(\Gamma\Theta) \leq s_{d+\mu-1}(\Gamma).$$

Therefore $d+\mu-1 < k+\mu$ and so $d \leq k$, which completes the proof of Theorem 6.1 in the compact case. $\square$

7. Schatten–von Neumann Classes S_p

In this section we study Hankel operators of Schatten–von Neumann class S_p. We state the main result, which describes the Hankel operators of class S_p, for $0 < p < \infty$, as those whose symbols belong to the Besov class $B_p^{1/p}$. However, we give the proof here only in the case $p = 1$.

DEFINITION. Let $\mathcal{H}_1$ and $\mathcal{H}_2$ be Hilbert spaces and let $T : \mathcal{H}_1 \to \mathcal{H}_2$ be a bounded linear operator. Given p with $0 < p < \infty$, we say that $T \in S_p(\mathcal{H}_1, \mathcal{H}_2)$ (or simply $T \in S_p$), if the sequence $\{s_j\}_{j \geq 0}$ of the singular values of T belongs to ℓ^p. We put

$$\|T\|_{S_p} \overset{\text{def}}{=} \left(\sum_{j \geq 0} s_j^p \right)^{1/p}. \tag{7–1}$$

For $1 \leq p < \infty$ the class $S_p(\mathcal{H}_1, \mathcal{H}_2)$ forms a Banach space with norm given by (7–1). If $\mathcal{H}$ is a Hilbert space and T is an operator on $\mathcal{H}$ of class S_1, one can define the trace of T by

$$\operatorname{trace} T \overset{\text{def}}{=} \sum_{j \geq 0} (Te_j, e_j), \tag{7–2}$$

where $\{e_j\}_{j \geq 0}$ is an orthonormal basis of $\mathcal{H}$. The right-hand side of (7–2) does not depend on the choice of the orthonormal basis. The trace is a linear functional on $\mathcal{H}$, and $|\operatorname{trace} T| \leq \|T\|_{S_1}$. The dual space of $S_1(\mathcal{H}_1, \mathcal{H}_2)$ can be identified with the space $\mathbb{B}(\mathcal{H}_2, \mathcal{H}_1)$ of bounded linear operators from $\mathcal{H}_2$ to $\mathcal{H}_1$ with respect to the pairing

$$\langle T, R \rangle \overset{\text{def}}{=} \operatorname{trace} TR \quad \text{for } T \in S_1(\mathcal{H}_1, \mathcal{H}_2) \text{ and } R \in \mathbb{B}(\mathcal{H}_2, \mathcal{H}_1).$$

We refer the reader to [Gohberg and Kreĭn 1965] for more detailed information about the classes S_p.

We now define the Besov classes B_p^s of functions on $\mathbb{T}$. They admit many different equivalent definitions; see [Peetre 1976], for example. We need regularized de la Vallée Poussin type kernels V_n, which can be defined as follows. Let v be an infinitely differentiable function on $\mathbb{R}$ such that $\operatorname{supp} v = [\frac{1}{2}, 2]$, $v \geq 0$, and

$$\sum_{j \geq 0} v \left(\frac{x}{2^j} \right) = 1 \quad \text{for } x \geq 1.$$

It is very easy to construct such a function v. We can now define V_n by

$$V_n = \begin{cases} \sum_{k \in \mathbb{Z}} v \left(\dfrac{k}{2^n} \right) z^k & \text{for } n \geq 1, \\ \overline{V_{-n}} & \text{for } n < 0, \end{cases}$$

$$V_0(z) = \bar{z} + 1 + z.$$

DEFINITION. Let $0 < p < \infty$. The *Besov space* B_p^s consists of the distributions f on $\mathbb{T}$ satisfying

$$\sum_{n \in \mathbb{Z}} \left(2^{s|n|} \| f * V_n \|_{L^p} \right)^p < \infty.$$

If $p \geq 1$, the function v does not have to be infinitely smooth. In particular, in this case we can replace v by the piecewise linear function w satisfying $w(1) = 1$ and $\operatorname{supp} w = [\frac{1}{2}, 2]$, and replace the V_n by the trigonometric polynomials

$$W_n = \begin{cases} \sum_{k \in \mathbb{Z}} w\left(\dfrac{k}{2^n} \right) z^k & \text{for } n \geq 1, \\ \overline{W_{-n}} & \text{for } n < 0, \end{cases}$$
$$W_0(z) = \bar{z} + 1 + z.$$

It is clear from this definition that $\mathbb{P}_+ B_p^s \subset B_p^s$. We can identify in a natural way a function f in $\mathbb{P}_+ B_p^s$ with the function $\sum_{j \geq 0} \hat{f}(j) z^j$, analytic in $\mathbb{D}$. A function f analytic in $\mathbb{D}$ belongs to $\mathbb{P}_+ B_p^s$ if and only if

$$\int_{\mathbb{D}} |f^{(n)}(\zeta)|^p (1 - |\zeta|)^{(n-s)p-1} \, d\boldsymbol{m}_2(\zeta) < \infty,$$

where $n \in \mathbb{Z}_+$ and $n > s$.

For $s > \max\{1/p - 1, 0\}$, the class B_p^s consists of the functions f on $\mathbb{T}$ for which

$$\int_{\mathbb{T}} \frac{\|\mathbb{D}_\tau^n f\|_{L^p}^p}{|\tau - 1|^{1+sp}} \, d\boldsymbol{m}(\tau) < \infty,$$

where $n > s$ is a positive integer and $(\mathbb{D}_\tau f)(\zeta) \overset{\text{def}}{=} f(\tau\zeta) - f(\zeta)$.

THEOREM 7.1. *Let φ be a function on $\mathbb{T}$ of class* BMO *and let $0 < p < \infty$. Then $H_\varphi \in \boldsymbol{S}_p$ if and only if $\mathbb{P}_- \varphi \in B_p^{1/p}$.*

For technical reasons it is more convenient to work with Hankel matrices $\Gamma_\varphi = \{\hat{\varphi}(j+k)\}_{j,k \geq 0}$, where $\varphi = \sum_{j \geq 0} \hat{\varphi}(j) z^j$ is a function analytic in the unit disk. We shall identify Hankel matrices Γ_φ with operators on the space ℓ^2.

Clearly, the following statement is equivalent to Theorem 7.1.

THEOREM 7.2. *Let φ be a function analytic in the unit disk and let $0 < p < \infty$. Then the Hankel operator Γ_φ belongs to the class $\boldsymbol{S}_p$ if and only if $\varphi \in B_p^{1/p}$.*

Theorem 7.1 was proved in [Peller 1980] for $p \geq 1$, and in [Semmes 1984] and [Peller 1983] for $p < 1$ (the proofs are quite different). Pekarskii's theorem [1985] on rational approximation also gives another proof of Theorem 7.2. Later other proofs were found; see, for example, [Coifman and Rochberg 1980] for $p = 1$, and [Rochberg 1982; Peetre and Svensson 1984] for $1 < p < \infty$.

We prove Theorem 7.2 only for $p = 1$. We present the original proof from [Peller 1980], which gives rather sharp estimates from above and from below for the norms $\|\Gamma_\varphi\|_{\boldsymbol{S}_1}$.

PROOF OF THEOREM 7.2 FOR $p = 1$.. We first prove that $\Gamma_\varphi \in \boldsymbol{S}_1$ if $\varphi \in B_1^1$. It is easy to see that

$$\varphi = \sum_{n \geq 0} \varphi * W_n.$$

We have

$$\sum_{n \geq 0} 2^n \|\varphi * W_n\|_{L^1} < \infty.$$

Clearly, $\varphi * W_n$ is a polynomial of degree at most $2^{n+1} - 1$. The following lemma gives sharp estimates of the trace norm of a Hankel operator with polynomial symbol.

LEMMA 7.3. *Let f be an analytic polynomial of degree m. Then*

$$\|\Gamma_f\|_{\boldsymbol{S}_1} \leq (m + 1)\|f\|_1.$$

PROOF. Given $\zeta \in \mathbb{T}$, we define elements x_ζ and y_ζ of ℓ^2 by

$$x_\zeta(j) = \begin{cases} \zeta^j & \text{if } 0 \leq j \leq m, \\ 0 & \text{if } j > m; \end{cases}$$

$$y_\zeta(k) = \begin{cases} f(\zeta)\bar{\zeta}^k & \text{if } 0 \leq k \leq m, \\ 0 & \text{if } k > m. \end{cases}$$

Define the rank-one operator A_ζ on ℓ^2 by setting $A_\zeta x = (x, x_\zeta)y_\zeta$ for $x \in \ell^2$. Then $A_\zeta \in \boldsymbol{S}_1$ and

$$\|A_\zeta\|_{\boldsymbol{S}_1} = \|x_\zeta\|_{\ell^2}\|y_\zeta\|_{\ell^2} = (m + 1)|f(\zeta)|.$$

We prove that

$$\Gamma_f = \int_{\mathbb{T}} A_\zeta \, d\boldsymbol{m}(\zeta) \tag{7–3}$$

(the function $\zeta \mapsto A_\zeta$ being continuous, the integral can be understood as the limit of integral sums). We have

$$(\Gamma_f e_j, e_k) = \hat{f}(j + k) = \int_{\mathbb{T}} f(\zeta)\bar{\zeta}^{j+k} \, d\boldsymbol{m}(\zeta),$$

$$(A_\zeta e_j, e_k) = f(\zeta)\bar{\zeta}^j \bar{\zeta}^k.$$

Therefore (7–3) holds and

$$\|\Gamma_f\|_{\boldsymbol{S}_1} \leq \int_{\mathbb{T}} \|A_\zeta\|_{\boldsymbol{S}_1} \, d\boldsymbol{m}(\zeta) \leq (m + 1) \int_{\mathbb{T}} |f(\zeta)| \, d\boldsymbol{m}(\zeta). \qquad \square$$

We now complete the proof of the sufficiency of the condition $\varphi \in B_1^1$. It follows from Lemma 7.3 that

$$\|\Gamma_\varphi\|_{\boldsymbol{S}_1} \leq \sum_{n \geq 0} \|\Gamma_{\varphi * W_n}\|_{\boldsymbol{S}_1} \leq \sum_{n \geq 0} 2^{n+1} \|\varphi * W_n\|_{L^1}.$$

Now suppose that $\Gamma_\varphi \in \boldsymbol{S}_1$. Define polynomials Q_n and R_n, for $n \geq 1$, by

$$\hat{Q}_n(k) = \begin{cases} 0 & \text{if } k \leq 2^{n-1}, \\ 1 - |k - 2^n|/2^{n-1} & \text{if } 2^{n-1} \leq k \leq 2^n + 2^{n-1}, \\ 0 & \text{if } k \geq 2^n + 2^{n-1}; \end{cases}$$

$$\hat{R}_n(k) = \begin{cases} 0 & \text{if } k \leq 2^n, \\ 1 - |k - 2^n - 2^{n-1}|/2^{n-1} & \text{if } 2^n \leq k \leq 2^{n+1}, \\ 0 & \text{if } k \geq 2^{n+1}. \end{cases}$$

Clearly, $W_n = Q_n + \frac{1}{2} R_n$ for $n \geq 1$.

We now show that

$$\sum_{n \geq 0} 2^{2n+1} \|\varphi * Q_{2n+1}\|_{L^1} < \infty. \tag{7-4}$$

One proves in exactly the same way that

$$\sum_{n \geq 1} 2^{2n} \|\varphi * Q_{2n}\|_{L^1} < \infty,$$

$$\sum_{n \geq 0} 2^{2n+1} \|\varphi * R_{2n+1}\|_{L^1} < \infty,$$

$$\sum_{n \geq 1} 2^{2n} \|\varphi * R_{2n}\| < \infty.$$

To prove (7–4), we construct an operator B on ℓ^2 such that $\|B\| \leq 1$ and $\langle \Gamma_\varphi, B \rangle = \sum_{n \geq 0} 2^{2n} \|f * Q_{2n+1}\|_{L^1}$.

Consider the squares $S_n = [2^{2n-1}, 2^{2n-1} + 2^{2n} - 1] \times [2^{2n-1} + 1, 2^{2n-1} + 2^{2n}]$, for $n \geq 1$, on the plane.

Let $\{\psi_n\}_{n \geq 1}$ be a sequence of functions in L^∞ such that $\|\psi_n\|_{L^\infty} \leq 1$. We define the matrix $\{b_{jk}\}_{j,k \geq 0}$ of B by

$$b_{jk} = \begin{cases} \hat{\psi}_n(j+k) & \text{if } (j,k) \in S_n \text{ for } n \geq 1, \\ 0 & \text{if } (j,k) \notin \bigcup_{n \geq 1} S_n. \end{cases}$$

We show that $\|B\| \leq 1$. Consider the subspaces

$$\mathcal{H}_n = \text{span}\{e_j : 2^{2n-1} \leq j \leq 2^{2n-1} + 2^{2n} - 1\},$$

$$\mathcal{H}'_n = \text{span}\{e_j : 2^{2n-1} + 1 \leq j \leq 2^{2n-1} + 2^{2n}\}.$$

It is easy to see that

$$B = \sum_{n \geq 1} P'_n \Gamma_{\psi_n} P_n,$$

where P_n and P'_n are the orthogonal projection onto $\mathcal{H}_n$ and $\mathcal{H}'_n$. Since the spaces $\{\mathcal{H}_n\}_{n \geq 1}$ are pairwise orthogonal as well as the spaces $\{\mathcal{H}'_n\}_{n \geq 1}$, we have

$$\|B\| = \sup_n \|P'_n \Gamma_{\psi_n} P_n\| \leq \sup_n \|\Gamma_{\psi_n}\| \leq \sup_n \|\psi_n\|_{L^\infty} \leq 1.$$

We show that

$$\langle \Gamma_\varphi, B \rangle = \sum_{n \geq 0} 2^{2n} \langle Q_{2n+1} * \varphi, \psi_n \rangle,$$

where $\langle g, h \rangle \overset{\text{def}}{=} \int_{\mathbb{T}} f(\zeta) h(\bar{\zeta}) \, d\boldsymbol{m}(\zeta)$, for $g \in L^1$ and $h \in L^\infty$. We have

$$\langle \Gamma_\varphi, B \rangle = \sum_{n \geq 1} \langle \Gamma_\varphi, P_n' \Gamma_{\psi_n} P_n \rangle$$

$$= \sum_{n \geq 1} \sum_{j=2^{2n}}^{2^{2n}+2^{2n+1}} \left(2^{2n} - |j - 2^{2n+1}| \right) \hat{\varphi}(j) \hat{\psi}_n(j)$$

$$= \sum_{n \geq 1} 2^{2n} \langle Q_{2n+1} * \varphi, \psi_n \rangle.$$

We can now pick a sequence $\{\psi_n\}_{n \geq 1}$ such that $\langle Q_{2n+1} * \varphi, \psi_n \rangle = \|Q_{2n+1} * \varphi\|_{L^1}$. Then

$$\langle \Gamma_\varphi, B \rangle = \sum_{n \geq 1} 2^{2n} \|Q_{2n+1} * \varphi\|_{L^1}.$$

Hence

$$\sum_{n \geq 1} 2^{2n+1} \|Q_{2n+1} * \varphi\|_{L^1} = 2 \langle \Gamma_\varphi, B \rangle \leq 2 \|\Gamma_\varphi\|_{\boldsymbol{S}_1} < \infty. \qquad \square$$

REMARK. This proof easily leads to the estimates

$$\frac{1}{6} \sum_{n \geq 1} 2^n \|\varphi * W_n\|_{L^1} \leq \|\Gamma_\varphi\|_{\boldsymbol{S}_1} \leq 2 \sum_{n \geq 0} 2^n \|\varphi * W_n\|_{L^1}.$$

8. Rational Approximation

Classical theorems on polynomial approximation, as found in [Akhiezer 1965], for example, describe classes of smooth functions in terms of the rate of polynomial approximation in one norm or another. The smoother the function, the more rapidly its deviations relative to the set of polynomials of degree n decay. However, it turns out that in the case of rational approximation the corresponding problems are considerably more complicated. The first sharp result was obtained in [Peller 1980]; it concerned rational approximation in the BMO norm and was deduced from the $\boldsymbol{S}_p$ criterion for Hankel operators given in Theorem 7.1. There were also earlier results [Gonchar 1968; Dolženko 1977; Brudnyĭ 1979], but there were gaps between the "direct" and "inverse" theorems.

In this section we describe the Besov spaces $B_p^{1/p}$ in terms of the rate of rational approximation in the norm of BMO. Then we obtain an improvement of Grigoryan's theorem which estimates the L^∞ norm of $\mathbb{P}_- f$ in terms of $\|f\|_{L^\infty}$ for functions f such that $\mathbb{P}_- f$ is a rational function of degree n. As a consequence we obtain a sharp result about rational approximation in the L^∞ norm.

There are many different natural norms on BMO. We can use, for example,

$$\|f\|_{\text{BMO}} \overset{\text{def}}{=} \inf\{\|\xi\|_{L^\infty} + \|\eta\|_{L^\infty} : f = \xi + \mathbb{P}_+ \eta \text{ for } \xi, \eta \in L^\infty\}.$$

Denote by $\mathcal{R}_n$, for $n \geq 0$, the set of rational functions of degree at most n with poles outside $\mathbb{T}$. For $f \in \mathrm{BMO}$ put

$$r_n(f) \overset{\text{def}}{=} \mathrm{dist}_{\mathrm{BMO}}\{f, \mathcal{R}_n\}.$$

The following theorem was proved in [Peller 1980] for $p \geq 1$ and in [Peller 1983; Semmes 1984; Pekarskiĭ 1985] for $p < 1$. Pekarskii [1985; 1987] also obtained similar results for rational approximation in the L^p norms. See also [Parfenov 1986] for other applications of Hankel operators in rational approximation.

THEOREM 8.1. *Let* $\varphi \in \mathrm{BMO}$ *and* $0 < p < \infty$. *Then* $\{r_n(\varphi)\}_{n \geq 0} \in \ell^p$ *if and only if* $\varphi \in B_p^{1/p}$.

PROOF. We have $\mathbb{P}_+ \mathrm{BMO} \subset \mathrm{BMO}$ (see the introduction), $\mathbb{P}_+ B_p^{1/p} \subset B_p^{1/p}$ (Section 7), and $\mathbb{P}_+ \mathcal{R}_n \subset \mathcal{R}_n$. Therefore it is sufficient to prove the theorem for $\mathbb{P}_- \varphi$ and $\mathbb{P}_+ \varphi$. We do it for $\mathbb{P}_- \varphi$; the corresponding result for $\mathbb{P}_+ \varphi$ follows by passing to complex conjugate.

It follows from Theorem 6.1 that

$$s_n(H_\varphi) = \inf\{\|H_\varphi - H_r\| : \mathrm{rank}\, H_r \leq n\}.$$

Without loss of generality we may assume that $r = \mathbb{P}_- r$. By Corollary 4.2, $\mathrm{rank}\, H_r \leq n$ if and only if $r \in \mathcal{R}_n$. Together with Theorem 2.3 this yields

$$c_1 s_n(H_\varphi) \leq \inf\{\|\varphi - r\|_{\mathrm{BMO}} : r \in \mathcal{R}_n\} \leq c_2 s_n(H_\varphi)$$

for some positive constants c_1 and c_2.

The result follows now from Theorem 7.1. $\qquad\square$

Denote by $\mathcal{R}_n^+$ the set of rational functions of degree at most n with poles outside the closed unit disk, and put

$$r_n^+(f) \overset{\text{def}}{=} \mathrm{dist}_{\mathrm{BMOA}}\{f, \mathcal{R}_n^+\}.$$

COROLLARY 8.2. *Let* $\varphi \in \mathrm{BMOA}$ *and* $0 < p < \infty$. *Then* $\{r_n^+(\varphi)\}_{n \geq 0} \in \ell^p$ *if and only if* $\varphi \in \mathbb{P}_+ B_p^{1/p}$.

We now prove an improvement of a theorem of Grigoryan [1976], which estimates the $\|\mathbb{P}_- \varphi\|_{L^\infty}$ in terms of $\|\varphi\|_{L^\infty}$ in the case $\mathbb{P}_- \varphi \in \mathcal{R}_n$. Clearly, the last condition is equivalent to the fact that φ is a boundary value function of a meromorphic function in $\mathbb{D}$ bounded near $\mathbb{T}$ and having at most n poles, counted with multiplicities. It is not obvious that such an estimate exists. If we consider the same question in the case where $\mathbb{P}_- \varphi$ is a polynomial of degree n, it is well known that $\|\mathbb{P}_- \varphi\|_{L^\infty} \leq \mathrm{const}\, \log(1 + n)$ (see [Zygmund 1968]; this follows immediately from the fact that $\|\sum_{j=0}^n z^j\|_{L^1} \leq \mathrm{const}\, \log(1 + n)$). Grigoryan's theorem claims that, if $\mathbb{P}_- \varphi \in \mathcal{R}_n$, then

$$\|\mathbb{P}_- \varphi\|_{L^\infty} \leq \mathrm{const}\, \cdot n \|\varphi\|_{L^\infty}. \tag{8-1}$$

The following result, obtained in [Peller 1983], improves this estimate. The proof is based on the S_1 criterion for Hankel operators given in Theorem 7.1.

THEOREM 8.3. *Let n be a positive integer and let φ be a function in L^∞ such that $\mathbb{P}_- \varphi \in \mathcal{R}_n$. Then*

$$\|\mathbb{P}_- \varphi\|_{B_1^1} \leq \text{const} \cdot n \|\varphi\|_{L^\infty}. \tag{8-2}$$

Observe first that (8–2) implies (8–1). Indeed, if $f \in B_1^1$, then $\sum_{n \geq 0} 2^n \|f *W_n\|_{L^1} \leq \text{const} \|f\|_{B_1^1}$ (see Section 7). It is easy to show that

$$\|\varphi\|_{L^\infty} \leq \sum_{j \geq 0} |\hat{f}(j)| \leq \text{const} \sum_{n \geq 0} 2^n \|f * W_n\|_{L^1},$$

which proves the claim.

PROOF OF THEOREM 8.3. Consider the Hankel operator H_φ. By Nehari's theorem, $\|H_\varphi\| \leq \|\varphi\|_{L^\infty}$. By Kronecker's theorem, rank $H_\varphi \leq n$. Therefore $\|H_\varphi\|_{S_1} \leq n \|H_\varphi\|$. The result now follows from Theorem 7.1, which guarantees that $\|\mathbb{P}_- \varphi\|_{B_1^1} \leq \text{const} \|H_\varphi\|_{S_1}$. $\square$

To conclude this section we obtain a result on rational approximation in the L^∞ norm [Peller 1983]. For $\varphi \in L^\infty$ we put

$$\rho_n(\varphi) \overset{\text{def}}{=} \text{dist}_{L^\infty}\{\varphi, \mathcal{R}_n\} \quad \text{for } n \in \mathbb{Z}_+.$$

THEOREM 8.4. *Let $\varphi \in L^\infty$. Then the $\rho_n(\varphi)$ decay more rapidly than any power of n if and only if $\varphi \in \bigcap_{p>0} B_p^{1/p}$.*

Pekarskii [1987] obtained a result similar to Theorem 8.1 for rational approximation in L^∞ in the case $0 < p < 1$.

LEMMA 8.5. *Let $r \in \mathcal{R}_n$. Then*

$$\|r\|_{L^\infty} \leq \text{const} \cdot n \, \|r\|_{\text{BMO}}.$$

PROOF. It suffices to prove the inequality for $\mathbb{P}_- r$ and $\mathbb{P}_+ r$; we do it for $\mathbb{P}_- r$. Let f be the symbol of H_r-minimal norm, that is, such that $\mathbb{P}_- r = \mathbb{P}_- f$ and $\|f\|_{L^\infty} = \|H_r\|$ (see Corollary 2.5). We have

$$\|\mathbb{P}_- r\|_{L^\infty} = \|\mathbb{P}_- f\|_{L^\infty} \leq \text{const} \cdot n \|f\|_{L^\infty} = \text{const} \cdot n \|H_r\| \leq \text{const} \cdot n \|\mathbb{P}_- r\|_{\text{BMO}},$$

by Theorems 8.3 and 2.3. $\square$

Theorem 5.8 is an easy consequence of the following lemma.

LEMMA 8.6. *Let $\lambda > 1$ and let φ be a function in L^∞ such that $r_n(\varphi) \leq \text{const} \cdot n^{-\lambda}$ for $n \geq 0$. Then*

$$\rho_n(\varphi) \leq \text{const} \cdot n^{-\lambda+1} \quad \text{for } n \geq 0.$$

PROOF. Suppose that $r_n \in \mathcal{R}_{2^n}$ and $\|\varphi - r_n\|_{BMO} \leq \text{const } 2^{-n\lambda}$. We have

$$\varphi - r_n = \sum_{j \geq 0} \left((\varphi - r_{n+j}) - (\varphi - r_{n+j+1})\right) = \sum_{j \geq 0} (r_{n+j+1} - r_{n+j}).$$

Under the hypotheses of the lemma ,

$$\|r_{n+j+1} - r_{n+j}\|_{BMO} \leq \text{const } 2^{-(n+j)\lambda},$$

and, since $r_{n+j+1} - r_{n+j} \in \mathcal{R}_{2^{n+j+2}}$, Lemma 8.5 gives

$$\|r_{n+j+1} - r_{n+j}\|_{L^\infty} \leq \text{const } 2^{-(n+j)(\lambda-1)}.$$

Therefore

$$\rho_{2^n}(\varphi) \leq \|\varphi - r_n\|_{L^\infty} \text{const } 2^{-n(\lambda-1)},$$

which implies the conclusion of the lemma. $\square$

9. The Operator of Best Approximation by Analytic Functions

Let φ be a function in VMO. By Corollary 2.5, there exists a unique function f in BMOA such that $\varphi - f$ is bounded on $\mathbb{T}$ and

$$\|\varphi - f\|_{L^\infty} = \inf\{\|\varphi - g\|_{L^\infty} : g \in \text{BMOA with } \varphi - g \in L^\infty(\mathbb{T})\} = \|H_\varphi\|.$$

We define the nonlinear *operator of best approximation by analytic functions* on the space VMO by setting $\mathcal{A}\varphi \stackrel{\text{def}}{=} f$. This operator is very important in applications such as control theory and prediction theory.

We are going to study hereditary properties of $\mathcal{A}$. This means the following: Suppose that $X \subset \text{VMO}$ is a space of functions on $\mathbb{T}$. For which X does the operator $\mathcal{A}$ maps X into itself? Certainly not for arbitrary X: for example, $\mathcal{A}C(\mathbb{T}) \not\subset C(\mathbb{T})$, as follows from the remark after Corollary 5.7.

Shapiro [1952] showed that $\mathcal{A}X \subset X$ if X is the space of functions analytic in a neighbourhood of $\mathbb{T}$. Carleson and Jacobs [1972] proved that $\mathcal{A}\Lambda_\alpha \subset \Lambda_\alpha$ if $\alpha > 0$ and $\alpha \notin \mathbb{Z}$, where the $\Lambda_\alpha \stackrel{\text{def}}{=} B^\alpha_\infty$ are the Hölder–Zygmund classes (see Section 7).

In [Peller and Khrushchëv 1982] three big classes of function spaces X were found for which $\mathcal{A}X \subset X$. The first consists of the so-called $\mathcal{R}$-spaces, which are, roughly speaking, function spaces that can be described in terms of rational approximation in the BMO norm. The Besov spaces $B_p^{1/p}$, for $0 < p < \infty$, and the space VMO are examples of $\mathcal{R}$-spaces. I will not give a precise definition here.

The second class consists of function spaces X that satisfy the following axioms:

(A1) If $f \in X$, then $\bar{f} \in X$ and $\mathbb{P}_+ f \in X$.

(A2) X is a Banach algebra with respect to pointwise multiplication.

(A3) The trigonometric polynomials are dense in X.

(A4) The maximal ideal space of X can be identified naturally with $\mathbb{T}$.

Many classical spaces of functions on the unit circle satisfy these axioms: the space of functions with absolutely convergent Fourier series, the Besov spaces B_p^s for $1 \le p < \infty$, and many others (see [Peller and Khrushchëv 1982]). However, the Hölder–Zygmund classes Λ_α do not satisfy axiom (A3).

The third class of function spaces described in [Peller and Khrushchëv 1982] contains many nonseparable Banach spaces. In particular, it contains the classes Λ_α, for $\alpha > 0$. I will not define the third class here; see [Peller and Khrushchëv 1982] for the definition and other examples.

Other function spaces satisfying the property $\mathcal{A}X \subset X$ are described in [Vol'berg and Tolokonnikov 1985; Tolokonnikov 1991].

Another related question, also important in applications, is the continuity problem. Merino [1989] and Papadimitrakis [1993] showed that the operator $\mathcal{A}$ is discontinuous at any function $\varphi \in C(\mathbb{T}) \setminus H^\infty$ in the L^∞ norm. For function spaces satisfying Axioms (A1)–(A4), continuity points of $\mathcal{A}$ in the norm of X were described in [Peller 1990b]: if $\varphi \in X \setminus H^\infty$, then $\mathcal{A}$ is continuous at φ if and only if the singular value $s_0(H_\varphi)$ of the Hankel operator $H_\varphi : H^2 \to H_-^2$ has multiplicity one.

In this section we prove that $\mathcal{A}$ preserves the spaces $B_p^{1/p}$, for $0 < p < \infty$, and the space VMO. Moreover, it turns out that the operator $\mathcal{A}$ is *bounded* on such spaces; that is,

$$\|\mathcal{A}\varphi\|_X \le \text{const}\,\|\varphi\|_X, \qquad (9\text{–}1)$$

for $X = B_p^{1/p}$ or $X = \text{VMO}$. Note, however, that this is a rather exceptional property. It was proved in [Peller 1992] that $\mathcal{A}$ is unbounded on X if $X = B_p^s$, with $s > 1/p$, and on Λ_a, with $\alpha > 0$. Then it was shown in [Papadimitrakis 1996] that $\mathcal{A}$ is unbounded on the space of functions with absolutely convergent Fourier series.

THEOREM 9.1. *Let* $X = B_p^{1/p}$, *with* $0 < p < \infty$, *or* $X = \text{VMO}$. *Then* $\mathcal{A}X \subset X$ *and* (9–1) *holds.*

To prove Theorem 9.1 we need a formula that relates the moduli of the Toeplitz operators T_u and $T_{\bar{u}}$ for a *unimodular* function u (one satisfying $|u(\zeta)| = 1$ a.e. on $\mathbb{T}$). This formula was found in [Peller and Khrushchëv 1982]:

$$H_{\bar{u}}^* H_{\bar{u}} T_u = T_u H_u^* H_u. \qquad (9\text{–}2)$$

It is an immediate consequence of the definitions of the Toeplitz and Hankel operators. Nonetheless, it has many important applications.

Recall that each bounded linear operator T on a Hilbert space $\mathcal{H}$ admits a polar decomposition $T = U(T^*T)^{1/2}$, where U is an operator such that $\text{Ker}\,U = \text{Ker}\,T$ and $U|\mathcal{H} \ominus \text{Ker}\,U$ is an isometry onto the closure of the range of T. The operator U is called the *partially isometric factor* of T.

We need the following well-known fact [Halmos 1967, Problem 152]. Let A and B be selfadjoint operators on Hilbert space and let T be an operator such that $AT = TB$. Then $AU = UB$, where U is the partially isometric factor of T.

We apply this to formula (9–2). Let u be a unimodular function on $\mathbb{T}$. Denote by $\boldsymbol{U}$ the partially isometric factor of T_u. Then

$$H_{\bar{u}}^* H_{\bar{u}} \boldsymbol{U} = \boldsymbol{U} H_u^* H_u. \qquad (9\text{–}3)$$

The following theorem was proved in [Peller and Khrushchëv 1982].

THEOREM 9.2. *Let u be a unimodular function on $\mathbb{T}$ such that T_u has dense range in H^2. Then $H_{\bar{u}}^* H_{\bar{u}}$ is unitarily equivalent to $H_u^* H_u | H^2 \ominus E$, where*

$$E = \operatorname{Ker} T_u = \{ f \in H^2 : H_u^* H_u f = f \}.$$

PROOF. Since $\boldsymbol{U}$ maps $H^2 \ominus E$ isometrically onto H^2, it follows from (9–3) that

$$H_{\bar{u}}^* H_{\bar{u}} = \boldsymbol{U} H_u^* H_u \boldsymbol{U}^* = \boldsymbol{U}(H_u^* H_u | H^2 \ominus E)\boldsymbol{U}^*,$$

which proves the result. $\square$

To prove Theorem 9.1 we need one more elementary fact [Peller and Khrushchëv 1982].

LEMMA 9.3. *Let h be an outer function in H^2, τ an inner function, and let $u = \bar{\tau}\bar{h}/h$. Then T_u has dense range in H^2.*

PROOF. Assume that $f \perp T_u H^2$ is nonzero. Then $(f, ug) = 0$ for any $g \in H^2$. We have $f = f_{(o)} f_{(i)}$, where $f_{(o)}$ is outer and $f_{(i)}$ is inner. Put $g = \tau f_{(i)} h$. Then

$$(f, ug) = (f_{(i)} f_{(o)}, \bar{\tau}\tau f_{(i)} \bar{h}) = (f_{(o)}, \bar{h}) = f_{(o)}(0)h(0) = 0,$$

which is impossible since both h and $f_{(o)}$ are outer. $\square$

PROOF OF THEOREM 9.1. We prove the theorem for $X = B_p^{1/p}$. The proof for $X = \mathrm{VMO}$ is exactly the same.

Without loss of generality we may assume that $\mathbb{P}_- \varphi \neq 0$. Multiplying φ, if necessary, by a suitable constant, we may also assume that $\|H_\varphi\| = 1$. Let $f = \mathcal{A}\varphi$. Put $u = \varphi - f$. By Corollary 2.5, u is unimodular and has the form $u = \bar{z}\bar{\vartheta}\bar{h}/h$, where ϑ is an inner function and h is an outer function in H^2. It follows from Lemma 9.3 that T_u has dense range in H^2.

Since $\mathbb{P}_- u = \mathbb{P}_- \varphi$, Theorem 7.1 implies that $H_u \in \boldsymbol{S}_p$ and $\|H_u\|_{\boldsymbol{S}_p}$ is equivalent to $\|\mathbb{P}_- \varphi\|_{B_p^{1/p}}$. We can now apply Theorem 9.2, which implies that

$$\|H_{\bar{u}}\|_{\boldsymbol{S}_p} \le \|H_u\|_{\boldsymbol{S}_p},$$

and so

$$\|\mathbb{P}_+ u\|_{B_p^{1/p}} \le \mathrm{const}\, \|\mathbb{P}_- u\|_{B_p^{1/p}}.$$

The result follows now from the obvious observation $f = \mathbb{P}_+ f = \mathbb{P}_+ \varphi - \mathbb{P}_+ u.$ $\square$

10. Hankel Operators and Prediction Theory

In this section we demonstrate how Hankel operators can be applied in prediction theory. By a discrete time stationary Gaussian process we mean a two-sided sequence $\{X_n\}_{n\in\mathbb{Z}}$ of random variables which belong to a Gaussian space (i.e., space of functions normally distributed) such that

$$EX_n = 0$$

and

$$EX_n X_k = c_{n-k}$$

for some sequence $\{c_n\}_{n\in\mathbb{Z}}$ of real numbers, where E is mathematical expectation.

It is easy to see that the sequence $\{c_n\}_{n\in\mathbb{Z}}$ is positive semi-definite, so by the Riesz–Herglotz theorem [Riesz and Sz.-Nagy 1965] there exists a finite positive measure μ on $\mathbb{T}$ such that $\hat{\mu}(n) = c_n$. The measure μ is called the *spectral measure of the process*.

We can now identify the closed linear span of $\{X_n\}_{n\in\mathbb{Z}}$ with the space $L^2(\mu)$ using the unitary map defined by

$$X_n \mapsto z^n \quad \text{for } n \in \mathbb{Z}.$$

This allows one to reduce problems of prediction theory to the corresponding problems in the space $L^2(\mu)$, and instead of the sequence $\{X_n\}_{n\in\mathbb{Z}}$ we can study the sequence $\{z^n\}_{n\in\mathbb{Z}}$. Note that if μ is the spectral measure of a stationary Gaussian process, its Fourier coefficients are real, so μ satisfies the condition

$$\mu(E) = \mu\{\zeta : \bar{\zeta} \in E\} \quad \text{for } E \in \mathbb{T}. \tag{10–1}$$

It can be shown that any finite positive measure satisfying (10–1) is the spectral measure of a stationary Gaussian process. However, to study regularity conditions in the space $L^2(\mu)$ we do not need (10–1). So from now on μ is an arbitrary positive finite Borel measure on $\mathbb{T}$, though if it does not satisfy (10–1), the results described below have no probabilistic interpretation.

With the process $\{z^n\}_{n\in\mathbb{Z}}$ we associate the following subspaces of $L^2(\mu)$:

$$\boldsymbol{G}^n \stackrel{\text{def}}{=} \operatorname*{span}_{L^2(\mu)}\{z^m : m \geq n\} = z^n H^2(\mu)$$

("future starting at the moment n") and

$$\boldsymbol{G}_n \stackrel{\text{def}}{=} \operatorname*{span}_{L^2(\mu)}\{z^m : m < n\} = z^n H^2_-(\mu)$$

("past till the moment n"). Here

$$H^2(\mu) \stackrel{\text{def}}{=} \operatorname*{span}_{L^2(\mu)}\{z^m : m \geq 0\}, \qquad H^2_-(\mu) \stackrel{\text{def}}{=} \operatorname*{span}_{L^2(\mu)}\{z^m : m < 0\},$$

and span means the closed linear span.

The process $\{z^n\}_{n\in\mathbb{Z}}$ is called *regular* if

$$\bigcap_{n\geq 0} \boldsymbol{G}^n = \{0\}.$$

We denote by $\mathcal{P}^n$ and $\mathcal{P}_n$ the orthogonal projections onto $\boldsymbol{G}^n$ and $\boldsymbol{G}_n$ respectively. It is easy to see that the process is regular if and only if $\lim_{n\to\infty} \mathcal{P}_0\mathcal{P}^n = 0$ in the strong operator topology.

By Szegö's theorem [Ibragimov and Rozanov 1970] the process is regular if an only if μ is absolutely continuous with respect to Lebesgue measure and its density w (called the *spectral density of the process*) satisfies $\log w \in L^1$.

In prediction theory it is important to study other regularity conditions (i.e., conditions expressing that the operators $\mathcal{P}_0\mathcal{P}^n$ are small in a certain sense) and characterize the processes satisfying such conditions in terms of the spectral densities.

A process $\{z^n\}_{n\in\mathbb{Z}}$ in $L^2(\mu)$ is called *completely regular* if

$$\rho_n \overset{\text{def}}{=} \lim_{n\to\infty} \|\mathcal{P}_0\mathcal{P}^n\| = 0;$$

this means that the spaces $\boldsymbol{G}^n$ and $\boldsymbol{G}_0$ become asymptotically orthogonal as $n \to \infty$, or the corresponding Gaussian subspaces become asymptotically independent.

The following results describe processes satisfying certain regularity conditions. See [Peller and Khrushchëv 1982] for other regularity conditions.

THEOREM 10.1. *The process $\{z^n\}_{n\in\mathbb{Z}}$ in $L^2(w)$ is completely regular if and only if w admits a representation*

$$w = |P|^2 e^\varphi, \tag{10-2}$$

where φ is a real function in VMO *and P is a polynomial with zeros on $\mathbb{T}$.*

Theorem 10.1 was proved in [Helson and Sarason 1967] and [Sarason 1972] (without mention of the space VMO, which was introduced later).

THEOREM 10.2. *The process $\{z^n\}_{n\in\mathbb{Z}}$ in $L^2(w)$ satisfies the condition*

$$\rho_n \leq \text{const}(1 + n)^{-\alpha}, \quad \text{for } \alpha > 0,$$

if and only if w admits a representation of the form (10–2)*, where φ is a real function in Λ^α and P is a polynomial with zeros on $\mathbb{T}$.*

Theorem 10.2 was obtained by Ibragimov; see [Ibragimov and Rozanov 1970].

THEOREM 10.3. *The process $\{z^n\}_{n\in\mathbb{Z}}$ in $L^2(w)$ satisfies the condition*

$$\mathcal{P}_0\mathcal{P}^0 \in \boldsymbol{S}_p, \quad \text{for } 0 < p < \infty,$$

if and only if w admits a representation of the form (10–2)*, where φ is a real function in $B_p^{1/p}$ and P is a polynomial with zeros on $\mathbb{T}$.*

For $p = 2$ Theorem 10.3 was proved by Ibragimov and Solev; see [Ibragimov and Rozanov 1970]. It was generalized for $1 \leq p < \infty$ in [Peller 1980; Peller and Khrushchëv 1982] and for $p < 1$ in [Peller 1983]. The Ibragimov–Solev proof works only for $p = 2$.

Original proofs were different for different regularity conditions; some of them (in particular, the original proof of Theorem 10.2) were technically very complicated. In [Peller and Khrushchëv 1982] a unified method was found that allowed one to prove all such results by the same method. The method involves Hankel and Toeplitz operators and it simplifies considerably many original proofs. In [Peller 1990a] the method was simplified further.

In this section we prove Theorem 10.1. The proofs of Theorems 10.2 and 10.3 are similar.

To prove Theorem 10.1 we need several well-known results from the theory of Toeplitz operators. We mention some elementary properties, which follow immediately from the definition:

$$T_\varphi^* = T_{\bar\varphi} \qquad \text{for } \varphi \in L^\infty,$$
$$T_{\bar\varphi f \psi} = T_{\bar\varphi} T_f T_\psi \quad \text{for } f \in L^\infty \text{ and } \varphi, \psi \in H^\infty.$$

An operator T on Hilbert space is called *Fredholm* if there exists an operator R such that $TR - I$ and $RT - I$ are compact. It is well-known that T is Fredholm if and only if $\dim \operatorname{Ker} T < \infty$, $\dim \operatorname{Ker} T^* < \infty$, and the range of T is closed. The *index* $\operatorname{ind} T$ of a Fredholm operator T is defined by

$$\operatorname{ind} T = \dim \operatorname{Ker} T - \dim \operatorname{Ker} T^*.$$

If T_1 and T_2 are Fredholm, then $\operatorname{ind} T_1 T_2 = \operatorname{ind} T_1 + \operatorname{ind} T_2$. The proofs of these facts can be found in [Douglas 1972].

Clearly, a Fredholm operator with zero index is not necessarily invertible. However, the following result of Coburn (see [Sarason 1978; Nikol'skiĭ 1986, Appendix 4, 43], for example) shows that a Fredholm Toeplitz operator with zero index must be invertible.

LEMMA 10.4. *Let $\varphi \in L^\infty$. Then $\operatorname{Ker} T_\varphi = \{0\}$ or $\operatorname{Ker} T_\varphi^* = \{0\}$.*

PROOF. Let $f \in \operatorname{Ker} T_\varphi$ and $g \in \operatorname{Ker} T_\varphi^*$. Then $\varphi f \in H_-^2$ and $\bar\varphi g \in H_-^2$. Consequently, $\varphi f \bar g \in H_-^1 \overset{\text{def}}{=} \{\psi \in L^1 : \hat\psi(n) = 0, \ n \leq 0\}$ and $\bar\varphi \bar f g \in H_-^1$. Thus the Fourier coefficients of $\varphi f \bar g$ are identically equal to zero, and so $\varphi f \bar g = 0$. Therefore if φ is a nonzero function, then either f or g must vanish on a set of positive measure which implies that $f = 0$ or $g = 0$. $\square$

We need one more well-known lemma of Devinatz and Widom; see, for example, [Douglas 1972; Nikol'skiĭ 1986, Appendix 4, 36].

LEMMA 10.5. *Let u be a unimodular function such that T_u is invertible. Then there exists an outer function η such that $\|u - \eta\|_{L^\infty} < 1$.*

PROOF. Clearly,

$$\|H_u f\|_{L^2}^2 + \|T_u f\|_{L^2}^2 = \|f\|_{L^2}^2 \quad \text{for } f \in H^2.$$

Since T_u is invertible, it follows that

$$\|H_u\| = \text{dist}_{L^\infty}\{u, H^\infty\} < 1.$$

Let η be a function in H^∞ such that $\|u - \eta\|_{L^\infty} < 1$. We show that η is outer. We have

$$\|I - T_{\bar{u}\eta}\| = \|\mathbf{1} - \bar{u}\eta\|_{L^\infty} = \|u - \eta\|_{L^\infty} < 1$$

(here $\mathbf{1}$ is the function identically equal to 1). Thus $T_{\bar{u}\eta} = T_u^* T_\eta$ is invertible. Hence T_η is invertible. Clearly T_η is multiplication by η on H^2, and so η must be invertible in H^∞ which implies that η is outer. □

Finally, we prove the theorem of Sarason [1978] that describes the unimodular functions in VMO. We give the proof from [Peller and Khrushchëv 1982], which is based on Toeplitz operators.

THEOREM 10.6. *A unimodular function u belongs to* VMO *if an only if u admits a representation*

$$u = z^n \exp i(\tilde{q} + r), \tag{10–3}$$

where $n \in \mathbb{Z}$ and q and r are real functions in $C(\mathbb{T})$.

In other words, u belongs to VMO if and only if $u = z^n e^{i\kappa}$, where κ is a real function in VMO.

PROOF. Suppose that u is given by (10–3). Then

$$u = z^n \exp(q + i\tilde{q}) \exp(-q + ir) \in H^\infty + C,$$

since $H^\infty + C$ is an algebra (see Theorem 5.1). Hence H_u is compact, and so $\mathbb{P}_- u \in$ VMO (see Theorems 5.5 and 5.8). Similarly, $\mathbb{P}_- \bar{u} \in$ VMO, and so $u \in$ VMO.

Now suppose that $u \in$ VMO. It follows immediately from the definitions of Hankel and Toeplitz operators, that

$$I - T_u T_{\bar{u}} = H_{\bar{u}}^* H_{\bar{u}} \quad \text{for } I - T_{\bar{u}} T_u = H_u^* H_u.$$

Since the Hankel operators H_u and $H_{\bar{u}}$ are compact, the operator T_u is Fredholm. Put $u = z^n v$, where $n = \text{ind } T_u$. If $n \geq 0$, then $T_u = T_v T_{z^n}$, whereas if $n \leq 0$, then $T_u = T_{z^n} T_v$. Therefore $\text{ind } T_u = \text{ind } T_v + \text{ind } T_{z^n} = \text{ind } T_v - n = -n$. Hence $\text{ind } T_v = 0$, and T_v is invertible by Lemma 10.4.

By Lemma 10.5 there exists an outer function η such that

$$\|v - \eta\|_{L^\infty} = \|\mathbf{1} - \bar{v}\eta\|_{L^\infty} < 1.$$

Hence $\bar{v}\eta$ has a logarithm in the Banach algebra $H^\infty + C$. Let $f \in C(\mathbb{T})$ and let $g \in H^\infty$ satisfy $(\bar{v}\eta)^{-1} = v/\eta = \exp(f + g)$. We have

$$v = \exp(ic + \log|\eta| + i\widetilde{\log|\eta|} + f + g),$$

where $c \in \mathbb{R}$. Since v is unimodular, it follows that $\log|\eta| + \mathrm{Re}(f + g) = 0$. Therefore, setting $q \stackrel{\text{def}}{=} \log|\eta| + \mathrm{Re}\,g$, we have $q \in C(\mathbb{T})$. Since $g \in H^\infty$, we get

$$\widetilde{\log|\eta|} + \mathrm{Im}\,g = \tilde{q} + \mathrm{Im}\,\hat{g}(0).$$

To complete the proof it remains to put $r \stackrel{\text{def}}{=} \mathrm{Im}\,f + c + \mathrm{Im}\,\hat{g}(0)$ and observe that u satisfies (10–3). $\qquad\square$

PROOF OF THEOREM 10.1. We first write the operator $\mathcal{P}_0\mathcal{P}^n$ in terms of a Hankel operator. Let h be an outer function in H^2 such that $|h|^2 = w$. Consider the unitary operators $\mathcal{U}$ and $\mathcal{V}$ from L^2 onto $L^2(w)$ defined by

$$\mathcal{U}f = f/h, \qquad \mathcal{V}f = f/\bar{h}, \qquad f \in L^2.$$

Since h is outer, it follows from Beurling's theorem (see [Nikol'skiĭ 1986], for example) that $\mathcal{U}H^2 = H^2(w)$ and $\mathcal{V}H^2_- = H^2_-(w)$. Therefore

$$\mathcal{P}_0 g = \mathcal{V}\mathbb{P}_-\mathcal{V}^{-1}g \quad \text{for } g \in L^2(w),$$

and

$$\mathcal{P}^n g = \mathcal{U}z^n\mathbb{P}_+\bar{z}^n\mathcal{U}^{-1}g \quad \text{for } g \in L^2(w).$$

Hence

$$\mathcal{P}_0\mathcal{P}^n g = \mathcal{V}\mathbb{P}_-\mathcal{V}^{-1}\mathcal{U}z^n\mathbb{P}_+\bar{z}^n\mathcal{U}^{-1}g = \mathcal{V}\mathbb{P}_-(\bar{h}/h)z^n\mathbb{P}_+\bar{z}^n\mathcal{U}^{-1}g \quad \text{for } g \in L^2(w).$$

It follows that

$$\rho_n = \|H_{z^n\bar{h}/h}\| \quad \text{for } n \geq 0. \tag{10–4}$$

LEMMA 10.7. *The process $\{z^n\}_{n\in\mathbb{Z}}$ is completely regular if and only if $\bar{h}/h \in$ VMO.*

PROOF. It follows from (10–4) that complete regularity is equivalent to the fact that $\|H_{z^n\bar{h}/h}\| \to 0$. We have

$$\|H_{z^n\bar{h}/h}\| = \mathrm{dist}_{L^\infty}\{z^n\bar{h}/h, H^\infty\}$$
$$= \mathrm{dist}_{L^\infty}\{\bar{h}/h, \bar{z}^n H^\infty\} \to \mathrm{dist}_{L^\infty}\{\bar{h}/h, H^\infty + C\},$$

so $\lim_{n\to\infty}\rho_n = 0$ if and only if $\bar{h}/h \in H^\infty + C$. The last condition means that $H_{\bar{h}/h}$ is compact, which is equivalent to the fact that $\mathbb{P}_-\bar{h}/h \in$ VMO (see Theorem 5.8). It remains to show that this is equivalent to the inclusion $\bar{h}/h \in$ VMO.

Put $u = \bar{h}/h$. By Lemma 9.3 the Toeplitz operator T_u has dense range in H^2, so by Theorem 9.2 the Hankel operator $H_{\bar{u}}$ is compact. The result now follows from Theorem 5.8. $\qquad\square$

We resume the proof of Theorem 10.1. It is easy to see that

$$\overline{\frac{z-\gamma}{z-\gamma}} = -\bar{\gamma}\bar{z} \quad \text{for } \gamma \in \mathbb{T}.$$

Therefore for a polynomial P of degree m with zeros on $\mathbb{T}$

$$\frac{\overline{P}}{P} = c\bar{z}^m$$

for $c \in \mathbb{T}$.

Suppose first that $w = |P|^2 e^\varphi$, where $\varphi \in \mathrm{VMO}$ and P is a polynomial of degree m. Consider the outer function $h_1 = \exp\frac{1}{2}(\varphi + i\tilde{f})$. Since Ph_1 is outer and $|Ph_1| = |h|$, we have $h = \omega Ph_1$, where $\omega \in \mathbb{T}$. Hence

$$\frac{\overline{h}}{h} = \bar{\omega}^2 \frac{\overline{P}}{P} \frac{\overline{h}_1}{h_1} = \lambda \exp(-i\tilde{\varphi}), \quad \lambda \in \mathbb{T}.$$

By Theorem 10.6, $\bar{h}/h \in \mathrm{VMO}$, and so by Lemma 10.7 the process is completely regular.

Conversely, suppose that the process is completely regular. By Lemma 10.7, $\bar{h}/h \in \mathrm{VMO}$, so by Theorem 10.6

$$\frac{\overline{h}}{h} = \bar{z}^m e^{i\psi}$$

for some $m \in \mathbb{Z}$ and $\omega \in \mathrm{VMO}$. By Lemma 9.3, $T_{\bar{h}/h}$ has dense range, so $\operatorname{ind} T_{\bar{h}/h} \geq 0$. It follows from the proof of Theorem 10.6 that $T_{e^{i\psi}}$ is invertible, which implies that $m \geq 0$. Now consider the outer function

$$h_1 \stackrel{\text{def}}{=} \exp(-\tilde{\psi}/2 + i\psi/2).$$

As in the remark after Corollary 5.7 we can conclude that $h_1 \in H^2$.

Consider the Toeplitz operator $T_{\bar{z}\bar{h}/h} = T_{\bar{z}^{m+1}\bar{h}_1/h_1}$. Its index equals $m+1$, so it has $(m+1)$-dimensional kernel. Obviously, $z^j h_1 \in \operatorname{Ker} T_{\bar{z}^{m+1}\bar{h}_1/h_1}$ for $0 \leq j \leq m$, and so the functions $z^j h_1$, for $0 \leq j \leq m$, form a basis in $\operatorname{Ker} T_{\bar{z}\bar{h}/h}$. It is also obvious that $h \in \operatorname{Ker} T_{\bar{z}\bar{h}/h}$. Hence $h = Ph_1$ for some polynomial P of degree at most m. Since h and h_1 are outer, so is P, which implies that P has no zeros outside the closed unit disk.

We show that P has degree m and has no zeros in $\mathbb{D}$. Let $P = P_1 P_2$, where P_1 has zeros on $\mathbb{T}$ and P_2 has zeros outside the closed unit disk. Let $k = \deg P_1$. Then

$$\frac{\overline{h}}{h} = \frac{\overline{P}_1}{P_1} \frac{\overline{P}_2}{P_2} \frac{\overline{h}_1}{h_1} = \omega \bar{z}^k \overline{P}_2 \frac{\overline{h}_1}{h_1} P_2^{-1} \quad \text{for } \omega \in \mathbb{T}.$$

Consequently,

$$\operatorname{ind} T_{\bar{h}/h} = \operatorname{ind} T_{\bar{z}^k \overline{P}_2 (\bar{h}_1/h_1) P_2^{-1}} = k + \operatorname{ind} T_{\overline{P}_2} + \operatorname{ind} T_{\bar{h}_1/h_1} + \operatorname{ind} T_{P_2^{-1}} = k,$$

since the operators $T_{\overline{P}_2}$, $T_{\bar{h}_1/h_1}$, and $T_{P_2^{-1}}$ are clearly invertible. Hence $k = m$ which completes the proof. $\qquad\square$

11. Spectral Properties of Hankel Operators with Lacunary Symbols

To speak about spectral properties we certainly have to realize Hankel operators as operators from a certain Hilbert space into itself. For $\varphi \in L^\infty$ we can consider the operator Γ_φ on ℓ^2 with Hankel matrix $\{\hat{\varphi}(j+k)\}_{j,k\geq 0}$ in the standard basis of ℓ^2. Not much is known about spectral properties of such operators in terms of φ. Power [1982] described the essential spectrum of Γ_φ for piecewise continuous functions φ. See also [Howland 1986; 1992a; 1992b], where spectral properties of self-adjoint operators Γ_φ with piecewise continuous φ are studied.

For a long time it was unknown whether there exists a nonzero quasinilpotent Hankel operator Γ, i.e., a Hankel operator Γ such that $\sigma(\Gamma) = \{0\}$ [Power 1984]. This question was answered affirmatively by Megretskii [1990], who considered Hankel operators with lacunary symbols and found an interesting approach to the description (in a sense) of their spectra. In particular, his method allows one to construct nonzero quasinilpotent Hankel operators.

In this section we describe the method of [Megretskii 1990]. In particular we prove that the operator with the following Hankel matrix is compact and quasinilpotent:

$$\Gamma_\# = \begin{pmatrix} i & \frac{1}{2} & 0 & \frac{1}{4} & 0 & \cdots \\ \frac{1}{2} & 0 & \frac{1}{4} & 0 & 0 & \cdots \\ 0 & \frac{1}{4} & 0 & 0 & 0 & \cdots \\ \frac{1}{4} & 0 & 0 & 0 & \frac{1}{8} & \cdots \\ 0 & 0 & 0 & \frac{1}{8} & 0 & \cdots \\ \vdots & \vdots & \vdots & \vdots & \vdots & \ddots \end{pmatrix} . \tag{11-1}$$

We consider a more general situation of Hankel operators of the form

$$\Gamma = \begin{pmatrix} \alpha_0 & \alpha_1 & 0 & \alpha_2 & 0 & \cdots \\ \alpha_1 & 0 & \alpha_2 & 0 & 0 & \cdots \\ 0 & \alpha_2 & 0 & 0 & 0 & \cdots \\ \alpha_2 & 0 & 0 & 0 & \alpha_3 & \cdots \\ 0 & 0 & 0 & \alpha_3 & 0 & \cdots \\ \vdots & \vdots & \vdots & \vdots & \vdots & \ddots \end{pmatrix} ,$$

where $\{\alpha_k\}_{k\geq 0}$ is a sequence of complex numbers. In other words, we set $\Gamma = \{\gamma_{j+k}\}_{j,k\geq 0}$, where

$$\gamma_j = \begin{cases} \alpha_k & \text{if } j = 2^k - 1 \text{ with } k \in \mathbb{Z}_+, \\ 0 & \text{otherwise.} \end{cases}$$

We evaluate the norm of Γ and give a certain description of its spectrum.

Since $\mathrm{BMOA} = (H^1)^*$ by Fefferman's theorem [Garnett 1981] with respect to the natural duality, it follows from Paley's theorem [Zygmund 1968] that $\sum_{k\geq 0} \alpha_k z^{2^k-1} \in \mathrm{BMOA}$ if and only if $\{\alpha_k\}_{k\geq 0} \in \ell^2$. Therefore, by Nehari's theorem, Γ is a matrix of a bounded operator if and only if $\{\alpha_k\}_{k\geq 0} \in \ell^2$;

moreover $\|\Gamma\|$ is equivalent to $\|\{\alpha_k\}_{k\geq 0}\|_{\ell^2}$. It is also clear that for $\{\alpha_k\}_{k\geq 0} \in \ell^2$ the function $\sum_{k\geq 0} \alpha_k z^{2^k-1}$ belongs to $VMOA$, so Γ is bounded if and only if it is compact.

We associate with Γ the sequence $\{\mu_k\}_{k\geq 0}$ defined by

$$\mu_0 = 0,$$
$$\mu_{k+1} = \tfrac{1}{2}\big(\mu_k + 2|\alpha_{k+1}|^2 + (\mu_k^2 + 4|\alpha_{k+1}|^2)^{1/2}\big) \quad \text{for } k \in \mathbb{Z}_+.$$
(11–2)

The following theorem evaluates the norm of Γ.

THEOREM 11.1. *If $\{\alpha_k\}_{k\geq 0} \in \ell^2$, the sequence $\{\mu_k\}_{k\geq 0}$ converges and*

$$\|\Gamma\|^2 = \lim_{k\to\infty} \mu_k.$$

To describe the spectrum of Γ consider the class Λ of sequences of complex numbers $\{\lambda_j\}_{j\geq 0}$ satisfying

$$\lambda_0 = \alpha_0,$$
$$(\lambda_j - \lambda_{j-1})\lambda_j = \alpha_j^2 \quad \text{for } j \geq 1.$$
(11–3)

THEOREM 11.2. *Suppose that $\{\alpha_j\}_{j\geq 0} \in \ell^2$. Any sequence $\{\lambda_j\}_{j\geq 0}$ in Λ converges. The spectrum $\sigma(\Gamma)$ consists of 0 and the limits of such sequences.*

To prove Theorems 11.1 and 11.2 we consider finite submatrices of Γ. Let $\mathcal{L}_k$ be the linear span of the basis vectors e_j, for $j = 0, 1, \ldots, 2^k - 1$, and let P_k be the orthogonal projection from ℓ^2 onto $\mathcal{L}_k$. Consider the operator $\Gamma_k \stackrel{\text{def}}{=} P_k\Gamma|\mathcal{L}_k$ and identify it with its $2^k \times 2^k$ matrix. Put

$$\tilde{\Gamma}_k \stackrel{\text{def}}{=} \Gamma_k P_k = \begin{pmatrix} \Gamma_k & 0 \\ 0 & 0 \end{pmatrix}.$$

It is easy to see that $\|\Gamma_k\| = \|\tilde{\Gamma}_k\|$ and $\sigma(\tilde{\Gamma}_k) = \sigma(\Gamma_k) \cup \{0\}$. Clearly,

$$\Gamma_{k+1} = \begin{pmatrix} \Gamma_k & \alpha_{k+1}J_k \\ \alpha_{k+1}J_k & 0 \end{pmatrix},$$

where J_k is the $2^k \times 2^k$ matrix given by

$$J_k = \begin{pmatrix} 0 & 0 & \cdots & 0 & 1 \\ 0 & 0 & \cdots & 1 & 0 \\ \vdots & \vdots & \ddots & \vdots & \vdots \\ 0 & 1 & \cdots & 0 & 0 \\ 1 & 0 & \cdots & 0 & 0 \end{pmatrix}.$$

We need a well-known fact from linear algebra: Let N be a block matrix of the form

$$N = \begin{pmatrix} A & B \\ C & D \end{pmatrix},$$

where A and D are square matrices and D is invertible. Then

$$\det N = \det D \det(A - BD^{-1}C).$$
(11–4)

See [Gantmakher 1988, Chapter 2, §5.3], for example.

PROOF OF THEOREM 11.1. Since $J_k^* = J_k$ and J_k^2 is the identity matrix of size $2^k \times 2^k$ (which we denote by I_k), we have

$$\Gamma_{k+1}^* \Gamma_{k+1} = \begin{pmatrix} \Gamma_k^* \Gamma_k + |\alpha_{k+1}|^2 I_k & \alpha_{k+1} \tilde{\Gamma}_k^* J_k \\ \bar{\alpha}_{k+1} J_k \tilde{\Gamma}_k & |\alpha_{k+1}|^2 I_k \end{pmatrix}. \tag{11-5}$$

Applying formula (11–4) to the matrix $\Gamma_{k+1}^* \Gamma_{k+1} - \lambda I_{k+1}$, where $\lambda \neq |\alpha_{k+1}|^2$, we obtain

$$\det(\Gamma_{k+1}^* \Gamma_{k+1} - \lambda I_{k+1}) = \rho^{2^k} \det\left(-\frac{\lambda}{\rho} \Gamma_k^* \Gamma_k + \rho I_k\right), \tag{11-6}$$

where $\rho \overset{\text{def}}{=} |\alpha_{k+1}|^2 - \lambda$.

Since Γ is a bounded operator, we have $\|\Gamma\| = \lim_{k\to\infty} \|\Gamma_k\|$. Therefore it is sufficient to show that $\mu_k = \|\Gamma_k\|^{1/2}$ or, which is the same, that μ_k is the largest eigenvalue of $\Gamma_k^* \Gamma_k$. We proceed by induction on k. For $k = 0$ the assertion is obvious.

If $\Gamma_k = 0$, the assertion is obvious. Otherwise, it follows easily from (11–5) that $\|\Gamma_{k+1}^* \Gamma_{k+1}\| > |\alpha_{k+1}|^2$.

It is easy to see from (11–6) that $\lambda \neq |\alpha_{k+1}|^2$ is an eigenvalue of $\Gamma_{k+1}^* \Gamma_{k+1}$ if and only if ρ^2/λ is an eigenvalue of $\Gamma_k^* \Gamma_k$. Put

$$\mu = \rho^2/\lambda = (|\alpha_{k+1}|^2 - \lambda)^2/\lambda.$$

If μ is an eigenvalue of $\Gamma_k^* \Gamma_k$, it generates two eigenvalues of $\Gamma_{k+1}^* \Gamma_{k+1}$:

$$\tfrac{1}{2}\left(\mu + 2|\alpha_{k+1}|^2 + (\mu^2 + 4|\alpha_{k+1}|^2)^{1/2}\right) \quad \text{and} \quad \tfrac{1}{2}\left(\mu + 2|\alpha_{k+1}|^2 - (\mu^2 + 4|\alpha_{k+1}|^2)^{1/2}\right).$$

Clearly, to get the largest eigenvalue of $\Gamma_{k+1}^* \Gamma_{k+1}$ we have to put $\mu = \mu_k$ and choose the first of the eigenvalues above. This proves that μ_{k+1} defined by (11–2) is the largest eigenvalue of $\Gamma_{k+1}^* \Gamma_{k+1}$. $\square$

To prove Theorem 11.2 we need two lemmas.

LEMMA 11.3. *Let Λ_k be the set of k-th terms of sequences in Λ; that is,*

$$\Lambda_k = \{\lambda_k : \{\lambda_j\}_{j\geq 0} \in \Lambda\}.$$

If $\{\zeta_j\}_{j\geq 0}$ is an arbitrary sequence satisfying $\zeta_j \in \Lambda_j$, then it converges if and only if $\lim_{j\to\infty} \zeta_j = 0$ or there exists a sequence $\{\lambda_j\}_{j\geq 0} \in \Lambda$ such that $\zeta_j = \lambda_j$ for sufficiently large j.

LEMMA 11.4. *Let A be a compact operator on Hilbert space and let $\{A_j\}_{j\geq 0}$ be a sequence of bounded linear operators such that $\lim_{j\to\infty} \|A - A_j\| = 0$. Then the spectrum $\sigma(A)$ consists of the limits of all convergent sequences $\{\nu_j\}_{j\geq 0}$ such that $\nu_j \in \sigma(A_j)$.*

Lemma 11.4 is well known [Newburgh 1951] and we don't prove it here. Note that we need Lemma 11.4 for compact operators A_j, in which case it is proved in [Gohberg and Kreĭn 1965, Theorem 4.2].

PROOF OF THEOREM 11.2 ASSUMING LEMMAS 11.3 AND 11.4. Since Γ is compact, $0 \in \sigma(\Gamma)$.

For $\lambda \in \mathbb{C} \setminus \{0\}$ we apply formula (11–4) to the matrix $\Gamma_{k+1} - \lambda I_{k+1}$ and obtain

$$\det(\Gamma_{k+1} - \lambda I_{k+1}) = (-\lambda)^{2^k} \det\left(\Gamma_k - \left(\lambda - \frac{\alpha_{k+1}^2}{\lambda}\right) I_k\right). \qquad (11\text{–}7)$$

Obviously,

$$0 \in \sigma(\Gamma_k) \quad \text{if and only if} \quad \alpha_k = 0.$$

Together with (11–7) this implies that $\lambda \in \sigma(\Gamma_k)$ if and only if there exists $\lambda' \in \sigma(\Gamma_{k-1})$ such that $(\lambda - \lambda')\lambda = \alpha_k^2$.

Let λ be a nonzero point in the spectrum of Γ. By Lemma 11.4, there exists a sequence $\{\nu_j\}_{j \geq 0}$ such that $\nu_j \to \lambda$ as $j \to \infty$ and $\nu_j \in \sigma(\tilde{\Gamma}_j)$. Since $\lambda \neq 0$ we may assume without loss of generality that $\nu_j \in \sigma(\Gamma_j)$. It follows now from Lemma 11.3 that there exists a sequence $\{\lambda_j\}_{j \geq 0}$ in Λ such that $\lambda_j = \nu_j$ for sufficiently large j, so $\lambda = \lim_{j \to \infty} \lambda_j$.

Conversely, let $\{\lambda_j\}_{j \geq 0} \in \Lambda$. By Lemma 11.3, $\{\lambda_j\}_{j \geq 0}$ converges to a point $\lambda \in \mathbb{C}$. As we have already observed, $\lambda_j \in \sigma(\Gamma_j)$, so Lemma 11.4 gives $\lambda \in \sigma(\Gamma)$. $\qquad \square$

PROOF OF LEMMA 11.3. Let $\{\lambda_j\}_{j \geq 0} \in \Lambda$. Then $|\lambda_j| \leq |\lambda_{j-1}| + |\alpha_j|^2/|\lambda_j|$ for $j \geq 1$. It follows that

$$|\lambda_j| \leq \max\{\varepsilon, |\lambda_{j-1}| + |\alpha_j|^2/\varepsilon\} \qquad (11\text{–}8)$$

for any $\varepsilon > 0$. We show that either $\lambda_j \to 0$ as $j \to \infty$ or $|\lambda_j| \geq \delta$ for some $\delta > 0$ for sufficiently large j.

To do this, we show first that, if $\varepsilon > 0$ and $\liminf_{j \to \infty} |\lambda_j| < \varepsilon$, then $\limsup_{j \to \infty} |\lambda_j| \leq 2\varepsilon$. Assume to the contrary that $\liminf_{j \to \infty} |\lambda_j| < \varepsilon$ and $\limsup_{j \to \infty} |\lambda_j| > 2\varepsilon$ for some $\varepsilon > 0$. It follows that for any $N \in \mathbb{Z}_+$ there exist positive integers m and n such that $N \leq m < n$, $|\lambda_{m-1}| < \varepsilon$, $|\lambda_j| \geq \varepsilon$ for $m \leq j \leq n$, and $|\lambda_n| \geq 2\varepsilon$. It follows from (11–8) that $|\lambda_j| \leq |\lambda_{j-1}| + |\alpha_j|^2/\varepsilon$ for $m \leq j \leq n$. Therefore

$$|\lambda_n| \leq |\lambda_{m-1}| + \frac{\sum_{j=m}^n |\alpha_j|^2}{\varepsilon}.$$

Since $\{\alpha_k\}_{k \geq 0} \in \ell^2$, we can choose N so large that $(\sum_{j=m}^\infty |\alpha_j|^2)/\varepsilon < \varepsilon$, which contradicts the inequality $|\lambda_n| \geq 2\varepsilon$.

If $|\lambda_j| \geq \delta > 0$ for large values of j, then by (11–3)

$$|\lambda_j - \lambda_{j-1}| \leq \frac{|\alpha_j|^2}{\delta}.$$

Therefore $\{\lambda_j\}_{j \geq 0}$ converges.

Now suppose that $\{\lambda_j\}_{j\geq 0}$ and $\{\nu_j\}_{j\geq 0}$ are sequences in Λ which have nonzero limits. Then, for sufficiently large j,

$$|\lambda_{j-1} - \nu_{j-1}| = \left|(\lambda_j - \nu_j)\left(1 + \frac{\alpha_j^2}{\lambda_j\nu_j}\right)\right| \leq |\lambda_j - \nu_j|(1 + d|\alpha_j|^2)$$

for some $d > 0$. Iterating this inequality, we obtain

$$|\lambda_{j-1} - \nu_{j-1}| \leq \left|\lim_{j\to\infty}\lambda_j - \lim_{j\to\infty}\nu_j\right| \prod_{m=j}^{\infty}(1 + d|\alpha_m|^2)$$

(the infinite product on the right-hand side converges since $\{\alpha_j\}_{j\geq 0} \in \ell^2$). Therefore if $\lim_{j\to\infty}\lambda_j = \lim_{j\to\infty}\nu_j$, then $\lambda_j = \nu_j$ for sufficiently large j.

For $\varepsilon > 0$ we consider the set of sequences $\{\lambda_j\}_{j\geq 0}$ in Λ such that $\limsup\{|\lambda_j| : j \geq k\} \geq \varepsilon$ for any positive integer k. We show that the number of such sequences is finite. Suppose that $\{\lambda_j\}_{j\geq 0} \in \Lambda$ is such a sequence. As we observed at the beginning of the proof, there exists $\delta > 0$ and $j_0 \in \mathbb{Z}_+$ such that $|\lambda_j| \geq \delta$ for sufficiently large j. Clearly, $|\alpha_j| < \delta$ for sufficiently large j. It follows that if j is sufficiently large, then λ_j is uniquely determined by λ_{j-1} and the conditions

$$(\lambda_j - \lambda_{j-1})\lambda_j = \alpha_j^2, \quad |\lambda_j| \geq \delta.$$

Hence there are only finitely many possibilities for such sequences.

Now let $\{\zeta_j\}_{j\geq 0}$ be a converging sequence such that $\zeta_j \in \Lambda_j$ for $j \geq 0$ and such that $\lim_{j\to\infty}\zeta_j \neq 0$. As already proved, there are sequences $\{\lambda_j^{(s)}\}_{j\geq 0} \in \Lambda$, for $s = 1, \ldots, m$, such that $\zeta_j \in \{\lambda_j^{(1)}, \ldots, \lambda_j^{(m)}\}$ for sufficiently large j and the sequences $\{\lambda_j^{(s)}\}_{j\geq 0}$ have distinct limits. It follows that there exists an s in the range $1 \leq s \leq m$ such that $\zeta_j = \lambda_j^{(s)}$ for sufficiently large j. $\qquad\square$

We now proceed to the operator $\Gamma_\#$ defined by (11–1). In other words, we consider the operator Γ with

$$\alpha_0 = i,$$
$$\alpha_j = 2^{-j} \quad \text{for } j \geq 1.$$

THEOREM 11.5. $\Gamma_\#$ is a compact quasinilpotent operator.

PROOF. It is easy to see by induction that if $\{\lambda_j\}_{j\geq 0}$ satisfies (11–3), then $\lambda_j = 2^{-j}i$, so Theorem 11.2 gives $\sigma(\Gamma_\#) = \{0\}$. We have already seen that bounded Hankel operators of this form are always compact. $\qquad\square$

REMARK. We can consider a more general situation where $\Gamma = \{\gamma_{j+k}\}_{j,k\geq 0}$, with

$$\gamma_j = \begin{cases} \alpha_k & \text{for } j = n_k - 1 \text{ with } k \in \mathbb{Z}_+, \\ 0 & \text{otherwise}, \end{cases}$$

where $\{n_k\}_{k\geq 0}$ is a sequence of natural numbers such that $n_{k+1} \geq 2n_k$ for $k \geq 0$ and $\{\alpha_k\}_{k\geq 0} \in \ell^2$. It is easy to see that the same results hold and the same proofs also work in this situation, which allows one to construct other quasinilpotent Hankel operators. In particular, the Hankel operator

$$\begin{pmatrix} 0 & i & 0 & \frac{1}{2} & 0 & \cdots \\ i & 0 & \frac{1}{2} & 0 & 0 & \cdots \\ 0 & \frac{1}{2} & 0 & 0 & 0 & \cdots \\ \frac{1}{2} & 0 & 0 & 0 & \frac{1}{4} & \cdots \\ 0 & 0 & 0 & \frac{1}{4} & 0 & \cdots \\ \vdots & \vdots & \vdots & \vdots & \vdots & \ddots \end{pmatrix}.$$

is quasinilpotent.

It is still unknown whether there exist noncompact quasinilpotent Hankel operators.

12. Recent Developments

In this section we discuss briefly three recent developments in Hankel operators and their applications. We describe the results without proofs but give references.

Self-adjoint operators unitarily equivalent to Hankel operators. In Section 11 we discussed some spectral properties of Hankel operators. Here we consider the problem of describing all possible spectral types of self-adjoint Hankel operators.

The problem can also be described as follows. Let R be a (bounded) self-adjoint operator on a Hilbert space $\mathcal{H}$. When is R unitarily equivalent to a Hankel operator? In other words, is there an orthonormal basis $\{e_j\}_{j \geq 0}$ in $\mathcal{H}$ in which R is represented by a Hankel matrix?

Let me first say a few words about another related problem, posed in [Khrushchëv and Peller 1984], which appeared while we were studying geometric features of prediction theory. Let $\mathcal{K}$ and $\mathcal{L}$ be subspaces of a Hilbert space $\mathcal{H}$. The problem is to find out under which conditions there exists a stationary process $\{x_j\}_{j \in \mathbb{Z}}$ in $\mathcal{H}$ (i.e., the inner products $(x_j, x_k)_{\mathcal{H}}$ depend only on $j - k$) such that

$$\operatorname{span}\{x_j : j < 0\} = \mathcal{K} \quad \text{and} \quad \operatorname{span}\{x_j : j \geq 0\} = \mathcal{L}.$$

It was shown in [Khrushchëv and Peller 1984] that this problem is equivalent to the following one. Let K be a nonnegative self-adjoint operator on Hilbert space. Under which conditions does there exist a Hankel operator Γ whose modulus $|\Gamma| \stackrel{\text{def}}{=} (\Gamma^*\Gamma)^{1/2}$ is unitarily equivalent to K? In the same paper the following two simple necessary conditions were found:

(i) $\operatorname{Ker} K$ is either trivial or infinite-dimensional;

(ii) K is noninvertible.

We asked whether these conditions together are also sufficient.

Partial results in this direction were obtained in [Treil' 1985b; Vasyunin and Treil' 1989; Ober 1987; 1990], where the case of operators K with discrete spectrum was considered. The last two of these papers suggested a very interesting approach to the problem, based on linear systems with continuous time. Using Ober's approach, Treil [1990] gave in a complete solution by proving that conditions (i) and (ii) are sufficient. In the same article he showed that under these conditions there exists a *self-adjoint* Hankel operator whose modulus is unitarily equivalent to K.

Let me explain why the problem of describing the self-adjoint operators that are unitarily equivalent to Hankel operators is considerably more delicate. Recall that by von Neumann's spectral theory each self-adjoint operator R on Hilbert space is unitarily equivalent to multiplication by the independent variable on a direct integral of Hilbert spaces $\int \oplus \mathcal{K}(t) \, d\mu(t)$ that consists of measurable functions f such that $f(t) \in \mathcal{K}(t)$ and

$$\int \|f(t)\|^2_{\mathcal{K}(t)} d\mu(t) < \infty$$

(μ is a positive Borel measure on $\mathbb{R}$, called a *scalar spectral measure* of R). The *spectral multiplicity function* ν_R is defined μ-a.e. by $\nu_R(t) \stackrel{\text{def}}{=} \dim \mathcal{K}(t)$. Two self-adjoint operators are unitary equivalent if and only if their scalar spectral measures are mutually absolutely continuous and their spectral multiplicity functions coincide almost everywhere. See [Birman and Solomyak 1980] for the theory of spectral multiplicity.

Conditions (i) and (ii) describe the spectral multiplicity function $\nu_{|\Gamma|}$ of the moduli of self-adjoint Hankel operators. Namely, (i) means that $\nu(0) = 0$ or $\nu(0) = \infty$, while (ii) means that $0 \in \operatorname{supp} \nu$. Clearly, $\nu_{|\Gamma|}(t) = \nu_\Gamma(t) + \nu_\Gamma(-t)$, for $t > 0$. So the problem of describing the self-adjoint operators that are unitarily equivalent to Hankel operators is equivalent to the problem of investigating how $\nu_{|\Gamma|}(t)$ can be distributed between $\nu_\Gamma(t)$ and $\nu_\Gamma(-t)$.

The problem was solved recently in [Megretskii et al. 1995]. The main result of that paper is the following theorem. As usual, μ_a and μ_s are the absolutely continuous and the singular parts of a measure μ.

THEOREM 12.1. *Let R be a selfadjoint operator on Hilbert space, μ a scalar spectral measure of R, and ν its spectral multiplicity function. Then R is unitarily equivalent to a Hankel operator if and only if the following conditions hold:*

(i) *Either* $\operatorname{Ker} R = \{0\}$ *or* $\dim \operatorname{Ker} R = \infty$.

(ii) *R is noninvertible.*

(iii) *$|\nu(t) - \nu(-t)| \leq 1$, μ_a-a.e., and $|\nu(t) - \nu(-t)| \leq 2$, μ_s-a.e.*

The necessity of (i) and (ii) is almost obvious. The necessity of (iii) is more complicated. To prove that (iii) is necessary certain commutation relations between Hankel operators, the shift operator, and the backward shift were used in [Megretskii et al. 1995].

However, the most difficult problem is to prove sufficiency. It would be natural to try the method of linear systems with continuous time. Unfortunately (or perhaps fortunately), it does not work. To be more precise, it works if we replace (iii) by the stronger condition: $|\nu(t) - \nu(-t)| \leq 1$, μ-a.e.

To prove sufficiency we used in [Megretskii et al. 1995] linear systems with discrete time (with scalar input and scalar output). Let A be a bounded linear operator on a Hilbert space $\mathcal{H}$, and let $b, c \in \mathcal{H}$. Consider the linear system

$$\begin{cases} x_{n+1} = Ax_n + u_n b, \\ y_n = (x_n, c), \end{cases} \qquad (12\text{--}1)$$

for $n \in \mathbb{Z}$. Here $u_n \in \mathbb{C}$ is the input, $x_n \in \mathcal{H}$, and $y_n \in \mathbb{C}$ is the output. We assume that $\sup_{n \geq 0} \|A^n\| < \infty$.

We can associate with (12–1) the Hankel matrix $\Gamma_\alpha = \{\alpha_{j+k}\}_{j,k \geq 0}$, where $\alpha_j \overset{\text{def}}{=} (A^j b, c)$.

The Hankel operator Γ_α is related to the system (12–1) in the following way. We can associate with a sequence $v = \{v_n\}_{n \geq 0} \in \ell^2$ the input sequence $u = \{u_n\}_{n \in \mathbb{Z}}$ defined by

$$u_n = \begin{cases} v_{-1-n} & \text{if } n < 0, \\ 0 & \text{if } n \geq 0. \end{cases}$$

It is easy to see that under the initial condition $\lim_{n \to -\infty} x_n = 0$ the output $y = \{y_n\}_{n \geq 0}$ of the system (12–1) with input u satisfies $y = \Gamma_\alpha v$.

It was shown in [Megretskii et al. 1995] that under conditions (i)–(iii) of Theorem 12.1 there exists a triple $\{A, b, c\}$ such that the Hankel operator Γ_α is unitarily equivalent to R. The proof is very complicated. The triple $\{A, b, c\}$ is found as a solution of certain Lyapunov–type equations. In addition to that, A must satisfy the asymptotic stability condition

$$\|A^n x\| \to 0 \quad \text{for } x \in \mathcal{H}.$$

The most complicated part of the proof is to construct a solution satisfying the asymptotic stability condition above.

Approximation by analytic matrix functions. As mentioned in the introduction, Hankel operators play an important role in control theory, and it is especially important in control theory to consider Hankel operators whose symbols are matrix functions or even operator functions. Let $\mathcal{H}$ and $\mathcal{K}$ be Hilbert spaces and let $\Phi \in L^\infty(\mathbb{B}(\mathcal{H}, \mathcal{K}))$, i.e., Φ is a bounded weakly measurable function taking values in the space $\mathbb{B}(\mathcal{H}, \mathcal{K})$ of bounded linear operators from $\mathcal{H}$ to $\mathcal{K}$. We can define the Hankel operator $H_\Phi : H^2(\mathcal{H}) \to H^2_-(\mathcal{K})$ by

$$H_\Phi f \overset{\text{def}}{=} \mathbb{P}_- \Phi f \quad \text{for } f \in H^2(\mathcal{H}),$$

where the spaces of vector functions $H^2(\mathcal{H})$ and $H^2_-(\mathcal{K})$ are defined as in the scalar case and $\mathbb{P}_-$ is the orthogonal projection onto $H^2_-(\mathcal{K})$. The analog of

Nehari's theorem says that

$$\|H_\Phi\| = \inf\{\|\Phi - F\|_{L^\infty(\mathbb{B}(\mathcal{H},\mathcal{K}))} : F \in H^\infty(\mathbb{B}(\mathcal{H},\mathcal{K}))\}. \qquad (12\text{--}2)$$

The operator H_Φ is compact if and only if $\Phi \in H^\infty(\mathbb{B}(\mathcal{H},\mathcal{K})) + C(\mathfrak{K}(\mathcal{H},\mathcal{K}))$, where $C(\mathfrak{K}(\mathcal{H},\mathcal{K}))$ is the space of continuous functions that take values in the space $\mathfrak{K}(\mathcal{H},\mathcal{K})$ of compact operators from $\mathcal{H}$ to $\mathcal{K}$. The proofs of these facts can be found in [Page 1970]. As in the scalar case, Nehari's problem is to find, for a given $\Phi \in L^\infty(\mathbb{M}_{m,n})$, a function $F \in H^\infty(\mathbb{B}(\mathcal{H},\mathcal{K}))$ that minimizes the right-hand side of (12–2).

If $\dim \mathcal{H} = n < \infty$ and $\dim \mathcal{K} = m < \infty$, then $\mathcal{H}$ can be identified with $\mathbb{C}^n$, $\mathcal{K}$ with $\mathbb{C}^m$, and $\mathbb{B}(\mathcal{H},\mathcal{K})$ with the space $\mathbb{M}_{m,n}$ of $m \times n$ matrices.

It is important in applications to be able to solve Nehari's problem for matrix functions (and for operator functions). However, unlike the scalar case, it is only exceptionally that the problem has a unique solution. Consider the matrix function

$$\Phi = \begin{pmatrix} \bar{z} & 0 \\ 0 & \frac{1}{2}\bar{z} \end{pmatrix}.$$

Since $\|H_{\bar{z}}\| = 1$, it follows that $\operatorname{dist}_{L^\infty}\{\bar{z}, H^\infty\} = 1$, and since $\|\Phi\|_{L^\infty(\mathbb{M}_{2,2})} = 1$, we have $\operatorname{dist}_{L^\infty(\mathbb{M}_{2,2})}\{\Phi, H^\infty(\mathbb{M}_{2,2})\} = 1$. On the other hand, if f is a scalar function in H^∞ and $\|f\|_{H^\infty} \le \frac{1}{2}$, it is obvious that

$$\left\| \Phi - \begin{pmatrix} 0 & 0 \\ 0 & f \end{pmatrix} \right\|_{L^\infty(\mathbb{M}_{2,2})} = 1,$$

so Φ has infinitely many best uniform approximants by bounded analytic functions. Intuitively, however, it is clear that the "very best" approximation is the zero function, since a nonzero $f \in H^\infty$ increases the L^∞-norm of the lower right entry.

This suggests the idea of imposing additional constraints on $\Phi - F$. Given a matrix function Φ, we put

$$\Omega_0 = \{F \in H^\infty(\mathbb{M}_{m,n}) : F \text{ minimizes } t_0 = \sup_{\zeta \in \mathbb{T}} \|\Phi(\zeta) - F(\zeta)\|\};$$

$$\Omega_j = \{F \in \Omega_{j-1} : F \text{ minimizes } t_j = \sup_{\zeta \in \mathbb{T}} s_j(\Phi(\zeta) - F(\zeta))\}.$$

Here s_j is the j-th singular value.

Functions in $F \in \Omega_{\min\{m,n\}-1}$ are called *superoptimal approximations of Φ by analytic functions*, or superoptimal solutions of Nehari's problem. The numbers t_j are called *superoptimal singular values* of Φ. The notion of superoptimal approximation was introduced in [Young 1986]; it is important in H^∞ control theory.

The following uniqueness theorem was obtained in [Peller and Young 1994a].

THEOREM 12.2. *Let* $\Phi \in H^\infty + C(\mathbb{M}_{m,n})$. *Then there exists a unique super-optimal approximation* $F \in H^\infty(\mathbb{M}_{m,n})$ *by bounded analytic functions. It satisfies the equalities*

$$s_j(\Phi(\zeta) - F(\zeta)) = t_j \quad a.e. \ on \ \mathbb{T} \quad for \ 0 \le j \le \min\{m,n\} - 1.$$

Here is briefly the method of the proof. Let $v \in H^2(\mathbb{C}^n)$ be a maximizing vector of H_Φ (which exists since H_Φ is compact). Consider the vector function $w = H_\Phi v \in H^2(\mathbb{C}^m)$. It can be shown that v and $\bar{z}\bar{w}$ admit the factorizations

$$v = \vartheta_1 h\boldsymbol{v} \quad \text{and} \quad \bar{z}\bar{w} = \vartheta_2 h\boldsymbol{w},$$

where h is a scalar outer function, ϑ_1 and ϑ_2 are scalar inner functions, and $\boldsymbol{v}$ and $\boldsymbol{w}$ are column functions which are inner and co-outer (this means that $\|\boldsymbol{v}(\zeta)\|_{\mathbb{C}^n} = \|\boldsymbol{w}(\zeta)\|_{\mathbb{C}^m} = 1$ a.e. on $\mathbb{T}$, and both $\boldsymbol{v}$ and $\boldsymbol{w}$ have coprime entries, i.e., they do not have a common nonconstant inner divisor).

It is proved in [Peller and Young 1994a] that $\boldsymbol{v}$ and $\boldsymbol{w}$ admit *thematic completions*; that is, there exist matrix functions $V_c \in H^\infty(\mathbb{M}_{n,n-1})$ and $W_c \in H^\infty(\mathbb{M}_{m,m-1})$ such that the matrix functions $V \overset{\text{def}}{=} (\boldsymbol{v} \ \ \overline{V}_c)$ and $W \overset{\text{def}}{=} (\boldsymbol{w} \ \ \overline{W}_c)$ have the following properties:

(i) V and W take unitary values.
(ii) all minors of V and W on the first column are in H^∞.

Let Q be an arbitrary best approximant in $H^\infty(\mathbb{M}_{m,n})$. It is shown in [Peller and Young 1994a] that

$$W^t(\Phi - Q)V = \begin{pmatrix} t_0 u_0 & 0 \\ 0 & \Phi^{(1)} \end{pmatrix}, \tag{12-3}$$

where $u_0 \overset{\text{def}}{=} \bar{z}\bar{\vartheta}_1\bar{\vartheta}_2\bar{h}/h$ and $\Phi^{(1)} \in H^\infty + C(\mathbb{M}_{m-1,n-1})$ (this inclusion is deduced in [Peller and Young 1994a] from the analyticity property (ii) of the minors). It is shown in the same article that the problem of finding a superoptimal approximation of Φ reduces to the problem of finding one for $\Phi^{(1)}$. Namely, if $F^{(1)}$ is a superoptimal approximation of $\Phi^{(1)}$, the formula

$$W^t(\Phi - F)V = \begin{pmatrix} t_0 u_0 & 0 \\ 0 & \Phi^{(1)} - F^{(1)} \end{pmatrix}$$

determines a superoptimal approximation F to Φ. This allows us to reduce the size of the matrix function. Uniqueness now follows from the uniqueness result in the case $n = 1$, whose proof is the same as that of Theorem 2.4.

The proof of Theorem 12.2 given on [Peller and Young 1994a] is constructive. Another (less constructive) method for proving the same result was given in [Treil' 1995].

The proof obtained in [Peller and Young 1994a] gives interesting factorizations (*thematic factorizations*) of the error functions $\Phi - F$. To describe such factorizations, assume for simplicity that $m = n$. We denote by $\boldsymbol{I}_j$ the constant $j \times j$ identity matrix function.

THEOREM 12.3. *Let Φ be an $n \times n$ function satisfying the hypotheses of Theorem 12.2 and let F be the unique superoptimal approximation of Φ by bounded analytic functions. Then $\Phi - F$ admits a factorization*

$$\Phi - F = \overline{W}_0 \overline{W}_1 \cdots \overline{W}_{n-2} D V_{n-2}^* \cdots V_1^* V_0^*, \qquad (12\text{--}4)$$

where

$$D = \begin{pmatrix} t_0 u_0 & 0 & \cdots & 0 \\ 0 & t_1 u_1 & \cdots & 0 \\ \vdots & \vdots & \ddots & \vdots \\ 0 & 0 & \cdots & t_{n-1} u_{n-1} \end{pmatrix},$$

$u_0, \ldots, u_{n-1}$ *are unimodular functions in* VMO *such that* $\operatorname{dist}_{L^\infty}\{u_j, H^\infty\} = 1$ *for* $0 \le j \le n-1$,

$$W_j = \begin{pmatrix} \boldsymbol{I}_j & 0 \\ 0 & \breve{W}_j \end{pmatrix},$$

$$V_j = \begin{pmatrix} \boldsymbol{I}_j & 0 \\ 0 & \breve{V}_j \end{pmatrix} \quad \text{for } 0 \le j \le n-2,$$

and $\breve{V}_j$ and $\breve{W}_j$, for $0 \le j \le n-2$, are thematic matrix functions.

We can associate with the factorization (12–4) the indices $k_j \overset{\text{def}}{=} \dim \operatorname{Ker} T_{u_j}$, for $0 \le j \le n-1$. Since $\|H_{u_j}\| = 1$ and H_{u_j} is compact, it follows that $k_j \ge 1$. It was shown in [Peller and Young 1994a] that the indices are not determined uniquely by the function Φ: they can depend on the choice of a thematic factorization. However, combining our earlier methods with those of [Treil' 1995], we showed in [Peller and Young 1994b] that the sums of the indices that correspond to equal superoptimal singular values are uniquely determined by Φ.

Another result obtained in [Peller and Young 1994b] is an inequality between the singular values of the Hankel operator H_Φ and the terms of the *extended t-sequence*, which is defined as follows:

$$\tilde{t}_0 \overset{\text{def}}{=} t_0, \ldots, \tilde{t}_{k_0-1} \overset{\text{def}}{=} t_0, \ \tilde{t}_{k_0} \overset{\text{def}}{=} t_1, \ldots, \tilde{t}_{k_0+k_1-1} \overset{\text{def}}{=} t_1, \ \tilde{t}_{k_0+k_1} \overset{\text{def}}{=} t_2, \ldots$$

(each term of the sequence $\{t_j\}_{0 \le j \le n-1}$ is repeated k_j times). The inequality is

$$\tilde{t}_j \le s_j(H_\Phi) \quad \text{for } 0 \le j \le k_0 + k_1 + \cdots + k_{n-1} - 1. \qquad (12\text{--}5)$$

A similar result holds for infinite matrix functions (or operator functions) Φ under the condition that H_Φ is compact [Treil' 1995; Peller 1995; Peller and Treil' 1995].

In [Peller and Treil' 1997] the preceding results were shown to be true in a more general context, when the matrix function Φ does not necessarily belong to $H^\infty + C$. It is shown there also that these results generalize to the case when the essential norm of H_Φ is less than the smallest nonzero superoptimal singular value. In fact, the paper deals with the so-called four-block problem, which is more general than Nehari's problem. Another result obtained in it is the following inequality, which is stronger than (12–5):

$$s_j(H_{\Phi^{(1)}}) \leq s_{j+k_0}(H_\Phi) \quad \text{for } j \geq 0,$$

where $\Phi^{(1)}$ is defined in (12–3).

It is also shown in [Peller and Young 1994a] that the nonlinear operator of superoptimal approximation has hereditary properties similar to those discussed in Section 9. Continuity properties of the operator of superoptimal approximation are studied in [Peller and Young 1997].

Similarity to a contraction. Here we consider one more application of Hankel operators, which has led recently to a solution of the famous problem of similarity to a contraction. Recall that operators T_1 and T_2 on Hilbert space are called *similar* if there exists an invertible linear operator V such that $T_2 = VT_1V^{-1}$. Clearly, similar operators have identical spectral properties. Sometimes one can prove that if an operator has the same properties as operators from a certain class, it is similar to an operator from that class. For example, it was proved in [Sz.-Nagy 1947] that if T is invertible and $\sup_{n\in\mathbb{Z}} \|T^n\| < \infty$, then T is similar to a unitary operator. However, if we know only that T satisfies $\sup_{n\geq 0} \|T^n\| < \infty$, it is not true that T must be similar to a contraction. The first example of such an operator was constructed in [Foguel 1964]; see [Davie 1974; Peller 1982; Bożejko 1987] for other examples).

It follows from von Neumann's inequality [von Neumann 1951] that any operator similar to a contraction is *polynomially bounded*, i.e.,

$$\|\varphi(T)\| \leq \text{const} \cdot \max_{|\zeta|\leq 1} |\varphi(\zeta)|$$

for any analytic polynomial φ.

The question of whether the converse is true was posed by Halmos [1970] and remained opened until recently.

Paulsen [1984] proved that T is similar to a contraction under the stronger condition of *complete polynomial boundedness*, which means that

$$\left\|\begin{pmatrix} \varphi_{11}(T) & \varphi_{12}(T) & \cdots & \varphi_{1n}(T) \\ \varphi_{21}(T) & \varphi_{22}(T) & \cdots & \varphi_{2n}(T) \\ \vdots & \vdots & \ddots & \vdots \\ \varphi_{n1}(T) & \varphi_{n2}(T) & \cdots & \varphi_{nn}(T) \end{pmatrix}\right\| \leq c \cdot \max_{|\zeta|\leq 1} \left\|\begin{pmatrix} \varphi_{11}(\zeta) & \varphi_{12}(\zeta) & \cdots & \varphi_{1n}(\zeta) \\ \varphi_{21}(\zeta) & \varphi_{22}(\zeta) & \cdots & \varphi_{2n}(\zeta) \\ \vdots & \vdots & \ddots & \vdots \\ \varphi_{n1}(\zeta) & \varphi_{n2}(\zeta) & \cdots & \varphi_{nn}(\zeta) \end{pmatrix}\right\|$$

for any positive integer n and any polynomial matrix $\{\varphi_{jk}\}_{1\leq j,k\leq n}$; the constant c in the inequality does not depend on n.

Now let f be a function analytic in the unit disk $\mathbb{D}$. Consider the operator R_f on $\ell^2 \oplus \ell^2$ defined by

$$R_f = \begin{pmatrix} S^* & \Gamma_f \\ 0 & S \end{pmatrix}, \tag{12–6}$$

where S is the shift operator on ℓ^2 and Γ_f is the Hankel operator on ℓ^2 with matrix $\{\hat{f}(j+k)\}_{j,k\geq 0}$ in the standard basis of ℓ^2. Such operators were introduced and used in [Peller 1982] to construct power bounded operators which are not similar to contractions. The operators R_f were considered independently by Foiaş and Williams; see [Carlson et al. 1994].

In [Peller 1984] the problem was posed of whether it is possible to find a function f for which R_f is polynomially bounded but not similar to a contraction. The reason why the operators R_f are convenient for this purpose is that functions of R_f can be evaluated explicitly: if φ is an analytic polynomial, then

$$\varphi(R_f) = \begin{pmatrix} \varphi(S^*) & \Gamma_{\varphi'(S^*)f} \\ 0 & \varphi(S) \end{pmatrix}. \tag{12--7}$$

It was shown in [Peller 1984] that, if $f' \in$ BMOA (see the definition on page 70), R_f is polynomially bounded. A stronger result was obtained later in by Bourgain [1986]: if $f' \in$ BMOA, then R_f is completely polynomially bounded, and so it is similar to a contraction. Another proof of Bourgain's result was obtained later in [Stafney 1994].

Recently Paulsen has shown that R_f is similar to a contraction if and only if the matrix $\{(j-k)\hat{f}(j+k)\}_{j,k\geq 0}$ determines a bounded operator on ℓ^2. It follows from results of [Janson and Peetre 1988] that the last condition is equivalent to the fact that $f' \in$ BMOA. (In the latter paper instead of matrices the authors study integral operators on $L^2(\mathbb{R})$, but their methods also work for matrices.) This implies that R_f is similar to a contraction if and only if $f' \in$ BMOA.

There was hope of finding a function f with $f' \notin$ BMOA such that R_f is polynomially bounded, which would solve the problem negatively. However, this was recently shown to be is impossible, in [Aleksandrov and Peller 1996], the main result of the paper being the following:

THEOREM 12.4. *Let f be a function analytic in $\mathbb{D}$. The following statements are equivalent*:

(i) *R_f is polynomially bounded.*
(ii) *R_f is similar to a contraction.*
(iii) *$f' \in$ BMOA.*

To prove Theorem 12.4 the following factorization result is established in [Aleksandrov and Peller 1996]. We denote by C_A the disk algebra of functions analytic in $\mathbb{D}$ and continuous in clos $\mathbb{D}$.

THEOREM 12.5. *Let f be a function analytic in $\mathbb{D}$. Then $f \in H^1$ if and only if its derivative f' admits a representation*

$$f' = \sum_{j\geq 0} g'_j h_j,$$

where $g_j \in C_A$, $h_j \in H^1$, and

$$\sum_{j \geq 0} \|g_j\|_{L^\infty} \|h_j\|_{H^1} < \infty.$$

However, this was not the end of the story. Pisier [1997] considered the operator R_f on the space of vector functions $H^2(\mathcal{H}) \oplus H^2(\mathcal{H})$, where $\mathcal{H}$ is a Hilbert space. The definition of R_f is exactly the same as given in (12–6). In this case S is the shift operator on $H^2(\mathcal{H})$, f is a function analytic in $\mathbb{D}$ and taking values in the space $\mathbb{B}(\mathcal{H})$ of bounded linear operators on $\mathcal{H}$, and Γ_f is the operator on $H^2(\mathcal{H})$ given by the block Hankel matrix $\{\hat{f}(j+k)\}_{j,k \geq 0}$. It is easy to see that (12–7) also holds in this setting for any scalar analytic polynomial φ. Pisier managed to construct a function f for which R_f is polynomially bounded but not similar to a contraction. To do that he used a sequence of operators $\{C_j\}_{j \geq 0}$ on $\mathcal{H}$ with the properties

$$\left\| \sum_{j \geq 0} \alpha_j C_j \right\| = \left(\sum_{j \geq 0} |\alpha_j|^2 \right)^{1/2} \quad \text{with } \alpha_j \in \mathbb{C}$$

and

$$\frac{1}{2} \sum_{j \geq 0} |\alpha_j| \leq \left\| \sum_{j \geq 0} \alpha_j C_j \otimes C_j \right\| \leq \sum_{j \geq 0} |\alpha_j| \quad \text{for } \alpha_j \in \mathbb{C}.$$

Such a sequence $\{C_j\}_{j \geq 0}$ always exists; see [Pisier 1996], for example.

The following result from [Pisier 1997] solves the problem of similarity to a contraction.

THEOREM 12.6. *Let $\{\alpha_j\}_{j \geq 0}$ be a sequence of complex numbers such that*

$$\sup_{k \geq 0} k^2 \sum_{j \geq k} |\alpha_j|^2 < \infty$$

and

$$\sum_{j \geq 1} j^2 |\alpha_j|^2 = \infty.$$

Then the operator R_f with $f = \sum_{j \geq 0} \alpha_j z^j C_j$ is polynomially bounded but not similar to a contraction.

It is very easy to construct such a sequence $\{\alpha_j\}_{j \geq 0}$. For example, one can put $\alpha_j = (j+1)^{-3/2}$.

Pisier's proof is rather complicated and involves martingales. Kislyakov [1996], Davidson and Paulsen [1997], and McCarthy [1996] have simplified the original argument and got rid of martingales.

References

[Adamyan et al. 1968a] V. M. Adamjan, D. Z. Arov, and M. G. Kreĭn, "Бесконечные ганкелевы матрицы и обобщённые задачи Каратеодори–Фейера и И. Шура", *Funkcional. Anal. i Priložen.* **2**:4 (1968), 1–17. Translated as "Infinite Hankel matrices and generalized Carathéodory–Fejér and I. Schur problems", *Functional Anal. Appl.* **2** (1968), 269–281.

[Adamyan et al. 1968b] V. M. Adamjan, D. Z. Arov, and M. G. Kreĭn, "О бесконечных ганкелевых матрицах и обобщённых задачах Каратеодори–Фейера и Ф. Рисса", *Funkcional. Anal. i Priložen.* **2**:1 (1968), 1–19. Translated as "Infinite Hankel matrices and generalized problems of Carathéodory–Fejér and F. Riesz", *Functional Anal. Appl.* **2** (1968), 1-18.

[Adamyan et al. 1971] V. M. Adamjan, V. Z. Arov, and M. G. Kreĭn, "Аналитические свойства пар Шмидта ганкелева оператора и обобщённая задача Шура–Такаги", *Mat. Sbornik* **86** (1971), 33–75. Translated as "Analytic properties of Schmidt pairs of a Hankel operator and generalized Schur–Takagi problem", *Math. USSR Sb.* **15** (1971), 31-73.

[Adamyan et al. 1984] V. M. Adamjan, D. Z. Arov, and M. G. Kreĭn, "Approximation of bounded functions by elements of $H^\infty + C$", pp. 254–258 in *Linear and complex analysis problem book: 199 research problems*, edited by V. P. Havin et al., Lecture Notes in Mathematics **1043**, Springer, Berlin, 1984. Russian original in Исследования по линеиным операторам и теорий функций: 99 нерешенных задач линейного и комплексного анализа, Записки научных семинаров ЛОМИ **81**, edited by N. K. Nikol'skii, V. P. Khavin, and S. V. Khrushchëv, Nauka, Leningrad, 1978.

[Akhiezer 1965] N. I. Akhiezer, Лекции по теории аппроксимаций, 2nd ed., Nauka, Moscow, 1965. Translated as *Theory of approximation*, Frederick Ungar Pub., New York, 1956; reprinted by Dover, New York, 1992.

[Aleksandrov and Peller 1996] A. B. Aleksandrov and V. V. Peller, "Hankel operators and similarity to a contraction", *Internat. Math. Res. Notices* **6** (1996), 263–275.

[Axler et al. 1979] S. Axler, I. D. Berg, N. Jewell, and A. Shields, "Approximation by compact operators and the space $H^\infty + C$", *Ann. of Math.* (2) **109**:3 (1979), 601–612.

[Birman and Solomyak 1980] M. S. Birman and M. Z. Solomyak, Спектральная теория самосопряжённых операторов в гильбертовом пространстве, 1980. Translated as *Spectral theory of selfadjoint operators in Hilbert space*, D. Reidel Pub., Dordrecht, 1987.

[Bourgain 1986] J. Bourgain, "On the similarity problem for polynomially bounded operators on Hilbert space", *Israel J. Math.* **54**:2 (1986), 227–241.

[Bożejko 1987] M. Bożejko, "Littlewood functions, Hankel multipliers and power bounded operators on a Hilbert space", *Colloq. Math.* **51** (1987), 35–42.

[Brudnyĭ 1979] J. A. Brudnyĭ, "Rational approximation and imbedding theorems", *Dokl. Akad. Nauk SSSR* **247**:2 (1979), 269–272. In Russian; translation in *Soviet Math. Dokl.* **20** (1979), 681-684.

[Carleson and Jacobs 1972] L. Carleson and S. Jacobs, "Best uniform approximation by analytic functions", *Ark. Mat.* **10** (1972), 219–229.

[Carlson et al. 1994] J. F. Carlson, D. N. Clark, C. Foiaş, and J. P. Williams, "Projective Hilbert $\mathbf{A(D)}$-modules", *New York J. Math.* **1** (1994), 26–38.

[Coifman and Rochberg 1980] R. R. Coifman and R. Rochberg, "Representation theorems for holomorphic and harmonic functions in L^p", pp. 11–66 in *Representation theorems for Hardy spaces*, Astérisque **77**, Soc. Math. France, Paris, 1980.

[Davidson and Paulsen 1997] K. R. Davidson and V. I. Paulsen, "Polynomially bounded operators", *J. Reine Angew. Math.* **487** (1997), 153–170.

[Davie 1974] A. M. Davie, "Power-bounded elements in a Q-algebra", *Bull. London Math. Soc.* **6** (1974), 61–65.

[Dolženko 1977] E. P. Dolženko, "The dependence of the boundary properties of an analytic function on the rate of its approximation by rational functions", *Mat. Sb. (N.S.)* **103(145)**:1 (1977), 131–142, 144. In Russian.

[Douglas 1972] R. G. Douglas, *Banach algebra techniques in operator theory*, Pure and Applied Mathematics **49**, Academic Press, New York, 1972.

[Foguel 1964] S. R. Foguel, "A counterexample to a problem of Sz.-Nagy", *Proc. Amer. Math. Soc.* **15** (1964), 788–790.

[Francis 1987] B. A. Francis, *A course in H_∞ control theory*, Lecture Notes in Control and Information Sciences **88**, Springer, Berlin, 1987.

[Fuhrmann 1981] P. A. Fuhrmann, *Linear systems and operators in Hilbert space*, McGraw-Hill, New York, 1981.

[Gantmakher 1988] F. R. Gantmakher, Теория матриц, 4 ed., Nauka, Moscow, 1988. First edition translated as *The theory of matrices* (2 vols.), Chelsea, New York, 1959.

[Garnett 1981] J. B. Garnett, *Bounded analytic functions*, Pure and Applied Mathematics **96**, Academic Press, New York, 1981.

[Glover 1984] K. Glover, "All optimal Hankel-norm approximations of linear multivariable systems and their L^∞-error bounds", *Internat. J. Control* **39**:6 (1984), 1115–1193.

[Gohberg and Kreĭn 1965] I. C. Gohberg and M. G. Kreĭn, Введение в теорию линейных несамосопряженных операторов в гильбертовом пространстве, Nauka, Moscow, 1965. Translated as *Introduction to the theory of linear nonselfadjoint operators*, Translations of Mathematical Monographs **18**, Amer. Math. Soc., Providence, 1969.

[Gonchar 1968] A. A. Gonchar, "Скорость приближения рациональными дробями и свойства функций", pp. 329–346 in *Proc. Internat. Congr. Math.* (Moscow, 1966), Mir, Moscow, 1968. Translated as "Rate of approximation by rational functions and properties of functions", *Amer. Math. Soc. Transl.* (2) **91** (1969).

[Grigoryan 1976] L. D. Grigoryan, "Estimates of the norm of holomorphic components of meromorphic functions in domains with a smooth boundary", *Mat. Sb. (N.S.)* **100(142)**:1 (1976), 156–164. In Russian; translation in *Math. USSR Sb.* **29** (1976).

[Halmos 1967] P. R. Halmos, *A Hilbert space problem book*, Van Nostrand, Princeton, NJ, and London, 1967. Second edition published by Springer, New York, 1982.

[Halmos 1970] P. R. Halmos, "Ten problems in Hilbert space", *Bull. Amer. Math. Soc.* **76** (1970), 887–933.

[Hamburger 1920] H. Hamburger, "Über eine Erweiterung des Stieltjesschen Moment-problems, I", *Math. Ann.* **81** (1920), 235–319.

[Hamburger 1921] H. Hamburger, "Über eine Erweiterung des Stieltjesschen Moment-problems, II and III", *Math. Ann.* **82** (1921), 120–164, 168–187.

[Hankel 1861] H. Hankel, *Über eine besondere Classe der symmetrischen Determinan-ten*, (Leipziger) dissertation, Göttingen, 1861.

[Hartman 1958] P. Hartman, "On completely continuous Hankel matrices", *Proc. Amer. Math. Soc.* **9** (1958), 862–866.

[Helson and Sarason 1967] H. Helson and D. Sarason, "Past and future", *Math. Scand.* **21** (1967), 5–16.

[Hoffman 1962] K. Hoffman, *Banach spaces of analytic functions*, Prentice-Hall Series in Modern Analysis, Prentice-Hall, Englewood Cliffs, NJ, 1962. Reprinted by Dover, New York, 1988.

[Howland 1986] J. S. Howland, "Spectral theory of selfadjoint Hankel matrices", *Michigan Math. J.* **33**:2 (1986), 145–153.

[Howland 1992a] J. S. Howland, "Spectral theory of operators of Hankel type, I", *Indiana Univ. Math. J.* **41**:2 (1992), 409–426.

[Howland 1992b] J. S. Howland, "Spectral theory of operators of Hankel type, II", *Indiana Univ. Math. J.* **41**:2 (1992), 427–434.

[Ibragimov and Rozanov 1970] I. A. Ibragimov and J. A. Rozanov, Гауссовские случайные процессы, Nauka, Moscow, 1970. Translated as *Gaussian random processes*, Applications of Mathematics, **9**, Springer, New York, 1978.

[Janson and Peetre 1988] S. Janson and J. Peetre, "Paracommutators—boundedness and Schatten – von Neumann properties", *Trans. Amer. Math. Soc.* **305**:2 (1988), 467–504.

[Khavinson 1951] S. Y. Khavinson, "О некоторых экстремальных задачах теории аналитических функций", *Moskov. Gos. Univ. Učenye Zapiski Matematika* **148(4)** (1951), 133–143. Translated as "On some extremal problems of the theory of analytic functions", *Amer. Math. Soc. Translations* (2) **32** (1963), 139-154.

[Khrushchëv and Peller 1984] S. V. Hruščëv and V. V. Peller, "Moduli of Hankel operators, past and future", pp. 92–97 in *Linear and complex analysis problem book: 199 research problems*, edited by V. P. Havin et al., Lecture Notes in Math. **1043**, Springer, Berlin, 1984.

[Kislyakov 1996] S. V. Kislyakov, "Operators (not) similar to a contraction: Pisier's counterexample via singular integrals", preprint 42, Université de Bordeaux, 1996.

[Kronecker 1881] L. Kronecker, "Zur Theorie der Elimination einer Variablen aus zwei algebraischen Gleichungen", *Monatsber. Königl. Preussischen Akad. Wies. (Berlin)* (1881), 535–600. Reprinted as pp. 113-192 in *Mathematische Werke*, vol. 2, B. G. Teubner, Leipzig, 1897 or Chelsea, New York, 1968.

[Luecking 1980] D. H. Luecking, "The compact Hankel operators form an *M*-ideal in the space of Hankel operators", *Proc. Amer. Math. Soc.* **79**:2 (1980), 222–224.

[Megretskii 1990] A. V. Megretskiĭ, "A quasinilpotent Hankel operator", *Algebra i Analiz* **2**:4 (1990), 201–212.

[Megretskii et al. 1995] A. V. Megretskiĭ, V. V. Peller, and S. R. Treil', "The inverse spectral problem for self-adjoint Hankel operators", *Acta Math.* **174**:2 (1995), 241–309.

[Merino 1989] O. Merino, "Stability of qualitative properties and continuity of solutions to problems of optimisation over spaces of analytic functions", preprint, 1989.

[McCarthy 1996] J. E. McCarthy, "On Pisier's construction of a polynomially bounded operator not similar to a contraction", preprint, MSRI, 1996. Available at http://www.msri.org/MSRI-preprints/online/1996-017.html.

[Nehari 1957] Z. Nehari, "On bounded bilinear forms", *Ann. of Math.* (2) **65** (1957), 153–162.

[von Neumann 1951] J. von Neumann, "Eine Spektraltheorie für allgemeine Operatoren eines unitären Raumes", *Math. Nachr.* **4** (1951), 258–281.

[Newburgh 1951] J. D. Newburgh, "The variation of spectra", *Duke Math. J.* **18** (1951), 165–176.

[Nikol'skiĭ 1986]   N. K. Nikol'skiĭ, *Treatise on the shift operator*, Grundlehren der mathematischen Wissenschaften **273**, Springer, Berlin, 1986. With an appendix by S. V. Hruščev and V. V. Peller. Translation of Лекции об операторе сдвига, Nauka, Moscow, 1980.

[Ober 1987] R. J. Ober, "A note on a system theoretic approach to a conjecture by Peller–Khrushchev", *Systems Control Lett.* **8**:4 (1987), 303–306.

[Ober 1990] R. Ober, "A note on a system-theoretic approach to a conjecture by Peller–Khrushchev: the general case", *IMA J. Math. Control Inform.* **7**:1 (1990), 35–45.

[Page 1970] L. B. Page, "Bounded and compact vectorial Hankel operators", *Trans. Amer. Math. Soc.* **150** (1970), 529–539.

[Papadimitrakis 1993] M. Papadimitrakis, "Continuity of the operator of best uniform approximation by bounded analytic functions", *Bull. London Math. Soc.* **25**:1 (1993), 44–48.

[Papadimitrakis 1996]   M. Papadimitrakis, "On best uniform approximation by bounded analytic functions", *Bull. London Math. Soc.* **28**:1 (1996), 15–18.

[Parfenov 1986]   O. G. Parfenov, "Estimates for singular numbers of the Carleson embedding operator", *Mat. Sb.* (*N.S.*) **131(173)**:4 (1986), 501–518. In Russian; translation in *Math. USSR Sb.* **59**:2 (1988), 497–514.

[Paulsen 1984] V. I. Paulsen, "Every completely polynomially bounded operator is similar to a contraction", *J. Funct. Anal.* **55**:1 (1984), 1–17.

[Peetre 1976] J. Peetre, *New thoughts on Besov spaces*, Duke University Mathematics Series **1**, Mathematics Department, Duke University, Durham, NC, 1976.

[Peetre and Svensson 1984] J. Peetre and E. Svensson, "On the generalized Hardy's inequality of McGehee, Pigno and Smith and the problem of interpolation between BMO and a Besov space", *Math. Scand.* **54**:2 (1984), 221–241.

[Pekarskiĭ 1985] A. A. Pekarskiĭ, "Classes of analytic functions defined by best rational approximations in H_p", *Mat. Sb.* (*N.S.*) **127(169)**:1 (1985), 3–20. In Russian; translation in *Math. USSR Sb.* **55** (1986), 1–18.

[Pekarskiĭ 1987]   A. A. Pekarskiĭ, "Chebyshev rational approximation in a disk, on a circle and on a segment", *Mat. Sb. (N.S.)* **133(175)**:1 (1987), 86–102, 144. In Russian.

[Peller 1980]   V. V. Peller, "Операторы Ганкеля класса $\mathfrak{S}_p$ и их приложения (рациональная аппроксимация, гауссовские процессы, проблема мажорации операторов)", *Mat. Sb. (N.S.)* **113(155)**:4 (1980), 538–581. Translated as "Hankel operators of class $\mathfrak{S}_p$ and their applications (rational approximation, Gaussian processes, the problem of majorization of operators)", *Math. USSR Sb.* **41** (1982), 443–479.

[Peller 1982]   V. V. Peller, "Estimates of functions of power bounded operators on Hilbert spaces", *J. Operator Theory* **7**:2 (1982), 341–372.

[Peller 1983]   V. V. Peller, "Описание операторов Ганкеля класса $\mathfrak{S}_p$ при $p > 0$, исследование скорости рациональной аппроксимации и другие приложения", *Mat. Sb. (N.S.)* **122(164)**:4 (1983), 481–510. Translated as "Description of Hankel operators of the class $\mathfrak{S}_p$ for $p > 0$, investigation of the rate of rational approximation and other applications", *Math. USSR Sb.* **50** (1985), 465-494.

[Peller 1984]   V. V. Peller, "Estimates of functions of Hilbert space operators, similarity to a contraction and related function algebras", pp. 199–204 in *Linear and complex analysis problem book: 199 research problems*, edited by V. P. Havin et al., Lecture Notes in Math. **1043**, Springer, Berlin, 1984.

[Peller 1985]   V. V. Peller, "Операторы Ганкеля в теории возмущений унитарных и самосопряжённых операторов", *Funktsional. Anal. i Prilozhen.* **19**:2 (1985), 37–51, 96. Translated as "Hankel operators in the theory of perturbations of unitary and selfadjoint operators", *Functional Anal. Appl.* **19**:2 (1985), 111–123.

[Peller 1990a]   V. V. Peller, "Hankel operators and multivariate stationary processes", pp. 357–371 in *Operator theory: operator algebras and applications* (Durham, NH, 1988), vol. 1, edited by W. B. Arveson and R. G. Douglas, Proc. Sympos. Pure Math. **51**, Amer. Math. Soc., Providence, RI, 1990.

[Peller 1990b]   V. V. Peller, "Операторы Ганкеля и свойства непрерывности операторов наилучшего приближения", *Algebra i Analiz* **2**:1 (1990), 163–189. Translated as "Hankel operators and continuity properties of best approximation operators", *Leningrad Math. J.*, **2** (1991), 139-160.

[Peller 1992]   V. V. Peller, "Boundedness properties of the operators of best approximation by analytic and meromorphic functions", *Ark. Mat.* **30**:2 (1992), 331–343.

[Peller 1995]   V. V. Peller, "Approximation by analytic operator-valued functions", pp. 431–448 in *Harmonic analysis and operator theory* (Caracas, 1994), Contemp. Math. **189**, Amer. Math. Soc., Providence, RI, 1995.

[Peller and Khrushchëv 1982]   V. V. Peller and S. V. Khrushchëv, "Операторы Ганкеля, наилучшие приближения и стационарные гауссовские процессы", *Uspehi Mat. Nauk (N.S.)* **37**:1 (1982), 53–124. Translated as "Hankel operators, best approximations and stationary Gaussian processes", *Russian Math. Surveys* **37**:1 (1982), 61-144.

[Peller and Treil' 1995]   V. V. Peller and S. R. Treil', "Superoptimal singular values and indices of infinite matrix functions", *Indiana Univ. Math. J.* **44**:1 (1995), 243–255.

[Peller and Treil' 1997] V. V. Peller and S. R. Treil', "Approximation by analytic matrix functions: the four block problem", *J. Funct. Anal.* **148**:1 (1997), 191–228.

[Peller and Young 1994a] V. V. Peller and N. J. Young, "Superoptimal analytic approximations of matrix functions", *J. Funct. Anal.* **120**:2 (1994), 300–343.

[Peller and Young 1994b] V. V. Peller and N. J. Young, "Superoptimal singular values and indices of matrix functions", *Integral Equations Operator Theory* **20**:3 (1994), 350–363.

[Peller and Young 1997] V. V. Peller and N. J. Young, "Continuity properties of best analytic approximation", *J. Reine Angew. Math.* **483** (1997), 1–22.

[Pisier 1996] G. Pisier, *Similarity problems and completely bounded maps*, Lecture Notes in Mathematics **1618**, Springer, Berlin, 1996.

[Pisier 1997] G. Pisier, "A polynomially bounded operator on Hilbert space which is not similar to a contraction", *J. Amer. Math. Soc.* **10**:2 (1997), 351–369.

[Power 1982] S. C. Power, *Hankel operators on Hilbert space*, Research Notes in Mathematics **64**, Pitman, Boston and London, 1982.

[Power 1984] S. C. Power, "Quasinilpotent Hankel operators", pp. 259–261 in *Linear and complex analysis problem book: 199 research problems*, edited by V. P. Havin et al., Lecture Notes in Mathematics **1043**, Springer, Berlin, 1984.

[Riesz and Sz.-Nagy 1965] F. Riesz and B. Sz.-Nagy, *Leçons d'analyse fonctionnelle*, 4th ed., Gauthier-Villars, Paris, 1965.

[Rochberg 1982] R. Rochberg, "Trace ideal criteria for Hankel operators and commutators", *Indiana Univ. Math. J.* **31**:6 (1982), 913–925.

[Rogosinski and Shapiro 1953] W. W. Rogosinski and H. S. Shapiro, "On certain extremum problems for analytic functions", *Acta Math.* **90** (1953), 287–318.

[Sarason 1967] D. Sarason, "Generalized interpolation in H^∞", *Trans. Amer. Math. Soc.* **127**:2 (1967), 179–203.

[Sarason 1972] D. Sarason, "An addendum to: 'Past and future' [Helson and Sarason 1967]", *Math. Scand.* **30** (1972), 62–64.

[Sarason 1978] D. Sarason, "Function theory on the unit circle", lecture notes, Virginia Polytechnic Inst. and State Univ., Blacksburg, VA, 1978.

[Semmes 1984] S. Semmes, "Trace ideal criteria for Hankel operators, and applications to Besov spaces", *Integral Equations Operator Theory* **7**:2 (1984), 241–281.

[Shapiro 1952] H. S. Shapiro, *Extremal problems for polynomials and power series*, Dissertation, MIT, Cambridge, MA, 1952.

[Stafney 1994] J. D. Stafney, "A class of operators and similarity to contractions", *Michigan Math. J.* **41**:3 (1994), 509–521.

[Sz.-Nagy 1947] B. de Sz. Nagy, "On uniformly bounded linear transformations in Hilbert space", *Acta Sci. Math. (Szeged)* **11** (1947), 152–157.

[Sz.-Nagy and Foiaş 1967] B. Sz.-Nagy and C. Foiaş, *Analyse harmonique des opérateurs de l'espace de Hilbert*, Masson, Paris, and Akadémiai Kiadó, Budapest, 1967. Translated as *Harmonic analysis of operators on Hilbert space*, North-Holland, Amsterdam, and Akadémiai Kiadó, Budapest, 1970.

[Tolokonnikov 1991] V. A. Tolokonnikov, "Generalized Douglas algebras", *Algebra i Analiz* **3**:2 (1991), 231–252. In Russian; translation in *St. Petersburg Math. J.* **3**:2 (1992), 455–476.

[Treil' 1985a] S. R. Treil', "The Adamyan–Arov–Kreĭn theorem: a vector version", *Zap. Nauchn. Sem. Leningrad. Otdel. Mat. Inst. Steklov.* **141** (1985), 56–71. In Russian.

[Treil' 1985b] S. R. Treil', "Moduli of Hankel operators and the V. V. Peller – S. V. Khrushchëv problem", *Dokl. Akad. Nauk SSSR* **283**:5 (1985), 1095–1099. In Russian; translation in *Soviet Math. Dokl.* **32** (1985), 293-297.

[Treil' 1990] S. R. Treil', "An inverse spectral problem for the modulus of the Hankel operator, and balanced realizations", *Algebra i Analiz* **2**:2 (1990), 158–182. In Russian; translation in *Leningrad Math. J.* **2**:2 (1991), 353–375.

[Treil' 1995] S. Treil', "On superoptimal approximation by analytic and meromorphic matrix-valued functions", *J. Funct. Anal.* **131**:2 (1995), 386–414.

[Vasyunin and Treil' 1989] V. I. Vasyunin and S. R. Treil', "The inverse spectral problem for the modulus of a Hankel operator", *Algebra i Analiz* **1**:4 (1989), 54–66. In Russian; translation in *Leningrad Math. J.* **1**:4 (1990), 859–870.

[Vol'berg and Tolokonnikov 1985] A. L. Vol'berg and V. A. Tolokonnikov, "Hankel operators and problems of best approximation of unbounded functions", *Zap. Nauchn. Sem. Leningrad. Otdel. Mat. Inst. Steklov. (LOMI)* **141** (1985), 5–17, 188. In Russian.

[Young 1986] N. J. Young, "The Nevanlinna–Pick problem for matrix-valued functions", *J. Operator Theory* **15**:2 (1986), 239–265.

[Zygmund 1968] A. Zygmund, *Trigonometric series: I, II*, 2nd ed., Cambridge University Press, London, 1968.

VLADIMIR V. PELLER
MATHEMATICS DEPARTMENT
KANSAS STATE UNIVERSITY
MANHATTAN, KS 66506
UNITED STATES
peller@math.ksu.edu

Holomorphic Spaces
MSRI Publications
Volume **33**, 1998

Hankel-Type Operators,
Bourgain Algebras,
and Uniform Algebras

PAMELA GORKIN

ABSTRACT. Let $H^\infty(D)$ denote the algebra of bounded analytic functions on the open unit disc in the complex plane. For a function $g \in L^\infty(D)$, the Hankel-type operator S_g is defined by $S_g(f) = gf + H^\infty(D)$. We give here an overview of the study of the symbol of the Hankel-type operator, with emphasis on those symbols for which the operator is compact, weakly compact, or completely continuous. We conclude with a look at this operator on more general domains and several open questions.

We look at a uniform algebra A on a compact Hausdorf space X. We let $M(A)$ denote the maximal ideal space of A. We will consider the Hankel-type operator $S_g : A \to C(X)/A$ with symbol $g \in C(X)$ defined by $S_g(f) = fg + A$ for all $f \in A$.

Even though the space L^∞ does not look like an algebra of continuous functions, it is possible to identify it with the space of continuous functions on its maximal ideal space X as follows: for f in L^∞ define the Gelfand transform of f by $\hat{f}(x) = x(f)$ for all $x \in X$. Since the topology on X is given by saying that a net x_α converges to x in X if and only if $x_\alpha(f)$ converges to $x(f)$ for all $f \in L^\infty$, we see that the Gelfand transform defines a continuous function on X.

We will be most interested in the case in which $A = H^\infty(U)$, the algebra of bounded analytic functions on a bounded domain U in the complex plane, and $C(X) = L^\infty(U)$ with respect to area measure. When the domain does not matter or when we think no confusion should arise, we will write simply H^∞ or L^∞.

The purpose of this article is to indicate why we look at such operators, what one can do with these Hankel-type operators, and some of what remains to be done in this area.

1. The Relationship to Classical Hankel Operators

Why are these called Hankel-type operators? To answer this question, we first consider the Hankel-type operator S_f defined for $f \in L^\infty(\partial D)$. Let H^2 denote the usual Hardy space of functions on the circle ∂D. Let P denote the orthogonal projection of $L^2(\partial D)$ onto H^2 and let $f \in L^\infty(\partial D)$. Recall that the multiplication operator with symbol f is defined on $L^2(\partial D)$ by $M_f(g) = fg$ for $g \in L^2(\partial D)$. The Toeplitz operator T_f is defined by $T_f(g) = P(fg)$ and the (classical) Hankel operator H_f with symbol f is defined as the operator from H^2 into $H^{2^\perp}$ such that

$$H_f(g) = (I - P)(fg) \quad \text{for } g \in H^2.$$

These operators have been studied over the years and there exist many good references about them. See [Power 1982; Zhu 1990], as well as [Peller 1998] in this volume, for more information about classical Hankel operators.

Now suppose that we replace the *Hilbert space* L^2 above by the *uniform algebra* L^∞, and the Hardy space H^2 by the algebra H^∞ of boundary values of bounded analytic functions on the open unit disc D. What is the appropriate replacement for the Hankel operator? If we look closely at the Hankel operator, we see that it is a multiplication operator followed by an operator with kernel equal to the space H^2. Thus the replacement should ideally be a multiplication operator followed by a map that annihilates H^∞ functions. Our Hankel-type operators are multiplication operators followed by the quotient map, a map that takes functions in H^∞ to zero.

When we work on the unit disc rather than the unit circle, our Hankel-type operators are a generalization of the Hankel operator on the Bergman space $L_a^2(D)$, the space of square-integrable analytic functions on the disc. For $f \in L^\infty(D)$ we define the Hankel operator acting on the Bergman space as above, replacing the Szegő projection with the Bergman projection. We will return to these Hankel operators frequently for comparison.

2. Why Should We Look at Hankel-Type Operators?

One reason for studying Hankel-type operators is that they are a natural generalization of classical Hankel operators to uniform algebras. Multiplication operators, Hankel operators, and Toeplitz operators are important in the study of closed subalgebras of $L^\infty(\partial D)$ and many interesting results in this area were a consequence of careful study of these operators. In what follows, we will look at when these operators are compact (that is, when the norm closure of the image of the closed unit ball under S_g is compact), weakly compact (the weak closure of the image of the closed unit ball under S_g is weakly compact), and completely continuous (S_g takes weakly null sequences to norm null sequences). Complete continuity and compactness are equivalent in reflexive spaces; this was Hilbert's original definition of compactness. However, as we will see below, these three types of compactness need not be the same. See [Dunford and Schwartz 1958, Chapter 5] for elementary information on the subject.

There are two more good reasons for studying the operators S_g, both connected with the properties of compactness. The first has a long history and began with work of Sarason in 1967. As usual, $C(\partial D)$ denotes the algebra of continuous functions on the unit circle. Sarason [1976] looked at the linear space $H^\infty(\partial D) + C(\partial D)$ and showed the following:

THEOREM 2.1. *The space $H^\infty(\partial D) + C(\partial D)$ is a closed subalgebra of $L^\infty(\partial D)$.*

In fact, $H^\infty(\partial D) + C(\partial D)$ is the closed algebra generated by H^∞ and the conjugate of the inner function z on the unit circle. Hartman's theorem [Power 1982] tells us that the classical Hankel operator H_g is compact if and only if the symbol g belongs to $H^\infty(\partial D) + C(\partial D)$. Douglas, in connection with the study of Toeplitz operators on the circle, asked whether every closed subalgebra B of $L^\infty(\partial D)$ containing $H^\infty(\partial D)$ was generated by $H^\infty(\partial D)$ together with the set of conjugates of inner functions invertible in B. Algebras with this property became known as Douglas algebras, and Sarason's theorem inaugurated the study of closed subalgebras of $L^\infty(\partial D)$ containing $H^\infty(\partial D)$. One of the most important theorems in this study is the Chang–Marshall theorem [Chang 1976; Marshall 1976], which answers Douglas's question affirmatively and gives a beautiful description of all closed subalgebras of $L^\infty(\partial D)$ containing $H^\infty(\partial D)$. This theory does not generalize well to spaces of bounded functions on other domains in the complex plane, but Sarason's theorem above does. Rudin [1975] showed that the same is true of algebras on the disc:

THEOREM 2.2. *The space $H^\infty(D) + C(\overline{D})$ is a closed subalgebra of $L^\infty(D)$.*

It turns out that whenever U is a bounded open subset of the complex plane and σ is area measure on U, the closure of $H^\infty(U) + C(\overline{U})$ is a closed subalgebra of $L^\infty(\sigma)$ [Dudziak et al. $\geq$ 1998]. So the interesting part of the question really is: When is $H^\infty(U) + C(\overline{U})$ closed? Looking at the proofs of Theorems 2.1 and 2.2 (see [Axler and Shields 1987; Garnett 1981, p. 137], for example), one observes that they share a common ingredient: both use approximation of functions by a related harmonic extension of the function via the Poisson kernel.

Zalcman [1969] extended Sarason's theorem to algebras of analytic functions on certain infinitely connected domains (to be studied later in this paper). Davie, Gamelin and Garnett continued work along this line [Davie et al. 1973] and looked at algebras on a bounded open subset U of the complex plane for which the functions in $A(U) = H^\infty(U) \cap C(\overline{U})$ are pointwise boundedly dense in $H^\infty(U)$; that is, every function in $H^\infty(U)$ can be approximated pointwise on U by a bounded sequence in $A(U)$. They asked the following question: if $H^\infty(U) + C(\overline{U})$ is a closed subspace of $L^\infty(U)$, is $A(U)$ pointwise boundedly dense in $H^\infty(U)$?

Their work contains many related results and seems to suggest that this must be correct, yet this problem remains open.

Study of sums of closed algebras and investigation of when the sum is again a closed algebra continued. Aytuna and Chollet [1976] extended Sarason's result to strictly pseudoconvex domains in C^n. Cole and Gamelin [1982] continued this work in a natural way and studied the problem of when the double dual A^{**} of a uniform algebra A has the property that $A^{**} + C(X)$ is a closed subalgebra of $C(X)^{**}$. One of their main results is that S_g is weakly compact for all $g \in C(X)$ if and only if $A^{**} + C(X)$ is a closed subalgebra of $C(X)^{**}$. Thus, knowing when S_g is weakly compact is connected to the question of when sums of uniform algebras are closed algebras. If S_g is not weakly compact for all $g \in C(X)$, can we determine the space of functions for which S_g is weakly compact?

As it turns out, Cole and Gamelin showed that many of the algebras had the seemingly stronger property that S_g is compact for all $g \in C(X)$. In the same paper, they defined the notion of tightness: a uniform algebra A is said to be tight if S_g is weakly compact for all $g \in C(X)$. They showed that under certain conditions on the domain, one can use the fact that $A(U)$ is tight to show that $H^\infty(U) + C(\overline{U})$ is a closed algebra of continuous functions on U. Cole and Gamelin's results are quite general, and they gave plenty of examples of tight algebras. Their work concentrated on looking at algebras for which S_g is weakly compact for all $g \in C(X)$. Saccone [1995] continued studying tight algebras as well as strong tightness; a uniform algebra A is strongly tight if S_g is compact for all $g \in C(X)$. He discusses in some depth properties of tight and strongly tight algebras as well as the problem of tightness versus strong tightness.

In all cases that have been studied, the operators S_g are weakly compact if and only if they are compact. This brings us to one more question: For which $g \in C(X)$ is S_g compact?

It is not difficult to see that every compact operator is weakly compact and completely continuous. However, one can give examples to show that the three properties may be different in a space. (Saccone's thesis [1995] and Diestel's work [1984] are excellent references for some of these examples.)

It is an interesting problem to try to discover spaces in which these types of compactness or complete continuity actually coincide. A well-known problem in this direction is to characterize Banach spaces that have the *Dunford–Pettis property*, that is, those on which any weakly compact operator is completely continuous. This property was named after N. Dunford and B. J. Pettis, who first introduced it [1940] and showed that it holds for L^1 spaces. Grothendieck [1953] showed that $C(X)$ spaces have the Dunford–Pettis property. Bourgain [1984] showed that $H^\infty(D)$ has the same property, using the theory of ultra-products. In studying his work, Cima and Timoney [1987] noted that if the operators S_g are completely continuous for all $g \in C(X)$, then A has the Dunford–Pettis property. Using this approach, these authors were able to show that certain spaces from rational approximation theory have the property. It was their hope that they could do the same for $H^\infty(D)$ by characterizing the subalgebra of $L^\infty(\partial D)$ consisting of those symbols g for which S_g is completely continuous. They defined the Bourgain algebra of an algebra B to be

$$B_{\mathrm{cc}} = \{f \in C(X) : S_f \text{ is completely continuous}\}.$$

Their work is another version of the question we looked at above in connection with Cole and Gamelin's work; that is, for which $g \in C(X)$ is S_g completely continuous?

Thus we are interested in knowing three things: for which $g \in C(X)$ is the Hankel-type operator S_g compact, weakly compact, or completely continuous.

3. Bourgain Algebras

Cima and Timoney [1987] showed that the Bourgain algebra of a uniform algebra B is a closed subalgebra of $C(X)$ and that $B \subset B_{\mathrm{cc}}$. A great deal of work on Bourgain algebras followed; the reader is referred to the exposition in [Yale 1992] for a description of early work on the subject. One can change the domain of definition of the functions (see, for example, [Cima et al. 1993; Dudziak et al. $\geq$ 1998]), the subalgebra B (as in [Gorkin et al. 1992]), the superalgebra $C(X)$ [Izuchi et al. 1994], or the space on which the continuous functions act [Ghatage et al. 1992]. Finally, one can try to work in as general a context as possible [Izuchi 1992].

Many related questions arose. For example, Izuchi, Stroethoff and Yale [Izuchi et al. 1994] looked at the Bourgain algebra of closed linear subspaces rather than closed algebras. Tonev and Yale [1996] study invariance of Hankel-type operators under isomorphisms. One can also ask how the second Bourgain algebra of an algebra is related to the first [Cima et al. 1993; Gorkin et al. 1992; Izuchi 1992]; sometimes they are the same, sometimes not. Another interesting question is when the Bourgain algebras are monotonic; that is, if $A \subset B$, when is $A_{\mathrm{cc}} \subset B_{\mathrm{cc}}$?

4. Compactness of Hankel-Type Operators on the Circle

From here on, unless otherwise stated, we will only be concerned with algebras $H^\infty(U)$ of bounded analytic functions on a domain U in the complex plane as subalgebras of $L^\infty(U)$ with respect to area measure. One can also look at algebras on the boundary of these domains [Dudziak et al. $\geq$ 1998], but we will do so only for $H^\infty(\partial D)$ as a subalgebra of $L^\infty(\partial D)$ on the unit circle. We now return to the question of compactness of Hankel type operators.

As we mentioned above, for classical Hankel operators on the Hardy space we have this result:

THEOREM 4.1. *Let $f \in L^\infty(\partial D)$. The Hankel operator H_f defined on H^2 is compact if and only if $f \in H^\infty(\partial D) + C(\partial D)$.*

Cima, Janson and Yale [Cima et al. 1989] proved the following.

THEOREM 4.2. *Let $f \in L^\infty(\partial D)$. Then the Hankel-type operator S_f defined on H^∞ is completely continuous if and only if $f \in H^\infty(\partial D) + C(\partial D)$.*

Their proof uses a theorem of P. Beurling, as well as the theory of BMO and the Chang–Marshall theorem. It turns out to be relatively easy to eliminate the BMO theory, and this allows one to study Bourgain algebras of closed subalgebras of $L^\infty(\partial D)$ containing $H^\infty(\partial D)$ [Gorkin et al. 1992]. The proof of the theorem as stated above also does not use the full strength of the Chang–Marshall theorem. However, the theorem of P. Beurling, or ideas therein, turn out to be essential to the study of Hankel-type operators. As we shall see, once one has all these ideas in place it is not too difficult to show that complete continuity, compactness, and weak compactness are equivalent for these operators in this context. This seems to have first been noticed in [Dudziak et al. $\geq$ 1998].

In order to prove any result in this direction, one needs examples of weakly convergent sequences. Because we are working in the uniform algebra context, the Lebesgue dominated convergence theorem shows that a sequence of bounded analytic functions $\{f_n\}$ on D converges weakly to zero if and only if it is uniformly bounded and its Gelfand transform tends to zero pointwise on X. This makes it a bit easier to think about weakly null sequences, but Beurling's theorem helps us to construct many more weakly null sequences. Beurling's theorem is actually much more general than the version stated as Theorem 4.3 below; see [Garnett 1981].

Recall that if z_n are points in the disc satisfying $\sum(1-|z_n|) < \infty$ the Blaschke product with zeroes z_n is given by

$$B(z) = \prod \frac{\bar{z}_n}{|z_n|} \frac{z_n - z}{1 - z\bar{z}_n}.$$

A sequence $\{z_n\}$ of points in D is called *interpolating* if it has the property that, whenever $\{w_n\}$ is a bounded sequence of complex numbers, there exists a function $f \in H^\infty$ with $f(z_n) = w_n$. While it is not clear that interpolating sequences exist in general, for the disc such sequences have been characterized; see, for example, [Garnett 1981, Chapter 7]. In fact, every sequence tending to the boundary of the disc has an interpolating subsequence. An infinite Blaschke product B is called an *interpolating Blaschke product* if the zero sequence of B forms an interpolating sequence. The following general result of P. Beurling [Garnett 1981, p. 298] is what we need to prove a version of Theorem 4.2; see Theorem 4.5 below.

THEOREM 4.3. *Let $\{z_n\}$ be an interpolating sequence in the disc. Then there are $H^\infty(D)$ functions $\{f_n\}$ such that*

$$f_n(z_n) = 1,$$
$$f_n(z_m) = 0 \quad \text{for } n \neq m,$$

and a constant M such that $\sum |f_n(z)| < M$ for all $z \in D$.

Note that any such sequence of functions must converge to zero weakly, since, for any element φ in the dual of H^∞ and any positive integer N, if we let $a_n = \text{sgn}\,\overline{(\varphi(f_n))}$, we have

$$\sum_1^N |\varphi(f_n)| = \sum_1^N a_n(\varphi(f_n)) = \varphi\left(\sum_1^N a_n f_n\right) \leq \|\varphi\|M.$$

Thus $\varphi(f_n) \to 0$ for all φ in the dual space of H^∞.

In the setting of this paper one can show [Dudziak et al. $\geq$ 1998] that, if S_g not compact, then it is an isomorphism (a bicontinuous operator onto its range) on a subspace J of $H^\infty(U)$ isomorphic to ℓ^∞. This implies that S_g cannot be completely continuous or weakly compact. That is why, in all situations presented here, these properties are all equivalent.

The following theorem, which is a special case of the Chang–Marshall theorem, can be used to give a quick proof of Theorem 4.5.

THEOREM 4.4. *Suppose that f is in $L^\infty(\partial D)$ but not in $H^\infty(\partial D) + C(\partial D)$. Then the closed subalgebra of $L^\infty(\partial D)$ generated by H^∞ and f contains the conjugate of an interpolating Blaschke product.*

This theorem has the advantage that it gives an easily understood proof of Cima, Janson and Yale's result containing the major ingredients of many of the proofs in this area. It has the disadvantage that it does not generalize easily to functions on general domains. Before turning to the proof, note that as long as $S_g(z^n)$ converges to zero in norm, we have $g \in H^\infty + C$, for

$$\text{dist}(g, H^\infty + C) \leq \text{dist}(g, \bar{z}^n H^\infty) = \text{dist}(gz^n, H^\infty) = \|S_g z^n\| \to 0.$$

Thus, the strength of the following theorem really lies in the final assertion.

THEOREM 4.5. *Let $g \in L^\infty(\partial D)$. Then the Hankel-type operator S_g defined on H^∞ is compact if and only if $g \in H^\infty(\partial D) + C(\partial D)$. Furthermore, if g is not in $H^\infty(\partial D) + C(\partial D)$, then S_g is neither compact, completely continuous, nor weakly compact.*

PROOF. First we need to show that S_g is compact if $g \in H^\infty(\partial D) + C(\partial D)$. Since the result is clear for functions in $H^\infty(\partial D)$, we only have to show it for continuous functions. In addition, one can check that the space of symbols for which S_g is compact is a closed algebra, so it suffices to show that S_g is compact if $g(z) = \bar{z}$.

Suppose that $\{f_n\}$ is a bounded sequence of $H^\infty(\partial D)$ functions. By Montel's theorem there is a subsequence of $\{f_n\}$ converging uniformly on compacta to an $H^\infty(\partial D)$ function. Thus we may assume that $f_n \to 0$ uniformly on compacta. Note that, since $|z| = 1$, we have for $f \in H^\infty(\partial D)$

$$\bar{z}f = \frac{f - f(0)}{z} + \bar{z}f(0).$$

Now $\bar{z} f_n(0) \to 0$ and

$$\bar{z} f + H^\infty = \frac{f - f(0)}{z} + \bar{z} f(0) + H^\infty = \bar{z} f(0) + H^\infty,$$

so

$$\|S_{\bar{z}}(f_n)\| = \|\bar{z} f_n + H^\infty(\partial D)\| \le \|\bar{z} f_n(0)\| \to 0.$$

Therefore, S_g is compact.

For the other direction, suppose that g is not in $H^\infty(\partial D) + C(\partial D)$. By Sarason's theorem (Theorem 4.4), there is an interpolating Blaschke product b with $\bar{b} \in H^\infty[g]$. Let $\{z_n\}$ denote the zero sequence of b. By Beurling's theorem, we can obtain a constant M and a sequence $\{f_n\}$ tending to zero weakly with $\sum |f_n| < M$ such that $f_n(z_m) = \delta_{nm}$. Let $\alpha = \{\alpha_n\}$ be an arbitrary sequence in ℓ^∞. Let $\tilde{g} = \sum \alpha_n f_n$. Then $\tilde{g} \in H^\infty$ and $\|\tilde{g}\| \le \|\alpha\| M$. Now $\|S_{\bar{b}} \tilde{g}\| = \|\bar{b} \tilde{g} + H^\infty\|$. But $|b| = 1$ almost everywhere on ∂D, so we see that

$$\|S_{\bar{b}} \tilde{g}\| = \|\tilde{g} + b H^\infty\| \ge \sup |\tilde{g}(z_n)| \ge \sup |\alpha_n| \ge \|\tilde{g}\|/M.$$

Now the map defined for each $\alpha \in \ell^\infty$ by $\alpha \to \sum \alpha_n f_n$ is an embedding of ℓ^∞ into H^∞, and therefore $S_{\bar{b}}$ is an isomorphism on a subspace J of H^∞ isomorphic to ℓ^∞. This implies that $S_{\bar{b}}$ is not compact, weakly compact, or completely continuous. Since $\bar{b} \in H^\infty[g]$ and the spaces of symbols f for which S_f is compact, weakly compact, or completely continuous are uniformly closed algebras containing H^∞, we see that S_g cannot be compact, weakly compact, or completely continuous. □

One can also obtain the isomorphism statement directly for the operator S_g by using the full strength of the Chang–Marshall theorem rather than Sarason's theorem.

5. Compact Hankel-Type Operators on General Domains

The next result that appeared in this context was by Cima, Stroethoff and Yale. They replaced the domain above by the disc, but they used the result for the circle to obtain their result in this new situation.

Let's try to guess what the result might be. Obviously S_g is completely continuous for g in $H^\infty(D)$. We probably expect that it would be completely continuous for continuous symbols. Is there any other symbol that might make this operator completely continuous? We are looking for a symbol g such that $f_n \to 0$ weakly implies that $\|g f_n + H^\infty(D)\| \to 0$. Now we know from the uniform boundedness principle that the $\|f_n\|$ are uniformly bounded. Thus $f_n \to 0$ on compact subsets of D. So if we have a symbol for which f_n converging uniformly to zero on compacta implies $\|g f_n\| \to 0$, the corresponding Hankel type operator would be completely continuous. Of course any $L^\infty(D)$ function that vanishes outside a compact subset of D will have this property. Since such a function could be discontinuous on D, we see that it need not be in the algebra $H^\infty(D) + C(\bar{D})$.

Following Cima, Stroethoff, and Yale, we define the space L_o^∞ to be the closure of the set of functions in L^∞ that vanish outside a compact set contained in D. This space of functions turns out to be precisely the space of functions for which the multiplication operator is compact. (See [Dudziak et al. $\geq$ 1998] for this result and a similar result for multiplication operators on general domains.)

Now we can state the result.

THEOREM 5.1. *Let* $g \in L^\infty(D)$. *Then* S_g *is completely continuous if and only if* $g \in H^\infty(D) + C(\overline{D}) + L_o^\infty(D)$.

The original proof of the above result used Cima, Janson, and Yale's theorem. The first obstacle to be overcome is that it is difficult to start with a function in L^∞ and pass in a natural way to a function on the boundary. Later proofs actually went the other way around, proving the result on an open subset of the plane rather than the boundary. Many of these proofs can be adapted to work on the boundary of the domain as well.

Cima, Stroethoff, and Yale's result appears to be quite different from the result on the circle. It was Izuchi who first noticed what the connection is. In order to state Izuchi's result here, we need to recall some definitions. Izuchi's result was stated for a general uniform algebra. However, we will continue to work on algebras of bounded analytic functions on domains in the plane.

Izuchi noticed that the results obtained depend on the circle having the property that every point of ∂D is what is called a peak point for the algebra $H^\infty(D)$. We can state this for more general domains (and it can be stated for more general algebras; see [Gamelin 1969]).

Let U be a domain in C and let $\lambda \in \overline{U}$. The fiber $M_\lambda(H^\infty)$ over λ is defined by

$$M_\lambda(H^\infty) = \{\varphi \in M(H^\infty) : \varphi(z) = \lambda\}.$$

We say that λ is a peak point for H^∞ if there exists a function $f \in H^\infty$ such that $f|M_\lambda(H^\infty) = 1$ (that is, $\varphi(f) = 1$ for all $\varphi \in M_\lambda(H^\infty)$) and $|f|M_\alpha(H^\infty)| < 1$ for all $\alpha \neq \lambda$. In the case of the unit circle, every point is a peak point, since a rotation of the function $(z + 1)/2$ produces a peaking function. However, no point of the open unit disc can be a peak point of $H^\infty(D)$, for this would violate the maximum principle.

Note that when our domain is the unit disc or unit circle the function $\bar{z}$ is constant on every fiber, so if f is continuous, then f restricted to each fiber is constant as well. Therefore, in both results above we see if we let

$$B_{cc} = \{g \in L^\infty : S_g \text{ is completely continuous}\},$$

then $B_{cc}|M_\lambda(H^\infty) = (H^\infty + C)|M_\lambda(H^\infty) = H^\infty|M_\lambda(H^\infty)$ over every peak point λ. Thus it seems that our algebra does not change over a peak point, while it may change over a nonpeak point. Izuchi's result deals with general uniform algebras, but an inspection of the proof shows the following:

THEOREM 5.2. *Suppose that every point of ∂U is a peak point for $H^\infty(U)$. If $g \in L^\infty(U)$ and S_g is completely continuous, then $g|M_\lambda(H^\infty) \in H^\infty|M_\lambda(H^\infty)$ for all $\lambda \in \partial U$.*

Izuchi's proof uses the peak functions to construct a sequence of functions that have properties similar to the P. Beurling functions. His proof is like an earlier proof of Gamelin and Garnett [1970], which shows that a point $\lambda \in \bar{U}$ is a peak point if and only if every sequence of points in U tending to λ has an interpolating subsequence.

If we set
$$B_c = \{g \in L^\infty : S_g \text{ is compact}\},$$
$$B_{wc} = \{g \in L^\infty : S_g \text{ is weakly compact}\},$$
the next theorem tells us that what one now expects to be true is in fact so; see [Dudziak et al. $\geq$ 1998].

THEOREM 5.3. *Let U be a domain and $\lambda \in \bar{U}$ be a peak point for $H^\infty(U)$. Then*
$$B_c|M_\lambda(H^\infty) = H^\infty|M_\lambda(H^\infty).$$

If, in addition, every point of ∂U is a peak point for $H^\infty(U)$, then S_g is compact if and only if g is in the uniform closure of
$$H^\infty(U) + C(\bar{U}) + L_o^\infty(U).$$

Finally, if S_g is not compact, there is a subspace J of $H^\infty(U)$ isomorphic to ℓ^∞ on which S_g is an isomorphism.

6. Nonpeak Points and Hankel-Type Operators

The last result along these lines is one in which not every point of the boundary of the domain is a peak point. The so-called L-domains or roadrunner domains are well-known examples of this behavior. Denote an open disc of radius r_n and center c_n by Δ_n. The roadrunner domain that we will work on consists of the unit disc minus the union of the origin and a sequence of disjoint closed discs $\bar{\Delta}_n$. We require that the centers c_n be positive real numbers decreasing to 0, and that the discs accumulate only at zero. Such domains were first studied by Zalcman [1969], who showed that 0 is a peak point if and only if $\sum r_n/c_n = \infty$. He also showed that, if this sum is finite, one can define a homomorphism φ_0 by setting
$$\varphi_0(f) = \frac{1}{2\pi i} \int_{\partial U} \frac{f(z)}{z} \, dz \quad \text{for } f \in H^\infty(D).$$
Such a homomorphism, called a distinguished homomorphism, is special in that it is the only weak-star continuous homomorphism in the fiber over 0. In fact, if we identify a point z with the linear functional that is evaluation at z, there exists a sequence of points $\{z_n\}$ of U converging to φ_0 in the norm of the dual space of H^∞! Now any point $z_0 \in U$ has a trivial fiber consisting only of

such a homomorphism, and in some sense φ_0 thinks of itself as one of these homomorphisms, living in U. For this reason, the results and proofs mentioned above do not extend to the roadrunner set when 0 is not a peak point. One has to replace Beurling's result by something more general.

One well-known general result in this direction is the Rosenthal–Dor theorem [Diestel 1984, p. 201; Dor 1975]: In order that each bounded sequence in a Banach space X have a weakly Cauchy subsequence, it is necessary and sufficient that X contain no isomorphic copy of ℓ^1. The next theorem is the version of the P. Beurling theorem that one needs in our situation [Dudziak et al. $\geq$ 1998].

THEOREM 6.1. *Let $\{\mu_n\}$ be a sequence of measures on U converging weak-star in the dual of $C(\overline{U})$ to the point mass at $z_0 \in \partial U$. Then either $\{\mu_n\}$ converges in norm (in $(H^\infty(U))^*$) or for every $\varepsilon > 0$ there is a subsequence $\{\mu_{n_j}\}$, a constant M, and a sequence of $H^\infty(U)$ functions $\{f_k\}$ such that $\sum |f_k(z)| < M$ for every $z \in U$ and $\int f_k d\mu_{n_j} = \delta_{jk}$ for all j and k.*

More can be said in this setting; see [Dudziak et al. $\geq$ 1998] for this information and a proof of the preceding theorem. What is important for us is that, when the sequence converges in the norm of the dual space, it converges to the distinguished homomorphism. When this does not happen, we are in a situation in which every sequence has an interpolating subsequence.

Unfortunately, results obtained thus far require that the roadrunner have one more property. We need to require that there exist disjoint closed discs D_n with center c_n and radius R_n containing $\overline{\Delta}_n$ and satisfying $\sum r_n/R_n < \infty$. Note that there is always a distinguished homomorphism in this kind of domain, since

$$\sum \frac{r_n}{c_n} < \sum \frac{r_n}{R_n} < \infty.$$

Such domains were studied by Behrens [1970], who discovered that they have the following property.

THEOREM 6.2. *Given $\varepsilon > 0$ and $M > 0$ there exists an integer N such that, if $f_n \in H^\infty(\Delta_n^c)$ satisfies $f_n(\infty) = 0$ and $\|f_n\| < M$, then*

$$\sum_{m \geq N} |f_m(z)| < \varepsilon \quad \text{for } z \notin \bigcup_{n \geq N} D_n,$$

$$\sum_{\substack{m \geq N \\ m \neq n}} |f_m(z)| < \varepsilon \quad \text{for } z \in D_n.$$

This property and the use of certain projections allowed Behrens to prove the Corona theorem in such domains. The projections are defined as follows: for $f \in H^\infty(U)$ and an integer $n > 0$, define

$$P_n(f)(z) = \frac{1}{2\pi i} \int_{\partial \Delta_n} \frac{f(\zeta)}{z - \zeta} \, d\zeta \quad \text{for } z \notin \Delta_n.$$

This allows us to take functions from $H^\infty(U)$ to functions in $H^\infty(\Delta_n^c)$ vanishing at ∞. One can then use Theorem 6.2 to sum the resulting functions. These same properties are used to describe the Bourgain algebra $H^\infty(U)_{cc}$, the algebra $H^\infty(U)_c$ (consisting of symbols of compact Hankel-type operators), and the algebra $H^\infty(U)_{wc}$ (consisting of symbols of weakly compact Hankel-type operators) for the algebra $H^\infty(U)$ when U is a Behrens roadrunner. A precise description of this result is given in [Dudziak et al. $\geq$ 1998]. We note here only that the three algebras turn out to be equal even in this general situation, but not equal to $H^\infty(U) + C(\bar{U}) + L_o^\infty(U)$.

7. Open Questions

We conclude by mentioning some open questions in this area, the first two of which arose in [Dudziak et al. $\geq$ 1998]. In order to help the reader, we write B_{cc} rather than the usual notation B_b for the Bourgain algebra.

QUESTION 7.1. *In every situation that we know of, $B_{cc} = B_c = B_{wc}$. Are any or all of these equalities true in $H^\infty(U)$ for an arbitrary domain U in the plane?*

As mentioned above, in the case of the Behrens roadrunner the algebra $H^\infty(U)_c$ turns out to be different from $H^\infty(U) + C(\bar{U}) + L_o^\infty(U)$. In fact, any function $g \in L^\infty(U)$ that is the constant value 1 or the constant 0 on $D_n \setminus \Delta_n$ for every n is the symbol of a compact Hankel-type operator. If we take such a g and require that it be identically 1 on $D_{2n} \setminus \Delta_{2n}$ and identically 0 on $D_{2n+1} \setminus \Delta_{2n+1}$, the operator S_g will be compact, but $g \notin H^\infty(U) + C(\bar{U}) + L_o^\infty(U)$. At the time of this writing the following question remains open:

QUESTION 7.2. *Let U be a domain in C. For which $g \in L^\infty(U)$ is the Hankel-type operator S_g compact? weakly compact? completely continuous?*

The answer in the case of the Behrens roadrunner depends on a description of points tending to the distinguished homomorphism. In the general case, one would expect the distance in the pseudohyperbolic metric to the distinguished homomorphism to play an important role.

QUESTION 7.3. *The study of Bourgain algebras arose in an effort to find a simpler proof of the fact that $H^\infty(D)$ has the Dunford–Pettis property. Does $H^\infty(U)$ have the Dunford–Pettis property?*

Acknowledgements

I would like to thank S. Axler, J. Dudziak, T. Gamelin, and R. Mortini for several very helpful suggestions, as well as the University of Bern and the Mathematical Sciences Research Institute for their support.

References

[Axler and Shields 1987] S. Axler and A. Shields, "Algebras generated by analytic and harmonic functions", *Indiana Univ. Math. J.* **36**:3 (1987), 631–638.

[Aytuna and Chollet 1976] A. Aytuna and A.-M. Chollet, "Une extension d'un résultat de W. Rudin", *Bull. Soc. Math. France* **104**:4 (1976), 383–388.

[Behrens 1970] M. Behrens, "The corona conjecture for a class of infinitely connected domains", *Bull. Amer. Math. Soc.* **76** (1970), 387–391.

[Bourgain 1984] J. Bourgain, "New Banach space properties of the disc algebra and H^∞", *Acta Math.* **152**:1-2 (1984), 1–48.

[Chang 1976] S. Y. A. Chang, "A characterization of Douglas subalgebras", *Acta Math.* **137**:2 (1976), 82–89.

[Cima and Timoney 1987] J. A. Cima and R. M. Timoney, "The Dunford–Pettis property for certain planar uniform algebras", *Michigan Math. J.* **34**:1 (1987), 99–104.

[Cima et al. 1989] J. A. Cima, S. Janson, and K. Yale, "Completely continuous Hankel operators on H^∞ and Bourgain algebras", *Proc. Amer. Math. Soc.* **105**:1 (1989), 121–125.

[Cima et al. 1993] J. A. Cima, K. Stroethoff, and K. Yale, "Bourgain algebras on the unit disk", *Pacific J. Math.* **160**:1 (1993), 27–41.

[Cole and Gamelin 1982] B. J. Cole and T. W. Gamelin, "Tight uniform algebras and algebras of analytic functions", *J. Funct. Anal.* **46**:2 (1982), 158–220.

[Davie et al. 1973] A. M. Davie, T. W. Gamelin, and J. Garnett, "Distance estimates and pointwise bounded density", *Trans. Amer. Math. Soc.* **175** (1973), 37–68.

[Diestel 1984] J. Diestel, *Sequences and series in Banach spaces*, Graduate Texts in Mathematics **92**, Springer, New York, 1984.

[Dor 1975] L. E. Dor, "On sequences spanning a complex l_1 space", *Proc. Amer. Math. Soc.* **47** (1975), 515–516.

[Dudziak et al. ≥ 1998] J. Dudziak, T. W. Gamelin, and P. Gorkin, "Hankel operators on bounded analytic functions". To appear in *Trans. Amer. Math. Soc.*

[Dunford and Pettis 1940] N. Dunford and B. J. Pettis, "Linear operations on summable functions", *Trans. Amer. Math. Soc.* **47** (1940), 323–392.

[Dunford and Schwartz 1958] N. Dunford and J. T. Schwartz, *Linear Operators, I: General Theory*, Pure and Applied Mathematics **7**, Interscience, New York, 1958. Reprinted by Wiley, New York, 1988.

[Gamelin 1969] T. W. Gamelin, *Uniform algebras*, Prentice-Hall, Englewood Cliffs, NJ, 1969. Second edition, Chelsea, New York, 1984.

[Gamelin and Garnett 1970] T. W. Gamelin and J. Garnett, "Distinguished homomorphisms and fiber algebras", *Amer. J. Math.* **92** (1970), 455–474.

[Garnett 1981] J. B. Garnett, *Bounded analytic functions*, Pure and Applied Mathematics **96**, Academic Press, New York and London, 1981.

[Ghatage et al. 1992] P. G. Ghatage, S. H. Sun, and D. C. Zheng, "A remark on Bourgain algebras on the disk", *Proc. Amer. Math. Soc.* **114**:2 (1992), 395–398.

[Gorkin et al. 1992] P. Gorkin, K. Izuchi, and R. Mortini, "Bourgain algebras of Douglas algebras", *Canad. J. Math.* **44**:4 (1992), 797–804.

[Grothendieck 1953] A. Grothendieck, "Sur les applications linéaires faiblement compactes d'espaces du type $C(K)$", *Canadian J. Math.* **5** (1953), 129–173.

[Izuchi 1992] K. Izuchi, "Bourgain algebras of the disk, polydisk, and ball algebras", *Duke Math. J.* **66**:3 (1992), 503–519.

[Izuchi et al. 1994] K. Izuchi, K. Stroethoff, and K. Yale, "Bourgain algebras of spaces of harmonic functions", *Michigan Math. J.* **41**:2 (1994), 309–321.

[Marshall 1976] D. E. Marshall, "Subalgebras of L^∞ containing H^∞", *Acta Math.* **137**:2 (1976), 91–98.

[Peller 1998] V. V. Peller, "An excursion into the theory of Hankel operators", pp. 65–120 in *Holomorphic Spaces*, edited by S. Axler et al., Math. Sci. Res. Inst. Publications **33**, Cambridge Univ. Press, 1998.

[Power 1982] S. C. Power, *Hankel operators on Hilbert space*, Research Notes in Mathematics **64**, Pitman, Boston, 1982.

[Rudin 1975] W. Rudin, "Spaces of type $H^\infty + C$", *Ann. Inst. Fourier (Grenoble)* **25**:1 (1975), 99–125.

[Saccone 1995] S. F. Saccone, "Banach space properties of strongly tight uniform algebras", *Studia Math.* **114**:2 (1995), 159–180.

[Sarason 1976] D. Sarason, "Algebras between L^∞ and H^∞", pp. 117–130 in *Spaces of analytic functions* (Kristiansand, 1975), edited by O. B. Bekken et al., Lecture Notes in Math. **512**, Springer, Berlin, 1976.

[Tonev and Yale 1996] T. Tonev and K. Yale, "Hankel type operators, Bourgain algebras, and isometries", pp. 413–418 in *Recent developments in operator theory and its applications* (Winnipeg, MB, 1994), edited by I. Gohberg et al., Oper. Theory Adv. Appl. **87**, Birkhäuser, Basel, 1996.

[Yale 1992] K. Yale, "Bourgain algebras", pp. 413–422 in *Function spaces* (Edwardsville, IL, 1990), edited by K. Jarosz, Lecture Notes in Pure and Appl. Math. **136**, Dekker, New York, 1992.

[Zalcman 1969] L. Zalcman, "Bounded analytic functions on domains of infinite connectivity", *Trans. Amer. Math. Soc.* **144** (1969), 241–269.

[Zhu 1990] K. H. Zhu, *Operator theory in function spaces*, Monographs and Textbooks in Pure and Applied Mathematics **139**, Marcel Dekker Inc., New York, 1990.

PAMELA GORKIN
DEPARTMENT OF MATHEMATICS
BUCKNELL UNIVERSITY
LEWISBURG, PA 17837
UNITED STATES
  pgorkin@bucknell.edu

Holomorphic Spaces
MSRI Publications
Volume **33**, 1998

Tight Uniform Algebras

SCOTT SACCONE

ABSTRACT. We discuss the relationships between a certain class of uniform
algebras, called tight uniform algebras, and various concepts from Banach
space theory, such as the Dunford–Pettis property, the Pełczyński property,
and weak sequential completeness. We also mention some connections with
the $\bar{\partial}$-problem, interpolation, pointwise bounded approximation, and inner
functions on strictly pseudoconvex domains.

1. Introduction

B. Cole and T. W. Gamelin [1982] introduced a generalized notion of analyticity, which they called tightness. If K is a compact space and $X \subset C(K)$ is a
closed subspace (in the uniform norm) we say X is a *tight subspace* if the Hankel-
type operator $S_g : X \to C(K)/X$ defined by $f \mapsto fg + X$ is weakly compact for
every $g \in C(K)$. Recall that a *uniform algebra* A on K is a closed, separating
subalgebra of $C(K)$ which contains the constants. We say a uniform algebra A
on K is a *tight uniform algebra* if it is a tight subspace of $C(K)$. The following prototypical example from [Cole and Gamelin 1982] illustrates how tightness
could be thought of as an abstract version of the solvability of a $\bar{\partial}$-problem with
a mild gain in smoothness.

Let D be a strictly pseudoconvex domain in $\mathbb{C}^n$ with C^2 boundary and let $A =
A(D)$ be the uniform algebra on $\bar{D}$ of functions analytic in D. Let $K^{\infty}_{(0,1)}$ be the
space of smooth $\bar{\partial}$-closed $(0,1)$-forms on D. Then there exists a compact linear
operator $R : K^{\infty}_{(0,1)} \to C(\bar{D})$ which solves the $\bar{\partial}$-problem in D; that is, $\bar{\partial} \circ R = I$.
The compactness follows from the fact that there exist Hölder estimates on the
solutions, hence the mild gain in smoothness. If $g \in C^{\infty}(\bar{D})$ then we claim S_g
can be factored through R and is therefore compact. If we set $T_g(f) = f\bar{\partial}g$ and
let $q : C(\bar{D}) \to C(\bar{D})/A$ be the natural quotient map, the diagram

$$
\begin{array}{ccc}
A & \xrightarrow{\ S_g\ } & C(\bar{D})/A \\
{\scriptstyle T_g}\big\downarrow & & \big\uparrow{\scriptstyle q} \\
K^{\infty}_{(0,1)} & \xrightarrow{\ R\ } & C(\bar{D})
\end{array}
$$

clearly commutes. It now follows from a density argument that S_g is compact and therefore weakly compact for every g in $C(\bar{D})$. Thus $A(D)$ is a tight uniform algebra.

Another example, also from [Cole and Gamelin 1982], is this. Let K be a compact planar set and let $A = R(K)$ be the uniform algebra of continuous functions on K which are uniform limits of rational functions with poles off K. A similar argument shows that the operators S_g on $R(K)$ are compact for every $g \in C(K)$. If we consider the Vitushkin localization operator $T_g : C(K) \to C(K)$ defined by

$$(T_g f)(\zeta) = f(\zeta)g(\zeta) + \frac{1}{\pi} \iint\limits_K \frac{1}{z - \zeta} \, \bar{\partial} g(z) f(z) \, dx \, dy(z)$$

where $g \in C_c^1(\mathbb{C})$, then T_g is a continuous linear operator under which $R(K)$ is invariant [Gamelin 1969]. If we define $V_g : A \to C(K)$ by

$$(V_g f)(\zeta) = \frac{1}{\pi} \iint\limits_K \frac{1}{z - \zeta} \, \bar{\partial} g(z) f(z) \, dx \, dy(z),$$

it can be seen that V_g is compact. Since the diagram

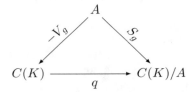

commutes (where q is the natural quotient map) it follows that S_g is compact for every $g \in C(K)$. In particular, $R(K)$ is tight for every compact K. If we note that the Cauchy transform solves the $\bar{\partial}$-problem in the plane (see [Gamelin 1969] or [Conway 1991]), then we see the method applied to $R(K)$ is exactly the same as the method applied to $A(D)$.

In all our examples thus far we have found that the operators S_g are compact. We say $X \subseteq C(K)$ is a *strongly tight subspace* if S_g is compact for every g, and similarly we define *strongly tight uniform algebras*.

Clearly any uniform algebra on a compact planar set K which is invariant under T_g for all smooth g is strongly tight. We say A is a *T-invariant uniform algebra* if A is a uniform algebra on a compact planar set K which contains $R(K)$ and is invariant under T_g. For example the algebra $A(K)$ of functions in $C(K)$ which are analytic in the interior of K is T-invariant [Gamelin 1969] and is therefore strongly tight.

One of the main results in [Cole and Gamelin 1982] is that if $C = C(K)$ then A is a tight uniform algebra if and only if $A^{**} + C$ is a (closed) subalgebra of C^{**} ($A^{**} + C$ is always a closed subspace of C^{**}). This result allowed the authors to extend Sarason's theorem about the Hardy space H^∞ on the unit circle to other domains. For example, let K be any compact planar set and let Q be the set of non-peak points for $R(K)$ (z is a *peak point* for A if there exists an $f \in A$ with $f(z) = 1$ and $|f| < 1$ elsewhere). Let λ_Q be Lebesgue measure restricted to Q and define $H^\infty(\lambda_Q)$ to be the weak-star closure of $R(K)$ in $L^\infty(\lambda_Q)$. Cole and Gamelin proved that $H^\infty(\lambda_Q) + C$ is a closed subalgebra of $L^\infty(\lambda_Q)$.

Tight uniform algebras have some Banach space properties that are typical of $C(K)$ spaces. We will discuss these properties and their relation to tightness, and also mention some applications of tightness to pointwise bounded approximation theory and inner functions on domains in $\mathbb{C}^n$.

2. The Dunford–Pettis Property

The following result is from [Dunford and Pettis 1940]. Recall that a continuous linear operator $T : X \to Y$ is *completely continuous* if T takes weakly null sequences in X to norm null sequences in Y.

THEOREM 2.1. *Let (Ω, Σ, μ) be any measure space. Then*:

(a) *If Y is a Banach space and $T : L^1(\mu) \to Y$ is a weakly compact operator then T is completely continuous.*

(b) *If Y is a separable dual space then any bounded linear operator $T : L^1(\mu) \to Y$ is completely continuous.*

Later, Grothendieck [1953] studied Banach spaces X that exhibited property (a) of the above theorem. Following Grothendieck, we say a Banach space X has the *Dunford–Pettis property* if whenever Y is a Banach space and $T : X \to Y$ is a weakly compact linear operator then T is completely continuous.

Part of Grothendieck's work was to provide various characterizations of the Dunford–Pettis property, some of which do not involve operators. It is not difficult to deduce from these characterizations, which we shall present shortly, that part (b) of Theorem 2.1 can be deduced from part (a). Furthermore, by using the result on the factorization of weakly compact operators in [Davis et al. 1974], it can be shown that (a) can also be deduced from (b). These ideas can also be found in [Diestel 1980].

Evidently $L^1(\mu)$ has the Dunford–Pettis property for every μ. From this it can be deduced that $C(K)$ has the Dunford–Pettis property for every compact space K. It has been shown that many uniform algebras, such as the disk algebra, have the Dunford–Pettis property as well. It is easy to see that any infinite dimensional reflexive space fails to have the Dunford–Pettis property. Also, the Hardy space H^1 on the unit circle fails to have the Dunford–Pettis property. To see this, consider the Paley operator $P : H^1 \to l^2$ defined by $f \mapsto (\hat{f}(2^k))_{k=1}^\infty$. It is well-known that P is a bounded linear operator [Pełczyński 1977] and is therefore weakly compact since it maps into a Hilbert space. However, if $f_n(\zeta) = \zeta^{2^n}$ then $\|Pf_n\| = 1$ while $f_n \xrightarrow{w} 0$ in H^1 by the Riemann–Lebesgue Lemma. Therefore P is not completely continuous.

We have mentioned that many uniform algebras have the Dunford–Pettis property. However, this is not true for all uniform algebras. It is easy to see that if X has the Dunford–Pettis property and Y is a complemented subspace of X then Y has the Dunford–Pettis property as well. It is a theorem of Milne [1972] (also, see [Wojtaszczyk 1991]) that every Banach space X is isomorphic to a complemented subspace of a uniform algebra A. The space A can be taken to be the uniform algebra on B_{X^*} (the unit ball in X^* with the weak-star topology) generated by X. If we let $X = l^2$ then A fails to have the Dunford–Pettis property. The author is not aware of any uniform algebra on a compact subset of $\mathbb{R}^n$ which fails the to have the Dunford–Pettis property.

Grothendieck proved the following popular characterizations of the Dunford–Pettis property. (See also [Diestel 1980].) Recall that a sequence $\{x\}$ in a Banach space X is a *weak-Cauchy sequence* if $\lim x^*(x_n)$ exists for every $x^* \in X^*$.

PROPOSITION 2.2. *The following statements are equivalent for any Banach space* X.

(a) X *has the Dunford–Pettis property.*
(b) *If* $T : X \to c_0$ *is a weakly compact linear operator then* T *is completely continuous.*
(c) *If* $x_n \xrightarrow{w} 0$ *in* X *and* $x_n^* \xrightarrow{w} 0$ *in* X^* *then* $x_n^*(x_n) \longrightarrow 0$.
(d) *If* $x_n \xrightarrow{w} 0$ *in* X *and* $\{x_n^*\}$ *is a weak Cauchy sequence in* X^* *then* $x_n^*(x_n) \longrightarrow 0$.
(e) *If* $x_n \xrightarrow{w} 0$ *in* X *and* $E \subset X^*$ *is relatively weakly compact then*

$$\lim_{n \to \infty} \sup_{x^* \in E} |x^*(x_n)| = 0.$$

It is well-known that, if K is a compact space, the dual of $C(K)$ is isomorphic to some L^1-space. The following corollary is now an immediate consequence of the proposition and Theorem 2.1.

COROLLARY 2.3. (a) *If* X *is a Banach space and* X^* *has the Dunford–Pettis property then* X *has the Dunford–Pettis property.*
(b) *If* K *is a compact space then* $C(K)$ *has the Dunford–Pettis property.*

We mentioned above that the two conclusions of the theorem of Dunford and Pettis are actually equivalent. This is not difficult to prove with the aid of the following important result from [Davis et al. 1974].

THEOREM 2.4. *Let* X *and* Y *be Banach spaces and suppose* $T : X \to Y$ *is a weakly compact linear operator. Then there exist bounded linear operators* S_1 *and* S_2 *and a reflexive Banach space* Z *such that the diagram*

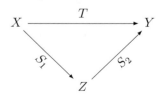

commutes.

This theorem can be used to form yet another characterization of the Dunford–Pettis property. We require the following result of Rosenthal and Dor [Diestel 1984]. We say a sequence $\{x_n\}$ in a Banach space X is a c_0-*sequence* if $\{x_n\}$ is equivalent to the unit vector basis of c_0; that is, there exists an isomorphic embedding $T : c_0 \to X$ such that $T(e_n) = x_n$. We define l^1-*sequences* in the same manner. The result is that if X is any Banach space and $\{x_n\}$ is a sequence in B_X which fails to have a weak-Cauchy subsequence then $\{x_n\}$ has an l^1-subsequence.

THEOREM 2.5. *The following statements are equivalent for a Banach space X.*

(a) *X has the Dunford–Pettis property.*

(b) *If $T : X \to Z^*$ is a bounded linear operator and Z contains no copy of l^1 then T is completely continuous.*

(c) *Every bounded linear operator $T : X \to Y$ from X to a separable dual space Y is completely continuous.*

(d) *Every bounded linear operator $T : X \to Y$ from X to a reflexive Banach space is completely continuous.*

PROOF. (a) $\Longrightarrow$ (b) Assume X has the Dunford–Pettis property and let $T : X \to Z^*$ be a bounded linear operator where Z contains no copy of l^1. Let $\{x_n\}$ be a weakly null sequence in X and assume $\{Tx_n\}$ fails to tend to zero in norm, so after passing to a subsequence we may assume $\|Tx_n\| > \varepsilon$ for some $\varepsilon > 0$ and for all n. Define $z_n^* = Tx_n$ and let $z_n \in B_Z$ be such that $z_n^*(z_n) > \varepsilon$ for every n. Since Z contains no copy of l^1, by the Rosenthal–Dor Theorem we may assume $\{z_n\}$ is a weak-Cauchy sequence. Let $x_n^* = T^*(z_n)$. Then by Proposition 2.2 we have $x_n^*(x_n) \longrightarrow 0$. However $x_n^*(x_n) = z_n^*(z_n)$, a contradiction.

Now (b) $\Longrightarrow$ (c) $\Longrightarrow$ (d) follows easily and the rest follows from Theorem 2.4. $\qquad\square$

Property (c) provides us with another way of showing the space H^1 fails to have the Dunford–Pettis property. This is because H^1 is separable and is easily seen to be a dual space by the F. and M. Riesz Theorem. If H^1 had the Dunford–Pettis property then by (c) the identity operator on H^1 would be completely continuous. This would say that weakly null sequences in H^1 are norm null. However the characters $f_n(\zeta) = \zeta^n$ are weakly null by the Riemann-Lebesgue Lemma. Thus, H^1 cannot have the Dunford–Pettis property.

We now present a well-known way of detecting when a bounded subset of the dual of a Banach space fails to be weakly compact. We say that a sequence $\{x_n\}$ in a Banach space X is a *weakly unconditionally Cauchy series* (w.u.C. series for short) if $\sum |x^*(x_n)| < \infty$ for every $x^* \in X$. For example, if X is a closed subspace of $C(K)$ then $\{f_n\}$ is a w.u.C. series if and only if $\sum |f_n(z)| \le C$ for every $z \in K$ for some constant C. We say a continuous linear operator $T : X \to Y$ is an *unconditionally converging operator* if T takes every weakly unconditionally Cauchy series to a series which converges unconditionally in norm. It is a theorem of Bessaga and Pełczyński [1958] that T is an unconditionally converging operator if and only if there does not exist a copy of c_0 in X on which T is an isomorphism.

PROPOSITION 2.6. *Let X be a Banach space and suppose $E \subset X^*$ is a bounded subset. Then the following statements are equivalent*:

(a) *There exists a weakly unconditionally Cauchy series $\sum x_n$ in X such that*

$$\varlimsup_{n \to \infty} \sup_{x^* \in E} |x^*(x_n)| > 0. \tag{2-1}$$

(b) *There exists a c_0-sequence $\{x_n\}$ in X such that (2-1) holds.*

If (a) *and* (b) *hold, E contains an l^1-sequence. In particular, E fails to be relatively weakly compact.*

PROOF. (a) $\Longrightarrow$(b) Assume $\{x_n\}$ is a w.u.C. series in X such that (2-1) holds. Then there exists some $\varepsilon > 0$ and a sequence $\{x_n^*\}$ in E such that, after passing to a subsequence of $\{x_n\}$ if necessary, we have $|x_n^*(x_n)| > \varepsilon$ for all n. Define $T : X \to l^\infty$ by $T(x) = (x_n^*(x))_{n=1}^\infty$, so $\|Tx_m\| > \varepsilon$ for $m \geq 1$. In particular, the series $\sum Tx_n$ does not converge and so by definition T fails to be an unconditionally converging operator. By the result of Bessaga and Pełczyński mentioned above there exists a subspace $X_0 \subseteq X$ such that X_0 is isomorphic to c_0 and $T|_{X_0}$ is an isomorphic embedding. The unit vector basis in X_0 is the desired c_0-sequence.

Part (a) follows trivially from (b) since the unit vector basis in c_0 is a weakly unconditionally Cauchy series.

To prove the final claim, assume E fails to contain an l^1-sequence. By the Rosenthal–Dor Theorem every sequence in E has a weak-Cauchy subsequence. Let $\{x_n\}$ be a w.u.C. series in X. We now claim that $\lim\sup_{x^* \in E} |x^*(x_n)| = 0$. $_{n\to\infty}$ To see this, define an operator $T : X^* \to l^1$ by $T(x^*) = (x^*(x_n))_{n=1}^\infty$. It follows from the Closed Graph Theorem that T is a bounded linear operator. Let $K = T(E)$. Recall that l^1 is a Schur space, i.e., the weak and norm convergence of sequences coincide. It follows from this that K has compact closure. In particular, K is totally bounded and it now follows easily that given an $\varepsilon > 0$ there exists an integer N such that $\sum_{k=N}^\infty |x^*(x_k)| \leq \varepsilon$ for every $x^* \in E$. This proves the claim and finishes the proposition. $\square$

An l^1-sequence $\{x_n^*\}$ in a dual space X^* cannot always be paired with a c_0-sequence in the way described above. For example, let $Y = C[0,1]$ and let $X = Y^*$. The Banach–Mazur Theorem states that every separable Banach space is isometrically isomorphic to a closed subspace of $C[0,1]$ (we say that $C[0,1]$ is a *universal space*; see [Wojtaszczyk 1991]). In particular X^* contains a copy of l^1, but X contains no copy of c_0. To see this, we recall the well-known fact that X is isomorphic to $L^1(\mu)$ for some abstract measure μ. We say a Banach space X is *weakly sequentially complete* if every weak-Cauchy sequence converges weakly in X. Every L^1-space is weakly sequentially complete [Dunford and Schwartz 1958]. However c_0 is not, which implies X contains no copy of c_0.

We will now mention some of the work of J. Bourgain, who showed that certain spaces of continuous functions, such as the ball-algebras and the polydisk algebras, have the Dunford–Pettis property. The following two theorems are easily deduced from the results in [Bourgain 1984a], as was observed in [Cima and Timoney 1987]. Recall the definition of the operators S_g from Section 1.

THEOREM 2.7 [Bourgain 1984a]. *Let X be a closed subspace of $C(K)$ and assume S_g is completely continuous for every $g \in C(K)$. Then:*

(a) *If $f_n \xrightarrow{w} 0$ in X and $E \subset X^*$ is a bounded subset with*

$$\varlimsup_{n \to \infty} \sup_{x^* \in E} |x^*(f_n)| > 0,$$

there exists a c_0-sequence $\{x_n\}$ in X failing to tend to zero uniformly on E.
(b) *X has the Dunford–Pettis property.*

The conclusion of (a) implies that E fails to be relatively weakly compact, by our previous proposition. Therefore any weakly null sequence must tend to zero uniformly on relatively weakly compact subsets of X^*. Hence, (a) implies (b) by Proposition 2.2.

THEOREM 2.8 [Bourgain 1984a]. *Let X be a closed subspace of $C(K)$ and assume $(S_g)^{**}$ is completely continuous for every $g \in C(K)$. Then X^* has the Dunford–Pettis property.*

We therefore have the following immediate consequence of Bourgain's work.

THEOREM 2.9. *If X is a strongly tight subspace of $C(K)$ then X and X^* have the Dunford–Pettis property.*

In fact, it follows from the technique mentioned in Theorem 2.7 that X has a property somewhat stronger than the Dunford–Pettis property. We will discuss this more in the next section.

It now follows immediately from [Cole and Gamelin 1982] that if A is any T-invariant uniform algebra (for example $R(K)$ or $A(K)$ for compact planar K) then A and A^* have the Dunford–Pettis property. Cima and Timoney, using different methods, also proved these results by showing S_g^{**} is completely continuous for every $g \in C(K)$ for a T-invariant uniform algebra A. It also follows from the work in [Cole and Gamelin 1982] that when $A = A(D)$ for D strictly pseudoconvex with C^2 boundary, then A and A^* have the Dunford–Pettis property (also, see [Li and Russo 1994]).

Incidentally, if X is a subspace of $C(K)$ we define X_b to be those $g \in C(K)$ such that S_g is completely continuous. Cima and Timoney [1987] showed that X_b is always an algebra, called the *Bourgain algebra* of X. The set of g for which S_g is weakly compact is also an algebra, and likewise the set of g for which S_g is compact. The weakly compact case is done in [Cole and Gamelin 1982] (for the case when $X = A$ itself is an algebra, but the result holds for subspaces) and the compact case is in [Saccone 1995a].

3. The Pełczyński Property

The Pełczyński property involves certain types of series in Banach spaces and how the convergence of these series is affected by linear operators. We begin with a result that follows from the work of Orlicz [1929]: if X and Y are Banach spaces and $T : X \to Y$ is a weakly compact linear operator then T takes weakly unconditionally Cauchy series to series that converge unconditionally in norm. In other words, weakly compact operators are necessarily unconditionally converging operators (defined above). For example, this implies any w.u.C. series in $L^2[0, 1]$ must be a norm convergent series.

Now we turn to the paper [Pełczyński 1962], entitled "Banach spaces on which every unconditionally converging operator is weakly compact," where the converse of Orlicz's result is studied. As the title suggests, we say a Banach space X has the *Pełczyński property* if whenever Y is a Banach space and $T : X \to Y$ is an unconditionally converging operator then T is weakly compact. Trivially, every reflexive space has the Pełczyński property. It follows from the result of Bessaga and Pełczyński mentioned in the previous section that X has the Pełczyński property if and only if every non-weakly compact linear operator from X fixes a copy of c_0 (i.e., is an isomorphism on a copy of c_0 in X).

Evidently, when studying the Pełczyński property it is important to know which Banach spaces contain copies of c_0. If (Ω, Σ, μ) is any measure space then $L^1(\mu)$ is weakly sequentially complete and therefore does not contain a copy of c_0. Assume $L^1(\mu)$ has the Pełczyński property. Then the identity operator cannot fix a copy of c_0. Therefore, the identity operator is weakly compact and so $L^1(\mu)$ is reflexive. Furthermore, $L^1(\mu)$ has the Dunford–Pettis property so the identity operator is also completely continuous which implies weakly convergent sequences in $L^1(\mu)$ are norm convergent. It now follows that every bounded sequence in $L^1(\mu)$ has a weakly convergent subsequence (by the reflexivity) and therefore a norm convergent subsequence. Therefore the unit ball in $L^1(\mu)$ is compact and $L^1(\mu)$ is finite-dimensional. Hence, $L^1(\mu)$ has the Pełczyński property if and only if it is reflexive which occurs if and only if it is finite-dimensional.

The Pełczyński property does not share the duality property of the Dunford–Pettis property. It was shown in Pełczyński's original paper [Pełczyński 1962] that if K is any compact space then $C(K)$ has the Pełczyński property. Therefore, any infinite-dimensional L^1-space fails to have the Pełczyński property in spite of the fact that its dual, which is isomorphic to $C(K)$ for some K, has the Pełczyński property. To complete this picture, suppose X is any Banach space such that X and X^* have the Pełczyński property. Then X^* is weakly sequentially complete and therefore contains no copy of c_0. This implies the identity operator on X^* is an unconditionally converging operator. Since X^* has the Pełczyński property, the identity operator must be weakly compact and so X^* is reflexive. It now follows that X and X^* have the Pełczyński property if and only if X is reflexive.

The following result presents some more or less well-known characterizations of the Pełczyński property.

THEOREM 3.1. *If X is a Banach space, the following conditions are equivalent.*

(a) *X has the Pełczyński property.*

(b) *If $T : X \to Y$ is a continuous linear operator which fails to be weakly compact then T is an isomorphism on some copy of c_0 in X.*

(c) *If $E \subseteq X^*$ is a bounded subset and the weak closure of E fails to be weakly compact then there exists a weakly unconditionally Cauchy series $\{x_n\}$ in X which fails to tend to zero uniformly on E.*

(d) (i) *X^* is weakly sequentially complete, and* (ii) *if $\{x_n^*\}$ is an l^1-sequence in X^* then there exists a c_0-sequence $\{x_k\}$ in X such that $|x_{n_k}^*(x_k)| > \delta > 0$ for all k for some sequence $\{n_k\}$.*

The equivalence of (a) and (b) follows from the remarks above. That of (a) and (c) is due to Pełczyński. The equivalence of (a) and (d) is less well-known, but can be deduced from (c) and the Rosenthal–Dor Theorem.

Note that (i) and (ii) of part (d) are distinct properties. Bourgain and Delbaen [1980] have constructed a Banach space X such that X^* is isomorphic to l^1, while X contains no copy of c_0; thus (i) holds while (ii) fails. R. C. James [1950] constructed a separable Banach space X such that X is nonreflexive and whose natural embedding into X^{**} has a codimension 1 image. In particular X^{**} is separable and therefore X^* contains no copy of l^1 and every sequence in X^* has a weak-Cauchy subsequence. Therefore X satisfies (ii) but fails (i).

As an illustration, consider the following theorem.

THEOREM 3.2 [Mooney 1972]. *Let m be Lebesgue measure on the unit circle Γ and let $H^\infty \subset L^\infty(m)$ be the Hardy space of boundary values of bounded analytic functions in the unit disk. Suppose $\{f_n\}$ is a bounded sequence in $L^1(m)$ such that $\lim_{n\to\infty} \int f_n h \, dm$ exists for every $h \in H^\infty$. Then there exists an element $f \in L^1(m)$ such that*

$$\lim_{n\to\infty} \int f_n h \, dm = \int f h \, dm$$

for every $h \in H^\infty$.

A proof of this can be found in [Garnett 1981]. It uses facts about peak sets in the maximal ideal space of H^∞. Mooney's theorem can easily be related to weak sequential completeness. Since $H^\infty = (L^1/H_0^1)^*$ where H_0^1 is the subspace of the Hardy space H^1 consisting of functions vanishing at the origin, Mooney's theorem is equivalent to the weak sequential completeness of L^1/H_0^1. Let A

be the disk algebra on the unit circle. By the F. and M. Riesz Theorem we have $A^\perp = \{f\,dm : f \in H_0^1\}$. It follows that if L is the space of measures singular to Lebesgue measure then A^* is isometrically isomorphic to $L^1/H_0^1 \oplus_{l^1} L$. Furthermore, L is isomorphic to $L^1(\mu)$ for some abstract measure μ. Since $L^1(\mu)$ is weakly sequentially complete, Mooney's theorem is equivalent to the weak sequential completeness of A^*. The disk algebra is an example of a tight uniform algebra and as we note below, all tight uniform algebras have weakly sequentially complete duals.

The Pełczyński property can be related to some ideas in interpolation. It is not hard to see that a bounded sequence $\{x_n\}$ in a Banach space is an l^1-sequence if and only if it interpolates X^*; i.e., for every bounded sequence of scalars $\{\beta_n\}$ there exists an x^* in X^* with $x^*(x_n) = \beta_n$. Consider the following result of P. Beurling (which can be found in [Garnett 1981]). Let D be the open unit disk. If $\{z_n\}$ is a sequence of points in D which interpolates H^∞ then there exists a sequence $\{h_n\}$ in H^∞ such that $\sum |h_n(z)| \leq C$ for all $z \in D$ and some constant C and $h_n(z_k) = \delta_{nk}$ where δ is the Kronecker delta function. Note that if A is the disk algebra then the point evaluations in A^* corresponding to the sequence $\{z_n\}$ form an l^1-sequence, and it is not hard to see that the sequence $\{h_n\}$ is a w.u.C. series in H^∞. The Pełczyński property for the disk algebra A offers a similar, but different, result. Given an arbitrary l^1-sequence $\{x_n^*\}$ in the dual of A (not necessarily point evaluations), there exists a c_0-sequence $\{x_n\}$ in A such that $|x_{n_k}^*(x_k)| > \delta > 0$ for some subsequence. This result applies to more general sequences in the dual, but the conclusion is weaker than that of Beurling.

Delbaen [1977] and Kisliakov [1975] independently showed that the disk algebra has the Pełczyński property. Delbaen [1979] extended these results to $R(K)$ for special classes of planar sets K, as did Wojtaszczyk [1979] (although the results of Delbaen were more extensive). It was shown in [Saccone 1995a] that $R(K)$ has the Pełczyński property for every compact planar set K, and also that every T-invariant uniform algebra on a compact planar set has the Pełczyński property. The T-invariant class includes $R(K)$ as well as $A(K)$ for all compact planar sets K. Bourgain [1983] showed that the ball-algebras and the polydisk-algebras have the Pełczyński property. This result was extended in [Saccone 1995a] to $A(D)$ for strictly pseudoconvex domains D in $\mathbb{C}^n$.

It follows from Milne's theorem, mentioned in the previous section, that there exist uniform algebras which fail to have the Pełczyński property. As in the case of the Dunford–Pettis property, the author is not aware of any uniform algebras on compact subsets of $\mathbb{R}^n$ which fail to have the Pełczyński property.

We will now elaborate on work from [Bourgain 1983]. If $X \subseteq C(K)$ we say that $m \in M(K)$ is a *weakly rich measure* for X if, whenever $\{f_n\}$ is a bounded sequence in X such that $\int |f_n|\,d|m| \longrightarrow 0$, the sequence $f_n g + X$ converges weakly to 0 for every $g \in C(K)$. If $f_n g + X$ converges to 0 in norm, we say that m is a *strongly rich measure*. This latter concept was introduced by Bourgain [1984a], who showed that X has the Pełczyński property if there exists a strongly rich measure for X. For example, it is shown that the surface-area measure on the unit sphere in $\mathbb{C}^n$ is a strongly rich measure for the ball-algebras. Note that weakly rich measures on strongly tight spaces (where the operators S_g are compact) are strongly rich.

As in the case of determining whether S_g is weakly compact, compact, or completely continuous, to show m is a weakly or strongly rich measure it suffices to check only a collection of those g which generate $C(K)$ as a uniform algebra. Given a measure $m \in M(K)$ and a closed subspace X of $C(K)$, define $(X, m)_{\text{wr}}$ and $(X, m)_{\text{sr}}$ to be the sets of those $g \in C(K)$ such that $f_n g + X \overset{w}{\longrightarrow} 0$ and $\|f_n g + X\| \longrightarrow 0$, respectively, whenever $\{f_n\}$ is a bounded sequence in X such that $\int |f_n| \, d|m| \longrightarrow 0$. Then $(X, m)_{\text{wr}}$ and $(X, m)_{\text{sr}}$ are closed subalgebras of $C(K)$ where $(X, m)_{\text{sr}}$ clearly contains X_b [Saccone 1995a]. Furthermore, it follows from Bourgain's result that if X possesses a strongly rich measure then X is a tight subspace.

It was shown in [Saccone 1995a] that every strongly tight uniform algebra on a compact metric space possesses a strongly rich measure and therefore has the Pełczyński property. (The proof actually works for strongly tight subspaces.) As noted in the same paper, it now follows from results in [Cole and Gamelin 1982] that $R(K)$ has the Pełczyński property for every compact planar set, and that $A(D)$ has the Pełczyński property for every strictly pseudoconvex domain D in $\mathbb{C}^n$ with C^2 boundary. It was also noted in [Saccone 1995a] that by examining Bourgain's proof it can be seen that indeed every strongly tight uniform algebra (or subspace) on an arbitrary compact space K has the Pełczyński property.

The following more general result is proved in [Saccone 1997].

THEOREM 3.3. *Let K be a compact space and let X be a tight subspace of $C(K)$. Then X has the Pełczyński property and X^* is weakly sequentially complete.*

If we only assume the operators S_g to be weakly compact instead of compact (that is, if we assume X is tight instead of strongly tight) then Bourgain's results no longer appear to be of use. The basic gliding hump construction used to prove the theorem remains essentially the same, however some calculations in Bourgain's original proof which involved Hilbert space geometry had to be replaced by more general arguments dealing with weak compactness in arbitrary Banach spaces.

COROLLARY 3.4 [Saccone 1995a]. (a) *If K is any compact planar set and A is a T-invariant uniform algebra on K then A has the Pełczyński property. In particular $R(K)$ and $A(K)$ have the Pełczyński property.*
(b) *If D is any strictly pseudoconvex domain in $\mathbb{C}^n$ with C^2 boundary then $A(D)$ has the Pełczyński property.*

Although it is now known that a large of class of planar uniform algebras, including $R(K)$ and $A(K)$, have such Banach space properties as the Dunford–Pettis property and the Pełczyński property, it is not known if any of these spaces, when they are not all of $C(K)$, fail to be isomorphic (as Banach spaces) to the disk algebra. For example, it is a theorem of Milutin that if K is any uncountable compact metric space then $C(K)$ is isomorphic to $C[0,1]$; see [Wojtaszczyk 1991]. On the other hand, the polydisk algebras were shown in [Henkin 1968] not to be isomorphic to the ball-algebras in higher dimensions.

It is not currently known whether $H^\infty(\mathbb{B}_n)$ has the Dunford–Pettis property or the Pełczyński property when $n > 1$. Astoundingly enough, Bourgain [1984b] has shown that if A is the disk algebra then all the duals of A have the Dunford–Pettis property and all the even duals of A have the Pełczyński property. The proof involves the theory of ultraproducts of Banach spaces. It follows from this that H^∞ on the unit circle has the Pełczyński property.

We say a Banach space X is a *Grothendieck space* if weak-star null sequences in X^* are weakly null. It is known that any dual space with the Pełczyński property is a Grothendieck space. It follows from Bourgain's work that all the even duals of the disk algebra are Grothendieck spaces, and in particular H^∞ is a Grothendieck space. It is not hard to see that this implies every continuous linear operator from H^∞ to a separable Banach space is weakly compact. In particular, if A is the disk algebra, every continuous linear operator $T : H^\infty \to A$ is weakly compact, and furthermore T^2 must be compact since A has the Dunford–Pettis property.

4. Band Theory

The theory of bands is useful for studying abstract properties of uniform algebras. Good sources for band theory are [Cole and Gamelin 1982] and [Conway 1991].

Let K be a compact Hausdorff space. If $\mathcal{B} \subseteq M(K)$ we say $\mathcal{B}$ is a *band of measures* if $\mathcal{B}$ is a closed subspace of $M(K)$ and has the property that when $\mu \in \mathcal{B}$, $\nu \in M(K)$, and $\nu \ll \mu$, then $\nu \in \mathcal{B}$. There is a Lebesgue decomposition theorem for bands which says that if $\mu \in M(K)$ then μ can be uniquely written as $\mu = \mu_a + \mu_s$ where $\mu_a \in \mathcal{B}$ and μ_s is singular to every element of $\mathcal{B}$. If $\mathcal{B}$ is a band the *complementary band* $\mathcal{B}'$ of $\mathcal{B}$ is the collection of measures singular to every measure in $\mathcal{B}$. It follows from the Lebesgue decomposition that $M(K) = \mathcal{B} \oplus_{l^1} \mathcal{B}'$. It is well known that if $\mathcal{B}$ is a band of measures then there exists some measure space (Ω, Σ, μ) such that $\mathcal{B}$ is isomorphic to $L^1(\mu)$.

If $\mathcal{B}$ is a band we define $L^\infty(\mathcal{B})$ to be the space of uniformly bounded families of functions $F = \{F_\nu\}_{\nu \in \mathcal{B}}$ where $F_\nu \in L^\infty(\nu)$ and $F_\nu = F_\mu$ a.e. with respect to $d\nu$ whenever we have $\nu \ll \mu$. The norm in $L^\infty(\mathcal{B})$ is given by $\|F\| = \sup_{\nu \in \mathcal{B}} \|F_\nu\|_{L^\infty(\nu)}$. The pairing $\langle \nu, F \rangle = \int F_\nu \, d\nu$ for $\nu \in \mathcal{B}$ and $F \in L^\infty(\mathcal{B})$ defines an isometric isomorphism between $L^\infty(\mathcal{B})$ and $\mathcal{B}^*$. If A is a uniform algebra on K we define $H^\infty(\mathcal{B})$ and $H^\infty(\mu)$ to be the weak-star closure of A in

$L^\infty(\mathcal{B})$ and $L^\infty(\mu)$ respectively. If $\mu \in \mathcal{B}$ there is a natural "projection"

$$H^\infty(\mathcal{B}) \xrightarrow{\tau} H^\infty(\mu)$$

defined by $F \mapsto F_\mu$. Since $H^\infty(m)$ is not identified with a subspace of $H^\infty(\mathcal{B})$ the map τ is not a projection in the usual sense. However, A is a closed subspace of $H^\infty(\mathcal{B})$ and we have $\tau(f) = f$ for every $f \in A$. The map τ is the adjoint of the natural injection

$$\frac{L^1(\mu)}{L^1(\mu) \cap A^\perp} \xrightarrow{\sigma} \frac{\mathcal{B}}{\mathcal{B} \cap A^\perp} \qquad (4\text{-}1)$$

defined by

$$f \, d\mu + L^1(\mu)/L^1(\mu) \cap A^\perp \mapsto f \, d\mu + A^\perp.$$

It is easy to see that the intersection of an arbitrary collection of bands is a band. If $\mathcal{C}$ is an arbitrary subset of $M(K)$ we define the *band generated by* $\mathcal{C}$ to be the smallest band containing $\mathcal{C}$. If $\mu \in M(K)$ then we identify the space $L^1(\mu)$ with the band of measures absolutely continuous with respect to μ. If K is a metric space and $\mathcal{C}$ is a separable subset of $M(K)$ then the band $\mathcal{B}$ generated by $\mathcal{C}$ is separable and a band $\mathcal{B}$ will be separable if and only if there exists some measure $\mu \in M(K)$ such that $\mathcal{B} = L^1(\mu)$.

If A is a uniform algebra on K we define $\mathcal{B}_{A^\perp}$ to be the band generated by the measures in $A^\perp$ and $\mathcal{S}$ to be the band complement of $\mathcal{B}_{A^\perp}$. It follows from the Lebesgue decomposition that

$$A^* \cong \frac{\mathcal{B}_{A^\perp}}{A^\perp} \oplus_{l^1} \mathcal{S}$$

and

$$A^{**} \cong H^\infty(\mathcal{B}_{A^\perp}) \oplus_{l^\infty} L^\infty(\mathcal{S}),$$

where the above isomorphisms are isometries.

We say a band $\mathcal{B}$ is a *reducing band* for A if for any measure $\nu \in A^\perp$ the projection ν_a of ν into $\mathcal{B}$ by the Lebesgue decomposition is also in $A^\perp$. We say $\mathcal{B}$ is a *minimal reducing band* if $\mathcal{B} \neq \{0\}$ while $\{0\}$ is the only reducing band properly contained in $\mathcal{B}$. It is easy to see that the intersection of two reducing bands is a reducing band. Therefore, any two minimal reducing bands are either identical or singular.

If A is a uniform algebra we denote the maximal ideal space of A by $\mathcal{M}_A$. The following version of the abstract F. and M. Riesz Theorem can be found in [Cole and Gamelin 1982].

THEOREM 4.1. *Let A be a uniform algebra and let $\varphi \in \mathcal{M}_A$. Then the band generated by the representing measures for φ is a minimal reducing band.*

We say a point $z \in K$ is a *peak point* for A if there exists an element $f \in A$ such that $f(z) = 1$ and $|f(w)| < 1$ for $w \neq z$. We say z is a *generalized peak point* if the only complex representing measure for z is the point mass at z. The *Choquet boundary* of A is the collection of all generalized peak points.

Now suppose $\mathcal{B}$ is a minimal reducing band such that $\mathcal{B} \subseteq \mathcal{S}$, the singular band to $\mathcal{B}_{A^\perp}$. Then every subband of $\mathcal{B}$ is reducing. By the minimality of $\mathcal{B}$ it can be seen that this implies $\mathcal{B}$ is all multiples of a point mass δ_z at some point $z \in K$. Theorem 4.1 now implies that z is a generalized peak point. Conversely, if z is a generalized peak point then it can be seen that the point mass δ_z at z lies in $\mathcal{S}$ and therefore all multiples of δ_z form a minimal reducing band contained in $\mathcal{S}$. We call these reducing bands *trivial minimal reducing bands* and the others *nontrivial minimal reducing bands*. Note that a minimal reducing band $\mathcal{B}$ is trivial if and only if $\mathcal{B} \cap A^\perp = 0$. Furthermore, since the intersection of two reducing bands is a reducing band, $\mathcal{B}$ is non-trivial if and only if $\mathcal{B} \subseteq \mathcal{B}_{A^\perp}$.

Let $\varphi \in \mathcal{M}_A$ and let $\mathcal{B}_\varphi$ be the band generated by the representing measures for φ. The *Gleason part* of φ is the collection of elements $\psi \in \mathcal{M}_A$ such that ψ has a representing measure in $\mathcal{B}_\varphi$. This is equivalent to saying that $\|\psi - \varphi\|_{A^*} < 2$. If φ and ψ lie in the same Gleason part then $\mathcal{B}_\varphi = \mathcal{B}_\psi$, otherwise $\mathcal{B}_\varphi$ and $\mathcal{B}_\psi$ are singular. We say a Gleason part is a *trivial Gleason part* if it corresponds to a point on the Choquet boundary and a *non-trivial Gleason part* otherwise. Note that a trivial Gleason part is a one-point part, but that there may be some one-point parts which are non-trivial. (This is not standard; usually a Gleason part is called trivial if it simply consists of one point. We therefore have more non-trivial Gleason parts than usual.)

Let z be a point in K and let $\varphi_z \in A^*$ be the point-evaluation at z. Let $\mathcal{B}_z$ be the minimal reducing band generated by the representing measures for z. If z lies off the Choquet boundary then $\mathcal{B}_z$ is non-trivial and so $\mathcal{B}_z \subseteq \mathcal{B}_{A^\perp}$. Therefore, every representing measure for z lies in $\mathcal{B}_{A^\perp}$ and we have $\varphi_z \in \mathcal{B}_{A^\perp}/A^\perp$. Similarly, if z lies on the Choquet boundary then $\varphi_z \in \mathcal{S}$.

If we let $\{\mathcal{B}_\alpha\}$ be the collection of all the non-trivial minimal reducing bands then $\bigoplus_{l^1} \mathcal{B}_\alpha$ is a reducing band contained in $\mathcal{B}_{A^\perp}$. However, this may not be all of $\mathcal{B}_{A^\perp}$. For more information, see [Cole and Gamelin 1982]. The sum $\bigoplus_{l^1} \mathcal{B}_\alpha/\mathcal{B}_\alpha \cap A^\perp$ is now isometric to a closed subspace of A^* which is contained in $\mathcal{B}_{A^\perp}/A^\perp$.

5. Pointwise Bounded Approximation and the Space $\mathcal{B}_{A^\perp}/A^\perp$

Given a uniform algebra A, the space $\mathcal{B}_{A^\perp}/A^\perp$ can be a useful object to study. It controls the uniform algebra in certain ways. For example, representing measures for points off the Choquet boundary lie in $\mathcal{B}_{A^\perp}$, and therefore their corresponding point evaluations lie in $\mathcal{B}_{A^\perp}/A^\perp$. Furthermore, since the dual of $\mathcal{B}_{A^\perp}/A^\perp$ is isometrically isomorphic to $H^\infty(\mathcal{B}_{A^\perp})$, and A is identified isometrically with a subspace of $H^\infty(\mathcal{B}_{A^\perp})$, $\mathcal{B}_{A^\perp}/A^\perp$ is a norming set for A.

It is proved in [Saccone 1997] that when A is a tight uniform algebra on a compact metric space K then $\mathcal{B}_{A^\perp}/A^\perp$ is separable. This separability property gives us even further control over the uniform algebra A. For example, it follows immediately that A has at most countably many non-trivial Gleason parts since if φ and ψ are elements of two different non-trivial Gleason parts, then they both lie in $\mathcal{B}_{A\perp}/A^{\perp}$ and $\|\varphi - \psi\|_{A^*} = 2$. The countability of the non-trivial Gleason parts can also be deduced from the fact that every such part corresponds to a distinct nontrivial minimal reducing band. If $\mathcal{B}_{A\perp}/A^{\perp}$ is separable then from the comments at the end of the last section we see that there can be at most countably many nontrivial minimal reducing bands.

This separability property can be used to construct special measures for the uniform algebra. We illustrate with the following result of A.M. Davie. Let K be a compact planar set and let $A = R(K)$. Let Q be the set of non-peak points for $R(K)$ and let λ_Q be Lebesgue measure restricted to Q. Davie's result [1972] (see also [Conway 1991]) is that if $f \in H^{\infty}(\lambda_Q)$ then there exists a sequence $\{f_n\}$ in A with $\|f_n\| \leq \|f\|$ that converges to f pointwise a.e. with respect to λ_Q. Such a phenomenon is sometimes referred to as pointwise bounded approximation with a reduction in norm. It is easy to see that there exists such a sequence converging pointwise to f, the hard part is to find a sequence that is bounded. It can be deduced from Davie's result, without much difficulty, that every point in Q has a representing measure absolutely continuous with respect to λ_Q.

We will now study Davie's result by considering the space $H^{\infty}(\mathcal{B}_{A\perp})$. The following proposition is not difficult to prove and can be found in [Saccone 1997]. Recall that a linear operator $T : X \to Y$ is a quotient map if the induced injection $S : X/Z \to Y$, where $Z = \ker T$, is an isometry.

PROPOSITION 5.1. *Let A be a uniform algebra on a compact space K and let $m \in \mathcal{B}_{A\perp}$. The following statements are equivalent:*

(a) *For every $f \in H^{\infty}(m)$ there exists a sequence $\{f_n\}$ in A with $\|f_n\| \leq \|f\|$ such that $f_n \longrightarrow f$ pointwise a.e. with respect to m.*

(b) *The natural projection $H^{\infty}(\mathcal{B}_{A\perp}) \overset{\tau}{\longrightarrow} H^{\infty}(m)$ is a quotient map.*

If A is a uniform algebra on a compact space K and $m \in \mathcal{B}_{A\perp}$ we say m is an *ordinary Davie measure* if τ is a quotient map and m is a *strong Davie measure* if τ is an isometry. In general, a linear operator between Banach spaces is an isometric embedding if and only if its dual is a quotient map. Therefore m is an ordinary Davie measure if and only if the map σ in (4–1) (where $\mathcal{B}$ should be taken to be $\mathcal{B}_{A\perp}$) is an isometric embedding and is a strong Davie measure if and only if σ is a surjective isometry. Since τ is an algebra homomorphism between uniform algebras, τ will be an isometry as soon as it is an isomorphism. Since σ is always injective it follows that m is a strong Davie measure if and only if σ is onto. Since the evaluations for the points off the Choquet boundary lie in $\mathcal{B}_{A\perp}/A^{\perp}$, it follows easily that when m is a strong Davie measure then every point off the Choquet boundary has a representing measure absolutely continuous with respect to m.

Interestingly enough, the injectivity of τ is closely related to Bourgain's rich measures, as the next proposition shows. If $m \in M(K)$ let $m = m_a + m_s$ be the Lebesgue decomposition of m with respect to $\mathcal{B}_{A\perp}$.

PROPOSITION 5.2 [Saccone 1995a]. *Let A be a uniform algebra on a compact space K and let m be an element of $M(K)$. Then the following statements are equivalent*:

(a) *The natural projection $H^\infty(\mathcal{B}_{A^\perp}) \xrightarrow{\tau} H^\infty(m_a)$ is one-to-one.*

(b) *If $\{f_n\}$ is a bounded sequence in A such that $\int |f_n|\,d|m| \longrightarrow 0$ then $f_n \xrightarrow{w^*} 0$ in $L^\infty(\mu)$ for every $\mu \in A^\perp$.*

(c) *m is a weakly rich measure for A.*

If these conditions hold, and K is metrizable, then $\mathcal{B}_{A^\perp}/A^\perp$ is separable.

We now have the following result from [Saccone 1997]. The fact that (b) holds for tight uniform algebras appeared in [Cole and Gamelin 1982], although the proof in [Saccone 1997] is more elementary.

THEOREM 5.3. *Let A be a tight uniform algebra on a compact metric space K. Then the following hold.*

(a) *$\mathcal{B}_{A^\perp}/A^\perp$ is separable.*

(b) *A has at most countably many non-trivial Gleason parts and at most countably many nontrivial minimal reducing bands.*

(c) *A has a strong Davie measure m. In particular, every non-peak point for A has a representing measure absolutely continuous with respect to m.*

Part (b) follows from a more general result which is proved in [Saccone 1997]. We say a Banach space X is a *separable distortion of an L^1-space* if $X = M \oplus_{l^1} L^1(\mu)$ where M is separable and μ is some measure. Since every band is isomorphic to $L^1(\mu)$ for some μ, A^* will be isomorphic to a separable distortion of an L^1-space whenever $\mathcal{B}_{A^\perp}/A^\perp$ is separable. The next result from [Saccone 1997] now generalizes the observation implicit in part (b).

THEOREM 5.4. *Let A be a uniform algebra and suppose A^* is isomorphic to a closed subspace of a separable distortion of an L^1-space. Then A has at most countably many non-trivial minimal reducing bands and therefore at most countably many non-trivial Gleason parts.*

6. Tightness Versus Strong Tightness

The following problem is open: does there exist a tight uniform algebra which fails to be strongly tight? We need not ask the question of tight subspaces, for if we let $X = l^2$ then X is tight (in any $C(K)$-space). However, from Theorem 2.7 we know that all strongly tight subspaces have the Dunford–Pettis property, a property which l^2 clearly fails to have. Thus l^2 is not strongly tight in any $C(K)$-space.

This problem has been studied in [Carne et al. 1989; Jaramillo and Prieto 1993]. The first of these papers dealt with the following uniform algebra. Let X be a Banach space and let A be the uniform algebra on B_{X^*} (with the weak-star topology) generated by X. Then A consists of analytic functions of possibly infinitely many variables. Recall the result of Milne which states that X is complemented in A (by a norm-one projection in fact). It now follows from Theorem 3.3 that X must have the Pełczyński property if A is to be tight. However, a much stronger result is proved in [Carne et al. 1989], namely that if A is tight then X is reflexive.

We now claim that A is strongly tight if and only if X is finite-dimensional (this result was also proved in [Carne et al. 1989] by a more direct means). Since the ball-algebras are known to be strongly tight (proved in [Cole and Gamelin 1982], for example), we need only prove necessity. If A is strongly tight then A has the Dunford–Pettis property by Corollary 2.3 and therefore X has the Dunford–Pettis property since it is complemented in A. However, by the result in [Carne et al. 1989] , X must also be reflexive, from which it follows that X must be finite-dimensional.

It was shown in [Carne et al. 1989] that if $X = l^2$ then A fails to be tight. However, it is still unknown if there exists an infinite-dimensional (reflexive) space X such that A is tight. In [Jaramillo and Prieto 1993], some strong versions of reflexivity were studied, but no example was produced.

Another example that is not well understood is the space $A(D)$ when D is a bounded domain in $\mathbb{C}^n$. It is known that $A(D)$ is strongly tight whenever the $\bar{\partial}$-problem can be solved in D with Hölder estimates on the solutions; however no satisfactory necessary conditions are known, although the following is proved in [Saccone 1995b].

PROPOSITION 6.1. *Suppose D is a bounded domain in $\mathbb{C}^n$ and let $A = A(D)$. Then the following statements are equivalent*:

(a) *A is strongly tight on $\bar{D}$.*
(b) *A has the property that when $\{f_n\}$ is a bounded sequence in D that tends to zero pointwise in D then we have $\|f_n g + A\| \longrightarrow 0$ for every $g \in C(\bar{D})$.*

7. Inner Functions

Recall the Chang–Marshall Theorem, which states that if m is Lebesgue measure on the unit circle then every closed subalgebra of $L^\infty(m)$ which contains H^∞ is generated by H^∞ and a collection of conjugates of inner functions. The following result shows how this phenomenon breaks down in higher dimensions.

THEOREM 7.1. *Let D be a strictly pseudoconvex domain with C^2 boundary in $\mathbb{C}^n$. Suppose f is an inner function in $H^\infty(\partial D)$. If $f(z_n) \longrightarrow 0$ for some sequence $\{z_n\}$ tending to ∂D then $\bar{f} \notin H^\infty + C$. In particular, if $n > 1$ then $\bar{f} \in H^\infty + C$ if and only if f is constant. If D is the unit disk then $\bar{f} \in H^\infty + C$ if and only if f is a finite Blaschke product.*

The proof is indirect and uses such tools as the Pełczyński property and tight uniform algebras. Given a function $g \in L^\infty(m)$, we define

$$S_{g,H^\infty} : H^\infty(m) \longrightarrow L^\infty(m)/H^\infty(m)$$

by $f \mapsto fg + H^\infty(m)$. It is shown that when $g \in H^\infty + C$ then S_{g,H^∞} is compact and if f is an inner function that tends to zero towards the boundary then $S_{\bar{f},H^\infty}$ is an isomorphism on a copy of c_0 in H^∞.

References

[Bessaga and Pełczyński 1958]   C. Bessaga and A. Pełczyński, "On bases and unconditional convergence of series in Banach spaces", *Studia Math.* **17** (1958), 151–164.

[Bourgain 1983] J. Bourgain, "On weak completeness of the dual of spaces of analytic and smooth functions", *Bull. Soc. Math. Belg. Sér. B* **35**:1 (1983), 111–118.

[Bourgain 1984a] J. Bourgain, "The Dunford–Pettis property for the ball-algebras, the polydisc-algebras and the Sobolev spaces", *Studia Math.* **77**:3 (1984), 245–253.

[Bourgain 1984b] J. Bourgain, "New Banach space properties of the disc algebra and H^∞", *Acta Math.* **152**:1-2 (1984), 1–48.

[Bourgain and Delbaen 1980] J. Bourgain and F. Delbaen, "A class of special $\mathcal{L}_\infty$ spaces", *Acta Math.* **145**:3-4 (1980), 155–176.

[Carne et al. 1989] T. K. Carne, B. Cole, and T. W. Gamelin, "A uniform algebra of analytic functions on a Banach space", *Trans. Amer. Math. Soc.* **314**:2 (1989), 639–659.

[Cima and Timoney 1987] J. A. Cima and R. M. Timoney, "The Dunford–Pettis property for certain planar uniform algebras", *Michigan Math. J.* **34**:1 (1987), 99–104.

[Cole and Gamelin 1982] B. J. Cole and T. W. Gamelin, "Tight uniform algebras and algebras of analytic functions", *J. Funct. Anal.* **46**:2 (1982), 158–220.

[Conway 1991] J. B. Conway, *The theory of subnormal operators*, Mathematical Surveys and Monographs **36**, Amer. Math. Soc., Providence, 1991.

[Davie 1972] A. M. Davie, "Bounded limits of analytic functions", *Proc. Amer. Math. Soc.* **32** (1972), 127–133.

[Davis et al. 1974] W. J. Davis, T. Figiel, W. B. Johnson, and A. Pełczyński, "Factoring weakly compact operators", *J. Functional Analysis* **17** (1974), 311–327.

[Delbaen 1977] F. Delbaen, "Weakly compact operators on the disc algebra", *J. Algebra* **45**:2 (1977), 284–294.

[Delbaen 1979] F. Delbaen, "The Pełczyński property for some uniform algebras", *Studia Math.* **64**:2 (1979), 117–125.

[Diestel 1980] J. Diestel, "A survey of results related to the Dunford–Pettis property", pp. 15–60 in *Proceedings of the Conference on Integration, Topology, and Geometry in Linear Spaces* (Chapel Hill, NC, 1979), edited by W. H. Graves, Contemp. Math. **2**, Amer. Math. Soc., Providence, RI, 1980.

[Diestel 1984] J. Diestel, *Sequences and series in Banach spaces*, Graduate Texts in Mathematics **92**, Springer, New York, 1984.

[Dunford and Pettis 1940] N. Dunford and B. J. Pettis, "Linear operations on summable functions", *Trans. Amer. Math. Soc.* **47** (1940), 323–392.

[Dunford and Schwartz 1958] N. Dunford and J. T. Schwartz, *Linear Operators, I: General Theory*, Pure and Applied Mathematics **7**, Interscience, New York, 1958. Reprinted by Wiley, New York, 1988.

[Gamelin 1969] T. W. Gamelin, *Uniform algebras*, Prentice-Hall, Englewood Cliffs, NJ, 1969. Second edition, Chelsea, New York, 1984.

[Garnett 1981] J. B. Garnett, *Bounded analytic functions*, Pure and Applied Mathematics **96**, Academic Press, New York, 1981.

[Grothendieck 1953] A. Grothendieck, "Sur les applications linéaires faiblement compactes d'espaces du type $C(K)$", *Canadian J. Math.* **5** (1953), 129–173.

[Henkin 1968] G. M. Henkin, "The Banach spaces of analytic functions in a ball and in a bicylinder are nonisomorphic", *Funkcional. Anal. i Priložen.* **2**:4 (1968), 82–91. In Russian; translation in *Functional Anal. Appl.* **2**:4 (1968), 334–341.

[James 1950] R. C. James, "Bases and reflexivity of Banach spaces", *Ann. of Math.* (2) **52** (1950), 518–527.

[Jaramillo and Prieto 1993] J. A. Jaramillo and A. Prieto, "Weak-polynomial convergence on a Banach space", *Proc. Amer. Math. Soc.* **118**:2 (1993), 463–468.

[Kisljakov 1975] S. V. Kisljakov, "On the conditions of Dunford–Pettis, Pełczyński and Grothendieck", *Dokl. Akad. Nauk SSSR* **225**:6 (1975), 1252–1255. In Russian; translation in *Soviet Math. Dokl.* **16** (1975), 1616–1621.

[Li and Russo 1994] S. Y. Li and B. Russo, "The Dunford-Pettis property for some function algebras in several complex variables", *J. London Math. Soc.* (*2*) **50**:2 (1994), 392–403.

[Milne 1972] H. Milne, "Banach space properties of uniform algebras", *Bull. London Math. Soc.* **4** (1972), 323–326.

[Mooney 1972] M. C. Mooney, "A theorem on bounded analytic functions", *Pacific J. Math.* **43** (1972), 457–463.

[Orlicz 1929] W. Orlicz, "Beiträge zur Theorie der Orthogonalentwicklungen, II", *Studia Math.* **11** (1929), 241–255.

[Pełczyński 1962] A. Pełczyński, "Banach spaces on which every unconditionally converging operator is weakly compact", *Bull. Acad. Polon. Sci. Sér. Sci. Math. Astronom. Phys.* **10** (1962), 641–648.

[Pełczyński 1977] A. Pełczyński, *Banach spaces of analytic functions and absolutely summing operators* (Kent, OH, 1976), CBMS Regional Conference Series in Mathematics **30**, Amer. Math. Soc., Providence, RI, 1977.

[Saccone 1995a] S. F. Saccone, "Banach space properties of strongly tight uniform algebras", *Studia Math.* **114**:2 (1995), 159–180.

[Saccone 1995b] S. F. Saccone, *A study of strongly tight uniform algebras*, Ph.D. thesis, Brown University, 1995. Available at http://www.math.wustl.edu/~sfs/papers/thesis.pdf.

[Saccone 1997] S. F. Saccone, "The Pełczyński property for tight subspaces", *J. Funct. Anal.* **148**:1 (1997), 86–116.

[Wojtaszczyk 1979] P. Wojtaszczyk, "On weakly compact operators from some uniform algebras", *Studia Math.* **64**:2 (1979), 105–116.

[Wojtaszczyk 1991] P. Wojtaszczyk, *Banach spaces for analysts*, Cambridge Studies in Advanced Mathematics **25**, Cambridge University Press, Cambridge, 1991.

SCOTT SACCONE
DEPARTMENT OF MATHEMATICS
CAMPUS BOX 1146
WASHINGTON UNIVERSITY
ST. LOUIS, MO 63130-4899
UNITED STATES
  sfs@math.wustl.edu

Holomorphic Spaces
MSRI Publications
Volume **33**, 1998

Higher-Order Hankel Forms and Commutators

RICHARD ROCHBERG

ABSTRACT. We discuss the algebraic structure of the spaces of higher-order Hankel forms and of the spaces of higher-order commutators. In both cases we find a close relationship between the space of order $n + 1$ and the derivations of the underlying algebra of functions into the space of order n.

1. Introduction and Summary

Let $\mathcal{H}^2$ be the Hardy space of the unit circle, Γ; that is, $\mathcal{H}^2$ is the space of functions, $f = \sum a_n z^n$, holomorphic on the unit disk for which

$$\|f\| = \left(\sum |a_n|^2 \right)^{1/2} < \infty.$$

Such an f will be identified with its boundary values on Γ. A *Hankel form* on $\mathcal{H}^2$ is a bilinear map $B = B_b : \mathcal{H}^2 \times \mathcal{H}^2 \to \mathbb{C}$ that has the characteristic form

$$B(f, g) = \int_\Gamma f g \bar{b}. \tag{1–1}$$

Here b is the (boundary value) of a holomorphic function, the *symbol function* of B. In terms of Taylor coefficients,

$$B(f, g) = \sum_{n,k \geq 0} \hat{f}(n) \hat{g}(k) \overline{\hat{b}(n + k)}.$$

When B is viewed as acting on functions, its characteristic property is that its value only depends on the product fg. When viewed as acting on the coefficients, the characteristic property of B is that its matrix,

$$\{\beta_{n,k}\} = \{\overline{\hat{b}(n + k)}\},$$

is a Hankel matrix; that is, a matrix on $\mathbb{Z}^+ \times \mathbb{Z}^+$ whose entries only depend on the sum of the indices. The associated Hankel operator is the linear map, $\mathbf{B}$, from $\mathcal{H}^2$ to its linear dual space which takes f to the linear functional $\mathbf{B}(f)(\cdot) = B(f, \cdot)$. The analytic properties of these forms and the associated operators have been

studied extensively, with much attention given to the relationship between the
properties of **B** and B, and those of b.

The idea of bilinear forms given by a representation such as (1–1), and thus
only depending on the product of the arguments, can certainly be extended
to other function spaces. One investigation of those more general forms is in
[Janson et al. 1987]. In the more general contexts operators based on expressions
such as (1–1) are sometimes called *small Hankel operators*. There is another
generalization, the *large Hankel operators*; the two types agree for the Hardy
space.

Recently there has also been consideration of more general classes, the Hankel
forms of higher type or order. For each nonnegative integer n there is a class,
H^n, of Hankel forms of type n. The elements of H^1 are the traditional Hankel
forms, and $H^n \subset H^{n+1}$ for each n.

An example of a Hankel form of type 2 on the Hardy space is

$$E(f,g) = \int_\Gamma f'g\bar{b}. \qquad (1\text{–}2)$$

The characteristic property of such a form is that for any polynomial, p, the new
bilinear form $C_p(f,g) = E(pf,g) - E(f,pg)$ is a Hankel form. On the coefficient
side the matrix is of the form

$$\{e_{n,k}\} = \{n\overline{\hat{b}(n+k)}\}.$$

Thus such forms are obtained by perturbing Hankel forms in a controlled way.
Higher-order forms were introduced in [Janson and Peetre 1987]. The point
of view there was that certain Lie groups have irreducible representations on
the Hardy and Bergman spaces. The representations on the function spaces
induce representations on the associated spaces of bilinear forms. However those
representations are not irreducible. An analysis shows that the Hankel forms are
the simplest irreducible component of the induced representation. The higher-
order Hankel forms are, by definition, the other irreducible components. Thus
the space of Hilbert–Schmidt bilinear forms on, say, $\mathcal{H}^2$ can be decomposed as
$\bigoplus_n (H^n \ominus H^{n-1})$. In [Janson and Peetre 1987] the basic analytic properties of
these forms (boundedness, Schatten ideal membership, and so on) are worked
out using a mixture of harmonic analysis and representation theory. That point
of view has been taken to other contexts; see [Rosengren 1996] and the references
there.

It is also possible to develop a theory of higher-order Hankel forms in the absence of a group action. That is done for general spaces of holomorphic functions
in [Peetre and Rochberg 1995], but those analytical results are in a much more
primitive state than the results for forms on the Hardy and Bergman spaces.
Here we continue to study higher-order forms, concentrating now on the algebraic structure of the spaces H^n. Even in the simplest case of the Hardy space

the analytical theory of the higher-order Hankel forms is a bit recalcitrant. However there are algebraic relations between the higher-order forms and the classical forms. It may be that the algebraic structure can be used to carry the analytical results from the classical forms to the higher-order forms. The only instance of this so far is a Kronecker theorem (Theorem 2.12 below), which perhaps doesn't really qualify as an analytical result. A more intriguing possibility of doing this is discussed in Remark 4.4 (page 176). Also, there appears to be a rich, but not well understood, relation between algebraic aspects of the theory of higher-order Hankel forms and algebraic aspects of the theory of higher-order commutators. This is particularly intriguing because the commutators considered need not be linear. Here we present analogous algebraic results for the two topics side by side (in Sections 2 and 3) as a step in developing this relation.

The results for Hankel forms are in Section 2. We will look at bilinear forms on a space, K, of holomorphic functions. We show that, roughly, for each n there is a natural identification of the quotient space H^{n+1}/H^n with the space of derivations mapping K into H^n/H^{n-1}. By iterating this result we find that, roughly, all elements H^n are built from elements of H^1 and differential operators of order at most $n-1$. This gives a new proof of the Kronecker theorem characterizing higher-order Hankel forms of finite rank which was first proved in [Rochberg 1995]. We will also be able to extend that theorem to new contexts.

In Section 3 we discuss commutators. A basic example involves the nonlinear operator Λ defined densely on $L^2(\mathbb{R})$ by $\Lambda f = f \ln |f|$. Let P be the Cauchy–Szegő projection, the orthogonal projection of $L^2(\mathbb{R})$ to $\mathcal{H}^2(\mathbb{R})$, and for any $b \in L^\infty(\mathbb{R})$ denote by M_b the operator of pointwise multiplication by b. It is easy to see that the nonlinear operator $[M_b, \Lambda] = M_b \Lambda - \Lambda M_b$ is bounded on $L^2(\mathbb{R})$. Less obvious, but also true, is that the operator $[P, \Lambda]$ is bounded on $L^2(\mathbb{R})$. This commutator is, in a sense discussed a bit more fully in Section 3, a nonlinear analog of a Hankel operator acting on $\mathcal{H}^2(\mathbb{R})$. In this analogy the boundedness of $[P, \Lambda]$ is analogous to the basic boundedness result for Hankel operators on $\mathcal{H}^2(\mathbb{R})$. Recently there has also been consideration of higher-order commutators such as $[M_b, [M_b, \Lambda]]$, $[M_b, [M_b, [M_b, \Lambda]]]$, etc., which we also consider here. These operators arise in the study of the internal structure of interpolation theory but they also have applications to classical analysis; see for instance [Cwikel et al. 1989; Iwaniec 1995; Pérez 1996]. In Section 3 we show that many of the algebraic results of Section 2 for higher-order Hankel forms have analogs for higher-order commutators. In my opinion the main conclusion of that section is not those relatively straightforward algebraic observations, rather it is the evidence of a possible systematic relation between higher-order Hankel forms and higher-order commutators. Another reason for interest in the algebraic structure of commutators is in the hope of extracting analytical information. The theory in [Milman and Rochberg 1995] proves, for instance, the boundedness on $L^2(\mathbb{R})$

of an operator whose main term is

$$R = [P, \Lambda^2] - \Lambda[P, \Lambda]. \tag{1-3}$$

This is a *higher-order* extension of the boundedness of $[P, \Lambda]$. Estimates on operators such as R have been less useful in analysis than the more elementary boundedness result for $[P, \Lambda]$. The interaction of nonlinearities makes it quite difficult to extract analytical information from estimates on R. It may be that the algebraic viewpoint can help.

The last section contains some brief further comments.

For a broader view of the topics of higher-order Hankel forms and higher-order commutators as well as for further references we refer to [Peetre and Rochberg 1995; Milman and Rochberg 1995]. For more recent work see [Rosengren 1996; Cwikel et al. $\geq$ 1997; Carro et al. 1995a; 1995b] and the references listed there.

2. Higher-Order Hankel Forms

Background and Notation. Let K be a Hilbert space of holomorphic functions defined on some domain D in $\mathbb{C}^N$. We assume that K contains a dense subalgebra, A, of bounded functions, and that for all a in A and k in K we have the norm estimate

$$\|ak\|_K \leq \|a\|_\infty \|k\|_K. \tag{2-1}$$

The choice of the Bergman space of the open unit disk as K and of $\mathcal{H}^\infty$, the algebra of bounded holomorphic functions on the disk, as A is a basic example. That is, K equals

$$A^2(\mathbb{D}, dx\, dy) = \left\{ f : f \in \mathrm{Hol}(\mathbb{D}) \text{ and } \|f\|^2 = \int_\mathbb{D} |f|^2 \, dx\, dy < \infty \right\}.$$

We emphasize that we do not assume that A, or even K, contains the polynomials or even the constant function. In particular, we want to include the Bergman space of the upper half-plane as an example, namely,

$$K = A^2(\mathbb{R}_+^2, dx\, dy) = \left\{ f : f \in \mathrm{Hol}(\mathbb{R}_+^2) \text{ and } \|f\|^2 = \int_{\mathbb{R}_+^2} |f|^2 dx\, dy < \infty \right\}.$$

In this case a convenient choice for A would be the polynomials in $(z + i)^{-1}$. The Bergman spaces of the disk and the half-plane and their standard weighted variants are the type of examples we have in mind. But many of the results have analogs on, for instance, the Fock space, where there is no natural choice for A that would have the norm estimates (2–1).

Let $\mathrm{Bilin}(K)$ be the space of continuous bilinear maps from $K \times K$ to $\mathbb{C}$.

DEFINITION 2.1. For $a \in A$ and $B \in \mathrm{Bilin}(A)$, define the elements aB and Ba of $\mathrm{Bilin}(A)$ by setting, for all $f, g \in A$,

$$(aB)(f, g) = B(af, g), \qquad (Ba)(f, g) = B(f, ag).$$

DEFINITION 2.2. For each $a \in A$, define δ_a to be the map of Bilin(A) to itself given by $\delta_a B = aB - Ba$. Thus

$$\delta_a B(f, g) = B(af, g) - B(f, ag).$$

We now collect some computational properties of these maps, which follow directly from the definition.

PROPOSITION 2.3. *For all $a, b \in A$ and all $B \in$ Bilin(A) we have*:

(1) $\delta_a \delta_b B = \delta_b \delta_a B$.
(2) $\delta_{ab} B = a(\delta_b B) + (\delta_a B)b = b(\delta_a B) + (\delta_b B)a$.
(3) $\delta_{ab} B = a(\delta_b B) + b(\delta_a B) - \delta_b \delta_a B$.

We now define Hankel forms and higher-order Hankel forms. In [Peetre and Rochberg 1995] several different definitions were offered; although it was clear that they agree on the standard examples, the general story is not clear. Here we use the definitions in [Peetre and Rochberg 1995] that are based on pairs (A, K).

DEFINITION 2.4. (1) $H^0 = \{0\}$.
(2) We say $B \in$ Bilin(K) is a *Hankel form* if $\delta_a B = 0$ for all $a \in A$. We denote the collection of all such forms by H^1.
(3) For $n = 2, 3, \ldots$, we define H^n, the set of Hankel forms of *type* (or *order*) n, to be the set of all $B \in$ Bilin(K) such that $\delta_a B \in H^{n-1}$ for all $a \in A$.
(4) For $n = 1, 2, \ldots$, set $J^n = H^n / H^{n-1}$.

Using part (1) of Proposition 2.3 and induction it is easy to check that

$$B \in H^n \text{ if and only if } (\delta_a)^n B = 0 \text{ for all } a \in A.$$

Here are some examples. Let K be the Bergman space of the upper half plane, U, and let A be the bounded analytic functions on U. Let μ be a finite measure supported on V, a compact subset of U. Define the bilinear forms B and C on K by $B(f, g) = \int fg \, d\mu$ and $C(f, g) = B(f', g) = \int f'g \, d\mu$. The fact that V is a compact subset of U insures that both B and C are continuous. It is then immediate that $B \in H^1$, and it follows from the product rule for differentiation that $C \in H^2$. It is also clear how to continue and construct elements of all the H^n, and that similar constructions will work as long as the functions in K and their derivatives have good bounds on compact subsets of D. Part of the content of Theorem 2.5 is that, in some sense, this method of constructing elements of H^2 from elements of H^1 gives all of H^2 and similarly for the higher-order forms.

Modules. Fix K and A. The elements of A multiply the elements of K on the left and on the right; thus K is a *bimodule* over A. For $a \in A$ we will use L_a or juxtaposition to denote left multiplication by a acting on K. Thus, for $B \in$ Bilin(K) and $f, g \in K$, we have $(L_a B)(f, g) = (aB)(f, g) = B(af, g)$. Similarly for right multiplication: $(R_a B)(f, g) = (Ba)(f, g) = B(f, ag)$.

For any $a, b, c \in A$, the three operations δ_a, L_b, and R_c on $\mathrm{Bilin}(K)$ all commute. Using this it is easy to check that, for each n, both left and right multiplication map each element of H^n to another element of H^n, and thus that H^n is an A-bimodule. It then follows that there is an induced bimodule structure on the quotients, J^n. Here however even more is true. We observe that, given $a \in A$ and $B \in H^n$, then $L_a B - R_a B = \delta_a B \in H^{n-1}$. Thus, using the same notation for the induced maps on J^n, we see that $L_a = R_a$ as operators on J^n; that is for, $j \in J^n$ and $a \in A$, we have $aj = ja$ as elements of J^n.

For normed spaces V and W we denote by $\mathrm{Map}(V, W)$ the space of continuous linear maps from V to W.

If X is an A-bimodule, an element $D \in \mathrm{Map}(A, X)$ is called a *derivation* if $D(ab) = aD(b) + D(a)b$ for all $a, b \in A$. We denote the space of such derivations by $\mathrm{Deriv}(A, X)$. For instance, $M = \mathrm{Map}(\mathrm{Bilin}(K), \mathrm{Bilin}(K))$ is an A-bimodule if we define left and right multiplication by

$$(am)(B) = a(mB) \quad \text{and} \quad (mb)(B) = (mB)b$$

for all $a, b \in A$, $m \in M$, and $B \in \mathrm{Bilin}(K)$. If we now define a map, D, of A into M by setting $D(a) = \delta_a$, part (2) of Proposition 2.3 implies that $D \in \mathrm{Deriv}(A, M)$.

Given $B \in \mathrm{Bilin}(K)$, we define $\Delta(B) \in \mathrm{Map}(A, \mathrm{Bilin}(K))$ by

$$\Delta(B)(a) = \delta_a B.$$

Now fix $\alpha \in A$. We define a mapping ∇_α that takes $\mathrm{Map}(A, \mathrm{Bilin}(K))$ to densely defined bilinear forms by the following rule: For $\tilde{\Delta} \in \mathrm{Map}(A, \mathrm{Bilin}(K))$,

$$\nabla_\alpha(\tilde{\Delta})(f, g) = \tilde{\Delta}(f)(\alpha, g).$$

Thus $\nabla_\alpha(\tilde{\Delta})$ is defined for $f \in A$ and $g \in K$; recall that A is dense in K. If $1 \in A$, the choice $\alpha = 1$ is a natural one to consider.

The next theorem concerns the properties of these two maps.

THEOREM 2.5. *For $n = 1, 2, \ldots$ we have:*

(1) $\Delta : J^{n+1} \to \mathrm{Deriv}(A, J^n)$.

(2) $\nabla_\alpha : \mathrm{Deriv}(A, J^n) \to J^{n+1}$.

(3) $\nabla_\alpha(\Delta(B)) = \alpha B$ *for any* $B \in J^{n+1}$, *and* $\Delta(\nabla_\alpha(D)) = \alpha D$ *for any* $D \in \mathrm{Deriv}(A, J^n)$.

REMARK 2.6. We are abusing the notation slightly when we use Δ and ∇_α for induced maps. This should cause no problem.

REMARK 2.7. Informally, and most clearly for the case $\alpha = 1$, the theorem says that $J^{n+1} = \mathrm{Deriv}(A, J^n)$.

PROOF OF THEOREM 2.5. (1) We know that $\Delta(B)$ takes A to bilinear forms. Pick $B \in H^{n+1}$. By (3) in Proposition 2.3 we know that, for all $a, b \in A$,

$$\Delta(B)(ab) = \delta_{ab}B = a\delta_b B + b\delta_a B - \delta_a \delta_b B$$
$$= a(\Delta(B)b) + b(\Delta(B)a) - \delta_a \delta_b B.$$

The second line shows that $\Delta(B)(ab)$ is in H^n. Now note that $\delta_a \delta_b B \in H^{n-1}$; and hence, as a map into the quotient J^n, $\Delta(B)$ satisfies

$$\Delta(B)(ab) = a\Delta(B)(b) + b\Delta(B)(a).$$

Recall that left and right multiplication by A agree on J^n. Hence the previous equation establishes that $\Delta(B)$ is a derivation. Finally note that, if we change B by an element of H^n, the range of $\Delta(B)$ changes by elements of H^{n-1}. Hence, as an element of J^n, the image is unchanged. Thus our map is well-defined on J^{n+1}.

(2) We start with $D \in \mathrm{Deriv}(A, J^n)$. Define B by

$$B(x, y) = \nabla_\alpha(D)(x, y) = D(x)(\alpha, y).$$

Certainly B is bilinear. Pick $a \in A$. We have

$$(\delta_a B)(x, y) = B(ax, y) - B(x, ay)$$
$$= D(ax)(\alpha, y) - D(x)(\alpha, ay)$$
$$= (aD(x) + xD(a))(\alpha, y) - D(x)(\alpha, ay).$$

Here we used the fact that D is a derivation and the fact that its range is J^n, a module on which left and right multiplication by A agree. We continue with

$$(\delta_a B)(x, y) = D(x)(a\alpha, y) + D(a)(\alpha x, y) - D(x)(\alpha, ay)$$
$$= \delta_a(D(x))(\alpha, y) + D(a)(\alpha x, y).$$

We want to show $(\delta_a B) \in H^n$. $D(a)$ is in H^n, hence so is the mapping from (x, y) to $D(a)(\alpha x, y)$. To finish we need to shown that C, defined by $C(x, y) = \delta_a(D(x))(\alpha, y)$, is in H^{n-1}. We do this by induction on n. First note that if $n = 1$ then $D(x)$ is in H^1, hence $\delta_a(D(x)) \equiv 0$ and we are fine. Suppose now that we are fine up to index $n - 1$. It is direct to check that $D \in \mathrm{Deriv}(A, J^n)$ implies that the map $\delta_a \circ D$ taking x to $\delta_a(D(x))$ is in $\mathrm{Deriv}(A, J^{n-1})$. Hence, by the computations in the proof of the case $n - 1$ of the theorem, we know that $\nabla_\alpha(\delta_a \circ D) \in H^{n-1}$. Unwinding the definition of ∇_α we find that this is what we needed. Finally note that, if the choice of the representative of $D(x)$ in J^n is changed, then $\delta_a(D(x))$ changes by an element of H^{n-1}, so we get the same image of $\nabla_\alpha D$ in J^{n+1}; that is, the choice of representative doesn't change the outcome of the computation.

(3) Suppose $B \in J^{n+1}$. We have

$$\nabla_\alpha(\Delta(B))(x, y) = \Delta(B)(x)(\alpha, y) = (\delta_x B)(\alpha, y) = B(\alpha x, y) - B(\alpha, xy).$$

The bilinear form that takes (x, y) to $B(\alpha, xy)$ is in H^1. Hence the previous computation shows that $\nabla_\alpha(\Delta(B)) = \alpha B$ as maps into J^{n+1}. As before, note that the outcome doesn't depend on choices made for the representative of $\Delta(B)(x)$.

Now select $D \in \mathrm{Deriv}(A, J^n)$, $a \in A$, and $x, y \in K$. We have

$$\Delta(\nabla_\alpha D)(a)(x, y) = \delta_a(\nabla_\alpha D)(x, y) = (\nabla_\alpha D)(ax, y) - (\nabla_\alpha D)(ax, y)$$
$$= (aD(x) + xD(a))(\alpha, y) - D(x)(\alpha, ay)$$
$$= D(x)(\alpha a, y) + D(a)(\alpha x, y) - D(x)(\alpha, ay)$$
$$= (\delta_a D(x))(\alpha, y) + \alpha D(a)(x, y).$$

The second term on the right is exactly what we wanted. We are working with derivations into J^n, so we are done if we show that the bilinear map of (x, y) to $(\delta_a D(x))(\alpha, y)$ is a map into H^{n-1}. As before, this follows by induction. If $n = 1$ then $\delta_a D(x)$ is the zero functional. Then we proceed as in the end of the proof of part (2), to see that $\nabla_\alpha(\delta_a \circ D) \in H^{n-1}$. $\square$

We now refine these calculations to develop structure theorems. When doing this we make further assumptions: that K is a space of functions of one variable and that the polynomials, P, are contained in A. The first assumption is for notational convenience. We return to the second later in the section.

THEOREM 2.8. *Suppose $P \subset A$ and $n \geq 1$. If $B \in H^n$ then*

$$B(p, g) = B(1, pg) + \sum_{j=1}^{n-1} \frac{(-1)^{j+1}}{j!}(\delta_z^n B)(p^{(j)}, g). \tag{2-2}$$

for all $p \in P$ and $g \in K$.

PROOF. First we develop a combinatoric formula for δ_f, where $f \in P$. We apply Proposition 2.3 (3) to the function z^2 and obtain $\delta_{z^2} = 2z\delta_z + \delta_z^2$. This is our starting point for an inductive proof that

$$\delta_f = \sum_{n=1}^{\infty} \frac{(-1)^{n+1}}{n!} f^{(n)} \delta_z^n. \tag{2-3}$$

Repeated application of Proposition 2.3 (3) clearly gives

$$\delta_{f(z)} = \sum_{n=1}^{\infty} \Lambda_n(f)\delta_z^n$$

for some linear operators Λ_n. By Proposition 2.3 (3) we have

$$\delta_{zf} = z\delta_f + f\delta_z - \delta_z \delta_f; \tag{2-4}$$

this insures that

$$\Lambda_1(zf) = z\Lambda_1(f) + f. \tag{2-5}$$

We already observed that $\Lambda_1(z^2) = 2z$. The required formula for Λ_1 for the remaining monomials follows from (2–5) by induction on degree and then for all of P by linearity. Equation (2–4) also implies that

$$\Lambda_n(zf) = z\Lambda_n(f) - \Lambda_{n-1}(f)$$

for $n > 1$. We now proceed by induction on n. Suppose the formula for Λ_{n-1} is established. Thus

$$\Lambda_n(zf) = z\Lambda_n(f) + \frac{(-1)^n}{(n-1)!} f^{(n-1)}. \tag{2–6}$$

It follows easily from (2–4) that $\Lambda_n(z^k) = 0$ for $k < n$ and that $\Lambda_n(z^n) = (-1)^{n+1}$. The formula for general monomials now follows by induction using (2–6) and, again, for general polynomials by linearity. This gives us (2–3).

   The theorem now follows by writing $B(p,g) = B(1,pg) - (\delta_p B)(1,g)$, applying (2–3) to $\delta_p B$, and noting that, because $B \in H^n$, the series ends after the term involving δ_z^{n-1}. $\square$

REMARK 2.9.   Equation (2–3) can be viewed as a formal Taylor series and summed, yielding $\delta_{f(z)} = f(z) - f(z - \delta_z)$. This formula can also be derived formally by writing $\delta_{f(z)} = L_{f(z)} - R_{f(z)} = f(L) - f(R) = f(z) - f(z - \delta_z)$.

REMARK 2.10.   The restriction to functions of a single variable was for notational convenience. The analog of (2–3), for instance, for polynomials in two variables is, symbolically,

$$\delta_{f(z,w)} = f(z,w) - f(z - \delta_z, w - \delta_w).$$

REMARK 2.11.   From an algebraic point of view, Theorem 2.8 is a complete structure theorem for Hankel forms. If we also consider topology, the situation is not clear. Suppose $n = 2$. If we restrict to polynomials, p, q, then for $B \in H^2$

$$B(p,q) = B(1,pq) + \frac{1}{2}(\delta_z B)(p',q).$$

Formally the map $C(p,q) = B(1,pq)$ is a Hankel form of type 1; that is, $\delta_f C = 0$ for any polynomial f. Likewise, $\delta_z B$ is a form of type 1; thus the form of type 2 has been represented as a linear combination of a form of type 1 and a form of type 1 composed with differentiation (which produces a form of type 2). For general n the conclusion is that every form of type n is built from forms of lower type by composing with differentiation in certain explicit ways. Algebraically this is the whole story. However, it is not clear how to obtain the continuity results for the new forms: are there estimates of the form $|B(1,pq)| \le c\,\|p\|_K\,\|q\|_K$ and $|(\delta_z B)(p',q)| \le c\,\|p\|_K\,\|q\|_K$? (Because each form is a bounded perturbation of the other, the two estimates are equivalent.) It is not clear if this issue can be settled algebraically or needs analytical work. We return to this issue in Remark 4.4, where we settle a minor variation of this question for $n = 2$.

We now turn to forms of finite rank. We will say a bilinear form, B, is of *rank* k if there are $2k$ continuous linear functionals, $h_1, \ldots, h_k, j_1, \ldots, j_k$, such that

$$B(f,g) = \sum_{i=1}^{k} h_i(f)k_i(g)$$

for any $f, g \in K$. We denote the set of all such forms by $\mathcal{F}_k$ and write $\mathcal{F} = \bigcup_n \mathcal{F}_n$. If ζ is a point in D for which all the functionals $h_i(f) = f^{(i)}(\zeta)$ are continuous then $B(f,g) = f^{(i)}(\zeta)g^{(n-i)}(\zeta)$ is an example of a form in $H^{n+1} \cap \mathcal{F}_1$. The content of the next theorem is, roughly, that these are the only examples.

We will use the following convexity and continuity hypothesis on K.

CONVEXITY HYPOTHESIS. For each point $\zeta \notin D$ and each $M > 0$ there is a polynomial $p \in K$ so that $\|p\|_K = 1$ and $|p(\zeta)| > M$.

CONTINUITY HYPOTHESIS. For each $\zeta \in D$ and each integer n the functional which takes $f \in K$ to $f^{(n)}(\zeta)$ is continuous.

THEOREM 2.12. *Suppose $P \subset K$ and that K satisfies the two preceding hypotheses. Given $B \in H^n \cap \mathcal{F}_k$ for some $n, k \geq 1$, it is possible to find constants $M = M(n,k)$ and $C = C(n,k)$, points $\zeta_1, \ldots, \zeta_M \in D$, and scalars $c_{i,j,m}$ such that*

$$B(f,g) = \sum_{\substack{i \leq n-1 \\ j \leq \bar{C}, \, m \leq M}} c_{i,j,m}(f^{(i)}g)^{(j)}(\zeta_m). \tag{2-7}$$

REMARK 2.13. The form that takes (f,g) to $(f^{(i)}g)^{(j)}(\zeta_m)$ is of type $i + 1$, independently of j. This can be verified by an elementary induction on i. Also note that it suffices to use only these asymmetric representations because, for example, $fg' = (fg)' - f'g$.

PROOF OF THEOREM 2.12. We work by induction on n. The case $n = 1$ is in [Janson et al. 1987, Section 14]. The basic point there is that by restricting from $K \times K$ to $P \times P$ we end up with a problem about ideals in polynomial rings, which can then be analyzed using tools from commutative algebra. We continue to restrict our attention to the forms acting on $P \times P$ and will next show that forms, B, on $P \times P$ that satisfy $\delta_a^n B = 0$ for every $a \in P$ have representations of the form (2–7). Suppose $B \in H^2 \cap \mathcal{F}$. By Theorem 2.8,

$$B(f,g) = B(1, fg) + (\delta_z B)(f', g).$$

The maps $(f,g) \mapsto B(f,g)$ and $(f,g) \mapsto (\delta_z B)(f', g)$ are both in $\mathcal{F}$, hence so is the form $C(f,g) = B(1, fg)$. Thus, in terms of their action on polynomials, both C and $\delta_z B$ are in $H^1 \cap \mathcal{F}$. (That is, they are annihilated by δ_z; we are not considering continuity.) Hence the result for $n = 1$, for forms acting on $P \times P$, can be applied to both of them. This gives the required form. The rank of C is at most the sum of the ranks of B and $\delta_z B$ and hence at most $3k$; thus we keep control of the ranks. Now we suppose that, as a form on $P \times P$, we have

$B \in H^3 \cap \mathcal{F}$; the argument here will make the general induction step clear. By Theorem 2.8,

$$B(f,g) = B(1,fg) + (\delta_z B)(f',g) + c(\delta_z^2 B)(f'',g)$$

(with $c = -\frac{1}{2}$). We can write $C(f,g) = B(1,fg)$ as a sum of three terms, all in $\mathcal{F}$; hence $C(f,g)$ is in $\mathcal{F}$ and we can estimate its rank in terms of k. We now apply the case $n = 1$ of the theorem to C and to $\delta_z^2 B$, and the case $n = 2$ to δB, to obtain the required form. Clearly this approach will also deal with the general induction step.

We now need to check that the ζ_i's are in D. Suppose $\zeta_1 \notin D$. Suppose N is the highest-order derivative of f that is evaluated at ζ_1 in the representation. Pick $r \in P$ so that r vanishes at $\zeta_2, \ldots, \zeta_M$ to order higher than the order of any derivative being evaluated at ζ_j, for $j \geq 2$, and vanishes to order $N - 1$ at ζ_1. Pick $g \in P$ so that neither g nor any of its first N derivatives vanish at ζ_1. Let p be the polynomial whose existence is insured by the Convexity Hypothesis, which has unit norm and is large at ζ_1. We have $B(rp, g) = c(r, g, B)p(\zeta_1)$. This isn't compatible with the boundedness estimate

$$|B(rp,g)| \leq c_B \|rp\| \|g\| \leq c_B' \|p\| \|g\| = c_B' \|g\|.$$

Now that we know that all the points are in D, the representation extends to all of K by continuity, using the Continutity Hypothesis. □

REMARK 2.14. Again the restriction to functions of a single variable is for notational convenience. The formulation of the more general result can be seen in [Rochberg 1995]. Those proofs are more computational than the ones here.

The requirement in Theorem 2.12 that the polynomials be dense precludes such basic examples as $K = A^2(\mathbb{R}_+^2)$, the Bergman space of the upper half-plane. We now show how to use the previous result (or rather its proof) to obtain results for $K = A^2(\mathbb{R}_+^2)$. Similar arguments could be used for other simple examples, for instance the spaces $A^2(\mathbb{R}_+^2; y^\alpha dx\, dy)$ with $\alpha > -1$, but the story for complicated choices of K is not clear.

THEOREM 2.15. Let $K = A^2(\mathbb{R}_+^2)$ and suppose $B \in \mathrm{Bilin}(K)$. If $B \in H^n \cap \mathcal{F}_k$ for some $n, k \geq 1$ then B has a representation of the type (2–7).

PROOF. B is of finite rank; hence we can find linear functionals h_i, k_i such that $B(f,g) = \sum_1^k h_i(f)k_i(g)$, and we may assume that the k_i are linearly independent. Set

$$V = \{f : B(f,g) = 0 \text{ for all } g \in K\}.$$

Because the k_i are linearly independent, $V = \bigcap \ker(h_i)$. Let α be a conformal map of the upper half-plane to the unit disk, and set

$$W = \{f : f, \alpha f, \ldots, \alpha^{n-1} f \in V\} = \bigcap_{i,j} \{f : h_i(\alpha^j f) = 0\}.$$

From this it is clear that W is a closed subspace of K of finite codimension, $\dim(K/W) \leq kn$. We now claim that W is invariant under multiplication by α. Pick $f \in W$; we need to show that $B(\alpha^n f, g) = 0$ for all $g \in K$. To see this note that $B \in H^n$, so $\delta_a^n(B) = 0$. Hence

$$B(\alpha^n f, g) = \sum_{k=0}^{n-1} c_k B\big(\alpha^{n-1-k} f, \alpha^{k+1} g\big).$$

Because $f \in W$ the right hand side is 0. This shows that $\alpha W \subset W$.

Let X be the closed subalgebra of $\mathcal{H}^\infty(\mathbb{R}_+^2)$ generated by 1 and α. Let

$$Y = \{\beta \in X : \beta f \in W \text{ for all } f \in K\}.$$

It is not clear at first glance that Y has any nonzero elements, but we now show that it is rather large. Certainly Y is a closed ideal in X. We will show that Y contains a polynomial in α. Pick $f \in K \setminus W$. Because W has codimension at most nk, there is a polynomial, p_1, of degree $n_1 \leq nk$, such that $p_1(\alpha)f \in W$. Let $W_1 = \operatorname{span}\{f, \alpha f, \ldots, , \alpha^{n_1-1} f\}$. If $W \oplus W_1 \neq K$ we continue, picking $g \in (W \oplus W_1) \setminus W$. As before there is a polynomial, now p_2 of degree n_2, such that $p_2(\alpha)g \in W \oplus W_1$. Set $W_2 = \operatorname{span}\{g, \alpha g, \ldots, , \alpha^{n_2-1} g\}$. This process must eventually fill K and we will have

$$K = W \oplus \bigoplus_i W_i. \qquad (2\text{--}8)$$

Set $Q_1(\alpha) = \prod p_i(\alpha)$. If $f \in K$ then $Q_1(\alpha)f \in W$. To see this, split f using the decomposition (2–8) as $f = g + g_1 + g_2 + \cdots + g_j$. It is enough to look at each summand, and g_2 is typical: multiplication by $p_2(\alpha)$ takes g_2 into $W \oplus W_1$, and hence further multiplication by $p_1(\alpha)$ takes the product into W. Because W is α-invariant, multiplication by the remaining factors of Q_1 does no harm. Y is an ideal and we have now seen that $Q_1 \in Y$. Hence $Q_1 X \subset Y$. The characteristic property of Q_1 is that $B(Q_1 f, g) = 0$ for all $f, g \in K$. In exactly the same way we can find Q_2 such that $B(f, Q_2 g) = 0$ for all $f, g \in K$. Set $Q = Q_1 Q_2$.

We now split K as an orthogonal direct sum

$$K = R \oplus Q(\alpha)K.$$

The fundamental property of this splitting is that, for any $r_1, r_2 \in R$ and f_1, f_2 in K,

$$B\big(r_1 + Q(\alpha)f_1,\ r_2 + Q(\alpha)f_2\big) = B(r_1,\ r_2). \qquad (2\text{--}9)$$

Suppose that $Q(\alpha)$ vanishes to order n_i at ζ_i in $\mathbb{R}_+^2$, for $i = 1, \ldots, N$, and that this is a complete listing of the zeros of $Q(\alpha)$ in $\mathbb{R}_+^2$. If $k \in K$ and K vanishes at each ζ_i to order at least n_i then $k = Q(\alpha)h$ for some $h \in K$. Hence for each polynomial S there is a unique $k(S)$ in R that agrees with S to order n_i at ζ_i, for $i = 1, \ldots, N$. We now define a bilinear form on polynomials by setting

$$\tilde{B}(S, T) = B\big(k(S), k(T)\big).$$

Let γ be an element of $\mathcal{H}^\infty(\mathbb{R}_+^2)$ that agrees with the monomial Z to order n_i at ζ_i, for $i = 1, \ldots, N$. For any polynomial U, we have

$$k(ZU) \equiv \gamma k(U) \pmod{Q(\alpha)K}.$$

Hence, taking into account (2–9),

$$\tilde{B}(ZS, T) - \tilde{B}(S, ZT) = B\big(\gamma k(S), k(T)\big) - B\big(k(S), \gamma k(T)\big).$$

Hence, for any polynomial U, and for all S, T,

$$(\delta_U^n \tilde{B})(S, T) = (\delta_{U(\gamma)}^n B)\big(k(S), k(T)\big).$$

Since $B \in H^n$, the expression on the right is always zero. This shows that $\tilde{B}$ is a Hankel form of type n on polynomials. The proof of Theorem 2.12 then shows that $\tilde{B}$ has the form (2.8), and certainly the ζ's that show up in (2.8) must be among our ζ_i's. Now note that if $k_i \in K$, for $i = 1, 2$, then $k_i = k(S_i) + Q(\alpha)h_i$ for some polynomials S_i and some $h_i \in K$. Also k_i, $k(S_i)$, and S_i all agree to order n_j at each ζ_j. Moreover,

$$B(k_1, k_2) = B\big(k(S_1), k(S_2)\big) = \tilde{B}(S_1, S_2).$$

We have seen that the expression on the right is a linear combination of values and derivatives of the S_i of a sort given by (2.8). Hence the expression on the left must be the same combination of values and derivatives of the k_i. That B has such a representation is what we wanted to show.

To finish we note that the degree of Q is controlled by n and k, and hence so is the number of terms in (2.8). $\square$

3. Commutators

In this section we develop analogs of the results in the previous section for certain nonlinear operators. That we worked with bilinear forms before and now work with operators is not a major change in point of view. Those results for bilinear forms could have been formulated as results about operators. Given a bilinear form B on K there is an induced linear map T_B from K to its dual space K', given by $T_B(f)(g) = B(f, g)$. (We don't want to identify the Hilbert space K with its dual using the inner product because that map is conjugate linear rather than linear.) If we have an algebra A of functions that act on K by multiplication, we can consider the corresponding multiplication operators $M_a(f) = af$. Everything in the previous section that was formulated in terms of operators that send B to aB, Ba, and $aB - Ba$ can be recast in terms of the operators that send T_B to $T_B M_a$, $M_a^* T_B$, and $T_B M_a - M_a^* T_B$.

Before going further we mention why one might look for such analogies between the results in the previous sections and commutators.

We'll work on $L^2(\mathbb{R})$. For $f \in L^2$ set $\Lambda_1(f) = f \ln |f|$. For any function a defined on $\mathbb{R}$ let M_a be the operator of multiplication by a. Pick an unbounded function b in BMO($\mathbb{R}$). (We need nothing here about BMO except that it contains unbounded functions—for instance, $b(x) = \ln|x|$). Let $\Lambda_2 = M_b$. Let P be the Cauchy–Szegő projection acting on $L^2(\mathbb{R})$. For any operators acting on functions, linear or not, we define the commutator $[A, B]$ of A and B by

$$[A, B](f) = (AB - BA)(f) = A(B(f)) - B(A(f)).$$

It is elementary that both Λ_1 and Λ_2 are unbounded, that if a is a bounded function then the operator $[M_a, \Lambda_1]$ is bounded on $L^2(\mathbb{R})$, and that $\big[M_a, [M_a, \Lambda_1]\big] = \big[M_b, \Lambda_2\big] = 0$. Also true, but not elementary, is that $[\Lambda_1, P]$, $[\Lambda_2, P]$, and $\big[\Lambda_2, [\Lambda_2, P]\big]$ are all bounded on $L^2(\mathbb{R})$. The boundedness of all three can be given a unified proof using interpolation theory. (I don't think that $\big[\Lambda_1, [\Lambda_1, P]\big]$ is bounded. The story there is complicated by the nonlinearities. Set $\Omega f = \frac{1}{2} f \big(\ln|f|\big)^2$. The operator that is bounded is $[P, \Omega] - \Lambda_1[P, \Lambda_1]$.) To see how this is related to the previous section we define Hankel forms on $\mathcal{H}^2(\mathbb{R}) = P\big(L^2(\mathbb{R})\big)$ by the natural analog of (1–1), namely

$$B(f, g) = \int_{\mathbb{R}} f g \bar{b}. \tag{3–1}$$

Now $f, g \in \mathcal{H}^2(\mathbb{R})$ and b is the boundary value of a function holomorphic in the upper half-plane. It is direct to check that

$$B(f, g) = \big\langle g, \overline{(I - P)(\bar{b}f)} \big\rangle.$$

Hence the properties of B are all to be found in the theory of the linear operator that takes f to $(I - P)(\bar{b}f)$. Now note that if we choose for b in the definition of Λ_2 the function $\bar{b}$ in (3–1), we have

$$[\Lambda_2, P]f = \bar{b}Pf - P(\bar{b}f) = \bar{b}f - P(\bar{b}f) = (I - P)(\bar{b}f).$$

Thus, from good information about $[\Lambda_2, P]$ we can derive equally good information about the Hankel operator with symbol b. Conversely, for any choice of b, the commutator $[\Lambda_2, P]$ can be written as an orthogonal direct sum of two operators, one a Hankel operator and the other unitarily equivalent to a Hankel operator. In sum, the theories of Hankel operators on $\mathcal{H}^2(\mathbb{R})$ and the theory of the operators $[\Lambda_2, P]$ are essentially equivalent. This suggests that the results for higher-order Hankel forms may have analogs for commutators (including possibly the nonlinear ones involving Λ_1, Ω, and related operators). The relations between the two topics, both those mentioned and those we develop in this section, suggest that there may be deeper connections.

(In another direction, the implication for Hankel operators and their generalizations of the boundedness of $\big[\Lambda_2, [\Lambda_2, P]\big]$, as well as higher-order commutators such as $\big[\Lambda_2, [\Lambda_2, [\Lambda_2, P]]\big]$ that are also bounded, is not clear.)

Another reason for suspecting that there may be a connection between the two topics is a similarity between some of the detailed computations that led to the results in [Janson and Peetre 1987] and those that led to the results in [Rochberg 1996].

Finally, the theory of Hankel forms is related to the study of bilinear forms on vector spaces that have additional structure; namely, there is a notion of pointwise products. The Hankel forms are those that only depend on the product of its arguments. The higher-order Hankel forms represent, in some sense, infinitesimal perturbations away from that situation PR. The commutators we consider arise in interpolation theory in ways related to infinitesimal changes in Banach space structure. The commutators that arise when working with the L^p scale involve the infinitesimal versions of changes which respect the multiplicative structure but not the linear structure. We return to this rather vague comment in Remark 4.5.

The setup. Our main interest here is in developing algebraic properties. We will be quite informal about the type of continuity possessed by the various maps considered.

Let X be a space of functions (on some space that we generally won't bother to specify) and let A be an algebra of functions with the property that, for $a \in A$ and $x \in X$, we have $ax \in X$ and $\|ax\|_X \leq C \|a\|_A \|x\|_X$. For instance, A could be the bounded holomorphic functions on the disk and X the Bergman space of the disk, or A could be $L^\infty(\mathbb{R})$ and $X = L^2(\mathbb{R})$. We also assume $A \cap X$ is dense in X. (We could get by with less, but for now we just want to capture the basic examples.)

Suppose Λ is a map from X to functions and that $a \in A$. We *do not* assume that Λ is linear. We define left and right multiplication as before: for all $a \in A$ and $f \in X$, we set

$$(L_a\Lambda)(f) = (a\Lambda)(f) = a(\Lambda f),$$
$$(R_a\Lambda)(f) = (\Lambda a)(f) = \Lambda(af),$$
$$\delta_a\Lambda = L_a\Lambda - R_a\Lambda.$$

Continuing the analogy with the previous section we say that $\Lambda \in L^1$ if $\delta_a\Lambda = 0$ for all $a \in A$, and for $n = 1, 2, \ldots$ we say that $\Lambda \in L^{n+1}$ if $\delta_a\Lambda \in L^n$ for all $a \in A$. If we are looking at linear operators this would be the setup for an operator version of the higher-order Hankel forms. The point now is that many of the algebraic results of the previous section still go through without the assumption of linearity.

Elementary properties. The following facts, valid for $a, b, c \in A$ and $\Lambda \in L^n$, follow immediately from the definitions:

(1) $L^n \subset L^{n+1}$.
(2) $L_a\Lambda$, $R_a\Lambda \in L^n$, and $\delta_a\Lambda \in L^{n-1}$.
(3) L_a, R_b, and δ_c commute.
(4) $\delta_a\delta_b = \delta_b\delta_a$.

(5) $\delta_{ab} = L_a \delta_b + R_b \delta_a = L_b \delta_a + R_a \delta_b$.

(6) $\delta_{ab} = a \delta_b + b \delta_a - \delta_b \delta_a$.

(7) $\Omega \in L^n$ if and only if $\delta_a^n \Omega = 0$ for all $a \in A$.

L^{n+1} and L^n are vector spaces and A-bimodules. Hence the quotient spaces, $I^{n+1} = L^{n+1}/L^n$, are A-bimodules and the induced action of A on the I^n is commutative; that is, for $a \in A$ and $i \in I^n$, we have $ai = ia$ as elements of I^n.

Having collected all of these facts, which are analogous to Proposition 2.3 and related facts in the proceeding section, we should note that the situation is not always that simple. Recall that a linear map D of A to itself is called a derivation if $D(ab) = aD(b) + bD(a)$ for all $a, b \in A$. Given such a D and given $\Lambda \in L^1$, is $\Lambda D \in L^2$? (This construction with the choice $D(f) = f'$ was used frequently in the previous section.) We compute

$$(\delta_a(\Lambda D))(f) = a(\Lambda D)(f) - (\Lambda D)(af) = a\Lambda(Df) - \Lambda(D(af))$$
$$= \Lambda(aDf) - \Lambda(fDa + aDf).$$

If Λ is linear we can continue with

$$(\delta_a(\Lambda D))(f) = \Lambda(aDf) - \Lambda(fDa) - \Lambda(aDf) = -Da\Lambda(f).$$

Thus $\delta_a(\Lambda D)$ is in L^1. Thus we see that the linearity of D is irrelevant but that this process will construct an element of L^2 from $\Lambda \in L^1$ only if Λ is linear. To go even further and show ΛD^2 is in L^3 we would also need D to be linear.

Examples. Having just seen that composition with derivations doesn't necessarily generate elements of L^n for $n > 1$, we now show how to generate elements of L^1 and, more generally, all the L^n.

PROPOSITION 3.1. *Suppose A is dense in X and suppose D is a derivation on A. For $n = 1, 2, \dots$ the map Λ_n given by*

$$\Lambda_n(f) = f\left(\frac{Df}{f}\right)^n$$

is an element of L^{n+1}.

REMARK 3.2. Of course Λ_n is only defined on the dense subspace A and for each f there is a problem on the set where $f = 0$. We put aside these issues and concentrate on formal structure.

REMARK 3.3. We *do not* assume that D is linear.

PROOF. The case $n = 0$ is trivial. For $n = 1$ we compute

$$(\delta_a \Lambda_1)(f) = (\delta_a D)(f) = D(af) - aD(f) = aD(f) + D(a)f - aD(f) = D(a)f.$$

As required, $\delta_a \Lambda_1$ is an element of L^1. It is straightforward to complete the proof by induction. However, with an eye to later discussion, we take a slightly less direct route.

For $a \in A$ and $t \in \mathbb{R}$ we define

$$T_t(a) = a \exp(t(\frac{Da}{a})).$$

LEMMA 3.4. *For all* $a, b \in A$ *we have* $T_t(ab) = T_t(a)T_t(b).$

PROOF.

$$T_t(ab) = ab \exp\left(t\left(\frac{D(ba)}{ab}\right)\right) = ab \exp\left(t\left(\frac{aDb + bDa}{ab}\right)\right)$$
$$= ab \exp\left(t\left(\frac{Db}{b} + \frac{Da}{a}\right)\right) = ab\left(\exp t\left(\frac{Da}{a}\right)\right)\left(\exp t\left(\frac{Db}{b}\right)\right)$$
$$= T_t(a)T_t(b),$$

as required. □

We write

$$T_t(a) = \sum \Gamma_n(a)t^n.$$

We now equate powers of t in the proposition. That gives

$$\Gamma_n(ab) = \sum c_k \Gamma_{n-k}(a)\Gamma_k(b) = a\Gamma_n(b) + \sum_{k=0}^{n-1} c_k \Gamma_{n-k}(a)\Gamma_k(b).$$

We now complete an inductive proof of the proposition. We need to show $\Gamma_n \in L^{n+1}$, and the case $n = 1$ is done. Suppose the cases up to $n - 1$ are done.

$$(\delta_a \Gamma_n)(b) = a\Gamma_n(b) - \Gamma_n(ab)$$
$$= a\Gamma_n(b) - \left(a\Gamma_n(b) + \sum_{k=0}^{n-1} c_k \Gamma_{n-k}(a)\Gamma_k(b)\right)$$
$$= \sum_{k=0}^{n-1} c_k \Gamma_{n-k}(a)\Gamma_k(b).$$

By the induction hypothesis this is of the required form, and the proposition is proved. □

To use the proposition to construct examples we need examples of operators D that are not assumed linear but satisfy

$$D(fg) = fD(g) - gD(f). \tag{3-2}$$

Suppose that D is of the form $D(f) = f\varphi(\ln(f))$ for some operator φ. To verify (3–2) we compute

$$D(fg) = fg\varphi(\ln(fg)) = fg\varphi(\ln(f) + \ln(g)).$$

Hence we will have the required form if φ is any linear operator. Similarly if $D(f) = f\varphi_1(\ln|f|)$ then (3–2) will be satisfied if φ_1 is linear.

Here are some examples of elements of L^n.

EXAMPLE 3.5. If D is a derivation, the proposition tells us that the map of f to $D(f)$ is in L^2.

EXAMPLE 3.6. Suppose A consists of holomorphic functions and set $\varphi(\ln f) = (\ln(f))'$. We have $\Lambda_1(f) = f'$. That example was central in the previous sections, as were the powers $\Lambda_1^n(f) = f^{(n)}$. However, the fact that the powers have properties which we can work with rests on the linearity of Λ_1.

For $n > 1$ the previous computations give us expressions that are in the L^{n+1}; we have $\Lambda_n(f) = f^{-n+1}(f')^n$. We don't know of other places where these operators arise.

EXAMPLE 3.7. Suppose $D(f) = f\varphi_1(\ln|f|) = f\ln|f|$. We have $\Lambda_n(f) = f(\ln|f|)^n$. This series of examples, starting with $\Lambda_1(f) = f\ln|f|$, is used to form some of the basic examples the nonlinear commutators that arise in interpolation theory. For instance, with $X = L^2(\mathbb{R})$ and P the Cauchy–Szegő projection, the operator

$$2[P, \Lambda_2] - \Lambda_1[P, \Lambda_1] \tag{3-3}$$

is bounded.

EXAMPLE 3.8. The previous two examples are related to the themes of this section and the preceding one. As soon as we move to other examples we encounter rather unfamiliar nonlinear operators. For instance, suppose $D(f) = f\varphi(\ln(f))$, with $\varphi(\ln f) = (\ln(f))''$; then

$$\Lambda_0(f) = f, \quad \Lambda_1(f) = \frac{ff'' - f'^2}{f}, \quad \dots$$

Derivations. The results in the previous section relating J^{n+1} to derivations into J^n used the fact that the J's are linear spaces but not the fact that the elements of the spaces were linear forms. Hence the results go through for I's. We need to change the details slightly because we are now dealing with operators rather than forms and we need to be attentive to the fact that the elements of the I's may be nonlinear. But the similarities are very strong, so we will be quick.

Let M denote the set of maps from X to functions. Given $\Lambda \in M$ we define a map from A to M by

$$\Delta(\Lambda)(a) = \delta_a\Lambda.$$

Pick and fix $\alpha \in A$. We define an operator, ∇_α, that takes a mapping, $\tilde{\Delta}$, of A into M to an element of M by

$$\nabla_\alpha(\tilde{\Delta})(f) = \tilde{\Delta}(f)(\alpha).$$

Again, if $1 \in A$ the choice $\alpha = 1$ is a natural one to consider.

We continue to denote by $\mathrm{Deriv}(A, M)$ the derivations of A into an A-module M.

THEOREM 3.9. *The following results hold for $n = 0, 1, 2, \dots$:*

(1) Δ maps I^{n+1} into $\text{Deriv}(A, I^n)$.

(2) ∇_α maps $\text{Deriv}(A, I^n)$ into I^{n+1}.

(3) $\nabla_\alpha(\Delta(\Lambda)) = -\alpha\Lambda$ for all $\Lambda \in I^{n+1}$, and $\Delta(\nabla_\alpha(D)) = -\alpha D$ for all $D \in \text{Deriv}(A, I^n)$.

REMARK 3.10. Again we abuse notation slightly, in using Δ and ∇_α for induced maps.

PROOF OF THEOREM 3.9. (1) Pick $\Lambda \in L^{n+1}$. For all $a, b \in A$, we have

$$\Delta(\Lambda)(ab) = \delta_{ab}\Lambda = a\delta_b\Lambda + b\delta_a\Lambda - \delta_a\delta_b\Lambda$$
$$= a(\Delta(\Lambda)b) + b(\Delta(\Lambda)a) - \delta_a\delta_b\Lambda.$$

The second equality shows that $\Delta(\Lambda)(ab)$ is in L^n. Now $\delta_a\delta_b\Lambda \in L^{n-1}$, so

$$\Delta(\Lambda)(ab) = a\Delta(\Lambda)(b) + b\Delta(\Lambda)(a)$$

as maps into I^n. Thus $\Delta(\Lambda)$ is a derivation. Finally, if we change Λ by an element of L^n, the range of $\Delta(\Lambda)$ changes by elements of L^{n-1}; hence, as an element of I^n, the image is unchanged. Thus our map is well-defined on I^{n+1}.

(2) We start with $D \in \text{Deriv}(A, I^n)$. Define Λ by

$$\Lambda(x) = \nabla_\alpha(D)(x) = D(x)(\alpha).$$

Pick $a, x \in A$. Then

$$(\delta_a\Lambda)(x) = a\Lambda(x) - \Lambda(ax) = aD(x)(\alpha) - D(ax)(\alpha)$$
$$= aD(x)(\alpha) - aD(x)(\alpha) - xD(a)(\alpha) = -xD(a)(\alpha).$$

We now recall that the product xD is the product in the A-module $\text{Deriv}(A, I^n)$. Thus $xD(a)(\alpha) = D(a)(\alpha x)$, which is an element of I^n applied to x. This shows that the map goes into L^{n+1}. The verification that the coset in I^{n+1} is unchanged if the choice of $D(x)$ is changed by an element of I^{n-1} is routine.

(3) Suppose $\Lambda \in I^{n+1}$. We have

$$\nabla_\alpha(\Delta(\Lambda))(x) = \Delta(\Lambda)(x)(\alpha) = (\delta_x\Lambda)(\alpha)$$
$$= x\Lambda(\alpha) - \Lambda(\alpha x) = x\Lambda(\alpha) - R_\alpha\Lambda(x).$$

The operator that takes x to $x\Lambda(\alpha)$ is in L^1. Hence the previous computation shows that $\nabla_\alpha(\Delta(\Lambda)) = -R_\alpha\Lambda$ as maps into I^{n+1}. As before, the outcome doesn't depend on choices made for the representative of $\Delta(\Lambda)(x)$. Finally recall that I^n is a commutative A-module and hence $R_\alpha\Lambda = L_\alpha\Lambda = \alpha\Lambda$, as required.

Now select $D \in \text{Deriv}(A, I^n)$ and $a, x \in A$. Recalling again that aD is the module product, we have

$$\Delta(\nabla_\alpha D)(a)(x) = \delta_a(\nabla_\alpha D)(x) = a(\nabla_\alpha D)(x) - (\nabla_\alpha D)(ax)$$
$$= a(D(x)(\alpha)) - D(ax)(\alpha) = D(x)(a\alpha) - (aD(x) + xD(a))(\alpha)$$
$$= (\delta_a D(x))(\alpha) - \alpha D(a)(x).$$

As in the previous section, an easy induction shows that the first term is of lower order and hence drops out when we pass to quotients. □

4. Further Remarks

REMARK 4.1. Some of the results in the previous sections involved derivations of an algebra of functions A into an A-bimodule. Such results can be reformulated in terms of module cohomology. At this point we don't see a direct use for that viewpoint, so we only offer the observation. Background can be found in [Ferguson 1996] and the references there.

REMARK 4.2. A *Foguel-type operator* is an operator of the form

$$R(X) = \begin{bmatrix} S^* & X \\ 0 & S \end{bmatrix},$$

where S is the unilateral shift on $l^2(\mathbb{Z}^+)$ and X is an operator on $l^2(\mathbb{Z}^+)$. Such operators have been considered in the investigation of polynomially bounded operators (see [Davidson and Paulsen 1997] for details and further references). One reason is that if $X = \Gamma_f$, the Hankel operator on $l^2(\mathbb{Z}^+)$ with symbol function f, then polynomials in $R(X)$ are particularly easy to compute. If p is a polynomial then

$$p(R(\Gamma_f)) = \begin{bmatrix} S^* & \Gamma_f \, p'(S) \\ 0 & S \end{bmatrix} = \begin{bmatrix} p(S^*) & \Gamma_{fp'} \\ 0 & p(S) \end{bmatrix}. \tag{4-1}$$

This follows from an elementary induction and the fact that Hankel operators satisfy $S^*\Gamma - \Gamma S = 0$. Let L_{S^*} denote multiplication on the left by S^* and R_S denote multiplication on the right by S. Set $\delta_z = L_{S^*} - R_S$. For any polynomial p we use the obvious extension of the notation and set $\delta_{p(z)} = L_{p(S^*)} - R_{p(S)}$. We are now in the notational set-up of Section 2, adapted to operators. The Hankel operators are exactly those X for which $\delta_{p(z)}(X) = 0$ for all polynomials p. (Equivalently, $\delta_z(X) = 0$.) To describe the general pattern, for any polynomial p in one variable define a new polynomial by

$$p^*(x, y) = \frac{p(x) - p(y)}{x - y}.$$

A quick induction on the degree of monomials shows that, for any p,

$$p(R(X)) = \begin{bmatrix} p(S^*) & p^*(L_{S^*}, R_S)(X) \\ 0 & p(S) \end{bmatrix}.$$

Writing $L_{S^*} = \delta_z + R_S$ we get

$$p(R(X)) = \begin{bmatrix} p(S^*) & \sum_{n=1}^{\infty} p^{(n)}(R_S)\delta_z^{n-1}(X) \\ 0 & p(S) \end{bmatrix}$$

$$= \begin{bmatrix} p(S^*) & \sum_{n=1}^{\infty} \delta_z^{n-1}(X)p^{(n)}(S) \\ 0 & p(S) \end{bmatrix}.$$

The linear operators induced by Hankel forms of order k are exactly those X for which $\delta_z^k(X) = 0$. For those X the infinite series ends and we have a slightly more complicated but still quite explicit formula for composition with polynomials. For instance, if X is a Hankel operator of order 2 we find

$$p(R(X)) = \begin{bmatrix} p(S^*) & Xp'(S) + \frac{1}{2}\delta_z(X)p''(S) \\ 0 & p(S) \end{bmatrix}.$$

REMARK 4.3. In Section 3 we defined classes of operators L^n by $\Lambda \in L^n$ if $\delta_u^n(\Lambda) = 0$ for all u in some class of functions. However that discussion was motivated by the study of commutators that arise in interpolation theory and the results in interpolation theory also suggest another, slightly more sophisticated point of view. We mention it here because it raises a number of interesting questions. We don't pursue it further here because it seems less amenable to formal algebraic analysis than the ideas in Section 3.

The commutator results which arise in interpolation theory involve operators Λ acting on a space X. The operators are generically unbounded and often nonlinear. The basic results are that for certain linear operators T which are bounded on X it is also true that the commutator $[T, \Lambda]$ is bounded on X. Often the class of T for which this holds includes all operators of the form M_u for bounded functions u. However, the construction of Λ involves a number of choices and in [Cwikel et al. $\geq$ 1997], for instance, the operators Λ are really viewed modulo bounded operators. Furthermore, even when it is true that $\delta_u(\Lambda) = [M_u, \Lambda]$ is bounded it is generally not true that $\delta_u^2(\Lambda) = 0$. This suggests considering bounded operators as being of type 1 and defining an operator to be of type n if $\delta_u^{n-1}(\Lambda)$ is a bounded operator for each bounded function u. This would come closer to the viewpoint of [Cwikel et al. $\geq$ 1997] and the results from interpolation theory generate a variety of type 2 operators that are not type 1. For example, if X is an L^p space and φ is any Lipschitz function, $\Lambda(f) = f\varphi(\ln|f|)$ will be of type 2 (in this sense) and not of type 1. However it is not generally true that $\delta_u^2(\Lambda) = 0$, although it is true in the special case $\varphi(x) = x$.

The difficulty with this approach is that it is not clear how to generate objects of type 3 other than with the ideas used in Section 3. That is, the examples $\Lambda_n(f) = f(\ln|f|)^n$ give examples of objects of arbitrary type in Section 3. It is not clear what other examples we would have for the variation just described.

In this context we should mention the very interesting results of Kalton [1988], which say, roughly, that in some circumstances if Λ has the property that $[M_u, \Lambda]$ is bounded for each bounded function u then $[T, \Lambda]$ is also bounded for a much larger of linear operators T.

A final word on interpolation. As we mentioned in the introduction it may be that the computations in Section 3 can be used to extract information from the boundedness of (3–3) and the more complicated combinations in [Milman and Rochberg 1995].

REMARK 4.4. We mentioned after Theorem 2.8 that a flaw in this representation is that we can't insure that the individual summands are bounded. For $n = 2$ we can do that if we accept a more complicated, but also more symmetrical, formulation. Suppose $B \in H^2$. By Theorem 2.8 we know that, for $p, q \in P$,

$$B(p,q) = B(1, pq) + \tfrac{1}{2}(\delta_z B)(p', q).$$

A similar argument in the second variable gives

$$B(p,q) = B(pq, 1) - \tfrac{1}{2}(\delta_z B)(p, q').$$

Adding gives

$$2B(p,q) = B(pq, 1) + B(1, pq) + (\delta_z B)(p', q) - (\delta_z B)(p, q'),$$

which we rewrite as

$$2B(p,q) = B(pq, 1) + B(1, pq) + (\delta_D(\delta_z B))(p, q).$$

Computing $(\delta_p \delta_q B)(1, 1)$ and recalling that $B \in H^2$ we get

$$B(pq, 1) + B(1, pq) = B(p, q) + B(q, p). \tag{4-2}$$

Writing $B_{\mathrm{sym}}(p, q) = B(p, q) + B(q, p)$ we have

$$2B(p,q) = B_{\mathrm{sym}}(p, q) + (\delta_D(\delta_z B))(p, q). \tag{4-3}$$

which is the representation we wanted. Clearly

$$|B_{\mathrm{sym}}(p, q)| \leq C\|p\|\,\|q\|$$

and hence

$$|(\delta_D(\delta_z B))(p, q)| \leq C\|p\|\,\|q\|.$$

Using (4–2) we see that $B_{\mathrm{sym}} \in H^1$. Thus (4–3) gives a representation of B as a type 1 form plus a form built from a type 1 form by composing with differentiation. All the forms are continuous.

Unfortunately it is not clear how to continue this analysis to higher n.

REMARK 4.5. We have mentioned that higher-order Hankel forms are related to deformation of multiplicative structure. One way to formulate this is to introduce the family of operators M_ε defined by $M_\varepsilon(f, g) = f(z + \varepsilon)g(z - \varepsilon)$. If we expand this in powers of ε we get

$$\sum B_n(f, g)\varepsilon^n = fg + (f'g - gf')\varepsilon + \cdots.$$

Hankel forms are those that only depend on $B_0(f, g) = fg$; Hankel forms of type 2 are linear functions of $B_0(f, g)$ and $B_1(f, g)$; and so on.

In Section 3 we introduced the operator $T_t(a) = a \exp(t(Da/a))$ and noted that $T_t(ab) = T_t(a)T_t(b)$ (Lemma 3.4). In the basic example from interpolation theory, when $D(f) = f \ln|f|$, this becomes $T_t(f) = f|f|^t$. The reason this operator plays a crucial role in interpolation theory is that

$$\|f\|_{L^p} = \|T_t(f)\|_{L^{p/(1+t)}}.$$

The crucial fact is that T_t moves functions through the scale of spaces without changing the norm. For general scales of spaces the operators Λ that arise are the derivative of a family of operators with similar properties. Thus, from the interpolation theoretic point of view, it is just an accident that the formula for $D(f) = f \ln|f|$ can be used to generate a multiplicative operator. However, this fact is basic to our algebraic computations.

QUESTION 4.6. Can the algebraic approach in Section 2 be extended to trilinear forms in a natural way? Cobos, Kühn, and Peetre have developed a theory of trilinear forms acting on a Hilbert space in [Cobos et al. 1992].

Acknowledgements

My thanks to Nigel Kalton and to Mario Milman for interesting comments and useful suggestions while I was developing these ideas.

References

[Carro et al. 1995a] M. J. Carro, J. Cerdà, and J. Soria, "Commutators and interpolation methods", *Ark. Mat.* **33**:2 (1995), 199–216.

[Carro et al. 1995b] M. J. Carro, J. Cerdà, and J. Soria, "Higher order commutators in interpolation theory", *Math. Scand.* **77**:2 (1995), 301–319.

[Cobos et al. 1992] F. Cobos, T. Kühn, and J. Peetre, "Schatten–von Neumann classes of multilinear forms", *Duke Math. J.* **65**:1 (1992), 121–156.

[Cwikel et al. 1989] M. Cwikel, B. Jawerth, M. Milman, and R. Rochberg, "Differential estimates and commutators in interpolation theory", pp. 170–220 in *Analysis at Urbana* (Urbana, IL, 1986–1987), vol. II, edited by E. Berkson et al., London Math. Soc. Lecture Note Ser. **138**, Cambridge Univ. Press, Cambridge, 1989.

[Cwikel et al. ≥ 1997] M. Cwikel, M. Milman, and R. Rochberg, "A unified approach to derivation mappings Ω for a class of interpolation methods". In preparation.

[Davidson and Paulsen 1997] K. R. Davidson and V. I. Paulsen, "Polynomially bounded operators", *J. Reine Angew. Math.* **487** (1997), 153–170.

[Ferguson 1996] S. H. Ferguson, "Polynomially bounded operators and Ext groups", *Proc. Amer. Math. Soc.* **124**:9 (1996), 2779–2785.

[Iwaniec 1995] T. Iwaniec, "Integrability theory of the Jacobians", lecture notes, Bonn University, 1995.

[Janson and Peetre 1987] S. Janson and J. Peetre, "A new generalization of Hankel operators (the case of higher weights)", *Math. Nachr.* **132** (1987), 313–328.

[Janson et al. 1987] S. Janson, J. Peetre, and R. Rochberg, "Hankel forms and the Fock space", *Rev. Mat. Iberoamericana* **3**:1 (1987), 61–138.

[Kalton 1988] N. J. Kalton, *Nonlinear commutators in interpolation theory*, Memoirs of the American Mathematical Society **385**, Amer. Math. Soc., Providence, 1988.

[Milman and Rochberg 1995] M. Milman and R. Rochberg, "The role of cancellation in interpolation theory", pp. 403–419 in *Harmonic analysis and operator theory* (Caracas, 1994), edited by S. A. M. Marcantognini et al., Contemp. Math. **189**, Amer. Math. Soc., Providence, RI, 1995.

[Peetre and Rochberg 1995] J. Peetre and R. Rochberg, "Higher order Hankel forms", pp. 283–306 in *Multivariable operator theory* (Seattle, WA, 1993), edited by R. E. Curto et al., Contemp. Math. **185**, Amer. Math. Soc., Providence, 1995.

[Pérez 1996]   C. Pérez, "Sharp estimates for commutators of singular integrals via iterations of the Hardy–Littlewood maximal function", preprint, Universidad Autónoma de Madrid, 1996.

[Rochberg 1995] R. Rochberg, "A Kronecker theorem for higher order Hankel forms", *Proc. Amer. Math. Soc.* **123**:10 (1995), 3113–3118.

[Rochberg 1996] R. Rochberg, "Higher order estimates in complex interpolation theory", *Pacific J. Math.* **174**:1 (1996), 247–267.

[Rosengren 1996]   H. Rosengren, "Multilinear Hankel forms of higher order and orthogonal polynomials", preprint, Lund University, 1996. Available at http://www.maths.lth.se/matematiklu/personal/hjalmar/engHR.html.

RICHARD ROCHBERG
DEPARTMENT OF MATHEMATICS
CAMPUS BOX 1146
WASHINGTON UNIVERSITY
ST. LOUIS MO, 63130
UNITED STATES
rr@math.wustl.edu

Holomorphic Spaces
MSRI Publications
Volume **33**, 1998

Function Theory and Operator Theory
on the Dirichlet Space

ZHIJIAN WU

ABSTRACT. We discuss some recent achievements in function theory and operator theory on the Dirichlet space, paying particular attention to invariant subspaces, interpolation and Hankel operators.

Introduction

In recent years the Dirichlet space has received a lot of attention from mathematicians in the areas of modern analysis, probability and statistical analysis. We intend to discuss some recent achievements in function theory and operator theory on the Dirichlet space. The key references are [Richter and Shields 1988; Richter and Sundberg 1992; Aleman 1992; Marshall and Sundberg 1993; Rochberg and Wu 1993; Wu 1993]. In this introductory section we state the basic results. Proofs will be discussed in the succeeding sections.

Denote by $\mathbb{D}$ the unit disk of the complex plane. For $\alpha \in \mathbb{R}$, the space $\mathcal{D}_\alpha$ consists of all analytic functions $f(z) = \sum_0^\infty a_n z^n$ defined on $\mathbb{D}$ with the norm

$$\|f\|_\alpha = \left(\sum_0^\infty (n+1)^\alpha |a_n|^2 \right)^{1/2}.$$

For $\alpha = -1$ one has $\mathcal{D}_{-1} = \mathcal{B}$, the Bergman space; for $\alpha = 0$, $\mathcal{D}_0 = \mathcal{H}^2$, the Hardy space; and for $\alpha = 1$, $\mathcal{D}_1 = \mathcal{D}$, the Dirichlet space. The space $\mathcal{D}_\alpha$ is referred to as a weighted Dirichlet space if $\alpha > 0$, and a weighted Bergman space if $\alpha < 0$. It is trivial that $\mathcal{D}_\alpha \subset \mathcal{D}_\beta$ if $\alpha > \beta$. In particular, the Dirichlet space is contained in the Hardy space.

For any $w \in \mathbb{D}$, the point evaluation at w is a bounded linear functional on $\mathcal{D}$. Therefore there is a corresponding reproducing kernel. It is given by

$$k_w(z) = k(z, w) = \frac{1}{\bar{w}z} \log \frac{1}{1 - \bar{w}z}.$$

Research supported in part by the NSF Grant DMS-9622890.

Let

$$D(f) = \int_{\mathbb{D}} |f'(z)|^2 \, dA(z) = \sum_{n=1}^{\infty} n \, |a_n|^2,$$

where $dA(z) = (1/\pi) \, dx \, dy$ is normalized Lebesgue measure on $\mathbb{D}$. The square of the norm of a function f in $\mathcal{D}$ can be also expressed as

$$\|f\|^2 = \|f\|_1^2 = \|f\|_0^2 + D(f).$$

Arazy, Fisher and Peetre proved in [Arazy et al. 1988] that the number $\sqrt{D(f)}$ is the Hilbert–Schmidt norm of the big Hankel operator on the Bergman space (introduced first in [Axler 1986]) with the analytic symbol f. Arazy and Fisher [1985] proved that the Dirichlet space $\mathcal{D}$ with the norm $\| \cdot \| = \sqrt{D(\cdot)}$ is the unique Möbius invariant Hilbert space on $\mathbb{D}$. A more general result in [Arazy et al. 1990] implies that for any Bergman type space $\mathcal{B}(\nu) = \mathcal{L}^2(\mathbb{D}, \nu) \cap \{\text{analytic functions on } \mathbb{D}\}$, which has a reproducing kernel $k_\nu(z, w)$, one has the formula

$$\int_{\mathbb{D}} \int_{\mathbb{D}} |f(z) - f(w)|^2 |k_\nu(z, w)|^2 \, d\nu(z) \, d\nu(w) = D(f) \quad \text{for all } f \in \mathcal{D}.$$

The operator M_z of multiplication by z on the Dirichlet space, denoted sometimes by $(M_z, \mathcal{D})$, is a bounded linear operator. Moreover it is an analytic 2-isometry; that is,

$$\left\| M_z^2 f \right\|^2 - 2 \left\| M_z f \right\|^2 + \|f\|^2 = 0 \quad \text{for all } f \in \mathcal{D}, \quad \text{and} \quad \bigcap_{n=0}^{\infty} M_z^n \mathcal{D} = \{0\}.$$

Richter [1991] proved that every cyclic analytic 2-isometry can be represented as multiplication by z on a Dirichlet-type space $\mathcal{D}(\mu)$ with the norm $\| \cdot \|_\mu$ defined by

$$\|f\|_\mu^2 = \|f\|_0^2 + D_\mu(f).$$

Here μ is a nonnegative finite Borel measure on the unit circle $\partial \mathbb{D}$; the number $D_\mu(f)$ is defined by

$$D_\mu(f) = \int_{\mathbb{D}} |f'(z)|^2 h_\mu(z) \, dA(z),$$

where $h_\mu(z)$ is the harmonic extension of the measure μ to $\mathbb{D}$, defined as the integral of the Poisson kernel $P_z(e^{i\theta}) = (1 - |z|^2)/|e^{i\theta} - z|^2$ against $d\mu$:

$$h_\mu(z) = \frac{1}{2\pi} \int_0^{2\pi} P_z(e^{i\theta}) \, d\mu(\theta).$$

It is not hard to see that $D_\mu(f) < \infty$ implies $f \in \mathcal{H}^2$. Therefore $\mathcal{D}(\mu) \subseteq \mathcal{H}^2$. Deep results involving $\mathcal{D}(\mu)$ can be found in [Richter and Sundberg 1991].

A closed subspace $\mathcal{N}$ of $\mathcal{D}$ is called invariant if M_z maps $\mathcal{N}$ into itself. We shall discuss the following two theorems for invariant subspaces.

THEOREM 0.1. *Let $\mathcal{N} \neq \{0\}$ be an invariant subspace for $(M_z, \mathcal{D})$. Then*

$$\dim(\mathcal{N} \ominus z\mathcal{N}) = 1,$$

that is, $z\mathcal{N}$ is a closed subspace of $\mathcal{N}$ of codimension one.

For $f \in \mathcal{D}$, denote by $[f]$ the smallest invariant subspace of $\mathcal{D}$ containing f. An analytic function φ defined on $\mathbb{D}$ is called a multiplier of $\mathcal{D}$ if $\varphi \mathcal{D} \subseteq \mathcal{D}$. The multiplier norm of φ is defined by

$$\|\varphi\|_{\mathcal{M}} = \sup\{\|\varphi f\| : f \in \mathcal{D}, \|f\| = 1\}.$$

THEOREM 0.2. *Every nonzero invariant subspace $\mathcal{N}$ of $(M_z, \mathcal{D})$ has the form*

$$\mathcal{N} = [\varphi] = \varphi \mathcal{D}(m_\varphi),$$

where $\varphi \in \mathcal{N} \ominus z\mathcal{N}$ is a multiplier of $\mathcal{D}$, and $dm_\varphi = |\varphi(e^{i\theta})|^2 \, d\theta/2\pi$.

The codimension-one property for invariant subspaces of the Dirichlet space was first proved in [Richter and Shields 1988]. Another proof that works for the more general operators $(M_z, \mathcal{D}(\mu))$ and gives more information can be found in [Richter and Sundberg 1992]. Aleman [1992] generalized the argument in [Richter and Sundberg 1992] so that it works for the weighted Dirichlet spaces $\mathcal{D}_\alpha$ for $0 < \alpha \leq 1$ ($\alpha > 1$ is trivial). Recently, Aleman, Richter, and Ross provided another approach to the codimension-one property which is good for a large class of weighted Dirichlet spaces and certain Banach spaces. Part of Theorem 0.2 was proved in [Richter 1991]. That the generator φ is a multiplier of $\mathcal{D}$ was proved in [Richter and Sundberg 1992]. (The result there is for the operator $(M_z, \mathcal{D}(\mu))$).

Carleson [1958] proved that, for a disjoint sequence $\{z_n\} \subset \mathbb{D}$, the interpolation problem

$$\varphi(z_n) = w_n, \quad \text{for } n = 1, 2, 3, \ldots \tag{0--1}$$

has a solution $\varphi \in \mathcal{H}^\infty$ for every given $\{w_n\} \in \ell^\infty$ if and only if there are a $\delta > 0$ and a $C < \infty$ such that

$$\left| \frac{z_n - z_m}{1 - \bar{z}_n z_m} \right| \geq \delta \quad \text{for all} \quad n \neq m$$

and

$$\sum_{z_n \in S(I)} (1 - |z_n|^2) \leq C\,|I| \quad \text{for all arcs } I \subset \partial\mathbb{D}.$$

Here $|I|$ is the arc length of I and $S(I)$ is the Carleson square based on I, defined as

$$S(I) = \{z \in \mathbb{D} : z/|z| \in I \text{ and } |z| > 1 - |I|/2\pi\}.$$

Let $\mathcal{M}_\mathcal{D}$ denote the space of multipliers of $\mathcal{D}$. It is clear that $\mathcal{M}_\mathcal{D}$ is an algebra and $\mathcal{M}_\mathcal{D} \subset \mathcal{H}^\infty$. We remark that $\mathcal{H}^\infty$ is in fact the space of all multipliers of $\mathcal{H}^2$.

A sequence $\{z_n\} \subset \mathbb{D}$ is called an interpolating sequence for $\mathcal{M}_\mathcal{D}$ if, for each bounded sequence of complex numbers $\{w_n\}$, the interpolation problem (0–1) has a solution φ in $\mathcal{M}_\mathcal{D}$. By the closed graph theorem, we know that if $\{z_n\}$ is an interpolating sequence for $\mathcal{M}_\mathcal{D}$ then there is a constant $C < \infty$ so that the interpolation can be done with a function $\varphi \in \mathcal{M}_\mathcal{D}$ satisfying $\|\varphi\|_{\mathcal{M}} \leq C \|\{w_n\}\|_{\ell^\infty}$.

Axler [1992] proved that any sequence $\{z_n\} \subset \mathbb{D}$ with $|z_n| \to 1$ contains a subsequence that is an interpolating sequence for $\mathcal{M}_\mathcal{D}$. Marshall and Sundberg [1993] gave the following necessary and sufficient conditions for an interpolating sequence for $\mathcal{M}_\mathcal{D}$.

THEOREM 0.3. *A sequence $\{z_n\}$ is an interpolating sequence for $\mathcal{M}_\mathcal{D}$ if and only if there exist a $\gamma > 0$ and a $C_0 < \infty$ such that*

$$1 - \left| \frac{z_n - z_m}{1 - \bar{z}_n z_m} \right|^2 \leq (1 - |z_n|^2)^\gamma \quad \text{for all } n \neq m \qquad (0\text{--}2)$$

and

$$\sum_{z_n \in \cup S(I_j)} \left(\log \frac{1}{1 - |z_n|^2} \right)^{-1} \leq C \left(\log \frac{1}{\text{Cap}(\bigcup I_j)} \right)^{-1}. \qquad (0\text{--}3)$$

Here $\{I_j\}$ is any finite collection of disjoint arcs on $\partial \mathbb{D}$.

Bishop also proved this theorem independently. Sundberg told me that a similar result for $\mathcal{M}_{\mathcal{D}_\alpha}$ with $0 < \alpha < 1$ is also true.

Condition (0–3) is a geometric condition for a Carleson measure for $\mathcal{D}$. Carleson measures for $\mathcal{D}_\alpha$ were first characterized by Stegenga [1980]. His result says that a nonnegative measure μ on $\mathbb{D}$ satisfies

$$\int_\mathbb{D} |g(z)|^2 \, d\mu(z) \leq C \|g\|_\alpha^2 \quad \text{for all } g \in \mathcal{D}_\alpha$$

(in other words, is a Carleson measure for $\mathcal{D}_\alpha$) if and only if

$$\mu\left(\bigcup S(I_j) \right) \leq C \, \text{Cap}_\alpha \left(\bigcup I_j \right)$$

for any finite collection of disjoint arcs $\{I_j\}$ on $\partial \mathbb{D}$. Here $\text{Cap}_\alpha(\cdot)$ denotes an appropriate Bessel capacity depending on α. When $\alpha = 1$, the usual logarithmic capacity may be used for $\text{Cap}_1(\cdot)$. Using Stegenga's theorem, condition (0–3) can be replaced by

$$\sum \left(\log \frac{1}{1 - |z_n|^2} \right)^{-1} \delta_{z_n} \quad \text{is a Carleson measure for } \mathcal{D}. \qquad (0\text{--}4)$$

Hankel operators (small and big) on the Hardy and the Bergman spaces have been studied intensively in the past fifteen years. We refer the reader to [Luecking 1992; Peller 1982; Rochberg 1985; Zhu 1990] and references therein for more information. Denote by $P_{\mathcal{H}^2}$ the orthogonal projection from $\mathcal{L}^2(\partial \mathbb{D})$ onto $\mathcal{H}^2$. On the Hardy space, the Hankel operator with symbol $b \in \mathcal{L}^2(\partial \mathbb{D})$ can be written as

$$(I - P_{\mathcal{H}^2})(\bar{b}g) \quad \text{for } g \in \mathcal{H}^\infty.$$

Another existing definition is

$$\overline{P_{\mathcal{H}^2}(b\bar{g})} \quad \text{for } g \in \mathcal{H}^{\infty}.$$

In fact, if $\{\bar{b}_n\}$ is the sequence of Fourier coefficients of b, the two operators above correspond to the Hankel matrices $\{b_1, b_2, b_3, \ldots\}$ and $\{b_0, b_1, b_2, \ldots\}$, respectively. In more general spaces, these expressions define two different operators, called big Hankel and small Hankel, respectively. One basic question in the study of Hankel operators is that of understanding the "size" of the operators (for example boundedness or compactness) via the "smoothness" of their symbols.

We can view the Dirichlet space as a subspace of the Sobolev space $\mathcal{L}^{2,1}(\mathbb{D})$, defined as the completion of $C^1(\mathbb{D})$ under the norm

$$\|f\| = \left\{ \left| \int_{\mathbb{D}} f \, dA \right|^2 + \int_{\mathbb{D}} (|\partial_z f|^2 + |\partial_{\bar{z}} f|^2) \, dA \right\}^{1/2}.$$

Note that the restriction of this norm to $\mathcal{D}$, which yields $\|f\|^2 = |f(0)|^2 + D(f)$ if $f \in \mathcal{D}$, is different from but equivalent to the norm of $\mathcal{D}$ introduced previously. Denote by $\mathcal{P}$ the set of all polynomials in z on $\mathbb{D}$. Clearly, $\mathcal{P}$ is dense in $\mathcal{D}_\alpha$. Let $P_{\mathcal{D}}$ be the orthogonal projection from $\mathcal{L}^{2,1}(\mathbb{D})$ onto $\mathcal{D}$. On the Dirichlet space, the small Hankel operator with symbol b is defined densely by

$$\mathcal{H}_b(g) = \overline{P_{\mathcal{D}}(b\bar{g})} \quad \text{for } g \in \mathcal{P}.$$

It turns out that the big Hankel operator on the Dirichlet space with an analytic symbol is easy to study; this is opposite to the situation on the Bergman space. Therefore we discuss here only the small Hankel operator. The following two theorems can be found in [Rochberg and Wu 1993; Wu 1993], where they are proved for weighted Dirichlet and Bergman spaces.

THEOREM 0.4. *Suppose b is analytic on $\mathbb{D}$.*

(a) *The Hankel operator $\mathcal{H}_b$ is bounded on $\mathcal{D}$ if and only if $|b'(z)|^2 \, dA(z)$ is a Carleson measure for $\mathcal{D}$.*

(b) *The Hankel operator $\mathcal{H}_b$ is compact on $\mathcal{D}$ if and only if $|b'(z)|^2 \, dA(z)$ is a Carleson measure for $\mathcal{D}$ and satisfies*

$$\int_{\bigcup S(I_j)} |b'(z)|^2 \, dA(z) = o\left(\left\{ \log \frac{1}{\mathrm{Cap}(\bigcup I_j)} \right\}^{-1} \right). \tag{0--5}$$

Denote by $\mathcal{W}_{\mathcal{D}}$ and $w_{\mathcal{D}}$, respectively, the sets of all analytic functions b that satisfy conditions (a) and (b) in Theorem 0.4. Let $\mathcal{X}_{\mathcal{D}}$ be the set of all analytic functions f on $\mathbb{D}$ that can be expressed as $f = \sum g_j h_j'$, where $g_j, h_j \in \mathcal{D}$ with $\sum \|g_j\| \|h_j\| < \infty$. Define the norm of f in $\mathcal{X}_{\mathcal{D}}$ as

$$\|f\|_{\mathcal{X}_{\mathcal{D}}} = \inf \left\{ \sum \|g_j\| \|h_j\| : f = \sum g_j h_j' \quad \text{with } g_j, h_j \in \mathcal{D} \right\}.$$

THEOREM 0.5. *The dual of* $\mathfrak{X}_{\mathcal{D}}$ *is* $\mathcal{W}_{\mathcal{D}}$, *realized by the pairing*

$$\langle f, b \rangle^* = \int_{\mathbb{D}} f(z)\overline{b'(z)}\, dA(z), \qquad \text{with } f \in \mathfrak{X}_{\mathcal{D}} \text{ and } b \in \mathcal{W}_{\mathcal{D}};$$

The dual of $w_{\mathcal{D}}$ *is* $\mathfrak{X}_{\mathcal{D}}$, *realized by the pairing*

$$^*\langle b, f \rangle = \int_{\mathbb{D}} b'(z)\overline{f(z)}\, dA(z), \qquad \text{with } b \in w_{\mathcal{D}} \text{ and } f \in \mathfrak{X}_{\mathcal{D}}.$$

The form of the definition of $\mathfrak{X}_{\mathcal{D}}$ is natural. For example, a result in [Coifman et al. 1976] suggests that the right way to look at a function f in the Hardy space $\mathcal{H}^1$ (on the unit ball) is as $f = \sum f_j g_j$, where $f_j, g_j \in \mathcal{H}^2$, and

$$\|f\|_{\mathcal{H}^1} = \inf\{\sum \|f_j\|_{\mathcal{H}^2} \|g_j\|_{\mathcal{H}^2}\}.$$

Corresponding to the weighted Dirichlet spaces, one can define

$$\mathcal{W}_{\mathcal{D}_\alpha} = \{b \in \mathcal{D}_\alpha : g \mapsto b'g \text{ is bounded from } \mathcal{D}_\alpha \text{ to } \mathcal{D}_{\alpha-2}\}$$

and

$$w_{\mathcal{D}_\alpha} = \{b \in \mathcal{D}_\alpha : g \mapsto b'g \text{ is compact from } \mathcal{D}_\alpha \text{ to } \mathcal{D}_{\alpha-2}\}.$$

It is not hard to see that $\mathcal{W}_{\mathcal{H}^2} = \text{BMOA}$ and $w_{\mathcal{H}^2} = \text{VMOA}$. Therefore Theorem 0.5 is similar to the analytic versions of Fefferman's and Sarason's well-known theorems, saying that $(\mathcal{H}^1)^* = \text{BMOA}$ and $(\text{VMOA})^* = \mathcal{H}^1$.

NOTATION. In the rest of the paper the letter C denotes a positive constant that many vary at each occurrence but is independent of the essential variables or quantities.

1. Invariant Subspaces

The codimension-one property for invariant subspaces of the Dirichlet space is related to the cellular indecomposibility of the operator $(M_z, \mathcal{D})$. This concept was first introduced and studied by Olin and Thomson [1984] for more general Hilbert spaces. Later Bourdon [1986] proved in that if the operator M_z is cellular indecomposable on a Hilbert space $\mathcal{H}$ of analytic functions on $\mathbb{D}$ with certain properties, then every nonzero invariant subspace for $(M_z, \mathcal{H})$ has the codimension-one property. The required properties for $\mathcal{H}$ in Bourdon's paper are: the polynomials are dense in $\mathcal{H}$; the operator M_z is a bounded linear operator on $\mathcal{H}$; if $zg \in \mathcal{H}$ and g is analytic on $\mathbb{D}$ then $g \in \mathcal{H}$; and for each point $w \in \mathbb{D}$ the point evaluation at w is a bounded linear functional on $\mathcal{H}$. Clearly our spaces satisfy these requirements. The operator $(M_z, \mathcal{H})$ is said to be cellular indecomposable if $\mathcal{N} \cap \mathcal{Q} \neq \{0\}$ for any two nonzero invariant subspaces $\mathcal{N}$ and $\mathcal{Q}$ of $(M_z, \mathcal{H})$. We note that $(M_z, \mathcal{H}^2)$ clearly is cellular indecomposable by Beurling's Theorem. However, from an example constructed by Horowitz [1974], we know that $(M_z, \mathcal{B})$ is not cellular indecomposable. We also know that $(M_z, \mathcal{B})$

does not have the codimension-one property; see, for example, [Bercovici et al. 1985] or [Seip 1993].

To prove Theorem 0.1, we show that the operator $(M_z, \mathcal{D})$ is cellular indecomposable. This is a consequence of the following result.

THEOREM 1.1. *If $f \in \mathcal{D}$, then $f = \varphi/\psi$, where φ and ψ are in $\mathcal{D} \cap \mathcal{H}^\infty$.*

This was proved in [Richter and Shields 1988] for the Dirichlet spaces on connected domains, in [Richter and Sundberg 1992] for $\mathcal{D}(\mu)$, and in [Aleman 1992] for weighted Dirichlet spaces. We discuss the proof later.

Let $\mathcal{N} \neq \{0\}$ be an invariant subspace for $(M_z, \mathcal{D})$ and $f \in \mathcal{N} \setminus \{0\}$. By Theorem 1.1, there are functions $\varphi, \psi \in \mathcal{D} \cap \mathcal{H}^\infty$ such that $f = \varphi/\psi$. We claim that $\varphi = \psi f$ is in $[f] \subseteq \mathcal{N}$. Let $\psi_r(z) = \psi(rz)$, for $0 < r < 1$. It is clear that we only need to show $\psi_r f \to \varphi$ in $\mathcal{D}$. Straightforward estimates show that

$$\|\psi_r f - \varphi\| \leq \|\psi_r(f - f_r)\| + \|\varphi_r - \varphi\|$$
$$\leq \|\psi_r'(f - f_r)\|_{\mathcal{B}} + \|\psi\|_{\mathcal{H}^\infty}\|f - f_r\| + \|\varphi_r - \varphi\|.$$

It suffices to show the first term on the right goes to zero as $r \to 1_-$. Recall that $\mathcal{D}$ is contained in the little Bloch space, which consists of all the analytic functions g on $\mathbb{D}$ such that $(1 - |z|^2)|g'(z)| = o(1)$ as $|z| \to 1_-$. We have therefore

$$(1 - |rz|^2)|\psi_r'(z)| \leq C \quad \text{for all } |rz| < 1.$$

Write $f(z) = \sum_0^\infty a_n z^n$. Then

$$\left\|\frac{f - f_r}{1 - |rz|^2}\right\|_{\mathcal{B}}^2 \leq \sum_0^\infty |a_n|^2 \int_0^1 \frac{t^{2n} - (rt)^{2n}}{(1 - r^2t^2)^2} t\,dt \leq C \sum_0^\infty n |a_n|^2.$$

This is enough for the claim. The discussion proves in fact the following result.

LEMMA 1.2. *Suppose $\psi \in \mathcal{D} \cap \mathcal{H}^\infty$ and $f \in \mathcal{D}$. If $\psi f \in \mathcal{D}$, then $\psi f \in [f]$.*

From Theorem 1.1, we see that there is always a nonzero bounded function $\varphi \in \mathcal{N}$ if $\mathcal{N}$ is a nonzero invariant subspace for $(M_z, \mathcal{D})$. If $\mathcal{N}$ and $\mathcal{Q}$ are two nonzero invariant subspaces for $(M_z, \mathcal{D})$, there are nonzero bounded functions $\varphi \in \mathcal{N}$ and $\psi \in \mathcal{Q}$. Clearly $\varphi\psi$ is also nonzero, bounded and in $\mathcal{D}$, therefore in both $\mathcal{N}$ and $\mathcal{Q}$ by Lemma 1.2. Thus $\mathcal{N} \cap \mathcal{Q} \neq \{0\}$; that is, $(M_z, \mathcal{D})$ is cellular indecomposable.

To prove Theorem 1.1, we need several lemmas.

LEMMA 1.3. *Suppose $a_n, b_n \geq 0$, for $n = 0, 1, 2, \ldots$. If*

$$\sum b_n(1 - r^n) \leq \sum a_n(1 - r^n)$$

for every r in the interval $(1-\delta, 1)$ with some $\delta \in (0, 1)$, then

$$\sum n b_n \leq \sum n a_n.$$

PROOF. Without loss of generality, we assume $\sum na_n < \infty$. Clearly we have

$$\sum_{n=1}^{N} nb_n = \lim_{r \to 1_-} \frac{1}{1-r} \sum_{n=1}^{N} b_n(1-r^n) \le \lim_{r \to 1_-} \frac{1}{1-r} \sum_{n=1}^{\infty} a_n(1-r^n) \le \sum_{n=1}^{\infty} na_n.$$

This is enough. $\qquad\square$

Recall the inner-outer factorization for a nonzero function $f \in \mathcal{H}^2$: $f(z) = I(z)F(z)$, where the inner and outer factors $I(z)$ and $F(z)$ satisfy $|I(e^{i\theta})| = 1$ a.e. on $\partial\mathbb{D}$ and

$$F(z) = \exp\left(\frac{1}{2\pi} \int_0^{2\pi} \frac{e^{i\theta} + z}{e^{i\theta} - z} \log|f(e^{i\theta})| d\theta\right) \quad \text{for all } z \in \mathbb{D}.$$

One application of Lemma 1.3 is the following.

COROLLARY 1.4. *Let $f \in \mathcal{D}$ and let F be an analytic function on $\mathbb{D}$ with $|F(z)| \ge |f(z)|$ in $\mathbb{D}$ and $|F(e^{i\theta})| = |f(e^{i\theta})|$ a.e. on $\partial\mathbb{D}$. Then $F \in \mathcal{D}$ and $\|F\| \le \|f\|$. In particular this estimate is true if F is the outer factor of $f \in \mathcal{D}$.*

PROOF. Write $f = \sum f_n z^n$ and $F = \sum F_n z^n$. We have clearly

$$\sum |f_n|^2 = \|f\|_0^2 = \|F\|_0^2 = \sum |F_n|^2 \quad \text{and}$$

$$\sum |f_n|^2 r^{2n} = \|f(r\,\cdot)\|_0^2 \le \|F(r\,\cdot)\|_0^2 = \sum |F_n|^2 r^{2n} \quad \text{for all } r \in (0,1).$$

Since $D(f) = \sum n|f_n|^2 < \infty$, applying Lemma 1.3 to the inequality

$$\sum |F_n|^2(1-r^{2n}) = \|F\|_0^2 - \|F(r\,\cdot)\|_0^2 \le \|f\|_0^2 - \|f(r\,\cdot)\|_0^2 = \sum |f_n|^2(1-r^{2n}),$$

we get $D(F) \le D(f) < \infty$. This yields $F \in \mathcal{D}$ and $\|F\| \le \|f\|$. $\qquad\square$

Let (X, μ) be a probability space and let $g \in \mathcal{L}^1(X, \mu)$ be positive μ-a.e. on X and satisfy $\log g \in \mathcal{L}^1(X, \mu)$. Jensen's inequality says that $\exp\left(\int_X \log g \, d\mu\right) \le \int_X g \, d\mu$. Set $E(g) = \int_X g \, d\mu - \exp\left(\int_X \log g \, d\mu\right)$.

Applying Jensen's inequality, Aleman [1992] proved the following inequality.

LEMMA 1.5. *Suppose that g is positive μ-a.e. and that $\log g \in \mathcal{L}^1(X, \mu)$. Then*

$$E(\min\{g, 1\}) \le E(g) \quad and \quad E(\max\{g, 1\}) \le E(g).$$

Let $F \in \mathcal{H}^2$ be a outer function. Define the outer functions F_- and F_+ by

$$F_-(z) = \exp\left(\frac{1}{2\pi} \int_0^{2\pi} \frac{e^{i\theta} + z}{e^{i\theta} - z} \log\left(\min\{|F(e^{i\theta})|, 1\}\right) d\theta\right)$$

$$F_+(z) = \exp\left(\frac{1}{2\pi} \int_0^{2\pi} \frac{e^{i\theta} + z}{e^{i\theta} - z} \log\left(\max\{|F(e^{i\theta})|, 1\}\right) d\theta\right).$$

It is clear that F_- and $1/F_+$ are in $\mathcal{H}^\infty$, with norms bounded by 1, and that

$$F_-(z)F_+(z) = F(z), \qquad |F_-(z)| \le |F(z)|.$$

THEOREM 1.6. *Let $f \in \mathcal{H}^2$ and let $f = IF$ be the inner-outer factorization. If $f \in \mathcal{D}$, then $D(F_+) \le D(f)$ and $D(IF_-) \le D(f)$. Moreover F_+, IF_-, and $1/F_+$ are in $\mathcal{D}$, with*

$$\|F_+\| \le 1 + \|f\|, \qquad \|1/F_+\| \le 1 + \|f\|, \qquad \|IF_-\| \le \|f\|.$$

PROOF. We note first that $\|IF_-\|_0 \le \|f\|_0$. Using the fact $|1/F_+(z)| \le 1$, we get

$$|(1/F_+(z))'| = |F_+'(z)| / |F_+(z)|^2 \le |F_+'(z)|.$$

This implies $D(1/F_+) \le D(F_+)$. Therefore we only need to show $D(F_+) \le D(f)$ and $D(IF_-)$.

Applying Lemma 1.5 with $X = [0, 2\pi]$, $d\mu = 1/(2\pi)P_z(e^{i\theta})\,d\theta$, and $g(\theta) = |F(e^{i\theta})|^2$, we have

$$\int_0^{2\pi} P_z(e^{i\theta}) \left|F_\pm(e^{i\theta})\right|^2 \frac{d\theta}{2\pi} - \left|F_\pm(z)\right|^2 \le \int_0^{2\pi} P_z(e^{i\theta}) \left|F(e^{i\theta})\right|^2 \frac{d\theta}{2\pi} - \left|F(z)\right|^2.$$

$$(1\text{--}1)$$

Integrating both sides of the inequality for F_+ over $|z| = r \in (0, 1)$ with respect to the measure $d\theta/2\pi$, we obtain

$$\|F_+\|_0^2 - \|F_+(r \cdot)\|_0^2 \le \|F\|_0^2 - \|F(r \cdot)\|_0^2.$$

Applying Lemma 1.3 (as in the proof of Corollary 1.4), we obtain $D(F_+) \le D(F)$. Corollary 1.4 yields therefore $D(F_+) \le D(f)$.

Since $|F_-(z)| \le |F(z)|$, we have

$$\left(1 - |I(z)|^2\right) \left|F_-(z)\right|^2 \le \left|F(z)\right|^2 - \left|f(z)\right|^2.$$

Adding this inequality to (1--1) for F_-, we get

$$\int_0^{2\pi} P_z(e^{i\theta}) \left|I(e^{i\theta})F_-(e^{i\theta})\right|^2 \frac{d\theta}{2\pi} - \left|I(z)F_-(z)\right|^2$$

$$\le \int_0^{2\pi} P_z(e^{i\theta}) \left|f(e^{i\theta})\right|^2 \frac{d\theta}{2\pi} - \left|f(z)\right|^2.$$

Reasoning as above, we obtain $D(IF_-) \le D(f)$. $\qquad\square$

REMARK 1.7. By Theorem 1.6, together with Lemma 1.2 and the identity $(1/F_+)f = IF_-$, we have $IF_- \in [f]$ if $f \in \mathcal{D}$.

PROOF. Proof of Theorem 1.1 Assume $f \ne 0$ and $f \in \mathcal{D}$. Let $f = IF$ be the inner-outer factorization. Since $F_-F_+ = F$, we have

$$f = \frac{IF_-}{1/F_+}.$$

Since IF_-, $1/F_+ \in \mathcal{H}^\infty$ and IF_-, $1/F_+ \in \mathcal{D}$ by Theorem 1.6, we obtain the desired decomposition by letting $\varphi = IF_-$ and $\psi = 1/F_+$. $\qquad\square$

PROOF OF THEOREM 0.2. Let $\varphi \in \mathcal{N} \ominus z\mathcal{N}$ and $\|\varphi\| = 1$. Note that the polynomials are dense in $\mathcal{D}(m_\varphi)$. By the codimension-one property, the polynomial multiples of φ are dense in $\mathcal{N}$ (see also [Richter 1988]). Thus, to see why $\mathcal{N} = [\varphi] = \varphi\mathcal{D}(m_\varphi)$, it is enough to show that $\|\varphi p\| = \|p\|_{m_\varphi}$ for every polynomial p. One can compute this directly by using the fact that $(M_z, \mathcal{D})$ is an analytic 2-isometry; see [Richter 1991] for details. To see why the function φ is a multiplier of $\mathcal{D}$, note that $\mathcal{D} \subseteq \mathcal{D}(m_\varphi)$ if φ is bounded. Hence $\varphi\mathcal{D} \subseteq \varphi\mathcal{D}(m_\varphi) = \mathcal{N} \subseteq \mathcal{D}$, and

$$\|\varphi f\| = \|f\|_{m_\varphi} = \left(\|f\|_0^2 + D_{m_\varphi}(f)\right)^{1/2}$$
$$\leq \left(\|f\|_0 + \|\varphi\|_{\mathcal{H}^\infty}^2 D(f)\right)^{1/2} \leq \max\{1, \|\varphi\|_{\mathcal{H}^\infty}\} \|f\|.$$

We show now φ is bounded. Let k be the order of the zero of φ at the origin; thus $\varphi = z^k\psi$ with $\psi \in \mathcal{D}$ and $\psi(0) \neq 0$. It is well known that $\varphi = z^k\psi$ is a solution of the extremal problem

$$\inf\left\{\frac{\|z^k f\|}{|f(0)|} : z^k f \in \mathcal{N}\right\}. \tag{1–2}$$

We shall show that any unbounded function $z^k f \in \mathcal{N}$ with $f(0) \neq 0$ is not a solution of (1–2); that is, we can construct a bounded function f_N so that $z^k f_N \in \mathcal{N}$ and $\|z^k f_N\| / |f_N(0)| < \|z^k f\| / |f(0)|$, or, equivalently,

$$\|z^k f\|^2 \left(1 - \frac{|f_N(0)|^2}{|f(0)|^2}\right) < \|z^k f\|^2 - \|z^k f_N\|^2. \tag{1–3}$$

Let $f = IF$ be the inner-outer factorization of f. Consider $F_N = N(F/N)_-$, the outer function defined by

$$F_N(z) = \exp\left(\frac{1}{2\pi}\int_0^{2\pi} \frac{e^{i\theta} + z}{e^{i\theta} - z} \log\left(\min\{|F(e^{i\theta})|, N\}\right) d\theta\right).$$

Let $f_N = IF_N$. Then $|f_N(z)| \leq N$ for every $z \in \mathbb{D}$ and $\|f_N\| \leq \|f\|$ by Theorem 1.6; and $z^k f_N = (z^k f)_N \in [z^k f] \subseteq \mathcal{N}$ by Remark 1.7. Straightforward computation yields

$$1 - \frac{|f_N(0)|^2}{|f(0)|^2} = 1 - \frac{|F_N(0)|^2}{|F(0)|^2} = 1 - \exp\left(-\frac{1}{\pi}\int_0^{2\pi} \log\frac{\max\{|f(e^{i\theta})|, N\}}{N} d\theta\right).$$

Using the inequalities $\log(1 + x) \leq x$ and $1 - e^{-x} \leq x$ for $x \geq 0$, we obtain

$$1 - \frac{|f_N(0)|^2}{|f(0)|^2} \leq 1 - \exp\left(-\frac{1}{\pi}\int_0^{2\pi} \frac{\max\{|f(e^{i\theta})|, N\} - N}{N} d\theta\right)$$
$$\leq \frac{1}{N\pi}\int_0^{2\pi}\left(\max\{|f(e^{i\theta})|, N\} - N\right) d\theta.$$

Since $D(z^k f_N) = D((z^k f)_N) \leq D(z^k f)$ by Theorem 1.6, we have

$$\left\| z^k f \right\|^2 - \left\| z^k f_N \right\|^2 = \|f\|_0^2 - \|f_N\|_0^2 + D(z^k f) - D(z^k f_N) \geq \|F\|_0^2 - \|F_N\|_0^2$$

$$= \frac{1}{2\pi} \int_0^{2\pi} \left(|F(e^{i\theta})|^2 - \min\{|F(e^{i\theta})|^2, N^2\} \right) d\theta$$

$$= \frac{1}{2\pi} \int_0^{2\pi} \left(\max\{|f(e^{i\theta})|^2, N^2\} - N^2 \right) d\theta$$

$$\geq \frac{N}{2\pi} \int_0^{2\pi} \left(\max\{|f(e^{i\theta})|, N\} - N \right) d\theta.$$

These estimates show that, if f is unbounded, inequality (1–3) holds for large enough N. $\qquad\square$

A set $Z \subset \mathbb{D}$ is a zero set of a space $\mathcal{H}$ of functions on $\mathbb{D}$ if $Z = \{z \in \mathbb{D} : f(z) = 0\}$ for some $f \in \mathcal{H}$. Theorem 0.2 implies the following result [Marshall and Sundberg 1993].

THEOREM 1.8. *A set $Z \subset \mathbb{D}$ is a zero set of $\mathcal{D}$ if and only if it is a zero set of $\mathcal{M}_\mathcal{D}$.*

PROOF. It is trivial that a zero set of $\mathcal{M}_\mathcal{D}$ is a zero set of $\mathcal{D}$. Assume Z is a zero set of $\mathcal{D}$. Consider the set of functions $\mathcal{Z} = \{f \in \mathcal{D} : f(Z) = 0\}$. It is clear that $\mathcal{Z}$ is a nonzero invariant subspace for $(M_z, \mathcal{D})$. By Theorem 0.2, we have a function $\varphi \in \mathcal{Z} \ominus z\mathcal{Z}$ so that $\mathcal{Z} = [\varphi]$ and φ is a multiplier of $\mathcal{D}$. Clearly this φ has zero set Z. $\qquad\square$

2. Interpolation

The connection between reproducing kernels and interpolation can be explained in terms of the adjoints of multiplication operators. If $\varphi \in \mathcal{M}_\mathcal{D}$ then for $f \in \mathcal{D}$ and $\zeta \in \mathbb{D}$, we have

$$M_\varphi^*(k_\zeta)(z) = \langle M_\varphi^*(k_\zeta), k_z \rangle = \overline{\langle \varphi k_z, k_\zeta \rangle} = \overline{\varphi(\zeta)}\, k_\zeta(z).$$

Suppose $\{z_j\}_1^n$ is a finite sequence of distinct points in $\mathbb{D}$. Let $\varphi \in \mathcal{M}_\mathcal{D}$ satisfy $\|\varphi\|_\mathcal{M} \leq 1$ and

$$\varphi(z_j) = w_j, \quad j = 1, 2, \ldots, n. \tag{2–1}$$

Then for any finite sequence $\{a_j\}_1^n$ of complex numbers we have, by straightforward computation,

$$0 \leq \left\| \sum a_j k_{z_j} \right\|^2 - \left\| M_\varphi^* \left(\sum a_j k_{z_j} \right) \right\|^2 = \sum (1 - w_j \bar{w}_k) k(z_j, z_k) a_k \bar{a}_j. \tag{2–2}$$

This shows that the positive semidefiniteness of the $n \times n$ matrix

$$\left\{ (1 - w_j \bar{w}_k) k(z_j, z_k) \right\}$$

is a necessary condition for the interpolation problem (2–1) to have a solution in $\mathcal{M}_\mathcal{D}$. This argument in fact works in any reproducing kernel Hilbert space. Agler [1986] proved that for the Dirichlet space the necessary condition is also sufficient.

THEOREM 2.1. *If $\{z_j\}_1^n \subset \mathbb{D}$ and $\{w_j\}_1^n \subset \mathbb{C}$ satisfy*

$$\{(1 - w_j \bar{w}_k) k(z_j, z_k)\} \geq 0,$$

there exists $\varphi \in \mathcal{M}_\mathcal{D}$ with $\|\varphi\|_\mathcal{M} \leq 1$ and $\varphi(z_i) = w_i$ for $i = 1, 2, \ldots, n$.

One says that the Dirichlet space has the Pick property, because Pick first established such a theorem for interpolation in $\mathcal{H}^\infty$. Note that, if the Pick property holds, it also applies to countable sequences.

REMARK 2.2. For $n = 2$, inequality (2–2) becomes

$$\frac{|k(z_1, z_2)|^2}{\|k_{z_1}\|^2 \|k_{z_2}\|^2} \leq \frac{(1 - |w_1|^2)(1 - |w_2|^2)}{|1 - w_1 \bar{w}_2|^2}. \qquad (2\text{–}3)$$

A sequence of vectors $\{x_n\}$ in a Hilbert space $\mathcal{H}$ is called independent if $x_n \notin$ Span$\{x_k : k \neq n\}$ for all n. A sequence of unit vectors $\{u_n\}$ in a Hilbert space $\mathcal{H}$ is called an interpolating sequence for $\mathcal{H}$ if the map $x \mapsto \{\langle x, u_n \rangle\}$ maps $\mathcal{H}$ onto ℓ^2. We cite the Köthe–Toeplitz theorem here, which can be found in [Nikolskii 1986].

THEOREM 2.3. *Let $\{u_n\}$ be a sequence of unit vectors contained in a Hilbert space $\mathcal{H}$. Let $\mathcal{K}$ be the smallest closed subspace of $\mathcal{H}$ containing $\{u_n\}$. Then the following statements are equivalent.*

(1) *The sequence $\{u_n\}$ is an interpolating sequence for $\mathcal{H}$.*
(2) *For all $x \in \mathcal{K}$ satisfying $\|x\|^2 \asymp \sum \|\langle x, u_n \rangle\|$, and $\{u_n\}$ is independent.*
(3) *$\left\|\sum a_n u_n\right\|^2 \asymp \sum |a_n|^2$ for all sequences $\{a_n\}$.*
(4) *$\left\|\sum b_n u_n\right\| \leq C \left\|\sum a_n u_n\right\|$ for all sequences $\{a_n\}$ and $\{b_n\}$ such that $|b_n| \leq |a_n|$ for all n.*

A unit vector sequence $\{u_n\}$ with property (3) above is called a Riesz sequence in $\mathcal{H}$, and one with property (4) is called an unconditional basic sequence in $\mathcal{H}$. The following result ties the Pick property and Köthe–Toeplitz theorem together.

THEOREM 2.4. *Let $\{z_n\} \subset \mathbb{D}$ and let $\tilde{k}_n = k_{z_n} / \|k_{z_n}\|$. The following statements are equivalent:*

(1) *$\{z_n\}$ is an interpolating sequence for $\mathcal{M}_\mathcal{D}$.*
(2) *$\{\tilde{k}_n\}$ is an interpolating sequence for $\mathcal{D}$.*
(3) *$\{\tilde{k}_n\}$ is an unconditional basic sequence in $\mathcal{D}$.*
(4) *$\{\tilde{k}_n\}$ is a Riesz sequence in $\mathcal{D}$.*

PROOF. Theorem 2.3 implies the equivalence of (2), (3), and (4). To prove that (1) implies (3), suppose $\{z_n\}$ is an interpolating sequence for $\mathcal{M}_{\mathcal{D}}$. Let $\{a_n\}$ and $\{b_n\}$ be sequences such that $|b_n| \leq |a_n|$ for all n. Then there is a $\varphi \in \mathcal{M}_{\mathcal{D}}$ with $\varphi(z_n) = \overline{b_n/a_n}$ for $n = 1, 2, \ldots$. Let $C = \|\varphi\|_{\mathcal{M}}$. We have

$$0 \leq \sum \left(C^2 - \varphi(z_j)\overline{\varphi(z_k)}\right) k(z_j, z_k) \frac{\bar{a}_j}{\|k_{z_j}\|} \frac{a_k}{\|k_{z_k}\|}$$

$$= C^2 \left\|\sum a_j \tilde{k}_j\right\|^2 - \left\|M_\varphi^*\left(\sum a_j \tilde{k}_j\right)\right\|^2$$

$$= C^2 \left\|\sum a_j \tilde{k}_j\right\|^2 - \left\|\sum b_j \tilde{k}_j\right\|^2.$$

This proves that $\{\tilde{k}_n\}$ is an unconditional basic sequence in $\mathcal{D}$.

Conversely, suppose (3) holds. To prove (1), by weak convergence of operators of the form M_φ, it suffices to show that each finite subsequence of $\{z_n\}$ is an interpolating sequence with the solutions in $\mathcal{M}_{\mathcal{D}}$ having uniformly bounded norms. This follows from the Pick property and the following inequality, which is equivalent to (3):

$$\sum \left(C^2 - \frac{\bar{b}_j}{\bar{a}_j} \frac{b_k}{a_k}\right) k(z_j, z_k) \frac{\bar{a}_j}{\|k_{z_j}\|} \frac{a_k}{\|k_{z_k}\|} = C^2 \left\|\sum a_j \tilde{k}_j\right\|^2 - \left\|\sum b_j \tilde{k}_j\right\|^2 \geq 0,$$

where $|b_j| \leq |a_j|$ for all j. $\qquad\square$

SKETCH OF PROOF OF THEOREM 0.3. The proof is found in [Marshall and Sundberg 1993]. Suppose $\{z_n\}$ is an interpolating sequence for $\mathcal{M}_{\mathcal{D}}$; then $\{\tilde{k}_n\}$ is an interpolating sequence for $\mathcal{D}$, by Theorem 2.4. This is equivalent to

$$\sum |f(z_n)|^2 \left(\log \frac{1}{1 - |z_n|^2}\right)^{-1} = \sum |\langle f, \tilde{k}_n \rangle|^2 \leq C \|f\|^2 \quad \text{for all } f \in \mathcal{D}.$$

Thus (0–4) holds, and hence so does (0–3).

Since $\{z_n\}$ is an interpolating sequence for $\mathcal{M}_{\mathcal{D}}$, we can find, for any distinct m and n, a $\varphi \in \mathcal{M}_{\mathcal{D}}$ so that $\varphi(z_n) = 1$, $\varphi(z_m) = 0$, and $\|\varphi\|_{\mathcal{M}} \leq C$. By Remark 2.2 we have

$$\frac{\left|\log \dfrac{1}{1 - \bar{z}_m z_n}\right|^2}{\log \dfrac{1}{1 - |z_n|^2} \log \dfrac{1}{1 - |z_m|^2}} = \frac{|k(z_n, z_m)|^2}{\|k_{z_n}\|^2 \|k_{z_m}\|^2} \leq 1 - \frac{1}{C^2}.$$

Marshall and Sundberg [1993] proved that this condition is equivalent to (0–2).

Now suppose that (0–2) and (0–3) hold. To show that $\{z_n\}$ is an interpolating sequence, we use the Pick property of the Dirichlet space, which allows us to convert the interpolation problem to an "$\mathcal{L}^2$" problem. That is, we prove that $\{\tilde{k}_n\}$ is a Riesz sequence in $\mathcal{D}$, then use Theorem 2.4. It can be shown that the sequence $\{\tilde{k}_n\}$ will be a Riesz sequence if the sequence $\{K_n = \text{Re}(\tilde{k}_n)\}$ is a Riesz sequence in the harmonic Dirichlet space $\mathcal{D}_h$, which consists of all harmonic functions in $\mathbb{D}$ with finite Dirichlet integral, the norm being given by $\|u\|^2 = \|u\|^2_{\mathcal{L}^2(\partial\mathbb{D})} + \|\nabla u\|^2_{\mathcal{L}^2(\mathbb{D})}$. By Theorem 2.3, the last condition holds if $\{K_n\}$ is an unconditional basic sequence in $\mathcal{D}_h$.

Suppose $|b_n| \leq |a_n|$ and set $t_n = a_n/b_n$ for all n. Then $|t_n| \leq 1$. To show that $\{K_n\}$ is an unconditional basic sequence in $\mathcal{D}_h$, we must show that

$$\left\|\sum t_n a_n K_n\right\| = \left\|\sum b_n K_n\right\| \leq C \left\|\sum a_n K_n\right\|.$$

Set $T(K_n) = t_n K_n$. It suffices to show that T can be extended to a bounded linear operator on the harmonic Dirichlet space $\mathcal{D}_h$. Note that

$$T^*(u)(z_n) = \left\langle T^*(u), \operatorname{Re}(k_{z_n})\right\rangle = \|k_{z_n}\| \left\langle u, T(K_n)\right\rangle = \bar{t}_n u(z_n).$$

Therefore it suffices to find a bounded linear map $u \mapsto v$ on $\mathcal{D}_h$ such that

$$v(z_n) = \bar{t}_n u(z_n) \quad \text{and} \quad \|v\| \leq C \|u\|.$$

The construction of the desired linear map in [Marshall and Sundberg 1993] requires deep and elegant estimates involving Stegenga's capacity condition for Carleson measures for $\mathcal{D}$. We shall sketch the idea. First, one uses conditions (0–2) and (0–3) to construct a bounded function φ on the disk $\mathbb{D}$ so that $\varphi(z_n) = t_n$ for $n = 1, 2, \ldots$ and such that

$$|\nabla\varphi(z)|^2 \, dA(z)$$

is a Carleson measure for $\mathcal{D}$. This implies, by the Dirichlet principle,

$$\int_{\mathbb{D}} |\nabla P(\varphi^* u^*)|^2 \, dA \leq \int_{\mathbb{D}} |\nabla(\varphi u)|^2 \, dA \leq C \|u\|^2 \quad \text{for all } u \in \mathcal{D}_h,$$

where $P(\psi^*)$ is the Poisson integral of the boundary function ψ^* of ψ. The desired linear map comes from correcting the function $P(\varphi^* u^*)$ by setting

$$v = P(\varphi^* u^*) + \sum \left(t_n u(z_n) - P(\varphi^* u^*)(z_n)\right) f_n,$$

where f_n is a harmonic function in $\mathcal{D}_h$ with $f_n(z_m) = \delta_{n,m}$. The existence of such functions, which can even be chosen to be analytic, follows from a result proved in [Shapiro and Shields 1962], which requires the condition

$$\sum \left(\log \frac{1}{1 - |z_n|^2}\right)^{-1} \leq M.$$

It is obvious that this condition follows from (0–3). Of course one needs to show that

$$\sum \left|P(\varphi^* u^*)(z_n) - t_n u(z_n)\right| \leq C \|u\| \quad \text{for all } u \in \mathcal{D}_h.$$

This again requires deep estimates related to capacity. $\qquad\square$

As noted in [Marshall and Sundberg 1993], the idea above provides also an easy proof of Carleson's interpolation theorem.

To end this section, we turn to another result proved in [Marshall and Sundberg 1993], which shows a different application of the Pick property.

Let $Z \subset \mathbb{D}$ be a zero set of $\mathcal{D}$ (or $\mathcal{M}_\mathcal{D}$), and take $z_0 \notin Z$. Consider the extremal problems

$$C_\mathcal{D} = \inf\{\|f\| : f(z_0) = 1 \text{ and } f(Z) = 0\} \tag{2–4}$$

and

$$C_{\mathcal{M}_\mathcal{D}} = \inf\{\|\varphi\|_\mathcal{M} : \varphi(z_0) = 1 \text{ and } \varphi(Z) = 0\}. \tag{2–5}$$

THEOREM 2.5. *The problems* (2–4) *and* (2–5) *have unique solutions* f_0 *and* φ_0, *and they satisfy*

$$\|f_0\| = \frac{\|\varphi_0\|_\mathcal{M}}{\|k_{z_0}\|} \quad \text{and} \quad f_0 = \varphi_0 \frac{k_{z_0}}{\|k_{z_0}\|^2}.$$

PROOF. Standard reasoning shows that solutions exist. A little elementary work shows that the solution of (2–4) is unique. Indeed, if f and g are distinct solutions to (2–4), then $\|f\| = \|g\|$, $f(z_0) = g(z_0)$, and $f(Z) = g(Z) = 0$, where we have set $Z = \{z_j : j = 1, 2, \ldots\}$. We claim that $\operatorname{Re}\langle f, g \rangle = \|f\|^2$, and therefore that

$$\|f - g\|^2 = \|f\|^2 + \|g\|^2 - 2\operatorname{Re}\langle f, g \rangle = 0.$$

If this is not the case, then $h = \frac{1}{2}(f + g)$ satisfies $h(z_0) = 1$, $h(Z) = 0$, and

$$\|h\|^2 = \frac{1}{4}\left(\|f\| + \|g\|^2 + 2\operatorname{Re}\langle f, g \rangle\right) < \|f\|^2.$$

This is impossible.

Let f_0 be the unique solution of (2–4) and let φ be any solution of (2–5). We show that

$$f_0 = \varphi \frac{k_{z_0}}{\|k_{z_0}\|^2},$$

so φ is unique. It is easy to see that

$$0 \le \left\|\sum a_j k_{z_j}\right\|^2 - \frac{1}{\|f_0\|^2}\left|\left\langle \sum a_j k_{z_j}, f_0 \right\rangle\right|^2 = \sum \left(1 - \frac{f_0(z_j)\overline{f_0(z_k)}}{\|f_0\|^2 \|k_{z_0}\|^2}\right) k(z_j, z_k)\, \bar{a}_j a_k.$$

By the Pick property, there is a $\psi \in \mathcal{M}_\mathcal{D}$ with $\|\psi\|_\mathcal{M} \le 1$, and

$$\psi(z_0) = \frac{1}{\|f_0\| \, \|k_{z_0}\|}, \quad \text{and} \quad \psi(z_j) = 0 \text{ for } j = 1, 2, \ldots.$$

Since φ is an extremal solution, we have

$$\|\varphi\|_\mathcal{M} \le \|\|f_0\| \, \|k_{z_0}\| \, \psi\|_\mathcal{M} \le \|f_0\| \, \|k_{z_0}\|.$$

On the other hand, the function

$$g = \varphi \frac{k_{z_0}}{\|k_{z_0}\|^2}$$

satisfies $g(z_0) = 1$, $g(Z) = 0$, and

$$\|g\| \le \|\varphi\|_{\mathcal{M}} / \|k_{z_0}\| \le \|f_0\|.$$

Therefore $f_0 = g$, as we wished to show.   □

It is not hard to show that f_0 and φ_0 in Theorem 2.5 have zero set exactly Z (counting multiplicities!).

3. Hankel Operators on the Dirichlet Space

In this section, we use the norm $\|f\| = \left(|f(0)|^2 + D(f)\right)^{1/2}$ for the Dirichlet space. In this case the reproducing kernel of $\mathcal{D}$ is

$$k_{\mathcal{D}}(z, w) = 1 + \log\left(\frac{1}{1 - \bar{w}z}\right).$$

We note that

$$k_{\mathcal{B}}(z, w) = \partial_z \partial_{\bar{w}} k_{\mathcal{D}}(z, w) = \frac{1}{(1 - \bar{w}z)^2}$$

is the reproducing kernel for the Bergman space $\mathcal{B}$.

Assume that the Hankel operator $\mathcal{H}_b$ is bounded on $\mathcal{D}$. For g in $\mathcal{P}$, we have

$$\mathcal{H}_b(g)(z) = \overline{\langle b\bar{g}, k_{\mathcal{D}}(\cdot, z)\rangle} = \int_{\mathbb{D}} \bar{b}g\, dA + \overline{\langle b'\bar{g}, \partial_w k_{\mathcal{D}}(\cdot, z)\rangle}_{\mathcal{L}^2(\mathbb{D})}.$$

For simplicity, we remove the rank-one operator $g \mapsto \int_{\mathbb{D}} \bar{b}g\, dA$, and take

$$\mathcal{H}_b(g)(z) = \overline{\langle b'\bar{g}, \partial_w k_{\mathcal{D}}(\cdot, z)\rangle}_{\mathcal{L}^2(\mathbb{D})} \quad \text{for all } g \in \mathcal{P}, \tag{3–1}$$

as the definition for our Hankel operator on $\mathcal{D}$. It is easy to compute that

$$\partial_z \overline{\mathcal{H}_b(g)(z)} = \overline{\langle b'\bar{g}, k_{\mathcal{B}}(\cdot, z)\rangle}_{\mathcal{L}^2(\mathbb{D})} = \overline{P_{\mathcal{B}}(b'\bar{g})(z)},$$

where $P_{\mathcal{B}}$ is the orthogonal projection from $\mathcal{L}^2 = \mathcal{L}^2(\mathbb{D})$ onto $\mathcal{B}$. We have then

$$\langle h, \overline{\mathcal{H}_b(g)}\rangle = \langle h', P_{\mathcal{B}}(b'\bar{g})\rangle_{L^2} = \langle h'g, b'\rangle_{L^2} \quad \text{for all } h,\, g \in \mathcal{P}. \tag{3–2}$$

We see that $b \in \mathcal{D}$ is a necessary condition for the Hankel operator $\mathcal{H}_b$ to be bounded on $\mathcal{D}$. From (3–2), we get also that the boundedness or compactness of $\mathcal{H}_b$ on $\mathcal{D}$ are equivalent, respectively, to the boundedness or compactness of the operator

$$g \mapsto P_{\mathcal{B}}(b'\bar{g}) \tag{3–3}$$

from $\overline{\mathcal{D}} = \{\bar{g} : g \in \mathcal{D}\}$ to $\mathcal{B}$.

The following result is standard; see [Arazy et al. 1990], for example.

LEMMA 3.1. *Suppose φ is a C^1 function on a neighborhood of $\mathbb{D}$. Then*

$$\varphi(z) = P_{\mathcal{B}}(\varphi)(z) + \int_{\mathbb{D}} \frac{\partial_{\bar{w}}\varphi(w)(1 - |w|^2)}{(z - w)(1 - \bar{w}z)}\, dA(w) \quad \text{for all } z \in \mathbb{D}.$$

LEMMA 3.2. *The linear operator*

$$f(z) \mapsto \int_{\mathbb{D}} \frac{f(w)(1 - |w|^2)}{|z - w||1 - \overline{w}z|} \, dA(w)$$

is bounded on $\mathcal{L}^2(\mathbb{D})$.

Lemma 3.2 can be proved easily by using Schur's test with the test function $u(z) = (1 - |z|^2)^{-1/4}$.

PROOF OF THEOREM 0.4. Suppose the measure $|b'(z)|^2 \, dA(z)$ is a Carleson measure for $\mathcal{D}$. Then $b'\bar{g}$ is in $\mathcal{L}^2$. This implies the map (3–3) is bounded.

Now suppose the Hankel operator $\mathcal{H}_b$ or, equivalently, the map (3–3) is bounded. We shall show that $b'\bar{g}$ is in $\mathcal{L}^2$, for every $g \in \mathcal{D}$, with norm bounded by $C \|g\|$. We note that $P_{\mathcal{B}}(b'\bar{g})$ is the best approximation to $b'\bar{g}$ in $\mathcal{B}$. Therefore by Lemmas 3.1 and 3.2, we have the following formula for the difference between $b'\bar{g}$ and $P_{\mathcal{B}}(b'\bar{g})$:

$$b'(z)\overline{g(z)} - P_{\mathcal{B}}(b'\bar{g})(z) = \int_{\mathbb{D}} \frac{b'(w)\overline{g'(w)}(1 - |w|^2)}{(z - w)(1 - \overline{w}z)} \, dA(w) \quad \text{for all } g \in \mathcal{P}. \quad (3\text{–}4)$$

Since $b \in \mathcal{D}$ and $\mathcal{D}$ is contained in the little Bloch space, we have

$$(1 - |w|^2)\big|b'(w)\big| \le C \|b\|.$$

Thus by Lemma 3.2 and the assumption, we obtain

$$\|b'\bar{g}\|_{\mathcal{L}^2}^2 = \big\|P_{\mathcal{B}}(b'\bar{g})\big\|_{\mathcal{L}^2}^2 + \big\|b'\bar{g} - P_{\mathcal{B}}(b'\bar{g})\big\|_{\mathcal{L}^2}^2 \le C \|b\|^2 \|g\|^2 \quad \text{for all } g \in \mathcal{P}.$$

This proves part (a) of the theorem.

The operator defined by (3–4) is in fact compact from $\bar{\mathcal{D}}$ to $\mathcal{L}^2$, because $\mathcal{D}$ is a subset of the little Bloch space. Therefore the compactness of $\mathcal{H}_b$ is equivalent to the compactness of the multiplier $M_{b'} : \mathcal{D} \to \mathcal{B}$. A result in [Rochberg and Wu 1992] implies therefore that condition (0–5) is necessary and sufficient for the compactness of $\mathcal{H}_b$. □

The proof of Theorem 0.5 requires a general result about pairing of operators. Suppose $\mathcal{H}$ and $\mathcal{K}$ are Hilbert spaces. The trace class of linear operators from $\mathcal{H}$ to $\mathcal{K}$, denoted by $\mathcal{S}_1 = \mathcal{S}_1(\mathcal{H}, \mathcal{K})$, is the set of all compact operators T from $\mathcal{H}$ to $\mathcal{K}$ for which the sequence of singular numbers

$$\big\{s_k(T) = \inf\{\|T - R\| : \text{rank}(R) < k\}\big\}_1^\infty$$

belongs to ℓ^1. The $\mathcal{S}_1$ norm of T is defined by

$$\|T\|_{\mathcal{S}_1} = \|\{s_k(T)\}_1^\infty\|_{\ell^1}.$$

We will use $\mathcal{S}_0 = \mathcal{S}_0(\mathcal{H}, \mathcal{K})$ and $\mathcal{S}_\infty = \mathcal{S}_\infty(\mathcal{H}, \mathcal{K})$ for the sets of compact operators and bounded operators from $\mathcal{H}$ to $\mathcal{K}$, respectively. Let T and S be bounded linear operators from $\mathcal{H}$ to $\mathcal{K}$ and from $\mathcal{K}$ to $\mathcal{H}$, respectively. The pairing of T and S is given by

$$\langle T, S \rangle = \text{trace}(TS).$$

The following standard theorem can be found in [Zhu 1990], for example.

THEOREM 3.3. (a) $(\mathcal{S}_0)^* = \mathcal{S}_1$ and $(\mathcal{S}_1)^* = \mathcal{S}_\infty$.

(b) (*Schmidt decomposition*) *Φ is a compact operator from $\mathcal{H}$ to $\mathcal{K}$ if and only if Φ can be written as*

$$\Phi = \sum \lambda_j \langle \, \cdot \, , f_j \rangle_{\mathcal{H}} e_j,$$

where $\{\lambda_j\}_1^\infty$ is a sequence of numbers tending to 0, and $\{f_j\}_1^\infty$ and $\{e_j\}_1^\infty$ are orthonormal sequences in $\mathcal{H}$ and $\mathcal{K}$, respectively. Moreover, if $\{\lambda_j\}_1^\infty$ is in ℓ^1, then

$$\|\Phi\|_{\mathcal{S}_1} = \sum |\lambda_j|.$$

PROOF OF THEOREM 0.5. It is easy to check that $\mathcal{W}_{\mathcal{D}} \subseteq (\mathcal{X}_{\mathcal{D}})^*$ and $\mathcal{X}_{\mathcal{D}} \subseteq (w_{\mathcal{D}})^*$ by using formula (3–2). We prove $(\mathcal{X}_{\mathcal{D}})^* \subseteq \mathcal{W}_{\mathcal{D}}$ next.

Suppose $T \in (\mathcal{X}_{\mathcal{D}})^*$. For any $g, h \in \mathcal{D}$, it is clear that $gh' \in \mathcal{X}_{\mathcal{D}}$ and $\|gh'\|_{\mathcal{X}_{\mathcal{D}}} \leq \|g\| \|h\|$. Hence if in addition $h(0) = 0$, then

$$|T(gh')| \leq \|T\| \|gh'\|_{\mathcal{X}_{\mathcal{D}}} \leq \|T\| \|g\| \|h\| = \|T\| \|g\| \|h'\|_{\mathcal{B}}.$$

This inequality shows that for fixed $g \in \mathcal{D}$ the linear functional $h' \mapsto T(gh')$ on $\mathcal{B}$ is bounded. Hence by the Riesz–Fischer Theorem there is a $T_g \in \mathcal{B}$ such that

$$T(gh') = \langle h', T_g \rangle_{L^2} \quad \text{for all } h' \in \mathcal{B}.$$

Clearly T_g is uniquely determined by g and the linear map $g \mapsto T_g$ from $\mathcal{D}$ to $\mathcal{B}$ is bounded with $\|T_g\|_{\mathcal{B}} \leq \|T\| \|g\|$.

Let $b(z) = b_T(z) = \int_0^z T_1(\zeta)\, d\zeta \in \mathcal{D}$. For any $g \in \mathcal{D}$ we have

$$T_g(w) = \langle T_g, k_{\mathcal{B}}(\,\cdot\,, w) \rangle_{L^2} = \overline{T(gk_{\mathcal{B}}(\,\cdot\,, w))}.$$

Since for fixed $w \in \mathbb{D}$, $gk_{\mathcal{B}}(\,\cdot\,, w)$ is always in $\mathcal{B}$, we have

$$T(gk_{\mathcal{B}}(\,\cdot\,, w)) = \langle gk_{\mathcal{B}}(\,\cdot\,, w), T_1 \rangle_{L^2} = \langle gk_{\mathcal{B}}(\,\cdot\,, w), b' \rangle_{L^2} = \partial_{\bar{w}} \langle g\partial_z k_{\mathcal{D}}(\,\cdot\,, w), b' \rangle_{L^2}.$$

This implies that $T_g(w) = \partial_w \overline{\mathcal{H}_{b_T}(g)(w)}$, for any $g \in \mathcal{P}$. We conclude therefore

$$\|\mathcal{H}_b(g)\| = \|T_g\|_{\mathcal{B}} \leq \|T\| \|g\|,$$

and hence $\|\mathcal{H}_b\| \leq \|T\|$. By Theorem 0.4, we have $b \in \mathcal{W}_{\mathcal{D}}$ and then

$$T_g(w) = \partial_w \overline{\mathcal{H}_{b_T}(g)(w)} \quad \text{for all } g \in \mathcal{D}. \tag{3–5}$$

This discussion also yields (since $b \in \mathcal{W}_{\mathcal{D}}$)

$$T(gh') = \langle gh', T_1 \rangle_{L^2} \quad \text{for all } g, h \in \mathcal{D}.$$

This implies that the map $T \mapsto b_T$ from $(\mathfrak{X}_\mathcal{D})^*$ to $\mathcal{W}_\mathcal{D}$ is bounded and one-to-one. To complete the proof, it remains to verify that

$$T(f) = \langle f, b_T \rangle^* \quad \text{for all } f \in \mathfrak{X}_\mathcal{D}.$$

This is easy to check by using (3–5) and (3–2).

To prove that $(w_\mathcal{D})^* \subseteq \mathfrak{X}_\mathcal{D}$, we consider the map $b \mapsto \partial \overline{\mathcal{H}}_b$ from $w_\mathcal{D}$ to $\mathcal{S}_0(\overline{\mathcal{D}}, \mathcal{B})$. This map is clearly one-to-one and maps $w_\mathcal{D}$ onto a closed subspace of $\mathcal{S}_0$. Take $L \in (w_\mathcal{D})^*$. Extend L to a bounded linear functional $\tilde{L}$ on $\mathcal{S}_0$ so that $\|\tilde{L}\| = \|L\|$. By Theorem 3.3(a), there is a Φ in $\mathcal{S}_1(\mathcal{B}, \overline{\mathcal{D}})$ such that $\|\Phi\|_{\mathcal{S}_1} = \|\tilde{L}\|$ and $\tilde{L}(T) = \langle T, \Phi \rangle$, for any $T \in \mathcal{S}_0$. Suppose the Schmidt decomposition of Φ given by Theorem 3.3(b) is

$$\Phi = \sum s_j \langle \, \cdot \, , f_j \rangle_{\mathcal{L}^2} \bar{g}_j,$$

where $\{s_j\}_1^\infty$ is the sequence of singular numbers of Φ, and $\{f_j\}_1^\infty$ and $\{\bar{g}_j\}_1^\infty$ are orthonormal sequences in $\mathcal{B}$ and $\overline{\mathcal{D}}$, respectively.

It is clear that $\left\{ h_j(z) = \int_0^z f_j(\zeta) \, d\zeta \right\}_0^\infty$ is an orthonormal sequence in $\mathcal{D}$. Set

$$f = f_L = \sum s_j g_j f_j = \sum s_j g_j h_j'. \tag{3–6}$$

Then clearly f is in $\mathfrak{X}_\mathcal{D}$, and

$$\|f\|_{\mathfrak{X}_\mathcal{D}} \leq \sum |s_j| = \|\Phi\|_{\mathcal{S}_1} = \|L\|.$$

For any $b \in w_\mathcal{D}$, we have

$$L(b) = \tilde{L}\left(\partial \overline{\mathcal{H}}_b\right) = \langle \partial \overline{\mathcal{H}}_b, \Phi \rangle = \text{trace}\left(\partial \overline{\mathcal{H}}_b \Phi\right) = \text{trace}\left(\Phi \partial \overline{\mathcal{H}}_b\right)$$
$$= \sum \langle \Phi \partial \overline{\mathcal{H}}_b(\bar{g}_j), \bar{g}_j \rangle = \sum s_j \langle \partial \overline{\mathcal{H}}_b(g_j), f_j \rangle_{\mathcal{L}^2} = \sum s_j \langle b', g_j h_j' \rangle_{\mathcal{L}^2}$$
$$= \langle b', f \rangle_{\mathcal{L}^2};$$

thus

$$L(b) = {}^* \langle b, f \rangle \quad \text{for all } b \in w_\mathcal{D}. \tag{3–7}$$

This implies $\|L\| \leq \|f\|_{\mathfrak{X}_\mathcal{D}}$, and hence $\|L\| = \|f\|_{\mathfrak{X}_\mathcal{D}}$.

To complete the proof it remains to show that the map $L \mapsto f_L$ defined by (3–6) is well defined and one-to-one.

In fact for any $\zeta \in \mathbb{D}$, let $b_\zeta(z) = \partial_{\bar\zeta} k_\mathcal{D}(z, \zeta)$. By formula (3–1) and the equality $b_\zeta'(z) = k_\mathcal{B}(z, \zeta)$, we get

$$\mathcal{H}_{b_\zeta}(g)(w) = \langle g \partial_z k_\mathcal{D}(\cdot, w), b_\zeta' \rangle_{\mathcal{L}^2} = \partial_\zeta k_\mathcal{D}(\zeta, w) g(\zeta) \quad \text{for all } g \in \mathcal{P}.$$

This shows that $\mathcal{H}_{b_\zeta}$ is a compact operator (of rank one!). Thus $b_\zeta \in w_\mathcal{D}$ and

$$L(b_\zeta) = {}^* \langle b_\zeta, f_L \rangle = \langle b_\zeta', f_L \rangle_{\mathcal{L}^2} = \overline{f_L(\zeta)}.$$

The "one-to-one" part is then an immediate consequence of the identity (3–7). $\square$

References

[Agler 1986] J. Agler, "Interpolation", preprint, 1986.

[Aleman 1992] A. Aleman, "Hilbert spaces of analytic functions between the Hardy and the Dirichlet space", *Proc. Amer. Math. Soc.* **115**:1 (1992), 97–104.

[Arazy and Fisher 1985] J. Arazy and S. D. Fisher, "The uniqueness of the Dirichlet space among Möbius-invariant Hilbert spaces", *Illinois J. Math.* **29**:3 (1985), 449–462.

[Arazy et al. 1988] J. Arazy, S. D. Fisher, and J. Peetre, "Hankel operators on weighted Bergman spaces", *Amer. J. Math.* **110**:6 (1988), 989–1053.

[Arazy et al. 1990] J. Arazy, S. D. Fisher, S. Janson, and J. Peetre, "An identity for reproducing kernels in a planar domain and Hilbert-Schmidt Hankel operators", *J. Reine Angew. Math.* **406** (1990), 179–199.

[Axler 1986] S. Axler, "The Bergman space, the Bloch space, and commutators of multiplication operators", *Duke Math. J.* **53**:2 (1986), 315–332.

[Axler 1992] S. Axler, "Interpolation by multipliers of the Dirichlet space", *Quart. J. Math. Oxford Ser.* (2) **43**:172 (1992), 409–419.

[Bercovici et al. 1985] H. Bercovici, C. Foiaş, and C. Pearcy, *Dual algebras with applications to invariant subspaces and dilation theory*, CBMS Regional Conference Series in Mathematics **56**, Amer. Math. Soc., Providence, 1985.

[Bourdon 1986] P. S. Bourdon, "Cellular-indecomposable operators and Beurling's theorem", *Michigan Math. J.* **33**:2 (1986), 187–193.

[Carleson 1958] L. Carleson, "An interpolation problem for bounded analytic functions", *Amer. J. Math.* **80** (1958), 921–930.

[Coifman et al. 1976] R. R. Coifman, R. Rochberg, and G. Weiss, "Factorization theorems for Hardy spaces in several variables", *Ann. of Math.* (2) **103**:3 (1976), 611–635.

[Horowitz 1974] C. Horowitz, "Zeros of functions in the Bergman spaces", *Duke Math. J.* **41** (1974), 693–710.

[Luecking 1992] D. H. Luecking, "Characterizations of certain classes of Hankel operators on the Bergman spaces of the unit disk", *J. Funct. Anal.* **110**:2 (1992), 247–271.

[Marshall and Sundberg 1993] D. E. Marshall and C. Sundberg, "Interpolating sequences for the multipliers of the Dirichlet space", preprint, 1993. Available at http://www.math.washington.edu/~marshall/preprints/preprints.html% posturlhook.

[Nikolskii 1986] N. K. Nikol'skiĭ, *Treatise on the shift operator*, Grundlehren der mathematischen Wissenschaften **273**, Springer, Berlin, 1986.

[Olin and Thomson 1984] R. F. Olin and J. E. Thomson, "Cellular-indecomposable subnormal operators", *Integral Equations Operator Theory* **7**:3 (1984), 392–430.

[Peller 1982] V. V. Peller, "Vectorial Hankel operators, commutators and related operators of the Schatten–von Neumann class γ_p", *Integral Equations Operator Theory* **5**:2 (1982), 244–272.

[Richter 1988] S. Richter, "Invariant subspaces of the Dirichlet shift", *J. Reine Angew. Math.* **386** (1988), 205–220.

[Richter 1991] S. Richter, "A representation theorem for cyclic analytic two-isometries", *Trans. Amer. Math. Soc.* **328**:1 (1991), 325–349.

[Richter and Shields 1988] S. Richter and A. Shields, "Bounded analytic functions in the Dirichlet space", *Math. Z.* **198**:2 (1988), 151–159.

[Richter and Sundberg 1991] S. Richter and C. Sundberg, "A formula for the local Dirichlet integral", *Michigan Math. J.* **38**:3 (1991), 355–379.

[Richter and Sundberg 1992] S. Richter and C. Sundberg, "Multipliers and invariant subspaces in the Dirichlet space", *J. Operator Theory* **28**:1 (1992), 167–186.

[Rochberg 1985] R. Rochberg, "Decomposition theorems for Bergman spaces and their applications", pp. 225–277 in *Operators and function theory* (Lancaster, 1984), edited by S. C. Power, NATO Adv. Sci. Inst. Ser. C: Math. Phys. Sci. **153**, Reidel, Dordrecht, 1985.

[Rochberg and Wu 1992] R. Rochberg and Z. Wu, "Toeplitz operators on Dirichlet spaces", *Integral Equations Operator Theory* **15**:2 (1992), 325–342.

[Rochberg and Wu 1993] R. Rochberg and Z. Wu, "A new characterization of Dirichlet type spaces and applications", *Illinois J. Math.* **37**:1 (1993), 101–122.

[Seip 1993] K. Seip, "Beurling type density theorems in the unit disk", *Invent. Math.* **113**:1 (1993), 21–39.

[Shapiro and Shields 1962] H. S. Shapiro and A. L. Shields, "On the zeros of functions with finite Dirichlet integral and some related function spaces", *Math. Z.* **80** (1962), 217–229.

[Stegenga 1980] D. A. Stegenga, "Multipliers of the Dirichlet space", *Illinois J. Math.* **24**:1 (1980), 113–139.

[Wu 1993] Z. Wu, "The predual and second predual of W_α", *J. Funct. Anal.* **116**:2 (1993), 314–334.

[Zhu 1990] K. H. Zhu, *Operator theory in function spaces*, Monographs and Textbooks in Pure and Applied Mathematics **139**, Marcel Dekker Inc., New York, 1990.

ZHIJIAN WU
DEPARTMENT OF MATHEMATICS
UNIVERSITY OF ALABAMA
TUSCALOOSA, AL 35487
UNITED STATES
zwu@euler.math.ua.edu

Holomorphic Spaces
MSRI Publications
Volume **33**, 1998

Some Open Problems in the Theory
of Subnormal Operators

JOHN B. CONWAY AND LIMING YANG

ABSTRACT. Subnormal operators arise naturally in complex function theory, differential geometry, potential theory, and approximation theory, and
their study has rich applications in many areas of applied sciences as well
as in pure mathematics. We discuss here some research problems concerning the structure of such operators: subnormal operators with finite-rank
self-commutator, connections with quadrature domains, invariant subspace
structure, and some approximation problems related to the theory. We also
present some possible approaches for the solution of these problems.

Introduction

A bounded linear operator S on a separable Hilbert space $\mathcal{H}$ is called *subnormal* if there exists a normal operator N on a Hilbert space $\mathcal{K}$ containing $\mathcal{H}$ such
that $N\mathcal{H} \subset \mathcal{H}$ and $N|_{\mathcal{H}} = S$. The operator S is called *cyclic* if there exists an
x in $\mathcal{H}$ such that

$$\mathcal{H} = \mathrm{clos}\{p(S)x : p \text{ is a polynomial}\},$$

and is called *rationally cyclic* if there exists an x such that

$$\mathcal{H} = \mathrm{clos}\{r(S)x : r \text{ is a rational function with poles off } \sigma(S)\}.$$

The operator S is *pure* if S has no normal summand and is *irreducible* if S is
not unitarily equivalent to a direct sum of two nonzero operators.

The theory of subnormal operators provides rich applications in many areas,
since many natural operators that arise in complex function theory, differential
geometry, potential theory, and approximation theory are subnormal operators.
Many deep results have been obtained since Halmos introduced the concept of
a subnormal operator. In particular, Thomson's solution of the long-standing
problem on the existence of bounded point evaluations reveals a structure theory
of cyclic subnormal operators. Thomson's work answers many questions that had

Yang was partially supported by NSF grant DMS-9401234.

been open for a long time and promises to enable researchers to answer many more; see [Thomson 1991] or [Conway 1991]. The latter is a general reference for the theory of subnormal operators.

Here we will present some research problems on subnormal operators and discuss some possibilities for their solution.

1. Subnormal Operators with Finite-Rank Self-Commutator

Let A denote area measure in the complex plane $\mathbb{C}$. A bounded domain G is a *quadrature domain* if there exist points $z_1, \ldots, z_N$ in G and constants $a_{m,n}$ such that

$$\int_G f(z)\, dA = \sum_{n=1}^{N} \sum_{m=0}^{N_n} a_{m,n} f^{(m)}(z_n)$$

for every function f analytic in G that is area-integrable. The theory of quadrature domains has been successfully studied by the techniques of compact Riemann surfaces, complex analysis and potential theory [Aharonov and Shapiro 1976; Gustafsson 1983; Sakai 1988].

The *self-commutator* of S is the operator $[S^*, S] = S^*S - SS^*$. The structure of subnormal operators with finite-rank self-commutator has been studied by many authors. Morrel [1973/74] showed that every subnormal operator with rank one self-commutator is a linear combination of the unilateral shift and the identity. Olin et al. [$\geq$ 1997] classified all cyclic subnormal operators with finite-rank self-commutator. D. Xia [1987a; 1987b] attempted to classify all subnormal operators with finite-rank self-commutator. His results, however, are incomplete. In [McCarthy and Yang 1997; 1995] a connection between a class of subnormal operators with finite-rank self-commutator and quadrature domains was established. However, the following problem still remains open.

PROBLEM 1.1. Classify all subnormal operators whose self-commutator has finite rank.

One can show that, if S is a pure subnormal operator with a finite rank self-commutator, the spectrum of the minimal normal extension N is contained in an algebraic curve. This is a modification of a result in [Xia 1987a]. If one assumes some additional conditions—for example, that the index of $S - \lambda$ is constant—then it turns out that S is unitarily equivalent to the direct sum of the bundle shifts over quadrature domains introduced in [Abrahamse and Douglas 1976]. (The argument is similar to that in [Putinar 1996], and uses the fact that the principle function is a constant on the spectrum minus the essential spectrum). For the general case, where the index of the subnormal operator $S - \lambda$ may change from one component to another, difficulties remain.

EXAMPLE 1.2. Set $r(z) = z(2z - 1)/(z + 2)$ and let $\Omega = r(\mathbb{D})$, where $\mathbb{D}$ is the open unit disk. If $\Gamma_1 = \partial\Omega$ and $\Gamma_2 = \text{clos}(r(\partial\mathbb{D}) \setminus \partial\Omega)$, then Γ_2 is a closed simple curve. Let Ω_0 denote the region bounded by Γ_2 and let T_r be the Toeplitz operator on H^2 with symbol r. It is easy to show that T_r is a subnormal operator with a finite-rank self-commutator and that $\text{ind}(T_r - \lambda) = -2$ for $\lambda \in \Omega_0$ and $\text{ind}(T_r - \lambda) = -1$ for $\lambda \in \Omega \setminus \bar{\Omega}_0$. Obviously, neither Ω nor Ω_0 is a quadrature domain. The techniques in [McCarthy and Yang 1997] do not work for regions that have inner curves like Γ_2.

In [McCarthy and Yang 1997] one sees that the spectral pictures of rationally cyclic subnormal operators determine their structure. Therefore, in general, the difficulty consists of using the spectral picture of a subnormal operator with finite-rank self-commutator to obtain its structure.

EXAMPLE 1.3. Use the notation of Example 1.2. Let S be an irreducible subnormal operator satisfying these properties:

(1) $\sigma(S) = \bar{\Omega}$.
(2) $\sigma_e(S) = \Gamma_1 \cup \Gamma_2$.
(3) $\text{ind}(S - \lambda) = -1$ for $\lambda \in \Omega \setminus \bar{\Omega}_0$ and $\text{ind}(S - \lambda) = -3$ for $\lambda \in \Omega_0$.

The existence of such an irreducible subnormal operator is established by a method in [Thomson and Yang 1995]. Such an operator should not have a finite-rank self-commutator. However, how can one prove it? Understanding these examples will give some ideas on how to solve the problem.

Recently M. Putinar [1996] found another connection between operator theory and the theory of quadrature domains, by studying hyponormal operators with rank one self-commutator. It is of interest to see how operator-theoretical methods can be applied to the theory of quadrature domains.

Let G be a quadrature domain and let ω be the harmonic measure of G with respect to a fixed point in G. Let H_ω be the operator of multiplication by z on the Hardy space $H^2(G)$. It is easy to see that H_ω is irreducible with a finite-rank self-commutator and that the spectrum of the minimal normal extension N_ω is ∂G. Therefore, one sees that ∂G is a subset of an irreducible algebraic curve (the spectrum of N_ω is a subset of an algebraic curve). This was proved by A. Aharanov and H. Shapiro [1976] using function theory in 1976. B. Gustafsson [1983] obtained even better results: he showed that, except for possibly finitely many points, the boundary of a quadrature domain is an irreducible algebraic curve. In order to prove Gustafsson's theorem by using operator theory, one needs to prove the following operator-theoretical conjecture.

CONJECTURE 1.4. If S is an irreducible subnormal operator with finite-rank self-commutator, then except for possibly a finite number of points, the spectrum of the minimal normal extension N of S is equal to an irreducible algebraic curve.

A solution to this should indicate some ideas for obtaining the structure of subnormal operators with finite rank self-commutator.

Let G be a quadrature domain of connectivity $t + 1$ and order n. Let W be a plane domain bounded by finitely many smooth curves and conformally equivalent to G. Let $\phi : W \to G$ be a conformal map. Define

$$C = \{z \in \partial G : z = \phi(w) \text{ for some } w \in \partial W \text{ with } \phi'(w) = 0\},$$
$$D = \{z \in \partial G : z = \phi(w_1) = \phi(w_2) \text{ for two different } w_j \in \partial W\}.$$

It was shown in [Aharonov and Shapiro 1976; Gustafsson 1983] that, if G is a quadrature domain, there is a meromorphic function $S(z)$ on G that is continuous on $\bar{G}$ except at its poles and such that $S(z) = \bar{z}$ on ∂G; moreover the equation $S(z) = \bar{z}$ has at most finitely many solutions in G. Let E denote the set of the solutions in G. Let c, d, e denote the cardinalities of C, D, E.

THEOREM 1.5 [Gustafsson 1988]. $t + c + 2d + e \leq (n - 1)^2$.

THEOREM 1.6 [Sakai 1988]. $c + e \geq t + n - 1$.

Using the model for rationally cyclic subnormal operators with finite rank self-commutator, it has only been possible to show that $t + e \leq (n - 1)^2$ [McCarthy and Yang 1997].

PROBLEM 1.7. Use operator-theoretical methods to prove Gustafsson's theorem and Sakai's theorem.

In order to solve the problem, one has to understand deeply the operator-theoretical meanings of the numbers t, n, c, d, e. Discovering such methods may improve the results in the theory of quadrature domains and will help us understand the connections between these two areas better.

2. Invariant Subspaces of Subnormal Operators

Let S be a cyclic subnormal operator. A standard result [Conway 1991, p. 52] says that S is unitarily equivalent to the operator S_μ of multiplication by z on the space $P^2(\mu)$, the closure of the polynomials in $L^2(\mu)$.

A point $\lambda \in \mathbb{C}$ is a *bounded point evaluation* for $P^2(\mu)$ if there exists $C > 0$ such that $|p(\lambda)| \leq C\|p\|$ for every polynomial p. In this section, we will assume that S_μ is an irreducible operator and that the set of bounded point evaluations for $P^2(\mu)$ is the open unit disk.

Apostol, Bercovici, Foiaş, and Pearcy showed in [Apostol et al. 1985] that, if the measure restricted to the unit circle is zero, there exists for each integer $1 \leq n \leq \infty$ an invariant subspace M of S_μ such that $\dim(M \ominus zM) = n$. It seems that the following problem has an affirmative answer.

QUESTION 2.1. Suppose that the measure μ restricted to the unit circle is non-zero. Does every invariant subspace M of S_μ have the codimension-one property (which means that $\dim(M \ominus zM) = 1$)?

Partial answers are known. Miller [1989] showed that, if μ is area measure on the unit disk plus the Lebesgue measure on an arc, the question has an affirmative answer. In [Yang 1995] it is shown that, if μ is the area measure on the unit disk and Lebesgue measure restricted to a compact subset of the unit circle having positive Lebesgue measure, then each invariant subspace has the codimension-one property. The following is an analogous problem.

PROBLEM 2.2. Characterize all subnormal operators S_μ such that the restriction of S to any invariant subspace M is cyclic.

It seems that, if the measure μ restricted to the unit circle is nonzero, then S_μ restricted to each invariant subspace is cyclic. However, Problem 2.2 is harder than 2.1.

Say that an invariant subspace M of S_μ has finite codimension if the dimension of $P^2(\mu) \ominus M$ is finite. Here is an interesting problem concerning invariant subspaces of S_μ.

PROBLEM 2.3. Characterize the cyclic subnormal operators S_μ that have the property that every invariant subspace is an intersection of invariant subspaces having finite codimension.

This property implies that every invariant subspace is hyperinvariant (because every invariant subspace having finite codimension is hyperinvariant), and therefore that $P^\infty(S_\mu) = \{S_\mu\}'$, where $P^\infty(S_\mu)$ is the w^*-closure of

$$\{p(S_\mu) : p \text{ is a polynomial}\} \text{ and } \{S_\mu\}' = \{A : AS_\mu = S_\mu A\}.$$

3. Mean and Uniform Approximation

Let S_μ be irreducible and let G be the set of bounded point evaluations for $P^t(\mu)$. In [Olin and Yang 1995], it was shown that the algebra $P^2(\mu) \cap C(\operatorname{spt} \mu)$ equals $A(G)$, the algebra of continuous functions on $\bar{G}$ that are also analytic inside G. By a modification of the proof, one can actually show that $P^t(\mu) \cap C(\operatorname{spt} \mu) = A(G)$ for $1 < t < \infty$. However, for mean rational approximation, the situation is more complicated, since the closure of the set of bounded point evaluations may be strictly smaller than K; see [Conway 1991], for example.

Let Ω be the interior of the bounded point evaluations for $R^t(K, \mu)$ and let $A(K, \Omega)$ be the set of continuous functions on K that are also analytic on Ω. Using a result in [Conway and Elias 1993], it is easy to construct a space $R^t(K, \mu)$ with the set of bounded point evaluations Ω such that $R^t(K, \mu)$ does not contain $A(K, \Omega)$. Therefore, the correct question about the analogous property would be the following.

QUESTION 3.1. If $f \in R^t(K, \mu) \cap C(\operatorname{spt} \mu)$, is the function f continuous on K?

The answer is affirmative if we assume that the boundary of K contains no bounded point evaluations. The proof is similar to that in [Olin and Yang 1995].

Therefore, it only needs to be shown that the function f is continuous at each bounded point evaluation on the boundary.

Let $R(K)$ be the uniform closure of the rational functions with poles off K. It is well known that a smooth function f is in $R(K)$ if and only if $\bar{\partial} f = 0$ on the set of nonpeak points for $R(K)$; see [Browder 1969, p. 166, Theorem 3.2.9]. The situation for the mean rational approximation is different. For example, in [McCarthy and Yang 1997] it is shown that there are a lot of nontrivial $R^t(K, \mu)$ spaces such that $\bar{z} p(z) \in R^t(K, \mu)$, where p is a polynomial. Notice that $\bar{\partial} \bar{z} p(z) = p(z) \neq 0$ except for finitely many points. Therefore, it is interesting to consider the following question.

QUESTION 3.2. Let $R^t(K, \mu)$ be an irreducible space and let f be a smooth function on $\mathbb{C}$. If $f \in R^t(K, \mu)$, what is the relation between f, K, and μ?

If $f = \bar{z} p(z)$, where p is a polynomial, the relation is well understood by the theorem in [McCarthy and Yang 1997]. That is, $K = \mathrm{clos}(\overset{\circ}{K})$, the set $\overset{\circ}{K}$ is a quadrature domain, and spt μ is the union of the boundary of K with finitely many points in $\overset{\circ}{K}$. For an arbitrary smooth function f, one may guess that the support of μ inside the set $\{z : \bar{\partial} f \neq 0\}$ should not be big. On the other hand, the boundary of K should not be arbitrary. The boundary should have some "generalized quadrature domain" properties. This investigation may raise some interesting function-theoretical problems. Further study of those problems may have rich applications in complex analysis and potential theory.

Next we give a problem concerning uniform polynomial approximation. For a compact subset K of $\mathbb{C}$, let $P(K)$ be the uniform closures in $C(K)$ of polynomials.

EXAMPLE 3.3. Set

$$K = \{z : |z| = 1 \text{ and } \mathrm{Im}\, z \geq 0\} \cup \{z : \mathrm{Im}\, z = 0 \text{ and } -1 \leq \mathrm{Re}\, z \leq 1\}.$$

Then $\bar{z} P(K) + P(K)$ is dense in $C(K)$, since $1 - |z|^2$ is in $\bar{z} P(K) + P(K)$; but $P(K) \neq C(K)$. On the other hand, if K is the unit circle, it is easy to show that $\bar{z} P(K) + P(K)$ is not dense in $C(K)$.

Naturally, we have the following problem.

PROBLEM 3.4. Characterize the compact subsets K of the complex plane such that $\bar{z} P(K) + P(K)$ is dense in $C(K)$.

Example 3.3 indicates the hard part of the problem. That is, there are two toplogically equivalent compact subsets K_1 and K_2, for which $\bar{z} P(K_1) + P(K_1)$ is dense in $C(K_1)$ but $\bar{z} P(K_2) + P(K_2)$ is not dense in $C(K_2)$. However, there might be a geometric solution to the problem.

4. The Existence of Bounded Point Evaluations

Thomson's proof of the existence of bounded point evaluations for $P^t(\mu)$ uses Davie's deep estimation of analytic capacity, S. Brown's technique, and

Vitushkin's localization for uniform rational approximation. The proof is excellent but complicated. We believe that a simpler proof may exist. In this section, we will present one approach and point out the difficult part that is left unsolved.

THEOREM 4.1 [Luecking 1981]. *Let D be an open disk and let G be a subset of D. Then*

$$\int_D |p|^t \, dA \leq C \int_G |p|^t \, dA$$

for all polynomials p if and only if there are positive numbers $\varepsilon, \delta > 0$ such that, for each disk O centered on the boundary of D with radius less than δ, we have

$$\text{Area}(O \cap G) \geq \varepsilon \, \text{Area}(O \cap D). \tag{4-1}$$

Equation (4–1) is called Luecking's condition.

Let ν be a complex Borel finite measure with $\int p \, d\nu = 0$ for each polynomial p. Let $\hat{\nu}$ be the Cauchy transform of ν, that is

$$\hat{\nu}(\lambda) = \int \frac{1}{z - \lambda} \, d\nu.$$

We say a point $\lambda \in \mathbb{C}$ is *heavy* if there exists a disk D centered at λ and there exists $\alpha > 0$ such that the set $G_\lambda = \{z : |\hat{\nu}(z)| \geq \alpha\} \cap D$ satisfies Luecking's condition.

PROPOSITION 4.2. *Suppose that λ is a heavy point. Then λ is a bounded point evaluation for $P^1(|\nu|)$.*

PROOF. From Luecking's theorem, we see that

$$\int_D |p| \, dA \leq C \int_{G_\lambda} |p| \, dA \leq \frac{C}{\alpha} \int_{G_\lambda} |p\hat{\nu}| \, dA = \frac{C}{\alpha} \int_{G_\lambda} |\widehat{p\nu}| \, dA \leq C' \int |p| \, d|\nu|.$$

Since $|p(\lambda)| \leq C \int_D |p| \, dA$, we see that λ is a bounded point evaluation for $P^1(|\nu|)$. $\square$

Therefore, in order to prove the existence of bounded point evaluations, for each non-heavy point we need to construct some kind of "peak functions". Those "peak functions" enable us to show that the measure μ is zero on the complement of the closure of heavy points. Hence, the set of heavy points will not be empty since μ is a nonzero measure. It is possible to construct such functions because for each non-heavy point every disk centered at the point doesn't satisfy Luecking's condition. That means that the Cauchy transform will be small in some sense near the point.

References

[Abrahamse and Douglas 1976] M. B. Abrahamse and R. G. Douglas, "A class of subnormal operators related to multiply-connected domains", *Advances in Math.* **19**:1 (1976), 106–148.

[Aharonov and Shapiro 1976] D. Aharonov and H. S. Shapiro, "Domains on which analytic functions satisfy quadrature identities", *J. Analyse Math.* **30** (1976), 39–73.

[Apostol et al. 1985] C. Apostol, H. Bercovici, C. Foiaş, and C. Pearcy, "Invariant subspaces, dilation theory, and the structure of the predual of a dual algebra, I", *J. Funct. Anal.* **63**:3 (1985), 369–404.

[Browder 1969] A. Browder, *Introduction to function algebras*, W. A. Benjamin, Inc., New York-Amsterdam, 1969.

[Conway 1991] J. B. Conway, *The theory of subnormal operators*, Mathematical Surveys and Monographs **36**, Amer. Math. Soc., Providence, 1991.

[Conway and Elias 1993] J. B. Conway and N. Elias, "Analytic bounded point evaluations for spaces of rational functions", *J. Funct. Anal.* **117**:1 (1993), 1–24.

[Gustafsson 1983] B. Gustafsson, "Quadrature identities and the Schottky double", *Acta Appl. Math.* **1**:3 (1983), 209–240.

[Gustafsson 1988] B. Gustafsson, "Singular and special points on quadrature domains from an algebraic geometric point of view", *J. Analyse Math.* **51** (1988), 91–117.

[Luecking 1981] D. H. Luecking, "Inequalities on Bergman spaces", *Illinois J. Math.* **25**:1 (1981), 1–11.

[McCarthy and Yang 1995] J. E. McCarthy and L. Yang, "Cyclic subnormal operators with finite rank self-commutators", *Proc. Royal Irish Acad. A* **95** (1995), 173–177.

[McCarthy and Yang 1997] J. E. McCarthy and L. Yang, "Subnormal operators and quadrature domains", *Adv. Math.* **127**:1 (1997), 52–72.

[Miller 1989] T. L. Miller, "Some subnormal operators not in $\mathbb{A}_2$", *J. Funct. Anal.* **82**:2 (1989), 296–302.

[Morrel 1973/74] B. B. Morrel, "A decomposition for some operators", *Indiana Univ. Math. J.* **23** (1973/74), 497–511.

[Olin and Yang 1995] R. F. Olin and L. M. Yang, "The commutant of multiplication by z on the closure of polynomials in $L^t(\mu)$", *J. Funct. Anal.* **134**:2 (1995), 297–320.

[Olin et al. ≥ 1997] R. Olin, J. Thomson, and T. Trent, "Subnormal operators with finite rank self-commutators", *Trans. Amer. Math. Soc.*.

[Putinar 1996] M. Putinar, "Extremal solutions of the two-dimensional L-problem of moments", *J. Funct. Anal.* **136**:2 (1996), 331–364.

[Sakai 1988] M. Sakai, "An index theorem on singular points and cusps of quadrature domains", pp. 119–131 in *Holomorphic functions and moduli* (Berkeley, CA, 1986), vol. I, edited by D. Drasin et al., Math. Sci. Res. Inst. Publ. **10**, Springer, New York, 1988.

[Thomson 1991] J. E. Thomson, "Approximation in the mean by polynomials", *Ann. of Math.* (2) **133**:3 (1991), 477–507.

[Thomson and Yang 1995] J. E. Thomson and L. Yang, "Invariant subspaces with the codimension one property in $L^t(\mu)$", *Indiana Univ. Math. J.* **44**:4 (1995), 1163–1173.

[Xia 1987a] D. X. Xia, "The analytic model of a subnormal operator", *Integral Equations Operator Theory* **10**:2 (1987), 258–289.

[Xia 1987b] D. X. Xia, "Analytic theory of subnormal operators", *Integral Equations Operator Theory* **10**:6 (1987), 880–903. Errata in **12**:6 (1989), 898–899.

[Yang 1995] L. M. Yang, "Invariant subspaces of the Bergman space and some subnormal operators in $\mathcal{A}_1 \setminus \mathcal{A}_2$", *Michigan Math. J.* **42**:2 (1995), 301–310.

JOHN B. CONWAY
DEPARTMENT OF MATHEMATICS
UNIVERSITY OF TENNESSEE
KNOXVILLE, TN 37996
UNITED STATES
conway@novell.math.utk.edu

LIMING YANG
DEPARTMENT OF MATHEMATICS
UNIVERSITY OF HAWAII
HONOLULU, HI 96822
UNITED STATES
yang@math.hawaii.edu

Holomorphic Spaces
MSRI Publications
Volume **33**, 1998

Elements of Spectral Theory
in Terms of the Free Function Model
Part I: Basic Constructions

NIKOLAI NIKOLSKI AND VASILY VASYUNIN

ABSTRACT. This is a survey of the function model approach to spectral
theory, including invariant subspaces, generalized spectral decompositions,
similarity to normal operators, stability problems for the continuous spectrum of unitary and selfadjoint operators, and scattering theory.

Part I contains a revised version of the coordinate-free function model
of a Hilbert space contraction, based on analysis of functional embeddings
related to the minimal unitary dilation of the operator. Using functional
embeddings, we introduce and study all other objects of model theory, including the characteristic function, various concrete forms (transcriptions)
of the model, one-sided resolvents, and so on. For the case of an inner
scalar characteristic function we develop the classical H^∞-calculus up to
a local function calculus on the level curves of the characteristic function.
The spectrum of operators commuting with the model operator, and in
particular functions of the latter, are described in terms of their liftings.
A simplified proof of the invariant subspace theorem is given, using the
functional embeddings and regular factorizations of the characteristic function. As examples, we consider some compact convolution-type integral
operators, and dissipative Schrödinger (Sturm–Liouville) operators on the
half-line.

Part II, which will appear elsewhere, will contain applications of the
model approach to such topics as angles between invariant subspaces and
operator corona equations, generalized spectral decompositions and free interpolation problems, resolvent criteria for similarity to a normal operator,
and weak generators of the commutant and the reflexivity property. Classical topics of stability of the continuous spectrum and scattering theory
will also be brought into the fold of the coordinate-free model approach.

Vasyunin was supported by INTAS grant no. 93-0249-ext and RFFI grant no. 95-01-00482.

CONTENTS

Part I: Basic Constructions

Part II: The Function Model in Action (to appear elsewhere)

Foreword

This text is a detailed and enlarged version of an introductory mini-course on model theory that we gave during the Fall of 1995 at the Mathematical Sciences Research Institute in Berkeley. We are indebted to the Institute for granting us a nice opportunity to work side by side with our colleagues, operator analysts, from throughout the world.

We were rewarded by the highly professional audience that attended our lectures, and are very grateful to our colleagues for making stimulating comments and for encouraging us to give a course to students whose knowledge of the subject was sometimes better than ours. Having no possibility to list all of you, we thankfully mention the initiators of the course, our friends Joseph Ball and Cora Sadosky.

We are extremely indebted to the editors of this volume for inexhaustible patience, to Maria Gamal and Vladimir Kapustin for helping to compile the list of references, and to Alexander Plotkin, David Sherman, and Donald Sarason for their careful reading the manuscript.

Introduction: A Brief Account

About the reader. We hope that most of the material in this paper will be accessible to nonexperts having general knowledge of operator theory and of basic analysis, including the spectral theorem for normal Hilbert space operators. Some more special facts and terminology are listed below in Section 0.7 of this Introduction.

0.1. A bit of philosophy. A *model* of an operator $T : H \to H$ is another operator, say $M : K \to K$, that is in some sense equivalent to T. There exist models up to *unitary equivalence*, $T = U^{-1}MU$ for a unitary operator $U : H \to K$ acting between Hilbert spaces; up to *similarity equivalence*, $T = V^{-1}MV$ for a linear isomorphism $V : H \to K$ acting between Banach spaces; up to quasi-similarity; pseudo-similarity; and other equivalences. So, in fact, these transformations U, V, etc., change the notation, reducing the operator to a form convenient for computations, especially for a functional calculus admitted by the operator. The calculus contains both algebraic and analytic features (in particular, norm estimates of expressions in the operator), and the requirements to simplify them lay the foundation of modern model theories based on *dilations* of the operator under question. As soon as such a dilation is established, any object related to the analysis of our operator can be the subject of a *lifting* up to the level of the dilation (the calculus, the commutant, spectral decompositions, etc.), the level where they can be treated more easily.

The function model of Livsic, de Branges, Sz.-Nagy, and Foiaş follows precisely this course. Starting with a Hilbert space contraction T, or a dissipative operator in the initial Livsic form $(\mathrm{Im}(Tx, x) \geq 0$, for $x \in \mathrm{Dom}\, T)$, it makes use of a unitary (selfadjoint) dilation $\mathcal{U}$ realized by use of the von Neumann spectral theorem as a multiplication operator $\mathcal{U} : f \to e^{it}f$. It then uses advanced trigonometric harmonic analysis on the circle (on the line, in the Livsic case), including Hardy classes, multiplicative Nevanlinna theory and other developed techniques. It is exactly these circumstances that gave B. Sz.-Nagy and C. Foiaş the chance for such an elegant expression as *harmonic analysis of operators* for the branch of operator theory based on the function model, and not the fact that "all is harmonious in the developed theory" as was suggested by M. Krein in his significant preface to the Russian translation of [Sz.-Nagy and Foiaş 1967]. We can even say that such a function model is a kind of noncommutative discrete Fourier transform of an operator, and mention with astonishment that a continuous version of this transform exists only nominally—which fact evidently retards applications to semigroup theory and scattering theory. We will have an opportunity to observe the absence of an "automatic translator" from the circle to the line when dealing with the Cayley transform and Schrödinger operators (Chapter 2).

0.2. What is the free function model? In constructing the free—that is, coordinate-free—function model of a Hilbert space contraction, we follow the Sz.-Nagy–Foiaş turnpike strategy by starting with a *unitary dilation*

$$\mathcal{U} : \mathcal{H} \longrightarrow \mathcal{H}$$

of a given contraction

$$T : H \longrightarrow H,$$

so that we get the dilation (calculus) property

$$p(T) = P_H p(\mathcal{U})|H$$

for all polynomials p. The usual way to continue is to realize the action of $\mathcal{U}$ by use of the spectral theorem, and so to represent T as a compression

$$f \longmapsto P_{\mathcal{K}} z f, \quad \text{for } f \in \mathcal{K},$$

of the multiplication operator $f \mapsto zf$ onto the orthogonal difference $\mathcal{K}$ of two z-invariant subspaces of a weighted space $L^2(\mathbb{T}, \mathcal{E}, W)$, with respect to an operator-valued weight $W(\zeta) : \mathcal{E} \to \mathcal{E}$, in a way similar to the way that the spectral theorem itself represents the unitary operator $\mathcal{U}$. (Here ζ is a complex number on the unit circle $\mathbb{T} = \{\zeta \in \mathbb{C} : |\zeta| = 1\}$.)

The idea of the free function model is to stop this construction halfway from the unitary dilation to the final formulae of the Sz.-Nagy–Foiaş model. In other words, we fix neither a concrete spectral representation of a unitary dilation nor the dilation itself, but work directly with an (abstract) dilation equipped with two "functional embeddings" that carry the information on the fine function structure of the operator and the dilation. So, we can say that the free function model of a contraction T having defect spaces E and E_* is a class of in some sense equivalent (see Chapter 1) isometric functional embeddings

$$\pi : L^2(E) \longrightarrow \mathcal{H} \quad \text{and} \quad \pi_* : L^2(E_*) \longrightarrow \mathcal{H}$$

of E- and E_*-valued L^2-spaces $(L^2(E) \overset{\text{def}}{=} L^2(\mathbb{T}, E))$ into a Hilbert space $\mathcal{H}$ satisfying a minimality property, in such a way that the product $\pi_*^* \pi$ is a multiplication operator by a function unitarily equivalent to the *characteristic function* Θ_T of T,

$$\Theta_T(z)h = (-T + z D_{T^*}(I - zT^*)^{-1} D_T)h \quad \text{for } h \in E_*, \ z \in \mathbb{D}$$

(where $\mathbb{D} \overset{\text{def}}{=} \{\zeta \in \mathbb{C} : |\zeta| < 1\}$). The minimal unitary dilation $\mathcal{U}$ can be defined by the splitting property $\Pi z = \mathcal{U}\Pi$, where

$$\Pi = \pi \oplus \pi_* : L^2(E) \oplus L^2(E_*) \to \mathcal{H}$$

and z stands for the shift operator $z(f \oplus f_*) = zf \oplus zf_*$ on this orthogonal sum of L^2-spaces.

At this stage, we have a great deal of freedom: we can write down or not write down the operator $\mathcal{U}$ using the spectral theorem, where the freedom is in a choice of spectral coordinates; and we can choose in very diverse ways the functional embeddings, keeping $\pi_*^*\pi$ unitarily equivalent to Θ_T (in fact, equivalent with respect to a special class of unitary equivalences). These transcription problems are considering in Chapter 3.

So, we can say that the free model is based on the intrinsic functional structures of T and the unitary dilation $\mathcal{U}$, and it is a kind of virtual function representation of a given operator: we can still stay in the same Hilbert space H where the initial contraction T is defined, but the pair $\{H,T\}$ is now equipped with two isometric embeddings π and π_* satisfying the above-mentioned conditions. Working with an operator on this free function model $\{H, T, \pi_*, \pi\}$ we can at any time transfer (by means of π, π_*) an arbitrary operator computation to functions in the spaces $L^2(E)$ and $L^2(E_*)$, whose elements have all of the advantages of concrete objects with concrete local structures. This is the idea, and in its realization we follow [Vasyunin 1977; Makarov and Vasyunin 1981; Nikolski and Vasyunin 1989; Nikolski and Khrushchev 1987; Nikolski 1994].

As concrete examples of model computations we consider in Chapter 2 two operators: the operator of indefinite integration on the space $L^2((0,1);\,\mu)$ with respect to an arbitrary finite measure μ, and Schrödinger operators with real potentials and dissipative boundary conditions.

0.3. What is spectral theory?

0.3. What is spectral theory? This is a really good question. We can say, having no intention to give a truism and rather looking for an explanation of the title, that for us it is the study of *intrinsic structures* of an operator. So, for instance, even speaking of such well-studied subjects as selfadjoint and more generally of normal operators, the attempts to identify the theory with the spectral theorem are unjustified: this is the main but not sole gist of the theory. In fact, the structures of invariant subspaces and of the restrictions of the operator under consideration to them (the parts of the operator) are far from being straightforward consequences of the spectral theorem, and these subjects should be considered as independent ingredients of spectral theory. All the more, the same is true for the interplay between functional calculus and geometric properties of the operator (recall for example, the problem of density of polynomials in different spaces related to the spectral measure, or the problem of the asymptotic behavior of evolutions defined by the operator).

Beyond normality, there is again the study of invariant subspaces as a path to the parts of an operator, and also the analysis of decompositions into invariant subspaces (series, integrals), and, of course, the functional calculus admitted by an operator, and related studies of its "space action", that is, properties of the evolution defined by the operator.

(By the way, this is why the famous polar decomposition of a Hilbert space operator, $T = VR$ with $R \geq 0$ and V a partial isometry, is not a crucial reduction of the problem to the selfadjoint and unitary cases: it does not respect any of basic properties and operations mentioned above. Another sentence we have the courage to put in these parentheses is that the previous paragraph explains why, in our opinion, spectral theory is something different from an algebraic approach (say, C^*-algebraic) of replacing an operator by a letter (a symbol) and then studying it: operator structures are lost in the kernel of such a morphism.)

So, in this paper, as a spectral theory, we consider the attempts

- to describe and to study invariant subspaces;
- to use them for decompositions of an operator into sums of its restrictions to such subspaces of a particular type respecting all other operator structures (so-called spectral subspaces);
- to distinguish operators enjoying the best possible spectral decompositions, that is, operators similar to a normal one;
- to study stability properties of fine spectral structures of an operator with respect to small perturbations and to study evolutions defined by it (scattering problems).

Of course, to construct such a spectral theory and to work with it using the function model technique, we need to develop some routine prerequisites, in particular to compute the spectrum and the resolvent of a contraction in terms of our free function model; this is done in Chapter 5 below.

Now we describe briefly how we intend to realize this program.

0.4. Conceptual value of invariant subspaces. The invariant subspaces of an operator, their classification and description, have played, explicitly or implicitly, the central role in operator theory for the last 50 years. The reader may ask, why?

The first and rather formal reason is that invariant subspaces are a direct analogue of the eigenvectors of linear algebra. Indeed, diagonal or Jordan form representations of a matrix are nothing but decompositions of the space an operator acts on into a direct sum of the operator's invariant subspaces of particular types. In the twenties and thirties, the theories of selfadjoint and unitary operators, and later that of normal operators, culminating in various forms of the spectral theorem and its applications, showed that this approach is fruitful. In the fifties and sixties, making use of distribution theory and advanced functional calculi, several generalizations followed this approach: the kernel theorem of Gelfand and L. Schwartz, the spectral operators of Dunford and J. Schwartz, Foiaş's decomposable operators, and some more specialized theories. See [Gel'fand and Vilenkin 1961; Dunford and Schwartz 1971; Dowson 1978; Colojoara and Foiaş 1968].

Another, and even double, reason to put invariant subspaces in the foundation of general spectral theory was provided by a real breakthrough that happened fifty years ago when M. Livsic [1946] discovered the notion of the characteristic function and A. Beurling [1949] established a one-to-one correspondence between the invariant subspaces of a Hilbert space isometry and the Nevanlinna inner functions. During the fifties and sixties, mainly due to efforts by M. Livsic and M. Brodski, and V. Potapov (for dissipative operators) and by B. Sz.-Nagy and C. Foiaş, and L. de Branges (for contractions), these observations together became the cornerstone of the theories linking invariant subspaces and factorizations (inner or not) of the characteristic function of an operator. These techniques make it possible to translate spectral decomposition problems for large classes of operators into the function language, and then to use methods of hard analysis (of complex analysis and harmonic analysis) to solve them.

The reader will find in Chapter 6 descriptions of the invariant subspaces of a contraction in terms of its functional embeddings and factorizations of the characteristic function, as well as some concrete examples of such descriptions.

In the Afterword (page 288), which functions as a preview of the Part II of this paper, we will briefly outline how this machinery works for several sample problems: spectral model operators and generalized free interpolation, rational tests for the similarity to a normal operator, stability of the continuous spectrum for trace class perturbations of unitary operators, and scattering theory. The Afterword also contains a brief discussion of factorizations of the characteristic function related to invariant subspaces of different types, and of operator Bezout ("corona") equations.

But first, in the next section, we sketch two of the main technical tools of the model theory: the commutant lifting theorem (CLT) and local functional calculi.

0.5. The commutant lifting theorem (CLT) and local calculi.

As we have already mentioned, the main trick of model theory is to consider a unitary dilation $\mathcal{U} : \mathcal{H} \to \mathcal{H}$ of a contractive operator $T : H \to H$, where $H \subset \mathcal{H}$:

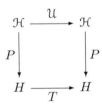

to try to "lift" all things related to T up to the level of $\mathcal{U}$, and then to work with these lifted functional objects. One of the main objects related to an operator T is its *commutant* $\{T\}'$, the set of $X : H \to H$ such that $XT = TX$. All functions of an operator, that is, operators included in a functional calculus, belong to $\{T\}'$; so do all projections on an invariant subspace parallel to a complementary invariant subspace (that is, maps P of the form $P(x_1 + x_2) = x_1$ for $x_1 \in H_1$ and $x_2 \in H_2$, where $H = H_1 + H_2$, with $TH_1 \subset H_1$, $TH_2 \subset H_2$); and hence so does the spectral measure of an operator if it exists.

The commutant lifting theorem (CLT) says that the commutant of T can be lifted up to the commutant of $\mathcal{U}$: for $X \in \{T\}'$ there exists an $Y \in \{\mathcal{U}\}'$, a "symbol of X", such that

$$X = P_H Y | H \quad \text{and} \quad \|X\| = \|Y\|.$$

This fundamental result was proved by B. Sz.-Nagy and C. Foiaş [1968], preceded by the important partial case of an inner scalar characteristic function Θ_T, discovered by D. Sarason [1967]. For more details we refer to Chapter 4 below, and for the history and more references refer to [Sz.-Nagy and Foiaş 1967].

The functional structure of Y, related to the functional embeddings of the model, is important, as are the links between this lifting process and the fundamental Nehari theorem on "Laurent extension" (lifting) of a Hankel type form. Without entering into these details now, we refer to [Nikolski 1986; Pták and Vrbová 1988] for the interplay between the CLT and Nehari type theorems and to [Cotlar and Sadosky 1984/85; 1986a; 1986b; 1988; 1992] for an extensive theory unifying these two techniques. To finish with our appreciation of the CLT, we mention its basic role for control theory and interpolation by analytic, rational, and other functions with constraints (of the Carathéodory or Nevanlinna–Pick type). A vast array of techniques have been developed for these applications: the Adamyan–Arov–Krein step-by-step extension process, choice sequences by Apostol, Foiaş et al., Schur parameters techniques, and more; for details we refer to [Foiaş and Frazho 1990; Ball et al. 1990; Bakonyi and Constantinescu 1992; Nikolski 1986; Alpay $\geq$ 1998].

However, we will avoid these deep theories and instead give, in Chapter 4, the simplest direct proof of the CLT based on S. Parrott's approach [1970] (see also [Davidson 1988]). This approach makes use of a version of the CLT: the Ando theorem on commuting unitary dilations of two commuting contractions [Ando 1963], which says that, if $T_1 : H \to H$ and $T_2 : H \to H$ satisfy $\|T_1\| \leq 1$, $\|T_2\| \leq 1$, and

$$T_1 T_2 = T_2 T_1,$$

there exist commuting unitary dilations $\mathfrak{U}_1$ and $\mathfrak{U}_2$:

$$\mathfrak{U}_1 \mathfrak{U}_2 = \mathfrak{U}_2 \mathfrak{U}_1.$$

We give a short new proof of Ando's theorem and derive the CLT and describe the admissible symbols Y in functional terms characterizing the functional (model) parameters of Y related to the functional embeddings π and π_*. Also, we give a formula expressing the norm of X as the distance

$$\|X\| = \inf\{\|Y + \pi \Gamma \pi_*^*\| : \Gamma \in H^\infty(E_* \to E)\},$$

where $H^\infty(E_* \to E)$ stands for the space of bounded holomorphic functions in the unit disc $\mathbb{D}$ taking as values operators $f(z) : E_* \to E$ from E_* to E (see Chapter 4).

For *two-sided inner* characteristic functions Θ, that is, for H^∞-functions whose boundary values $\Theta(\zeta)$ (where $\zeta \in \mathbb{T}$) are unitary almost everywhere on

$\mathbb{T}$, the CLT is a partial case of the vector-valued version of the Nehari theorem: if $X \in \{T\}'$ and $\mathbf{H} = \pi^* X P_H \pi_*$, then $\mathbf{H}$ is a Hankel operator (that is, $\mathbf{H} : H^2(E_*) \to H^2_-(E)$ and $\mathbf{H}z = P_- z \mathbf{H}$), and the function $\pi^* Y \pi_*$ is a symbol of $\mathbf{H}$ (that is, $\mathbf{H}f = P_-(\pi^* Y \pi_*)f$ for $f \in H^2(E_*)$). The formula for the norm of X is the usual Nehari formula

$$\|X\| = \mathrm{dist}_{L^\infty(E_* \to E)}(\pi^* Y \pi_*, H^\infty(E_* \to E));$$

see [Nikolski 1986].

The H^∞-calculus exists for every completely nonunitary contraction already by the initial unitary dilation theorem. It takes the following simplified form for two-sided inner characteristic functions:

$$f \longmapsto f(T) \overset{\mathrm{def}}{=} P_H \pi_* f \pi_*^* | H,$$
$$\|f(T)\| = \mathrm{dist}_{L^\infty(E_* \to E)}(\Theta^* f, H^\infty(E_* \to E)).$$

Dealing with the calculus, and even with the whole commutant, it is important to localize, if possible, the expressions for $f(T)$ and for the norm $\|f(T)\|$, so as to make visible their dependence on the local behavior of f near the spectrum of T (whereas the definition and aforementioned formulas depend on the global behavior of f in the disc $\mathbb{D}$ and on the torus $\mathbb{T}$). Following [Nikolski and Khrushchev 1987] we give such expressions for operators T with scalar inner characteristic functions Θ, showing, for instance, an estimate on level sets of Θ:

$$\|f(T)\| \le C_\varepsilon \sup\{|f(z)| : z \in \mathbb{D}, |\Theta(z)| < \varepsilon\}, \quad \text{for } f \in H^\infty.$$

Later on we use these estimates for our treatment of free interpolation problems (Chapter 8).

That is all we include in Part I of the article.

0.6. Prehistory. When writing an expository paper, one automatically accepts an extra assignment to comment on the history of the subject. Frankly speaking, we would like to avoid this obligation and to restrict ourselves to sporadic remarks. In fact, model theory is a quite recent subject, and we hope that careful references will be enough. Nonetheless, we will recall a kind of prehistory of the theory. This consists of four fundamental results by V. I. Smirnov (1928), A. I. Plessner (1939), M. A. Naimark (1943), and G. Julia (1944), which anticipated the subsequent developments of the field and ought to been have its clear landmarks, but were not identified properly and are still vaguely cited.

Smirnov's results. The following result, usually attributed to A. Beurling, is Theorem 2 of [Smirnov 1928a]: If $f \in H^2$ is an outer function (see Section 0.7 below), then

$$\mathrm{span}_{H^2}(z^n f : n \ge 0) = H^2.$$

In [Smirnov 1932, Section 10] there is another proof of this fact—essentially the same that Beurling published in 1949. Another fundamental result, the inner-outer factorization $f = f_{\text{inn}} \cdot f_{\text{out}}$ of H^p functions (see Section 0.7 for definitions), appears in [1928b, Section 5]. Together, these two results immediately imply the well-known formula for the invariant subspace generated by a function $f \in H^2$:

$$\operatorname*{span}_{H^2}(z^n f : n \geq 0) = f_{\text{inn}} H^2.$$

This is the main ingredient of the famous Beurling paper [1949] on the shift operator, which stimulated the study of invariant subspaces. Unfortunately, in Smirnov's time the very notion of a bounded operator was not completely formulated—Banach's book dates from 1932—and, what is more, during the next ten years nobody thought about the shift operator, until H. Wold and A. I. Plessner.

Plessner's results. In 1939, A. I. Plessner constructed the H^∞-calculus for isometric operators allowing a unitary extension with an absolutely continuous spectrum. His paper [1939a] contains a construction of L^∞-functions of a maximal symmetric operator $A : H \to H$ having an absolutely continuous selfadjoint extension $\mathcal{A} : \mathcal{H} \to \mathcal{H}$:

$$f(A) = P_H f(\mathcal{A})|H;$$

the multiplicativity is proved for H^∞ functions.

Plessner [1939b] also proved that any Hilbert space isometry $V : H \to H$ is unitarily equivalent to an orthogonal sum of a unitary operator U and a number of copies of the (now standard!) H^2-shift operator $S : H^2 \to H^2$, given by $Sf = zf$:

$$V \simeq U \oplus \left(\sum_i \in I \oplus S \right), \quad \text{with } \operatorname{card}(I) = \dim(H \ominus VH).$$

To apply to isometries the calculus from the first paper, Plessner proved that an isometry V having an absolutely continuous unitary part U is an inner function of a maximal symmetric operator of the previous type: $V = f(A)$.

Clearly, the missing link to complete the construction of the H^∞-calculus for contractions would be the existence of isometric dilations (and their absolute continuity). The first step to such a dilation was accomplished by G. Julia in 1944, but his approach remained overlooked until the mid-fifties.

Julia's result. Julia [1944a; 1944b; 1944c] discovered a "one-step isometric dilation" of a Hilbert space contraction $T : H \to H$. Precisely, he observed that the operator defined on $H \oplus H$ by the block matrix

$$V = \begin{pmatrix} T & D_{T^*} \\ -D_T & T^* \end{pmatrix}$$

is an isometry, and that $T = P_H V|H$.

Of course, V is not necessarily a "degree-two dilation" $T^2 = P_H V^2 | H$, nor a full polynomial dilation $T^n = P_H V^n | H$, for $n \geq 0$. We know now that to get such a dilation one has to extend the isometry V by the shift operator of multiplicity H, or simply take

$$\overline{V}(x, x_1, x_2, \ldots) = (Tx, D_T x, x_1, x_2, \ldots)$$

for $(x, x_1, \ldots) \in H \oplus H \oplus \cdots$. But Julia did not make by this observation.

In fact, isometric and even unitary dilations were discovered ten years later by B. Sz.-Nagy [1953], ignoring Julia's step but using Naimark's dilation theorem.

Naimark's result. Historically, this was the last latent cornerstone of model theory. M. A. Naimark [1943] proved that every operator-valued positive measure $\mathcal{E}(\sigma) : H \to H$ taking contractive values, $0 \leq \mathcal{E}(\sigma) \leq I$, has a dilation that is a spectral measure: there exists an orthoprojection-valued measure $E(\sigma) : \mathcal{H} \to \mathcal{H}$, where $\mathcal{H} \supset H$, such that $\mathcal{E}(\sigma) = P_H E(\sigma) | H$ for all σ.

Later, Sz.-Nagy [1953] derived from this result the existence of the minimal unitary dilation for an arbitrary contraction, and the history of model theory started. In fact, as already mentioned, it started seven years earlier, with Livsic [1946], motivated rather by the theory of selfadjoint extensions of symmetric operators (developed by Krein, Naimark, and many others) than by the dilation philosophy.

0.7. Prerequisites. For the reader's convenience, we collect here some more specialized facts and terminology used throughout the paper.

General terminology. All vector spaces are considered over the complex numbers $\mathbb{C}$; a subspace of a normed space means a closed vector subspace. For a subset A of a normed space X, we denote by

$$\mathrm{span}(A) \overset{\mathrm{def}}{=} \mathrm{clos}(\mathrm{Lin}(A))$$

the closed linear hull of A, and by

$$\mathrm{dist}_X(x, A) \overset{\mathrm{def}}{=} \inf\{\|x - y\| : y \in A\}$$

the distance of a vector $x \in X$ from A. For a subspace $E \subset H$, the orthogonal complement is denoted by $E^\perp$ or $H \ominus E$, and the orthogonal projection on E by P_E. The orthogonal sum of a family $\{E_n\}$ of subspaces of a Hilbert space is denoted by

$$\sum_n \oplus E_n \overset{\mathrm{def}}{=} \Big\{ \textstyle\sum_n x_n : x_n \in E_n \text{ and } \sum_n \|x_n\|^2 < \infty \Big\}.$$

Operators on a Hilbert space. "Operator" means a bounded linear operator unless otherwise indicated; $L(E \to F)$ stands for the space of all operators from E to F. A subspace $E \subset H$ is called *invariant* with respect to an operator $T : H \to H$ if $TE \subset E$; the restriction of T to E is denoted by $T|E$ and T is called an *extension* of $T|E$. The set of all T-invariant subspaces is denoted by

Lat T (or Lat(T)). *Reducing* subspaces are subspaces from Lat $T \cap$ Lat T^*; T^* stands for the (hermitian) adjoint of T. We denote by Range(T) the range of T, that is, the linear set $\{Th : h \in H\}$, and by Ker $T \stackrel{\text{def}}{=} \{x : Tx = 0\}$ for the kernel of T.

A *partial isometry* $V : H \to K$ is an operator between Hilbert spaces such that $V|(\text{Ker } V)^\perp$ is an isometry; the subspaces $(\text{Ker } V)^\perp$ and Range V are called, respectively, the *initial* and *final* subspaces of V. The final subspace of a partial isometry V is the initial subspace of the adjoint V^*, and vice versa; moreover $V^*V = P_{(\text{Ker } V)^\perp}$.

The *polar decomposition* of an operator $T : H \to K$ is the representation $T = VR$, where $R = |T| \stackrel{\text{def}}{=} (T^*T)^{1/2} \geq 0$ is the positive square root of T^*T (the modulus of T) and $V : H \to K$ a partial isometry with the initial space Range $R = (\text{Ker } T)^\perp$ and the final space Range T. The equation $A^*A = B^*B$ is equivalent to saying that $B = VA$, where V stands for a partial isometry with the initial space Range A and the final space Range B.

The *spectral theorem* (in the von Neumann form) says that a *normal operator* $N : H \to H$ (that is, one such that $N^*N = NN^*$) is unitarily equivalent to the multiplication operator

$$f(z) \longmapsto zf(z)$$

on the space

$$\{f \in L^2(H, \mu_N) : f(z) \in E_N(z)H \text{ a.e. with respect to } \mu_N\},$$

where $E_N(\cdot)$ stands for a mesurable family of orthoprojections on H, and μ_N stands for a so-called scalar spectral measure of N carried by the spectrum $\sigma(N)$; the class of measures equivalent to μ_N and the dimension function $z \mapsto \dim H(z)$ defined μ_N-a.e. are complete unitary invariants of N.

Isometries and co-isometries. The Wold–Kolmogorov theorem says that an isometry $V : H \to H$ gives rise to an orthogonal decomposition

$$H = H_\infty \oplus \left(\sum_n \geq 0 \oplus V^n E \right),$$

where $H_\infty = \bigcap_{n \geq 0} V^n H$ and $E = H \ominus VH = \text{Ker } V^*$. The subspace H_∞ reduces V and the restriction $V|H_\infty$ is unitary; any other subspace with these properties is contained in H_∞. An isometry V with $H_\infty = \{0\}$ is called *pure* (or an *abstract shift operator*), and a subspace $L \subset H$ satisfying $V^n L \perp V^m L$ for $n \neq m \geq 0$ is called a *wandering subspace* of V; in particular, $E = H \ominus VH$ is a wandering subspace. A pure isometry V is unitarily equivalent to the *shift operator* of multiplicity $\dim E = \dim \text{Ker } V^*$, defined by

$$f \longmapsto zf \quad \text{for } f \in H^2(E),$$

where

$$H^2(E) = \{f = \sum_{n \geq 0} \hat{f}(n)z^n : \hat{f}(n) \in E \text{ and } \sum_{n \geq 0} \|\hat{f}(n)\|^2 < \infty\}$$

stands for E-valued *Hardy space* (see also below).

A *co-isometry* $V : H \to K$ is an operator (a partial isometry) whose adjoint V^* is an isometry; it is called *pure* if V^* is pure.

Contractions. A *contraction* $T : H \to K$ is an operator with $\|T\| \leq 1$. The *defect operators* of T are $D_T = (I - T^*T)^{1/2}$ and $D_{T^*} = (I - TT^*)^{1/2}$; the closures of their ranges $\mathcal{D}_T \overset{\text{def}}{=} \text{clos Range } D_T$ and $\mathcal{D}_{T^*}$ are called the *defect subspaces*, and $\dim \mathcal{D}_T = \text{rank}(I - T^*T)$ and $\dim \mathcal{D}_{T^*}$ are the *defect numbers*. The defect operators are intertwined by the contraction, $TD_T = D_{T^*}T$, and hence the restriction $T|\mathcal{D}_T^{\perp}$ is an isometry from $\mathcal{D}_T^{\perp}$ to $\mathcal{D}_{T^*}^{\perp}$. A contraction T is called *completely nonunitary* if there exists no reducing subspace (or invariant subspace) where T acts as a unitary operator.

Function spaces and their operators. Lebesgue spaces of vector-valued functions are defined in the standard way: the notation $L^p(\Omega, E, \mu)$ means the L^p-space of E-valued functions on a mesure space (Ω, μ). We can abbreviate $L^p(\Omega, E, \mu)$ as $L^p(E, \mu)$, $L^p(E)$, $L^p(\Omega, E)$, or $L^p(\Omega)$ (this last when $E = \mathbb{C}$).

Our standard case is $\Omega = \mathbb{T}$, endowed with normalized Lebesgue measure $\mu = m$. The *Hardy subspace* of $L^p(E) = L^p(\mathbb{T}, E)$ is defined in the usual way,

$$H^p(E) = \{f \in L^p(E) : \hat{f}(n) = 0 \text{ for } n < 0\},$$

where the $\hat{f}(n)$, for $n \in \mathbb{Z}$, are the Fourier coefficients. Obviously, $L^2(E \oplus F) = L^2(E) \oplus L^2(F)$. Also as usual, $H^p(E)$ is identified with the space of boundary functions of the corresponding space of functions holomorphic in the unit disc $\mathbb{D}$ (for $p = 2$, see above). The *reproducing kernel* of the space $H^2(E)$ is defined by the equality

$$\left(f, \frac{e}{1 - \bar{\lambda}z}\right)_{H^2(E)} = (f(\lambda), e)_E \quad \text{for } f \in H^2(E) \text{ and } e \in E.$$

The *Riesz projections* $P_{H^2} = P_+$ and $P_- = I - P_+$ are defined on $L^2(E)$ by

$$P_+\left(\sum_{n \in \mathbb{Z}} \hat{f}(n)z^n\right) = \sum_{n \geq 0} \hat{f}(n)z^n, \quad \text{for } |z| = 1.$$

Clearly, $L^2(E) = H^2(E) \oplus H^2_-(E)$, where $H^2_-(E) = P_- L^2(E)$.

The L^p and H^p spaces with values in the space $L(E \to F)$ of all operators from E to F are denoted by $L^p(E \to F)$ and $H^p(E \to F)$ respectively. Any operator A acting from $L^2(E_1)$ to $L^2(E_2)$ and intertwining multiplication by z on these spaces (which means that $Az = zA$) is the multiplication operator induced by a function $\Theta \in L^\infty(E_1 \to E_2)$, namely $(Af)(z) = \Theta(z)f(z)$; moreover, $\|A\| = \|A|H^2(E_1)\| = \|\Theta\|_{L^\infty(E_1 \to E_2)}$. We often identify A and Θ; we have $\Theta \in H^\infty(E_1 \to E_2)$ if and only if $\Theta \in L^\infty(E_1 \to E_2)$ and $\Theta H^2(E_1) \subset H^2(E_2)$. A contractive-valued function $\Theta \in H^\infty(E_1 \to E_2)$ is called *pure* if it does not reduce to a constant isometry on any subspace of E_1, or, equivalently, if $\|\Theta(0)e\|_{E_2} < \|e\|_{E_1}$ for all nonzero $e \in E_1$.

The last point is to recall the *Nevanlinna–Riesz–Smirnov canonical factorization* of scalar H^1-functions:

$$\Theta = BS \cdot F = \Theta_{\text{inn}} \cdot \Theta_{\text{out}},$$

where B is a Blaschke product, S a singular inner function, and F an outer function. We have

$$B(z) = \prod_{\lambda \in \mathbb{D}} b_\lambda^{k(\lambda)}(z),$$

where

$$b_\lambda(z) = \frac{|\lambda|}{\lambda} \frac{\lambda - z}{1 - \bar{\lambda}z}, \quad \text{with } |\lambda| < 1,$$

is a Blaschke factor and $k(\lambda)$ is the divisor of zero multiplicities, satisfying the Blaschke condition $\sum_{\lambda \in \mathbb{D}} k(\lambda)(1 - |\lambda|) < \infty$. We also have

$$S(z) = \exp\left(-\int_{\mathbb{T}} \frac{\zeta + z}{\zeta - z} \, d\mu_s(\zeta)\right),$$

μ_s being a positive measure singular with respect to Lebesgue measure m, and

$$F(z) = \exp\left(-\int_{\mathbb{T}} \frac{\zeta + z}{\zeta - z} \log \frac{1}{|\Theta(\zeta)|} \, dm(\zeta)\right).$$

Let δ_λ be the measure of mass 1 concentrated at the point λ (Dirac measure). We associate with the function Θ the measure on the closed unit disc defined by

$$d\mu_\Theta = \log \frac{1}{|\Theta|} \, dm + d\mu_s + \sum_{\lambda \in \mathbb{D}} k(\lambda)(1 - |\lambda|^2)\delta_\lambda.$$

This is often called the *representing measure* of Θ.

Chapter 1. Construction of the Function Model

1.1. Unitary dilation. We start constructing the function model for a contraction on a Hilbert space with an explicit description of its minimal unitary dilation. An operator $\mathcal{U}$ acting on a Hilbert space $\mathcal{H}$ is called a *dilation* of an operator T acting on H, where $H \subset \mathcal{H}$, if

$$T^n = P_H \mathcal{U}^n | H \quad \text{for all } n \geq 1.$$

The dilation is called *minimal* if $\text{span}\{\mathcal{U}^n H : n \in \mathbb{Z}\} = \mathcal{H}$. It is called *unitary* if $\mathcal{U}$ is a unitary operator.

The structure of the space $\mathcal{H}$ where a dilation of an operator acts is given by the following lemma of Sarason [1965].

1.2. Lemma. $\mathcal{U} : \mathcal{H} \to \mathcal{H}$ *is a dilation of* $T : H \to H$ *if and only if*

$$\mathcal{H} = G_* \oplus H \oplus G, \qquad (1.2.1)$$

with

$$\mathcal{U}G \subset G, \quad \mathcal{U}^*G_* \subset G_*, \quad \text{and} \quad P_H \mathcal{U}|H = T.$$

G *and* G_* *are called the* outgoing *and the* incoming *subspace, respectively.*

PROOF. To prove the "only if" part, put

$$G = \operatorname{span}\{\mathcal{U}^n H : n \geq 0\} \ominus H,$$
$$G_* = \mathcal{H} \ominus \operatorname{span}\{\mathcal{U}^n H : n \geq 0\}.$$

If $g = \lim p_n(\mathcal{U})h_n$ is orthogonal to H, $h_n \in H$, and $f \in H$, then

$$(\mathcal{U}g, f) = \lim(\mathcal{U}p_n(\mathcal{U})h_n, f) = \lim(Tp_n(T)h_n, f) = \lim(p_n(T)h_n, T^*f)$$
$$= \lim(p_n(\mathcal{U})h_n, T^*f) = (g, T^*f) = 0,$$

that is, $\mathcal{U}G \subset G$. The inclusion $\mathcal{U}^*G_* \subset G_*$ is obvious.

For the converse, we use the block matrix representation of $\mathcal{U}$ with respect to the decomposition (1.2.1). We get

$$\mathcal{U} = \begin{pmatrix} * & 0 & 0 \\ * & T & 0 \\ * & * & * \end{pmatrix} \implies \mathcal{U}^n = \begin{pmatrix} * & 0 & 0 \\ * & T^n & 0 \\ * & * & * \end{pmatrix} \implies P_H \mathcal{U}^n|H = T^n,$$

for all $n \geq 1$. $\qquad\square$

1.3. The matrix of a unitary dilation. We need a unitary operator $\mathcal{U} : \mathcal{H} \to \mathcal{H}$, where $\mathcal{H} = G_* \oplus H \oplus G$, of the form:

$$\mathcal{U} = \begin{pmatrix} \mathcal{E}_* & 0 & 0 \\ A & T & 0 \\ C & B & \mathcal{E} \end{pmatrix}.$$

In matrix form, the conditions $\mathcal{U}^*\mathcal{U} = I$ and $\mathcal{U}\mathcal{U}^* = I$ become

$$\begin{pmatrix} \mathcal{E}_*^*\mathcal{E}_* + A^*A + C^*C & A^*T + C^*B & C^*\mathcal{E} \\ T^*A + B^*C & T^*T + B^*B & B^*\mathcal{E} \\ \mathcal{E}^*C & \mathcal{E}^*B & \mathcal{E}^*\mathcal{E} \end{pmatrix} = I,$$

$$\begin{pmatrix} \mathcal{E}_*\mathcal{E}_*^* & \mathcal{E}_*A^* & \mathcal{E}_*C^* \\ A\mathcal{E}_*^* & AA^* + TT^* & AC^* + TB^* \\ C\mathcal{E}_*^* & CA^* + BT^* & CC^* + BB^* + \mathcal{E}\mathcal{E}^* \end{pmatrix} = I.$$

We put $D_T \overset{\text{def}}{=} (I - T^*T)^{1/2}$ and $\mathcal{D}_T \overset{\text{def}}{=} \operatorname{clos} D_T H$. Twelve different entries give twelve different equations. Here are ten of them:

$$
\left\{
\begin{array}{ll}
\mathcal{E}^*\mathcal{E} = I & \Longrightarrow \mathcal{E} \text{ is an isometry;} \\
\mathcal{E}_*\mathcal{E}_*^* = I & \Longrightarrow \mathcal{E}_* \text{ is a co-isometry;} \\
AA^* + TT^* = I & \Longrightarrow A^* = V_* D_{T^*} \text{ (polar decomposition);} \\
\mathcal{E}_* A^* = 0 & \Longrightarrow V_* : \mathcal{D}_{T^*} \to \operatorname{Ker} \mathcal{E}_*; \\
T^*T + B^*B = I & \Longrightarrow B = V D_T \text{ (polar decomposition);} \\
\mathcal{E}^* B = 0 & \Longrightarrow V : \mathcal{D}_T \to \operatorname{Ker} \mathcal{E}^*; \\
CA^* + BT^* = 0 & \Longrightarrow CV_* + VT^* = 0 \Longrightarrow C = -VT^*V_*^* + C_0, \\
 & \qquad\qquad\qquad\qquad\qquad C_0 | \operatorname{Range} V_* = 0; \\
A^*T + C^*B = 0 & \Longrightarrow C_0^* V D_T = 0 \Longrightarrow C_0^* | \operatorname{Range} V = 0; \\
\mathcal{E}_*^*\mathcal{E}_* + A^*A + C^*C = I & \Longrightarrow V_*V_*^* + C_0^*C_0 = P_{\operatorname{Ker} \mathcal{E}_*}; \\
CC^* + BB^* + \mathcal{E}\mathcal{E}^* = I & \Longrightarrow VV^* + C_0 C_0^* = P_{\operatorname{Ker} \mathcal{E}^*}.
\end{array}
\right.
$$

The last two of these identities mean that C_0 is a partial isometry with initial space $\operatorname{Ker} \mathcal{E}_* \ominus \operatorname{Range} V_*$ and final space $\operatorname{Ker} \mathcal{E}^* \ominus \operatorname{Range} V$; therefore, the two other identities $\mathcal{E}^* C = 0$ and $C\mathcal{E}_*^* = 0$ are fulfilled automatically.

Now, if we introduce

$$
G^{(1)} = \sum_n \geq 0 \oplus \mathcal{E}^n V \mathcal{D}_T, \qquad G_*^{(1)} = \sum_n \geq 0 \oplus \mathcal{E}_*^{*n} V_* \mathcal{D}_{T^*},
$$

$$
G^{(2)} = G \ominus G^{(1)}, \qquad\qquad G_*^{(2)} = G_* \ominus G_*^{(1)},
$$

then $G_*^{(2)} \oplus G^{(2)}$ is a reducing subspace of $\mathcal{U}$:

$$
\mathcal{U} | G_*^{(2)} \oplus G^{(2)} = \begin{pmatrix} \mathcal{E}_* | G_*^{(2)} & 0 \\ C_0 & \mathcal{E} | G^{(2)} \end{pmatrix}.
$$

Therefore, the operator

$$
\mathcal{U} | G_*^{(1)} \oplus H \oplus G^{(1)} = \begin{pmatrix} \mathcal{E}_* | G_*^{(1)} & 0 & 0 \\ D_{T^*} V_*^* & T & 0 \\ -VT^*V_*^* & V D_T & \mathcal{E} | G^{(1)} \end{pmatrix}
$$

is also a unitary dilation of T. Moreover, the dilation is minimal. Indeed, since

$$
\mathcal{U} H = \begin{pmatrix} 0 \\ T \\ V D_T \end{pmatrix} H,
$$

we see that $\operatorname{span}\{\mathcal{U}H, H\} = H \oplus V\mathcal{D}_T$, whence $\operatorname{span}\{\mathcal{U}^n H : n \geq 0\} = H \oplus G^{(1)}$. In a similar way we obtain $\operatorname{span}\{\mathcal{U}^n H : n \leq 0\} = G_*^{(1)} \oplus H$. So the dilation is minimal if and only if $\mathcal{E}$ and $\mathcal{E}_*$ are pure isometries and $\operatorname{Range} V = \operatorname{Ker} \mathcal{E}^*$, $\operatorname{Range} V_* = \operatorname{Ker} \mathcal{E}_*$.

Thus, we arrive at the following theorem.

1.4.   Theorem. *An operator* $\mathcal{U} : \mathcal{H} \to \mathcal{H}$ *is a minimal unitary dilation of* $T : H \to H$ *if and only if there exist subspaces* G *and* G_* *of* $\mathcal{H}$ *such that*

$$\mathcal{H} = G_* \oplus H \oplus G,$$

and, with respect to this decomposition, $\mathcal{U}$ *has matrix*

$$\mathcal{U} = \begin{pmatrix} \mathcal{E}_* & 0 & 0 \\ D_{T^*}V_*^* & T & 0 \\ -VT^*V_*^* & VD_T & \mathcal{E} \end{pmatrix},$$

where $\mathcal{E} = \mathcal{U}|G$ *and* $\mathcal{E}_*^* = \mathcal{U}^*|G_*$ *are pure isometries,* V *is a partial isometry with initial space* $\mathcal{D}_T$ *and final space* $\operatorname{Ker} \mathcal{E}^*$, *and* V_* *is a partial isometry with initial space* $\mathcal{D}_{T^*}$ *and final space* $\operatorname{Ker} \mathcal{E}_*$. $\square$

1.5. First glimpse into the function model. In principle, the construction of the minimal unitary dilation could be split into two steps: first we can construct a minimal isometric dilation (or a minimal co-isometric extension), and then apply the same procedure to the adjoint of the operator obtained. As a result, we get a minimal unitary dilation. It is worth mentioning that in our terms the restriction of $\mathcal{U}$ to $H \oplus G$ is a minimal isometric dilation of T and the compression of $\mathcal{U}$ onto $G_* \oplus H$ is a minimal co-isometric extension of T.

A function model of a contraction T arises whenever we realize $\mathcal{E}$ and $\mathcal{E}_*^*$, which are abstract shift operators, as the operators of multiplication by z on appropriate Hardy spaces. Such a realization provides us with two functional embeddings of the corresponding Lebesgue spaces in the space of the minimal unitary dilation. All other objects needed for handling such a model operator (the characteristic function, formulas for the projection onto H, for the dilation, and for the operator itself, etc.) are computed in terms of these embeddings. We present the necessary constructions in the rest of this chapter. In Chapter 3 we give some explicit formulas for the model as examples of the above-mentioned functional realizations of the coordinate-free model in terms of multiplication by z on certain L^2-spaces.

1.6.   Functional embeddings. As already mentioned, after fixing a minimal unitary dilation of a given contraction T, we begin with a spectral representation of the pure isometry $\mathcal{E}$ and the pure co-isometry $\mathcal{E}_*$ using the Wold decomposition. Let E and E_* be two auxiliary Hilbert spaces such that

$$\dim E = \dim \mathcal{D}_T = \dim \operatorname{Ker} \mathcal{E}^*,$$
$$\dim E_* = \dim \mathcal{D}_{T^*} = \dim \operatorname{Ker} \mathcal{E}_*.$$

Since $\operatorname{Ker} \mathcal{E}^* = G \ominus \mathcal{E}G = G \ominus \mathcal{U}G$ and $\operatorname{Ker} \mathcal{E}_* = G_* \ominus \mathcal{E}_*^*G = G_* \ominus \mathcal{U}^*G_*$, there exist unitary mappings

$$v : E \to G \ominus \mathcal{U}G \quad \text{and} \quad v_* : E_* \to G_* \ominus \mathcal{U}^*G_* \tag{1.6.1}$$

identifying these spaces. Now we can define a mapping

$$\Pi = (\pi_*, \pi) : L^2(E_*) \oplus L^2(E) \to \mathcal{H} \qquad (1.6.2)$$

by the formulas

$$\pi\left(\sum_{k \in \mathbb{Z}} z^k e_k\right) = \sum_{k \in \mathbb{Z}} \mathcal{U}^k v e_k \qquad \text{for } e_k \in E,$$

$$\pi_*\left(\sum_{k \in \mathbb{Z}} z^k e_{*k}\right) = \sum_{k \in \mathbb{Z}} \mathcal{U}^{k+1} v_* e_{*k} \qquad \text{for } e_{*k} \in E_*,$$

$$\Pi(f_* \oplus f) = \pi_* f_* + \pi f.$$

The operators Π, π, and π_* are called *functional mappings* or *functional embeddings*. Immediately from the above definition we deduce the following properties of the embeddings π and π_*:

(i) $\pi^* \pi = I_{L^2(E)}$ and $\pi_*^* \pi_* = I_{L^2(E_*)}$.
(ii) $\pi z = \mathcal{U}\pi$ and $\pi_* z = \mathcal{U}\pi_*$.
(iii) $\pi H^2(E) = G$ and $\pi_* H_-^2(E_*) = G_*$.

Property (i) is a consequence of the fact that $G \ominus \mathcal{U}G$ and $G_* \ominus \mathcal{U}^* G_*$ are wandering subspaces for $\mathcal{U}$; that is,

$$\mathcal{U}^n(G \ominus \mathcal{U}G) \perp \mathcal{U}^m(G \ominus \mathcal{U}G) \quad \text{and} \quad \mathcal{U}^n(G_* \ominus \mathcal{U}^* G_*) \perp \mathcal{U}^m(G_* \ominus \mathcal{U}^* G_*)$$

for distinct $n, m \in \mathbb{Z}$. Property (ii) is obvious, as are the relations $\pi z^n = \mathcal{U}^n \pi$ and $\pi_* z^n = \mathcal{U}^n \pi_*$ for all $n \in \mathbb{Z}$. To check (iii) we write

$$\pi H^2(E) = \left\{\pi \sum_{k \geq 0} z^k e_k : e_k \in E, \ \sum_{n \geq 0} \|e_k\|^2 < \infty\right\}$$

$$= \left\{\sum_{k \geq 0} \mathcal{U}^k v e_k : e_k \in E, \ \sum_{n \geq 0} \|e_k\|^2 < \infty\right\}$$

$$= \sum_k \geq 0 \oplus \mathcal{U}^k(G \ominus \mathcal{U}G) = G,$$

$$\pi_* H_-^2(E_*) = \left\{\pi_* \sum_{k < 0} z^k e_{*k} : e_{*k} \in E_*, \ \sum_{n \geq 0} \|e_{*k}\|^2 < \infty\right\}$$

$$= \left\{\sum_{k < 0} \mathcal{U}^{k+1} v_* e_{*k} : e_{*k} \in E_*, \ \sum_{n \geq 0} \|e_{*k}\|^2 < \infty\right\}$$

$$= \sum_k \geq 0 \oplus \mathcal{U}^{*k}(G_* \ominus \mathcal{U}^* G_*) = G_*.$$

In general, the space $\operatorname{clos} \operatorname{Range} \Pi$ may be different from the entire space $\mathcal{H}$ of the minimal unitary dilation of T. Now we describe this range as follows.

1.7. Lemma. *The orthogonal complement*

$$H_u \overset{\text{def}}{=} (\operatorname{Range} \Pi)^\perp = \left(\pi L^2(E) + \pi_* L^2(E_*)\right)^\perp$$

is contained in H and is a reducing subspace of T. Moreover, H_u is maximal among all T-invariant subspaces $H' \subset H$ for which the restriction $T|H'$ is unitary.

PROOF. Since $G = \pi H^2(E) \subset \text{Range}\,\Pi$ and $G_* = \pi_* H_-^2(E_*) \subset \text{Range}\,\Pi$, we have $H_u \subset H$. The intertwining property (ii) implies that H_u reduces $\mathcal{U}$. Therefore, the operator $T|H_u = \mathcal{U}|H_u$ is unitary.

It remains to prove the maximality of H_u. Let $TH' \subset H' \subset H$, and let $T|H'$ be unitary. Then for $h \in H'$ and $n \geq 0$ we have

$$\|h\| = \|T^n h\| = \|P_H \mathcal{U}^n h\| \leq \|\mathcal{U}^n h\| = \|h\|.$$

Therefore, $\mathcal{U}^n h = P_H \mathcal{U}^n h \in H$, whence $\mathcal{U}^n H' \subset H$, $n \geq 0$. Thus, $\mathcal{U}^n H' \perp \pi H^2(E)$ and $H' \perp \mathcal{U}^{*n} \pi H^2(E)$, which yields $H' \perp \pi L^2(E)$ because

$$\text{span}\{\mathcal{U}^{*n} \pi H^2(E) : n \geq 0\} = \text{span}\{\pi \bar{z}^n H^2(E) : n \geq 0\} = \pi L^2(E).$$

Similarly, the relations

$$\|h\| = \|T^{*n} h\| = \|P_H \mathcal{U}^{*n} h\| \leq \|\mathcal{U}^{*n} h\| = \|h\|$$

imply that $H' \perp \pi_* L^2(E_*)$, whence $H' \subset H_u$.   □

The lemma yields the well-known decomposition of a contraction into a unitary and a completely nonunitary part. We recall that a contraction is called *completely nonunitary* if $H_u = \{0\}$, that is, if the restriction of it to any nontrivial invariant subspace is not a unitary operator.

1.8. Corollary. *A contraction T on H can be uniquely represented in the form $T = T_u \oplus T_0$, where $T_u = T|H_u$ is unitary and $T_0 \overset{\text{def}}{=} T|H_0$, for $H_0 \overset{\text{def}}{=} H \ominus H_u$, is completely nonunitary. Therefore, a contraction is completely nonunitary if and only if*

(iv) $\text{clos}\,\text{Range}\,\Pi = \mathcal{H}$.

In what follows we shall deal with completely nonunitary contractions only.

1.9. Characteristic function. We continue the construction of the function model of a given completely nonunitary contraction T. We use the functional embeddings to introduce the central object of model theory, the *characteristic function* of an operator.

We define

$$\Theta \overset{\text{def}}{=} \pi_*^* \pi : L^2(E) \to L^2(E_*). \tag{1.9.1}$$

By property (ii) on page 228, the mapping Θ intertwines the operators of multiplication by z in the two L^2-spaces above:

$$\Theta z = z \Theta.$$

We saw in Section 0.7 that such a Θ is the operator of multiplication by a bounded operator-valued function, which we denote by the same symbol:

$$(\Theta f)(\zeta) = \Theta(\zeta) f(\zeta) \quad \text{for } \zeta \in \mathbb{T}.$$

We shall write $\Theta \in L^\infty(E \to E_*)$, which means that Θ is a measurable function defined and bounded almost everywhere on the unit circle whose values are operators acting from E to E_*. Moreover, this function is contractive-valued ($\|\Theta\| \leq \|\pi_*^*\| \cdot \|\pi\| = 1$) and analytic. Indeed, we have

$$\pi H^2(E) = G \perp G_* = \pi_* H_-^2(E_*),$$

whence

$$\Theta H^2(E) = \pi_*^* \pi H^2(E) \subset H^2(E_*);$$

that is, $\Theta \in H^\infty(E \to E_*)$.

Definition. The function Θ defined by (1.9.1) is called the *characteristic function* of the given completely nonunitary contraction T.

Thus, formally, the "characteristic function" of a contraction T is a family of functions (depending on $\mathcal{U}$ and Π). Our next goal is to describe the family of characteristic functions corresponding to all operators unitarily equivalent to a given operator. Later, in Remark 1.12, we shall make somewhat more precise the notion of a characteristic function.

1.10. Equivalence relations. The definition of the functional embeddings, and therefore that of the characteristic function, involve two arbitrary Hilbert spaces E and E_*, only their dimensions being essential. So, it is natural to regard all the objects obtained as equivalent if a pair of spaces E, E_* is replaced by another one, say E', E'_*, of the same dimension. *By definition*, two functions $\Theta \in L^\infty(E \to E_*)$ and $\Theta' \in L^\infty(E' \to E'_*)$ are said to be *equivalent* if there exist unitary mappings

$$u : E \to E' \quad \text{and} \quad u_* : E_* \to E'_* \tag{1.10.1}$$

such that

$$\Theta' u = u_* \Theta. \tag{1.10.2}$$

In what follows, being in the framework of Hilbert space theory, we shall often view our initial object, a Hilbert space contraction, up to unitary equivalence. The minimal unitary dilation of such a contraction is also defined up to the corresponding unitary equivalence $W : \mathcal{H} \to \mathcal{H}'$ intertwining $\mathcal{U}$ and $\mathcal{U}'$ and preserving the structure (1.2.1), that is, satisfying

$$WG = G', \quad WG_* = G'_*, \quad WH = H'. \tag{1.10.3}$$

Therefore, the following definition is natural. Two functional embeddings Π and Π' are said to be *equivalent* if there exist unitary mappings (1.10.1) and a unitary operator (1.10.3) such that

$$\Pi' \begin{pmatrix} u_* & 0 \\ 0 & u \end{pmatrix} = W\Pi.$$

1.11. Theorem. *Let T, T' be two completely nonunitary contractions; we denote by Π, Π' their functional embeddings and by Θ, Θ' their characteristic functions. The following assertions are equivalent.*

(i) T and T' are unitarily equivalent.

(ii) Π and Π' are equivalent.

(iii) Θ and Θ' are equivalent.

PROOF. (1) $\implies$ (2). Let W_0 be a unitary operator, $W_0 T = T' W_0$. We define W on finite sums $\sum \mathcal{U}^n h_n$, where $h_n \in H$, by the formula

$$W\left(\sum \mathcal{U}^n h_n\right) = \sum \mathcal{U}'^n W_0 h_n.$$

We check that W is norm-preserving; this will imply that W is well-defined and can be extended to an isometry acting on the whole of $\mathcal{H}$. Since the finite sums $\sum \mathcal{U}'^n W_0 h_n$ are dense in $\mathcal{H}'$, such an extension will be surjective and, therefore, unitary. We have

$$\left\| W\left(\sum_n \mathcal{U}'^n h_n\right) \right\|^2 = \sum_n \| \mathcal{U}'^n W_0 h_n \|^2 + 2 \operatorname{Re} \sum_{k>l} (\mathcal{U}'^{k-l} W_0 h_k, W_0 h_l)$$

$$= \sum_n \| h_n \|^2 + 2 \operatorname{Re} \sum_{k>l} (T'^{k-l} W_0 h_k, W_0 h_l)$$

$$= \sum_n \| \mathcal{U}^n h_n \|^2 + 2 \operatorname{Re} \sum_{k>l} (\mathcal{U}^{k-l} h_k, h_l) = \left\| \sum_n \mathcal{U}^n h_n \right\|^2.$$

From the definition we see that $W\mathcal{U} = \mathcal{U}'W$ and $WH = W_0 H = H'$. Moreover,

$$W(G \oplus H) = W \operatorname{span}\{\mathcal{U}^n H : n \geq 0\} = \operatorname{span}\{\mathcal{U}'^n W_0 H : n \geq 0\}$$

$$= \operatorname{span}\{\mathcal{U}'^n H' : n \geq 0\} = G' \oplus H';$$

that is, $WG = G'$, whence $W(G \ominus \mathcal{U}G) = G' \ominus \mathcal{U}'G'$. Similarly, $WG_* = G'_*$ and $W(G_* \ominus \mathcal{U}^* G_*) = G'_* \ominus \mathcal{U}'^* G'_*$. Using the unitary mappings v and v_* of (1.6.1), we can define

$$u = v'^* W v : E \to E', \quad u_* = v'^*_* W v_* : E_* \to E'_*.$$

Now we show that these operators provide an equivalence between Π and Π':

$$W\pi\left(\sum z^k e_k\right) = W \sum \mathcal{U}^k v e_k = \sum \mathcal{U}'^k W v e_k$$

$$= \sum \mathcal{U}'^k v' u e_k = \pi'\left(\sum z^k u e_k\right) = \pi' u\left(\sum z^k e_k\right).$$

Thus $W\pi = \pi' u$. That $W\pi_* = \pi'_* u_*$ can be checked similarly.

(2) $\implies$ (3). We have

$$\Theta' u = \pi'^*_* \pi' u = \pi'^*_* W \pi = (W^* \pi'_*)^* \pi = (\pi_* u^*_*)^* \pi = u_* \pi^*_* \pi = u_* \Theta.$$

(3) $\implies$ (2). Defining an operator W on the dense set $\Pi L^2(E_* \oplus E)$ by the identity

$$W\Pi = \Pi' \begin{pmatrix} u_* & 0 \\ 0 & u \end{pmatrix}$$

we check that W is norm-preserving. As in the first part of the proof, this will imply that W is well-defined and extends to a unitary operator.

First, we note that property (i) of page 228, combined with the definition of Θ, can be written as

$$\Pi^*\Pi = \begin{pmatrix} I & \Theta \\ \Theta^* & I \end{pmatrix}.$$

Using this relation and the identity $\Theta'u = u_*\Theta$, we obtain

$$\|W\Pi x\|^2 = \left\| \Pi' \begin{pmatrix} u_* & 0 \\ 0 & u \end{pmatrix} x \right\|^2 = \left(\begin{pmatrix} u_*^* & 0 \\ 0 & u^* \end{pmatrix} \begin{pmatrix} I & \Theta' \\ \Theta'^* & I \end{pmatrix} \begin{pmatrix} u_* & 0 \\ 0 & u \end{pmatrix} x,\, x \right)$$

$$= \left(\begin{pmatrix} I & \Theta \\ \Theta^* & I \end{pmatrix} x,\, x \right) = \|\Pi x\|^2,$$

which completes the proof that Π and Π' are equivalent.

(2) $\implies$ (1). We put

$$\Pi' \begin{pmatrix} u_* & 0 \\ 0 & u \end{pmatrix} = W\Pi.$$

Then $WG = W\pi H^2(E) = \pi'uH^2(E) = \pi'H^2(E') = G'$ and, similarly, $WG_* = G'_*$, whence $WH = H'$. Thus, $W_0 \stackrel{\text{def}}{=} W|H$ defines a unitary operator from H to H'. Since

$$W\mathcal{U}\Pi = W\Pi z = \Pi' \begin{pmatrix} u_* & 0 \\ 0 & u \end{pmatrix} z = \mathcal{U}'\Pi' \begin{pmatrix} u_* & 0 \\ 0 & u \end{pmatrix} = \mathcal{U}'W\Pi$$

and Range Π is dence in $\mathcal{H}$, we have $\mathcal{U}'W = W\mathcal{U}$. This yields

$$W_0 T = W_0 P_H \mathcal{U}|H = P_{H'} W\mathcal{U}|H = P_{H'}\mathcal{U}'W|H = T'W_0;$$

that is, T and T' are unitarily equivalent. $\square$

1.12. The dilations and the characteristic function of a given contraction.

The same arguments show that a minimal unitary dilation of a given contraction T is unique up to a unitary equivalence $W\mathcal{U} = \mathcal{U}'W$ such that $WG = G'$, $WG_* = G'_*$, and $W|H = I|H$.

Returning to the main definition 1.9 and taking Theorem 1.11 into account, we see that it is natural that *the characteristic function* of a contraction should mean any function Θ defined by (1.9.1) for any operator T unitarily equivalent to the given one.

Now we find an expression for the characteristic function of a completely nonunitary contraction in terms of the contraction itself.

1.13.   Theorem. *The characteristic function* Θ *of a completely nonunitary contraction T is equivalent to the function in $H^\infty(\mathcal{D}_T \to \mathcal{D}_{T^*})$ defined on the unit disc by the formula*

$$\Theta_T(\lambda) = \left(-T + \lambda D_{T^*}(I - \lambda T^*)^{-1} D_T\right)|\mathcal{D}_T \quad \text{for } \lambda \in \mathbb{D}. \tag{1.13.1}$$

PROOF. Let Θ be the function (1.9.1) defined by the embeddings (1.6.2) related to the unitary dilation described in Theorem 1.4. We shall prove that

$$\Theta(\lambda) = \Omega_*^*\left(-T + \lambda D_{T^*}(I - \lambda T^*)^{-1} D_T\right)\Omega$$

where

$$\Omega = V^* v : E \to H \quad \text{and} \quad \Omega_* = V_*^* v_* : E_* \to H,$$

V, V_* being defined in Theorem 1.4 and v, v_* in (1.6.1). The above Ω and Ω_* map E and E_* isometrically onto $\mathcal{D}_T$ and $\mathcal{D}_{T^*}$, respectively. In particular, choosing $E = \mathcal{D}_T$, $E_* = \mathcal{D}_{T^*}$ and $v = V|\mathcal{D}_T$, $v_* = V_*|\mathcal{D}_{T^*}$, we get the function Θ_T as one of the possible choices among equivalent representations of the characteristic function.

Let $e \in E$ and $e_* \in E_*$. Then, for $|\lambda| < 1$,

$$(\Theta(\lambda)e, e_*)_{E_*} = \left(\Theta e, \frac{e_*}{1 - \bar{\lambda}z}\right)_{L^2(E_*)} = \left(\pi e, \pi_* \frac{e_*}{1 - \bar{\lambda}z}\right)_{\mathcal{H}}$$

$$= \left(ve, \pi_* \sum_{k\geq 0} \bar{\lambda}^k z^k e_*\right)_{\mathcal{H}} = \left(ve, \sum_{k\geq 0} \bar{\lambda}^k \mathcal{U}^{k+1} v_* e_*\right)_{\mathcal{H}}$$

$$= \left(v_*^* \sum_{k\geq 0} \lambda^k \mathcal{U}^{*(k+1)} ve, e_*\right)_{E_*}.$$

Therefore,

$$\Theta(\lambda) = v_*^* \sum_{k\geq 0} \lambda^k \mathcal{U}^{*(k+1)} v = v_*^* \frac{(I - \lambda \mathcal{U}^*)^{-1} - I}{\lambda} v = \frac{1}{\lambda} v_*^*(I - \lambda \mathcal{U}^*)^{-1} v,$$

because $\text{Range}\, v \perp \text{Range}\, v_*$ (the latter fact follows from the relations $G \perp G_*$ and $\text{Range}\, v \subset G$, $\text{Range}\, v_* \subset G_*$).

The proof of the following formula for the inverse of a block matrix operator is left to the reader:

$$\begin{pmatrix} X & A & B \\ 0 & Y & C \\ 0 & 0 & Z \end{pmatrix}^{-1} = \begin{pmatrix} X^{-1} & -X^{-1}AY^{-1} & X^{-1}(-B + AY^{-1}C)Z^{-1} \\ 0 & Y^{-1} & -Y^{-1}CZ^{-1} \\ 0 & 0 & Z^{-1} \end{pmatrix}.$$

We apply this formula to the operator

$$I - \lambda \mathcal{U}^* = \begin{pmatrix} I - \lambda \mathcal{E}_*^* & -\lambda V_* D_{T^*} & \lambda V_* T V^* \\ 0 & I - \lambda T^* & -\lambda D_T V^* \\ 0 & 0 & I - \lambda \mathcal{E}^* \end{pmatrix}$$

(compare the expression for $\mathcal{U}$ in Theorem 1.4). This results in the relation

$$
\Theta(\lambda) = \frac{1}{\lambda}(v_*^*, 0, 0)(I - \lambda \mathcal{U}^*)^{-1} \begin{pmatrix} 0 \\ 0 \\ v \end{pmatrix}
$$

$$
= \frac{1}{\lambda} v_*^*(I - \lambda \mathcal{E}_*^*)^{-1}\big(-\lambda V_* T V^* + (-\lambda V_* D_{T^*})(I - \lambda T^*)^{-1}(-\lambda D_T V^*)\big)
$$
$$
\times (I - \lambda \mathcal{E}^*)^{-1} v
$$
$$
= \Omega_*^*\big(-T + \lambda D_{T^*}(I - \lambda T^*)^{-1} D_T\big)\Omega,
$$

because $v_*^* \mathcal{E}_*^* = 0$ and $\mathcal{E}^* v = 0$. $\qquad\square$

1.14. Two more expressions for Θ_T. Multiplying (1.13.1) from the right by D_T we get the following useful formula

$$
\Theta_T(\lambda) D_T = D_{T^*}(I - \lambda T^*)^{-1}(\lambda I - T); \tag{1.14.1}
$$

and multiplying from the left by D_{T_*} we get

$$
D_{T^*}\Theta_T(\lambda) = (\lambda I - T)(I - \lambda T^*)^{-1} D_T|\mathcal{D}_T. \tag{1.14.2}
$$

1.15. Corollary. *The characteristic function is a pure contractive-valued function; that is,*

$$
\|\Theta(0)e\|_{E_*} < \|e\|_E \quad \text{for any nonzero } e \in E.
$$

Indeed, if $\|\Theta(0)e\|_{E_*} = \|e\|_E$, then

$$
\|\Omega e\| = \|e\| = \|\Theta(0)e\| = \|\Omega_*^* T \Omega e\| = \|T\Omega e\|,
$$

that is, $\Omega e \in \operatorname{Ker} D_T$. However, since $\Omega e \in \mathcal{D}_T$, we have $\Omega e = 0$, whence $e = 0$.

1.16. Coordinate-free function model. Now we are ready to construct the coordinate-free function model for a contraction on a Hilbert space. This will help us to solve the following *inverse problem*: given a purely contractive-valued function Θ analytic in the unit disc, find a completely nonunitary contraction whose characteristic function is equivalent to Θ.

To this end, the following steps can be taken.

(i) We take a function $\Theta \in H^\infty(E \to E_*)$ such that $\|\Theta\|_\infty \le 1$ and $\|\Theta(0)e\|_{E_*} < \|e\|_E$ for all nonzero $e \in E$.

(ii) We take any functional embedding $\Pi = \pi_* \oplus \pi$ acting from $L^2(E_* \oplus E)$ to an arbitrary Hilbert space with the prescribed modulus

$$
\Pi^*\Pi = \begin{pmatrix} I & \Theta \\ \Theta^* & I \end{pmatrix}. \tag{1.16.1}
$$

(iii) We put $\mathcal{H} = \operatorname{clos} \operatorname{Range} \Pi$.

(iv) We introduce a unitary operator $\mathcal{U}$ on $\mathcal{H}$ by the relation

$$
\mathcal{U}\Pi = \Pi z.
$$

(v) We introduce the *model subspace*

$$\mathcal{K}_\Theta \overset{\text{def}}{=} \mathcal{H} \ominus \left(\pi H^2(E) \oplus \pi_* H^2_-(E_*) \right),$$

where, as before, $\pi = \Pi | L^2(E)$ and $\pi_* = \Pi | L^2(E_*)$.

(vi) Finally we define a *model operator* by the formula

$$\mathcal{M}_\Theta \overset{\text{def}}{=} P_\Theta \mathcal{U} | \mathcal{K}_\Theta,$$

where P_Θ stands for the orthogonal projection from $\mathcal{H}$ onto $\mathcal{K}_\Theta$.

The next theorem shows that these six steps really solve the inverse problem in question.

1.17. Theorem. $\mathcal{M}_\Theta$ *is a completely nonunitary contraction with the characteristic function* Θ. *The operator* $\mathcal{U}$ *is the minimal unitary dilation of* $\mathcal{M}_\Theta$.

PROOF. First of all, it is worth mentioning that the operator $\mathcal{U}$ described in step (4) above is well-defined and unitary. Clearly, the norm is preserved, since

$$\|\mathcal{U}\Pi x\|^2 = \|\Pi zx\|^2 = \left(\begin{pmatrix} I & \Theta \\ \Theta^* & I \end{pmatrix} zx, \, zx \right) = \left(\begin{pmatrix} I & \Theta \\ \Theta^* & I \end{pmatrix} x, \, x \right) = \|\Pi x\|^2;$$

therefore, $\mathcal{U}$ is well-defined and isometric. By step (3), $\mathcal{U}$ is densely defined and possesses a dense range; hence, it admits a unitary extension to $\mathcal{H}$.

Now, we prove that $\mathcal{U}$ is a minimal unitary dilation of $\mathcal{M} = \mathcal{M}_\Theta$. From the definition it is clear that $\mathcal{U}$ is a dilation. More precisely, this follows from the fact that the subspaces $G \overset{\text{def}}{=} \pi H^2(E)$ and $G_* \overset{\text{def}}{=} \pi_* H^2_-(E_*)$ are invariant under $\mathcal{U}$ and $\mathcal{U}^*$, respectively, which shows that $\mathcal{H} = G_* \oplus \mathcal{K}_\Theta \oplus G$ is the decomposition from Lemma 1.2. Thus, all we need to verify is minimality.

We check the identity

$$\mathcal{U} P_\Theta \pi \bar{z} e + P_\Theta \pi_* \Theta_0 e = \pi(I - \Theta_0^* \Theta_0) e \quad \text{for } e \in E, \tag{1.17.1}$$

where $\Theta_0 \overset{\text{def}}{=} \Theta(0)$. Using step (v), we get for the orthogonal projection onto $\mathcal{K}_\Theta$ the formula

$$P_\Theta = I - \pi P_+ \pi^* - \pi_* P_- \pi_*^*,$$

where P_+ and P_- stand for the Riesz projections (see Section 0.7). Thus, we obtain

$$\mathcal{U} P_\Theta \pi \bar{z} e = \mathcal{U}(\pi P_- - \pi_* P_- \Theta) \bar{z} e = \mathcal{U}(\pi \bar{z} e - \pi_* \Theta_0 \bar{z} e) = \pi e - \pi_* \Theta_0 e,$$

$$P_\Theta \pi_* \Theta_0 e = (\pi_* P_+ - \pi P_+ \Theta^*) \Theta_0 e = \pi_* \Theta_0 e - \pi \Theta_0^* \Theta_0 e,$$

which implies (1.17.1).

Since Θ is pure, the operator $I - \Theta_0^* \Theta_0$ has a dense range, whence

$$\pi E \subset \text{span}\{\mathcal{K}_\Theta, \mathcal{U} \mathcal{K}_\Theta\},$$

and

$$G = \pi H^2(E) = \text{span}\{\mathcal{U}^n \pi E : n \geq 0\} \subset \text{span}\{\mathcal{U}^n \mathcal{K}_\Theta : n \geq 0\}.$$

In a similar way, using the identity

$$P_\Theta \pi_* e_* + \mathcal{U} P_\Theta \pi \bar{z} \Theta_0^* e_* = \pi_* (I - \Theta_0 \Theta_0^*) e_*,$$

we get

$$G_* = \pi_* H_-^2(E_*) = \text{span}\{\mathcal{U}^n \pi_* E_* : n < 0\} \subset \text{span}\{\mathcal{U}^n \mathcal{K}_\Theta : n \leq 0\}.$$

Hence, $\mathcal{H} = G_* \oplus \mathcal{K}_\Theta \oplus G \subset \text{span}\{\mathcal{U}^n \mathcal{K}_\Theta : n \in \mathbb{Z}\}$, so $\mathcal{U}$ is a minimal dilation.

By construction, π and π_* are functional embeddings with $v = \pi|E$ and $v_* = \pi_* \bar{z}|E_*$. Therefore, Θ is the characteristic function of $\mathcal{M}_\Theta$.   □

Now comes the main result of the Chapter, to complete the construction of the function model.

1.18.   Theorem. *If $\Theta = \Theta_T$ is the characteristic function of a completely nonunitary contraction T, then T is unitarily equivalent to the model operator $\mathcal{M}_\Theta$ constructed with the help of the six steps in 1.16.*

PROOF. Obvious from Theorems 1.11 and 1.17.   □

Let T be a completely nonunitary contraction and let $\Theta = \Theta_T$ be the characteristic function of T. We say that $\mathcal{M}_\Theta$ is a *coordinate-free model* of T acting on the model space $\mathcal{K}_\Theta$. Some explicit formulas for the function model will be obtained as transcriptions of this free model by specifying a representation of the Hilbert space $\mathcal{H}$ and a solution Π of equation (1.16.1). Several examples of such transcriptions are given in Chapter 3.

1.19.   Residual subspaces. In this section we introduce and briefly discuss two more functional mappings related to the function model of a contraction; these are quite useful for the study of the commutant lifting (see Chapter 4) and invariant subspaces (see Chapter 6). These mappings, denoted below by τ and τ_*, arise necessarily when one studies the absolutely continuous spectrum of a contraction, because they give spectral representations of the restrictions of the unitary dilation $\mathcal{U}$ to the so called *residual* and *∗-residual* parts of $\mathcal{H}$, that is, to

$$\mathcal{R} \stackrel{\text{def}}{=} \mathcal{H} \ominus \pi_* L^2(E_*) \quad \text{and} \quad \mathcal{R}_* \stackrel{\text{def}}{=} \mathcal{H} \ominus \pi L^2(E).$$

Clearly,

$$\mathcal{R} = (I - \pi_* \pi_*^*)\mathcal{H} = \text{clos}(I - \pi_* \pi_*^*)\pi L^2(E) = \text{clos}(\pi - \pi_* \Theta)L^2(E).$$

Since

$$(\pi - \pi_* \Theta)^*(\pi - \pi_* \Theta) = (\pi^* - \Theta^* \pi_*^*)(\pi - \pi_* \Theta) = I - \Theta^* \Theta = \Delta^2,$$

the polar decomposition

$$\pi - \pi_* \Theta = \tau \Delta \tag{1.19.1}$$

provides us with a partial isometry τ acting from $L^2(E)$ to $\mathcal{H}$, which is isometric on $L^2(\Delta E) \stackrel{\text{def}}{=} \text{clos}\, \Delta L^2(E)$ and whose range is $\mathcal{R}$. It turns out to be more convenient to view τ as defined only on $L^2(\Delta E)$. Then τ is an isometry that intertwines z on $L^2(\Delta E)$ with $\mathcal{U}|\mathcal{R}$; that is, it provides a unitary equivalence of these operators. Algebraically, these properties can be written as follows:

$$\tau^*\tau = I,$$
$$\tau\tau^* = I - \pi_*\pi_*^*,$$
$$\tau z = \mathcal{U}\tau.$$

In a similar way,

$$\mathcal{R}_* = \tau_* L^2(\Delta_* E_*),$$

where τ_* is the partial isometry occurring in the polar decomposition

$$\pi_* - \pi\Theta^* = \tau_*\Delta_*. \tag{1.19.2}$$

Then

$$\tau_*^*\tau_* = I,$$
$$\tau_*\tau_*^* = I - \pi\pi^*,$$
$$\tau_* z = \mathcal{U}\tau_*.$$

Below we list some more relations for the embeddings τ, τ_* and π, π_*, which are simple consequences of the definitions:

$$\tau^*\pi = \Delta, \qquad \tau_*^*\pi_* = \Delta_*,$$
$$\tau^*\pi_* = 0, \qquad \tau_*^*\pi = 0,$$
$$\tau^*\tau_* = -\Theta^*, \qquad \tau_*^*\tau = -\Theta.$$

Chapter 2. Examples

In this chapter, we give two examples where the characteristic function is computed. Both deal with dissipative operators, rather than contractions. However, as is well known, the theories of these two classes are related by the Cayley transform, which allows us to transfer information obtained for dissipative operators to contractions and vice versa. This transfer is not completely automatic, and we start with some prerequisites.

2.1. Definition. A densely defined operator A is said to be *dissipative* if

$$\operatorname{Im}(Ax, x) \geq 0 \quad \text{for all } x \in \operatorname{Dom} A.$$

A dissipative operator is *maximal* if it has no proper dissipative extension. A dissipative operator is *completely nonselfadjoint* if it has no selfadjoint restriction on a nonzero invariant subspace.

The next lemma collects some properties of dissipative operators and their relations to contractions. The proof of the lemma is classical; see [Sz.-Nagy and Foiaş 1967], for example.

2.2. Lemma. 1. *If A is a dissipative operator, the operator*

$$T = \mathcal{C}(A) \stackrel{\text{def}}{=} (A - iI)(A + iI)^{-1}$$

is a contraction acting from $(A + iI)\operatorname{Dom} A$ to $(A - iI)\operatorname{Dom} A$ and such that $1 \notin \sigma_p(T)$. Conversely, if T is a contraction such that 1 is not an eigenvalue of T, the operator

$$A = \mathcal{C}^{-1}(T) = i(I + T)(I - T)^{-1}$$

is well-defined on $\operatorname{Dom} A = (I - T)\operatorname{Dom} T$ and is dissipative. The operator $\mathcal{C}(A)$ is called the Cayley transform of A, and $\mathcal{C}^{-1}(T)$ is called the Cayley transform of T.

2. *Every dissipative operator has a maximal dissipative extension. A maximal dissipative operator is closed. A dissipative operator A is maximal if and only if $\operatorname{Dom} \mathcal{C}(A) = H$.*
3. *The operator A is selfadjoint if and only if $\mathcal{C}(A)$ is unitary.*
4. *Two dissipative operators A_1 and A_2 are unitarily equivalent if and only if are $\mathcal{C}(A_1)$ and $\mathcal{C}(A_2)$ so.*

A bounded operator A is maximal dissipative if and only if it is defined on the whole space ($\operatorname{Dom} A = H$) and its imaginary part is nonnegative ($\operatorname{Im} A = \frac{1}{2i}(A - A^*) \geq 0$). In this simplest case, we compute the defect operators and the characteristic function of the Cayley transform.

2.3. Lemma. *Let A be a bounded dissipative operator and let $T = \mathcal{C}(A)$ be the Cayley transform of A. Then*

$$D_T^2 = 2i(A^* - iI)^{-1}(A^* - A)(A + iI)^{-1},$$
$$D_{T^*}^2 = 2i(A + iI)^{-1}(A^* - A)(A^* - iI)^{-1}.$$

Moreover, there exist partial isometries V and V_* with initial space

$$\operatorname{clos} \operatorname{Range}(\operatorname{Im} A)$$

and final spaces $\mathcal{D}_T$ and $\mathcal{D}_{T^*}$, respectively, such that

$$D_T = V2(\operatorname{Im} A)^{1/2}(A + iI)^{-1} = (A^* - iI)^{-1}2(\operatorname{Im} A)^{1/2}V^*, \quad (2.3.1)$$
$$D_{T^*} = V_*2(\operatorname{Im} A)^{1/2}(A^* - iI)^{-1} = (A + iI)^{-1}2(\operatorname{Im} A)^{1/2}V_*^*. \quad (2.3.2)$$

In particular, $\operatorname{rank} D_T = \operatorname{rank} D_{T^*} = \operatorname{rank}(\operatorname{Im} A)$.

PROOF. The first two formulas are straightforward consequences of the definitions of D_T, D_{T^*}, and $\mathcal{C}(A)$. The second two relations follow from the polar decomposition; see Section 0.7. □

2.4. Lemma. *The characteristic function $\Theta_T(z)$ is equivalent to the function*

$$\mathcal{S}_A\left(i \frac{1 + z}{1 - z}\right),$$

where $z \in \mathbb{D}$ and

$$S_A(\zeta) \stackrel{\text{def}}{=} \left(I + i(2\operatorname{Im} A)^{1/2}(A^* - \zeta I)^{-1}(2\operatorname{Im} A)^{1/2}\right)\big|\operatorname{Range}(\operatorname{Im} A), \quad \text{for } \operatorname{Im}\zeta > 0. \tag{2.4.1}$$

PROOF. Using the the expressions for $T = \mathcal{C}(A)$ and formulas (2.3.1)–(2.3.2) we can rewrite formula (1.14.1) for Θ_T as follows:

$$V_*^* \Theta_T(z) D_T$$

$$= V_*^* D_{T^*}(I - zT^*)^{-1}(zI - T)$$

$$= 2(\operatorname{Im} A)^{1/2}(A^* - iI)^{-1}\left(I - z(A^* + iI)(A^* - iI)^{-1}\right)^{-1}\left(zI - (A - iI)(A + iI)^{-1}\right)$$

$$= 2(\operatorname{Im} A)^{1/2}\left((A^* - iI) - z(A^* + iI)\right)^{-1}\left(z(A + iI) - (A - iI)\right)(A + iI)^{-1}$$

$$= -2(\operatorname{Im} A)^{1/2}\left(A^* - i\frac{1+z}{1-z}I\right)^{-1}\left(A - i\frac{1+z}{1-z}I\right)(A + iI)^{-1}$$

$$= -2(\operatorname{Im} A)^{1/2}(A^* - \zeta I)^{-1}(A - \zeta I)(A + iI)^{-1}$$

$$= -2(\operatorname{Im} A)^{1/2}(A^* - \zeta I)^{-1}(2i\operatorname{Im} A + A^* - \zeta I)(A + iI)^{-1}$$

$$= -\left(I + 2i(\operatorname{Im} A)^{1/2}(A^* - \zeta I)^{-1}(\operatorname{Im} A)^{1/2}\right)2(\operatorname{Im} A)^{1/2}(A + iI)^{-1}.$$

Taking into account (2.3.1), we get for Θ_T an expression equivalent to (2.4.1). $\square$

The function (2.4.1) is called *the characteristic function of the dissipative operator A*.

2.5. Example: The dissipative integration operator. Let μ be a positive finite Borel measure on the interval [0,1]. We consider the integration operator A acting on the space $L^2(\mu)$ and defined by

$$(Af)(x) = i\int_{[0,x\}} f(t)\,d\mu(t) \stackrel{\text{def}}{=} i\int_{[0,x)} f(t)\,d\mu(t) + \frac{i}{2}\mu(\{x\})f(x) \quad \text{for } x \in [0,1].$$

Obviously, A is well-defined and bounded (even compact) on $L^2(\mu)$, and its adjoint operator is given by

$$(A^* f)(x) = -i\int_{\{x,1]} f(t)\,d\mu(t).$$

Therefore, A is a dissipative operator with rank-one imaginary part:

$$(2\operatorname{Im} A)f = \int_{[0,1]} f(t)\,d\mu(t) = (f,\mathbf{1})\mathbf{1},$$

$$((2\operatorname{Im} A)f, f) = |(f,\mathbf{1})|^2 \geq 0 \quad \text{for } f \in L^2(\mu).$$

Our goal is to construct the model for the Cayley transform $\mathcal{C}(A)$ of the operator A. In accordance with the general theory of Chapter 1, the only thing we need is to find the completely nonunitary part of $\mathcal{C}(A)$ and to compute its characteristic function. However, it turns out that A is completely nonselfadjoint, so $\mathcal{C}(A)$ is completely nonunitary (see part 3 of Lemma 2.2). Taking into account Lemma 2.4, we will compute not the characteristic function of $\mathcal{C}(A)$, but the characteristic function of A itself.

2.6. Theorem. *The operator A is a completely nonselfadjoint dissipative operator with characteristic function*

$$S_A(\zeta) = \left(\prod_{0 \le t \le 1} \frac{\zeta - \frac{i}{2}\mu(\{t\})}{\zeta + \frac{i}{2}\mu(\{t\})} \right) \exp\left(-i \frac{\mu_c([0,1])}{\zeta} \right). \tag{2.6.1}$$

To check that A is a completely nonselfadjoint operator and to compute its characteristic function S_A, we need to compute the resolvent $(A^* - \zeta I)^{-1}$, that is, to find a solution f of the equation

$$(A^* - \zeta I)f = h, \tag{2.6.2}$$

or

$$-i \int_{\{x,1\}} f(t)\, d\mu(t) - \zeta f(x) = h(x).$$

Putting $M = \mu([0,1])$, we introduce the map $\varphi : [0,1] \to [0, M]$ given by

$$\varphi(x) = \begin{cases} \frac{1}{2}\mu(\{0\}) & \text{if } x = 0, \\ \mu([0,x)) + \frac{1}{2}\mu(\{x\}) & \text{if } x > 0, \end{cases}$$

and the map $\psi : [0, M] \to [0,1]$ given by

$$\psi(t) = \begin{cases} \inf\{x : \mu([0,x)) > t\} & \text{if } t < \mu([0,1)), \\ 1 & \text{if } t \ge \mu([0,1)). \end{cases}$$

Then the solution of (2.6.2) can be described as follows.

2.7. Lemma. *If $\operatorname{Re}\zeta \ne 0$ and $h \in L^2(\mu)$, equation (2.6.2) has a unique solution $f \in L^2(\mu)$, which can be recovered from the solution g of the differential equation*

$$g'(\tau) = \frac{g(\tau) - h(\psi(\tau))}{\tau - \varphi(\psi(\tau)) - i\zeta} \tag{2.7.1}$$

satisfying the initial condition $g(M) = 0$. Namely, the function

$$f \overset{\text{def}}{=} \frac{1}{\zeta}(g \circ \varphi - h)$$

solves (2.6.2).

PROOF. For ζ away from the imaginary line, the function in the denominator of (2.7.1) is bounded away from zero; hence equation (2.7.1) has an absolutely continuous solution g. Thus we need only to check that the function $f(x) = \frac{1}{\zeta}\big(g(\varphi(x)) - h(x)\big)$ solves equation (2.6.2).

Further, we note that φ is continuous at every point x of the interval $[0,1]$ for which $\mu(\{x\}) = 0$, and that each jump of φ, that is, each point mass of the measure μ, corresponds to an interval $(\varphi(x_{-0}), \varphi(x_{+0})) = (\mu([0,x)), \mu([0,x]))$ where the function ψ is constant. In turn, the function ψ is continuous everywhere off the set of points t such that $\varphi(x_1) = \varphi(x_2) = t$ for at least two different points $x_1 \neq x_2$. If x is not a mass point for μ and $x_1 = \inf\{x : \varphi(x) = t\}$, $x_2 = \sup\{x : \varphi(x) = t\}$, then ψ has a jump at t with $\psi(t_{-0}) = x_1$ and $\psi(t_{+0}) = x_2$.

We introduce functions f_1, h_1, and g on $[0, M]$ by the formulas

$$f_1(t) = f(\psi(t)), \quad h_1(t) = h(\psi(t)), \quad g(t) = -i \int_t^M f_1(s)\, ds.$$

From the definition we see that g is an absolutely continuous function, piecewise linear on the intervals where ψ is constant. We prove that g coincides with $h_1 + \zeta f_1$ on the image of φ.

Changing the variable, we get the relations

$$\int_0^{\varphi(x)} f_1(s)\, ds = \int_{[0,x\}} f(t)\, d\mu(t) \quad \text{and} \quad \int_{\varphi(x)}^M f_1(s)\, ds = \int_{\{x,1]} f(t)\, d\mu(t).$$

In particular,

$$g(\varphi(x)) = -i \int_{\{x,1]} f(t)\, d\mu(t) = h(x) + \zeta f(x). \tag{2.7.2}$$

If τ is an arbitrary point in $[0, M]$ and $x = \psi(\tau)$, then $f_1(s) = f_1(\tau) = f(x)$ on the interval $s \in (\varphi(x_{-0}), \varphi(x_{+0}))$, and

$$g(\tau) = -i \int_\tau^M f_1(s)\, ds = -i \int_{\varphi(x)}^M f_1(s)\, ds - i \int_\tau^{\varphi(x)} f_1(s)\, ds$$

$$= -i \int_{\{x,1]} f(t)\, d\mu(t) - i(\varphi(x) - \tau)f_1(\tau) = h(x) + \zeta f(x) - i(\varphi(x) - \tau)f_1(\tau)$$

$$= h_1(\tau) + i(\tau - \varphi(\psi(\tau)) - i\zeta)f_1(\tau),$$

which yields

$$if_1(\tau) = \frac{g(\tau) - h_1(\tau)}{\tau - \varphi(\psi(\tau)) - i\zeta}.$$

Since the definition of g is equivalent to the equation $g' = if_1$ with the initial condition $g(M) = 0$, the latter relation implies that g is a solution of (2.7.1). Therefore, the conclusion of the lemma follows from (2.7.2). $\qquad\square$

2.8. Corollary. *The operator A is completely nonselfadjoint.*

PROOF. Let H_0 be the maximal invariant subspace such that the restriction $A|H_0$ is selfadjoint. Then H_0 is a reducing subspace for A (see Lemma 2.2 and Corollary 1.8), hence $\sigma(A|H_0) \subset \sigma(A^*)$. By Lemma 2.7, equation (2.6.2) has an L^2-solution for all h and for all ζ, $\operatorname{Re}\zeta \neq 0$; that is, A^* has no real spectrum, except, maybe, the point $\zeta = 0$. Thus, to check that A is completely nonselfadjoint it suffices to check that the kernel of A^* is trivial.

Let $A^* f = 0$, that is,

$$\frac{1}{2}\mu(\{x\})f(x) + \int\limits_{(x,1]} f(t)\,d\mu(t) = 0 \quad \forall x \in [0,1].$$

Putting $x + \varepsilon$ in place of x and letting ε tend to zero, we obtain

$$\int\limits_{(x,1]} f(t)\,d\mu(t) = 0 \quad \text{for all } x \in [0,1],$$

or

$$\int\limits_{(x,y]} f(t)\,d\mu(t) = 0 \quad \text{for all } x,y \in [0,1],$$

which implies that f vanishes μ-a.e. $\qquad\square$

We mention that T. Kriete [1972] found a criterion for complete nonselfadjointness for a class of dissipative operators with rank-one imaginary part. Our operator A corresponding to an absolutely continuous measure μ is contained in this class.

2.9. Proof of Theorem 2.6. That A is completely nonselfadjoint has already been proved. Now we compute the characteristic function $\mathcal{S}_A$:

$$\mathcal{S}_A(\zeta) = \left((I + i\sqrt{2\operatorname{Im}A}(A^* - \zeta I)^{-1}\sqrt{2\operatorname{Im}A})\mathbf{1},\,\mathbf{1}\right)\|\mathbf{1}\|^{-2}$$
$$= 1 + i\|\mathbf{1}\|^{-2}\left((A^* - \zeta I)^{-1}\sqrt{2\operatorname{Im}A}\,\mathbf{1},\,\sqrt{2\operatorname{Im}A}\,\mathbf{1}\right)$$
$$= 1 + i\left((A^* - \zeta I)^{-1}\mathbf{1},\,\mathbf{1}\right),$$

because $\sqrt{2\operatorname{Im}A}\,f = \|\mathbf{1}\|^{-1}(f,\mathbf{1})\mathbf{1}$.

So, for computing $\mathcal{S}_A(\zeta)$ we need the solution f of (2.6.2) with $h = \mathbf{1}$. Actually, we need not f itself, but

$$\int\limits_{[0,1]} f(t)\,d\mu(t) = \int\limits_0^M f_1(s)\,ds = ig(0),$$

where g is the solution of

$$g'(\tau) = \frac{g(\tau) - 1}{\tau - \varphi(\psi(\tau)) - i\zeta} \quad \text{with } g(M) = 0.$$

Putting $g_1 = 1 - g$ and $\omega(\tau) = (\tau - \varphi(\psi(\tau)) - i\zeta)^{-1}$, we see that $\mathcal{S}_A(\zeta) = 1 - g(0) = g_1(0)$ and that

$$g_1'(\tau) = \omega(\tau)g_1(\tau), \quad g_1(M) = 1.$$

Thus

$$g_1(\tau) = \exp\left(-\int_\tau^M \omega(t)\, dt\right),$$

and

$$\mathcal{S}_A(\zeta) = \exp\left(-\int_0^M \omega(t)\, dt\right). \tag{2.9.1}$$

As mentioned in Section 2.7, ψ is constant on the intervals

$$\Omega_x \overset{\text{def}}{=} \big(\varphi(x_{-0}),\, \varphi(x_{+0})\big) = \big(\mu([0, x)),\, \mu([0, x])\big),$$

where its value is x; that is, $\omega(t) = (t - \varphi(x) - i\zeta)^{-1}$ for $t \in \Omega_x$. Set $\Omega \overset{\text{def}}{=} \bigcap\{\Omega_x : \mu(\{x\}) > 0\}$. For almost all $t \notin \Omega$ (in fact for all t except for the ends of Ω_x) we have $\varphi(\psi(t)) = t$, that is, $\omega(t) = \text{const} = -1/(i\zeta)$ for $t \in [0, M] \setminus \Omega$, whence

$$\int_0^M \omega(t)\, dt = -\frac{M}{i\zeta} + \int_\Omega \left(\omega(t) + \frac{1}{i\zeta}\right) dt = -\frac{\mu([0,1])}{i\zeta} + \sum_{x:\mu(\{x\})>0} \int_{\Omega_x} \left(\omega(t) + \frac{1}{i\zeta}\right) dt.$$

The integral over Ω can naturally be split into the sum of the integrals over the intervals $\Omega_x = \big(\varphi(x_{-0}),\, \varphi(x_{+0})\big)$ on which $\omega(t) = (t - \varphi(x) - i\zeta)^{-1}$, and

$$\int_{\varphi(x_{-0})}^{\varphi(x_{+0})} \left(\omega(t) + \frac{1}{i\zeta}\right) dt = \int_{\varphi(x_{-0})}^{\varphi(x_{+0})} \frac{ds}{s - \varphi(x) - i\zeta} + \frac{1}{i\zeta}\big(\varphi(x_{+0}) - \varphi(x_{-0})\big)$$

$$= \log\frac{\varphi(x_{+0}) - \varphi(x) - i\zeta}{\varphi(x_{-0}) - \varphi(x) - i\zeta} + \frac{1}{i\zeta}\big(\varphi(x_{+0}) - \varphi(x_{-0})\big)$$

$$= \log\frac{1 - i\mu(\{x\})/(2\zeta)}{1 + i\mu(\{x\})/(2\zeta)} + \frac{\mu(\{x\})}{i\zeta},$$

because $\varphi(x_{+0}) - \varphi(x) = \varphi(x) - \varphi(x_{-0}) = \frac{1}{2}\mu(\{x\})$. Finally, substituting this in (2.9.1), we obtain

$$\mathcal{S}_A(\zeta) = \left(\prod_{0 \le t \le 1} \frac{\zeta - \frac{i}{2}\mu(\{t\})}{\zeta + \frac{i}{2}\mu(\{t\})}\right) \exp\left(-i\,\frac{\mu_c([0,1])}{\zeta}\right),$$

where μ_c is the continuous part of μ:

$$\mu_c([0,1]) = \mu([0,1]) - \sum \mu(\{t\}). \qquad \square$$

2.10. Unitary classification. Thus, the characteristic function of A is a scalar inner function whose singular part has only one singularity, at $\zeta = 0$ (with mass $\mu_c([0,1])$) and whose zeros on the imaginary axis are at $\zeta = \frac{i}{2}\mu(\{t\})$ and have multiplicity equal to the number of points on $[0,1]$ with the same mass $\mu(\{t\})$. We see that S_A is independent of the distribution of μ on the interval $[0,1]$, depending only on the total continuous mass $\mu_c([0,1])$ and on the values of the point masses $\mu(\{t\})$ (regardless of their location). Therefore, two operators A determined by measures μ_1 and μ_2 are unitarily equivalent if and only if

$$\mu_{1c}([0,1]) = \mu_{2c}([0,1])$$

and

$$\operatorname{card}\{t : \mu_1(\{t\}) = \lambda\} = \operatorname{card}\{t : \mu_2(\{t\}) = \lambda\} \quad \text{for all } \lambda > 0.$$

We know the characteristic function of the operator, so we can describe its spectrum (see Chapter 5). In our case the spectrum of the operator A consists of the eigenvalues of multiplicity one at the points $i\lambda$ such that

$$k(\lambda) \stackrel{\text{def}}{=} \operatorname{card}\{t : \mu(\{t\}) = 2\lambda\} > 0.$$

Moreover, $k(\lambda)$ is the size of the corresponding Jordan block. If A is not a finite rank operator—that is, if the support of μ is not a finite set—the point $\lambda = 0$ is the sole point of the essential spectrum of A.

2.11. The matrix case. As an illustration to the previous computations, we rewrite a partial case of the operator A in a matrix form.

Set $\mu = \sum_{k \geq 1} \mu_k \delta_{t_k}$, where $\mu_k > 0$, $t_k > 0$ and $\sum_{k \geq 1} \mu_k < \infty$. Then our operator A is unitarily equivalent to the operator $\mathcal{A} : \ell^2(\mu_k) \to \ell^2(\mu_k)$ defined by the formula

$$\mathcal{A}f = \left\{ i\left(\sum_{t_j < t_k} f_j \mu_j + \tfrac{1}{2} f_k \mu_k \right) : k \geq 1 \right\}$$

on the sequence space

$$\ell^2(\mu_k) = \left\{ f = (f_k)_{k \geq 1} : \sum_{k \geq 1} |f_k|^2 \mu_k < \infty \right\}.$$

Taking the unitary transformation $V : \ell^2(\mu_k) \to \ell^2$ given by

$$Vf = (a_k f_k)_{k \geq 1},$$

where the a_k are complex numbers satisfying $|a_k|^2 = \mu_k$ for $k \geq 1$, we get a unitarily equivalent operator $\mathcal{J}_a : \ell^2 \to \ell^2$ given by

$$\mathcal{J}_a x = \left\{ i\left(\sum_{t_j < t_k} a_k \bar{a}_j x_j + \tfrac{1}{2} |a_k|^2 x_k \right) : k \geq 1 \right\} \quad \text{for } x \in \ell^2.$$

From the preceding discussion we know that two operators $\mathcal{J}_a$ and $\mathcal{J}_b$, with $a, b \in \ell^2$, are unitarily equivalent if and only if the decreasing rearrangements of $|a|$ and $|b|$ coincide, and that the spectrum of $\mathcal{J}_a$ is $\sigma(\mathcal{J}_a) = \frac{i}{2} \operatorname{Range} |a|^2 = \frac{i}{2}\{0, |a_1|^2, |a_2|^2, \ldots\}$. Every invariant subspace of $\mathcal{J}_a$ is generated by the eigenvectors and root vectors it contains; see Section 6.20.

For two special distributions of t_k the operators $\mathcal{J}_a$ are related to the triangular truncation of matrices on the space ℓ^2. If we take an increasing sequence, $t_k < t_{k+1}$ for $k \geq 1$ the corresponding operator $\frac{1}{i}\mathcal{J}_a$ is a kind of lower truncation of a selfadjoint matrix $\mathbf{a} = (a_k\bar{a}_j)_{k,j\geq 1}$:

$$\mathcal{J}_a^{(l)} = i \begin{pmatrix} \frac{1}{2}|a_1|^2 & 0 & 0 & 0 & \cdots \\ a_2\bar{a}_1 & \frac{1}{2}|a_2|^2 & 0 & 0 & \cdots \\ a_3\bar{a}_1 & a_3\bar{a}_2 & \frac{1}{2}|a_3|^2 & 0 & \cdots \\ a_4\bar{a}_1 & a_4\bar{a}_2 & a_4\bar{a}_3 & \frac{1}{2}|a_4|^2 & \cdots \\ \cdots & \cdots & \cdots & \cdots & \cdots \end{pmatrix}.$$

Likewise, if we take a decreasing sequence, $\frac{1}{i}\mathcal{J}_a$ acts as upper truncation:

$$\mathcal{J}_a^{(u)} = i \begin{pmatrix} \frac{1}{2}|a_1|^2 & a_1\bar{a}_2 & a_1\bar{a}_3 & a_1\bar{a}_4 & \cdots \\ 0 & \frac{1}{2}|a_2|^2 & a_2\bar{a}_3 & a_2\bar{a}_4 & \cdots \\ 0 & 0 & \frac{1}{2}|a_3|^2 & a_3\bar{a}_4 & \cdots \\ 0 & 0 & 0 & \frac{1}{2}|a_4|^2 & \cdots \\ \cdots & \cdots & \cdots & \cdots & \cdots \end{pmatrix}.$$

It is clear that the matrix $\mathbf{a}$ represents the rank-one operator $(\cdot, a)a$, and that the operators $\mathcal{J}_a^{(u)}$ and $\mathcal{J}_a^{(l)}$ are unitarily equivalent.

2.12. Example 2: The dissipative Sturm–Liouville operator. We consider the differential operator ℓ_h arising from the differential expression

$$\ell y = -y'' + qy,$$

where q is a real function, and by the boundary condition $y'(0) = hy(0)$, where h is a complex number. The operator ℓ_h has domain $\mathrm{Dom}(\ell_h) = \{y \in W_{2,\mathrm{loc}}^2(\mathbb{R}_+) : \ell y \in L^2(\mathbb{R}_+),\ y'(0) = hy(0)\}$, and is defined by

$$\ell_h(y) = \ell y \in L^2(\mathbb{R}_+). \tag{2.12.1}$$

The general facts about the operators ℓ_h can be found in [Reed and Simon 1975] or [Atkinson 1964]. In particular,

$$\ell_h^* = \ell_{\bar{h}}.$$

Moreover, ℓ_h is a rank-one perturbation of the operator ℓ_∞ defined by the differential expression ℓ and by the boundary condition $y(0) = 0$. The operator ℓ_∞ is selfadjoint. The operator ℓ_h is dissipative if and only if $\mathrm{Im}\, h > 0$.

Our goal is to compute the characteristic function S_{ℓ_h} of ℓ_h and, thus, to include the study of ℓ_h in the model theory.

To this end, we consider two solutions φ_ζ and ψ_ζ of the equation $\ell y = \zeta y$, satisfying the boundary conditions

$$\begin{cases} \varphi_\zeta(0) = 0, \\ \varphi_\zeta'(0) = 1 \end{cases} \quad \text{and} \quad \begin{cases} \psi_\zeta(0) = -1, \\ \psi_\zeta'(0) = 0. \end{cases}$$

It is well known that for every ζ with $\mathrm{Re}\,\zeta \neq 0$ there exists a unique L^2-solution y_ζ of our equation $\ell y = \zeta y$ representable as a linear combination of φ_ζ and ψ_ζ:

$$y_\zeta = \psi_\zeta + m(\zeta)\varphi_\zeta.$$

The function $m(\zeta)$ determined by this condition for all $\zeta \neq \bar\zeta$ is called the *Weyl function*.

Following B. Pavlov [1976], we can compute the characteristic function of ℓ_h in terms of the Weyl function m.

An important remark is that now we cannot use the formulas from Lemmas 2.3 and 2.4 for the defect operators and for the characteristic function, because our operator $A = \ell_h$ is unbounded, its domain is different from the domain of the adjoint, and the imaginary part is not well-defined. This difficulty can be overcome; we refer the reader to [Solomyak 1989], for example.

If $T = \mathcal{C}(A) = (A - iI)(A + iI)^{-1}$, then $T^* = (A^* + iI)(A^* - iI)^{-1}$. Taking an arbitrary vector $f \in \mathrm{Dom}\,A$ and putting $x = (A + iI)f$, we have $Tx = (A - iI)f$, whence

$$D_T^2 x = (A + iI)f - T^*(A - iI)f = (A + iI)f - (A^* + iI)(A^* - iI)^{-1}(A - iI)f.$$

Set $g \overset{\mathrm{def}}{=} (A^* - iI)^{-1}(A - iI)f$. Then

$$(A - iI)f = (A^* - iI)g,$$

or

$$\ell(f - g) = i(f - g),$$

which implies that $f - g = cy_i$, and

$$D_T^2 x = (A + iI)f - (A^* + iI)g = (\ell + iI)(f - g) = 2icy_i.$$

Thus, we have checked that the defect subspace $\mathcal{D}_T$ of the Cayley transform $T = \mathcal{C}(\ell_h)$ is the one-dimensional subspace

$$\mathcal{D}_T = \mathrm{span}\{y_i\}$$

generated by the solutions y_ζ for $\zeta = i$. In a similar way, the defect subspace of the adjoint operator T^* is also one-dimensional:

$$\mathcal{D}_{T^*} = \mathrm{span}\{y_{-i}\}.$$

Next, we note that $Ty_i = 0$ and $T^*y_{-i} = 0$, whence $D_T y_i = y_i$ and $D_{T^*} y_{-i} = y_{-i}$. Now, instead of (2.4.1), we use expression (1.14.2), namely,

$$D_{T^*}\Theta_T(z) = (zI - T)(I - zT^*)^{-1}D_T,$$

which can be rewritten in terms of $A = \ell_h$ as follows:

$$D_{T^*}\Theta_T(z) = -(A - \zeta I)(A + iI)^{-1}(A^* - iI)(A^* - \zeta I)^{-1}D_T,$$

where, as before, $\zeta = i(1 + z)/(1 - z)$. We apply this operator to the vector y_i and employ the formula

$$(\ell_h - \lambda I)^{-1} y_\mu = \frac{y_\mu + c y_\lambda}{\mu - \lambda},$$

(it will be used twice); here c is chosen so as to ensure that the vector above belongs to $\mathrm{Dom}(\ell_h)$, namely,

$$c = c(\lambda, \mu, h) = -\frac{m(\mu) + h}{m(\lambda) + h}.$$

As a result, we obtain

$$
\begin{aligned}
D_{T^*} \Theta_T(z) y_i &= -(\ell_h - \zeta I)(\ell_h + iI)^{-1}(\ell_{\bar{h}} - iI)(\ell_h - \zeta I)^{-1} y_i \\
&= -(\ell_h - \zeta I)(\ell_h + iI)^{-1}(\ell_{\bar{h}} - iI) \frac{y_i + c(\zeta, i, \bar{h}) y_\zeta}{i - \zeta} \\
&= (\ell_h - \zeta I)(\ell_h + iI)^{-1} c(\zeta, i, \bar{h}) y_\zeta \\
&= (\ell_h - \zeta I) c(\zeta, i, \bar{h}) \frac{y_\zeta + c(-i, \zeta, h) y_{-i}}{\zeta + i} \\
&= -c(\zeta, i, \bar{h}) c(-i, \zeta, h) y_{-i} = c_0 \frac{m(\zeta) + h}{m(\zeta) + \bar{h}},
\end{aligned}
$$

where $c_0 = -\big(m(i) + \bar{h}\big)/\big(m(-i) + h\big)$ is a unimodular constant. Since $y_{-i} = \bar{y}_i$ and, therefore, $\|y_i\| = \|y_{-i}\|$, we see that the characteristic function is equivalent to the simple expression

$$\mathcal{S}_{\ell_h}(\zeta) = \frac{m(\zeta) + h}{m(\zeta) + \bar{h}}.$$

Thus, we have proved the following theorem.

2.13. Theorem. *The characteristic function* $\Theta_{\mathcal{C}(\ell_h)}$ *of the Cayley transform of a dissipative Sturm–Liouville operator* (2.12.1) *is*

$$\Theta_{\mathcal{C}(\ell_h)}(z) = \mathcal{S}_{\ell_h}(\zeta) = \frac{m(\zeta) + h}{m(\zeta) + \bar{h}},$$

where $\zeta = i \dfrac{1 + z}{1 - z}$.

Chapter 3. Transcriptions of the Model

A coordinate transcription of the function model arises whenever we choose a spectral representation of the minimal unitary dilation $\mathcal{U}$ and a solution Π of the equation

$$\Pi^* \Pi = W_\Theta \overset{\mathrm{def}}{=} \begin{pmatrix} I & \Theta \\ \Theta^* & I \end{pmatrix}. \tag{3.0}$$

3.1. Multiplicity of the minimal dilation. To begin with, we mention that for a given completely nonunitary contraction T the spectral measure $E_{\mathcal{U}}$ of the minimal unitary dilation $\mathcal{U}$ and Lebesgue measure on the unit circle are mutually absolutely continuous. Moreover, the local spectral multiplicity of $E_{\mathcal{U}}$ is at least $\max(\partial, \partial_*)$ and at most $\partial + \partial_*$, where $\partial = \dim \mathcal{D}_T$ and $\partial_* = \dim \mathcal{D}_{T^*}$. This is an immediate consequence of the existence of the embeddings π and π_* intertwining $\mathcal{U}$ and z on $L^2(E)$ and $L^2(E_*)$, respectively, and the completeness property $\mathcal{H} = \operatorname{clos} \operatorname{Range} \Pi$.

3.2. Choosing a space of the minimal dilation. Thus, the minimal unitary dilation $\mathcal{U}$ is unitarily equivalent to the operator of multiplication by z on any weighted space

$$L^2(E_* \oplus E, W) \overset{\text{def}}{=} \left\{ f : \int_{\mathbb{T}} (W(\zeta)f(\zeta), f(\zeta))\, dm(\zeta) < \infty \right\}$$

whenever the operator-valued weight $W(\zeta) : E_* \oplus E \to E_* \oplus E$ satisfies the spectral multiplicity condition $\operatorname{rank} W(\zeta) = \operatorname{rank} E_{\mathcal{U}}(\zeta)$ for a.e. $\zeta \in \mathbb{T}$.

We could choose another coefficient space instead of $E_* \oplus E$; however, the latter space is natural and minimal among those ensuring that $\dim(E_* \oplus E) = \partial_* + \partial \geq \operatorname{rank} E_{\mathcal{U}}$.

3.3. Intermediate space. In order to separate the role of the weight and to facilitate computation we assume that the embeddings π and π_* are continuous as mappings into the nonweighted spaces $L^2(E) \to L^2(E_* \oplus E)$ and $L^2(E_*) \to L^2(E_* \oplus E)$, respectively. The corresponding adjoint mappings will be denoted by the symbols π^+ and π_*^+. To start with, we also assume that the weight W is bounded. The operator adjoint to the natural embedding $L^2(E_* \oplus E) \to L^2(E_* \oplus E, W)$ is the operator of multiplication by W. Thus, we have $\pi^* = \pi^+ W$ and $\pi_*^* = \pi_*^+ W$, that is, $\Pi^* = \Pi^+ W$.

3.4. Choosing functional embeddings. So, we must choose an operator Π satisfying (3.0). To solve this equation we rewrite it in terms of Π^+:

$$\Pi^* \Pi = W_\Theta \quad \Longleftrightarrow \quad \begin{cases} \pi^+ W \pi = I, \\ \pi_*^+ W \pi_* = I, \\ \pi_*^+ W \pi = \Theta. \end{cases} \qquad (3.4.1)$$

Here we shall not dwell on the description of all solutions of this system, nor do we discuss in detail the possible preferences in choosing a specific transcription; this can be found in [Nikolski and Vasyunin 1989]. We only present some ideas concerning what could be required from a transcription and describe three most popular transcriptions that are used in numerous papers of many authors.

3.5. The Szőkefalvi-Nagy–Foiaş transcription. First of all, we can prefer to work in a nonweighted L^2-space. Though it is not always possible to put $W = I$, in any case as $\mathcal{H}$ we can choose a subspace of $L^2(E_* \oplus E)$. This means the chosen W is a projection: $W(\zeta) = P_{\text{Range }\Pi(\zeta)}$, where $\zeta \in \mathbb{T}$. Moreover, we can try to take π_* (or π) to be the natural embedding.

Solving (3.4.1) under the assumptions

$$\pi_* = \begin{pmatrix} I \\ 0 \end{pmatrix}, \quad \pi = \begin{pmatrix} X \\ Y \end{pmatrix}, \quad W = W^2 = \begin{pmatrix} A & B \\ B^* & C \end{pmatrix},$$

we obtain

$$\pi_*^+ W \pi_* = I \implies A = I,$$
$$W^2 = W \implies B = 0 \text{ and } C^2 = C,$$
$$\pi_*^+ W \pi = \Theta \implies X = \Theta,$$
$$\pi^+ W \pi = I \implies Y^* C Y = \Delta^2 \overset{\text{def}}{=} I - \Theta^* \Theta.$$

The usual choice is

$$Y = \Delta, \quad C(\zeta) = P_{\text{Range }\Delta(\zeta)}.$$

In this way we arrive at the *Szőkefalvi-Nagy–Foiaş incoming transcription* of the model:

$$\pi = \begin{pmatrix} \Theta \\ \Delta \end{pmatrix}, \quad \pi_* = \begin{pmatrix} I \\ 0 \end{pmatrix}, \quad W = \begin{pmatrix} I & 0 \\ 0 & P_{\text{Range }\Delta} \end{pmatrix}, \quad \mathcal{H} = \begin{pmatrix} L^2(E_*) \\ L^2(\Delta E) \end{pmatrix}, \quad (3.5.1)$$

where $L^2(\Delta E) \overset{\text{def}}{=} \text{clos } \Delta L^2(E)$;

$$G = \begin{pmatrix} \Theta \\ \Delta \end{pmatrix} H^2(E), \quad G_* = \begin{pmatrix} H_-^2(E_*) \\ 0 \end{pmatrix}, \quad \mathcal{K}_\Theta = \begin{pmatrix} H^2(E_*) \\ L^2(\Delta E) \end{pmatrix} \ominus \begin{pmatrix} \Theta \\ \Delta \end{pmatrix} H^2(E),$$

$$\mathcal{M}_\Theta \begin{pmatrix} f \\ g \end{pmatrix} = \begin{pmatrix} zf - \Theta[z(\Theta^* f + \Delta g)]\widehat{}(0) \\ zg - \Delta[z(\Theta^* f + \Delta g)]\widehat{}(0) \end{pmatrix}, \quad \mathcal{M}_\Theta^* \begin{pmatrix} f \\ g \end{pmatrix} = \begin{pmatrix} \dfrac{f - f(0)}{z} \\ \bar{z}g \end{pmatrix}.$$

Choosing π (rather than π_*) to be the natural embedding, we obtain the *Szőkefalvi-Nagy–Foiaş outgoing transcription* of the model:

$$\pi = \begin{pmatrix} 0 \\ I \end{pmatrix}, \quad \pi_* = \begin{pmatrix} \Delta_* \\ \Theta^* \end{pmatrix}, \quad W = \begin{pmatrix} P_{\text{Range }\Delta_*} & 0 \\ 0 & I \end{pmatrix}, \quad \mathcal{H} = \begin{pmatrix} L^2(\Delta_* E_*) \\ L^2(E) \end{pmatrix},$$

$$G = \begin{pmatrix} 0 \\ H^2(E) \end{pmatrix}, \quad G_* = \begin{pmatrix} \Delta_* \\ \Theta^* \end{pmatrix} H_-^2(E_*), \quad \mathcal{K}_\Theta = \begin{pmatrix} L^2(\Delta_* E_*) \\ H_-^2(E) \end{pmatrix} \ominus \begin{pmatrix} \Delta_* \\ \Theta^* \end{pmatrix} H_-^2(E_*),$$

$$\mathcal{M}_\Theta \begin{pmatrix} f \\ g \end{pmatrix} = \begin{pmatrix} zf \\ zg - [zg]\widehat{}(0) \end{pmatrix}, \quad \mathcal{M}_\Theta^* \begin{pmatrix} f \\ g \end{pmatrix} = \begin{pmatrix} \bar{z}f - \bar{z}\Delta_*[\Delta_* f + \Theta g]\widehat{}(0) \\ \bar{z}g - \bar{z}\Theta^*[\Delta_* f + \Theta g]\widehat{}(0) \end{pmatrix}.$$

In the case of an inner characteristic function Θ (that is, $\Delta = 0$) the first transcription becomes especially simple:

$$\mathcal{K}_\Theta = H^2(E_*) \ominus \Theta H^2(E),$$

and the operator adjoint to the model operator is the restriction on $\mathcal{K}_\Theta$ of the backward shift

$$\mathcal{M}_\Theta^* f = \frac{f - f(0)}{z}.$$

If the characteristic function is $*$-inner (that is, $\Delta_* = 0$), then for the second transcription we have

$$\mathcal{K}_\Theta = H^2_-(E) \ominus \Theta^* H^2_-(E_*), \quad \mathcal{M}_\Theta g = zg - \widehat{[zg]}(0);$$

that is, now the model operator $\mathcal{M}_\Theta$ itself is a restriction of the backward shift. However, if we wish to work in the more usual space consisting of analytic functions rather than of anti-analytic functions, we can apply the transformation J: $(Jh)(z) = \bar{z}h(\bar{z})$, obtaining

$$\mathcal{K}_\Theta = H^2(E) \ominus \widetilde{\Theta} H^2(E_*), \quad \text{where} \quad \widetilde{\Theta} = J\Theta J, \quad \text{that is,} \quad \widetilde{\Theta}(z) = \Theta^*(\bar{z}),$$

and

$$\mathcal{M}_\Theta f = \frac{f - f(0)}{z}.$$

In this representation, the minimal unitary dilation of $\mathcal{M}_\Theta$ is the operator of multiplication by $\bar{z}$ on $\mathcal{H} = L^2(E)$.

3.6. The Pavlov transcription. It seems that the most natural way of choosing our embeddings π and π_* is to decide that both of them are the identity embeddings $\pi : L^2(E) \to L^2(E_* \oplus E)$ and $\pi_* : L^2(E_*) \to L^2(E_* \oplus E)$. However, in this case we cannot avoid some complications related to the weight W. Indeed, if we put $\Pi = \mathrm{id}$, then $\Pi^* = \Pi^+ W = W$, and (3.0) implies that $W = W_\Theta$,

$$\mathcal{H} = L^2\left(E_* \oplus E, \begin{pmatrix} I & \Theta \\ \Theta^* & I \end{pmatrix} \right), \quad G = \begin{pmatrix} 0 \\ H^2(E) \end{pmatrix}, \quad G_* = \begin{pmatrix} H^2_-(E_*) \\ 0 \end{pmatrix},$$

$$\mathcal{M}_\Theta \begin{pmatrix} f \\ g \end{pmatrix} = \begin{pmatrix} zf \\ zg - \widehat{[z(\Theta^* f + g)]}(0) \end{pmatrix}, \quad \mathcal{M}_\Theta^* \begin{pmatrix} f \\ g \end{pmatrix} = \begin{pmatrix} \bar{z}f - \bar{z}\widehat{[f + \Theta g]}(0) \\ \bar{z}g \end{pmatrix}.$$

This version of the model was proposed by B. Pavlov [1975] for the investigation of the problems of scattering theory, where the incoming and outgoing subspaces play an essential role. In this representation, these subspaces have the simplest possible form. However, this must be paid for with the complexity of computations in $\mathcal{K}_\Theta$. The vectors in $\mathcal{H}$ are no longer pairs of L^2-functions (this is due to possible degeneracy of the weight and the necessity of completion); moreover, it may happen that the model space contains no vector representable as a pair of L^2-functions, except, of course, the zero vector.

3.7. The de Branges–Rovnyak transcription. If we prefer the model subspace to consist of analytic functions only, we can choose $\Pi^* = \mathrm{id}$. Then (3.0) implies $\Pi = W_\Theta$, which yields $W = W_\Theta^{[-1]}$ (for a selfadjoint operator A, we denote by $A^{[-1]}$ the operator equal to zero on $\mathrm{Ker}\, A$ and to the left inverse of A on $\mathrm{Range}\, A$). Thus,

$$\mathcal{H} = L^2\left(E_* \oplus E, W_\Theta^{[-1]} \right), \quad G = \begin{pmatrix} \Theta \\ I \end{pmatrix} H^2(E), \quad G_* = \begin{pmatrix} I \\ \Theta^* \end{pmatrix} H^2_-(E_*).$$

Now the model space consists of pairs of analytic and anti-analytic functions:

$$\mathcal{K}_\Theta = \left\{ \begin{pmatrix} f \\ g \end{pmatrix} : f \in H^2(E),\ g \in H^2_-(E_*),\ g - \Theta^* f \in \Delta L^2(E) \right\}.$$

The action of the model operator is not more involved than in other transcriptions; we have

$$\mathcal{M}_\Theta \begin{pmatrix} f \\ g \end{pmatrix} = \begin{pmatrix} zf - \Theta[zg]\widehat{}\,(0) \\ zg - [zg]\widehat{}\,(0) \end{pmatrix}, \qquad \mathcal{M}_\Theta^* \begin{pmatrix} f \\ g \end{pmatrix} = \begin{pmatrix} \frac{f - f(0)}{z} \\ \bar{z}g - \Theta^* \bar{z} f(0) \end{pmatrix}.$$

However, the verification that a given pair of functions belongs to $\mathcal{K}_\Theta$ and the computation of the norm become rather difficult in this representation.

To identify this transcription with the original de Branges–Rovnyak form of the model we need the following description of the model space given in [de Branges and Rovnyak 1966]:

$$\mathcal{H}(\Theta) \overset{\text{def}}{=} (I - \Theta P_+ \Theta^*)^{1/2} H^2(E_*)$$

(this space endowed with the *range norm*),

$$\mathcal{D}(\Theta) \overset{\text{def}}{=} \left\{ \begin{pmatrix} f \\ g \end{pmatrix} : f \in \mathcal{H}(\Theta),\ g \in H^2(E),\ \text{and}\ z^n f - \Theta P_+ z^n Jg \in \mathcal{H}(\Theta)\ \text{for}\ n \geq 0 \right\},$$

where $(Jh)(\bar{z}) \overset{\text{def}}{=} \bar{z}h(\bar{z})$, as in Section 3.5. The norm on $\mathcal{D}(\Theta)$ is defined by

$$\left\| \begin{pmatrix} f \\ g \end{pmatrix} \right\|^2 \overset{\text{def}}{=} \lim \left(\| z^n f - \Theta P_+ z^n Jg \|^2_{\mathcal{H}(\Theta)} + \| P_+ z^n Jg \|^2_{H^2(E)} \right).$$

The original de Branges–Rovnyak model operator is

$$\mathrm{BR} \begin{pmatrix} f \\ g \end{pmatrix} \overset{\text{def}}{=} \begin{pmatrix} \frac{f - f(0)}{z} \\ zg - \Theta(\bar{z})^* f(0) \end{pmatrix}, \qquad \text{with}\ \begin{pmatrix} f \\ g \end{pmatrix} \in \mathcal{D}(\Theta).$$

3.8. Proposition.

$$\mathcal{K}_\Theta = \mathcal{J}\mathcal{D}(\Theta) \quad \textit{and} \quad \mathcal{J}\mathcal{M}_\Theta^* \mathcal{J} = \mathrm{BR},$$

where

$$\mathcal{J} = \begin{pmatrix} I & 0 \\ 0 & J \end{pmatrix} : \mathcal{H}(\Theta) \oplus L^2(E) \to \mathcal{H}(\Theta) \oplus L^2(E).$$

The proof can be found in [Nikolski and Vasyunin 1989].

Chapter 4. The Commutant Lifting Theorem and Calculi

Our first goal in this chapter is to give a simple proof of the Sz.-Nagy–Foiaş commutant lifting theorem (CLT) and to describe the parametrizations of the lifting operators in terms of the coordinate-free functional model. The second theme is the classical H^∞-function calculus for a completely nonunitary contraction, along with refinements pertaining to locally defined versions of the calculus.

We derive the CLT from the following theorem of T. Ando.

4.1. Theorem. *Any two commuting contractions have commuting unitary dilations.*

The idea for proving the commutant lifting theorem via the Ando theorem goes back to S. Parrott [1970]. Here we present it with all details, including a new simple proof of the Ando theorem. This simplicity makes the approach quite attractive.

We start by introducing the necessary terminology and proving an "abstract" version of the lifting theorem.

4.2. Definition. Let T be a contraction on H; suppose that $X : H \to H$ belongs to the commutant of T:

$$X \in \{T\}' \overset{\text{def}}{=} \{A : AT = TA\}.$$

Next, let $\mathcal{H} = G_* \oplus H \oplus G$ be the space of the minimal unitary dilation $\mathcal{U}$ of T. An operator Y acting on $\mathcal{H}$ is called a *lifting* of X if Y commutes with $\mathcal{U}$, $YG \subset G$, $Y^*G_* \subset G_*$, and $X = P_H Y|H$.

In other words, that Y is a lifting of X means that $Y \in \{\mathcal{U}\}'$ and Y is a dilation of X, that is, that the operator Y has the following matrix representation with respect to the decomposition $\mathcal{H} = G_* \oplus H \oplus G$:

$$Y = \begin{pmatrix} * & 0 & 0 \\ * & X & 0 \\ * & * & * \end{pmatrix} \begin{pmatrix} G_* \\ H \\ G \end{pmatrix} \longrightarrow \begin{pmatrix} G_* \\ H \\ G \end{pmatrix}. \tag{4.2.1}$$

4.3. Commutant Lifting Theorem. *Let T be a contraction on H. Then an operator X on H is a contraction commuting with T if and only if there exists a contractive lifting Y of X.*

4.4. Parametrization Theorem. *Let T be a contraction on a Hilbert space H. Then Y is a lifting of an operator commuting with T if and only if there exist two bounded analytic functions $A \in H^\infty(E \to E)$, $A_* \in H^\infty(E_* \to E_*)$ that are intertwined by the characteristic function of T,*

$$\Theta A = A_* \Theta, \tag{4.4.1}$$

and a bounded function $B \in L^\infty(\Delta_* E_* \to \Delta E)$ such that

$$Y = \pi_* A_* \pi_*^* + \tau \Delta A \pi^* + \tau B \tau_*^*. \tag{4.4.2}$$

Here π, π_* and τ, τ_* are the functional embeddings defined in 1.6 and 1.19, respectively, and $\Delta = (I - \Theta^*\Theta)^{1/2}$. Furthermore, if $X = P_H Y | H$, then

$$\{Y + \pi\Gamma\pi_*^* : \Gamma \in H^\infty(E_* \to E)\}$$

is the set of all liftings of X, and

$$\|X\|_H = \inf\{\|Y + \pi\Gamma\pi_*^*\|_{\mathcal{H}} : \Gamma \in H^\infty(E_* \to E)\}$$
$$= \mathrm{dist}_{\mathcal{H}}\big(Y, \pi H^\infty(E_* \to E)\pi_*^*\big). \tag{4.4.3}$$

The infimum is attained at an H^∞-function Γ.

This parametrization differs from the original one given by B. Sz.-Nagy and C. Foiaş [1973] in one point: here we have one free parameter B instead of two matrix entries subject to a certain relation in the Sz.-Nagy–Foiaş parametrization.

We would like to underline here that (4.4.2) describes *liftings* of operators commuting with T, rather than all operators commuting with $\mathcal{U}$ whose compressions to H commute with T. It is very essential that liftings are not arbitrary operators commuting with $\mathcal{U}$, but those respecting the triangular structure of $\mathcal{U}$ given by (1.2.1), that is, the operators described in (4.4.2) leave invariant the subspaces G and $H \oplus G$.

Now we begin to realize the program outlined above by proving the Ando theorem. As already mentioned, the theorem on the existence of a unitary dilation can be proved in two steps: first, we construct a co-isometric extension of a given contraction, and second, we apply the same step to the adjoint of this extension (we recall that an operator $A : \mathcal{H} \to \mathcal{H}$ is called an *extension* of $B : H \to H$ if $H \subset \mathcal{H}$ is invariant subspace of A and $B = A|H$).

We use this approach to prove the Ando theorem (Section 4.6) and the commutant lifting theorem (Section 4.8). The first step is as follows.

4.5. Lemma. *Any two commuting contractions have commuting co-isometric extensions.*

PROOF. Let T_1, T_2 be two commuting contractions on a Hilbert space H. It is always possible to find co-isometric extensions V_1, V_2 of T_1, T_2, respectively, acting on one and the same space $\mathcal{H}$. Indeed, if V_1 acts on a space $H \oplus H_1$ and V_2 on $H \oplus H_2$, we can take $\mathcal{H} = H \oplus H_1 \oplus H_2$, defining $V_1|H_2$ and $V_2|H_1$ as the identity operators.

Furthermore, we may assume that the operators $V_1 V_2$ and $V_2 V_1$ are unitarily equivalent. Indeed, $V_1 V_2$ and $V_2 V_1$ are two co-isometric extensions of the operator $T_1 T_2 = T_2 T_1$. Let V_0 be a minimal co-isometric extension of $T_1 T_2$ (minimality means that the space where V_0 acts is the smallest subspace reducing V_0 and containing H). Then the operators $V_1 V_2$ and $V_2 V_1$ are unitarily equivalent to certain orthogonal sums:

$$V_1 V_2 \simeq V_0 \oplus V_{12}, \quad V_2 V_1 \simeq V_0 \oplus V_{21},$$

where V_{12}, V_{21} are co-isometries acting on the corresponding spaces H_{12} and H_{21}. Now, we extend the operators V_1 and V_2 from $\mathcal{H}$ to $\mathcal{H} \oplus \sum_1^\infty \oplus (H_{12} \oplus H_{21})$, one of them by the identity operator on $\sum_1^\infty \oplus (H_{12} \oplus H_{21})$, and another by the infinite orthogonal sum of the operators $V_{12} \oplus V_{21}$. The operators obtained are also co-isometric extensions of T_i, but their products are unitarily equivalent, namely, they are equivalent to V_0 plus the infinite orthogonal sum of the operators $V_{12} \oplus V_{21}$.

Thus, we assume that $V_1 V_2$ and $V_2 V_1$ are unitarily equivalent, that is, there exists a unitary operator U such that

$$V_1 V_2 U = U V_2 V_1.$$

Put $W_1 = V_1 U^*$, $W_2 = U V_2$. Then W_1 and W_2 commute. Observing that U intertwines the minimal parts of $V_1 V_2$ and $V_2 V_1$ and that a unitary operator intertwining any two minimal co-isometric extensions of a contraction on H can be chosen as the identity on H, we conclude that the restriction $U|H$ is the identity operator, whence $W_i|H = V_i|H = T_i$, that is, the W_i are the required co-isometric extensions of T_i, for $i = 1, 2$. $\square$

Now the Ando theorem follows easily.

4.6. Proof of Theorem 4.1. If the operators T_i of Lemma 4.5 are isometries, then the extensions V_i can be taken unitary. (It is easily seen that the minimal co-isometric extension of an isometry is unitary.) Then the final commuting isometries W_i will also be unitary. So the theorem is proved for co-isometries.

As for arbitrary contractions, we can apply the result obtained above to the commuting isometries W_1^*, W_2^*. Their commuting unitary extensions are the desired dilations of T_1^*, T_2^*. $\square$

We start proving Theorem 4.3 with the following lemma.

4.7. Lemma. *Let T and X be commuting contractions on H, and let V be the minimal co-isometric extension of T acting on $\mathcal{H}_+$. Then there exists a contraction Y_+ on $\mathcal{H}_+$ commuting with V and extending the operator X.*

PROOF. Let V_T and V_X be arbitrary commuting co-isometric extensions of T and X, respectively. (Such extensions exist by Lemma 4.5). Putting $\mathcal{H}_+ = \text{span}\{V_T^{*n} H : n \geq 0\}$, we see that $V_T \mathcal{H}_+ \subset \mathcal{H}_+$ and $V = V_T|\mathcal{H}_+$ is a minimal co-isometric extension of T. Let $Y_+ = P_{\mathcal{H}_+} V_X|\mathcal{H}_+$. Since $\mathcal{H}_+$ is a reducing subspace for V_T, we have

$$V Y_+ = V_T P_{\mathcal{H}_+} V_X|\mathcal{H}_+ = P_{\mathcal{H}_+} V_T V_X|\mathcal{H}_+ = P_{\mathcal{H}_+} V_X V_T|\mathcal{H}_+ = Y_+ V$$

and, moreover,

$$Y_+|H = P_{\mathcal{H}_+} V_X|H = P_{\mathcal{H}_+} X = X. \qquad \square$$

4.8. Proof of Theorem 4.3. By Lemma 4.7, given $X \in \{T\}'$, we can find a contractive extension Y_+ of X^*, commuting with the minimal co-isometric extension V of T^*. We recall that $V = P_{H \oplus G} \mathcal{U}^* | H \oplus G$ (see Remark 1.5). Applying Lemma 4.7 once again, now to Y_+^* and V^*, we get the required dilation Y. Indeed, since the minimal unitary dilation $\mathcal{U}$ of T is the minimal co-isometric extension of V^*, we have $Y\mathcal{U} = \mathcal{U}Y$. Furthermore, being an extension of X^*, the operator Y_+ has the matrix structure

$$Y_+ = \begin{pmatrix} * & 0 \\ * & X^* \end{pmatrix} : \begin{pmatrix} G \\ H \end{pmatrix} \longrightarrow \begin{pmatrix} G \\ H \end{pmatrix},$$

or

$$Y_+^* = \begin{pmatrix} X & 0 \\ * & * \end{pmatrix} : \begin{pmatrix} H \\ G \end{pmatrix} \longrightarrow \begin{pmatrix} H \\ G \end{pmatrix}.$$

The operator Y, being an extension of Y_+^*, is of the form

$$Y = \begin{pmatrix} * & 0 & 0 \\ * & X & 0 \\ * & * & * \end{pmatrix} : \begin{pmatrix} G_* \\ H \\ G \end{pmatrix} \longrightarrow \begin{pmatrix} G_* \\ H \\ G \end{pmatrix},$$

which means, in accordance with Definition 4.2, that Y is a lifting of X.

The converse is obvious: if Y is a contractive lifting of $X = P_H Y | H$, then X is a contraction and

$$
\begin{aligned}
XT = P_H Y T &= P_H Y P_H \mathcal{U} | H \\
&= P_H Y (I - P_G - P_{G_*}) \mathcal{U} | H \quad \text{(since } \mathcal{U}H \subset H \oplus G \text{ and } YG \subset G) \\
&= P_H Y \mathcal{U} | H = P_H \mathcal{U} Y | H \\
&= P_H \mathcal{U} (P_H + P_G + P_{G_*}) Y | H \quad \text{(since } YH \subset H \oplus G \text{ and } \mathcal{U}G \subset G) \\
&= P_H \mathcal{U} P_H Y | H = TX. \qquad \qquad \square
\end{aligned}
$$

Now we employ the function model to give a functional parametrization of the commutant.

4.9.   Proof of Theorem 4.4. Since the functional embeddings π, π_*, τ, and τ_* all intertwine the dilation $\mathcal{U}$ and the operator of multiplication by z, expression (4.4.2) provides an operator commuting with $\mathcal{U}$ for every triple of operator-valued functions (A, A_*, B). Furthermore, (4.4.2) implies that $\pi_*^* Y = A_* \pi_*^*$; that is, $Y^* G_* = Y^* \pi_* H_-^2(E_*) = \pi_* A_*^* H_-^2(E_*) \subset \pi_* H_-^2(E_*) = G_*$, and $Y\pi = \pi_* A_* \Theta + \tau \Delta A$. By (4.4.1), the latter relation can be rewritten as $Y\pi = (\pi_* \Theta + \tau \Delta) A = \pi A$. Then, obviously, $YG = Y\pi H^2(E) = \pi A H^2(E) \subset \pi H^2(E) = G$.

So, we have proved that any operator Y of the form (4.4.2) is a lifting of an operator belonging to the commutant of T. Now we prove the converse.

Let Y be a lifting of an operator belonging to the commutant of T; this means that $Y\mathcal{U} = \mathcal{U}Y$, $YG \subset G$, and $Y^* G_* \subset G_*$. First, we use the inclusion $YG \subset G$, which we rewrite in the form $Y\pi H^2(E) \subset \pi H^2(E)$. The relations $\pi^*\pi = I$, $\pi\pi_*^* + \tau_*\tau_*^* = I$, and $\tau_*^*\pi = 0$ show that the latter inclusion is equivalent to the inclusion $\pi^*Y\pi H^2(E) \subset H^2(E)$ together with the identity $\tau_*^*Y\pi|H^2(E) = 0$. Since Y commutes with $\mathcal{U}$, the operators $\pi^*Y\pi$ and $\tau_*^*Y\pi$ commute with multiplication by z; that is, they are operators of multiplication by certain operator-valued functions, which we denote by the same symbols. Moreover, the inclusion $\pi^*Y\pi H^2(E) \subset H^2(E)$ means that $A \overset{\text{def}}{=} \pi^*Y\pi \in H^\infty(E \to E)$, and the identity $\tau_*^*Y\pi|H^2(E) = 0$ shows that $\tau_*^*Y\pi = 0$ everywhere. This yields $Y\pi = (\pi\pi^* + \tau_*\tau_*^*)Y\pi = \pi\pi^*Y\pi = \pi A$; that is, we have the intertwining relation

$$\pi A = Y\pi. \tag{4.9.1}$$

Similarly, introducing $A_* \overset{\text{def}}{=} \pi_*^*Y\pi_*$, we deduce from the inclusion $Y^*G_* \subset G_*$ that $A_* \in H^\infty(E_* \to E_*)$ and that

$$\pi_*^*Y = A_*\pi_*^*. \tag{4.9.2}$$

Multiplying (4.9.1) by π_*^* from the left and (4.9.2) by π from the right, we arrive at (4.4.1). Furthermore,

$$\begin{aligned}
Y &= (\pi_*\pi_*^* + \tau\tau^*)Y = \pi_*A_*\pi_*^* + \tau\tau^*Y(\pi\pi^* + \tau_*\tau_*^*) \\
&= \pi_*A_*\pi_*^* + \tau\tau^*\pi A\pi^* + \tau(\tau^*Y\tau_*)\tau_*^* = \pi_*A_*\pi_*^* + \tau\Delta A\pi^* + \tau B\tau_*^*, \quad (4.9.3)
\end{aligned}$$

where we have put $B \overset{\text{def}}{=} \tau^*Y\tau_*$. Since B intertwines the operators of multiplication by z on the spaces $L^2(\Delta_*E_*)$ and $L^2(\Delta E)$, B is the operator of multiplication by a function belonging to $L^\infty(\Delta_*E_* \to \Delta E)$. For convenience, B may be regarded as a function in $L^\infty(E_* \to E)$ equal to zero on $\text{Ker}\,\Delta_*$.

To complete the proof we must describe all liftings of a given operator and check the formula for the norm. First, we note that if Y is a lifting of X, then, clearly, $\|X\| \leq \|Y\|$. By Theorem 4.3, there exists a contractive lifting of the operator $X\|X\|^{-1}$. Multiplying it by $\|X\|$, we get a lifting Y of X with norm at most $\|X\|$. Therefore,

$$\|X\| = \inf\{\|Y\| : Y \text{ is a lifting of } X\}$$

and to prove formula (4.4.3) it suffices to check that the set of all liftings of X is of the form

$$\{Y + \pi\Gamma\pi_*^* : \Gamma \in H^\infty(E_* \to E)\},$$

where Y is an arbitrary lifting. In other words, we need to check that the set

$$\{\pi\Gamma\pi_*^* : \Gamma \in H^\infty(E_* \to E)\}$$

is merely the set of liftings of the zero operator. The latter assertion is proved in Lemma 4.10 below.

4.10. Lemma. *The following assertions are equivalent.*

(1) *Y is a lifting of the zero operator.*
(2) *$Y = \pi\Gamma\pi_*^*$ for some Γ, $\Gamma \in H^\infty(E_* \to E)$.*

(3) *For some function $\Gamma \in H^\infty(E_* \to E)$, the operator Y is representable as in (4.4.2), with*

$$A = \Gamma\Theta, \quad A_* = \Theta\Gamma, \quad B = \Delta\Gamma\Delta_*. \tag{4.10.1}$$

Here π and π_* are the functional embeddings defined in Section 1.6, $\Delta = (I - \Theta^*\Theta)^{1/2}$, and $\Delta_* = (I - \Theta\Theta^*)^{1/2}$.

PROOF. (1) $\Longrightarrow$ (2). If $P_H Y | H = 0$, then $Y(G \oplus H) \subset G$. Since $G = \pi H^2(E)$ and $G \oplus H = \mathcal{H} \ominus G_* = \pi_* H^2(E_*) \oplus \tau L^2(\Delta E)$, we can rewrite the latter inclusion in the form

$$Y(\pi_* H^2(E_*) \oplus \tau L^2(\Delta E)) \subset \pi H^2(E). \tag{4.10.2}$$

In particular,

$$Y\pi_* H^2(E_*) \subset \pi H^2(E),$$

which implies that $\Gamma \stackrel{\text{def}}{=} \pi^* Y \pi_* \in H^\infty(E_* \to E)$. Furthermore, applying the operators $\mathcal{U}^{*n}$ to the both sides of (4.10.2) and then letting n tend to infinity, we get

$$Y\mathcal{H} \subset \pi L^2(E);$$

that is, $\tau_*^* Y = 0$. Similarly, applying the operators $\mathcal{U}^n$ to the same inclusion and then letting n tend to infinity, we get

$$Y\tau L^2(\Delta E) = \{0\};$$

that is, $Y\tau = 0$. Therefore,

$$Y = (\pi\pi^* + \tau_*\tau_*^*)Y(\pi_*\pi_*^* + \tau\tau^*) = \pi(\pi^* Y \pi_*)\pi_*^* = \pi\Gamma\pi_*^*.$$

(2) $\Longleftrightarrow$ (3). Since $\pi = \pi_*\Theta + \tau\Delta$ and $\pi_* = \pi\Theta^* + \tau_*\Delta_*$, we have

$$\pi\Gamma\pi_*^* = (\pi_*\Theta + \tau\Delta)\Gamma\pi_*^* = \pi_*\Theta\Gamma\pi_* + \tau\Delta\Gamma(\Theta\pi^* + \Delta_*\tau_*^*)$$
$$= \pi_*(\Theta\Gamma)\pi_*^* + \tau\Delta(\Gamma\Theta)\pi^* + \tau(\Delta\Gamma\Delta_*)\tau_*^*;$$

that is, the identity $Y = \pi\Gamma\pi_*^*$ is equivalent to (4.4.2) with $A = \Gamma\Theta$, $A_* = \Theta\Gamma$, $B = \Delta\Gamma\Delta_*$.

(2) + (3) $\Longrightarrow$ (1). By the already proved part of Theorem 4.4 the operator Y defined by formula (4.4.2) is a lifting of the operator $X = P_H Y | H$. To compute X, take a vector $h \in H$. Then $\pi_*^* h \in H^2(E_*)$, whence

$$Xh = P_H Yh = P_H \pi\Gamma\pi_*^* h \in P_H \pi H^2(E) = \{0\},$$

so that $X = 0$. $\qquad\qquad\qquad\qquad\qquad\qquad\qquad\qquad\qquad\qquad\qquad\qquad\square$

4.11. Another expression for a lifting. The representation (4.4.2) can be rewritten in the form

$$Y = \pi A \pi^* + \pi_* A_* \Delta_* \tau_*^* + \tau B \tau_*^*. \tag{4.11.1}$$

The existence of the two representations (4.4.2) and (4.11.1) of the same operator is founded on the duality between the operators and their adjoints. We can prove this formula in the same way as (4.9.3):

$$Y = Y(\pi\pi^* + \tau_*\tau_*^*) = \pi A\pi^* + (\pi_*\pi_*^* + \tau\tau^*)Y\tau_*\tau_*^*$$
$$= \pi A\pi^* + \pi_* A_* \pi_*^* \tau_* \tau_*^* + \tau(\tau^* Y \tau_*)\tau_*^* = \pi A\pi^* + \pi_* A_* \Delta_* \tau_*^* + \tau B\tau_*^*.$$

Moreover, we note that formulas (4.4.2) and (4.11.1) represent one and the same operator if and only if relation (4.4.1) is fulfilled. Indeed, the difference of (4.4.2) and (4.11.1) is equal to

$$(\pi_* A_* \pi_*^* + \tau\Delta A\pi^*) - (\pi A\pi^* + \pi_* A_* \Delta_* \tau_*^*) = \pi_* A_* (\pi_*^* - \Delta_* \tau_*^*) - (\pi - \tau\Delta)A\pi^*$$
$$= \pi_*(A_*\Theta - \Theta A)\pi^*.$$

4.12. More function parameters. We introduce two more functions related to a lifting operator:

$$C \overset{\text{def}}{=} \tau^* Y \tau \in L^\infty(\Delta E \to \Delta E)$$
$$C_* \overset{\text{def}}{=} \tau_*^* Y \tau_* \in L^\infty(\Delta_* E \to \Delta_* E).$$

These functions satisfy the relations

$$C = \Delta A \Delta - B\Theta, \tag{4.12.1}$$
$$C_* = \Delta_* A_* \Delta_* - \Theta B, \tag{4.12.2}$$
$$Y\tau = \tau C, \tag{4.12.3}$$
$$\tau_*^* Y = C_* \tau_*^*. \tag{4.12.4}$$

Indeed, multiplying (4.4.2) by τ from the right, we obtain

$$Y\tau = (\pi_* A_* \pi_*^* + \tau\Delta A\pi^* + \tau B\tau_*^*)\tau = \tau(\Delta A\Delta - B\Theta),$$

which yields (4.12.1) and (4.12.3). Similarly, multiplying (4.11.1) by τ_*^* from the left, we get (4.12.2) and (4.12.4).   □

It should be noted that the parameters C and C_* of a lifting Y of an operator X are *uniquely determined by* X. Indeed, if $X = 0$, then $Y = \pi\Gamma\pi_*^*$, whence

$$C = \tau^* Y \tau = 0, \quad C_* = \tau_*^* Y \tau_* = 0.$$

4.13. Multiplication Theorem. *Let X_1 and X_2 be operators commuting with T. For $i = 1, 2$, let Y_i be the lifting of X_i with parameters $A_i, A_{*i}, B_i, C_i, C_{*i}$.*

Then the operator $Y \overset{\text{def}}{=} Y_2 Y_1$ *is a lifting of the product* $X_2 X_1$, *and the parameters of* Y *are*

$$A = A_2 A_1, \qquad A_* = A_{*2} A_{*1}, \tag{4.13.1}$$

$$B = \Delta A_2 \Theta^* A_{*1} \Delta_* + B_2 \Delta_* A_{*1} \Delta_* + \Delta A_2 \Delta B_1 - B_2 \Theta B_1, \tag{4.13.2}$$

$$C = C_2 C_1, \qquad C_* = C_{*2} C_{*1}. \tag{4.13.3}$$

PROOF. Relation (4.9.1) yields

$$Y\pi = Y_2 Y_1 \pi = Y_2 \pi A_1 = \pi A_2 A_1,$$

so that $A = A_2 A_1$. Similarly, $A_* = A_{*2} A_{*1}$ is a consequence of (4.9.2). In the same way, (4.12.3) and (4.12.4) imply (4.13.3). To check (4.13.2) we combine (4.9.1) and (4.9.2) with (4.12.1) and (4.12.3):

$$\begin{aligned}
B &= \tau^* Y \tau_* = \tau^* Y_2 (\pi_* \pi_*^* + \tau \tau^*) Y_1 \tau_* = \tau^* Y_2 (\pi_* A_{*1} \pi_*^* \tau_* + \tau B_1) \\
&= \tau^* Y_2 (\pi \Theta^* + \tau_* \Delta_*) A_{*1} \Delta_* + \tau^* \tau C_2 B_1 \\
&= (\Delta A_2 \Theta^* + B_2 \Delta_*) A_{*1} \Delta_* + (\Delta A_2 \Delta - B_2 \Theta) B_1.
\end{aligned}$$

$\square$

4.14. Intertwining two contractions. From the very beginning we could have considered the liftings of the operators intertwining two arbitrary contractions $T_1 : H_1 \to H_1$ and $T_2 : H_2 \to H_2$ instead of those of the operators intertwining a contraction with itself. More precisely, we mean the operators $X_{21} : H_1 \to H_2$ satisfying

$$T_2 X_{21} = X_{21} T_1. \tag{4.14.1}$$

An operator $Y_{21} : \mathcal{H}_1 \to \mathcal{H}_2$ is said to be a lifting of X_{21} if $X_{21} = P_{H_2} Y_{21} | H_1$, $Y_{21} G_1 \subset G_2$, and $Y_{21}^* G_{*2} \subset G_{*1}$. For such a lifting problem all results would be the same. The only difference is that in this more general situation we have different spaces and different function models from the right and from the left. For example, instead of (4.4.2), for a lifting Y_{21} of the intertwining operator X_{21} we have the following formula

$$Y_{21} = \pi_{2*} A_{21*} \pi_{1*}^* + \tau_2 \Delta_2 A_{21} \pi_1^* + \tau_2 B_{21} \tau_{1*}^* \tag{4.14.2}$$

acting between the corresponding spaces $\mathcal{H}_1$ and $\mathcal{H}_2$ of the minimal unitary dilations of T_1 and T_2.

Moreover, such a generalization is an immediate consequence of the lifting theorem for the commutant. Indeed, having the intertwining relation (4.14.1), we can introduce on the space $H = H_1 \oplus H_2$ the commuting operators

$$T = \begin{pmatrix} T_1 & 0 \\ 0 & T_2 \end{pmatrix} : \begin{pmatrix} H_1 \\ H_2 \end{pmatrix} \to \begin{pmatrix} H_1 \\ H_2 \end{pmatrix},$$

$$X = \begin{pmatrix} 0 & 0 \\ X_{21} & 0 \end{pmatrix} : \begin{pmatrix} H_1 \\ H_2 \end{pmatrix} \to \begin{pmatrix} H_1 \\ H_2 \end{pmatrix}.$$

Let Y be a lifting of X. Then the operator $Y_{21} \stackrel{\text{def}}{=} P_{\mathcal{H}_2} Y | \mathcal{H}_1$ is a lifting of X_{21}. Indeed,

$$Y_{21} G_1 = P_{\mathcal{H}_2} Y G_1 \subset P_{\mathcal{H}_2} Y (G_1 \oplus G_2) \subset P_{\mathcal{H}_2} (G_1 \oplus G_2) = G_2,$$

and, similarly,

$$Y_{21}^* G_{*2} = P_{\mathcal{H}_1} Y^* G_{*2} \subset P_{\mathcal{H}_1} Y^* (G_{*1} \oplus G_{*2}) \subset P_{\mathcal{H}_1} (G_{*1} \oplus G_{*2}) = G_{*1}.$$

The relations $X_{21} = P_{H_2} X | H_1 = P_{H_2} Y | H_1 = P_{H_2} Y_{21} | H_1$ are clear.

4.15. The special case of an inner characteristic function. All formulas become simpler under the assumption that $\Theta = \Theta_T$ is an inner or $*$-inner function. We recall that this is equivalent to saying that the imbedding π_* (or, respectively, π) is onto. Indeed, Θ is inner means $\Delta = 0$, which is equivalent to saying that $\tau = 0$, and the latter is possible (see Section 1.19) if and only if π_* is a co-isometry, that is, unitary. Similarly,

$$\Theta \text{ is } *\text{-inner} \iff \Delta_* = 0 \iff \tau_* = 0 \iff \pi \pi^* = I \iff \pi \text{ is unitary.}$$

In this section we assume that π_* is unitary, that is, that Θ is an inner function. In this case, formula (4.4.2) for a lifting becomes

$$Y = \pi_* A_* \pi_*^*,$$

where the sole free parameter A_* runs over all functions in $H^\infty (E_* \to E_*)$ satisfying $A \stackrel{\text{def}}{=} \Theta^* A_* \Theta \in H^\infty (E \to E)$.

The distance formula (4.4.3) becomes

$$\|X\| = \inf \{ \|Y + \pi \Gamma \pi_*^*\| : \Gamma \in H^\infty (E_* \to E) \}$$
$$= \inf \{ \|A_* + \Theta \Gamma\| : \Gamma \in H^\infty (E_* \to E) \}$$
$$= \text{dist} \big(A_*, \Theta H^\infty (E_* \to E) \big).$$

4.16. Lifting for the Sz.-Nagy–Foiaş model. Let

$$H = \mathcal{K}_\Theta = \begin{pmatrix} H^2(E_*) \\ L^2(\Delta E) \end{pmatrix} \ominus \begin{pmatrix} \Theta \\ \Delta \end{pmatrix} H^2(E)$$

be the Sz.-Nagy–Foiaş transcription of the model, with

$$T = \mathcal{M}_\Theta = P_\Theta z | \mathcal{K}_\Theta.$$

In this model, the liftings take the form of the operator of multiplication by the matrix function

$$Y = \begin{pmatrix} A_* & 0 \\ \Delta A \Theta^* + B \Delta_* & C \end{pmatrix},$$

where C is defined by (4.12.1). Indeed, in the Sz.-Nagy–Foiaş transcription the functional embeddings are

$$\pi_* = \begin{pmatrix} I \\ 0 \end{pmatrix}, \quad \pi = \begin{pmatrix} \Theta \\ \Delta \end{pmatrix}, \quad \tau = \begin{pmatrix} 0 \\ I \end{pmatrix}, \quad \tau_* = \begin{pmatrix} \Delta_* \\ -\Theta^* \end{pmatrix}.$$

Therefore, (4.4.2) turns into

$$Y = \begin{pmatrix} I \\ 0 \end{pmatrix} A_*(I,0) + \begin{pmatrix} 0 \\ I \end{pmatrix} \Delta A(\Theta^*, \Delta) + \begin{pmatrix} 0 \\ I \end{pmatrix} B(\Delta_*, -\Theta)$$

$$= \begin{pmatrix} A_* & 0 \\ \Delta A \Theta^* + B \Delta_* & \Delta A \Delta - B \Theta \end{pmatrix},$$

as claimed (see (4.12.1)).

In the case of an inner characteristic function we have $\mathcal{H} = L^2(E_*)$, $\pi_* = I$, and

$$H = \mathcal{K}_\Theta = H^2(E_*) \ominus \Theta H^2(E).$$

Then, the lifting Y is simply the operator of multiplication by A_*, and we get

$$X = P_\Theta Y | \mathcal{K}_\Theta = (I - \Theta P_+ \Theta^*) A_* | \mathcal{K}_\Theta.$$

4.17.   The special case of a scalar characteristic function. Now we consider a scalar characteristic function Θ; that is, we assume that $\dim E = \dim E_* = 1$; we identify $E = E_* = \mathbb{C}$. We put $a = A = A_*$. In this case, as a second parameter it is convenient to take the right bottom entry $c = \Delta^2 a - \Theta B$ satisfying the condition $a - c \in \Theta L^\infty(\Delta)$, where under the symbol $L^\infty(\Delta)$ we mean the space of all essentially bounded functions with respect to the measure $\Delta \, dm$. Then

$$Y = \begin{pmatrix} a & 0 \\ \Delta(a-c)\Theta^{-1} & c \end{pmatrix}, \tag{4.17.1}$$

where $a \in H^\infty$ and $c \in a + \Theta L^\infty(\Delta)$. Now the formula for the norm can be rewritten as

$$\|X\| = \|P_\Theta Y | \mathcal{K}_\Theta\| = \inf \left\{ \left\| \begin{pmatrix} a + \Theta \Gamma & 0 \\ \Delta(a-c)\Theta^{-1} + \Delta \Gamma & c \end{pmatrix} \right\|_\infty : \quad \Gamma \in H^\infty \right\}.$$

This expression can be estimated as follows:

$$\max \left\{ \operatorname{dist}\left(\frac{a - \Delta^2 c}{\Theta}, H^\infty \right), \|c\|_{L^\infty(\Delta)} \right\} \le \|X\| \le \operatorname{dist}\left(\frac{a - \Delta^2 c}{\Theta}, H^\infty \right) + 2\|c\|_{L^\infty(\Delta)}.$$

For functions of the model operator, that is, in the case where $c = a$, the estimate above takes the form

$$\max\{\operatorname{dist}(a\bar{\Theta}, H^\infty), \|a\|_{L^\infty(\Delta)}\} \le \|a(\mathcal{M}_\Theta)\| \le \operatorname{dist}(a\bar{\Theta}, H^\infty) + 2\|a\|_{L^\infty(\Delta)}.$$

For details, refer to [Nikolski and Khrushchev 1987].

4.18. The special case of a scalar inner characteristic function. For a scalar inner function Θ_T we have $|\Theta_T| = 1$ and $\Delta = 0$ a.e. on $\mathbb{T}$, and the Sz.-Nagy–Foiaş model reduces to

$$\mathcal{K}_\Theta = H^2 \ominus \Theta H^2,$$

$$\mathcal{M}_\Theta f = P_\Theta z f \quad \text{for } f \in \mathcal{K}_\Theta.$$

The commutant lifting theorem for this case was discovered by D. Sarason [1967] and served as a model for proving the general result [Sz.-Nagy and Foiaş 1968] presented in this chapter. The Sarason theorem says that whatever is $X \in \{\mathcal{M}_\Theta\}'$ there exists a function $Y = a \in H^\infty$ such that

$$X = P_\Theta a | \mathcal{K}_\Theta = a(\mathcal{M}_\Theta), \quad \|X\| = \|a\|_\infty.$$

In general, for $\varphi \in H^\infty$ we have

$$\|\varphi(\mathcal{M}_\Theta)\| = \mathrm{dist}_{L^\infty}(\varphi\bar{\Theta}, H^\infty) \le \|\varphi\|_\infty.$$

One can observe that this scalar version of the CLT says nothing but that the commutant $\{\mathcal{M}_\Theta\}'$ is reduced to functions of the model operator defined by the H^∞-calculus.

Now, we enter in some more details of the H^∞-calculus.

4.19. H^∞-calculus for completely nonunitary contractions. Roughly speaking, a calculus for an operator $T : H \to H$ is an algebra homomorphism extending the standard polynomial calculus $p \mapsto p(T)$. More precisely, let $\mathcal{A}$ be a topological function algebra containing the complex polynomials; an $\mathcal{A}$-*calculus* for a Hilbert space operator $T : H \to H$ is a continuous algebra homomorphism $f \mapsto f(T) \in L(H \to H)$, where $f \in \mathcal{A}$, such that $z^n(T) = T^n$ for $n \ge 0$.

The existence of unitary dilations for completely nonunitary contractions gives an easy possibility to define a rich functional calculus. Indeed, let $\mathcal{U}$ be the minimal unitary dilation of a completely nonunitary contraction $T : H \to H$. For any $f \in L^\infty$, we put

$$f(T) = P_H f(\mathcal{U})|H, \tag{4.19.1}$$

where $f(\mathcal{U})$ is well-defined because of absolute continuity of $\mathcal{U}$. Observe that $z^n(T) = T^n$ for $n \ge 0$, and that

$$\|f(T)\| \le \|f\|_\infty \tag{4.19.2}$$

for every f. We note that the mapping $f \mapsto f(\mathcal{U})$ is a calculus, but $f \mapsto f(T)$ is not. However, the restriction of this mapping to H^∞ is a calculus, called *the H^∞- or Sz.-Nagy–Foiaş calculus* for the completely nonunitary contraction T.

One of the easiest ways to check the multiplicativity in (4.19.1) for H^∞-functions is to observe that, due to the von Neumann spectral theorem for $\mathcal{U}$ (see Section 0.7), we have

$$(f(T)x, y) = (f(\mathcal{U})x, y) = \int_{\mathbb{T}} f(\zeta)\,\big(x(\zeta), y(\zeta)\big)\,dm(\zeta)$$

for all $x, y \in H$ and $f \in L^\infty$, where $(x(\,\cdot\,), y(\,\cdot\,)) \in L^1$. Therefore, the mapping $f \mapsto f(T)$ is continuous with respect to the w^*-topology of L^∞ and the weak operator topology of $L(H \to H)$. Clearly, having $(pq)(T) = p(T)q(T)$ for all complex polynomials p, q, we get first $(pf)(T) = p(T)f(T)$ for $f \in H^\infty$ (for example, using the Fejér sums approximations for f), whence $(gf)(T) = g(T)f(T)$ for all pairs $f, g \in H^\infty$.

It can be proved that for a function f different from the zero function the equality $f(\mathcal{M}_\Theta) = 0$ occurs if and only if Θ is two-sided inner and $f \cdot I \in \Theta H^\infty(E)$.

Further, from the asserted property of w^*-continuity, it is clear that the H^∞ calculus is compatible with any other calculus, continuous in a stronger sense (for instance, with the classical Riesz–Dunford holomorphic calculus). However, for many purposes (e. g., for applications to free interpolation, or for similarity problems; see Part II) we need calculi for functions defined locally, on a kind of neighborhood of the spectrum (ideally, on the spectrum itself, as for normal operators) and not on the entire unit disc $\mathbb{D}$. Clearly, such a local calculus requires a stronger upper estimate for the norms $\|f(T)\|$ than is given by (4.19.2). We give an example of such an estimate in the next section.

4.20. Level curves estimate. In this section, following [Nikolski and Khrushchev 1987], we obtain an estimate of $\|f(T)\|$ for the case of a scalar inner characteristic function $\Theta = \Theta_T$. This estimate depends on the values of f on the level sets $L(\Theta, \varepsilon)$ of Θ. The latter are defined as follows:

$$L(\Theta, \varepsilon) \overset{\text{def}}{=} \{z \in \mathbb{D} : |\Theta(z)| < \varepsilon\} \quad \text{for } 0 < \varepsilon < 1.$$

It is well known (and will be proved in Chapter 5) that the spectrum of the operator under consideration, $T \simeq \mathcal{M}_\Theta$, coincides with "zeros of Θ" in the sense that

$$\sigma(\mathcal{M}_\Theta) = \{\zeta \in \bar{\mathbb{D}} : \lim_{\substack{z \to \zeta \\ z \in \mathbb{D}}} |\Theta(z)| = 0\}.$$

Therefore, it is natural to consider the level sets $L(\Theta, \varepsilon)$ as "fine neighborhoods" of $\sigma(\mathcal{M}_\Theta)$ (it is clear that whatever is a neighborhood V of the spectrum, one has $L(\Theta, \varepsilon) \subset V$ for ε small enough). The following theorem [Nikolski and Khrushchev 1987] gives us an estimate of $\text{dist}_{L^\infty}(f\bar{\Theta}, H^\infty)$ in terms of the smallness of f on the level curves $\partial L(\Theta, \varepsilon)$.

4.21. Theorem. *Let Θ be a scalar inner function and $0 < \varepsilon < 1$. There exists a constant $A = A(\varepsilon)$ such that*

$$\|f(\mathcal{M}_\Theta)\| \leq A \cdot \sup\{|f(z)| : z \in L(\Theta, \varepsilon)\}, \quad f \in H^\infty.$$

PROOF. The proof is based on the existence of so-called Carleson contours [Carleson 1962; Garnett 1981, Chapter 8, section 5]. There exists a constant $p \geq 1$ such that for every $\varepsilon \in (0, 1)$ and for every H^∞-function Θ, one can find a contour γ_ε splitting the disc $\mathbb{D}$ into two parts: one, call it Ω, is contained in $\mathbb{D} \setminus L(\Theta, \varepsilon^p)$, and the other (the complement of clos Ω) is contained in $L(\Theta, \varepsilon)$, and such that the arclength on γ_ε is a Carleson measure with embedding norm C, $H^1|\gamma_\varepsilon \subset L^1(\gamma_\varepsilon, |dz|)$, depending only on ε. (Recall, that a measure μ on $\mathbb{D}$ is called a Carleson measure if the restriction on $\operatorname{supp}\mu$ is a continuous embedding of H^1 in $L^2(\mu)$.)

Let $\Omega_r = \Omega \cap \{|z| < \varepsilon\}$; the boundary $\partial\Omega_r$ consists of a part $\gamma_{\varepsilon,r}$ of γ_ε and a part $\Gamma_r = \Omega_r \cap \mathbb{T}_r$ of the circle $\mathbb{T}_r = r\mathbb{T}$.

Let g be a polynomial. Then

$$\int_{\gamma_{\varepsilon,r}} \frac{fg}{\Theta}\, dz = -\int_{\Gamma_r} \frac{fg}{\Theta}\, dz,$$

and hence

$$\left|\frac{1}{2\pi i}\int_{\Gamma_r} \frac{fg}{\Theta}\, dz\right| \leq \varepsilon^{-p}\sup_{\gamma_{\varepsilon,r}}|f| \cdot \frac{1}{2\pi}\int_{\gamma_{\varepsilon,r}}|g|\,|dz| \leq \varepsilon^{-p}C\|g\|_1 \cdot \sup_{L(\Theta,\varepsilon)}|f|.$$

Let E_r be the radial projection of the set Γ_r on the circle $\mathbb{T}$. Then

$$\frac{1}{2\pi i}\int_{\Gamma_r} \frac{fg}{\Theta}\, dz = \frac{r}{2\pi i}\int_{\mathbb{T}} \frac{f(r\zeta)g(r\zeta)}{\Theta}(r\zeta)\chi_{E_r}(\zeta)\, d\zeta.$$

It is clear that this integral tends to

$$\int_{\mathbb{T}} \frac{f}{\Theta} gz\, dm$$

as $r \to 1$. (Indeed, $|\Theta(r\zeta)| \geq \varepsilon^p$ for $\zeta \in E_r$ and $\chi_{E_r} \to 1$ a.e. on $\mathbb{T}$, because $\lim_{r\to 1}|\Theta(r\zeta)| = 1$ for almost all $\zeta \in \mathbb{T}$.) But, it is clear from duality arguments that

$$\operatorname{dist}_{L^\infty}(f\bar\Theta, H^\infty) = \sup\left\{\left|\int_{\mathbb{T}} f\bar\Theta gz\, dm\right| : \|g\|_1 \leq 1\right\},$$

and we get the required inequality with $A = \varepsilon^{-p}C$. $\qquad\square$

4.22. Local functional calculi. The estimates we have obtained actually allow us to extend considerably the functional calculus (4.19.1) from the algebra H^∞ to the algebra $H^\infty\big(L(\Theta,\varepsilon)\big)$ of bounded holomorphic functions on a level set $L(\Theta,\varepsilon)$, $\varepsilon > 0$. This is done in the following theorem [Nikolski and Khrushchev 1987].

4.23. Theorem. *Let Θ be a scalar inner function and $0 < \varepsilon < 1$. For every $f \in H^\infty\big(L(\Theta,\varepsilon)\big)$ there exists a function $\varphi \in H^\infty$ such that $f - \varphi \in \Theta H^\infty\big(L(\Theta,\varepsilon)\big)$ and $\|\varphi\|_\infty \leq A\|f\|_{H^\infty(L(\Theta,\varepsilon))}$ (with the same constant A as in Theorem 4.21). The mappping*

$$f \mapsto \varphi(\mathcal{M}_\Theta) \stackrel{\text{def}}{=} [f](\mathcal{M}_\Theta) \quad \text{for } f \in H^\infty\big(L(\Theta,\varepsilon)\big),$$

is a well-defined calculus. Its kernel consists (precisely) of the functions of the form Θh, $h \in H^\infty\big(L(\Theta,\varepsilon)\big)$. For the functions in H^∞ this calculus coincides with the Sz.-Nagy–Foiaş one, $[f](\mathcal{M}_\Theta) = f(\mathcal{M}_\Theta)$, and, considered on $\bigcup_{\varepsilon > 0} H^\infty\big(L(\Theta,\varepsilon)\big)$, it contains also the Riesz–Dunford calculus.

For the proof, we need the following remarkable lemma by L. Carleson [1962].

4.24. Lemma. *Let B be a finite Blaschke product with simple zeros $\lambda_1, \ldots, \lambda_n$, and let f be a function from $H^\infty(L(B, \varepsilon))$, where $\varepsilon > 0$. Then there exists a function $\varphi \in H^\infty$ such that $\varphi(\lambda_i) = f(\lambda_i)$ for $1 \le i \le n$ and such that $\|\varphi\|_\infty \le C\varepsilon^{-p}\|f\|_{H^\infty(L(B,\varepsilon))}$, where C is an absolute constant and p is the exponent in the definition of Carleson contours.*

SKETCH OF PROOF. (See [Carleson 1962] for the full version.) The general form of H^∞-functions φ interpolating $f(\lambda_i)$ is given by $\varphi = \varphi_0 + Bh$, with $h \in H^\infty$, so that

$$\inf\{\|\varphi\|_\infty : \varphi(\lambda_i) = f(\lambda_i) \text{ for } 1 \le i \le n\}$$

$$= \sup\left\{\left|\frac{1}{2\pi i}\int_{\mathbb{T}} \bar{B}\varphi_0 g\, dz\right| : g \in H^1 \text{ with } \|g\|_1 \le 1\right\}$$

$$= \sup\left\{\left|\frac{1}{2\pi i}\int_{\gamma_\varepsilon} \frac{\varphi_0 g}{B}\, dz\right| : g \in H^1 \text{ with } \|g\|_1 \le 1\right\}$$

$$= \sup\left\{\left|\frac{1}{2\pi i}\int_{\gamma_\varepsilon} \frac{fg}{B}\, dz\right| : g \in H^1 \text{ with } \|g\|_1 \le 1\right\}$$

$$\le C\varepsilon^{-p}\|f\|_{H^\infty(L(B,\varepsilon))}. \qquad \square$$

4.25. Proof of Theorem 4.23. The existence of the requsted function φ can be derived from Lemma 4.24 in a standard way by using the fact that the constants C and p are independent of the function B. First, Θ is assumed to be a Blaschke product B with simple zeros. We approximate B by its partial products B_n (so that, $|B| \le |B_n|$ in $\mathbb{D}$). Then we pass to the limit in the equalities $f - \varphi_n = B_n h_n$ (which follow from Lemma 4.24) using the compactness principle.

Next, an arbitrary inner function Θ can be uniformly approximated by Blaschke products with simple zeros—for instance, by its "Frostman shifts" $B = (\Theta - \lambda_n)(1 - \bar{\lambda}_n \Theta)^{-1}$, where the λ_n converge to zero and $\{z : \Theta(z) = \lambda_n, \Theta'(z) = 0\} = \varnothing$ for each n; see, for example, [Nikolski 1986, Chapter 2, Section 5]. Repeated application of the compactness principle proves the existence of the requested φ.

Moreover, if $\psi \in H^\infty$ is another function corresponding to the same f, then $(\varphi - \psi)/\Theta$ is analytic and bounded on $L(\Theta, \varepsilon)$ by the definition of φ and ψ, and on $\mathbb{D} \setminus L(\Theta, \varepsilon)$ by the definition of $L(\Theta, \varepsilon)$. Hence, $(\varphi - \psi)/\Theta \in H^\infty$ and $\varphi(\mathcal{M}_\Theta) = \psi(\mathcal{M}_\Theta)$. Thus, the mapping $f \mapsto \varphi(\mathcal{M}_\Theta) = [f](\mathcal{M}_\Theta)$ is well-defined and bounded.

The multiplicativity of $f \mapsto [f](\mathcal{M}_\Theta)$ and other properties are obvious.

4.26. Explicit formula. In fact, one can prove [Nikolski and Khrushchev 1987] the following explicit formula for the local calculus of Theorem 4.23:

$$[f](\mathcal{M}_\Theta)x = \frac{\Theta}{2\pi i}\int_{\gamma_\varepsilon} \frac{f(\zeta)x(\zeta)}{\Theta(\zeta)(\zeta - z)}\, d\zeta \quad \text{for } x \in \mathcal{K}_\Theta,$$

where γ_ε is the same contour as in the proof of Theorem 4.21.

The spectral mapping theorem for the H^∞-calculus and for our local calculi will be proved in Chapter 5.

Chapter 5. Spectrum and Resolvent

Now we use the description of the commutant to obtain a formula for the resolvent of a given contraction in its model representation. The regular points of a contraction T will be characterized by the existence and some analytic continuation properties of the inverse $\Theta(z)^{-1}$ of the characteristic function $\Theta = \Theta_T$.

The notation is the same as in the previous chapters, namely, $T : H \to H$ is a completely nonunitary contraction, $\mathcal{U} : \mathcal{H} \to \mathcal{H}$ its minimal unitary dilation, π and π_* the corresponding functional embeddings, $\Theta = \Theta_T$ the characteristic function of T, etc.

5.1. Theorem. *A point λ with $|\lambda| < 1$ belongs to the spectrum of T if and only if $\Theta(\lambda)$ is not invertible. Moreover, if $\Theta(\lambda)$ is invertible, the resolvent $R_\lambda \stackrel{\text{def}}{=} (T - \lambda I)^{-1}$ of T at the point λ commutes with T and has a lifting Y with the following parameters:*

$$A = \frac{I - \Theta(\lambda)^{-1}\Theta}{z - \lambda}, \qquad (5.1.1)$$

$$A_* = \frac{I - \Theta\Theta(\lambda)^{-1}}{z - \lambda}, \qquad (5.1.2)$$

$$B = -\frac{\Theta^* + \Delta\Theta(\lambda)^{-1}\Delta_*}{z - \lambda}. \qquad (5.1.3)$$

The resolvent can be written in the form

$$R_\lambda = P_H(\mathcal{U} - \lambda I)^{-1}(I - \pi\Theta(\lambda)^{-1}\pi_*^*)|H. \qquad (5.1.4)$$

PROOF. Assume that λ is a regular point of the contraction T, so there exists an operator R_λ such that

$$R_\lambda(T - \lambda I) = (T - \lambda I)R_\lambda = I.$$

Since $R_\lambda \in \{T\}'$, Theorems 4.3 and 4.4 imply that $R_\lambda = P_H Y|H$, where Y is a lifting of R_λ determined by certain parameters A, A_*, B (see formula (4.4.2)). The operator $T - \lambda I$ also belongs to the commutant of T and has a lifting with the parameters $A = (z-\lambda)I_E$, $A_* = (z-\lambda)I_{E_*}$, and $B = -(z-\lambda)\Theta^*$. Therefore, by the multiplication theorem (Theorem 4.13), the formulas

$A_0 = I - A(z - \lambda), \qquad A_{*0} = I - A_*(z - \lambda),$

$B_0 = -\Theta^* - (\Delta A\Theta^*(z-\lambda)\Delta_* + B\Delta_*(z-\lambda)\Delta_* - \Delta A\Delta\Theta^*(z-\lambda) + B\Theta\Theta^*(z-\lambda))$

$\quad = -\Theta^* - B(z - \lambda)$

provide us with the parameters of a lifting of the zero operator $I - R_\lambda(T - \lambda)$. Hence, by Lemma 4.11, there exists a function $\Gamma \in H^\infty(E_* \to E)$ such that

$$I - A(z - \lambda) = \Gamma\Theta, \qquad (5.1.5)$$

$$I - A_*(z - \lambda) = \Theta\Gamma, \qquad (5.1.6)$$

$$-\Theta^* - B(z - \lambda) = \Delta\Gamma\Delta_*. \qquad (5.1.7)$$

Evaluating (5.1.5) and (5.1.6) at the point λ, we conclude that the operator $\Theta(\lambda)$ is invertible and $\Gamma(\lambda) = \Theta(\lambda)^{-1}$.

Conversely, assuming that $\Theta(\lambda)$ is invertible, we prove the existence of the resolvent at the point λ. To this end we need to choose an analytic function Γ such that equations (5.1.5)–(5.1.7) are solvable with respect to A, A_*, B. The simplest choice is the constant function Γ: $\Gamma(z) = \Theta(\lambda)^{-1}$, $|z| < 1$. In fact, clearly, any function $\Gamma \in H^\infty(E_* \to E)$ satisfying $\Gamma(\lambda) = \Theta(\lambda)^{-1}$ gives a desired solution

$$A = \frac{I - \Gamma\Theta}{z - \lambda} \in H^\infty(E \to E),$$

$$A_* = \frac{I - \Theta\Gamma}{z - \lambda} \in H^\infty(E_* \to E_*),$$

$$B = -\frac{\Theta^* + \Delta\Gamma\Delta_*}{z - \lambda} \in L^\infty(\Delta_* E_* \to \Delta E).$$

To obtain formula (5.1.4) we plug the parameters (5.1.1)–(5.1.3) into formula (4.4.2), that is, we consider the following lifting of R_λ:

$$
\begin{aligned}
Y &= \pi_* \frac{I - \Theta\Theta(\lambda)^{-1}}{z - \lambda} \pi_*^* + \tau\Delta \frac{I - \Theta(\lambda)^{-1}\Theta}{z - \lambda} \pi^* - \tau \frac{\Theta^* + \Delta\Theta(\lambda)^{-1}\Delta_*}{z - \lambda} \tau_*^* \\
&= (\mathcal{U} - \lambda I)^{-1}(\pi_*\pi_*^* + \tau(\Delta\pi^* - \Theta^*\tau_*^*) - \pi_*\Theta\Theta(\lambda)^{-1}\pi_*^* - \tau\Delta\Theta(\lambda)^{-1}(\Theta\pi^* + \Delta_*\tau_*^*)) \\
&= (\mathcal{U} - \lambda I)^{-1}(\pi_*\pi_*^* + \tau\tau^* - (\pi_*\Theta + \tau\Delta)\Theta(\lambda)^{-1}\pi_*^*) \\
&= (\mathcal{U} - \lambda I)^{-1}(I - \pi\Theta(\lambda)^{-1}\pi_*^*).
\end{aligned}
$$

This yields

$$R_\lambda = P_H(\mathcal{U} - \lambda I)^{-1}(I - \pi\Theta(\lambda)^{-1}\pi_*^*)|H,$$

as claimed. $\square$

5.2. One-sided spectrum.

Actually, formula (5.1.4) contains more information than is asserted in Theorem 5.1. Indeed, if $\Theta(\lambda)^{-1}$ means only a left or a right inverse to $\Theta(\lambda)$, denoted in what follows by $[\Theta(\lambda)]_l^{-1}$ and $[\Theta(\lambda)]_r^{-1}$, respectively, then (5.1.4) yields a left or a right inverse to $T - \lambda I$, respectively. More precisely:

5.3. Theorem. *For $|\lambda| < 1$, the operator $T - \lambda I$ has a left inverse if and only if $\Theta(\lambda)$ is left invertible, and $T - \lambda I$ has a right inverse if and only if $\Theta(\lambda)$ is right* *invertible. In both cases the corresponding one-sided inverses can be expressed by the same formulas*:

$$[\Theta(\lambda)]_{l,r}^{-1} = \left(-T^* - D_T[T - \lambda I]_{l,r}^{-1}D_{T^*}\right)|\mathcal{D}_{T^*}, \tag{5.3.1}$$

$$[T - \lambda I]_{l,r}^{-1} = P_H(\mathcal{U} - \lambda I)^{-1}(I - \pi[\Theta(\lambda)]_{l,r}^{-1}\pi_*^*)|H. \tag{5.3.2}$$

PROOF. For example, assume that $\Theta(\lambda)$ is left invertible. Then, denoting by $\mathcal{R}_{l,r}$ the right-hand side of (5.3.2), we have

$$\begin{aligned}
\mathcal{R}_l(T - \lambda I) &= P_H(\mathcal{U} - \lambda I)^{-1}(I - \pi[\Theta(\lambda)]_l^{-1}\pi_*^*)P_H(\mathcal{U} - \lambda I)|H \\
&= P_H(I - \pi[\Theta(\lambda)]_l^{-1}\pi_*^*)|H \\
&\quad - P_H(\mathcal{U} - \lambda I)^{-1}(I - \pi[\Theta(\lambda)]_l^{-1}\pi_*^*)P_G(\mathcal{U} - \lambda I)|H \\
&\quad - P_H(\mathcal{U} - \lambda I)^{-1}(I - \pi[\Theta(\lambda)]_l^{-1}\pi_*^*)P_{G_*}(\mathcal{U} - \lambda I)|H.
\end{aligned}$$

We compute the three terms of the above sum separately. The first one is the identity operator. Indeed,

$$\pi[\Theta(\lambda)]_l^{-1}\pi_*^*H \subset \pi[\Theta(\lambda)]_l^{-1}H^2(E_*) \subset \pi H^2(E) = G \perp H. \tag{5.3.3}$$

The two other summands are zero operators, because

$$\begin{aligned}
(\mathcal{U} - \lambda I)^{-1}(I - \pi[\Theta(\lambda)]_l^{-1}\pi_*^*)G &= (\mathcal{U} - \lambda I)^{-1}\pi(I - [\Theta(\lambda)]_l^{-1}\Theta)H^2(E) \\
&= \pi\frac{I - [\Theta(\lambda)]_l^{-1}\Theta}{z - \lambda}H^2(E) \subset \pi H^2(E) = G \perp H
\end{aligned}$$

and

$$(\mathcal{U} - \lambda I)H \subset G \oplus H \perp G_*.$$

So $\mathcal{R}_l(T - \lambda I) = I$, as desired.

Conversely, let $T - \lambda I$ be left invertible, and let $[T - \lambda I]_l^{-1}$ be a left inverse to $T - \lambda I$ (this inverse is not necessarily equal to (5.3.2)). We check that the operator

$$\Lambda = \left(-T^* - D_T[T - \lambda I]_l^{-1}D_{T^*}\right)|\mathcal{D}_{T^*}, \tag{5.3.4}$$

is a left inverse to $\Theta(\lambda)$.

Since the operator (5.3.4) acts from $\mathcal{D}_{T^*}$ to $\mathcal{D}_T$, we can use formula (1.13.1) for the characteristic function Θ_T. This yields

$$\begin{aligned}
\Lambda\Theta_T(\lambda) &= \left(-T^* - D_T[T - \lambda I]_l^{-1}D_{T^*}\right)\left(-T + \lambda D_{T^*}(I - \lambda T^*)^{-1}D_T\right) \\
&= I - D_T^2 - \lambda T^*D_{T^*}(I - \lambda T^*)^{-1}D_T + D_T[T - \lambda I]_l^{-1}D_{T^*}T \\
&\quad - \lambda D_T[T - \lambda I]_l^{-1}D_{T^*}^2(I - \lambda T^*)^{-1}D_T \\
&= I - D_T[T - \lambda I]_l^{-1}\big\{(T - \lambda I)(I - \lambda T^*) + (T - \lambda I)\lambda T^* \\
&\quad - T(I - \lambda T^*) + \lambda(I - T^*T)\big\}(I - \lambda T^*)^{-1}D_T \\
&= I,
\end{aligned}$$

whence Λ is a left inverse to $\Theta_T(\lambda)$. $\qquad\square$

5.4. Resolvent and characteristic function off the unit disc. First, we observe that formula (1.13.1) allows us to define $\Theta(z)$ when $|z| > 1$ and $1/z \notin \sigma(T^*)$, by the same expression:

$$\Theta_T(z) \overset{\text{def}}{=} \left(-T + zD_{T^*}(I - zT^*)^{-1}D_T\right)|\mathcal{D}_T \quad \text{for } |z| > 1. \tag{5.4.1}$$

Also, it is clear that

$$\Theta_T(z)^* = \Theta_{T^*}(\bar{z}), \tag{5.4.2}$$

for $|z| < 1$ as well as for $|z| > 1$. Comparing these formulas with (5.3.1) for $|\lambda| < 1$, $\lambda \notin \sigma(T)$, we get

$$\Theta_T(\lambda)^{-1} = \Theta_T(\lambda^*)^* = \Theta_{T^*}(1/\lambda), \tag{5.4.3}$$

where $\lambda^* = 1/\bar{\lambda}$.

As for the resolvent, if $|\lambda| > 1$, then, clearly,

$$(T - \lambda I)^{-1} = -\frac{1}{\lambda}\sum_{n \geq 0}\left(\frac{T}{\lambda}\right)^n = -\frac{1}{\lambda}P_H\sum_{n \geq 0}\left(\frac{\mathcal{U}}{\lambda}\right)^n|H = P_H(\mathcal{U} - \lambda I)^{-1}|H, \tag{5.4.4}$$

which formally coincides with (5.1.4). Indeed, taking (5.4.3) into account, as in (5.3.3) we have

$$\pi\Theta(\lambda^*)^*\pi_*^*H \subset \pi\Theta(\lambda^*)^*H^2(E_*) \subset \pi H^2(E) = G,$$

and $(\mathcal{U} - \lambda I)^{-1}G \subset G$, whence $P_H(\mathcal{U} - \lambda I)^{-1}\pi\Theta(\lambda^*)^*\pi_*^*)|H = 0$ and

$$P_H(\mathcal{U} - \lambda I)^{-1}(I - \pi\Theta(\lambda^*)^*\pi_*^*)|H = P_H(\mathcal{U} - \lambda I)^{-1}|H.$$

It is worth mentioning that (5.4.1) and (1.13.1), formally coinciding, can give two different analytic functions. For example, let $\Theta = \frac{1}{2}(1 - z)$. According to (5.4.2) this is the characteristic function of a contraction unitarily equivalent to its adjoint; that is, the adjoint operator has the same characteristic function. However, if we substitute this expression in (5.4.3), we obtain

$$\Theta(\lambda)^{-1} = \frac{2}{1 - \lambda} \neq \frac{\lambda - 1}{2\lambda} = \Theta\left(\frac{1}{\lambda}\right).$$

But for $|z| > 1$ formula (5.4.1) gives us another expression, namely, $\Theta_T(z) = 2z/(z - 1)$. Inversion of this expression as the left-hand term in (5.4.3) gives a correct equality.

In what follows we shall see that (1.13.1) and (5.4.1) define the same analytic function if there exists a regular point of T on the unit circle.

5.5. Boundary spectrum. Now we consider the spectrum on the unit circle $\mathbb{T}$. Formulas (5.3.2) and (5.4.4) cannot be directly extended to $|\lambda| = 1$, because the entire unit circle is the spectrum of $\mathcal{U}$. However, rewriting these formulas in an appropriate form, we can describe the boundary spectrum $\sigma(T) \cap \mathbb{T}$ in terms of analytic continuation of Θ_T across the boundary.

A new local form of the resolvent is as follows.

5.6. Lemma. *Let $x \in H$. Then*

$$(T - \lambda I)^{-1}x = \begin{cases} (\mathcal{U} - \lambda I)^{-1}\big(x - \pi\Theta(\lambda)^{-1}(\pi_*^*x)(\lambda)\big) & \text{if } |\lambda| < 1, \ \lambda \notin \sigma(T), \\ (\mathcal{U} - \lambda I)^{-1}\big(x - \pi(\pi^*x)(\lambda)\big) & \text{if } |\lambda| > 1. \end{cases}$$

$$(5.6.1)$$

PROOF. For $|\lambda| < 1$, the inclusion

$$(\mathcal{U} - \lambda I)^{-1}(I - \pi\Theta(\lambda)^{-1}\pi_*^*)H \subset H \oplus G$$

can be verified by straightforward computation; also, it follows from the definition of lifting (Definition 4.2), because the operator on the left in the above inclusion is a lifting of $(T - \lambda I)^{-1}$. Therefore, for $|\lambda| < 1$ we have

$$R_\lambda x = (I - P_G)(\mathcal{U} - \lambda I)^{-1}\big(x - \pi\Theta(\lambda)^{-1}\pi_*^*x\big)$$

$$= (\mathcal{U} - \lambda I)^{-1}\big(x - \pi\Theta(\lambda)^{-1}\pi_*^*x\big) - \pi P_+ \frac{\pi^*x - \Theta(\lambda)^{-1}\pi_*^*x}{z - \lambda}$$

$$= (\mathcal{U} - \lambda I)^{-1}\big(x - \pi\Theta(\lambda)^{-1}\pi_*^*x\big) + \pi\Theta(\lambda)^{-1}\frac{\pi_*^*x - (\pi_*^*x)(\lambda)}{z - \lambda}$$

$$= (\mathcal{U} - \lambda I)^{-1}\big(x - \pi\Theta(\lambda)^{-1}(\pi_*^*x)(\lambda)\big).$$

Similarly, for $|\lambda| > 1$,

$$R_\lambda x = (I - P_G)(\mathcal{U} - \lambda I)^{-1}x = (\mathcal{U} - \lambda I)^{-1}x - \pi P_+ \frac{\pi^*x}{z - \lambda}$$

$$= (\mathcal{U} - \lambda I)^{-1}x + \pi\frac{(\pi^*x)(\lambda)}{z - \lambda} = (\mathcal{U} - \lambda I)^{-1}\big(x - \pi(\pi^*x)(\lambda)\big). \qquad \square$$

To show that formulas (5.6.1) can be extended to regular points of the operator T on the unit circle we need the following property.

5.7. Lemma. *Assume that there exists a neighbourhood $\mathcal{O}_\lambda$ of a point $\lambda \in \Theta$ such that Θ admits analytic continuation to $\mathcal{O}_\lambda$ and that there exists a left (right) inverse $[\Theta]_l^{-1}$ (respectively, $[\Theta]_r^{-1}$), coinciding with Θ^* on $\mathcal{O}_\lambda \cap \mathbb{T}$. Then for every vector $x \in H$ the function π^*x (respectively, π_*^*x) admits analytic continuation to $\mathcal{O}_\lambda$.*

PROOF. The hypotheses of the lemma imply $\Delta(\zeta) = 0$ on $\mathcal{O}_\lambda \cap \mathbb{T}$. Therefore, since $\pi = \pi_*\Theta + \tau\Delta$, on $\mathcal{O}_\lambda$ we have

$$(\pi^*x)(\zeta) = \Theta(\zeta)^*(\pi_*^*x)(\zeta) + \Delta(\zeta)(\tau^*x)(\zeta) = \Theta(\zeta)^*(\pi_*^*x)(\zeta)$$

$$= [\Theta(\zeta)]_l^{-1}(\pi_*^*x)(\zeta),$$

$$(5.7.1)$$

and the right-hand side in (5.7.1) gives an analytic continuation of $\pi^* x$ to $\mathcal{O}_\lambda$. If $[\Theta(\lambda)]_r^{-1}$ is a right inverse to $\Theta(\lambda)$, then $\Delta_*^2(\zeta) = I - \Theta(\zeta)\Theta(\zeta)^* = I - \Theta(\zeta)[\Theta(\zeta)]_r^{-1} = 0$ on $\mathcal{O}_\lambda \cap \mathbb{T}$, that is,

$$(\pi_*^* x)(\zeta) = \Theta(\zeta)(\pi^* x)(\zeta) + \Delta_*(\zeta)(\tau_*^* x)(\zeta) = \Theta(\zeta)(\pi^* x)(\zeta), \qquad (5.7.2)$$

and again the right-hand side gives an analytic continuation of $\pi_*^* x$ to $\mathcal{O}_\lambda$.   □

5.8. Theorem. *A point λ, $|\lambda| = 1$, belongs to the resolvent set of the operator T if and only if the characteristic function $\Theta = \Theta_T$ admits analytic continuation to a neighbourhood $\mathcal{O}_\lambda$ of the point λ, and $\Theta(\zeta)^{-1} = \Theta(\zeta)^*$ for $\zeta \in \mathcal{O}_\lambda \cap \mathbb{T}$.*

PROOF. If λ with $|\lambda| = 1$ is a regular point for T, then $\bar{\lambda}$ is a regular point for T^*, and formula (1.13.1) determines an analytic continuation to a neighbourhood of λ; moreover, identity (5.4.3) turns into $\Theta(\zeta)^{-1} = \Theta(\zeta)^*$ for $\zeta \in \mathcal{O}_\lambda \cap \mathbb{T}$.

Conversely, suppose that Θ is analytic in a neighbourhood of a point λ and that $\Theta(\zeta)^{-1} = \Theta(\zeta)^*$ for $\zeta \in \mathcal{O}_\lambda \cap \mathbb{T}$. Then, as before, each expression on the right in (5.6.1) is the resolvent at the point λ. Indeed, for every $x \in H$ evaluation at the point λ is well defined by Lemma 5.7, and by (5.7.1)–(5.7.2) we have $\Theta(\lambda)(\pi^* x)(\lambda) = (\pi_*^* x)(\lambda)$, that is, the two expressions on the right in (5.6.1) coincide. We shall check that $y \overset{\text{def}}{=} x - \pi(\pi^* x)(\lambda) \in (\mathcal{U} - \lambda I)H$; this will imply that (5.6.1) is well-defined at such a point λ and, after analytic continuation, gives us a formula for the resolvent at that point.

We have

$$\pi^* y = \pi^* x - (\pi^* x)(\lambda) \in (z - \lambda)H_-^2(E).$$

Since $\Delta_*(\zeta) = 0$ for $\zeta \in \mathcal{O}_\lambda \cap \mathbb{T}$, the operator of multiplication by $1/(z - \lambda)$ is bounded on $L^2(\Delta_* E_*)$, whence

$$y = (\mathcal{U} - \lambda I)y', \quad \text{where } y' = \pi \frac{\pi^* x - (\pi^* x)(\lambda)}{z - \lambda} + \tau_* \frac{\tau_*^* y}{z - \lambda},$$

and

$$\pi^* y' = \frac{\pi^* x - (\pi^* x)(\lambda)}{z - \lambda} \in H_-^2(E),$$

$$\pi_*^* y' = \Theta \frac{\pi^* y}{z - \lambda} + \Delta_* \frac{\tau_*^* y}{z - \lambda} = \frac{\pi_*^* y}{z - \lambda} = \frac{\pi_*^* x - \Theta\Theta(\lambda)^{-1}(\pi_*^* x)(\lambda)}{z - \lambda} \in H^2(E_*).$$

Consequently, $y' \in H$ and $y \in (\mathcal{U} - \lambda I)H$.   □

Now we consider the discrete spectrum of T and the corresponding eigen- and root-subspaces.

5.9. Theorem. *A point λ is an eigenvalue of T if and only if $|\lambda| < 1$ and $\operatorname{Ker}\Theta(\lambda) \neq \{0\}$. Moreover,*

$$\operatorname{Ker}(T - \lambda I) = \pi \frac{\operatorname{Ker}\Theta(\lambda)}{z - \lambda}.$$

PROOF. Let $x \in H$. Then $x \in \text{Ker}(T - \lambda I)$ if and only if $(\mathcal{U} - \lambda I)x \in G$, that is, there exists a function $e \in H^2(E)$ such that $(\mathcal{U} - \lambda I)x = \pi e$, whence

$$(z - \lambda)\pi^* x = e,$$
$$(z - \lambda)\pi_*^* x = \Theta e.$$

Since $\pi^* x \in H^2_-(E)$, the function e is a constant vector in E. And since $\pi_*^* x \in H^2(E)$, putting $z = \lambda$ in the second of the above relations, we get $e \in \text{Ker }\Theta(\lambda)$. All arguments can be reversed; that is, if $e \in \text{Ker }\Theta(\lambda)$, then the vector $x = \pi e/(z - \lambda)$ is in H and, moreover, $x \in \text{Ker}(T - \lambda I)$.   □

As for the root subspaces of T, they will be described in the next chapter, after we introduce the notion of regular factorization.

5.10. The spectral mapping theorem. Now, we prove a spectral mapping theorem for the H^∞ functional calculus in the case of a scalar characteristic function. Moreover, we compute the spectrum of an arbitrary operator in the commutant (in terms of the corresponding functional parameters of its lifting, see Section 4.17 above). To this end, given a function $\Theta \in H^\infty$, let us define the Θ-*range of a function* $f \in H^\infty$ by the formula

$$\text{Range}_\Theta f = \{\lambda \in \mathbb{C} : \inf_{z \in \mathbb{D}} (|f(z) - \lambda| + |\Theta(z)|) = 0\}.$$

Also, we recall the notion of the *essential range* of a measurable function g on a measure space:

$$\text{Range}(g) = \{\lambda \in \mathbb{C} : \text{essinf } |g - \lambda| = 0\}.$$

5.11.   Theorem. *Let* $\Theta \not\equiv 0$ *be a scalar characteristic function.   Take* $X \in \{\mathcal{M}_\Theta\}'$ *and let* Y *be the lifting of* X *written in the form* (4.17.1):

$$Y = \begin{pmatrix} a & 0 \\ \Delta(a - c)\Theta^{-1} & c \end{pmatrix}, \tag{5.11.1}$$

where $a \in H^\infty$ *and the entries on the bottom row lie in* $L^\infty(\Delta)$, *with* $\Delta^2 = 1 - |\Theta|^2$ *and* $L^\infty(\Delta)$ *the space of all essentially bounded functions with respect to the measure* $\Delta\,dm$ (m *being Lebesgue measure on* $\mathbb{T}$). *Then*

$$\sigma(X) = (\text{Range}_\Theta a) \cup (\text{Range}(c)),$$

where $\text{Range}(c)$ *is taken with respect to the measure* $\Delta\,dm$.

PROOF. Essentially, we repeat the scheme of the proof of Theorem 5.1. If X is invertible, then $X^{-1} \in \{\mathcal{M}_\Theta\}'$, and, by the commutant lifting theorem, $X^{-1} = P_\Theta Y'$, where $P_\Theta = P_{\mathcal{K}_\Theta}$ and the matrix Y' is of the form (5.11.1) with $a' \in H^\infty$ and $c' \in L^\infty(\Delta)$. By direct computation (or by using the Multiplication Theorem 4.13) we see that the matrix $I - YY'$ is of the form (5.11.1) with the parameters aa' and cc' instead of a and c, respectively. Since $I - YY'$ is a lifting of the zero operator, by Theorem 4.10 we have

$$I - YY' = \begin{pmatrix} \Theta\Gamma & 0 \\ \Delta\Gamma & 0 \end{pmatrix}, \tag{5.11.2}$$

for some $\Gamma \in H^\infty$. Thus, we have $1 - aa' = \Theta\Gamma$ and $cc' = 1$ a.e. with respect to $\Delta \, dm$. Hence, $0 \notin \mathrm{Range}_\Theta \cup \mathrm{Range}(c)$.

Conversely, if the equations $aa' + \Theta\Gamma = 1$ and $cc' = 1$ are solvable with a', $\Gamma \in H^\infty$, and $c' \in L^\infty(\Delta)$, the matrix

$$Y' = \begin{pmatrix} a' & 0 \\ \Delta(a' - c')/\Theta & c' \end{pmatrix} \tag{5.11.3}$$

determines a lifting of the operator inverse to X. First we have to check that Y' is a lifting of an operator in $\{\mathcal{M}_\Theta\}'$. To this end, we need to verify that $a' - c' \in \Theta L^\infty(\Delta)$ (see Section 4.17). Since

$$ac(a' - c') = (1 - \Theta\Gamma)c - a = (c - a) - \Theta\Gamma \in \Theta L^\infty(\Delta),$$

we need to check that the functions a and c are bounded away from zero on a subset of Θ where $\mathbb{T}$ is small, say, $|\Theta(\zeta)| < \frac{1}{2}\|\Gamma\|_\infty$. For such a ζ we have

$$|a| \geq \frac{|aa'|}{\|a'\|_\infty} = \frac{|1 - \Theta\Gamma|}{\|a'\|_\infty} \geq \frac{1 - |\Theta| \cdot \|\Gamma\|}{\|a'\|_\infty} \geq \frac{1}{2\|a'\|_\infty},$$

$$|c| \geq \frac{1}{\|c'\|_{L^\infty(\Delta)}},$$

whence $a' - c' \in \Theta L^\infty(\Delta)$ and (5.11.3) defines a lifting of an operator. To check that this operator is inverse of X, we multiply the matrices (5.11.1) and (5.11.3), obtaining (5.11.2).

To conclude we recall that, by the Carleson corona theorem (see [Garnett 1981] or [Nikolski 1986], for example), the existence of the needed solutions a' and Γ is equivalent to $\inf_{\mathbb{D}}(|a| + |\Theta|) > 0$. □

5.12. The spectrum of $\mathcal{M}_\Theta$. Clearly, Theorem 5.11 contains also the well-known formula for the spectrum of the model operator itself:

$$\sigma(\mathcal{M}_\Theta) = \{\lambda \in \overline{\mathbb{D}} : \lim_{\substack{z \in \mathbb{D} \\ z \to \lambda}} |\Theta(z)| = 0\} \cup \mathrm{supp}(\Delta) = \mathrm{supp}(\mu_\Theta),$$

μ_Θ being the representing measure of Θ in the Nevanlinna–Riesz–Smirnov sense (for the definition see Section 0.7).

This should be compared with the general case described in Theorems 5.1 and 5.8.

5.13. Corollary. *For $X = P_\Theta Y | \mathcal{K}_\Theta \in \{\mathcal{M}_\Theta\}'$, we have*

$$\|X\| \geq \sup\{|\lambda| : \lambda \in \sigma(X)\} = \max\left(\|c\|_{L^\infty(\Delta)}, \max\{|\lambda| : \lambda \in \mathrm{Range}_\Theta \, a\}\right).$$

An upper estimate of the same type as Corollary 5.13 (that is, in terms of the values of the functions a and c on the spectrum $\sigma(\mathcal{M}_\Theta)$) does not seem possible. Recall, that in Section 4.21, in the case $a = c = f$, we already presented an upper estimate depending on the values $a(\zeta)$ in some narrow "neighborhood" $L(\Theta, \varepsilon)$ of the spectrum $\sigma(\mathcal{M}_\Theta)$.

Chapter 6. Invariant Subspaces

In this chapter we describe the lattice of invariant subspaces of a completely nonunitary contraction in terms of certain functional embeddings and the characteristic function. In fact, the description that we are going to discuss is a direct consequence of the classical Wold–Kolmogorov formula, as is the case for the invariant subspaces of unitary operators. We recall the following well-known description of the lattice Lat $\mathcal{U}$ of a unitary operator.

6.1. Lemma. *Let $\mathcal{U}$ be a unitary operator on a Hilbert space $\mathcal{H}$, and let $L \in$ Lat $\mathcal{U}$. Then*
$$L = R \oplus \eta H^2(F),$$
where $F = L \ominus \mathcal{U}L$, $R = \bigcap_{n \geq 0} \mathcal{U}^n L$, and $\eta : L^2(F) \to \mathcal{H}$ is a "functional embedding" intertwining z and $\mathcal{U}$; that is, $\eta z = \mathcal{U}\eta$. Moreover, R is a reducing subspace of $\mathcal{U}$.

PROOF. The Wold–Kolmogorov lemma implies that $L = R \oplus \left(\sum_n \geq 0 \oplus \mathcal{U}^n F \right)$; of course, this is equivalent to the assertion claimed, with the natural definition of η, namely,
$$\eta \left(\sum_{k \in \mathbb{Z}} a_k e^{ikt} \right) = \sum_{k \in \mathbb{Z}} \mathcal{U}^k a_k, \quad \text{with } a_k \in F. \qquad \square$$

Also, it is worth recalling that reducing subspaces are well-behaved with respect to the spectral decomposition of $\mathcal{U}$. In particular, the description of such subspaces becomes explicit when one uses the von Neumann spectral theorem: L is a reducing subspace of a unitary operator $\mathcal{U}$ if and only if $L = \{f \in L^2(\mathcal{H}, \mu_\mathcal{U}) : f(\zeta) \in P_L(\zeta)\mathcal{H} \ \mu_\mathcal{U}\text{-a.e.}\}$, where P_L stands for a projection-valued function on $\mathbb{T}$ subordinate to $E_\mathcal{U}$, that is, $P_L(\zeta) \leq E_\mathcal{U}(\zeta) \ \mu_\mathcal{U}\text{-a.e.}$

Now, turning to the invariant subspaces of completely nonunitary contractions, we can benefit from the coordinate-free function model approach, because using the language of this approach we need only follow precisely the same lines as in Lemma 6.1. In order to make this point even more transparent, we start with an outline of the description of Lat T, paralleling the steps of this description with the proof of the leading partial case, Beurling's theorem [1949].

6.2 The description of the invariant subspaces: the first draft. The steps of the proof, which will be formalized later, will be listed in a table. On the left-hand side, we deal with a completely nonunitary contraction T and its function model $\{H, T, \mathcal{H}, \mathcal{U}, \pi, \pi_*, \ldots\}$, and on the right-hand side we have the standard shift operator S on the (scalar) Hardy space H^2, given by $Sf = zf$ for $f \in H^2$.

1. $T: H \to H$ a completely nonunitary contraction; $\mathcal{U}: \mathcal{H} \to \mathcal{H}$ the minimal unitary dilation; π, π_* the functional embeddings; $\Theta_T = \pi_*^* \pi$	1. $S: H^2 \to H^2$, $Sf = fz$ for $f \in H^2$; $\mathcal{H} = L^2(\mathbb{T})$, $\mathcal{U}f = zf$ for $f \in L^2$; $E = \{0\}$, $\pi = 0$; $E_* = \mathbb{C}$, $G_* = L^2$, $\pi_* = I$; $\Theta_T = 0_{1 \times 0}: \mathbb{C} \to \{0\}$
2. Let $L \in \operatorname{Lat} T$. Then $L \oplus G \in \operatorname{Lat} \mathcal{U}$ (Lemma 6.3). Take F: $\dim F = \dim((L \oplus G) \ominus \mathcal{U}(L \oplus G))$; $\eta: L^2(F) \to \mathcal{H}$, $\eta F = ((L \oplus G) \ominus \mathcal{U}(L \oplus G))$, $\eta z = \mathcal{U} \eta$	2. Let $L \in \operatorname{Lat} S$. Let $F = L \ominus SL$, $\dim F = 1$; $\eta: L^2 \to L^2$ as in Lemma 6.1, $\eta z = \mathcal{U} \eta$
3. Apply the Wold–Kolmogorov lemma: $L \oplus G = \mathcal{R} \oplus \eta H^2(F)$	3. $L = \mathcal{R} \oplus \eta H^2$, $\mathcal{R} \subset H^2 \implies \mathcal{R} = \{0\}$
4. The mapping $\eta^* \pi: L^2(E) \to L^2(F)$ commutes with z and is analytic, because $\eta^* \pi H^2(E) \subset H^2(F)$ (Section 6.5)	4. $\pi = 0$, hence $\eta^* \pi = 0$
5. The same for $\pi_*^* \eta: L^2(F) \to L^2(E_*)$	5. $\pi_*^* \eta = \eta: L^2 \to L^2$
6. Put $\Theta_1 = \eta^* \pi \in H^\infty(E \to F)$, $\Theta_2 = \pi_*^* \eta \in H^\infty(F \to E_*)$, $\Theta_T = \pi_*^* \pi = \pi_*^* (\eta \eta^*) \pi = \Theta_2 \Theta_1$ (cf. (6.5.3)), $L = \operatorname{span}(\pi L^2(E), \eta L^2(F)) \ominus$ $\quad (\pi H^2(E) \oplus \eta H_-^2(F))$ (cf. (6.7.1))	6. $\Theta_1 = 0_{1 \times 0}: \mathbb{C} \to \{0\}$, $\Theta_2 = \eta \in H^\infty$, $\Theta_T = \Theta_2 \Theta_1 = 0_{1 \times 0}: \mathbb{C} \to \{0\}$, $L = \Theta_2 H^2$

Now we check the proposed program step by step and describe the factorizations $\Theta_T = \Theta_2 \Theta_1$ of the characteristic function appearing in this way.

6.3. Lemma. $L \in \operatorname{Lat} T$ *if and only if* $L \oplus G \in \operatorname{Lat} \mathcal{U}$.

PROOF. Let $L \in \operatorname{Lat} T$. Since $G \in \operatorname{Lat} \mathcal{U}$, it suffices to check that $\mathcal{U}L \subset L \oplus G$. The inclusions $\mathcal{U}L \subset \mathcal{U}(H \oplus G) \subset H \oplus G$ imply that

$$\mathcal{U}L \subset P_H \mathcal{U}L \oplus P_G \mathcal{U}L \subset TL \oplus G \subset L \oplus G.$$

The converse is also obvious: $TL = P_H \mathcal{U}L \subset P_H(L \oplus G) = L$. $\qquad \square$

6.4. An additional functional embedding arises if the completely nonunitary (pure) part of the isometry $\mathcal{U}|L \oplus G$ is realized as the multiplication by z in a suitable H^2-space. More precisely, we can take an auxiliary Hilbert space F such that

$$\dim F = \dim\big((L \oplus G) \ominus \mathcal{U}(L \oplus G)\big)$$

and an isometry $\eta_L : L^2(F) \to \mathcal{H}$ with the following properties:

η_L intertwines $\mathcal{U}$ and z, that is, $\eta_L z = \mathcal{U}\eta_L$;

$\eta_L H^2(F)$ is the subspace of $L \oplus G$ where the pure part of $\mathcal{U}|(L \oplus G)$ acts.

Therefore, the reducing part of $\mathcal{U}|(L \oplus G)$ acts on the subspace

$$\mathcal{R}_L \overset{\text{def}}{=} (L \oplus G) \ominus \eta_L H^2(F); \qquad (6.4.1)$$

in other words, the formula

$$L \oplus G = \mathcal{R}_L \oplus \eta_L H^2(F) \qquad (6.4.2)$$

provides the Wold–Kolmogorov decomposition for the isometry $\mathcal{U}|(L\oplus G)$. Since $\mathcal{R}_L$ is reducing, it is orthogonal not only to $\eta_L H^2(F)$, but to the whole of $\eta_L L^2(F)$, that is, $\eta_L^* \mathcal{R}_L = \{0\}$.

In what follows we omit the index L and write simply η if we have no need to emphasize the fact that this embedding is generated by an invariant subspace L.

6.5. Properties of the additional embedding. It is clear that the operator

$$\Theta_1 \overset{\text{def}}{=} \eta^* \pi \qquad (6.5.1)$$

intertwines the multiplications by z in $L^2(E)$ and $L^2(F)$, that is, $\Theta_1 \in L^\infty(E \to F)$ (as usual, we identify an operator of multiplication with the corresponding function). Moreover, the function Θ_1 is analytic, that is, $\Theta_1 \in H^\infty(E \to F)$, because

$$\Theta_1 H^2(E) = \eta^* \pi H^2(E) = \eta^* G \subset \eta^*(\eta H^2(F) \oplus \mathcal{R}_L) = H^2(F).$$

Similarly, the operator

$$\Theta_2 \overset{\text{def}}{=} \pi_*^* \eta \qquad (6.5.2)$$

is also (an operator of multiplication by) a bounded analytic function:

$$\Theta_2^* H_-^2(E_*) = \eta^* \pi_* H_-^2(E^*) = \eta^* G_* \subset H_-^2(F),$$

because $G_* \perp H \oplus G \supset L \oplus G \supset \eta H^2(F)$.

As the third property of η we mention the identity

$$\pi_*^*(I - \eta\eta^*)\pi = 0. \qquad (6.5.3)$$

For the proof, note that $\pi L^2(E) \subset \eta L^2(F) \oplus \mathcal{R}_L$, that is, $(I - \eta\eta^*)\pi L^2(E) \subset \mathcal{R}_L$. It suffices to show that $\pi_*^* \mathcal{R}_L = \{0\}$. Since $\mathcal{R}_L$ reduces $\mathcal{U}$, the relation $\mathcal{R}_L \perp G_*$ implies that $\mathcal{R}_L \perp \pi_* L^2(E_*)$, that is, $\pi_*^* \mathcal{R}_L = \{0\}$.

6.6. Definition. An isometry η from $L^2(F)$ to $\mathcal{H} = \mathrm{span}\{\pi L^2(E), \pi_* L^2(E_*)\}$ is said to be *compatible with π and π_** if the following four conditions are fulfilled:

(a) $\eta z = \mathcal{U}\eta$ (that is, η is a $\mathcal{U}$-functional embedding);
(b) $\pi H^2(E) \perp \eta H_-^2(F)$;
(c) $\eta H^2(F) \perp \pi_* H_-^2(E_*)$;
(d) $\pi_*^*(I - \eta\eta^*)\pi = 0$.

In these terms, in Section 6.5 we proved that any T-invariant subspace L generates an embedding $\eta = \eta_L$ compatible with π and π_*.

6.7. Theorem. *Let T be a completely nonunitary contraction, π and π_* its functional embeddings. The mapping $L \mapsto \eta_L$ is a bijection of* $\mathrm{Lat}\, T$ *onto the set of isometries compatible with π and π_*. Moreover, the inverse mapping $\eta \mapsto L_\eta$ can be defined by the formula*

$$L_\eta = \mathrm{span}\{\pi L^2(E), \eta L^2(F)\} \ominus \left(\pi H^2(E) \oplus \eta H_-^2(F)\right). \qquad (6.7.1)$$

PROOF. First, we check that $L_\eta \in \mathrm{Lat}\, T$ for any isometry η compatible with π and π_*. To this end, we rewrite (6.7.1) in the form

$$L_\eta = \left(\eta H^2(F) \oplus \mathrm{clos}(I - \eta\eta^*)\pi L^2(E)\right) \ominus \pi H^2(E). \qquad (6.7.2)$$

Clearly, this representation implies that L_η is T-invariant. Indeed, since

$$\mathcal{U}L_\eta \subset \eta H^2(F) \oplus \mathrm{clos}(I - \eta\eta^*)\pi L^2(E),$$

after projecting onto H we find ourselves in the orthogonal complement of $G = \pi H^2(E)$, that is, in L_η.

Now we check that $L = L_{\eta_L}$ for every invariant subspace L. Comparing (6.7.2) with the definition of the subspace $\mathcal{R}_L$ in (6.4.1), we see that we must verify the formula

$$\mathcal{R}_L = \mathrm{clos}(I - \eta\eta^*)\pi L^2(E),$$

where $\eta = \eta_L$. The inclusion $\mathrm{clos}(I - \eta\eta^*)\pi L^2(E) \subset \mathcal{R}_L$ was already proved in Section 6.5. Now we consider the $\mathcal{U}$-reducing subspace $\mathcal{R}_L \ominus \mathrm{clos}(I - \eta\eta^*)\pi L^2(E)$ and verify that it is contained in L. If this is done, the complete nonunitarity of T will imply that the above subspace is, in fact, the zero subspace.

Let $x \in \mathcal{R}_L \ominus \mathrm{clos}(I - \eta\eta^*)\pi L^2(E)$. Since $x \in \mathcal{R}_L$, we have $\eta^* x = 0$. The property $x \perp (I - \eta\eta^*)\pi L^2(E)$ means $\pi^*(I - \eta\eta^*)x = 0$, so $\pi^* x = \pi^* \eta\eta^* x = 0$. Therefore, $x \perp G$; since $x \in L \oplus G$, this is equivalent to $x \in L$.

If we start with a compatible embedding η and construct the corresponding invariant subspace L_η as in (6.7.2), then, as is easily seen, the additional embedding constructed by L_η coincides with (is equivalent to) the initial η; that is, $\eta_{L_\eta} = \eta$. Indeed, by (6.7.2),

$$L_\eta \oplus G = \eta H^2(F) \oplus \mathrm{clos}(I - \eta\eta^*)\pi L^2(E).$$

This representation is the Wold–Kolmogorov decomposition (6.4.2) with $\eta_L = \eta$.

$\square$

6.8.   Regular factorizations. Using definitions (6.5.1) and (6.5.2) we can rewrite identity (6.5.3) in the form

$$\Theta = \Theta_2 \Theta_1. \tag{6.8.1}$$

This means that any compatible embedding η induces a factorization of the characteristic function into two contractive H^∞-factors. Among all possible contractive H^∞-factorizations of Θ, the factorizations induced by compatible embeddings are distinguished by the particular property of being *regular factorizations*. We can define regular factorization as a factorization (6.8.1) generated by a compatible embedding η by formulas (6.5.1) and (6.5.2). However, it is important that there exists another description of such factorizations that does not use the notion of functional embedding and, moreover, can be applied to a factorization of an arbitrary contraction.

We start by restating the invariant subspace theorem (Theorem 6.7) in terms dating back to Sz.-Nagy and Foiaş [1967]; then we present the above-mentioned characterization of regular factorizations (Section 6.10). To distinguish the definition of regularity via an additional embedding and the definition given in Section 6.10 we shall call the latter one sometimes a *locally regular factorization*. After some preparation (Lemmas 6.11–6.14) we shall be able to prove (Theorem 6.15) that these two definitions of regularity coincide.

6.9.   Theorem. *Let T be a completely nonunitary contraction, and let π, π_*, $\Theta = \Theta_T$, etc., have the usual meanings. There exists a bijection between the lattice $\operatorname{Lat} T$ and the set of all regular factorizations*

$$\Theta = \Theta_2 \Theta_1$$

of the characteristic function Θ. This bijection is given by the formula

$$L_{\Theta_2,\Theta_1} = \left(\eta H^2(F) \oplus \operatorname{clos}(I - \eta\eta^*)\pi L^2(E)\right) \ominus \pi H^2(E), \tag{6.9.1}$$

where η is the compatible embedding producing the functions Θ_i via (6.5.1) and (6.5.2). In the original Sz.-Nagy–Foiaş transcription the representation given by (6.9.1) takes the form

$$L_{\Theta_2,\Theta_1} = \begin{pmatrix} \Theta_2 \\ Z_2^* \Delta_2 \end{pmatrix} H^2(F) \oplus \begin{pmatrix} 0 \\ Z_1^* \end{pmatrix} L^2(\Delta_1 E) \ominus \begin{pmatrix} \Theta \\ \Delta \end{pmatrix} H^2(E),$$

where the Z_i and Δ_i are certain operator-valued functions to be defined in the next section.

PROOF. The bijection is the same as in Theorem 6.7, namely, $\eta \leftrightarrow L_\eta$. The formula in the Sz.-Nagy–Foiaş transcription follows from expression (6.13.1) for η combined with the definition of τ (1.19.1) and formulas (3.5.1) for the embeddings π and π_* in the Sz.-Nagy–Foiaş representation.   □

Now we introduce the language for the local description of the regular factorizations.

6.10. Definition. Let A and B be two Hilbert space contractions such that A acts into the space where B is defined, that is, the product BA is well defined. Let $\mathcal{D}_A$, $\mathcal{D}_B$, and $\mathcal{D}_{BA}$ be the defect spaces of these operators. We define an operator

$$Z : \mathcal{D}_{BA} \to \begin{pmatrix} \mathcal{D}_A \\ \mathcal{D}_B \end{pmatrix}$$

by the identity

$$Z D_{BA} \overset{\text{def}}{=} \begin{pmatrix} Z_1 \\ Z_2 \end{pmatrix} D_{BA} = \begin{pmatrix} D_A \\ D_B A \end{pmatrix}. \tag{6.10.1}$$

The factorization BA is called *regular* if Z is unitary.

For a factorization $\Theta = \Theta_2 \Theta_1$, the definition of Z becomes

$$Z\Delta = \begin{pmatrix} Z_1 \\ Z_2 \end{pmatrix} \Delta = \begin{pmatrix} \Delta_1 \\ \Delta_2 \Theta_1 \end{pmatrix} \tag{6.10.2}$$

where $\Delta_i = (I - \Theta_i^* \Theta_i)^{1/2}$, and all functions are regarded as multiplication operators on the corresponding spaces. It should be mentioned at once that a function $Z \in L^\infty(E_1 \to E_2)$ is unitary as an operator from $L^2(E_1)$ to $L^2(E_2)$ if and only if the operators $Z(\zeta) : E_1 \to E_2$ are unitary for almost all $\zeta \in \mathbb{T}$. This implies that *a factorization $\Theta = \Theta_2 \Theta_1$ is regular if and only if the factorizations $\Theta(\zeta) = \Theta_2(\zeta)\Theta_1(\zeta)$ are regular for almost all $\zeta \in \mathbb{T}$.*

The above observation motivates the use of the adverb *locally* when talking about regular function factorizations $\Theta = \Theta_2 \Theta_1$. In what follows, we shall often use the term *locally regular factorization* instead of "regular factorization", emphasizing the fact that we deal with functions; also, this will allow us to distinguish (temporarily) between this new wording and the previous one defined in terms of compatible embeddings.

Our next observation is that Z in (6.10.1) is *always* an isometry.

6.11. Lemma. *The operator Z defined by (6.10.1) (and, therefore, by (6.10.2)) is an isometry.*

PROOF. It suffices to check that Z is norm-preserving on a dense set of vectors. By the definition of $\mathcal{D}_{BA}$, the set of vectors $\{D_{BA}x : x \in H\}$ is dense there. On this dense set we have

$$\|Z D_{BA} x\|^2 = \|D_A x\|^2 + \|D_B A x\|^2$$
$$= \|x\|^2 - \|Ax\|^2 + \|Ax\|^2 - \|BAx\|^2 = \|D_{BA}x\|^2. \qquad \square$$

Thus, to prove that Z is unitary it suffices to check that this operator is a co-isometry, that is, that Z_1 and Z_2 are co-isometries. Moreover, the following assertion is true.

6.12. Lemma. Z *is a co-isometry if and only if Z_2 is.*

PROOF. If Z_2 is a co-isometry, then

$$0 < \left\| \begin{pmatrix} x_1 \\ x_2 \end{pmatrix} \right\|^2 - \left\| Z^* \begin{pmatrix} x_1 \\ x_2 \end{pmatrix} \right\|^2$$

$$= \|x_1\|^2 + \|x_2\|^2 - \|Z_1^* x_1\|^2 - 2 \operatorname{Re}(Z_1^* x_1, Z_2^* x_2) - \|Z_2^* x_2\|^2$$

$$= \|x_1\|^2 - \|Z_1^* x_1\|^2 - 2 \operatorname{Re}(Z_1^* x_1, Z_2^* x_2),$$

which is possible if and only if $Z_1 Z_2^* = 0$. Hence,

$$\begin{pmatrix} 0 \\ \mathcal{D}_B \end{pmatrix} = \begin{pmatrix} 0 \\ \operatorname{Range} Z_2 \end{pmatrix} = \operatorname{Range} Z Z_2^* Z_2 \subset \operatorname{Range} Z,$$

whence

$$\operatorname{Range} Z = \begin{pmatrix} \operatorname{clos} \operatorname{Range} Z_1 \\ \mathcal{D}_B \end{pmatrix} = \begin{pmatrix} \mathcal{D}_A \\ \mathcal{D}_B \end{pmatrix}.$$

Thus, Z is unitary.   □

6.13. Lemma. *Let η be an embedding compatible with π and π_*, and let $\Theta = \Theta_2 \Theta_1$ be the corresponding factorization of the characteristic function (that is, Θ_1 is defined by (6.5.1) and Θ_2 by (6.5.2)). Then*

$$\eta = \pi_* \Theta_2 + \tau Z_2^* \Delta_2, \tag{6.13.1}$$

where Z_2 is defined by (6.10.2) and $\Delta_2 \overset{\text{def}}{=} (I - \Theta_2^* \Theta_2)^{1/2}$, or symmetrically,

$$\eta = \pi \Theta_1^* + \tau_* Z_{*1}^* \Delta_{*1}, \tag{6.13.2}$$

where $\Delta_{*1} \overset{\text{def}}{=} (I - \Theta_1 \Theta_1^*)^{1/2}$, and Z_{*1} is the component of the operator Z_* corresponding to the factorization $\Theta^* = \Theta_1^* \Theta_2^*$ in accordance with (6.10.1):

$$Z_* \Delta_* \overset{\text{def}}{=} \begin{pmatrix} Z_{*2} \\ Z_{*1} \end{pmatrix} \Delta_* = \begin{pmatrix} \Delta_{*2} \\ \Delta_{*1} \Theta_2^* \end{pmatrix}. \tag{6.13.3}$$

PROOF. Since

$$\eta = (\pi_* \pi_*^* + \tau \tau^*) \eta = \pi_* \Theta_2 + \tau (\tau^* \eta),$$

relation (6.13.1) will be proved if we check that $\tau^* \eta = Z_2^* \Delta_2$, or $\eta^* \tau = \Delta_2 Z_2$. We have

$$\eta^* \tau \Delta = \eta^* (\pi - \pi_* \Theta) = \Theta_1 - \Theta_2^* \Theta = \Delta_2^2 \Theta_1 = \Delta_2 Z_2 \Delta.$$

Applying $I = \pi \pi^* + \tau_* \tau_*^*$ to η, we obtain the symmetric formula (6.13.2).   □

Now we prove that, conversely, a contractive H^∞-factorization $\Theta = \Theta_2 \Theta_1$ (not necessarily locally regular) gives rise to a functional embedding η (which is not necessarily isometric).

6.14. Lemma. *Let $\Theta = \Theta_2 \Theta_1$ be a contractive H^∞-factorization. Then the operators defined by (6.13.1) and (6.13.2) coincide and possess properties a)–d) of Definition 6.6.*

PROOF. First, let us check that formulas (6.13.1) and (6.13.2) are equivalent. Assuming, for example, that (6.13.2) defines an operator η, we check that (6.13.1) is also true, with the same η. We have

$$\pi_*^*\eta = \pi_*^*(\pi\Theta_1^* + \tau_* Z_{*1}^* \Delta_{*1}) = \Theta\Theta_1^* + \Delta_* Z_{*1}^* \Delta_{*1}$$
$$= \Theta\Theta_1^* + \Theta_2\Delta_{*1}\Delta_{*1} = \Theta_2(\Theta_1\Theta_1^* + \Delta_{*1}^2) = \Theta_2.$$

Here we have used definition (6.13.3). Now, using the formula (to be proved below)

$$Z_{*1}\Theta = \Theta_1 Z_1 - \Delta_{*1}\Delta_2 Z_2, \tag{6.14.1}$$

we get

$$\tau^*\eta = \tau^*(\pi\Theta_1^* + \tau_* Z_{*1}^* \Delta_{*1}) = \Delta\Theta_1^* - \Theta^* Z_{*1}^* \Delta_{*1}$$
$$= \Delta\Theta_1^* - (Z_1^*\Theta_1^* - Z_2^*\Delta_2\Delta_{*1})\Delta_{*1} = (I - Z_1^* Z_1)\Delta\Theta_1^* + Z_2^*\Delta_2\Delta_{*1}^2$$
$$= Z_2^* Z_2\Delta\Theta_1^* + Z_2^*\Delta_2\Delta_{*1}^2 = Z_2^*\Delta_2(\Theta_1\Theta_1^* + \Delta_{*1}^2) = Z_2^*\Delta_2.$$

Therefore,

$$\eta = (\pi_*\pi_*^* + \tau\tau^*)\eta = \pi_*\Theta_2 + \tau Z_2^*\Delta_2.$$

We complete the proof that (6.13.1) and (6.13.2) are equivalent by checking formula (6.14.1):

$$(\Theta_1 Z_1 - \Delta_{*1}\Delta_2 Z_2)\Delta = \Theta_1\Delta_1 - \Delta_{*1}\Delta_2^2\Theta_1 = \Delta_{*1}(I - \Delta_2^2)\Theta_1$$
$$= \Delta_{*1}\Theta_2^*\Theta_2\Theta_1 = Z_{*1}\Delta_*\Theta = Z_{*1}\Theta\Delta.$$

Now, if a contractive factorization $\Theta = \Theta_2\Theta_1$ is given and if η is defined by (6.13.1)–(6.13.2), then, clearly, $\pi^*\eta = \Theta_1^*$ and $\pi_*^*\eta = \Theta_2$; that is, formulas (6.13.1)–(6.13.2) determine the bijection inverse to that given by (6.5.1)–(6.5.2). Property (a) is obvious; properties (b) and (c) of Definition 6.6 are equivalent to the analyticity of Θ_1 and Θ_2, respectively. Property d) means merely that $\Theta = \Theta_2\Theta_1$.   □

So, in talking about a contractive H^∞-factorization of Θ, we deal with an embedding η satisfying conditions a)–d) of Definition 6.6. Thus, to get a compatible embedding we need only to check that η is an isometry. This property of η is equivalent to the property that the factorization $\Theta = \Theta_2\Theta_1$ is locally regular.

6.15. Theorem. *If $\Theta = \Theta_2\Theta_1$ is a locally regular factorization, formula (6.13.1) (or (6.13.2)) defines an embedding η compatible with π and π_*. Conversely, if η is a compatible embedding, then the factorization $\Theta = \Theta_1\Theta_2$ with the factors defined by (6.5.1), (6.5.2) is locally regular. Moreover, formulas (6.5.1), (6.5.2), and (6.13.1) (or (6.13.2)) determine a one-to-one correspondence between the set of all compatible embeddings and the set of all regular factorizations.*

PROOF. As already mentioned, after establishing Lemmas 6.13 and 6.14, it only remains to prove that η is an isometry if and only if the corresponding factorization is locally regular.

We have

$$I - \eta^*\eta = I - (\Theta_2^*\pi_*^* + \Delta_2 Z_2 \tau^*)(\pi_*\Theta_2 + \tau Z_2^*\Delta_2)$$
$$= I - \Theta_2^*\Theta_2 - \Delta_2 Z_2 Z_2^*\Delta_2 = \Delta_2(I - Z_2 Z_2^*)\Delta_2.$$

Hence, η is an isometry if and only if Z_2 is a co-isometry; by Lemma 6.12, the latter is equivalent to the fact that Z is a co-isometry. We conclude that η is an isometry if and only if $\Theta = \Theta_2\Theta_1$ is a locally regular factorization. $\square$

The factors in a regular factorization of a characteristic function Θ_T are very deeply related with the parts of the operator T induced by the corresponding invariant subspace $L = L_{\Theta_2, \Theta_1}$, namely, with the restriction $T_1 = T|L$ and the compression to the orthogonal complement $T_2 = P_{L^\perp} T|L^\perp$.

6.16. Theorem. *Let T be a contraction on a Hilbert space, $\Theta = \Theta_T$ the characteristic function of T. If L is a T-invariant subspace and $\Theta = \Theta_2\Theta_1$ is the corresponding regular factorization of Θ, then the pure part of Θ_1 is the characteristic function of the operator $T_1 = T|L$ and the pure part of Θ_2 is the characteristic function of the operator $T_2 = P_{L^\perp} T|L^\perp$.*

OUTLINE OF THE PROOF. Let η be the additional functional embedding corresponding to the invariant subspace L (see Section 6.4). Putting

$$E_1 \stackrel{\text{def}}{=} E \ominus \pi^*(\pi E \cap \eta F),$$
$$E_{*1} \stackrel{\text{def}}{=} F \ominus \eta^*(\pi E \cap \eta F),$$

we introduce two functional embeddings $\pi_1 \stackrel{\text{def}}{=} \pi|L^2(E_1)$ and $\pi_{*1} \stackrel{\text{def}}{=} \eta|L^2(E_{*1})$. It is not very difficult to verify that π_1 and π_{*1} form the pair of "canonical" functional embeddings related with the operator T_1 and, therefore, the characteristic function Θ_{T_1} of T_1 is equal to $\pi_{*1}^*\pi_1$, however the latter operator is the pure part of Θ_1.

In a similar way, for T_2 we put

$$E_2 \stackrel{\text{def}}{=} F \ominus \eta^*(\pi_* E_* \cap \eta F),$$
$$E_{*2} \stackrel{\text{def}}{=} E_* \ominus \pi_*^*(\pi_* E_* \cap \eta F),$$

and the corresponding embeddings are $\pi_2 \stackrel{\text{def}}{=} \eta|L^2(E_2)$ and $\pi_{*2} \stackrel{\text{def}}{=} \pi_*|L^2(E_{*2})$. $\square$

As an application of the above description of the invariant subspaces we find the factorizations corresponding to the root subspaces, in particular, to the eigenspaces, described in Theorem 5.10.

6.17. Theorem. *For every $\lambda \in \mathbb{D}$ we have*

$$\operatorname{Ker}(T - \lambda I)^n = \pi \sum_{k}^{n-1} = 0 \oplus \frac{\vartheta_k^* \operatorname{Ker} \Theta_k(\lambda)}{z - \lambda}, \qquad (6.17.1)$$

where $\Theta_T = \Theta_k \vartheta_k$ is the (locally) regular factorization corresponding to the invariant subspace $\operatorname{Ker}(T - \lambda I)^k$, $0 \leq k \leq n - 1$ (for $k = 0$ we take $\Theta_0 = \Theta_T$, $\vartheta_0 = I$).

OUTLINE OF THE PROOF. We work by induction on n. The case $n = 1$ is contained in Theorem 5.10 with $\Theta_0 \overset{\text{def}}{=} \Theta$ and $\vartheta_0 \overset{\text{def}}{=} I$. If $H_1 \overset{\text{def}}{=} \operatorname{Ker}(T - \lambda I)$ and $P_1 \overset{\text{def}}{=} P_{\operatorname{Ker}\Theta(\lambda)}$, then

$$H_1 \oplus G = \pi \frac{\operatorname{Ker} \Theta(\lambda)}{z - \lambda} \oplus \pi H^2(E) = \pi\left(\bar{b}_\lambda H^2(P_1 E) \oplus H^2((I - P_1)E)\right).$$

Hence, operating as in steps 2 and 3 of Section 6.2, we can take $F = E$ and $\eta = \pi\left(\bar{b}_\lambda P_1 + (I - P_1)\right)$. In accordance with (6.5.1) the factor ϑ_1 occurring in the corresponding regular factorization $\Theta = \Theta_1 \vartheta_1$ is as follows:

$$\vartheta_1 = \eta^* \pi = b_\lambda P_1 + (I - P_1).$$

In order to describe $\operatorname{Ker}(T - \lambda I)^2$ we consider the triangulation (that is, the block matrix form) of T with respect to the orthogonal decomposition $H = H_1^\perp \oplus H_1$. Putting $T_1 \overset{\text{def}}{=} (I - P_1)T|H_1^\perp$, we get $\operatorname{Ker}(T - \lambda I)^2 = \operatorname{Ker}(T_1 - \lambda I) \oplus H_1$; now we can apply the same Theorem 5.10 to the operator T_1:

$$\operatorname{Ker}(T_1 - \lambda I) = \pi_1 \frac{\operatorname{Ker} \Theta_1(\lambda)}{z - \lambda}.$$

The next observation is that, up to passage to the pure parts, the characteristic function of T_1 is the factor Θ_1, and the role of the canonical embedding π_1 is played by the embedding $\eta = \pi \vartheta_1^*$. Therefore, we have

$$\operatorname{Ker}(T_1 - \lambda I) = \pi \frac{\vartheta_1^* \operatorname{Ker} \Theta_1(\lambda)}{z - \lambda}.$$

By induction, if the subspace $H_k = \operatorname{Ker}(T - \lambda I)^k$ and the corresponding factorization $\Theta = \Theta_k \vartheta_k$ has been described, then we can write

$$H_{k+1} \overset{\text{def}}{=} \operatorname{Ker}(T - \lambda I)^{k+1} = H_k \oplus \pi \frac{\vartheta_k^* \operatorname{Ker} \Theta_k(\lambda)}{z - \lambda},$$

and compute the next regular factor $\vartheta_k = (b_\lambda P_k + (I - P_k))$, where $P_k \overset{\text{def}}{=} P_{\operatorname{Ker}\Theta_{k-1}(\lambda)}$. This gives us a recursive procedure for computing the regular factors ϑ_k for $0 \leq k \leq n - 1$, and, finally, to prove formula (6.17.1).

We omit some details of the proof, because they require more information on regular factorizations than we have prepared in this chapter. $\qquad \square$

6.18. Example: Scalar characteristic function. We recall that this term refers to the case where $\dim E = \operatorname{rank}(I - T^*T) = 1$ and $\dim E_* = \operatorname{rank}(I - TT^*) = 1$; thus, we can make the identifications $E = E_* = \mathbb{C}$ and $H^\infty(\mathbb{C} \to \mathbb{C}) = H^\infty$.

Let $\Theta_1 \in H^\infty(\mathbb{C}^1 \to F)$ and $\Theta_2 \in H^\infty(F \to \mathbb{C}^1)$ be contractive functions. The definition of regular factorizations (Section 6.10) and Lemma 6.11 imply that in the finite-dimensional case a factorization $\Theta = \Theta_2\Theta_1$ is regular if and only if

$$\operatorname{rank} \Delta(\zeta) = \operatorname{rank} \Delta_1(\zeta) + \operatorname{rank} \Delta_2(\zeta) \quad \text{for a.e. } \zeta \in \mathbb{T}. \tag{6.18.1}$$

In this and only in this case the final space of the isometry $Z(\zeta)$ has the same dimension as its initial space, so that Z is unitary. Whence the condition $0 \leq \dim F \leq 2$ is necessary for the factorization $\Theta = \Theta_2\Theta_1$ to be regular. We consider these three possibilities separately.

$\underline{\dim F = 0}$.

This case can occur for the function $\Theta = 0 : \mathbb{C}^1 \to \mathbb{C}^1$ only. Indeed, if $\dim F = 0$, then $\Theta_1 = 0 : \mathbb{C}^1 \to \{0\}$, $\Theta_2 = 0 : \{0\} \to \mathbb{C}^1$ and $\Theta = \Theta_2\Theta_1 = 0$. The factorization

$$0_{1\times 1} = 0_{1\times 0} 0_{0\times 1}.$$

is clearly regular, because here we have

$$\operatorname{rank} \Delta(\zeta) = \operatorname{rank} \Delta_1(\zeta) = 1, \quad \operatorname{rank} \Delta_2(\zeta) = 0.$$

The operator corresponding to this characteristic function is

$$\mathcal{M}_\Theta = S \oplus S^* \stackrel{\text{def}}{=} z|H^2 \oplus P_- z|H_-^2.$$

In this case, the underlying geometry is very transparent:

$$\mathcal{H} = \begin{pmatrix} L^2 \\ L^2 \end{pmatrix}, \quad \mathcal{K}_\Theta = \begin{pmatrix} H^2 \\ H_-^2 \end{pmatrix}, \quad \pi = \begin{pmatrix} 0 \\ 1 \end{pmatrix}, \quad G = \begin{pmatrix} 0 \\ H^2 \end{pmatrix}, \quad \pi_* = \begin{pmatrix} 1 \\ 0 \end{pmatrix}, \quad G_* = \begin{pmatrix} H^2 \\ 0 \end{pmatrix}.$$

Since $F = \{0\}$, from (6.7.1) we have

$$L = \pi L^2 \ominus \pi H^2 = \pi H_-^2 = \begin{pmatrix} 0 \\ H_-^2 \end{pmatrix},$$

whence we see that $L = \{0\} \oplus H_-^2$ is the reducing subspace where the operator $S^* = P_- z|H_-^2$ acts.

$\underline{\dim F = 1}$.

Now Θ_1 and Θ_2 are ordinary scalar functions belonging to the unit ball of H^∞. The factorization $\Theta = \Theta_2\Theta_1$ is regular if and only if

$$\max\{|\Theta_1(\zeta)|, |\Theta_2(\zeta)|\} = 1 \quad \text{for a.e. } \zeta \in \mathbb{T}. \tag{6.18.2}$$

Indeed, since $\operatorname{rank} \Delta(\zeta) \leq 1$, (6.18.1) implies that $\min\{\Delta_1(\zeta), \Delta_2(\zeta)\} = 0$, which is equivalent to (6.18.2). Conversely, if, for example, $|\Theta_1(\zeta)| = 1$, then

$|\Theta(\zeta)| = |\Theta_2(\zeta)|$, $\Delta(\zeta) = \Delta_2(\zeta)$, $\Delta_1(\zeta) = 0$, and (6.18.1) follows. Similarly for $|\Theta_2(\zeta)| = 1$.

Condition (6.18.2) means that the inner part of Θ can be arbitrarily factored in two inner factors, and the factorizations of the outer part are determined by the Borel subsets $\sigma \subset \mathbb{T}$ in such a way that

$$|\Theta_1(\zeta)| = \begin{cases} |\Theta(\zeta)| & \text{if } \zeta \in \sigma, \\ 1 & \text{if } \zeta \in \mathbb{T} \setminus \sigma, \end{cases} \qquad |\Theta_2(\zeta)| = \begin{cases} |\Theta(\zeta)| & \text{if } \zeta \in \mathbb{T} \setminus \sigma, \\ 1 & \text{if } \zeta \in \sigma. \end{cases}$$

Particular case: $\Theta = 0$. If $\Theta = 0$, then the only possibilities are $0 = 0 \cdot \vartheta$ or $0 = \vartheta \cdot 0$, with an inner ϑ. The corresponding invariant subspaces are as follows

$$0 = 0 \cdot \vartheta \Longleftrightarrow L = \begin{pmatrix} 0 \\ H^2_- \ominus \vartheta^* H^2_- \end{pmatrix}, \qquad 0 = \vartheta \cdot 0 \Longleftrightarrow L = \begin{pmatrix} \vartheta H^2 \\ H^2_- \end{pmatrix}.$$

We recall that in this case

$$\mathcal{K}_\Theta = \begin{pmatrix} H^2 \\ H^2_- \end{pmatrix}, \qquad \mathcal{M}_\Theta = S \oplus S^* = \begin{pmatrix} z \\ P_- z \end{pmatrix}.$$

$\underline{\dim F = 2}$.

In this case the factorizations under study are of the form

$$\Theta = \Theta_2 \Theta_1 = (f_1, \ f_2) \begin{pmatrix} g_1 \\ g_2 \end{pmatrix} = f_1 g_1 + f_2 g_2.$$

Such a factorization is regular if and only if $|\Theta(\zeta)| < 1$ for a.e. $\zeta \in \mathbb{T}$, Θ_2 is co-inner, and Θ_1 is inner, that is, $|f_1(\zeta)|^2 + |f_2(\zeta)|^2 = 1$ and $|g_1(\zeta)|^2 + |g_2(\zeta)|^2 = 1$ for a.e. $\zeta \in \mathbb{T}$. Indeed, since $\operatorname{rank} \Theta_2(\zeta) \leq \dim E_* = 1$, we have $\operatorname{rank} \Delta_2(\zeta) \geq \dim F - \operatorname{rank} \Theta_2(\zeta) \geq 1$. Hence, (6.18.1) implies that the case $\dim F = 2$ is possible only if $\operatorname{rank} \Delta(\zeta) \geq 1$, that is, we neccessarily have $|\Theta(\zeta)| < 1$ and $\operatorname{rank} \Delta(\zeta) = 1$ for a.e. ζ. Whence (6.18.1) is equivalent to the conditions $\operatorname{rank} \Delta_2(\zeta) = 1$ and $\operatorname{rank} \Delta_1(\zeta) = 0$, which means that Θ_2 is co-inner and Θ_1 is inner, correspondingly.

It should be noted that even the existence of factorizations of this sort are not completely obvious. However, they do exist for any Θ satisfying $|\Theta(\zeta)| < 1$ a.e.; moreover, there exist infinitely many such factorizations. To see that, we take any function φ on $\mathbb{T}$ whose values are ± 1 (that is, this is a function satisfying the condition $\varphi^2(\zeta) = 1$ a.e.). Then, assuming that $\Theta \not\equiv 0$, we take two outer functions f_1, f_2 such that

$$|f_1|^2 = \frac{1 - \varphi \Delta}{2}, \qquad |f_2|^2 = \frac{1 + \varphi \Delta}{2}.$$

This is indeed possible, because

$$\log \frac{1 \pm \varphi \Delta}{2} \geq \log \frac{1 - \Delta}{2} \geq \log \frac{1 - \Delta^2}{4} = \log \frac{|\Theta|^2}{4} \in L^1.$$

Since

$$|f_1|^2 |f_2|^2 = \frac{1 - \varphi^2 \Delta^2}{4} = \frac{1 - \Delta^2}{4} = \frac{|\Theta|^2}{4},$$

for the outer part of Θ we have $\Theta_{\text{out}} = 2f_1 f_2$. Hence, taking $g_1 = f_2 \Theta_{\text{inn}}$, $g_2 = f_1 \Theta_{\text{inn}}$, we get a solution; here Θ_{inn} stands for the inner part of Θ. The case $\Theta = 0$ is treated later on.

6.19. Problem. How can one describe all factorizations of the latter type, that is, the (1×2)–(2×1) factorizations of a scalar function Θ satisfying $|\Theta| < 1$ a.e.? We make some comments on this problem.

(a) Let a function Θ with $|\Theta(\zeta)| < 1$ a.e. be given. The problem consists in describing all regular factorizations of Θ with $\dim F = 2$ (that is, in completing the description of the invariant subspaces of a given operator with the characteristic function $\Theta_T = \Theta$). In detail, the problem is as follows:

Find all pairs of H^∞-functions f_1, f_2 and g_1, g_2 such that $f_1 g_1 + f_2 g_2 = \Theta$ and

$$|f_1(\zeta)|^2 + |f_2(\zeta)|^2 = |g_1(\zeta)|^2 + |g_2(\zeta)|^2 = 1 \quad \text{for a.e. } \zeta \in \mathbb{T}.$$

(b) Let a (2×1)-inner function Θ_1 be given. The problem consists in the description of all Θ such that $\Theta_1 = \begin{pmatrix} g_1 \\ g_2 \end{pmatrix}$ is a right regular factor of Θ (that is, in obtaining a description of the operators having a scalar characteristic function and an invariant subspace restriction that is unitarily equivalent to the given operator $\mathfrak{M}_{\Theta_1}$). In detail:

Find all H^∞-functions Θ, $|\Theta(\zeta)| < 1$ a.e., for which there exists a pair of H^∞-functions f_1, f_2 such that $f_1 g_1 + f_2 g_2 = \Theta$ and $|f_1(\zeta)|^2 + |f_2(\zeta)|^2 = 1$ a.e.

A particular case: $\Theta = 0$. Again, let $\Theta = 0$. Then it can be shown that all regular factorizations of $\dim F = 2$ type admit the following parametrization: if $\vartheta = \begin{pmatrix} \alpha \\ \beta \end{pmatrix}$ is an arbitrary (2×1)-function which is inner and co-outer (that is, the inner parts of α and β are coprime), and ϑ_1, ϑ_2 are two arbitrary scalar inner functions, then the representation

$$0 = (\vartheta_2 \beta, \, -\vartheta_2 \alpha) \begin{pmatrix} \vartheta_1 \alpha \\ \vartheta_1 \beta \end{pmatrix}$$

is a required factorization. All (1×2)–(2×1) regular factorizations of $\Theta = 0$ are of this type.

Moreover, for $\Theta = 0$ we can answer question (b) as well: every operator with a (2×1)-inner characteristic function can be realized as the restriction of the operator $S \oplus S^*$ to an invariant subspace.

The latter statement is not true for general functions Θ. For example, an operator with a nonzero scalar characteristic function possesses an invariant subspace where it acts as the unilateral shift S if and only if $\log \Delta \in L^1$. Indeed, to get such an invariant subspace we need a regular factorization of the form

$$\Theta = (f_1, \, f_2) \begin{pmatrix} 1 \\ 0 \end{pmatrix},$$

that is, we need $f_1 = \Theta$ and $|f_2| = (1 - |f_1|^2)^{1/2} = \Delta$, $f_2 \in H^\infty$, and this is possible if and only if $\log \Delta \in L^1$.

6.20. Example: Indefinite integration operator. We now return to the operator

$$A : f \mapsto i \int\limits_{[0,x\}} f(t)\, d\mu(t);$$

the corresponding characteristic function $\mathcal{S}_A$, computed in Theorem 2.6 is a scalar inner function. A description of the lattice of the invariant subspaces of the operator A was given by Leggett [1973]. We derive a description of this lattice as an immediate consequence of the description of invariant subspaces obtained in this chapter.

As is clear from Section 6.18, all regular factorizations of a scalar inner function are factorizations in two scalar inner factors. Assume that the continuous part μ_c of μ is nonzero. Then there exists a continuous nest of invariant subspaces corresponding to the factors

$$\mathcal{S}_\alpha = \exp\left(-\frac{i\alpha}{\zeta}\right) \quad \text{for } 0 \le \alpha \le \mu_c([0,1]).$$

They are the only invariant subspaces of A if and only if μ is a continuous measure ($\mu = \mu_c$). Clearly, in this case the lattice Lat A is completely ordered by inclusion, or, in other words, A is a unicellular operator.

If μ has a nontrivial discrete part, then A has isolated eigenvalues on the imaginary axis: $\lambda = \frac{i}{2}\mu(\{t\})$. For every such λ there exists a unique eigenvector, and the corresponding Jordan block has dimension card$\{t : \frac{i}{2}\mu(\{t\}) = \lambda\}$.

If μ is a purely discrete measure ($\mu_c = 0$), then $\mathcal{S}_A$ is a Blaschke product, and hence every inner factorization of $\mathcal{S}_A$, $\mathcal{S}_A = \mathcal{S}_2 \mathcal{S}_1$, is a factorization in two Blaschke products. Therefore, the characteristic function of the restriction $A|L$ to an invariant subspace L is always a Blaschke product. We recall that an operator with a scalar characteristic function Θ has a complete family of eigen- and root-vectors if and only if Θ is a Blaschke product (see [Nikolski 1986]). Thus, for the operator A we have the following spectral synthesis theorem: *if μ is a discrete measure, then every invariant subspace of A is generated by the eigen- and root-vectors it contains.*

If the measure μ has infinitely many point masses together with a nonzero continuous part μ_c, then the angle between any "continuous" invariant subspace and any infinite dimensional "discrete" subspace is equal to zero, because

$$\inf_{\operatorname{Im}\zeta > 0}(|\mathcal{S}_\alpha(\zeta)| + |B_\sigma(\zeta)|) = 0,$$

where

$$B_\sigma \stackrel{\text{def}}{=} \prod_{t \in \sigma} \frac{\zeta - \frac{i}{2}\mu(\{t\})}{\zeta + \frac{i}{2}\mu(\{t\})}.$$

Why this infimum determines the angle between the invariant subspaces will be explained in Chapter 7, in Part II of this article.

Afterword: Outline of Part II

The second part of this paper, "The Function Model in Action", will be published elsewhere. For the reader who expected to find applications of the function model here, now we describe briefly how the model works in spectral theory.

A.1. Angles between invariant subspaces, and operator Bezout equations. As is well known, angles between invariant subspaces are the key tool for studying all kinds of spectral decompositions. Recall that the *angle* $\alpha = \alpha(K, K')$ between two subspaces $K, K' \subset H$ of a Hilbert space H is, by definition, the number $\alpha \in [0, \pi/2]$ satisfying

$$\cos \alpha = \sup\{|(x, y)| : x \in K,\, y \in K',\, \|x\| = \|y\| = 1\}.$$

For instance, eigenvectors $(x_n)_{n \geq 1}$, or any other vectors of a Hilbert space, form a basis if and only if the angles between the subspaces $K_n = \mathrm{span}(x_k : 1 \leq k \leq n)$ and $K'_n = \mathrm{span}(x_k : k > n)$ are uniformly bounded away from zero. This is equivalent to saying that the *projections* $\mathcal{P}_n$ on K_n *parallel to* K'_n, defined by

$$\mathcal{P}_n(x + x') = x \quad \text{for } x \in K_n,\ x' \in K'_n,$$

are uniformly bounded: $\sup_n \|\mathcal{P}_n\| < \infty$. The same equivalence exists for unconditional bases, if we replace K_n by $K_\sigma = \mathrm{span}(x_k : k \in \sigma)$ and K'_n by $K'_\sigma = \mathrm{span}(x_k : k \notin \sigma)$, and finally $\mathcal{P}_n$ by $\mathcal{P}_\sigma$; the basis condition is

$$\sup\{\|\mathcal{P}_\sigma\| : \sigma \subset \mathbb{N}\} < \infty.$$

The same is valid for bases of subspaces, standard or unconditional. Taking spectral subspaces of a given operator (see below), we obtain a kind of spectral decomposition, and considering all families of spectral subspaces we arrive at Dunford spectral operators; see [Dunford and Schwartz 1971; Dowson 1978; Nikolski 1986] for details. This scheme explains why angles are important.

As was mentioned above, invariant subspaces of a contraction T correspond to regular factorizations of its characteristic function Θ_T, see Chapter 6 for details. Hence, to work with spectral decompositions in terms of the function model, we need to know which factorizations $\Theta_T = \Theta_2 \Theta_1$ and $\Theta_T = \Theta'_2 \Theta'_1$ correspond to invariant subspaces $TK \subset K$ and $TK' \subset K'$ having a positive angle between them. In some partial cases—for inner factorizations, for instance—this problem was solved by P. Fuhrmann [1981] and Teodorescu [1975]. A criterion that holds in full generality was proved in [Vasyunin 1994], using the CLT: the projection $\mathcal{P}$ is bounded if and only if *two* Bezout equations have bounded solutions: the "standard" one

$$W\Theta_1 + W'\Theta'_1 = I,$$

where W, W' are H^∞ operator-valued functions; and another "complementary" Bezout equation to be solved in L^∞ operator-valued functions. In Chapter 7 we will discuss this in detail; here we only mention that the obvious necessary condition, namely, the uniform local left invertibility condition

$$\|\Theta_1(z)x\| + \|\Theta_1'(z)x\| \geq \varepsilon\|x\| \quad \text{for all } z \in \mathbb{D} \text{ and all } x,$$

is, in general, not sufficient [Treil 1989].

A.2. Spectral subspaces and generalized free interpolation. The *free interpolation problem* of complex analysis, stated for a class X of functions holomorphic on a domain Ω, consists of a description of those subsets $\Lambda \subset \Omega$ for which the restriction space $X|\Lambda$ is free of traces of holomorphy, that is, is an ideal space of functions on Λ, in the sense that $a \in X|\Lambda$ and $|b(z)| \leq |a(z)|$, for $z \in \Lambda$, imply $b \in X|\Lambda$.

The interplay between this problem and operator theory goes back to the late sixties, when it was understood that the freedom property of a restriction space is adequate for the unconditional convergence of a spectral decomposition, or to the existence of a spectral measure [Nikolski and Pavlov 1968; 1970]. In fact, both of these properties are equivalent to the boundedness of the corresponding L^∞-calculus; for instance, it is clear that the space $X|\Lambda$ is an ideal if and only if every $\ell^\infty(\Lambda)$-function is a multiplier of $X|\Lambda$: $a \in X|\Lambda$ and $m \in \ell^\infty(\Lambda)$ imply $ma \in X|\Lambda$. The joint studies in interpolation theory and the function model have led to solutions of several concrete problems in both subjects, and to a considerable evolution of the very notion of interpolation, transforming it to what we call now generalized free interpolation. We present these results in Chapter 8, and now, to summarize them briefly, we start with the operator theory part of the problem, namely spectral subspaces.

Originally, spectral subspaces were introduced as a substitute for the spectral measure of a normal operator to serve more general classes of operators appearing in perturbation theory. Given a closed set $\sigma \subset \mathbb{C}$ and an operator T, the *spectral subspace* $K(\sigma)$ over σ can be defined as the set of x such that the local resolvent $\lambda \mapsto (\lambda I - T)^{-1}x$ admits an analytic extension to $\mathbb{C} \setminus \sigma$.

Another way to define such a subspace is to postulate its invariance and maximality properties: a subspace satisfying $TK(\sigma) \subset K(\sigma)$ and the corresponding spectrum inclusion $\sigma(T|K(\sigma)) \subset \sigma$, and such that $TE \subset E$ and $\sigma(T|E) \subset \sigma$ imply $E \subset K(\sigma)$.

Obviously, for normal operators, the subspaces $K(\sigma)$ coincide with the ranges of the spectral measure, $K(\sigma) = \text{Range } E(\sigma)$, and therefore they exist for σ taken from the whole Borel σ-algebra $\mathfrak{B}$. However, for general operators, it is not clear how to define subspaces $K(\sigma)$ for σ from some more or less rich σ-algebra of subsets of the complex plane $\mathbb{C}$ (any complete substitute of the spectral measure would be defined, of course, on the entire algebra $\mathfrak{B}$). This obstruction limits the use and the significance of this notion to general operators, endowed with no additional structure.

At this stage, the function model shows some of its advantages: using the multiplicative structure of H^∞- and H^2-functions, we can explicitly define an analogue of spectral subspaces for every Borel set σ and for every contraction whose characteristic function allows a "scalar multiple". For example, for the simplest case of a scalar characteristic function $\Theta_T \not\equiv 0$, when $I - T^*T$ and $I - TT^*$ both have rank 1, one can show, first, that the spectral subspaces $K(\sigma)$ over *closed* subsets σ coincide with invariant subspaces corresponding to factorizations

$$\Theta = \Theta_{\sigma'}\Theta_\sigma,$$

where Θ_σ and $\Theta_{\sigma'}$ stand for the *parts of Θ_T over σ and $\sigma' \overset{\text{def}}{=} \mathbb{C} \setminus \sigma$*, respectively. To define Θ_σ, we use the canonical *Nevanlinna–Riesz–Smirnov representation*,

$$\Theta_T(z) = \left(\prod_{\lambda \in \mathbb{D}} b_\lambda^{k(\lambda)}(z) \right) \exp\left(-\int_\mathbb{T} \frac{\zeta + z}{\zeta - z}\, d\mu(\zeta) \right),$$

putting

$$\Theta_\sigma(z) = \left(\prod_{\lambda \in \sigma} b_\lambda^{k(\lambda)}(z) \right) \exp\left(-\int_{\sigma \cap \mathbb{T}} \frac{\zeta + z}{\zeta - z}\, d\mu(\zeta) \right), \text{with} \quad |z| < 1.$$

And then, we can *define* an analogue of spectral subspaces, called *prespectral*, as invariant subspaces corresponding to factorizations $\Theta = \Theta_{\sigma'}\Theta_\sigma$ with an arbitrary Borel set σ. This is a simple but principal step, because it supplies us with a "space-valued analogue" of spectral measure. Now, to develop a kind of spectral decomposition, nothing remains but to check the angles between $K(\sigma)$ and $K(\sigma')$ for σ from a given σ-algebra.

One can join spectral decompositions and complex interpolation in at least two ways. The first way uses the commutant lifting theorem and H^∞-calculus mentioned above. The key observation is the following: if an operator A, induced on an algebraic sum $K(\sigma) + K(\sigma')$ by two restrictions $f_\sigma(T)|K(\sigma)$ and $f_{\sigma'}(T)|K(\sigma')$, where $f_\sigma, f_{\sigma'} \in H^\infty$, is well-defined and bounded, then, due to the CLT, there exists a function $f \in H^\infty$ "interpolating" the operator A in the sense $A = f(T)$. This equality is equivalent to saying that

$$f - f_\sigma \in \Theta_{\sigma'} H^\infty \quad \text{and} \quad f - f_{\sigma'} \in \Theta_\sigma H^\infty.$$

For the case when Θ_T is a Blaschke product with simple zeros, these inclusions can be interpreted in terms of classical interpolation theory. In more general situations they lead to what we call *generalized interpolation*, namely, to interpolation by germs of H^∞- and H^2-functions, for an arbitrary Blaschke product Θ_T; to a kind of asymptotic interpolation, for a point mass singular inner function Θ_σ; and to tangential interpolation on the boundary of the unit disc, for outer Θ_σ.

Continuing in this way and using the Carleson corona theorem [Garnett 1981; Nikolski 1986], we prove in Chapter 8 the following *generalized free interpolation theorem*, which is essentially contained in [Vasyunin 1978; Nikolski 1978b]. Given a bounded sequence $(\Theta_k)_{k\geq 1}$ of H^∞-functions, the following conditions are equivalent:

(1) For every $0-1$ sequence $(\varepsilon_k)_{k\geq 1}$ there exists an H^∞-function f such that $f - \varepsilon_k \in \Theta_k H^\infty$ for $k \geq 1$.

(2) For every sequence of functions $f_k \in H^\infty$, where $k \geq 1$, satisfying

$$\sup_k \operatorname{dist}_{L^\infty}(f_k \bar{\Theta}_k, H^\infty) < \infty,$$

there exists an H^∞-function f such that $f - f_k \in \Theta_k H^\infty$ for $k \geq 1$ and

$$\sup_k \left\| \frac{f - f_k}{\Theta_k} \right\|_\infty < \infty.$$

(3) There exists a uniformly equivalent sequence $(\theta_k)_{k\geq 1}$ (that is, one satisfying

$$0 < \varepsilon < \left| \frac{\Theta_k(z)}{\theta_k(z)} \right| < \varepsilon^{-1}$$

for all $k \geq 1$, $z \in \mathbb{D}$, and a fixed $\varepsilon > 0$), making the product $\theta = \prod_{k\geq 1} \theta_k$ convergent and satisfying the *generalized Carleson condition*

$$|\theta(z)| \geq \varepsilon \cdot \inf_k |\theta_k(z)| \quad \text{for } z \in \mathbb{D}.$$

As an operator-theoretic consequence of this theory we can mention a criterion for the Dunford spectrality (see [Dunford and Schwartz 1971] for definitions) of a contraction with defect numbers $(1, 1)$: it is necessary and sufficient that $\inf_{\mathbb{T}} |\Theta_T|$ be positive, that the singular measure μ_s be purely atomic, that $\mu_s = \sum_{n\geq 1} a_n \delta_{\zeta_n}$ for $\zeta_n \in \mathbb{T}$, and that the sequence $(\theta_k)_{k\geq 1}$ consisting of the corresponding point mass singular inner functions $\exp\big(-a_n(\zeta_n + z)(\zeta_n - z)\big)$ and all Blaschke factors $b_\lambda^{k(\lambda)}$ satisfy the generalized Carleson condition.

Another way to join spectral decompositions and interpolation is to go down from the calculus level to the individual vector level and to use N. Bari's characterization of unconditional bases of Hilbert space. Namely, a sum of subspaces $K(\sigma_1) + K(\sigma_2) + \cdots$ is an unconditionally convergent decomposition of the subspace $\operatorname{span}(K(\sigma_i) : i \geq 1)$ if and only if an approximate Parseval identity holds: there exists a constant $c > 0$ such that

$$c \sum_i \|x_i\|^2 \leq \left\| \sum_i x_i \right\|^2 \leq c^{-1} \sum_i \|x_i\|^2$$

for all sequences $x_i \in K(\sigma_i)$, $i \geq 1$. Now, at least for the case of *inner* characteristic functions Θ_T, one can use the interpolation meaning of the spectral projections f_i, where $f = \sum_i f_i$, corresponding to the *dual* biorthogonal decomposition $K(\sigma_1)' + K(\sigma_2)' + \ldots$, and add the following assertion, equivalent to (1)–(3) listed above:

(4) The functions $f_i \in H^2$ satisfy $\sum_i \|f_i\|_{H^2/\Theta_i H^2}^2 < \infty$ if and only if there exists a function $f \in H^2$ interpolating the f_i, in the sense that $f - f_i \in \Theta_i H^2$ for $i \geq 1$.

In particular, we get a *generalized embedding theorem*: the generalized Carleson condition implies the convergence

$$\sum_i \|f\|_{H^2/\Theta_i H^2}^2 < \infty$$

for every $f \in H^2$.

These results, in which we follow [Nikolski 1978a; 1987; Nikolski and Khrushchev 1987], were generalized in [Hartmann 1996] for all H^p with $1 < p < \infty$.

The last subject of interpolation theory that we will treat in Chapter 8 is a local estimate of interpolation data and locally defined data. Generally speaking, the idea is that the distances $\mathrm{dist}_{L^\infty}(f_i \bar{\Theta}_i, H^\infty)$ from the theorem above, in fact, coincide with operator norms of the restrictions $\|f_i(T)|K(\sigma_i)\|$ and hence should be expressed in local terms related to the behavior of f_i near the spectrum of $T|K(\sigma_i)$ (that is, $\mathrm{clos}(\sigma(T) \cap \sigma_i)$). For the inner function case, an appropriate technical tool is a local function calculus developed in Chapter 4 and especially an estimate of the mentioned operator norm by the local norm $\sup\{|f_i(z)| : z \in \Omega(\Theta_i, \varepsilon)\}$, where $\Omega(\Theta_i, \varepsilon) = \{z \in \mathbb{D} : |\Theta_i(z)| < \varepsilon\}$ stands for a level set of the function Θ_i; the latter essentially coincides with the characteristic function of the restriction $T|K(\sigma_i)$. This leads directly to the following *local interpolation* assertion which is stated to be equivalent to assertions (1)–(4) above:

(5) For all sequences of functions $f_i \in H^\infty(\Omega(\Theta_i, \varepsilon))$, where $i \geq 1$, such that $\sup_i \|f_i\|_{H^\infty(\Omega(\Theta_i, \varepsilon))} < \infty$, there exists an $H^\infty(\mathbb{D})$-function f *interpolating* the f_i in the sense that

$$f - f_i \in \Theta_i H^\infty(\Omega(\Theta_i, \varepsilon)) \quad \text{for } i \geq 1,$$

and

$$\sup_i \left\| \frac{f - f_i}{\Theta_i} \right\|_{H^\infty(\Omega(\Theta_i, \varepsilon))} < \infty.$$

It is curious that, for the Blaschke product case, where $\Theta_T = \prod b_\lambda^{k(\lambda)}$ and $\Theta_i = b_\lambda^{k(\lambda)}$, $\lambda = \lambda_i$, the corresponding level sets $\Omega(\Theta_i, \varepsilon)$ are noneuclidean discs $\{z : |b_\lambda(z)| < \varepsilon^{1/k(\lambda)}\}$, and that for this case it is well known [Vinogradov and Rukshin 1982] that the considered sets $\Omega(\Theta_i, \varepsilon)$ cannot be replaced by smaller ones. Namely, for discs $\{z : |b_{\lambda_i}(z)| < r_i\}$ with $r_i = o(\varepsilon^{1/k(\lambda_i)})$ for $i \to \infty$, the generalized Carleson condition does not imply the conclusion of assertion 5).

A.3. Similarity to a normal operator. It is clear from the previous discussion that we believe that spectral decompositions are one of the most powerful tools of spectral theory. Therefore, we rank operators admitting the same quality spectral decompositions as normal ones, as the best possible operators on

Hilbert space. As to normals, they can be characterized by any of the following properties, referring not only to decompositions but also to the calculus.

(1) An operator N on a Hilbert space H is normal if and only if there exists a projection-valued *contractive* measure $E(\,\cdot\,)$ such that $Nx = \int_\sigma z\, dE(z)x$ (Riemann convergent integral for every $x \in H$).

(2) An operator N on a Hilbert space H is normal if and only if there exists an *isometric* functional calculus $f \mapsto f(N)$ defined on the algebra $C(\sigma)$ of all continuous complex functions on a suitable compact set $\sigma \subset \mathbb{C}$, that is $\|f(T)\| = \|f\|_{C(\sigma)} \overset{\text{def}}{=} \sup_\sigma |f|$ for all $f \in C(\sigma)$.

John Wermer [1954] has discovered the remarkable fact that the passage to *operators similar to a normal one*, $N \mapsto T = V^{-1}NV$, transforms the picture in such a way that, to keep the criteria, we need simply to replace all equalities by norm inequalities. Precisely, the following theorems hold true.

(1′) An operator T on a Hilbert space H is similar to a normal one if and only if there exists a *bounded* projection-valued measure $\mathcal{E}(\,\cdot\,)$ such that $Tx = \int_\sigma z\, d\mathcal{E}(z)x$ (the Riemann convergent integral for every $x \in H$).

(2′) An operator T on a Hilbert space H is similar to a normal one if and only if there exists a *bounded* functional calculus $f \mapsto f(T)$ defined on the algebra $C(\sigma)$ of a suitable compact set $\sigma \subset \mathbb{C}$, that is

$$\|f(T)\| \leq C\|f\|_{C(\sigma)} \quad \text{for } f \in C(\sigma). \tag{A.3.1}$$

Nevertheless, an inconvenience of (A.3.1) as a sufficient condition for similarity to a normal operator is obvious. Indeed, what we have at hand when dealing with an operator T, are *rational* expressions in T, that is the resolvent

$$R(\lambda, T) = (\lambda I - T)^{-1} \quad \text{for } \lambda \in \mathbb{C} \setminus \sigma(T),$$

and hence $f(T)$ for f in the set $\mathrm{Rat}(\mathbb{C} \setminus \sigma(T))$ of rational functions having poles on $\mathbb{C} \setminus \sigma(T)$. The knowledge of other continous functions of T, required by (A.3.1), is still mostly implicit and the very definition of them is hiding in the existence of the functional calculus. For all that, for operators with a "thin" spectrum, rational functions are sufficient. For instance, if the set $\mathrm{Rat}(\mathbb{C} \setminus \sigma)$ is norm dense in the space $C(\sigma)$, then it suffices to require (A.3.1) for $f \in \mathrm{Rat}(\mathbb{C} \setminus \sigma)$ in order to guarantee similarity to a normal operator and the inclusion $\sigma(T) \subset \sigma$. However, for most concrete situations, for example those coming from perturbation theory, there are too many rational functions in $\mathrm{Rat}(\mathbb{C} \setminus \sigma)$ to consider even this restricted version of (A.3.1) as a practical test for similarity.

The idea of *rational tests* for similarity, presented in Chapter 9 following [Benamara and Nikolski 1997], is to reduce the number of test functions $\mathrm{Rat}(\mathbb{C} \setminus \sigma)$ to a reasonable part of it. A natural choice for such a reduction is to consider estimates (A.3.1) for $f \in \mathrm{Rat}_n(\mathbb{C} \setminus \sigma)$, the set of all rational functions of degree at most $n = 1, 2, \ldots$ and with poles in $\mathbb{C} \setminus \sigma$. The case $n = 1$, also called the *resolvent criterion*, is the most popular and consists of testing for (A.3.1) rational functions f of degree 1 only: thus $f = 1/(\lambda - z)$, with

$$\left\| \frac{1}{\lambda - z} \right\|_{C(\sigma)} = \frac{1}{\text{dist}(\lambda, \sigma)}.$$

In this case, the problem can be stated as follows: for which classes of operators is the *linear growth of the resolvent* (LGR), when approaching to the spectrum,

$$\|R(\lambda, T)\| \leq \frac{\text{const}}{\text{dist}(\lambda, \sigma(T))} \quad \text{for } \lambda \in \mathbb{C} \setminus \sigma(T), \tag{A.3.2}$$

sufficient for T to be similar to a normal operator? It is well known (A. Markus) that in general this is not the case even for operators with the real spectrum, $\sigma(T) \subset \mathbb{R}$; in [Benamara and Nikolski 1997] it is shown that no spectral restrictions, excepting only a finite spectrum $\sigma(T)$, together with the LGR, implies similarity to a normal operator. On the other hand, B. Sz.-Nagy and C. Foiaş [1967] proved that for *contractions* T with a unitary spectrum, $\sigma(T) \subset \mathbb{T}$, the LGR implies similarity to a normal, and hence to a unitary operator.

In Chapter 9, following [Benamara and Nikolski 1997], we explain why the resolvent criterion is still true for contractions T with $\sigma(T) \neq \overline{\mathbb{D}}$ and $\text{rank}(I - T^*T) < \infty$, $\text{rank}(I - TT^*) < \infty$, and fails for T with $\sigma(T) \neq \overline{\mathbb{D}}$ and $(I - T^*T) \in \bigcap_{p>1} \mathfrak{S}_p$, $(I - TT^*) \in \bigcap_{p>1} \mathfrak{S}_p$, where $\mathfrak{S}_p$ stands for the Schatten–von Neumann ideals of compact operators. The case of $p = 1$, trace class perturbations of unitary operators, is still open. Our technique is based on free interpolation results (Chapter 8) and on estimates of angles between invariant subspaces (Chapter 7).

Some higher rational tests, that is, estimates (A.3.1) for $f \in \text{Rat}_n(\mathbb{C} \setminus \sigma)$, $n > 1$, are also considered (and also following [Nikolski $\geq$ 1998]). For this case, a kind of Bernstein inequality is proved for operators with unitary or real spectra: it turns out that (A.3.2) implies inequality (A.3.1) for all $n = 2, 3, \ldots$ with constants const $= c_n$ depending on n. So, one can say that the $\text{Rat}_n(\mathbb{C} \setminus \mathbb{T})$ and $\text{Rat}_n(\mathbb{C} \setminus \mathbb{R})$ tests provide us with no new cases of similarity to a normal operator with respect to the simplest resolvent test. Of course, if the constants are uniformly bounded, $\sup_n c_n < \infty$, the operator is similar to a unitary (for $\sigma(T) \subset \mathbb{T}$) or a selfadjoint operator (for $\sigma(T) \subset \mathbb{R}$) by Wermer's theorem.

A.4. Stability of the continuous spectrum.

The problem of stability of the continuous spectrum goes back to H. Weyl (1909) and J. von Neumann (1935) for the selfadjoint case, and to A. Weinstein (1937), N. Aronszajn and A. Weinstein (the forties) and S. Kuroda (the fifties and sixties) for more general settings; see [Akhiezer and Glazman 1966; Kato 1967] for references and initial results. The *continuous spectrum* $\sigma_c(T)$ of an operator T is the spectrum with isolated eigenvalues of finite algebraic multiplicity removed, that is, eigenvalues whose Riesz projection is of finite rank. It is well known that if the resolvent set

$\mathbb{C} \setminus \sigma(T)$ is connected, the continuous spectrum is *stable* with respect to compact perturbations:

$$\sigma_c(T + K) = \sigma_c(T)$$

for all $K \in \mathfrak{S}_\infty$. If $\mathbb{C} \setminus \sigma(T)$ contains nontrivial bounded components, it may happen that $\sigma_c(T+K) \neq \sigma_c(T)$ even for finite rank operators K. In this case, at least one bounded component of $\mathbb{C} \setminus \sigma(T)$ is filled in by a layer of eigenvalues of perturbed operator $T + K$ (the so-called *singular case* of Fredholm perturbation theory, see [Kato 1967]). The problem is to distinguish operators T allowing the singular case from those whose continuous spectrum is stable with respect to perturbations from a given class of operators.

One of the classic examples of instability [Kato 1967] is a rank-one perturbation of the standard shift operator $\mathcal{S} : L^2(\mathbb{T}) \to L^2(\mathbb{T})$:

$$\sigma_c(\mathcal{S} + K) = \overline{\mathbb{D}},$$

where $Kf = -(f, \bar{z})1$ for $f \in L^2(\mathbb{T})$ (this contrasts with $\sigma_c(\mathcal{S}) = \mathbb{T}$). Clearly, the same construction works for any unitary operator U (instead of $\mathcal{S}$) whose spectral measure E_U dominates Lebesgue measure m of the unit circle $\mathbb{T}$ ($E_U \succeq m$). The inverse is also true, but is not so obvious. More precisely, it is proved in [Nikolski 1969] that these statements are equivalent for a given unitary operator U:

(1) $\sigma_c(U + K) = \sigma_c(U)$ for all rank-one operators K;
(2) $\sigma_c(U + K) = \sigma_c(U)$ for all $K \in \mathfrak{S}_1$;
(3) E_U does not dominate m, that is there exists a spectral gap $\sigma \subset \mathbb{T}$, $\sigma \in \mathfrak{B}$, of positive Lebesgue measure ($E_U(\sigma) = 0$ but $m\sigma > 0$);
(4) U belongs to the weak closed operator algebra generated by U^{-1};
(5) $\operatorname{Lat} U^{-1} \subset \operatorname{Lat} U$, where $\operatorname{Lat} U$ stands for the lattice of invariant subspaces of U.

The stability problem for bigger perturbations leads to a rougher picture:

$$\sigma_c(U + K) = \sigma_c(U) \quad \text{for all } K \in \mathfrak{S}_p \text{ with } p > 1$$

if and only if $\sigma(U) \neq \mathbb{T}$.

In the framework of perturbation theory, a natural problem was to find a stability criterion for trace class perturbations T of unitary operators, $T = U + C$, $C \in \mathfrak{S}_1$. The solution was found in [Makarov and Vasyunin 1981] by using the function model. The first problem here was to insert a noncontraction $T = U + C$ into model theory. This is done by considering the function model for an auxiliary "nearest to T" contraction T_0 and realizing T on this model. Another problem was to find a proper analogue of condition (3) above to express a spectral gap of the operator under question. The language of the function model helps once more, and the true expression for a completely nonunitary operator turns out to be the following:

(3′) the characteristic function Θ_T is J_T-unitary on a set $\sigma \subset \mathbb{T}$ of positive Lebesgue measure, where

$$\Theta_T(z) \overset{\text{def}}{=} (-TJ_T + zD_{T^*}(I - zT^*)^{-1}D_T)| \operatorname{Range} D_T,$$

and $D_A \overset{\text{def}}{=} |I - A^*A|^{1/2}$, $J_A \overset{\text{def}}{=} \operatorname{sign}(I - A^*A)$ (the square root $|\cdot|^{1/2}$ and the sign function $\operatorname{sign}(t) = t/|t|$ applied to the selfadjoint operator $I - A^*A$).

With these refinements, it will be proved in Chapter 10 that assertions (1)–(5) are still equivalent for an invertible operator $T = U + C$, $C \in \mathfrak{S}_1$ instead of U, if we add to condition 3) for the unitary part of T condition (3′) for its completely nonunitary part.

A.5. Scattering and other subjects. The quick development and the very rise of mathematical scattering theory in the late fifties was motivated by influences of both physical scattering theory and several mathematical theories. Among the latter, a leading role is playing by both of the subjects considered above, namely, by stability problems of the continuous spectrum and similarity problems. Of course, as we are speaking here about predecessors of scattering theory, we have in mind the classical framework of selfadjoint and unitary operators, in which domain the foundation of both theories was laid by H. Weyl (1909), J. von Neumann and K. Friedrichs (in the thirties through the fifties). Without entering into technical details, we trace here an approach to scattering problems that is adjustable to the use of the function model.

The main goal of scattering theory is to compare the asymptotic behavior for $t \to \pm\infty$ of two continuous groups on a Hilbert space: the "perturbed" group $S(t)$ and the "nonperturbed" group $S_0(t)$ ("free evolution"). Both continuous, $t \in \mathbb{R}$, and discrete, $t \in \mathbb{Z}$, times are considered. Being motivated by quantum physics scattering phenomena, mathematical scattering theory was started with the so-called *nonstationary approach* which consists of the following. Let S, S_0 be unitary groups on a Hilbert space H whose selfadjoint generators, A and A_0 respectively, differ by a "small", say of finite rank, operator, $A = A_0 + \Delta$ where the spectral measures of A, A_0 are absolutely continuous with respect to Lebesgue measure (E_A, $E_{A_0} \preceq m$). It turns out that under these hypotheses the asymptotic behavior of the groups are the same: for every $x \in H$ there exist unique vectors $x_0^{\pm} \in H$ such that $\lim_{t\to\pm\infty} \|S(t)x - S_0(t)x_0^{\pm}\| = 0$. The operators

$$x = \lim_{t\to\pm\infty} S(t)^{-1}S_0(t)x_0^{\pm} \overset{\text{def}}{=} W_{\pm}(A, A_0)x_0^{\pm}$$

are called the *wave operators* of the pair A, A_0 and $\mathbf{S} = W_+^*W_-$ is called the *scattering operator*. Under the above hypotheses, all three are unitary operators. The operator $\mathbf{S}$ links the "free asymptotics" for $t \to -\infty$ and $t \to \infty$, that is, $\mathbf{S}x_0^- = x_0^+$, and it commutes with A_0, whereas $W_{\pm}$ establish the similarity—in fact, the unitary equivalence—of A and A_0,

$$AW_{\pm} = W_{\pm}A_0,$$

and the semigroups themselves (the so-called *intertwining properties*).

In general, for instance for trace class selfadjoint perturbations Δ, the wave limits $W_\pm$ exist only on the absolutely continuous subspace of A_0, $H_0^{ac} = E_{A_0}^{ac}(H)$, and map it onto $H^{ac} = E_A^{ac}(H)$, establishing a unitary equivalence between the absolutely continuous parts $A|H^{ac}$, $A_0|H_0^{ac}$ and, *a fortiori*, the stability of the absolutely continuous spectrum with respect to nuclear perturbations. For the case of discrete time, all is similar, except we have no generators and deal directly with the unitary operators $U^n = S(1)^n$ and $U_0^n = S_0(1)^n$, for $n \in \mathbb{Z}$, and their absolutely continuous parts.

There exist several realizations of the second, *stationary approach* to scattering theory, that is, to find the wave operators $W_\pm$ avoiding the wave limits of the initial definition. The idea, coming back to K. Friedrichs, is to define these operators as solutions of some operator equations and then prove the above intertwining properties and the existence of the wave limits. The initial Friedrichs observation is that, for operators with absolutely continuous spectra,

$$W_\pm = I + \Gamma^\pm(\Delta W_\pm)$$

where $\Gamma^\pm(X) = \int_0^{\pm\infty} S_0(-t)X S_0(t)\,dt$. For methods of solving these Friedrichs (Γ)-equations we refer to [Kato 1967; Dunford and Schwartz 1971]. For other stationary approaches, making use of some resolvent equations (in a sense, the Fourier–Laplace transform of the Friedrichs ones) instead of the (Γ)-equations, see [Reed and Simon 1979; Yafaev 1992].

Function models are ideally adapted to stationary methods of scattering theory, because, as before, using the local function structure, one can guess explicit formulas conjecturally solving the needed operator equations, and then prove that they really provide the solutions. Another advantage is that the model approach is also well-adapted for an important passage from scattering for unitary (semi)groups to scattering for contractive semigroups. The pioneering role in applying function models to scattering problems was played by L. de Branges [de Branges 1962; de Branges and Rovnyak 1966; de Branges and Shulman 1968]. In Part II of this paper, we analize the de Branges and other model approaches to scattering theory, deriving from them the main facts of the theory. Here we simply mention some other sources: [Lax and Phillips 1967; Adamyan and Arov 1966; Naboko 1980; 1987].

Among other subjects that we plan to include in Part II of the paper, we can mention some properties of the weak star closed algebra alg M_Θ generated by the model operator M_Θ. For instance, we consider the problem of weak generators of this algebra, closely related to properties of the lattice Lat M_Θ of invariant subspaces of M_Θ; the problem, solved in [Kapustin 1992], of the reflexivity of M_Θ; and the problem of the "invisible spectrum" for alg M_Θ.

References

[Adamyan and Arov 1966]   V. M. Adamyan and D. Z. Arov, "Unitary couplings of semi-unitary operators", *Mat. Issled.* **1**:2 (1966), 3–64. In Russian.

[Akhiezer and Glazman 1966]   N. I. Akhiezer and I. M. Glazman, Теория линейных операторов в гильбертовом пространстве, 2nd ed., Nauka, Moscow, 1966. Translated as *Theory of linear operators in Hilbert space*, Pitman, Boston and London, 1981.

[Alpay ≥ 1998]   D. Alpay, "Algorithme de Schur, espaces à noyau reproduisant et théorie des systèmes". To appear.

[Ando 1963]   T. Andô, "On a pair of commutative contractions", *Acta Sci. Math. (Szeged)* **24** (1963), 88–90.

[Atkinson 1964]   F. V. Atkinson, *Discrete and continuous boundary problems*, Mathematics in Science and Engineering **8**, Academic Press, New York, 1964.

[Bakonyi and Constantinescu 1992]   M. Bakonyi and T. Constantinescu, *Schur's algorithm and several applications*, Pitman Research Notes in Mathematics Series **261**, Longman, Harlow, and Wiley, New York, 1992.

[Ball et al. 1990]   J. A. Ball, I. Gohberg, and L. Rodman, *Interpolation of rational matrix functions*, Operator Theory: Advances and Applications **45**, Birkhäuser, Basel, 1990.

[Benamara and Nikolski 1997]   N. Benamara and N. Nikolski, "Resolvent test for similarity to a normal operator", prépublication, Univ. Bordeaux I, 1997. To appear in *Proc. London Math. Soc.*

[Beurling 1949]   A. Beurling, "On two problems concerning linear transformations in Hilbert space", *Acta Math.* **81** (1949), 239–255.

[Carleson 1962]   L. Carleson, "Interpolations by bounded analytic functions and the corona problem", *Ann. of Math.* (2) **76**:3 (1962), 547–559.

[Colojoara and Foiaş 1968]   I. Colojoara and C. Foiaş, *The theory of generelized spectral operators*, Gordon and Breach, New York, 1968.

[Cotlar and Sadosky 1984/85]   M. Cotlar and C. Sadosky, "Generalized Toeplitz kernels, stationarity and harmonizability", *J. Analyse Math.* **44** (1984/85), 117–133.

[Cotlar and Sadosky 1986a]   M. Cotlar and C. Sadosky, "Lifting properties, Nehari theorem and Paley lacunary inequality", *Rev. Matem. Iberoamericana* **2** (1986), 55–71.

[Cotlar and Sadosky 1986b]   M. Cotlar and C. Sadosky, "A lifting theorem for subordinated invariant kernels", *J. Func. Anal.* **67** (1986), 345–359.

[Cotlar and Sadosky 1988]   M. Cotlar and C. Sadosky, "Toeplitz liftings of Hankel forms", pp. 22–43 in *Function spaces and applications* (Lund, 1986), Lecture Notes in Math. **1302**, Springer, Berlin and New York, 1988.

[Cotlar and Sadosky 1992]   M. Cotlar and C. Sadosky, "Weakly positive matrix measures, generalized Toeplitz forms, and their applications to Hankel and Hilbert transform operators", pp. 93–120 in *Continuous and discrete Fourier transforms, extension problems and Wiener–Hopf equations*, edited by I. Gohberg, Oper. Theory Adv. Appl. **58**, Birkhäuser, Basel, 1992.

[Davidson 1988]   K. R. Davidson, *Nest algebras*, Pitman Longman Sci. Tech., 1988.

[de Branges 1962]   L. de Branges, "Perturbations of self-adjoint operators", *Amer. J. Math.* **84**:4 (1962), 543–560.

[de Branges and Rovnyak 1966] L. de Branges and J. Rovnyak, "Canonical models in quantum scattering theory", pp. 295–392 in *Perturbation theory and its application in quantum mechanics* (Madison, 1965), edited by C. H. Wilcox, Wiley, New York, 1966.

[de Branges and Shulman 1968] L. de Branges and L. Shulman, "Perturbations of unitary transformations", *J. Math. Anal. Appl.* **23** (1968), 294–326.

[Dowson 1978] H. R. Dowson, *Spectral theory of linear operators*, Academic Press, 1978.

[Dunford and Schwartz 1971] N. Dunford and J. T. Schwartz, *Linear operators, III: Spectral operators*, Pure and Applied Mathematics **7**, Interscience, New York, 1971. Reprinted by Wiley, New York, 1988.

[Foiaş and Frazho 1990] C. Foiaş and A. E. Frazho, *The commutant lifting approach to interpolation problems*, Oper. Theory Adv. Appl. **44**, Birkhäuser, Basel, 1990.

[Fuhrmann 1981] P. A. Fuhrmann, *Linear systems and operators in Hilbert space*, McGraw-Hill, New York, 1981.

[Garnett 1981] J. B. Garnett, *Bounded analytic functions*, Pure and Applied Mathematics **96**, Academic Press, New York and London, 1981.

[Gel'fand and Vilenkin 1961] I. M. Gel'fand and N. Y. Vilenkin, Обобщенные функции, 4: Некоторые применения гармонического анализа. Оснащенные гильбертовы пространства, Gosudarstv. Izdat. Fiz.-Mat. Lit., Moscow, 1961. Translated as *Generalized functions, 4: Applications of harmonic analysis*, Academic Press, New York and London, 1964.

[Hartmann 1996] A. Hartmann, "Une approche de l'interpolation libre généralisée par la théorie des opérateurs et caractérisation des traces $H^p|_\Lambda$", *J. Operator Theory* **35**:2 (1996), 281–316.

[Julia 1944a] G. Julia, "Sur les projections des systèmes orthonormaux de l'espace hilbertien", *C. R. Acad. Sci. Paris* **218** (1944), 892–895.

[Julia 1944b] G. Julia, "Les projections des systèmes orthonormaux de l'espace hilbertien et les opérateurs bornés", *C. R. Acad. Sci. Paris* **219** (1944), 8–11.

[Julia 1944c] G. Julia, "Sur la représentation analytique des opérateurs bornés ou fermés de l'espace hilbertien", *C. R. Acad. Sci. Paris* **219** (1944), 225–227.

[Kapustin 1992] V. V. Kapustin, "Reflexivity of operators: general methods and a criterion for almost isometric contractions", *Algebra i Analiz* **4**:2 (1992), 141–160. In Russian. Translation in *St. Petersburg Math. J.*, **4**:2 (1993), 319–335.

[Kato 1967] T. Kato, *Perturbation theory for linear operators*, Springer, New York, 1967.

[Kriete 1972] T. L. Kriete, "Complete non-selfadjointness of almost selfadjoint operators", *Pacific J. Math.* **42** (1972), 413–437.

[Lax and Phillips 1967] P. D. Lax and R. S. Phillips, *Scattering theory*, Pure and Applied Mathematics **26**, Academic Press, New York and London, 1967.

[Leggett 1973] R. Leggett, "On the invariant subspace structure of compact, dissipative operators", *Indiana Univ. Math. J.* **22**:10 (1973), 919–928.

[Livšic 1946] M. S. Livšic, "On a class of linear operators on Hilbert space", *Mat. Sb.* **19** (1946), 239–260. In Russian.

[Makarov and Vasyunin 1981]   N. G. Makarov and V. I. Vasyunin, "A model for noncontractions and stability of the continuous spectrum", pp. 365–412 in *Complex analysis and spectral theory seminar* (Leningrad, 1979/80), edited by V. P. Havin [Khavin] and N. K. Nikolskii, Lecture Notes in Math. **864**, Springer, Berlin and New York, 1981.

[Naboko 1980]   S. N. Naboko, "Function model of perturbation theory and its applications to scattering theory", *Trudy MIAN* **147** (1980), 86–114. In Russian; translation in *Proc. Steklov Inst. Math.*, **147**:2 (1981).

[Naboko 1987]   S. N. Naboko, "Conditions for the existence of wave operators in the nonselfadjoint case", pp. 132–155 in Распространение волн. Теория рассеяния [*Wave propagation. Scattering theory*], edited by M. S. Birman, Проблемы математической физики [Problems in Mathematical Physics] **12**, Leningrad. Univ., Leningrad, 1987. In Russian.

[Naimark 1943]   M. A. Naimark, "On a representation of additive operator functions of sets", *Dokl. Akad. Nauk SSSR* **41** (1943), 373–375. In Russian.

[Nikolski 1969]   N. Nikolski, "The spectrum perturbations of unitary operators", *Mat. Zametki* **5** (1969), 341–349. In Russian; translation in *Math. Notes* **5** (1969), 207–211.

[Nikolski 1978a]   N. Nikolski, "Bases of invariant subspaces and operator interpolation", *Trudy Mat. Inst. Steklov* **130** (1978), 50–123. In Russian; translation in *Proc. Steklov Inst. Math.* **130** (1979).

[Nikolski 1978b]   N. Nikolski, "What spectral theory and complex analysis can do for each other", pp. 341–345 in *Proc. Intern. Congr. Math.* (Helsinki, 1978), Helsinki, 1978.

[Nikolski 1986]   N. K. Nikol'skiĭ, *Treatise on the shift operator*, Grundlehren der mathematischen Wissenschaften **273**, Springer, Berlin, 1986. With an appendix by S. V. Hruščev and V. V. Peller. Translation of Лекции об операторе сдвига, Nauka, Moscow, 1980.

[Nikolski 1987]   N. Nikolski, "Interpolation libre dans l'espace de Hardy", *C. R. Acad. Sci. Paris Sér. I* **304**:15 (1987), 451–454.

[Nikolski 1994]   N. Nikolski, "Modèles fonctionnels des opérateurs linéaires et interpolation libre", Notes de cours de D.E.A. 1993/94, École Doctorale, Univ. Bordeaux I, 1994.

[Nikolski ≥ 1998]   N. Nikolski, "Similarity to a normal operator and rational calculus". To appear.

[Nikolski and Vasyunin 1989]   N. K. Nikolskiĭ and V. I. Vasyunin, "A unified approach to function models, and the transcription problem", pp. 405–434 in *The Gohberg anniversary collection* (Calgary, AB, 1988), vol. 2, edited by H. Dym et al., Oper. Theory Adv. Appl. **41**, Birkhäuser, Basel, 1989.

[Nikolski and Khrushchev 1987]   N. Nikolski and S. V. Khrushchev, "A function model and some problems in the spectral function theory", *Trudy Math. Inst. Steklov* **176** (1987), 97–210. In Russian; translation in *Proc. Steklov Inst. Math.* **176**:3 (1988), 111–214.

[Nikolski and Pavlov 1968]   N. Nikolski and B. S. Pavlov, "Eigenvector expansions of nonunitary operators and the characteristic function", pp. 150–203 in Краевые задачи математической физики и смежные вопросы теорий функций, edited by O. A. Ladyzhen skaya, Zap. Nauchn. Semin. LOMI **11**, Nauka, Leningrad, 1968.

In Russian; translation on pp. 54–72 in *Boundary value problems of mathematical physics and related aspects of function theory, Part III*, Consultants Bureau, New York and London, 1970.

[Nikolski and Pavlov 1970]   N. Nikolski and B. S. Pavlov, "Eigenvector bases of completely nonunitary contractions and the characteristic function", *Izv. Akad. Nauk SSSR, Ser. Mat.* **34**:1 (1970), 90–133. In Russian; translation in *Math. USSR Izvestiya* **4** (1970).

[Parrott 1970]   S. Parrott, "Unitary dilations for commuting contractions", *Pacific Math. J.* **34**:2 (1970), 481–490.

[Pavlov 1975]   B. S. Pavlov, "On conditions for separation of the spectral components of a dissipative operator", *Izv. Akad. Nauk SSSR, Ser. Mat.* **39** (1975), 123–148. In Russian; translation in *Math. USSR Izvestiya* **9** (1975).

[Pavlov 1976]   B. S. Pavlov, "Theory of dilations and spectral analysis of nonselfadjoint differential operators", pp. 3–69 in *Theory of operators in linear spaces: Proceedings of the 7th winter school* (Drogobych, 1974), Moscow, 1976. In Russian; translation in *Amer. Math. Soc. Transl.* (2) **115** (1980).

[Plessner 1939a]   A. I. Plessner, "Functions of the maximal operator", *Dokl. Akad. Nauk SSSR* **23** (1939), 327–330. In Russian.

[Plessner 1939b]   A. I. Plessner, "On semi-unitary operators", *Dokl. Akad. Nauk SSSR* **25** (1939), 708–710. In Russian.

[Pták and Vrbová 1988]   V. Pták and P. Vrbová, "Lifting intertwining relations", *Int. Eq. Operator Theory* **11** (1988), 128–147.

[Reed and Simon 1975]   M. Reed and B. Simon, *Methods of modern mathematical physics, II: Fourier analysis, self-adjointness*, Academic Press, New York, 1975.

[Reed and Simon 1979]   M. Reed and B. Simon, *Methods of modern mathematical physics, III: Scattering theory*, Academic Press, New York, 1979.

[Sarason 1965]   D. Sarason, "On spectral sets having connected complement", *Acta Sci. Math. (Szeged)* **26** (1965), 289–299.

[Sarason 1967]   D. Sarason, "Generalized interpolation in H^{∞}", *Trans. Amer. Math. Soc.* **127**:2 (1967), 179–203.

[Smirnov 1928a]   V. I. Smirnov, "Sur la théorie des polynomes orthogonaux à une variable complexe", *J. Leningrad Fiz.-Mat. Obsch.* **2**:1 (1928), 155–179.

[Smirnov 1928b]   V. I. Smirnov, "Sur les valeurs limites des fonctions régulières à l'interieur d'un circle", *J. Leningrad Fiz.-Mat. Obsch.* **2**:2 (1928), 22–37.

[Smirnov 1932]   V. I. Smirnov, "Sur les formules de Cauchy et de Green et quelques problèmes qui s'y r'attachent", *Izv. Akad. Nauk SSSR, ser. fiz.-mat.* **3** (1932), 338–372.

[Solomyak 1989]   B. M. Solomyak, "A functional model for dissipative operators. A coordinate-free approach", *Zap. Nauchn. Semin. LOMI* **178** (1989), 57–91. In Russian; translation in *J. Soviet Math.* **61**:2 (1992), 1981–2002.

[Sz.-Nagy 1953]   B. Sz.-Nagy, "Sur les contractions de l'espace de Hilbert", *Acta Sci. Math. (Szeged)* **15** (1953), 87–92.

[Sz.-Nagy and Foiaş 1967]   B. Sz.-Nagy and C. Foiaş, *Analyse harmonique des opérateurs de l'espace de Hilbert*, Masson, Paris, and Akadémiai Kiadó, Budapest, 1967. Translated as *Harmonic analysis of operators on Hilbert space*, North-Holland, Amsterdam, and Akadémiai Kiadó, Budapest, 1970.

[Sz.-Nagy and Foiaş 1968] B. Sz.-Nagy and C. Foiaş, "Dilatation des commutants d'opérateurs", *C. R. Acad. Sci. Paris Sér. A-B* **266** (1968), A493–A495.

[Sz.-Nagy and Foiaş 1973] B. Sz.-Nagy and C. Foiaş, "On the structure of intertwining operators", *Acta Sci. Math. (Szeged)* **35** (1973), 225–254.

[Teodorescu 1975] R. I. Teodorescu, "Sur les decompositions directes des contractions de l'espace de Hilbert", *J. Funct. Anal.* **18**:4 (1975), 414–428.

[Treil 1989] S. R. Treil, "Geometric methods in spectral theory of vector-valued functions: some recent results", pp. 209–280 in *Toeplitz operators and spectral function theory*, edited by N. Nikolski, Operator Theory: Adv. and Appl. **42**, Birkhäuser, Basel, 1989.

[Vasyunin 1977] V. I. Vasyunin, "Construction of the functional model of B. Sz.-Nagy and C. Foiaş", *Zap. Nauchn. Semin. LOMI* **73** (1977), 6–23, 229. In Russian; translation in *J. Soviet Math.* **34**:6 (1986), 2028–2033.

[Vasyunin 1978] V. I. Vasyunin, "Unconditionally convergent spectral decompositions and interpolation problems", *Trudy Mat. Inst. Steklov* **130** (1978), 5–49, 223. In Russian; translation in *Proc. Steklov Inst. Math.* **4** (1979), 1–53.

[Vasyunin 1994] V. I. Vasyunin, "The corona problem and the angles between invariant subspaces", *Algebra i Analiz* **6**:1 (1994), 5–109. In Russian; translation in *St. Petersburg Math. J.* **6**:1 (1995), 77–88.

[Vinogradov and Rukshin 1982] S. A. Vinogradov and S. E. Rukshin, "Free interpolation of germs of analytic functions in Hardy spaces", *Zap. Nauchn. Semin. LOMI* **107** (1982), 36–45. In Russian; translation in *J. Soviet Math.* **36**:3 (1987), 319–325.

[Wermer 1954] J. Wermer, "Commuting spectral operators on Hilbert space", *Pacific J. Math.* **4** (1954), 335–361.

[Yafaev 1992] D. R. Yafaev, *Mathematical scattering theory: general theory*, Translations of Mathematical Monographs **105**, Amer. Math. Soc., Providence, 1992.

NIKOLAI NIKOLSKI
UFR MATHÉMATIQUES ET INFORMATIQUE
UNIVERSITÉ DE BORDEAUX-I
351, COURS DE LA LIBÉRATION
33405 TALENCE CEDEX
FRANCE
nikolski@math.u-bordeaux.fr

VASILY VASYUNIN
STEKLOV MATH. INSTITUTE
ST. PETERSBURG DIVISION FONTANKA 27
191011, ST. PETERSBURG
RUSSIA
vasyunin@pdmi.ras.ru

Liftings of Kernels Shift-Invariant
in Scattering Systems

CORA SADOSKY

To Daniel J. Goldstein, without whom this story could not have unfolded.

ABSTRACT. The Generalized Bochner Theorem (GBT) provides both integral representations and extensions of forms and kernels invariant under the shift operator. Even in the simplest setting of trigonometric polynomials, it allows a unified approach encompassing the Nehari approximation theorem and the Helson–Szegő and Helson–Sarason prediction theorems. It also gives results on weighted Lebesgue spaces that had been out of reach of classical methods.

The GBT's lifting approach is valid in abstract algebraic and hilbertian scattering systems, with one or several evolution groups (not necessarily commuting), and integral representations of Toeplitz extensions of Hankel forms are obtained in many such systems. These integral representations lead to applications to harmonic analysis in product spaces, such as the polydisk, and in symplectic spaces. In a different direction, a noncommutative extension of the GBT is given for kernels defined in terms of completely positive maps.

Introduction

The study of the generalized Toeplitz kernels and forms started as an attempt to apply Kreĭn's moment theory methods to the Hilbert transform. In particular, a generalization of the classical Herglotz–Bochner theorem, the GBT, yields a characterization of the pairs of measures for which the Hilbert transform operator is continuous in the corresponding weighted L^2 spaces. Yet the GBT, unlike the Bochner theorem, provides not only integral representations of bounded forms, but invariant extensions of them without norm increase. The GBT, therefore, is

1991 *Mathematics Subject Classification.* 47B35, 42B20.

This paper is an expanded version of lectures delivered during the Program on Holomorphic Spaces held at MSRI in the 1995 Fall semester. The author was partially supported by NSF grants DMS-9205926, INT-9204043 and GER-9550373. Her visit to MSRI was supported by NSF grant DMS-9022140 to MSRI.

closely related with the far-reaching lifting theory of Sz.-Nagy and Foiaş: their lifting theorem for intertwining contractions can be obtained as a corollary of its abstract generalization to scattering systems (see Section 2 below). Moreover, immediate applications to harmonic analysis follow through appropriate integral representations of the extended forms.

This paper is a self-contained exposition of research, joint with Mischa Cotlar, extending over two decades. (The abbreviation C-S is used in references to our papers.) Shortly after initiating this program, we collaborated with Rodrigo Arocena, who later carried on this lifting approach in his significant work in a somewhat different direction. Developments parallel to ours occurred during the same period in the work of other schools, some of which is represented in this volume. In many cases the relation between these other developments and ours still remains obscure, and should be investigated further.

This exposition concentrates on the basic lifting results, presents sketches of their proofs, and outlines some applications. It traces the historical development of our approach, starting with the concrete example of kernels on the integers, followed by the abstraction of that example in the setting of scattering systems. For background material, the reader can consult [Nikol'skiĭ 1986].

Section 1 centers on the GBT, introducing it as a result on integral representation of positive generalized Toeplitz kernels, and then as an extension property for bounded Hankel forms. The Helson–Szegő theorem for the Hilbert transform and the Nehari theorem for Hankel operators are given as corollaries, as well as our results on boundedness for those operators acting in two different weighted spaces. It is observed that the GBT holds for matrix- and operator-valued kernels, with essentially the same proof as presented here.

In Section 2, algebraic and hilbertian scattering structures appear as the natural settings for bounded Hankel forms. Thus the lifting theorems 2.1 and 2.2 emerge as a natural extension of the GBT. A constructive proof is sketched for them, based on the Wold–Kolmogorov decomposition. Satisfactory integral representations of the Hankel forms are obtained in the special cases of Adamyan–Arov and Lax–Phillips scattering systems. Section 2 also includes a lifting theorem for forms defined in general semi-invariant subspaces, such as the internal state space of a hilbertian scattering system, under a condition of "essential invariance." The Sarason representation theorem for contractions commuting with a compression of the shift, and the Sz.-Nagy–Foiaş lifting theorem for intertwining contractions are given as corollaries. Section 2 ends with a conditional lifting theorem from which follows the abstract Adamyan–Arov–Kreĭn (AAK) theorem for singular numbers of Hankel operators, and some of its applications.

In Section 3 the Lifting Theorem is extended to bounded Hankel forms in pairs of scattering systems with several evolution groups (not necessarily commuting). This result is one of the most significant features of our program. Although a physical interpretation of this setting is not yet clear, it allows many applications to function spaces in several variables. In particular, it yields a noncommutative Nehari theorem for forms acting in the space of Hilbert–Schmidt operators.

Section 4 shows another sense in which noncommutativity can usefully be introduced into our theory: the GBT can be extended from positive definite numerical-valued kernels to completely positive kernels whose values are operator-valued sesquilinear forms on a C^*-algebra. A Nehari theorem for sequences defined in a C^*-algebra is given as an application. The results of this section are being published here for the first time.

This whole paper is focused on liftings in discrete scattering systems, where the evolution group is a unitary representation of $\mathbb{Z}$. Alternatively, continuous scattering systems can be considered, where the evolution group is a unitary representation of $\mathbb{R}$. The lifting theorems of Section 2, as well as those of Section 3 for several evolution groups, also hold in the continuous case [C-S 1988; 1990a; 1994b]. The continuous version of the theory is invoked in the present paper only at the end of Section 3, when we consider operators in the symplectic plane.

The object of this paper is to show that invariant (Hankel) forms, bounded with respect to quadratic invariant norms, have invariant (Toeplitz) liftings. Two significant extensions of the theory — left out of this exposition for the sake of brevity — treat the cases when those norms are invariant but not quadratic (e.g., L^p norms for $p \neq 2$), or quadratic but not invariant (e.g., Sobolev norms). For the first case, see [C-S 1989a], where the pertinent previous papers are summarized, and [C-S 1990b], where the Hilbert transform in weighted $L^p(\mathbb{T}^2)$ is studied. For the second, see [C-S 1991], where the problem is related to unitary extensions in Kreĭn spaces and to scattering systems with evolution operators unitary with respect to an indefinite metric.

Other significant extensions not presented here are the (local) nonlinear theorems of [C-S 1989b], and the study of stationary, harmonizable and generalized stationary processes in scattering systems [C-S 1984/85; 1988; 1989b].

1. The General Bochner Theorem and Some of Its Applications

Positive definite functions admit integral representation as Fourier transforms of positive measures. This classical result, fertile in applications, is the starting point of our exposition.

In the simplest case, a *positive definite* sequence, that is, a sequence $s : \mathbb{Z} \to \mathbb{C}$ satisfying

$$\sum_{m,n} s(m-n)\, \lambda(m)\, \overline{\lambda(n)} \geq 0 \quad \text{for all } \lambda : \mathbb{Z} \to \mathbb{C} \text{ finitely supported,} \qquad (1\text{--}1)$$

is characterized as the Fourier coefficient sequence of a finite positive measure μ defined on $\mathbb{T}$:

$$s(n) = \hat{\mu}(n) := \int_{\mathbb{T}} e^{-int} \, d\mu \quad \text{for all } n \in \mathbb{Z}. \tag{1–2}$$

Note that we do not require strict inequality in (1–1).

Since (1–1) stands for the positive definiteness of the kernel $K \colon \mathbb{Z} \times \mathbb{Z} \to \mathbb{C}$, defined by $K(m,n) := s(m-n)$, and thus satisfying the *Toeplitz condition*

$$K(m+1, \, n+1) = K(m,n) \quad \text{for all } n \in \mathbb{Z}, \tag{1–3}$$

the result can be stated as

THEOREM 1.1 (HERGLOTZ–BOCHNER). *A kernel $K : \mathbb{Z} \times \mathbb{Z} \to \mathbb{C}$ is positive definite and Toeplitz if and only if there exists a uniquely determined finite measure $\mu \geq 0$ on $\mathbb{T}$ such that*

$$K(m,n) = \hat{\mu}(m-n) \quad \text{for all } m, n \in \mathbb{Z}. \tag{1–4}$$

In this section, this integral representation of positive kernels is extended to a class that includes the Toeplitz kernels and more. In this process a lifting property appears, having as corollaries classical theorems such as those of Helson–Szegő and Nehari — seemingly unrelated to the Herglotz–Bochner theorem — as well as new results.

Examples of numerical positive definite Toeplitz kernels are the autocorrelation kernels of stationary discrete (Hilbert space valued) stochastic processes X : $\mathbb{Z} \to H$, for H a Hilbert space, with autocorrelation $K(m,n) = \langle X(m), X(n) \rangle$. Not only is this kernel given by the Fourier coefficients of a positive measure μ (Herglotz–Bochner), but the process itself can be identified with the Fourier sequence of an orthogonally scattered bounded H-valued measure ν (Bochner–Khinchine):

$$X(n) = \hat{\nu}(n) \quad \text{for all } n \in \mathbb{Z}. \tag{1–5}$$

Among the simplest nonstationary processes (i.e., those whose kernels are not Toeplitz) are the harmonizable and the generalized stationary processes. Harmonizable processes admit a weakened version of representation (1–5) in terms of bounded (but not necessarily orthogonally scattered) ν. Alternatively, their kernels can be represented by an integration formally similar to (1–4), but involving not measures on $\mathbb{T}$ but bimeasures on $\mathbb{T} \times \mathbb{T}$. On the other hand, generalized stationary processes (those stationary except for one point) admit both a representation (1–4) in terms of positive numerical matrix-valued vector measures and a representation (1–5) in terms of pairs of bounded (mutually orthogonally scattered) vector-valued vector measures. This follows from Theorem 1.2 below.

DEFINITION. A kernel $K : \mathbb{Z} \times \mathbb{Z} \to \mathbb{C}$ is a *generalized Toeplitz kernel* (GTK) if

$$K(m+1, \, n+1) = K(m,n) \quad \text{except possibly when } m = -1 \text{ or } n = -1. \tag{1–6}$$

The autocorrelation kernel of a generalized stationary process is a GTK. The analog of the Herglotz–Bochner theorem for these kernels is the following result, whose proof will be sketched later in the section.

THEOREM 1.2 (THE GENERALIZED BOCHNER THEOREM [C-S 1979]). *A kernel* $K : \mathbb{Z} \times \mathbb{Z} \to \mathbb{C}$ *is positive definite and a GTK if and only if there exists a positive* 2×2 *matrix* $\mu = (\mu_{ij})$ *of (complex) measures defined on* $\mathbb{T}$ *such that*

$$K(m,n) = \hat{\mu}_{ij}(m - n) \quad \text{for } (m,n) \in \mathbb{Z}_i \times \mathbb{Z}_j, \quad i,j = 1,2, \qquad (1\text{--}7)$$

where $\mathbb{Z}_1 = \{k \in \mathbb{Z} : k \geq 0\}$ *and* $\mathbb{Z}_2 = \{k \in \mathbb{Z} : k < 0\}$.

For $\mu = (\mu_{ij})$ a 2×2 matrix of complex measures on $\mathbb{T}$, saying that μ is positive (denoted $\mu \geq 0$) is equivalent to saying that

$$\mu_{11} \geq 0, \quad \mu_{22} \geq 0, \quad \mu_{21} = \overline{\mu_{12}}, \quad \text{and}$$
$$|\mu_{12}(D)|^2 \leq \mu_{11}(D)\mu_{22}(D) \quad \text{for every Borel set } D \text{ in } \mathbb{T}. \qquad (1\text{--}8)$$

Since every Toeplitz kernel is obviously a GTK, the GBT (Theorem 1.2) includes the Herglotz–Bochner representation of the kernel, but here μ is *not* unique. This fact is at the heart of what makes the GBT not only a result on integral representation but on extension of forms. To show this it is helpful to rewrite the result as follows.

Let $\mathcal{P}$ be the set of trigonometric polynomials on $\mathbb{T}$. That is, $\mathcal{P}$ consists of finite sums of the form $f = \sum_n c_n e^{int}$. Given a hermitian kernel K, a sesquilinear form $B : \mathcal{P} \times \mathcal{P} \to \mathbb{C}$ can be defined by setting

$$B(e^{imt}, e^{int}) = K(m,n) \quad \text{for all } m,n \in \mathbb{Z},$$

and extending by linearity.

A form $B : \mathcal{P} \times \mathcal{P} \to \mathbb{C}$ is positive (that is, $B(f,f) \geq 0$ for all $f \in \mathcal{P}$) if and only if the corresponding kernel K is positive definite. If B is positive we write $B \geq 0$.

A kernel K is Toeplitz if and only if the corresponding form B is invariant under the shift operator $S : f \mapsto e^{it}f$, that is,

$$B(Sf, Sg) = B(f,g) \quad \text{for all } (f,g) \in \mathcal{P} \times \mathcal{P}. \qquad (1\text{--}9)$$

The forms B satisfying (1–9) are called *Toeplitz* or *S-invariant in* $\mathcal{P} \times \mathcal{P}$.

In this setting, the Herglotz–Bochner theorem translates to: *B is positive and S-invariant in* $\mathcal{P} \times \mathcal{P}$ *if and only if there exists* $\mu \geq 0$ *such that*

$$B(f,g) = \int f\bar{g} \, d\mu \quad \text{for all } f,g \in \mathcal{P}. \qquad (1\text{--}10)$$

Setting $\mathcal{P}_1 = \{f \in \mathcal{P} : f \text{ analytic}\}$ and $\mathcal{P}_2 = \{f \in \mathcal{P} : f \text{ antianalytic}\}$, we have $\mathcal{P} = \mathcal{P}_1 \dotplus \mathcal{P}_2$, and the domain of the form B splits into the four pieces

$\mathcal{P}_i \times \mathcal{P}_j$, for $i, j = 1, 2$. A weaker concept than S-invariance in $\mathcal{P} \times \mathcal{P}$ as in (1–9) is S-invariance in each $\mathcal{P}_i \times \mathcal{P}_j$; the form B has this property if

$$B(Sf, Sg) = B(f, g) \quad \text{for all } (f, g) \in (\mathcal{P}_1 \times \mathcal{P}_1) \cup (S^{-1}\mathcal{P}_2 \times S^{-1}\mathcal{P}_2) \cup (P_1 \times S^{-1}\mathcal{P}_2).$$

Then the GBT asserts that B *is positive in* $\mathcal{P} \times \mathcal{P}$ *and S-invariant in each* $\mathcal{P}_i \times \mathcal{P}_j$ *if and only if there exists* $\mu = (\mu_{ij}) \geq 0$ *such that*

$$B(f_1 + f_2, g_1 + g_2) = \sum_{i,j=1,2} \int f_i \bar{g}_j \, d\mu_{ij} \quad \text{for all } f_1, g_1 \in \mathcal{P}_1, \ f_2, g_2 \in \mathcal{P}_2. \quad (1\text{--}11)$$

SKETCH OF THE PROOF OF THE GBT (FOR EITHER FORMS OR KERNELS). $B \geq 0$ in $\mathcal{P} \times \mathcal{P}$ (or K positive definite) gives rise to a (possibly degenerate) scalar product in $\mathcal{P}$. Under the usual procedure, there is a Hilbert space H and a linear operator $J : \mathcal{P} \to H$ such that $J\mathcal{P}$ is dense in H and $B(f, g) = \langle Jf, Jg \rangle$ for $f, g \in \mathcal{P}$. Consider the operator V defined by $V(Jf) := J(Sf)$ for $f \in \mathcal{P}$. If B were S-invariant in $\mathcal{P} \times \mathcal{P}$ (or K Toeplitz), V would extend to a unitary operator in H. As B is S-invariant only in each $\mathcal{P}_i \times \mathcal{P}_j$ (K is a GTK), V extends to an isometry in H, with domain $J\mathcal{P}_1 + J(S^{-1}\mathcal{P}_2)$ and range $J(S\mathcal{P}_1) + J(\mathcal{P}_2)$. As is well known, such an isometry extends to a unitary operator U in a larger Hilbert space $\mathcal{H} \supset H$. The cyclic pair $\xi_1 = J(1)$, $\xi_2 = J(e^{-it})$ in $\mathcal{H}$ gives

$$B(e^{imt}, e^{int}) = K(m, n) = \langle U^m U^{(i-1)} \xi_i, U^n U^{(j-1)} \xi_j \rangle$$

if $m \in \mathbb{Z}_i, n \in \mathbb{Z}_j$, for $i, j = 1, 2$. Now define $\mu = (\mu_{ij})$ by

$$\mu_{ij}(D) = \langle E(D)\xi_i, \xi_j \rangle \quad \text{for any Borel set } D \text{ in } \mathbb{T},$$

where E is the spectral measure of U. This gives the representation (1–7), (1–11). □

This proof holds also for operator-valued kernels $K : \mathbb{Z} \times \mathbb{Z} \to \mathcal{L}(N)$, where N is a Hilbert space. Then $B : \mathcal{P}(N) \times \mathcal{P}(N) \to \mathbb{C}$ is the corresponding form, for $\mathcal{P}(N)$ the set of vector-valued trigonometric polynomials, $f = \sum_n \xi_n e^{int}$ a finite sum with $\xi_n \in N$, and $B(f, g) = \sum_{m,n} \langle K(m, n)\xi_m, \xi_n \rangle$.

Observe here that if B is S-invariant in $\mathcal{P} \times \mathcal{P}$ we have

$$B \geq 0 \iff \mu \geq 0,$$

whereas if B is S-invariant in each $\mathcal{P}_i \times \mathcal{P}_j$ we have

$$B \geq 0 \iff \sum_{i,j} \int f_i \bar{f}_j \, d\mu_{ij} \geq 0 \text{ only for } f_1 \in \mathcal{P}_1 \text{ and } f_2 \in \mathcal{P}_2, \quad (1\text{--}12)$$

which is far less than

$$\mu \geq 0 \iff \sum_{i,j} \int f_i \bar{f}_j \, d\mu_{ij} \geq 0 \text{ for all } f_1, f_2 \in \mathcal{P}, \quad (1\text{--}13)$$

To unveil what this discrepancy means, we look at the restrictions of $B \geq 0$ to $\mathcal{P}_i \times \mathcal{P}_j$. The restrictions $B_1 = B|(\mathcal{P}_1 \times \mathcal{P}_1)$ and $B_2 = B|(\mathcal{P}_2 \times \mathcal{P}_2)$ are also ≥ 0, and thus define (possibly degenerate) scalar products; in contrast, $B_0 = B|(\mathcal{P}_1 \times \mathcal{P}_2)$ is not positive, but it is bounded in the sense that

$$|B_0(f_1, f_2)| \leq B_1(f_1, f_1)^{1/2} B_2(f_2, f_2)^{1/2} \quad \text{for all } f_1 \in \mathcal{P}_1,\ f_2 \in \mathcal{P}_2. \qquad (1\text{-}14)$$

Conversely, if $B_0 : \mathcal{P}_1 \times \mathcal{P}_2 \to \mathbb{C}$, $B_1 : \mathcal{P}_1 \times \mathcal{P}_1 \to \mathbb{C}$, and $B_2 : \mathcal{P}_2 \times \mathcal{P}_2 \to \mathbb{C}$ satisfy (1–14), the form $B : \mathcal{P} \times \mathcal{P} \to \mathbb{C}$ coinciding with each of them in the respective domain is positive.

If $B \geq 0$ is S-invariant in $\mathcal{P}_i \times \mathcal{P}_j$, by the GBT, the positive forms B_1 and B_2 are given by the positive measures μ_{11} and μ_{22}, while

$$B_0(f_1, f_2) = \int f_1 \bar{f}_2 \, d\mu_{12} \quad \text{only for } f_1 \in \mathcal{P}_1,\ f_2 \in \mathcal{P}_2.$$

Furthermore, by (1–14) B_0 is bounded on $L^2(\mu_{11}) \times L^2(\mu_{22})$, with $\|B_0\| \leq 1$. But the complex measure μ_{12}, which is bounded by μ_{11} and μ_{22} in the sense of (1–8), defines a form B' in all of $\mathcal{P} \times \mathcal{P}$:

$$B'(f_1, f_2) := \int f_1 \bar{f}_2 \, d\mu_{12} \quad \text{for all } f_1, f_2 \in \mathcal{P}. \qquad (1\text{-}15)$$

Thus the GBT not only gives an integral representation of B, but extends B_0 to all of $\mathcal{P} \times \mathcal{P}$, without increasing its norm!

COROLLARY 1.3 (EXTENSION PROPERTY FOR BOUNDED S-INVARIANT FORMS IN $\mathcal{P}_1 \times \mathcal{P}_2$). *Given two positive measures μ_{11} and μ_{22} and a form $B_0 : \mathcal{P}_1 \times \mathcal{P}_2 \to \mathbb{C}$ satisfying*

$$B_0(Sf_1, f_2) = B_0(f_1, S^{-1}f_2) \qquad (1\text{-}16)$$

and

$$|B_0(f_1, f_2)| \leq \|f_1\|_{L^2(\mu_{11})} \|f_2\|_{L^2(\mu_{22})}, \qquad (1\text{-}17)$$

there exists an S-invariant form $B' : \mathcal{P} \times \mathcal{P} \to \mathbb{C}$ such that $B'|(\mathcal{P}_1 \times \mathcal{P}_2) = B_0$ and $\|B'\| = \|B_0\|$.

Furthermore, B' has the integral representation (1–15) in terms of a complex measure μ_{12} satisfying inequality (1–8) with respect to the given μ_{11} and μ_{22}.

The value of the GBT as an *extension* result is highlighted when the forms are already given by an integral representation through measures. Such is the case for the first application of the GBT [C-S 1979], still the most striking, since, as we shall show, it provides a direct link between the lifting theory of Sz.-Nagy and Foiaş and the continuity of the Hilbert transform in weighted spaces.

Let H be the *Hilbert transform operator*, defined in $\mathcal{P} = \mathcal{P}_1 \dotplus \mathcal{P}_2$ by

$$H(f_1 + f_2) = -if_1 + if_2 \quad \text{for } f_1 \in \mathcal{P}_1,\ f_2 \in \mathcal{P}_2. \qquad (1\text{-}18)$$

The problem now is to characterize the positive measures μ and ν on $\mathbb{T}$ for which the *weighted norm inequality*

$$\int |Hf|^2 \, d\mu \leq M^2 \int |f|^2 \, d\nu \qquad (1\text{-}19)$$

holds for all $f \in \mathcal{P}$. By (1–18), this can be rewritten as

$$\int |f_1 - f_2|^2 \, d\mu \leq M^2 \int |f_1 + f_2|^2 \, d\nu \quad \text{for } f_1 \in \mathcal{P}_1, \, f_2 \in \mathcal{P}_2,$$

or, equivalently, as

$$\sum_{i,j=1,2} \int f_i \bar{f}_j \, d\rho_{ij} \geq 0 \quad \text{for } f_1 \in \mathcal{P}_1, \, f_2 \in \mathcal{P}_2, \tag{1–20}$$

where

$$\rho_{11} = \rho_{22} = M^2 \nu - \mu, \quad \rho_{12} = \rho_{21} = M^2 \nu + \mu \tag{1–21}$$

are positive measures on $\mathbb{T}$.

Define $B : \mathcal{P} \times \mathcal{P} \to \mathbb{C}$ by

$$B(f,g) = B(f_1 + f_2, g_1 + g_2) = \sum_{i,j=1,2} \int f_i \bar{g}_j \, d\rho_{ij}.$$

Then B is S-invariant in $\mathcal{P}_i \times \mathcal{P}_j$ by definition. Furthermore, if the ρ_{ij}'s are related with μ, ν, M via (1–21), condition (1–19) is equivalent to $B \geq 0$. Then, by the GBT, there exist measures μ_{ij}, for $i, j = 1, 2$, satisfying (1–8) and such that $\hat{\rho}_{11}(n) = \hat{\mu}_{11}(n)$, $\hat{\rho}_{22}(n) = \hat{\mu}_{22}(n)$ for all $n \in \mathbb{Z}$, while $\hat{\rho}_{12}(n) = \hat{\mu}_{12}(n)$ only for $n < 0$. By the uniqueness of the Fourier transform and the theorem of F. and M. Riesz for analytic measures, this is equivalent to

$$\mu_{11} = \mu_{22} = M^2 \nu - \mu \text{ and } \mu_{12} = M^2 \nu + \mu - h \text{ with } h \in H^1(\mathbb{T}). \tag{1–22}$$

Therefore, (1–8) implies this result:

THEOREM 1.4 (HELSON–SZEGŐ THEOREM FOR TWO MEASURES [C-S 1979]). *Given $\mu \geq 0$ and $\nu \geq 0$, the Hilbert transform H is a bounded operator from $L^2(\nu)$ to $L^2(\mu)$ with norm M if and only if there exists $h \in H^1(\mathbb{T})$ such that*

$$\left| (M^2 \nu + \mu)(D) - \int_D h \, dt \right| \leq (M^2 \nu - \mu)(D) \quad \text{for all Borel sets } D \subset \mathbb{T}. \tag{1–23}$$

In particular, μ is an absolutely continuous measure: $d\mu = w \, dt$ for some $w \in L^1$ satisfying $w \geq 0$.

In the case $\mu = \nu$, we have:

COROLLARY 1.5 ([C-S 1979]; see also [C-S 1983]). *For weights $\omega \in L^1(\mathbb{T})$ satisfying $\omega \geq 0$, the following conditions are equivalent:*

(i) *The Hilbert transform H is a bounded operator in $L^2(\omega)$ with norm M.*

(ii) *There is a positive constant M and an $h \in H^1(\mathbb{T})$ for which*

$$\left| (M^2 + 1)w(t) - h(t) \right| \leq (M^2 - 1)\,\omega(t) \quad \text{a.e. in } \mathbb{T}.$$

(iii) *There is some $h \in H^1(\mathbb{T})$ such that, for appropriate constants c, C, ε,*

$$\text{Re}\, h(t) \geq c\omega(t), \quad |h(t)| \leq C\omega(t), \quad \left| \arg h(t) \right| \leq \pi/2 - \varepsilon, \quad \text{a.e. in } \mathbb{T}.$$

(iv) *There exist real-valued bounded functions* u, v *such that, for appropriate constants* C *and* ε, *we have* $\|u\|_\infty \leq C$, $\|v\|_\infty \leq \pi/2 - \varepsilon$, *and* $\omega = \exp(u + Hv)$.

(v) *There exists a real-valued function* w *such that, for appropriate constants* c, C, M', *we have* $c\omega(t) \leq w(t) \leq C\omega(t)$ *and*

$$|Hw(t)| \leq M'w(t) \quad \text{a.e. in } \mathbb{T}.$$

The constants in (i)–(v) *are related; in particular, the same* M *can be taken in* (i) *and* (ii).

REMARK. The equivalence of (iv) with (i) for *some* M is the Helson–Szegő theorem [Helson and Szegő 1960]. The sufficiency of condition (v) had already appeared in work of Gaposhkin in the late 1950's, as noted in [Helson and Szegő 1960]; its direct equivalence with (i) is the most significant sharpening of the Helson–Szegő theorem provided by the GBT. In fact, it opened the way for the characterization of weights for which the Hilbert transform is bounded in $L^p(\omega)$, for $1 < p < \infty$, equivalent to the A_p condition (see [C-S 1983], and also [C-S 1990b] for further results in product spaces).

REMARK. In the same context, another corollary of the GBT is the Helson–Sarason theorem on past and future [Helson and Sarason 1967], obtained by replacing $\mathcal{P}_1$ by $e^{ikt}\mathcal{P}_1$ for a positive integer k [Arocena, Cotlar, and Sadosky 1981]. The flexibility for making this change is an essential feature of this approach, as will be seen in Section 2.

On the other hand, the *integral representation* provided by the GBT for the extension of the S-invariant form $B_0 : \mathcal{P}_1 \times \mathcal{P}_2 \to \mathbb{C}$ bounded in $L^2(\mu_1) \times L^2(\mu_2)$ is essential for obtaining another significant corollary, the Nehari theorem for Hankel operators.

Let H^2 be the closure of $\mathcal{P}_1$ in the norm of L^2, and H^2_-, that of $\mathcal{P}_2$. A linear operator $\Gamma : H^2 \to H^2_-$ is *Hankel* if

$$\Gamma S = (1 - P)S\,\Gamma \quad \text{for } P : L^2 \to H^2 \text{ the orthoprojector.} \tag{1–24}$$

An example of a bounded Hankel operator is given by a bounded *symbol* φ, i.e.,

$$\Gamma_\varphi : f \mapsto (1 - P)\varphi f, \quad \text{where } \varphi \in L^\infty(\mathbb{T}). \tag{1–25}$$

In fact,

$$\left|\langle \Gamma_\varphi f, g\rangle\right| = \left|\int \varphi f \bar{g}\, dt\right| \leq \|\varphi\|_\infty \|f\|_2 \|g\|_2 \quad \text{for all } f \in \mathcal{P}_1,\ g \in \mathcal{P}_2,$$

and $\|\Gamma_\varphi\| \leq \|\varphi\|_\infty$. Hankel operators may have many symbols, and

$$\Gamma_\varphi = \Gamma_\psi \iff \varphi - \psi = h \quad \text{for } h \text{ an analytic function.} \tag{1–26}$$

Set $B_\varphi(f, g) = \langle \Gamma_\varphi f, g\rangle$. Then $B_\varphi : \mathcal{P}_1 \times \mathcal{P}_2 \to \mathbb{C}$ is S-invariant and bounded, with $\|B_\varphi\| = \|\Gamma_\varphi\|$. Conversely, if $B : \mathcal{P}_1 \times \mathcal{P}_2 \to \mathbb{C}$ is bounded in $L^2 \times L^2$ and

S-invariant, the associated bounded operator Γ such that $\langle \Gamma f, g \rangle = B(f,g)$ is Hankel. This justifies calling the S-invariant forms in $\mathcal{P}_1 \times \mathcal{P}_2$ *Hankel forms*.

The Nehari theorem shows that all bounded Hankel operators are as those in the example above. This important result, key to the solution of many interpolation and moment problems, and essential in modern H^∞-control theory, follows immediate from Corollary 1.3.

THEOREM 1.6 [Nehari 1957]. *Let $\Gamma : H^2 \to H^2_-$ be a Hankel operator. Then Γ is bounded if and only if there exists $\varphi \in L^\infty$ such that $\Gamma = \Gamma_\varphi$ and $\|\varphi\|_\infty = \|\Gamma\|$.*

PROOF. Let $B_0 : \mathcal{P}_1 \times \mathcal{P}_2 \to \mathbb{C}$ be the bounded Hankel form corresponding to Γ, and consider the absolutely continuous measures $d\mu_{11} = d\mu_{22} = \|\Gamma\| \, dt$. From Corollary 1.3, there exists a complex measure μ_{12} such that

$$\langle \Gamma f, g \rangle = \int f \bar{g} \, d\mu_{12} \quad \text{for all } f \in \mathcal{P}_1, g \in \mathcal{P}_2,$$

satisfying

$$|\mu_{12}(D)| \le \|\Gamma\| \, |D| \quad \text{for all Borel sets } D \subset \mathbb{T}.$$

Therefore, $d\mu_{12} = \varphi \, dt$, with $\|\varphi\|_\infty \le \|\Gamma\|$, while $\Gamma = \Gamma_\varphi$. $\qquad \square$

Thus, Corollary 1.3 is the *Nehari theorem for Hankel operators in weighted Hardy spaces*, $\Gamma : H^2(\mu_{11}) \to H^2_-(\mu_{22})$, where $H^2(\mu_{11})$ is the closure of $\mathcal{P}_1$ in the norm of $L^2(\mu_{11})$, while $H^2_-(\mu_{22})$ is the closure of $\mathcal{P}_2$ in the norm of $L^2(\mu_{22})$.

In [C-S 1993b] it was shown that when $L^2(\mu_{22}) = H^2(\mu_{22})$ every finite-rank Hankel operator is zero, while if $L^2(\mu_{22}) \ne H^2(\mu_{22})$ such operators admit a Kronecker-type representation of their symbols, in terms of the reproducing kernel of $H^2(\mu_{22})$.

In the case of absolutely continuous measures, we have the following characterization of the symbols of Hankel operators in two different weighted spaces.

COROLLARY 1.7 (NEHARI THEOREM IN TWO WEIGHTED SPACES [C-S 1993b]). *Let $\Gamma : H^2(w_1) \to H^2_-(w_2)$ be a Hankel operator, where w_1, w_2 are weights. The following conditions are equivalent:*

(a) $\|\Gamma\| = \|\Gamma\|_{H^2(w_1) \to H^2_-(w_2)} = 1$.

(b) *There exists $\varphi \in L^\infty$ such that $\|\varphi\|_\infty = 1$ and $\varphi \sqrt{w_1/w_2}$ is a symbol of Γ, i.e., $\Gamma f = P_2 \varphi \sqrt{w_1/w_2} f$ for all $f \in H^2(w_1)$, where $P_2 : L^2(w_2) \to H^2_-(w_2)$ is the orthoprojector.*

(c) *$\Gamma 1$ is the unique symbol in $H^2_-(w_2)$.*

(d) *If ψ is a symbol of Γ, then $\psi = \Gamma 1 - h/w_2$, for some analytic function h.*

To summarize, even in the most elementary setting of trigonometric polynomials in $\mathbb{T}$, the GBT

- unifies the solution of problems from different areas,
- sharpens known results, and

- solves *new* problems, with the *same proofs* with which it solves old ones, while those new problems were not approachable with the classical methods.

The extensions and integral representations given by the GBT are valid (with essentially the same proofs) for *vector-valued* functions (i.e., for *operator-valued* kernels), as noted above. Thus, results such as the Nehari–Page theorem for Hankel operators $\Gamma : H^2(\mathcal{H}) \to H^2_-(\mathcal{H})$, for $\mathcal{H}$ a Hibert space [Page 1970], are part of the theory.

2. Lifting Theorems in Scattering Systems and Integral Representations

Results analogous to those in Section 1 are valid in abstract settings, provided they have an underlying "scattering structure." The Lax–Phillips scattering theory considers systems defined in a Hilbert space H, with outgoing and incoming spaces being closed subspaces of H, and evolutions given by one-parameter groups of unitary operators in $\mathcal{L}(H)$. In classical mechanics, however, one sometimes deals with groups of linear isomorphisms in other vector spaces, thus it is natural to consider also algebraic scattering systems, defined as follows.

A (discrete) *algebraic scattering system* $[V; W^+, W^-; \sigma]$ consists of a vector space V, two subspaces W^+, W^- of V, and a linear isomorphism $\sigma : V \to V$ such that the discrete group $\{\sigma^n : n \in \mathbb{Z}\}$ satisfies the scattering property

$$\sigma^n W^+ \subset W^+ \quad \text{and} \quad \sigma^{-n} W^- \subset W^- \quad \text{for all } n \geq 0. \tag{2–1}$$

W^+ and W^- are called, respectively, the *outgoing* and the *incoming* spaces of the system.

The trigonometric polynomials of Section 1 (with scalar or vector-valued coefficients) are an example of such a system for $V = \mathcal{P}$, $W^+ = \mathcal{P}_1$, $W^- = \mathcal{P}_2$, and $\sigma = S$. A more general example is a function system $[V(E); W^+(E), W^-(E); T]$, defined by an arbitrary set E, two subsets $E_1 \subset E$ and $E_2 \subset E$, and a bijection $T : E \to E$ such that $TE_1 \subset E_1$, $T^{-1}E_2 \subset E_2$, with

$$V(E) := \{f : E \to \mathbb{C} \text{ finitely supported}\},$$
$$W^+(E) := \{f \in V : \operatorname{supp} f \subset E_1\},$$
$$W^-(E) := \{f \in V : \operatorname{supp} f \subset E_2\},$$
$$Tf(x) := f(Tx),$$

for all $f \in V$ and $x \in E$. Setting $E = \mathbb{Z}$, $E_1 = \mathbb{Z}_1$, $E_2 = \mathbb{Z}_2$, and $T : n \mapsto n + 1$, we get back to the previous example by identifying the trigonometric polynomials with the finite sequences of their Fourier coefficients. In this example we have

$$E_1 \cup E_2 = E, \quad E_1 \cap E_2 = \varnothing. \tag{2–2}$$

In general, neither of the equalities in (2–2) need hold. But if both hold for some E, E_1, E_2, then for every kernel $K : E \times E \to \mathbb{C}$ there exists a unique sesquilinear form $B = B_K : V(E) \times V(E) \to \mathbb{C}$ satisfying $B(1_x, 1_y) = K(x, y)$, the correspondence $K \mapsto B$ is bijective, and statements concerning kernels translate to statements on forms and vice versa.

If V is a Hilbert space, the outgoing and incoming spaces are closed subspaces of it, and σ is a unitary operator acting in V, then $[V; W^+, W^-; \sigma]$ is called a *hilbertian scattering system*. This name is justified by the fact that, under the additional conditions

$$\bigcap_{n \geq 0} \sigma^n W^+ = \{0\} = \bigcap_{n \geq 0} \sigma^{-n} W^- \tag{2-3}$$

and

$$V = V^1 \vee V^2, \quad \text{where } V^1 := \bigvee_{n \in \mathbb{Z}} \sigma^n W^+, \ V^2 := \bigvee_{n \in \mathbb{Z}} \sigma^n W^-, \tag{2-4}$$

$[V; W^+, W^-, \sigma]$ is an *Adamyan–Arov* (A-A) *scattering system*, with evolution group $\{\sigma^n : n \in \mathbb{Z}\}$. If, furthermore,

$$W^+ \perp W^- \tag{2-5}$$

and

$$V = V^1 = V^2, \tag{2-6}$$

then $[V; W^+, W^-; \sigma]$ is a *Lax–Phillips* (L-P) *scattering system*. Condition (2–3) implies that the trajectory $\{\sigma^n f : n \in \mathbb{Z}\}$ of every $f \in W^+$ contains some element of the complement of W^+, and similarly for the trajectory of every $f \in W^-$, while (2–4) means that V is spanned by the trajectories.

Let $[V; W^+, W^-; \sigma]$ be any scattering system. By analogy with the trigonometric case, a sesquilinear form $B : V \times V \to \mathbb{C}$ is called *Toeplitz* in the system if

$$B(\sigma f, \sigma g) = B(f, g) \quad \text{for all } f, g \in V, \tag{2-7}$$

while a form $B_0 : W^+ \times W^- \to \mathbb{C}$ is called *Hankel* in it if

$$B_0(\sigma f, g) = B_0(f, \sigma^{-1} g) \quad \text{for all } f \in W^+, \ g \in W^-. \tag{2-8}$$

This is the analog to Corollary 1.3 in algebraic scattering systems:

THEOREM 2.1 (LIFTINGS OF HANKEL FORMS BOUNDED WITH RESPECT TO TOEPLITZ FORMS IN ALGEBRAIC SCATTERING SYSTEMS [C-S 1987]). *Given an algebraic scattering system* $[V; W^+, W^-; \sigma]$ *and two positive Toeplitz forms* B_1 *and* B_2 *in it, for every Hankel form* $B_0 : W^+ \times W^- \to \mathbb{C}$ *bounded by* B_1 *and* B_2 *in the sense that*

$$|B_0(f, g)|^2 \leq B_1(f, f) B_2(g, g) \quad \text{for all } f \in W^+, \ g \in W^-, \tag{2-9}$$

there exists a Toeplitz lifting $B : V \times V \to \mathbb{C}$ *such that* $B|(W^+ \times W^-) = B_0$ *and*

$$|B(f, g)|^2 \leq B_1(f, f) B_2(g, g) \quad \text{for all } f, g \in V. \tag{2-10}$$

A short simple proof of this result is sketched in [C-S 1987]; see also [C-S 1990a] for details. Alternatively, Theorem 2.1 can be seen as a special case of Theorem 2.2 for hilbertian scattering systems, as follows.

Observe that, given an algebraic scattering system $[V; W^+, W^-; \sigma]$, a positive Toeplitz form $B_1 : V \times V \to \mathbb{C}$ can provide an inner product for a hilbertian system $[H_1; W_1^+, W_1^-; \sigma_1]$, where H_1 is the Hilbert space in which V can be identified as a dense subspace, W_1^+ and W_1^- are the closures in H_1 of W^+ and W^-, and σ extends to $\sigma_1 \in \mathcal{L}(H_1)$ as a unitary operator.

Thus, given an algebraic scattering system $[V; W^+, W^-; \sigma]$ and two positive Toeplitz forms B_1 and B_2, for any Hankel form B_0 satisfying (2–9), we can consider $B_0 : W_1^+ \times W_2^- \to \mathbb{C}$ as bounded in $H_1 \times H_2$, where, for $i = 1, 2$, H_i, W_i^+, and W_i^- are the closures of V, W^+, and W^-, respectively, in the norm induced by B_i.

A form $B_0 : W_1^+ \times W_2^- \to \mathbb{C}$ bounded in $H_1 \times H_2$ is called *bounded in the pair of scattering systems* $[H_1; W_1^+, W_1^-; \sigma_1]$ and $[H_2; W_2^+, W_2^-; \sigma_2]$.

In the next theorem, the hilbertian scattering systems satisfy only the defining condition (2–1), which is sufficient to insure the existence of Toeplitz liftings, and even to provide some form of integral representation for them. The case of A–A systems will be treated separately because the functional realization of these systems [Adamyan and Arov 1966] permits simplified integral representations.

THEOREM 2.2 (LIFTING THEOREM FOR HANKEL FORMS BOUNDED IN A PAIR OF HILBERTIAN SCATTERING SYSTEMS [C-S 1988; 1993a]). *Given two hilbertian scattering systems* $[H_1; W_1^+, W_1^-; \sigma_1]$ *and* $[H_2; W_2^+, W_2^-; \sigma_2]$, *every Hankel form* $B_0 : W_1^+ \times W_2^- \to \mathbb{C}$, *bounded in the pair has a Toeplitz lifting* $B : H_1 \times H_2 \to \mathbb{C}$, $B|(W_1^+ \times W_2^-) = B_0$, *such that* $\|B\| = \|B_0\|$.

SKETCH OF PROOF. Since $B_0 : W_1^+ \times W_2^- \to \mathbb{C}$ is bounded in $H_1 \times H_2$, the hermitian form

$$\langle (f_1, f_2), (g_1, g_2) \rangle := \langle f_1, g_1 \rangle_{H_1} + \langle f_2, g_2 \rangle_{H_2} + B_0(f_1, g_2) + \overline{B_0(g_1, f_2)} \quad (2\text{–}11)$$

is positive and gives a pre-Hilbert structure to $W_1^+ \times W_2^-$, where $\sigma : (f_1, f_2) \mapsto (\sigma_1 f_1, \sigma_2 f_2)$ is an isometry, with domain $W_1^+ \times \sigma_2^{-1} W_2^-$ and range $\sigma_1 W_1^+ \times W_2^-$. Then there is a unitary operator $T \in \mathcal{L}(N)$ in a larger Hilbert space $N \supset W_1^+ \times W_2^-$, such that $T = \sigma$ on its domain and $T^{-1} = \sigma^{-1}$ on the range. Identifying W_1^+ with $W_1^+ \times \{0\}$ and W_2^- with $\{0\} \times W_2^-$, we can consider W_1^+ and W_2^- as subspaces of N. Set

$$R_1 = W_1^+ \ominus \sigma_1 W_1^+ \quad \text{and} \quad R_2 = W_2^- \ominus \sigma_2^{-1} W_2^-. \quad (2\text{–}12)$$

As a consequence of the Wold–Kolmogorov decomposition, H_1 and H_2 can be expressed with respect to σ_1 and σ_2, respectively, as

$$H_1 = \bigoplus_{n \in \mathbb{Z}} \sigma_1^n R_1 \oplus H_1^0 \oplus H_1^1, \qquad H_2 = \bigoplus_{n \in \mathbb{Z}} \sigma_2^{-n} R_2 \oplus H_2^0 \oplus H_2^1, \quad (2\text{–}13)$$

where $H_1^1 = \bigcap_{n \geq 0} \sigma_1^n W_1^+$ and $H_2^1 = \bigcap_{n \geq 0} \sigma_2^{-n} W_2^-$, so that

$$\sigma_1 H_1^0 = H_1^0, \quad \sigma_1 H_1^1 = H_1^1, \quad \sigma_2^{-1} H_2^0 = H_2^0, \quad \sigma_2^{-1} H_2^1 = H_2^1,$$

and every $f_i \in H_i$ can be written as

$$f_i = \bigoplus_{n \in \mathbb{Z}} \sigma_i^{\pm n} f_{n,i} + f_i^0 + f_i^1 \text{ with } f_{n,i} \in R_i,\ f_i^0 \in H_i^0,\ f_i^1 \in H_i^1, \quad \text{for } i = 1, 2.$$

$$(2\text{–}14)$$

For f_i as in (2–14), set

$$[f_1] := \bigoplus_n T^n f_{n,1} + f_1^1, \qquad [f_2] := \bigoplus_n T^{-n} f_{n,2} + f_2^1,$$

and, for $i = 1, 2$, set $J_i : H_i \to N$ by $J_i f_i = [f_i]$.

Observing that $\|[f_i]\|_N \leq \|f_i\|_{H_i}$, for $i = 1, 2$, define a scalar product in $H_1 \times H_2$ by

$$\langle (f_1, f_2), (g_1, g_2) \rangle := \langle f_1, g_1 \rangle_{H_1} + \langle f_2, g_2 \rangle_{H_2} + \langle [f_1], [g_2] \rangle_N + \overline{\langle [g_1], [f_2] \rangle_N} \quad (2\text{–}15)$$

and, as before, call H the Hilbert space obtained from it. Hence, H contains $H_1 \equiv H_1 \times \{0\}$ and $H_2 \equiv \{0\} \times H_2$ as subspaces, and $U \in \mathcal{L}(H)$ is the unitary operator such that $U(f_1, f_2) = (\sigma_1 f_1, \sigma_2 f_2)$ for $(f_1, f_2) \in H_1 \times H_2$.

Define $B : H_1 \times H_2 \to \mathbb{C}$ by

$$B(f_1, f_2) = \langle f_1, f_2 \rangle_H \quad \text{for all } f_1 \in H_1,\ f_2 \in H_2. \qquad (2\text{–}16)$$

Then $\|B\| \leq 1$, and it is not difficult to check that B is Toeplitz since

$$\langle [\sigma_1 f_1], [\sigma_2 f_2] \rangle_N = \langle [f_1], [f_2] \rangle_N, \quad \text{for all } f_1 \in H_1,\ f_2 \in H_2. \qquad (2\text{–}17)$$

Furthermore, for any $f_1 \in W_1^+$ and $f_2 \in W_2^-$, we have $[f_1] = f_1$ and $[f_2] = f_2$. Therefore $B|(W_1^+ \times W_2^-) = B_0$, since

$$B(f_1, f_2) = \langle f_1, f_2 \rangle_H = \langle [f_1], [f_2] \rangle_N = \langle f_1, f_2 \rangle_N = B_0(f_1, f_2).$$

This completes the proof that B is the desired lifting of B_0. $\qquad\qquad \square$

REMARK. The spaces $R_1 = W_1^+ \ominus \sigma_1 W_1^+$ and $R_2 = W_2^- \ominus \sigma_2^{-1} W_2^-$ play in this proof the role played in the proof of the GBT (Section 1) by $\mathcal{P}_1 \ominus S\mathcal{P}_1 = \{c1\}$ and $\mathcal{P}_2 \ominus S^{-1}\mathcal{P}_2 = \{ce^{-it}\}$, where $J(1)$ and $J(e^{-it})$ were a cyclic pair. Here R_1 and R_2 need not be one-dimensional, but they are still cyclic sets.

The essence of the proof above is that there are two metrics defined in $W_1^+ \times W_2^- \subset H_1 \times H_2$, one by B_0 in (2–11), and the other induced in $H_1 \times H_2$ by (2–15). The maps $J_1 : H_1 \to N$ and $J_2 : H_2 \to N$ allow the transference of the metric in $W_1^+ \times W_2^-$ to the whole of $H_1 \times H_2$, providing the lifting.

From the proof of Theorem 2.2 it follows that, if E is the spectral measure of the unitary operator $T \in \mathcal{L}(N)$, for each pair of elements $(f_1, f_2) \in H_1 \times H_2$

there is a numerical measure μ_{f_1,f_2} defined on $\mathbb{T}$ by its values on the Borel sets $D \subset \mathbb{T}$:

$$\mu_{f_1,f_2}(D) := \langle E(D) J_1 f_1, J_2 f_2 \rangle_N. \tag{2–18}$$

On the pair $(f_1, f_2) \in H_1 \times H_2$, equation (2–14) gives the representation of the lifting $B : H_1 \times H_2 \to \mathbb{C}$ as

$$B(f_1, f_2) = \sum_{m,n} \int e^{i(m-n)t} \, d\mu_{f_{m,1}, f_{n,2}}$$

$$+ \sum_n \int e^{int} \, d\mu_{f_1^1, f_{n,2}} + \sum_m \int e^{imt} \, d\mu_{f_{m,1}, f_2^1} + \int d\mu_{f_1^1, f_2^1}. \tag{2–19}$$

In the particular case of forms acting in A–A scattering systems (including the Lax–Phillips type), the number of measures necessary for the integral representation can be substantially reduced (compare [C-S 1986; 1988]), which is not surprising since, for starts, $H_1^1 = H_2^1 = \{0\}$ in that situation.

A more economical integral representation of the lifting $B : H_1 \times H_2 \to \mathbb{C}$ in A–A scattering systems can be given by just *one* operator-valued measure, as done in [C-S 1988]. This is obtained through the functional realization of the systems given by Adamyan and Arov [1966]. For a scattering system $[H; W^+, W^-; \sigma]$ satisfying (2–3) and (2–4), their realization provides a scattering function $s : \mathbb{T} \to \mathcal{L}(R_1, R_2)$, where $R_1 = W^+ \ominus \sigma W^+$ and $R_2 = W^- \ominus \sigma^{-1} W^-$, with $\|s(t)\| \leq 1$ for all $t \in \mathbb{T}$, and an isometric mapping $\mathcal{F}$ of H onto $L^2(R_2) \oplus L_\Delta^2(R_1)$ for which $\mathcal{F}(\sigma f)(t) = e^{it}\mathcal{F}(f)(t)$ for all $f \in H$ and $t \in \mathbb{T}$. Here $L^2(R_i) = L^2(\mathbb{T}; R_i)$, for $i = 1, 2$, and $L_\Delta^2(R_1)$ is the closure of the space

$$\{\phi \in L^2(R_1) : \phi = \Delta h \text{ for } h \in L^2(R_1)\},$$

where $\Delta(t) = (I_{R_1} - s^*(t)s(t))^{1/2}$.

Furthermore, there are two isometries, $j_1 : W^+ \to H^2(R_1)$ and $j_2 : W^- \to H_-^2(R_2)$, such that, for $(f_1, f_2) \in W^+ \times W^-$, we have

$$\mathcal{F}(f_2) = j_2 f_2 \oplus \{0\},$$

$$\mathcal{F}(f_1) = s j_1 f_1 \oplus \Delta j_1 f_1,$$

$$j_1(\sigma f_1)(t) = e^{it}(j_1 f_1)(t),$$

$$j_2(\sigma^{-1} f_2)(t) = e^{-it}(j_2 f_2)(t).$$

Let us consider this in the simpler case of a pair of scattering systems consisting of two copies of the same $[H; W^+, W^-; \sigma]$. Then $B_0 : W^+ \times W^- \to \mathbb{C}$ is a Hankel form bounded in $[H; W^+, W^-; \sigma]$, an A–A scattering system. Now, through the functional realization of the system, the Toeplitz extension $B : H \times H \to \mathbb{C}$ of B_0 can be written explicitly in all of $H \times H$ as

$$B(\sigma^m f_1, \sigma^n f_2) = \int \langle d\mu \, e^{imt}\varphi_1, e^{int}\varphi_2 \rangle \quad \text{for all } m, n \in \mathbb{Z}, \tag{2–20}$$

for $\varphi_1 = j_1 f_1$, $\varphi_2 = j_2 f_2$, $f_1 \in W^+$, $f_2 \in W^-$, where μ is an $\mathcal{L}(R_1, R_2)$-valued measure on $\mathbb{T}$. Since $\mathcal{F}$ respects scalar products, we have

$$B_0(f_1, f_2) = \langle (f_1, 0), (0, f_2) \rangle_H = \langle \mathcal{F}(f_1, 0), \mathcal{F}(0, f_2) \rangle_{L^2(R_2) \oplus L^2_{\Delta}(R_1)}$$

$$= \int_{\mathbb{T}} \langle s(t) \varphi_1(t), \varphi_2(t) \rangle \, dt,$$

so that $d\mu = s(t) \, dt$ is given by the scattering function of the system. In the case of a Lax–Phillips scattering system, s coincides with the Heisenberg scattering function as defined in [Lax and Phillips 1967]. Details are given in [C-S 1988].

REMARK. The role of the operator-valued scattering function in the integral representation of the bounded Hankel form is played, in the trigonometric case, by the BMO functions. In fact, for $\phi \in L^2(\mathbb{T})$, a real-valued function, B_ϕ : $\mathcal{P}_1 \times \mathcal{P}_2 \to \mathbb{C}$ is defined by $B_\phi(f, g) = \int f \bar{g} \phi$. Then B_ϕ is bounded in $L^2 \times L^2$ if and only if $B_\phi = B_\varphi$ for some $\varphi \in L^\infty$ with $\|\varphi\|_\infty \leq \|B_\phi\|$ and $\phi - \varphi = h$, an analytic function. Then $\phi = \operatorname{Re} \phi = \operatorname{Re} \varphi + \operatorname{Re} h = \operatorname{Re} \varphi - H(\operatorname{Im} h) = \operatorname{Re} \varphi + H(\operatorname{Im} \varphi) \in L^\infty + HL^\infty \equiv \text{BMO}(\mathbb{T})$.

Liftings of forms defined on general semi-invariant spaces. Whereas in the trigonometric example $V = \mathcal{P} = \mathcal{P}_1 \dotplus \mathcal{P}_2 = W^+ \dotplus W^-$, in the general case of a hilbertian scattering system $[H; W^+, W^-; \sigma]$ in which $W^+ \perp W^-$, the subspace

$$W = H \ominus (W^+ \oplus W^-) \tag{2-21}$$

will be nontrivial, and will deserve study in itself. In major applications, it is normal to call this subspace the *internal state space*.

A subspace $V \subset H$ is called *semi-invariant* relative to a unitary $\sigma \in \mathcal{L}(H)$ if there exist V_1 and $V_2 \subset V_1$, both invariant under σ, such that

$$V = V_1 \ominus V_2,$$

or, equivalently, if there exist V_1' and $V_2' \subset V_1'$, both invariant under σ^{-1}, such that

$$V = V_1' \ominus V_2'. \tag{2-22}$$

Observe that in the setting of $[H; W^+, W^-; \sigma]$ the internal state space W defined in (2–21) is semi-invariant. For that matter, so are W^+, W^-, and their orthogonal complements.

The forms $B_0 : W_1 \times W_2 \to \mathbb{C}$ defined in a pair of (semi-invariant) internal state spaces are of special interest. Given a pair of hilbertian scattering systems $[H_1; W_1^+, W_1^-; \sigma_1]$ and $[H_2; W_2^+, W_2^-; \sigma_2]$, the condition under which the lifting for bounded forms $B_0 : W_1^+ \times W_2^- \to \mathbb{C}$ was obtained is that of *invariance* with respect to σ_1 and σ_2, that is,

$$B_0(\sigma_1 f_1, f_2) = B_0(f_1, \sigma_2^{-1} f_2) \quad \text{for } f_1 \in W_1^+, \ f_2 \in W_2^-. \tag{2-23}$$

This condition makes sense since, by the scattering property (2–1), the domain $W_1^+ \times W_2^-$ is invariant under $(\sigma_1, \sigma_2^{-1})$.

This need not be true for $B_0 : W_1 \times W_2 \to \mathbb{C}$, where $W_i = H_i \ominus (W_i^+ \oplus W_i^-)$, for $i = 1, 2$, are only semi-invariant subspaces.

Thus we consider an invariance condition for $B_0 : W_1 \times W_2 \to \mathbb{C}$ that, although weaker than the Hankel condition (2–23), is still sufficient for the existence of invariant liftings in $H_1 \times H_2$.

A form $B_0 : W_1 \times W_2 \to \mathbb{C}$ is called *essentially invariant* in a pair of scattering systems if

$$B_0(P_1 \sigma_1 f_1, f_2) = B_0(f_1, P_2 \sigma_2^{-1} f_2) \quad \text{for all } f_1 \in W_1,\ f_2 \in W_2, \qquad (2\text{–}24)$$

where $P_1 : H_1 \to W_1$ and $P_2 : H_2 \to W_2$ are the orthoprojectors. In the special case when $W_1^+ = \{0\} = W_2^-$, we have $\sigma_1 W_1 \subset W_1$ and $\sigma_2^{-1} W_2 \subset W_2$, and the condition (2–24) reduces to $B_0 : W_1 \times W_2 \to \mathbb{C}$ being Hankel. Conversely, for $[H_i; W_i^+, W_i^-; \sigma_i]$ and $W_i = H_i \ominus (W_i^+ \oplus W_i^-)$, setting

$$V_1 = W_1 \oplus W_1^+ \quad \text{and} \quad V_2 = W_2 \oplus W_2^-,$$

it is easy to see that

$$\sigma_1 V_1 \subset V_1 \quad \text{and} \quad \sigma_2^{-1} V_2 \subset V_2. \qquad (2\text{–}25)$$

Let a bounded essentially invariant form $B_0 : W_1 \times W_2 \to \mathbb{C}$ be given. If we define $B_0^{\#} : V_1 \times V_2 \to \mathbb{C}$ by

$$B_0^{\#}(f_1 + f_1^+, f_2 + f_2^-) := B_0(f_1, f_2), \qquad (2\text{–}26)$$

it is not difficult to check that $\|B_0^{\#}\| = \|B_0\|$, and that $B_0^{\#}$ is Hankel in $V_1 \times V_2$.

From (2–26) and Theorem 2.2 we obtain:

COROLLARY 2.3 (LIFTINGS FOR BOUNDED ESSENTIALLY HANKEL FORMS DEFINED IN SEMI-INVARIANT SUBSPACES IN SCATTERING SYSTEMS [C-S 1993a]). *Consider two scattering systems $[H_1; W_1^+, W_1^-; \sigma_1]$ and $[H_2; W_2^+, W_2^-; \sigma_2]$ with internal spaces $W_i = H_i \ominus (W_i^+ \oplus W_i^-)$, for $i = 1, 2$, and a bounded form $B_0 : W_1 \times W_2 \to \mathbb{C}$, essentially Hankel in the sense of (2–24). Then:*

(a) *There exists a Toeplitz form $B : H_1 \times H_2 \to \mathbb{C}$ such that $B|(W_1 \times W_2) = B_0$ and $\|B\| = \|B_0\|$.*

(b) *Furthermore, $B = 0$ on $W_1^+ \times W_2^-$, $W_1 \times W_2^-$, and $W_1^+ \times W_2$.*

Corollary 2.3 has important applications. For example, we now use both parts of it to provide a simple proof of Sarason's interpolation theorem, without relying on Beurling's characterization of the invariant subspaces of $H^2(\mathbb{T})$.

THEOREM 2.4 [Sarason 1967]. *Let $W \subset H^2(\mathbb{T})$ be a subspace invariant under the shift, S, let $K = H^2 \ominus W$ be the model space, and let $T = P_K S | K$ be the compression of S to K. For each contraction $A \in \mathcal{L}(K)$ commuting with T, there exists a bounded holomorphic function a satisfying $\|a\|_\infty \le 1$ and*

$$Af = P_K(af) \quad \text{for all } f \in K. \qquad (2\text{–}27)$$

PROOF. In Corollary 2.3, take $H_1 = H_2 = L^2(\mathbb{T})$, $W_1 = W_2 = K$, $W_1^+ = W_2^+ = W$, $W_1^- = W_2^- = L^2 \ominus H^2 = H_-^2$, $\sigma_1 = \sigma_2 = S$, and define $B_0 : W_1 \times W_2 \to \mathbb{C}$ by

$$B_0(f,g) := \langle Af, g \rangle \quad \text{for all } f, g \in K. \tag{2-28}$$

Then $\|A\| \leq 1$ is equivalent to $\|B_0\| \leq 1$, and $AT = TA$ is equivalent to B_0 being essentially Hankel as in (2–24). Then, by Corollary 2.3, there is a bounded Toeplitz lifting $B : L^2 \times L^2 \to \mathbb{C}$ of B_0, with $\|B\| \leq 1$. The restriction to trigonometric polynomials of the Toeplitz lifting B is given by $B(f,g) = L(f\bar{g})$, for L a linear functional. Since $|B(f,g)| \leq \|f\|_2 \|g\|_2$, we have $|L(f)| \leq \|f\|_1$, and there exists $a \in L^\infty$ with $\|a\|_\infty \leq 1$ and such that $B(f,g) = \int a f \bar{g} \, dt$. From $B = B_0$ in $K \times K$ follows the representation (2–27), while from $B = 0$ in $(K \oplus W) \times H_-^2 = H^2 \times H_-^2$ it follows that $a1 \perp H_-^2$, which means $a \in H^2 \cap L^\infty = H^\infty$. $\qquad\square$

The most important result for which Corollary 2.3 gives an immediate proof is the Lifting Theorem for intertwining contractions of Sz.-Nagy and Foiaş:

THEOREM 2.5 [Sz.-Nagy and Foiaş 1967]. *Let* $T_1 \in \mathcal{L}(H_1)$ *and* $T_2 \in \mathcal{L}(H_2)$ *be two contractions, with strong unitary dilations* $U_1 \in \mathcal{L}(\mathcal{H}_1)$ *and* $U_2 \in \mathcal{L}(\mathcal{H}_2)$. *If* $X : H_1 \to H_2$ *is a contraction intertwining* T_1 *and* T_2, *that is, such that* $XT_1 = T_2X$, *then there exists a contraction* $Y : \mathcal{H}_1 \to \mathcal{H}_2$ *such that* Y *intertwines* U_1, U_2 *that is,* $YU_1 = U_2Y$, *and*

$$X = P_{H_2} Y | H_1, \tag{2-29}$$

where $P_{H_2} : \mathcal{H}_2 \to H_2$ *is the orthoprojector.*

PROOF. By Sarason's Lemma [Sarason 1965], for $i = 1, 2$, the fact that U_i is a strong dilation of T_i means that $\mathcal{H}_i = H_i \oplus H_i^+ \oplus H_i^-$, where $U_i H_i^+ \subset H_i^+$ and $U_i^{-1} H_i^- \subset H_i^-$, and $T_i = P_{H_i} U_i | H_i$. Defining $B_0 : H_1 \times H_2 \to \mathbb{C}$ by

$$B_0(f_1, f_2) = \langle Xf_1, f_2 \rangle_{H_2} \quad \text{for all } f_1 \in H_1, \ f_2 \in H_2,$$

the intertwining condition for X is equivalent to B_0 being essentially Hankel in U_1, U_2. From $B : \mathcal{H}_1 \times \mathcal{H}_2 \to \mathbb{C}$, the lifting of B_0 in Corollary 2.3, define $Y : \mathcal{H}_1 \to \mathcal{H}_2$ by

$$\langle Yf_1, f_2 \rangle_{\mathcal{H}_2} := B(f_1, f_2) \quad \text{for all } f_1 \in \mathcal{H}_1, \ f_2 \in \mathcal{H}_2.$$

Then $\|B\| \leq 1$ is equivalent to $\|Y\| \leq 1$, and the property that B is Toeplitz for U_1, U_2 is equivalent to $YU_1 = U_2Y$. Finally, $B = B_0$ in $H_1 \times H_2$ is equivalent to (2–29). $\qquad\square$

Conditional liftings. To every form $B_0 : W_1^+ \times W_2^- \to \mathbb{C}$ bounded in a pair of hilbertian scattering systems we associate a bounded operator $\Gamma : W_1^+ \to W_2^-$, by setting $\langle \Gamma f, g \rangle = B_0(f, g)$. Then $\|\Gamma\| = \|B\|$. If the form B_0 is Hankel, the operator satisfies the Hankel condition

$$\Gamma \sigma_1 = P_- \sigma_2 \Gamma \quad \text{for } P_- : H_2 \to W_2^- \text{ the orthoprojector.} \tag{2--30}$$

The singular numbers s_n of B_0, for $n \geq 0$, are

$$s_n(B_0) := s_n(\Gamma) = \inf\{\|\Gamma - T_n\| : \operatorname{rank} T_n \leq n\}. \tag{2--31}$$

A form $B^{(n)} : H_1 \times H_2 \to \mathbb{C}$ is called an n-*conditional lifting* of B_0 if there exists a subspace $M_n \subset W_1^+$ of codimension at most n such that

$$B^{(n)}|(M_n \times W_2^-) = B_0 \quad \text{and} \quad \|B^{(n)}\| \leq s_n(B_0). \tag{2--32}$$

THEOREM 2.6 (EXISTENCE OF n-CONDITIONAL LIFTINGS FOR HANKEL FORMS BOUNDED IN HILBERTIAN SCATTERING SYSTEMS [C-S 1993b]). *Given a Hankel form* $B_0 : W_1^+ \times W_2^- \to \mathbb{C}$, *bounded in a pair of hilbertian scattering systems* $[H_1; W_1^+, W_1^-; \sigma_1]$ *and* $[H_2; W_2^+, W_2^-; \sigma_2]$, *there exists for every integer* $n \geq 0$ *a Toeplitz* n-*conditional lifting of* B_0.

SKETCH OF PROOF. Given n, let $s_n = s_n(B_0)$, and let Γ be the bounded Hankel operator associated with B_0. Considering the set

$$K = \{f \in W_1^+ : \|\Gamma f\| \leq s_n\|f\|\},$$

it is easy to check that $\sigma_1 K \subset K$. By a particular case of [Treil' 1985, Theorem 2], there exists a subspace $M_n \subset W_1^+$ such that $\operatorname{codim} M_n \leq n$, $M_n \subset K$, and $\sigma_1 M_n \subset M_n$. Since $\sigma_1 M_n \subset M_n$ and $\sigma_2^{-1} W_2^- \subset W_2^-$, and since $M_n \subset K$ implies that the restriction $B_0|(M_n \times W_2^-)$ has norm bounded by s_n, we can apply Theorem 2.2 to the Hankel form $B_0|(M_n \times W_2^-)$, bounded in the systems $[H_1; M_n, W_1^-; \sigma_1]$ and $[H_2; W_2^+, W_2^-; \sigma_2]$, to obtain a Toeplitz form $B^{(n)} : H_1 \times H_2 \to \mathbb{C}$ such that $B^{(n)}|(M_n \times W_2^-) = B_0$ and $\|B^{(n)}\| \leq s_n$. $\square$

COROLLARY 2.7 (ABSTRACT ADAMYAN–AROV–KREĬN THEOREM [C-S 1993b]). *Given a pair of hilbertian scattering systems* $[H_1; W_1^+, W_1^-; \sigma_1]$ *and* $[H_2; W_2^+, W_2^-; \sigma_2]$, *and a bounded Hankel operator* $\Gamma : W_1^+ \to W_2^-$, *there exists for each integer* $n \geq 0$ *a Hankel operator* Γ_n *of finite rank at most* n *and such that*

$$\|\Gamma - \Gamma_n\| = s_n(\Gamma). \tag{2--33}$$

PROOF. For $B^{(n)}$ as in Theorem 2.6, let $\tilde{\Gamma} : W_1^+ \to W_2^-$ be the operator associated to the form $B^{(n)}|(W_1^+ \times W_2^-)$. Setting $\Gamma_n := \Gamma - \tilde{\Gamma}$, we have $\|\Gamma - \Gamma_n\| = \|\tilde{\Gamma}\| \leq s_n = s_n(\Gamma)$. Furthermore, Γ_n is Hankel, and since by definition it vanishes on M_n, its rank is at most n. $\square$

In the case when, for $i = 1, 2$, $H_i = L^2(\mathbb{T}; \mu_i)$ for μ_i a positive measure on $\mathbb{T}$, while $W_1^+ = H^2(\mathbb{T}; \mu_1)$ and $W_2^- = H_-^2(\mathbb{T}; \mu_2)$, we have the following consequence of Theorem 2.6 together with Corollary 1.7:

COROLLARY 2.8 (WEIGHTED AAK THEOREM [C-S 1993b]). *Given a bounded Hankel form* $B : H^2(\mathbb{T}; \mu_1) \times H^2_-(\mathbb{T}; \mu_2) \to \mathbb{C}$, *where* μ_1 *and* μ_2 *are positive measures on* $\mathbb{T}$, *there exist for every integer* $n \geq 0$ *a complex measure* μ *on* $\mathbb{T}$ *and a subspace* $M_n \subset H^2(\mathbb{T}; \mu_1)$, *of codimension at most* n, *such that*

$$B(f,g) = \int f\bar{g}\,d\mu \quad \text{for all } f \in M_n, g \in H^2_-(\mathbb{T}; \mu_2), \qquad (2\text{--}34)$$

while

$$s_n(B) \geq \sup_D \frac{|\mu(D)|}{\mu_1(D)^{1/2}\mu_2(D)^{1/2}},$$

where the supremum is taken over all Borel sets $D \subset \mathbb{T}$.

REMARK. When both μ_1 and μ_2 are the Lebesgue measure on $\mathbb{T}$, Corollary 2.8 is the classical AAK theorem [Adamyan, Arov and Kreĭn 1971]. If, moreover, $n = 0$, we recover the classical Nehari theorem [1957]. Furthermore, if μ_2 is a deterministic measure, i.e., if $L^2(\mu_2) = H^2(\mu_2)$, then every Hankel form of finite rank in $H^2(\mu_1) \times H^2_-(\mu_2)$ is zero, while in the opposite case such a form can be represented in terms of the reproducing kernel of $H^2(\mu_2)$ [C-S 1993b].

3. Lifting of Forms Invariant with Respect to Several Evolution Groups. Some Applications to Analysis in Product Spaces

The lifting theorems of the preceding section extend to forms invariant in scattering systems having several evolution groups. In order to avoid notational complications, here we present only the case of two evolutions, but all concepts and results are general.

For simplicity, we consider only hilbertian scattering systems of the form $[H; W^+, W^-; \sigma, \tau]$, where both $\sigma \in \mathcal{L}(H)$ and $\tau \in \mathcal{L}(H)$ are unitary operators satisfying

$$\sigma\tau = e^{ia}\tau\sigma \text{ for some } a \in \mathbb{R}, \qquad (3\text{--}1)$$

and such that W^+ is invariant with respect to both σ and τ, while W^- is invariant with respect to both σ^{-1} and τ^{-1}. Compare (2–1).

We are now concerned with forms invariant with respect to both evolution groups $\{\sigma^n : n \in \mathbb{Z}\}$ and $\{\tau^n : n \in \mathbb{Z}\}$. More precisely, given a pair of scattering systems $[H_1; W_1^+, W_1^-; \sigma_1, \tau_1]$ and $[H_2; W_2^+, W_2^-; \sigma_2, \tau_2]$ as described above, a form $B : H_1 \times H_2 \to \mathbb{C}$ is *Toeplitz* in the pair if

$$B(\sigma_1 f_1, \sigma_2 f_2) = B(f_1, f_2) = B(\tau_1 f_1, \tau_2 f_2) \quad \text{for all } f_1 \in H_1, \ f_2 \in H_2. \quad (3\text{--}2)$$

A form $B_0 : W_1^+ \times W_2^- \to \mathbb{C}$ is *Hankel* in the pair if

$$B_0(\sigma_1 f_1, f_2) = B_0(f_1, \sigma_2^{-1} f_2) \quad \text{and} \quad B_0(\tau_1 f_1, f_2) = B_0(f_1, \tau_2^{-1} f_2)$$
$$\text{for all } f_1 \in W_1^+, \ f_2 \in W_2^-. \quad (3\text{--}3)$$

Theorem 2.2 applied to a Hankel form in a scattering system with two evolutions σ and τ provides two liftings of the form, one invariant with respect to $\{\sigma^n\}$ and the other invariant with respect to $\{\tau^n\}$, but it does not provide liftings invariant with respect to both groups.

A full extension of Theorem 2.2 providing lifting with respect to all evolutions of the scattering systems cannot hold in general. This follows from the relation between Theorem 2.2 and the Sz.-Nagy–Foiaş Lifting Theorem (compare Section 2, as well as [C-S 1987; 1993a]), since the latter does not have a full extension to two pairs of intertwining operators. What we can obtain is partial liftings, in the following sense.

Given a Hankel form $B_0 : W_1^+ \times W_2^- \to \mathbb{C}$, two Toeplitz forms $B' : H_1 \times H_2 \to \mathbb{C}$ and $B'' : H_1 \times H_2 \to \mathbb{C}$ form a *Toeplitz lifting pair* (B', B'') for B_0 if

$$\|B'\| \leq \|B_0\|, \quad \|B''\| \leq \|B_0\|, \tag{3-4}$$

$$B'|(W_1^+ \times W_2^\sigma) = B_0, \text{ and } B''|(W_1^+ \times W_2^\tau) = B_0, \tag{3-5}$$

where

$$W_2^\sigma = \{f \in W_2^- : \sigma_2^k f \in W_2^- \quad \text{for all } k \in \mathbb{Z}\},$$
$$W_2^\tau = \{f \in W_2^- : \tau_2^k f \in W_2^- \quad \text{for all } k \in \mathbb{Z}\}. \tag{3-6}$$

The following result provides Toeplitz lifting pairs for bounded Hankel forms in a pair of scattering systems with two evolutions. Its proof does not follow directly from a repeated use of Theorem 2.2, where a system with two evolutions $[H; W^+, W^-; \sigma, \tau]$ is taken as two single-evolution systems $[H; W^+, W^-; \sigma]$ and $[H; W^+, W^-; \tau]$, but requires an argument based on Banach limits.

THEOREM 3.1 [C-S 1990a; 1993a]. *Given a pair of scattering systems with two evolutions,* $[H_1; W_1^+, W_1^-; \sigma_1, \tau_1]$ *and* $[H_2; W_2^+, W_2^-; \sigma_2, \tau_2]$, *every Hankel form* $B_0 : W_1^+ \times W_2^- \to \mathbb{C}$ *bounded in the pair has a Toeplitz lifting pair* (B', B'') *in the sense of* (3–4)–(3–6).

REMARK. Theorem 3.1 holds equally when W_2^σ, W_2^τ are replaced by W_1^σ, W_1^τ, or by W_1^σ, W_2^τ, or by W_2^σ, W_1^τ, defined similarly to (3–6).

SKETCH OF PROOF. Consider B_0 as a Hankel form in the pair $[H_i; W_i^+, W_i^-; \sigma_i]$, $i = 1, 2$. Then Theorem 2.2 gives a lifting $B^\sigma : H_1 \times H_2 \to \mathbb{C}$ satisfying $\|B^\sigma\| = \|B_0\|$, $B^\sigma|(W_1^+ \times W_2^-) = B_0$, and $B^\sigma(\sigma_1 f_1, \sigma_2 f_2) = B^\sigma(f_1, f_2)$ for all $f_1 \in H_1, f_2 \in H_2$. Similarly, Theorem 2.2 gives another lifting $B^\tau : H_1 \times H_2 \to \mathbb{C}$ satisfying $\|B^\tau\| = \|B_0\|$, $B^\tau|(W_1^+ \times W_2^-) = B_0$, and $B^\tau(\tau_1 f_1, \tau_2 f_2) = B^\tau(f_1, f_2)$ for all $f_1 \in H_1, f_2 \in H_2$.

For a fixed pair $(f_1, f_2) \in H_1 \times H_2$ and any positive integer k, we have

$$|B^\tau(\sigma_1^k f_1, \sigma_2^k f_2)| \leq \|B^\tau\| \, \|\sigma_1^k f_1\|_{H_1} \, \|\sigma_2^k f_2\|_{H_2} = \|B^\tau\| \, \|f_1\|_{H_1} \, \|f_2\|_{H_2}, \tag{3-7}$$

and $\{B^\tau(\sigma_1^k f_1, \sigma_2^k f_2)\}$ is a bounded numerical sequence, for which there is a Banach–Mazur limit. Thus, define $B' : H_1 \times H_2 \to \mathbb{C}$ by

$$B'(f_1, f_2) := \mathrm{LIM}_k \, B^\tau(\sigma_1^k f_1, \sigma_2^k f_2) \quad \text{for all } f_1 \in H_1, f_2 \in H_2. \tag{3-8}$$

The form B' is sesquilinear by the properties of the Banach–Mazur limits, and $\|B'\| \leq \|B^\tau\| = \|B_0\|$ by (3–7). Also, for $f_1 \in H_1, f_2 \in H_2$, we have

$$B'(\sigma_1 f_1, \sigma_2 f_2) = \mathrm{LIM}_k\, B^\tau(\sigma_1^{k+1} f_1, \sigma_2^{k+1} f_2) = \mathrm{LIM}_k\, B^\tau(\sigma_1^k f_1, \sigma_2^k f_2) = B'(f_1, f_2).$$

Furthermore, (3–1), the sesquilinearity of B^τ, and its invariance with respect to τ_1, τ_2 yield

$$B'(\tau_1 f_1, \tau_2 f_1) = \mathrm{LIM}_k\, B^\tau(\sigma_1^k \tau_1 f_1, \sigma_2^k \tau_2 f_2) = \mathrm{LIM}_k\, B^\tau(\tau_1 \sigma_1^k f_1, \tau_2 \sigma_2^k f_2)$$
$$= \mathrm{LIM}_k\, B^\tau(\sigma_1^k f_1, \sigma_2^k f_2) = B'(f_1, f_2).$$

Finally, since $B^\tau = B_0$ in $W_1^+ \times W_2^-$, and since $(f_1, f_2) \in W_1^+ \times W_2^\sigma$ implies $(\sigma_1^k f_1, \sigma_2^k f_2) \in W_1^+ \times W_2^-$ for all $k \geq 0$, we have

$$B^\tau(\sigma_1^k f_1, \sigma_2^k f_2) = B_0(\sigma_1^k f_1, \sigma_2^k f_2) = B_0(f_1, f_2) \quad \text{for all } k \geq 0,$$

and $B' = B_0$ in $W_1^+ \times W_2^\sigma$.

Defining the form $B'' : H_1 \times H_2 \to \mathbb{C}$ by

$$B''(f_1, f_2) = \mathrm{LIM}_k\, B^\sigma(\tau_1^k f_1, \tau_2^k f_2) \quad \text{for all } f_1 \in H_1, f_2 \in H_2,$$

we obtain the lifting pair (B', B'') for B_0. $\qquad\square$

Observe that while the lifting B for a Hankel form in a pair of scattering systems with one evolution group determines the form B_0 in the whole of its domain $W_1^+ \times W_2^-$, the lifting pair (B', B'') in the case of systems with two evolution groups determines B_0 only in the two subspaces $W_1^+ \times W_2^\sigma$ and $W_1^+ \times W_2^\tau$ of the domain. Thus, the value of Theorem 3.1 in applications is determined by the relation between these subspaces. When, for instance,

$$W_2^- = W_2^\sigma + W_2^\tau \tag{3–9}$$

holds, in the sense that for each $f \in W_2^-$ there are $g \in W_2^\sigma$ and $h \in W_2^\tau$ such that $f = g + h$, the lifting pair determines B_0. This is the case in many examples in analysis, as shown below. In particular, Theorem 3.1 provides as corollaries multidimensional analogs of the one-dimensional Helson–Szegő and Nehari–Adamyan–Arov–Kreĭn theorems that were given in Section 1 as corollaries of the GBT, as well as related results in symplectic spaces and the Heisenberg group [C-S 1990a].

Applications to some classical operators in the two-dimensional torus. In what follows, fix $H_1 = H_2 = L^2(\mathbb{T}^2)$,

$$W_1^+ = H^2(\mathbb{T}^2) = \{f \in L^2 : \hat{f}(m, n) = 0 \text{ for } m < 0 \text{ or } n < 0\},$$

and $W_2^- = L^2 \ominus H^2 = H^{2\perp}$. Let $\sigma = S_x$ and $\tau = S_y$ be the shifts in each variable of $\mathbb{T}^2$, that is, $S_x : f(x, y) \mapsto e^{ix} f(x, y)$ and $S_y : f(x, y) \mapsto e^{iy} f(x, y)$. Then

$$W_2^\tau = H_{-x}^2 = \{f \in L^2 : \hat{f}(m, n) = 0 \text{ for } m \geq 0\},$$
$$W_2^\sigma = H_{-y}^2 = \{f \in L^2 : \hat{f}(m, n) = 0 \text{ for } n \geq 0\}.$$

Therefore (3–9) holds in this case, since $H^{2\perp} = H^2_{-x} + H^2_{-y}$.

REMARK. If, with the same H_1, H_2, W_1^+ and the same two shifts, we define $W_2^- = \{f \in L^2 : \hat{f}(m, n) = 0 \text{ for } m \geq 0\}$, then $W_2^\sigma = \{0\}$, while $W_2^\tau = W_2^-$. This example has applications that will not be explored here. On the other hand, choosing $W_2^- = \overline{H^2} = \{f \in L^2 : \hat{f}(m, n) = 0 \text{ for } m \geq 0 \text{ or } n \geq 0\}$, we have $W_2^\sigma = W_2^\tau = \{0\}$, and no lifting is obtained.

In this setting, since the Bochner theorem is valid in $\mathbb{T}^d$, for any $d \geq 1$, every positive Toeplitz form $B : L^2(\mathbb{T}^2) \times L^2(\mathbb{T}^2) \to \mathbb{C}$ can be represented in terms of a positive measure μ on $\mathbb{T}^2$, by $B(f, g) = \int f\bar{g}\,d\mu$.

By Theorem 3.1, any bounded Hankel form $B_0 : H^2 \times H^{2\perp} \to \mathbb{C}$ has a lifting pair of Toeplitz forms B' and B''. If μ' and μ'' are the measures on $\mathbb{T}^2$ corresponding to B' and B'', respectively, B_0 has an integral representation

$$B_0(f, g) = \begin{cases} B'(f, g) = \int f\bar{g}\,d\mu' & \text{for } f \in H^2, \ g \in H^2_{-y}, \\ B''(f, g) = \int f\bar{g}\,d\mu'' & \text{for } f \in H^2, \ g \in H^2_{-x}, \end{cases} \qquad (3\text{–}10)$$

which implies $\hat{\mu}'(m, n) = \hat{\mu}''(m, n)$ for $m < 0$ and $n < 0$. Since both forms B' and B'' are bounded in $L^2 \times L^2$, (3–10) implies that

$$d\mu' = \varphi_1\,dx\,dy, \ d\mu'' = \varphi_2\,dx\,dy, \text{ with } \|\varphi_1\|_\infty \leq \|B_0\|, \ \|\varphi_2\|_\infty \leq \|B_0\|. \quad (3\text{–}11)$$

As in the one-dimensional case, to the bounded form B_0 is associated a bounded operator $\Gamma : H^2 \to H^{2\perp}$, called a *big Hankel operator*, which satisfies

$$\Gamma S_x = (1-P)S_x\Gamma, \ \Gamma S_y = (1-P)S_y\Gamma, \text{ for } P : L^2 \to H^2 \text{ orthoprojector.} \quad (3\text{–}12)$$

Again, a function $\phi \in L^2(\mathbb{T}^2)$ is called a *symbol* of Γ if $\Gamma = \Gamma_\phi$ for $\langle \Gamma_\phi f, g \rangle = \int f\bar{g}\phi$.

COROLLARY 3.2 (NEHARI THEOREM FOR BIG HANKEL OPERATORS IN THE TORUS [C-S 1993b]). *Let* $\Gamma : H^2(\mathbb{T}^2) \to H^2(\mathbb{T}^2)^\perp$ *be a big Hankel operator. Then* $\|\Gamma\| \leq 1$ *if and only if there exist two bounded functions* $\varphi_1, \varphi_2 \in L^\infty(\mathbb{T}^2)$ *satisfying* $\|\varphi_1\|_\infty \leq 1$, $\|\varphi_2\|_\infty \leq 1$, $\hat{\varphi}_1(m, n) = \hat{\varphi}_2(m, n)$ *for* $m < 0$ *and* $n < 0$, *and*

$$\langle \Gamma f, g \rangle = \begin{cases} \int f\bar{g}\varphi_1 & \text{for } f \in H^2, \ g \in H^2_{-y}, \\ \int f\bar{g}\varphi_2 & \text{for } f \in H^2, \ g \in H^2_{-x}. \end{cases}$$

Hence, we fail to assign a bounded symbol to Γ, but we get a "pair of partial symbols" φ_1, φ_2.

In [C-S 1994a] we introduced the space BMOr of functions $\phi \in L^2(\mathbb{T}^2)$ that can be expressed as

$$\phi = \varphi_1 + h_x = \varphi_2 + h_y = \varphi_0 + h^\perp \qquad (3\text{–}13)$$

for some $\varphi_1, \varphi_2, \varphi_0 \in L^\infty(\mathbb{T}^2)$ and some $h_x \in H^2_x$, $h_y \in H^2_y$, and $h^\perp \in H^{2\perp}$. BMOr is a normed space under

$$\|\phi\|_{\text{BMOr}} := \inf\{\max\{\|\varphi_1\|_\infty, \|\varphi_2\|_\infty, \|\varphi_0\|_\infty\} : \text{all decompositions } (3\text{–}13)\}.$$

Then Corollary 3.2 can be rewritten as: A big Hankel operator Γ is bounded if and only if there exists $\phi \in$ BMOr such that $\Gamma = \Gamma_\phi$ and

$$\|\phi\|_{\mathrm{BMOr}} \leq \|\Gamma\| \leq \sqrt{2}\,\|\phi\|_{\mathrm{BMOr}}.$$

The notion of BMOr and this formulation of the Nehari theorem in $\mathbb{T}^2$ have been fully explored in [C-S 1996].

Here we want only to underline that the Lifting Theorem 3.1 is the appropriate tool for obtaining a new description of the bounded Hankel operators in several-dimensional spaces, which presents important differences with the classical one-dimensional theory [C-S 1993b; 1994a; 1996]. To give the flavor of these fundamental differences, we state two results, omitting the proofs, which are based on the Lifting Theorems 3.1 and 2.6 and their corollaries.

THEOREM 3.3 [C-S 1993b]. *Given a bounded big Hankel operator* $\Gamma : H^2(\mathbb{T}^2) \to H^2(\mathbb{T}^2)^\perp$, *its singular numbers* $s_n(\Gamma)$ *satisfy*

$$s_n(\Gamma) \geq 2^{-1/2}\|\Gamma\| \quad \text{for all } n \geq 0.$$

An immediate consequence of this result is that there are no nonzero Hankel operators in $\mathbb{T}^2$ either of finite rank or compact; thus no AAK theory of approximation, in the sense of [Adamyan, Arov and Kreĭn 1971], can be developed. (For a substitute approach based on so-called "sigma numbers," see [C-S 1996].)

The structure of the space BMOr provides the next result:

THEOREM 3.4 [C-S 1996]. *There are bounded big Hankel operators in* $\mathbb{T}^2$ *without bounded symbols.*

Obviously, the big Hankel operators defined by bounded symbols are themselves bounded. The surprising result of Theorem 3.4 leaves open the question: Given the bounded big Hankel operator Γ_φ defined by $\varphi \in L^\infty(\mathbb{T}^2)$, is there $\psi \in L^\infty(\mathbb{T}^2)$ such that $\Gamma_\psi = \Gamma_\varphi$ and $\|\psi\|_\infty$ is equivalent to $\|\Gamma_\varphi\|$? This question was posed in lectures and in [C-S 1996], with the suggestion that a positive answer was unlikely. Recently, two ingenious constructive negative answers have been given independently by Ferguson [1997] and by Bakonyi and Timotin [1997]. This inequivalence establishes the essential role of BMOr.

Another interesting consequence of the Lifting Theorem 3.1 is the following characterization of the weights ω for which the product Hilbert transform $H = H_x H_y$ is continuous in $L^2(\omega)$.

THEOREM 3.5 (HELSON–SZEGŐ THEOREM IN $\mathbb{T}^2$ [C-S 1990b]). *For a weight* $0 \leq \omega \in L^1(\mathbb{T}^2)$ *the following conditions are equivalent* (*with related constants*):

(i) *The product Hilbert transform* $H = H_x H_y$ *is continuous in* $L^2(\omega)$ *with norm* M.

(ii) *There exist real-valued functions* $u_1, u_2, v_1, v_2 \in L^\infty(\mathbb{T}^2)$ *and constants* C, ε *such that* $\|u_i\|_\infty \leq C$ *and* $\|v_i\|_\infty \leq \pi/2 - \varepsilon$ *for* $i = 1, 2$, *and*

$$\log \omega = u_1 + H_x v_1 = u_2 + H_y v_2.$$

(iii) *There exist real-valued functions w_1 and w_2 and constants C, c, M', such that $c\omega \le w_i \le C\omega$ for $i = 1, 2$, and*

$$|H_x w_1| \le M' w_1, |H_y w_2| \le M' w_2 \text{ a.e. in } \mathbb{T}^2.$$

For the details of the proof see [C-S 1990b], where a general result is also given for two weights, and for H acting in $L^p(\omega)$, $p \ne 2$.

It is noteworthy that condition (ii) in Theorem 3.5 is equivalent to $\varphi = \log w \in$ bmo, the proper subspace of product $\mathrm{BMO}(\mathbb{T}^2)$ consisting of functions of *bounded mean oscillation on rectangles*. For the relations and properties of $\mathrm{bmo}(\mathbb{T}^2) \subsetneq \mathrm{BMOr}(\mathbb{T}^2) \subsetneq \mathrm{BMO}(\mathbb{T}^2)$, see [C-S 1996].

The applications above followed from Theorem 3.1 through the integral representations of the lifting pairs provided by the classical Bochner theorem in $\mathbb{T}^2$. Applications to operators acting on the symplectic plane come from the continuous analog of Theorem 3.1 as follows (see [C-S 1990a]).

Consider $(\mathbb{C}, [\,,\,])$, the symplectic plane under the symplectic form $[z_1, z_2] = -\mathrm{Im}\, z_1 \overline{z_2}$, for $z_1, z_2 \in \mathbb{C}$. A unitary representation of the symplectic plane on a Hilbert space H is a function $z \mapsto W(z)$ assigning to each point z a unitary operator $W(z) \in \mathcal{L}(H)$, with $W(0) = I$ and satisfying the Weyl–Segal relation

$$W(z_1)W(z_2) = \exp(\pi_i[z_1, z_2])W(z_1 + z_2).$$

All irreducible representations are unitarily equivalent to the Schrödinger representation on $H = L^2(\mathbb{R})$, defined by $z \mapsto w(z)$, with

$$(w(\zeta)\varphi)(z) = \exp(-\pi i[z, \zeta])\,\varphi(z + \zeta) \quad \text{for } \zeta = (s, t), z = (x, y) \in \mathbb{C}.$$

For every $f \in L^1(\mathbb{R}^2)$ its *Weyl transform* is the bounded operator in $L^2(\mathbb{R})$ defined by

$$W(f) := \int_{\mathbb{R}^2} f(x, y)\, w(-x, -y)\, dx\, dy. \tag{3-14}$$

The Weyl transform establishes an isometric isomorphism between $L^2(\mathbb{R}^2)$ and $\mathcal{L}^2 = \mathcal{L}^2(L^2(\mathbb{R}))$, the space of Hilbert–Schmidt operators, which is a Hilbert space with scalar product $\langle A_1, A_2 \rangle = \mathrm{tr}\, A_2^* A_1$, for $A_1, A_2 \in \mathcal{L}^2$. Moreover, the product of operators corresponds to the twisted convolution in the symplectic plane,

$$W(f)W(g) = W(f \natural g) \quad \text{for } f, g \in L^2(\mathbb{R}^2),$$

where

$$f \natural g(z) = \int f(z)\, g(z - \zeta) \exp(i\pi[z, \zeta])\, d\zeta.$$

Under the Weyl isomorphism the regular representation $\{W(z) : z \in \mathbb{C}\}$ of the symplectic plane in $L^2(\mathbb{R}^2)$ passes into the unitary representation $\{w(z) : z \in \mathbb{C}\}$ of the symplectic plane in $\mathcal{L}^2(L^2(\mathbb{R}))$.

Set $\Delta = \{z = x + iy \in \mathbb{C} : x \geq 0, y \geq 0\}$. For $H = \mathcal{L}^2(L^2(\mathbb{R}))$, considering the subspaces

$$W^+ = \{A \in \mathcal{L}^2 : A = W(f), \operatorname{supp} f \subset \Delta\},$$
$$W^- = \{A \in \mathcal{L}^2 : A = W(f), \operatorname{supp} f \subset \Delta^c\}, \tag{3-15}$$

and defining the evolution groups $\{\sigma_t\}$ and $\{\tau_t\}$ by

$$\sigma_t A = w(t + i0)A, \quad \tau_t A = w(0 + it)A, \quad \text{for all } A \in \mathcal{L}^2, \, t \in \mathbb{R}, \tag{3-16}$$

we obtain a *continuous scattering system* $[H; W^+, W^-; \sigma_t, \tau_t, t \in \mathbb{R}]$ since $(3\text{--}1)$ is satisfied (although σ_s and τ_t do not commute, we have $\sigma_s \tau_t = \exp(i\pi st)\tau_t \sigma_s$, for all $t, s \in \mathbb{R}$).

A sesquilinear form $B : \mathcal{L}^2 \times \mathcal{L}^2 \to \mathbb{C}$ is Toeplitz in this scattering system if

$$B(\sigma_t A_1, \sigma_t A_2) = B(A_1, A_2) = B(\tau_t A_1, \tau_t A_2) \quad \text{for all } t \in \mathbb{R} \text{ and all } A_1, A_2 \in \mathcal{L}^2. \tag{3-17}$$

In $\mathcal{L}^2$ *measures* are replaced by their quantized analogs, *states* or trace class operators. A trace class operator $S \in \mathcal{L}^1(L^2(\mathbb{R}))$ satisfying $S \geq 0$ defines a form $B_S : \mathcal{L}^2 \times \mathcal{L}^2 \to \mathbb{C}$ by

$$B_S(A_1, A_2) = \operatorname{tr} S A_2^* A_1. \tag{3-18}$$

This definition keeps its sense when S is a bounded operator. The form B_S is Toeplitz and, for S bounded, B_S is continuous in the $\mathcal{L}^2$-topology, while for S trace class, B_S is bounded in the $\mathcal{L}^\infty$-topology of compact operators.

The following result of N. Wallach (compare [C-S 1990a]), later extended in [C-S 1990c], provides the representation for Toeplitz forms in this setting:

THEOREM 3.6 (BOCHNER THEOREM FOR THE UNITARY REPRESENTATION OF THE SYMPLECTIC PLANE). *Given a Toeplitz form* $B : \mathcal{L}^2 \times \mathcal{L}^2 \to \mathbb{C}$, *continuous in* $\mathcal{L}^2$, *there exists a bounded operator* S *in* $L^2(\mathbb{R})$ *such that*

$$B(A_1, A_2) = B_S(A_1, A_2) = \operatorname{tr} S A_2^* A_1 \quad \text{for all } A_1, A_2 \in \mathcal{L}^2. \tag{3-19}$$

Furthermore, if $B \geq 0$, *then* $S \geq 0$, *and if* B *is continuous in* $\mathcal{L}^\infty$, *then* S *is a trace class operator.*

Since the norm in $\mathcal{L}^2$ is given by $\|A\|_{\mathcal{L}^2} = \operatorname{tr} A^* A$, the expression $\operatorname{tr} S A^* A =: \|A\|_{\mathcal{L}^2(S)}$, for a bounded operator $S \geq 0$, can be considered as a "weighted" norm in $\mathcal{L}^2(S)$.

The continuous version of the Lifting Theorem 3.1 and the Representation Theorem 3.6 together imply:

COROLLARY 3.7 (QUANTIZED NEHARI THEOREM IN WEIGHTED $\mathcal{L}^2$). *Consider a pair of quantized scattering systems* $[\mathcal{L}^2(S_i); W_i^+, W_i^-; \sigma_t, \tau_t, t \in \mathbb{R}]$, *for* $i = 1, 2$, *where the* $S_i \geq 0$ *are bounded operators, the* W_i^+, W_i^- *are as in* $(3\text{--}15)$, *and*

σ_t, τ_t as in (3–16). *A Hankel form* $B_0 : W_1^+ \times W_2^- \to \mathbb{C}$ *in the pair is bounded,* $\|B_0\| \leq 1$, *if and only if there exist two bounded operators* S' *and* S'' *satisfying*

$$\max\{|\operatorname{tr} S' A_2^* A_1|, |\operatorname{tr} S'' A_2^* A_1|\} \leq \|A_1\|_{\mathcal{L}^2(S_1)} \|A_2\|_{\mathcal{L}^2(S_2)}, \qquad (3\text{–}20)$$

for all $A_1 \in \mathcal{L}^2(S_1)$, $A_2 \in \mathcal{L}^2(S_2)$, *and representing* B_0 *in the sense that*

$$B_0(A_1, A_2) = \begin{cases} \operatorname{tr} S' A_2^* A_1 & for \ A_1 \in W_1^+, A_2 \in W_2^\sigma, \\ \operatorname{tr} S'' A_2^* A_1 & for \ A_1 \in W_1^+, A_2 \in W_2^\tau, \end{cases} \qquad (3\text{–}21)$$

where the spaces

$$W_2^\sigma = \{A \in W_2^- : \sigma_t A \in W_2^- \text{ for all } t \in \mathbb{R}\},$$
$$W_2^\tau = \{A \in W_2^- : \tau_t A \in W_2^- \text{ for all } t \in \mathbb{R}\}$$

satisfy $W_2^\sigma + W_2^\tau = W_2^-$.

For $S \in \mathcal{L}^1(L^2(\mathbb{R}))$ a trace class operator, and for $s \in L^2(\mathbb{C})$, the equation $(W^{-1}S)(z) = s(z)$ in $\operatorname{Im} z > 0$ (or in $\operatorname{Re} z > 0$) is equivalent to

$$\operatorname{tr} SW(g)^* W(f) = \int s(z)(f \natural g^*)(z) \, dz = s(f \natural g^*)$$

$$\text{for } (W(f), W(g)) \in W^+ \times W^\sigma \text{ (or } W^+ \times W^\tau). \qquad (3\text{–}22)$$

For S a bounded operator, $W^{-1}S$ is not defined, but the expression (3–22) still makes sense.

THEOREM 3.8 (NEHARI THEOREM FOR OPERATORS IN $L^2(\mathbb{R})$ [C-S 1990a]). *For a given linear functional* s *in* $L^2(\mathbb{R}^2)$, *the following conditions are equivalent:*

(i) *There exist two bounded operators* S' *and* S'' *in* $L^2(\mathbb{R})$, *of norm at most* 1, *satisfying*

$$W^{-1}S' = s \text{ in } \operatorname{Im} z > 0, \quad W^{-1}S'' = s \text{ in } \operatorname{Re} z > 0$$

in the sense of (3–22).

(ii) *We have* $|s(f \natural g^*)| \leq \|f\|_{L^2} \|g\|_{L^2}$ *whenever* $\operatorname{supp} f \subset \Delta$, *and* $\operatorname{supp} g \subset \Delta^c$. *(Recall that* Δ *is the closed first quadrant of the complex plane.)*

For other results in the symplectic plane and in the dual of the Heisenberg group, see [C-S 1990a; 1990c; 1994b].

4. Lifting Theorem
for Completely Positive Definite Kernels

Section 3 concluded with a noncommutative application of the lifting theorem in scattering systems with several evolution groups. This section deals with an entirely different way in which noncommutative objects can usefully be brought into the scope of our lifting theory.

The idea here is to replace the commutative algebra $C(X)$ of continuous functions by a general C^*-algebra, and the positive linear forms on $C(X)$ (i.e., measures), by completely positive maps on the C^*-algebra.

The framework and some results from [C-S 1994c] on lifting and integral representation of kernels defined through completely positive maps are summarized below, concluding with a Nehari theorem for sequences in C^*-algebras. These results are being published here for the first time.

First, let us look again at the commutative set-up to be generalized.

If X is a group and H is a Hilbert space, every function $f : X \to \mathcal{L}(H)$ gives rise to a kernel $K : X \times X \to \mathcal{L}(H)$ defined by $K(x, y) = f(y^{-1}x)$, and K is *positive definite* if and only if f is positive definite, i.e., if

$$\sum_{j,k} \langle f(x_k^{-1}x_j)\xi_j, \, \xi_k \rangle \geq 0 \qquad (4\text{--}1)$$

for all finite sets $\{x_j\} \subset X$, $\{\xi_j\} \subset H$.

A kernel K defined by a function is invariant under the group action σ_z : $x \mapsto zx$, that is, $K(\sigma_z x, \sigma_z y) = K(x, y)$ for all $z, x, y \in X$. Conversely, every invariant kernel is defined by a function, as above.

How does this translate when functions are replaced by maps in a C^*-algebra?

If $\mathcal{A} \subset \mathcal{L}(H)$ is a unital C^*-algebra of operators, every linear map $\varphi : \mathcal{A} \to \mathcal{L}(H)$ gives rise to a kernel $K : \mathcal{A} \times \mathcal{A} \to \mathcal{L}(H)$ defined by $K(A, B) = \varphi(B^*A)$, and K is positive definite if and only if φ is *completely positive*, i.e., if

$$\sum_{j,k} \langle \varphi(A_k^*A_j)\xi_j, \, \xi_k \rangle \geq 0 \qquad (4\text{--}2)$$

for all finite sets $\{A_j\} \subset \mathcal{A}$, $\{\xi_j\} \subset \mathcal{H}$.

In this setting, a kernel defined by a map is invariant under every unitary operator U, that is, $K(UA, UB) = K(A, B)$ for all $U, A, B \in \mathcal{A}$, but the converse need not hold.

In his study of quantum probability problems, Holevo [1988] combined the notions of positive definiteness and complete positivity by considering kernels Φ defined by a function f whose values $f(x)$, for $x \in X$, are linear maps, $\mathcal{A} \to \mathcal{L}(H)$. The kernel Φ is defined by $\Phi(x, y) = f(y^{-1}x)$, and is positive definite if and only if each $f(y^{-1}x)$ is completely positive, i.e., if

$$\sum_{j,k} \langle f(x_k^{-1}x_j)(A_k^*A_j)\xi_j, \, \xi_k \rangle \geq 0 \qquad (4\text{--}3)$$

for all finite sets $\{x_j\} \subset X$, $\{A_j\} \subset \mathcal{A}$, $\{\xi_j\} \subset H$.

When the group $X = \{e\}$ consists of a single point, Φ is positive definite if and only if $\Phi(e)$ is completely positive. When $\mathcal{A} = \{cI\}$ is one-dimensional, Φ is positive definite if and only if f is positive definite. Observe that these kernels are invariant both under the group action σ_z, for $z \in X$, and under the left multiplication by unitary $U \in \mathcal{A}$. In the simplest case, when $X = \mathbb{Z}$, $H \cong \mathbb{C}$, and $\mathcal{A} \cong \mathbb{C}$, the map $\sigma = \sigma_1$ is the shift in $\mathbb{Z}$, $\sigma_1(n) = n + 1$ for all $n \in \mathbb{Z}$, and Φ has domain in $\mathbb{Z} \times \mathbb{Z}$ and numerical values. These are the kernels of Section 1, and it is natural to seek the generalization of the GBT for general kernels Φ under suitable invariance conditions.

For this purpose we consider a setting somewhat more general than that of Holevo. Instead of assuming that the values $\Phi(x, y)$ are already defined as invariant under unitary operators, by $\Phi(x, y)(A, B) = \phi(x, y)(B^*A)$, we assume $\Phi(x, y)(A, B)$ to be any sesquilinear form in $\mathcal{A} \times \mathcal{A}$ with values in $\mathcal{L}(H)$, on which we impose invariance conditions under certain groups of automorphisms of the algebra $\mathcal{A}$. Furthermore, we assume X to be any set, and not necessarily a group. This more general setting for the kernels is chosen for two reasons. In the first place, positive definite kernels $\Phi : X \times X \to \mathcal{L}(\mathcal{A} \times \mathcal{A}, \mathcal{L}(H))$ can be considered as Hilbert space reproducing kernels. Secondly, in order to obtain invariant liftings, it is necessary to restrict the invariance condition with respect to $\mathcal{A}$ imposed to the kernels. Among the more general kernels having as values sesquilinear forms in $\mathcal{A} \times \mathcal{A}$ to $\mathcal{L}(H)$, those satisfying

$$\Phi(x, y)(A, B) = \Phi(x, y)(B^*A, I) \quad \text{for all } A, B \in \mathcal{A} \text{ and } x, y \in X, \qquad (4\text{--}4)$$

are called of *Holevo type*.

For X a set, and $\mathcal{A} \subset \mathcal{L}(H)$ a unital C^*-algebra of operators, a kernel $\Phi : X \times X \to \mathcal{L}(\mathcal{A} \times \mathcal{A}, \mathcal{L}(H))$ is positive definite if and only if

$$\sum_{jk} \langle \Phi(x_j, x_k)(A_j, A_k)\xi_j, \xi_k \rangle \geq 0 \qquad (4\text{--}5)$$

for all finite sets $\{x_j\} \subset X$, $\{A_j\} \subset \mathcal{A}$, $\{\xi_j\} \subset H$. In what follows, in abuse of language, such a Φ is referred to as *completely positive definite*.

Fix a bijection $\sigma : X \to X$. The kernel Φ is called σ-*invariant* or *Toeplitz* if

$$\Phi(\sigma x, \sigma y) = \Phi(x, y) \quad \text{for all } x, y \in X. \qquad (4\text{--}6)$$

For two fixed subsets X_1, X_2 satisfying $X_1 \cup X_2 = X$, $X_1 \cap X_2 = \varnothing$, and

$$\sigma X_1 \subset X_1, \quad \sigma^{-1} X_2 \subset X_2, \qquad (4\text{--}7)$$

a kernel $\Phi_0 : X_1 \times X_2 \to \mathcal{L}(\mathcal{A} \times \mathcal{A}, \mathcal{L}(H))$ is σ-*invariant* in $X_1 \times X_2$ or *Hankel* if

$$\Phi_0(\sigma x, y) = \Phi_0(x, \sigma^{-1} y) \quad \text{for all } x \in X_1, y \in X_2. \qquad (4\text{--}8)$$

Just as in Section 1, to give three kernels Φ_1, Φ_2, Φ_0, where $\Phi_1 : X_1 \times X_1 \to \mathcal{L}(\mathcal{A} \times \mathcal{A}, \mathcal{L}(H))$ and $\Phi_2 : X_2 \times X_2 \to \mathcal{L}(\mathcal{A} \times \mathcal{A}, \mathcal{L}(H))$ are Toeplitz and $\Phi_0 : X_1 \times X_2 \to \mathcal{L}(\mathcal{A} \times \mathcal{A}, \mathcal{L}(H))$ is Hankel, is the same as to give a GTK $\Phi : X \times X \to \mathcal{L}(\mathcal{A} \times \mathcal{A}, \mathcal{L}(H))$ such that $\Phi|(X_1 \times X_1) = \Phi_1$, $\Phi|(X_2 \times X_2) = \Phi_2$, and $\Phi|(X_1 \times X_2) = \Phi_0$. In this situation we write $\Phi \sim (\Phi_1, \Phi_2, \Phi_0)$. Such a GTK

Φ is completely positive definite if and only if Φ_1 and Φ_2 are completely positive definite and $\Phi_0 \leq (\Phi_1, \Phi_2)$ in $X_1 \times X_2$, that is,

$$\left| \sum_{j,k} \langle \Phi_0(x_j, y_k)(A_j, B_k)\xi_j, \eta_k \rangle \right|^2$$
$$\leq \left(\sum_{j,k} \langle \Phi_1(x_j, x_k)(A_j, A_k)\xi_j, \xi_k \rangle \right) \left(\sum_{j,k} \langle \Phi_2(y_j, y_k)(B_j, B_k)\eta_j, \eta_k \rangle \right) \quad (4\text{--}9)$$

for all finite sets $\{x_j\} \subset X_1, \{y_j\} \subset X_2, \{A_j\}, \{B_j\} \subset \mathcal{A}, \{\xi_j\}, \{\eta_j\} \subset H$.

For G a group and $\alpha : G \to \mathcal{L}(\mathcal{A})$ a representation of G by linear maps $\alpha_\gamma : \mathcal{A} \to \mathcal{A}$, for $\gamma \in G$, we call $\{\mathcal{A}, G, \alpha\}$ a *linear dynamical system*. Here we restrict ourselves to the case when G is an amenable group and $\alpha_\gamma(A) = U_\gamma A$, for $\gamma \mapsto U_\gamma$, a unitary representation of G satisfying $U_\gamma(\mathcal{A}) \subset \mathcal{A}$ for all $\gamma \in G$.

A kernel Φ is called *invariant with respect to such linear dynamical system*, or, simply, α-*invariant* if

$$\Phi(x,y)\big(\alpha_\gamma(A), \alpha_\gamma(B)\big) = \Phi(x,y)(A,B) \quad \text{for all } \gamma \in G, A, B \in \mathcal{A}, x, y \in X.$$
$$(4\text{--}10)$$

For instance, if Φ is of Holevo type, it is α-invariant for $\alpha_\gamma : A \mapsto U_\gamma A$, for all $U_\gamma \in \mathcal{A}$ unitary. But such invariance does not, in general, imply that Φ must be of Holevo type.

There is a large class of linear dynamical systems $\{\mathcal{A}, G, \alpha\}$, with $\alpha_\gamma(A) = U_\gamma A$, such that α-invariance of a kernel Φ *is* equivalent to Φ being of Holevo type. We call systems of this class *reducing*. Examples of reducing systems and their representations are (i) the Schrödinger representation, when $\mathcal{A}$ is the algebra of Hilbert–Schmidt operators acting in $L^2(\mathbb{R})$, and (ii) a unitary representation of a nilpotent Lie group G, when $\mathcal{A}$ is the Schwartz algebra of operators $\{A_f : f \in \mathcal{S}(G)\}$, for $A_f = \int_G f(\gamma) U_\gamma d\gamma$, and $\mathcal{S}(G)$ the Schwartz class of functions in G (compare [du Cloux 1989]).

Given a completely positive definite kernel Φ, set $Y = X \otimes \mathcal{A} \otimes H$ and define a numerical kernel $K : Y \times Y \to \mathbb{C}$ by

$$K((x, A, \xi), (y, B, \eta)) := \langle \xi, \Phi(x,y)(A,B)\eta \rangle \quad (4\text{--}11)$$

for all $(x, A, \xi), (y, B, \eta) \in Y$. Let $\mathcal{H}$ be the Hilbert space spanned by the functions $K_{xA\xi} : Y \to \mathbb{C}$ given by

$$K_{xA\xi}(y, B, \eta) = K((x, A, \xi), (y, B, \eta)) \quad (4\text{--}12)$$

with the scalar product

$$\langle K_{xA\xi}, K_{yB\eta} \rangle = K_{xA\xi}(y, B, \eta).$$

A bijection $\sigma : X \to X$ gives rise to a unitary operator $\sigma : \mathcal{H} \to \mathcal{H}$, defined by

$$\sigma(K_{xA\xi}) := K_{\sigma(x)A\xi} \quad \text{for all } x \in X, A \in \mathcal{A}, \xi \in H. \quad (4\text{--}13)$$

If the completely positive definite Φ is a GTK, that is, if $\Phi \sim (\Phi_1, \Phi_2, \Phi_0)$ for two Toeplitz kernels and a Hankel kernel in $X = X_1 \cup X_2$, then it defines a pair of scattering systems $[\mathcal{H}_1; W_1^+, W_1^-; \sigma_1]$ and $[\mathcal{H}_2; W_2^+, W_2^-; \sigma_2]$, where Φ_1, Φ_2 are the reproducing kernels for $\mathcal{H}_1$ and $\mathcal{H}_2$, σ_1 and σ_2 are the corresponding unitary operators, for which the forms $B_i : \mathcal{H}_i \times \mathcal{H}_i \to \mathbb{C}$ defined by

$$B_i(K_{xA\xi}^i, K_{yB\eta}^i) := \langle \xi, \Phi_i(x, y)(A, B)\eta \rangle \tag{4-14}$$

are Toeplitz. It is not difficult to check that for W_i^+, W_i^- corresponding to the subspaces of $\mathcal{H}_i$ spanned by $\{K_{xA\xi}^i \in \mathcal{H}_i : x \in X_i\}$, for $i = 1, 2$, a scattering system is obtained in which the form $B_0 : W_1^+ \times W_2^- \to \mathbb{C}$ defined by K^1 and K^2 in a way similar to (4–14) is Hankel. Then the Lifting Theorem 2.2 yields a lifting theorem for α-invariant completely positive definite GTKs:

THEOREM 4.1 [C-S 1994c]. *Given a linear dynamical system* $\{A, G, \alpha\}$ *and an α-invariant completely positive definite GTK* $\Phi \sim (\Phi_1, \Phi_2, \Phi_0) : X \times X \to \mathcal{L}(A \times A, \mathcal{L}(H))$, *there exists an α-invariant Toeplitz* $\Phi' : X \times X \to \mathcal{L}(A \times A, \mathcal{L}(H))$ *such that* $\Phi'|(X_1 \times X_2) = \Phi_0$ *and* $\Phi' \le (\Phi_1, \Phi_2)$ *in all of* $X \times X$. *Furthermore, if* $\{A, G, \alpha\}$ *is reducing and* Φ *is of Holevo type,* Φ' *is also of Holevo type.*

In the special case when $X = \mathbb{Z}$, $X_1 = \mathbb{Z}_1$, $X_2 = \mathbb{Z}_2$, there is a precise integral representation, closely related to the GBT.

THEOREM 4.2 (INTEGRAL REPRESENTATION FOR α-INVARIANT COMPLETELY POSITIVE DEFINITE GTK DEFINED IN $\mathbb{Z} \times \mathbb{Z}$ [C-S 1994c]). *Set* $\Omega = \{1, 2\}$. *Given a completely positive definite GTK* $\Phi \sim (\Phi_1, \Phi_2, \Phi_0) : \mathbb{Z} \times \mathbb{Z} \to \mathcal{L}(A \times A, \mathcal{L}(H))$, *there exists a 2×2 matrix of measures* $\mu = (\mu_{jk})$ *defined on* $\mathbb{T}$ *such that, for every Borel set* $D \subset \mathbb{T}$, $\mu(D) : \Omega \times \Omega \to \mathcal{L}(A \times A, \mathcal{L}(H))$ *is a completely positive definite kernel, satisfying*

$$\Phi_i(m, n)(A, B) = \hat{\mu}_{ii}(m, n)(A, B) = \int_{\mathbb{T}} e^{i(m-n)t} d\mu_{ii}(A, B)$$

for all $m, n \in \mathbb{Z}$ *and* $i = 1, 2$, *and*

$$\Phi_0(m, n)(A, B) = \hat{\mu}_{12}(m, n)(A, B) \quad \text{for } m \in \mathbb{Z}_1, \, n \in \mathbb{Z}_2.$$

Furthermore, the measures μ_{11} *and* μ_{22} *are α-invariant.*

These lifting theorems provide results in dilation theory, including one on intertwining contractions coupled with ∗-representations generalizing the Sz.-Nagy–Foiaş theorem. These results from [C-S 1994c] will not be stated here.

The last result of this section reduces to the Nehari–Page theorem [Page 1970] in the case dim $A = 1$.

THEOREM 4.3 (NEHARI THEOREM FOR SEQUENCES DEFINED IN C^*-ALGEBRAS [C-S 1994c]). *Let* $A \subset \mathcal{L}(H)$ *be the C^*-algebra of operators defined in the*

Schwartz space $\mathcal{S}(G)$, *for* G *a nilpotent Lie group, through a unitary representation. Given a sequence of linear maps,* $s_n : \mathcal{A} \to \mathcal{L}(H)$, *for* $n = 1, 2, \ldots$, *and a* $*$-*representation* $\theta : \mathcal{A} \to \mathcal{L}(H)$ *of* $\mathcal{A}$, *the following conditions are equivalent*:

(i) *There exists a function* $f : \mathbb{T} \to \mathcal{L}(\mathcal{A}, \mathcal{L}(H))$ *completely contractive with respect to* θ, *that is, satisfying*

$$\left| \sum_{j,k} \langle f(t)(A_k^* A_j)\, \xi_k, \xi_j \rangle \right| \leq \sum_j \|\theta(A_j)\xi_j\|^2,$$

and such that $\hat{f}(-n) = s_n$ *for* $n = 1, 2, \ldots$.

(ii) *The Hankel kernel* $\{s(m + n) : m, n > 0\}$ *satisfies*

$$\left| \sum_{m,n>0} \langle s(m + n)(A_n^* A_m)\eta_n, \, \xi_m \rangle \right|^2 \leq \left(\sum_{m>0} \|\theta(A_m)\xi_m\|^2 \right) \left(\sum_{n>0} \|\theta(A_n)\eta_n\|^2 \right).$$

Acknowledgements

I want to thank Chandler Davis for helpful discussions on the organization of this paper, and Silvio Levy for his excellent editing. The strong support of the directors and all staff members of MSRI, and of Don Sarason, co-chair of the program on Holomorphic Spaces, is gratefully acknowledged.

References

[Adamyan and Arov 1966] V. M. Adamyan and V. Z. Arov, "On unitary couplings of semiunitary operators", *Matem. Issledovanyia* **1** (1966), 3–64. In Russian; translation in *Amer. Math. Soc. Transl. Ser.* 2 **95** (1970), 75–169.

[Adamyan, Arov and Kreĭn 1971] V. M. Adamyan, V. Z. Arov, and M. G. Kreĭn, "Аналитические свойства пар Шмидта ганкелева оператора и обобщённая задача Шура–Такаги", *Mat. Sbornik* **86** (1971), 33–75. Translated as "Analytic properties of Schmidt pairs of a Hankel operator and generalized Schur–Takagi problem", *Math. USSR Sb.* **15** (1971), 31-73.

[Arocena, Cotlar, and Sadosky 1981] R. Arocena, M. Cotlar, and C. Sadosky, "Weighted inequalities in L^2 and lifting properties", pp. 95–128 in *Mathematical analysis and applications*, part A, edited by L. Nachbin, Adv. in Math. Suppl. Stud. **7**, Academic Press, New York and London, 1981.

[Bakonyi and Timotin 1997] M. Bakonyi and D. Timotin, "On a conjecture of Cotlar and Sadosky on multidimensional Hankel operators", preprint, 1997.

[C-S 1979] M. Cotlar and C. Sadosky, "On the Helson–Szegő theorem and a related class of modified Toeplitz kernels", pp. 383–407 in *Harmonic analysis in Euclidean spaces* (Williamstown, MA, 1978), part 1, edited by G. Weiss and S. Wainger, Proc. Sympos. Pure Math. **35**, Amer. Math. Soc., Providence, 1979.

[C-S 1983] M. Cotlar and C. Sadosky, "On some L^p versions of the Helson–Szegő theorem", pp. 306–317 in *Conference on harmonic analysis in honor of Antoni Zygmund* (Chicago, 1981), vol. 1, edited by W. Beckner et al., Wadsworth, Belmont, CA, 1983.

[C-S 1984/85] M. Cotlar and C. Sadosky, "Generalized Toeplitz kernels, stationarity and harmonizability", *J. Analyse Math.* **44** (1984/85), 117–133.

[C-S 1986] M. Cotlar and C. Sadosky, "A lifting theorem for subordinated invariant kernels", *J. Funct. Anal.* **67**:3 (1986), 345–359.

[C-S 1987] M. Cotlar and C. Sadosky, "Prolongements des formes de Hankel généralisées en formes de Toeplitz", *C. R. Acad. Sci. Paris Sér. I Math.* **305**:5 (1987), 167–170.

[C-S 1988] M. Cotlar and C. Sadosky, "Integral representations of bounded Hankel forms defined in scattering systems with a multiparametric evolution group", pp. 357–375 in *Contributions to operator theory and its applications* (Mesa, AZ, 1987), Oper. Theory Adv. Appl. **35**, Birkhäuser, Basel and Boston, 1988.

[C-S 1989a] M. Cotlar and C. Sadosky, "Generalized Bochner theorem in algebraic scattering systems", pp. 144–169 in *Analysis at Urbana* (Urbana, IL, 1986–1987), vol. 2, London Math. Soc. Lecture Note Ser. **138**, Cambridge Univ. Press, Cambridge and New York, 1989.

[C-S 1989b] M. Cotlar and C. Sadosky, "Nonlinear lifting theorems, integral representations and stationary processes in algebraic scattering systems", pp. 97–123 in *The Gohberg anniversary collection* (Calgary, AB, 1988), vol. 2, edited by H. Dym et al., Oper. Theory Adv. Appl. **41**, Birkhäuser, Basel and Boston, 1989.

[C-S 1990a] M. Cotlar and C. Sadosky, "Two-parameter lifting theorems and double Hilbert transforms in commutative and noncommutative settings", *J. Math. Anal. Appl.* **150**:2 (1990), 439–480.

[C-S 1990b] M. Cotlar and C. Sadosky, "The Helson–Szegő theorem in L^p of the bidimensional torus", pp. 19–37 in *Harmonic analysis and partial differential equations* (Boca Raton, FL, 1988), edited by M. Milman and T. Schonbek, Contemp. Math. **107**, Amer. Math. Soc., Providence, 1990.

[C-S 1990c] M. Cotlar and C. Sadosky, "Toeplitz and Hankel forms related to unitary representations of the symplectic plane", *Colloq. Math.* **60/61**:2 (1990), 693–708.

[C-S 1991] M. Cotlar and C. Sadosky, "Toeplitz liftings of Hankel forms bounded by non-Toeplitz norms", *Integral Equations Operator Theory* **14**:4 (1991), 501–532.

[C-S 1993a] M. Cotlar and C. Sadosky, "Transference of metrics induced by unitary couplings, a Sarason theorem for the bidimensional torus, and a Sz.-Nagy–Foiaş theorem for two pairs of dilations", *J. Funct. Anal.* **111**:2 (1993), 473–488.

[C-S 1993b] M. Cotlar and C. Sadosky, "Abstract, weighted, and multidimensional Adamjan–Arov–Kreĭn theorems, and the singular numbers of Sarason commutants", *Integral Equations Operator Theory* **17**:2 (1993), 169–201.

[C-S 1994a] M. Cotlar and C. Sadosky, "Nehari and Nevanlinna–Pick problems and holomorphic extensions in the polydisk in terms of restricted BMO", *J. Funct. Anal.* **124**:1 (1994), 205–210.

[C-S 1994b] M. Cotlar and C. Sadosky, "The Adamjan–Arov–Kreĭn theorem in general and regular representations of $\mathbb{R}^2$ and the symplectic plane", pp. 54–78 in *Toeplitz operators and related topics* (Santa Cruz, CA, 1992), edited by E. L. Basor and I. Gohberg, Oper. Theory Adv. Appl. **71**, Birkhäuser, Basel, 1994.

[C-S 1994c] M. Cotlar and C. Sadosky, "Completely positive kernels invariant with respect to a dynamical system", preprint, 1994. Presented at the International

Conference in honor of B. Sz.-Nagy (Szeged, August 1993) and at the International Conference in honor of M. Cotlar (Caracas, January 1994).

[C-S 1996] M. Cotlar and C. Sadosky, "Two distinguished subspaces of product BMO and the Nehari–AAK theory for Hankel operators on the torus", *Integral Equations Operator Theory* **26** (1996), 273–304.

[du Cloux 1989]   F. du Cloux, "Représentations tempérées des groupes de Lie nilpotents", *J. Funct. Anal.* **85**:2 (1989), 420–457.

[Ferguson 1997] S. H. Ferguson, "The Nehari Problem for the Hardy space of the torus", preprint, 1997.

[Helson and Sarason 1967] H. Helson and D. Sarason, "Past and future", *Math. Scand.* **21** (1967), 5–16.

[Helson and Szegő 1960] H. Helson and G. Szegő, "A problem in prediction theory", *Ann. Mat. Pura Appl.* (4) **51** (1960), 107–138.

[Holevo 1988]   A. S. Holevo, "A noncommutative generalization of conditionally positive definite functions", pp. 128–148 in *Quantum probability and applications, III* (Oberwolfach, 1987), Lecture Notes in Math. **1303**, Springer, Berlin and New York, 1988.

[Lax and Phillips 1967] P. D. Lax and R. S. Phillips, *Scattering theory*, Pure and Applied Mathematics **26**, Academic Press, New York and London, 1967.

[Nehari 1957] Z. Nehari, "On bounded bilinear forms", *Ann. of Math.* (2) **65** (1957), 153–162.

[Nikol'skiĭ 1986] N. K. Nikol'skiĭ, *Treatise on the shift operator*, Grundlehren der mathematischen Wissenschaften **273**, Springer, Berlin, 1986. With an appendix by S. V. Hruščev and V. V. Peller. Translation of Лекции об операторе сдвига, Nauka, Moscow, 1980.

[Page 1970] L. B. Page, "Bounded and compact vectorial Hankel operators", *Trans. Amer. Math. Soc.* **150** (1970), 529–539.

[Sarason 1965] D. Sarason, "On spectral sets having connected complement", *Acta Sci. Math.* (*Szeged*) **26** (1965), 289–299.

[Sarason 1967] D. Sarason, "Generalized interpolation in H^∞", *Trans. Amer. Math. Soc.* **127**:2 (1967), 179–203.

[Sz.-Nagy and Foiaş 1967]   B. Sz.-Nagy and C. Foiaş, *Analyse harmonique des opérateurs de l'espace de Hilbert*, Masson, Paris, and Akadémiai Kiadó, Budapest, 1967. Translated as *Harmonic analysis of operators on Hilbert space*, North-Holland, Amsterdam, and Akadémiai Kiadó, Budapest, 1970.

[Treil' 1985] S. R. Treil', "The Adamyan–Arov–Kreĭn theorem: a vector version", *Zap. Nauchn. Sem. Leningrad. Otdel. Mat. Inst. Steklov.* **141** (1985), 56–71. In Russian.

CORA SADOSKY
DEPARTMENT OF MATHEMATICS
HOWARD UNIVERSITY
WASHINGTON, DC 20059
UNITED STATES
cs@scs.howard.edu

Holomorphic Spaces
MSRI Publications
Volume **33**, 1998

Some Function-Theoretic Issues in Feedback Stabilisation

NICHOLAS YOUNG

ABSTRACT. This article aims to present for a mathematical audience some interesting function theory elaborated over recent years by control engineers in connection with the problem of robustly stabilising an imperfectly modelled physical device.

The theory of feedback extends over a broad field, from rarefied differential geometry to down-to-earth nuts-and-bolts engineering. Function theory enters into the study of a class of problems of great practical importance, those relating to linear time-invariant systems. It is still the case that the great majority of engineering devices are modelled by such systems, and function theory remains an important strand in recent engineering researches on the stabilisation of uncertain systems. The connection with H^∞ and Hankel operators is widely known by now, but the extent to which engineers have developed the mathematical theory along novel lines deserves publicity. Challenges of an engineering nature have given rise to some beautiful ideas and results in function theory, and the purpose of this expository article is to present some notions arising from studies of robust stabilisation which deserve the attention of mathematicians. These notions relate to certain spaces of functions on the real line or subsets of the complex plane and to sundry metrics on these spaces which measure closeness of functions from the point of view of stabilisability. Many engineers have contributed to these developments, notably M. Vidyasagar, T. T. Georgiou, M. C. Smith, K. Glover, D. C. McFarlane, and G. Vinnicombe. Virtually everything in this paper is from [Vidyasagar 1984; Georgiou and Smith 1990; 1993; McFarlane and Glover 1990; Vinnicombe 1993; Curtain and Zwart 1995]. A useful textbook covering the elements of H^∞-control is [Doyle et al. 1992]. However, these sources assume

1991 *Mathematics Subject Classification.* Primary 93D21, Secondary 30E10.

Key words and phrases. robust stabilisation, graph topology, gap metric, stability margin.

The author thanks the Mathematical Sciences Research Institute for hospitality while this work was carried out. His research at MSRI was supported in part by NSF grant DMS-9022140.

knowledge of control theory, and my hope here is to provide function-theorists with a way into some fine and relevant mathematics in places they would not normally look. I am grateful to Keith Glover and Malcolm Smith for helpful comments.

1. Robust Stabilisation

Consider this standard feedback configuration:

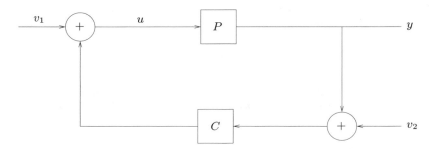

Here, as usual, P is the *plant*—that is, a function representing the object to be controlled—and C is the controller. We work in the frequency domain, so u and y represent the Laplace transforms of input and output signals to the plant. Subject to zero initial conditions, u and y are related by $y = Pu$, where u, y, and P are functions of the frequency variable s. We describe P as the *transfer function* of the system. The signal v_2 can be thought of as noise in the sensors that measure y. If the plant is modelled by a system of linear constant-coefficient differential equations then P will be a rational function, but if the model contains delays then P will have some exponential terms. Although much of the engineering literature concentrates on rational P there is a substantial body of work addressing the question of the most appropriate spaces of functions for the analysis of stablisation of general (not necessarily rational) plants. We shall return to this issue below, but to begin with let us think of rational plants P, and let us suppose that P is unstable, i.e., has at least one pole in the closed right half plane $\{s \in \mathbb{C} : \operatorname{Re} s \geq 0\}$. We wish to stabilise P with a rational controller C. In the feedback loop of Figure 1 we have

$$y = Pu = P(v_1 + Cy + Cv_2)$$

and so

$$y = (1 - PC)^{-1}Pv_1 + (1 - PC)^{-1}PCv_2$$

Thus for the loop to be stable we need the rational functions $(1 - PC)^{-1}P$ and $(1 - PC)^{-1}PC$ to be analytic in the closed right half-plane. For true peace of mind, though, we need somewhat more, since if there is instability in any of the connections of the loop then that connection will be liable to burn out.

Note that
$$u = v_1 + Cy + Cv_2 = v_1 + CPu + Cv_2$$
and so
$$u = (1 - CP)^{-1}v_1 + (1 - CP)^{-1}Cv_2.$$
Thus
$$\begin{bmatrix} y \\ u \end{bmatrix} = \begin{bmatrix} (1 - PC)^{-1}P & (1 - PC)^{-1}PC \\ (1 - CP)^{-1} & (1 - CP)^{-1}C \end{bmatrix} \begin{bmatrix} v_1 \\ v_2 \end{bmatrix} = \begin{bmatrix} P \\ 1 \end{bmatrix} (1 - CP)^{-1} \begin{bmatrix} 1 & C \end{bmatrix} \begin{bmatrix} v_1 \\ v_2 \end{bmatrix}$$
We shall say that the system (P, C) is *stable* or that C *stabilises* P if

$$H(P, C) \stackrel{\text{def}}{=} \begin{bmatrix} P \\ 1 \end{bmatrix} (1 - CP)^{-1} \begin{bmatrix} 1 & C \end{bmatrix} \tag{1--1}$$

is bounded and analytic in the closed right half-plane (this property is sometimes also called *internal stability*). The simplest form of the *stabilisation problem* is:

Given a plant P, find a controller C such that (P, C) is stable.

Classical control is full of recipes for constructing such Cs and simultaneously achieving various desirable performance characteristics [Rohrs et al. 1993]. This simple formulation, however, leaves out of account one of the most important aspects of control system design: the fact that models of plants are only approximate. Control engineers have long had ways of addressing this difficulty, but the development of a theory that meets it head-on is relatively recent. "H^∞ control" began around 1980 and is based on the notion that instead of finding C to stabilise a single plant P one should be looking for a C that stabilises not only the chosen model P but also all sufficiently close plants. The idea is natural: one constructs a model P_0 of a physical device by making several idealisations, simplifications, approximations and estimates. This P_0 is called the "nominal plant". One posits that there is a "true plant" P that is close to P_0, and it is P that one really wants to stabilise. Since P is unknown one should try to find a C that simultaneously stabilises as large a neighbourhood of P_0 as possible. To put this approach into effect for any given application one must decide

(1) What is the appropriate space of functions P?

(2) What is the appropriate notion of closeness?

The most natural interpretation of the second question is that one should give a metric on the chosen space of transfer functions; then one could try to stabilise the ball of greatest possible radius about P_0. However a weaker version is also of interest: what is the appropriate topology on the space of transfer functions [Vidyasagar 1984; Zames and El-Sakkary 1980]? Answering these questions requires significant mathematical as well as engineering considerations.

Corresponding to any answer to questions (1) and (2), then, we have a version of the *robust stabilisation problem*:

Given a nominal plant P_0 find the largest possible neighbourhood U of P_0 such that there is a controller C that stabilises every element of U.

In the case that closeness of plants is measurable by a metric then it is natural to seek the neighbourhood U that is the open ball of greatest possible radius.

The simplest version of the theory arises from choosing the space of rational functions without purely imaginary poles and the metric to be the L^∞ norm on the imaginary axis. This norm is physically well motivated, since it is roughly speaking the square root of the maximum ratio of the energies of output to input signals. Thus two plants are close in the L^∞ norm if, for the same input of unit energy they produce uniformly close outputs. This form of the robust stabilisation problem has an elegant solution in which an important ingredient is some classical analysis: Nevanlinna-Pick theory or Nehari's theorem. An account can be found in [Francis 1987, Chapter 6], and a simplified one in [Young 1988, Chapter 14]. The solution has, however, a slightly undesirable feature. Suppose we are given a strictly proper rational plant P_0 (a rational function is *strictly proper* if it vanishes at infinity). For $\varepsilon > 0$ we denote by $V(P_0, \varepsilon)$ the set of strictly proper rational plants P, analytic on the imaginary axis, such that $\|P - P_0\|_\infty < \varepsilon$ and *P has the same number of poles as P_0 in the right half-plane*. The theory tells us that the largest value of ε for which all members of $V(P_0, \varepsilon)$ can be simultaneously stabilised by a single controller is the reciprocal of the norm of a certain Hankel operator. This is an elegant robust stabilisation result, but it is not precisely the answer to the problem as posed. The restriction on the number of right-half-plane poles of the perturbed plant P is a requirement of the method of solution rather than a natural engineering assumption: the nominal plant P_0 and the nearby "true plant" P may perfectly easily have different numbers of poles in the right half-plane. Indeed, there is no reason to rule out plants with poles *on* the imaginary axis. These considerations led engineers to seek alternative approaches even within the framework of rational functions. One that has been particularly successful is a different representation of rational functions that allows us to use the L^∞ norm even in the presence of poles on the imaginary axis.

2. Graphs and Metrics

Stable plants have transfer functions that are bounded and analytic in the closed right half-plane, and so the H^∞ norm immediately gives a natural metric on the set of stable plants. The analysis of robust stabilization demands a metric for *unstable* plants, and it is less clear how that should be defined. An approach that has been successful is to focus not so much on the concrete rational transfer function but rather on the operator from inputs to outputs, and more particularly on its graph. One could after all argue that this is a more fundamental entity than the transfer function. This approach leads to a single natural topology on rational plants, the *graph topology*, which seems to have gained acceptance among engineers as the appropriate one for the robust stabilisation problem. However, there are several different metrics that give rise to this topology, or what is the same thing, several inequivalent ways of representing modelling uncertainty. Each corresponds to a version of the robust stabilisation problem; some have elegant solutions, some are as yet unsolved. The operator-theoretic view also has the merit of giving a lead towards generalisation of the theory to non-rational and even nonlinear plants. I particularly recommend [Georgiou and Smith 1993] as enjoyable reading for operator-theorists, since the paper gives a concise and elegant account of the basic notions of stabilisation from a congenial perspective.

For control applications one has to study plants with several inputs and outputs, so that transfer functions are matrix-valued. This makes problems significantly more complex and more interesting, but for the present purpose it is enough to consider plants with a single input and output, so that transfer functions will be scalar-valued. Consider a rational plant P. It determines a possibly unbounded closed linear operator M_P on the Hardy space H^2 of the right half-plane: $M_P u = Pu$ for $u \in H^2$ such that $Pu \in H^2$. Note that the domain of M_P,

$$\mathcal{D}(M_P) \overset{\text{def}}{=} \{u \in H^2 : Pu \in H^2\},$$

may be the whole of H^2 (if $P \in H^\infty$), a proper dense subspace (e.g., $P(s) = s^{-1}$), a proper closed subspace (e.g., $P(s) = (s-1)^{-1}$), or a subspace that is neither closed nor dense (e.g., $P(s) = s^{-1}(s-1)^{-1}$). Define the *graph of P*, denoted by $\mathcal{G}_P$, to be the graph of M_P:

$$\mathcal{G}_P = \left\{ \begin{bmatrix} u \\ Pu \end{bmatrix} : u \in \mathcal{D}(M_P) \right\}.$$

$\mathcal{G}_P$ is a closed subspace of $H^2 \oplus H^2$, and moreover it is invariant under multiplication by e^{-as} for all $a > 0$. By the Lax–Beurling theorem there is an inner $2 \times r$ function θ_P such that $\mathcal{G}_P = \theta_P H^2(\mathbb{C}^r)$ for some $r \in \mathbb{N}$. Obviously $r \leq 2$, and in fact we must have $r = 1$. For suppose $\theta_P = \begin{bmatrix} M \\ N \end{bmatrix}$ with M, N of type $1 \times r$ and let $Mx = 0$ for some $x \in H^2(\mathbb{C}^r)$. Then

$$\begin{bmatrix} 0 \\ Nx \end{bmatrix} = \begin{bmatrix} Mx \\ Nx \end{bmatrix} = \theta_P x \in \mathcal{G}_P,$$

and hence $Nx = 0$. Thus $\theta_P x = 0$, and since θ_P is inner, $x = 0$. That is, $Mx = 0$ implies $x = 0$, from which it follows that $r = 1$. Thus θ_P is a 2×1 inner function, which is called a *graph symbol* for P or a *normalised coprime factor representation* for P. The point of the latter terminology is that if $\theta_P = \begin{bmatrix} M \\ N \end{bmatrix}$ then $P = NM^{-1}$ is an expression of P as a ratio of two stable rational plants that are coprime elements of the algebra $\mathcal{S}$ of stable rational functions ($\mathcal{S} = \mathbb{C}(s) \cap H^\infty$ where $\mathbb{C}(s)$ denotes the field of rational functions in the variable s over $\mathbb{C}$), and that are normalised in the sense that $|N|^2 + |M|^2 = 1$ on the imaginary axis. Consider for example the unstable plant $P(s) = (s-1)/(s-2)$. For any choice of a in the left half plane

$$P(s) = \frac{s-1}{s-a} \bigg/ \frac{s-2}{s-a}$$

is an expression of P as a ratio of coprime elements of $\mathcal{S}$. A little calculation determines the unique a for which normalisation is achieved, and in fact

$$\theta_P(s) = \frac{1}{\sqrt{2}s + \sqrt{5}} \begin{bmatrix} s-1 \\ s-2 \end{bmatrix}.$$

Of course θ_P is only unique up to multiplication by complex numbers of unit modulus.

The correspondences

Plants $P \longleftrightarrow$ Closed subspaces $\mathcal{G}_P$ of $H^2(\mathbb{C}^2) \longleftrightarrow 2 \times 1$ inner functions θ_P

provide a useful conceptual and analytic framework for the study of robust stabilisation. Topologies and metrics on the family of closed subspaces of a Hilbert space induce the corresponding objects on the set of plants and the inner functions θ_P provide a tool for detailed analysis and computation.

The simplest metric on the set of closed subspaces of a Hilbert space is the *gap metric*. This is the metric induced by the operator norm via the identification of a closed subspace with the corresponding orthogonal projection. That is, if Π_K denotes the orthogonal projection operator in $\mathcal{L}(H)$ with range K then the gap between closed subspaces K_1, K_2 of H is

$$\mathrm{gap}(K_1, K_2) = \|\Pi_{K_1} - \Pi_{K_2}\|$$

Accordingly we may define a metric on rational plants by

$$\mathrm{gap}(P_1, P_2) = \mathrm{gap}(\mathcal{G}_{P_1}, \mathcal{G}_{P_2})$$

This metric was introduced in the present context in [Zames and El-Sakkary 1980] and argued to be an appropriate measure of closeness. Computing the gap is a (generalised) Nevanlinna-Pick interpolation problem [Georgiou 1988]:

$$\mathrm{gap}(P_1, P_2) = \max\{\delta_1(P_1, P_2),\, \delta_1(P_2, P_1)\}$$

where

$$\delta_1(P_1, P_2) = \inf_{Q \in H^\infty} \|\theta_{P_1} - \theta_{P_2} Q\|_{H^\infty}.$$

A variant is the *graph metric*

$$\mathrm{graph}(P_1, P_2) = \max\{\delta_2(P_1, P_2), \delta_2(P_2, P_1)\}$$

where

$$\delta_2(P_1, P_2) = \inf_{\|Q\|_{H^\infty} \leq 1} \|\theta_{P_1} - \theta_{P_2} Q\|_{H^\infty}.$$

This metric was proposed by Vidyasagar [1984], who gives examples and motivation. The graph and gap metrics generate the same topology on the space of rational functions. Vidyasagar calls this the *graph topology*, and advances convincing arguments for the thesis that it is the appropriate one for robust stabilisation. One is that the graph topology is the weak topology generated by the functions $H(\cdot, C)$ on $\mathbb{C}(s)$. That is, for any $P_0 \in \mathbb{C}(s)$, a neighbourhood sub-base of P_0 in the graph topology is furnished by the sets of the form

$$\left\{ P \in \mathbb{C}(s) : H(P, C) \in \mathcal{S} \text{ and } \left\| H(P, C) - H(P_0, C) \right\|_{H^\infty} < \varepsilon \right\}$$

where ε ranges over the positive reals and C ranges over the rational functions that stabilise P_0 (here the notation $H(P, C) \in \mathcal{S}$ means that each entry of the $2{\times}2$ matrix function $H(P, C)$ belongs to $\mathcal{S}$). One could express this characterisation by saying that the graph topology is the coarsest topology for which stabilisation by any controller is a robust property. The control community has evidently accepted that the graph topology is indeed the relevant one for discussion of the robust-stabilisation problem, but there are several rival metrics that induce this topology and it is perhaps not yet finally resolved which of them is best suited to engineering applications. There is a very strong candidate, due to G. Vinnicombe [1993], which is comparatively easy to compute and admits some sharp robustness results; we shall discuss this metric in the next section. By way of preparation we need two further notions. The first is that of the L^2-*gap metric* δ_{L^2}. This is just like the gap metric, except that we identify a plant P with the operator it induces on L^2 (of the imaginary axis) rather than H^2. More precisely, we define the possibly unbounded operator L_P on L^2 by

$$L_P u = P u \quad \text{for } u \in L^2 \text{ such that } P u \in L^2,$$

and we define $\delta_{L^2}(P_1, P_2)$ to be the gap between the closed subspaces of $L^2 \oplus L^2$ that are the graphs of L_{P_1} and L_{P_2}. The L^2-gap metric is much easier to compute than the gap metric—in fact, we have [Georgiou and Smith 1990]

$$\delta_{L^2}(P_1, P_2) = \left\| (1 + P_2 P_2^*)^{-1/2} (P_2 - P_1)(1 + P_1^* P_1)^{-1/2} \right\|_{L^\infty}. \tag{2–1}$$

To see this note that if a rational plant P has graph symbol $\theta \overset{\text{def}}{=} \begin{bmatrix} M \\ N \end{bmatrix}$ then $P = NM^{-1}$ and the L^2-graph of P is θL^2. Since θ is inner, the orthogonal projection on θL^2 is the multiplication operator $L_{\theta \theta^*}$ on L^2, and the projection on its orthogonal complement is $L_{1-\theta\theta^*}$, which equals $L_{\bar\theta_c \theta_c^T}$ where θ_c is the "complementary inner function",

$$\theta_c = \begin{bmatrix} -N \\ M \end{bmatrix},$$

so that $[\theta \;\; \bar\theta_c]$ is unitary-valued on the imaginary axis. The *directed gap*

$$\vec{\delta}(\theta_1 L^2, \theta_2 L^2)$$

between the closed subspaces $\theta_1 L^2$ and $\theta_2 L^2$ of L^2 is defined to be the norm of the orthogonal projection from $\theta_1 L^2$ to the orthogonal complement of $\theta_2 L^2$.

Hence, if θ_j is the symbol of P_j,

$$\vec{\delta}(\theta_1 L^2, \theta_2 L^2) = \|L_{\bar{\theta}_{2c}\theta_{2c}^T} L_{\theta_1 \theta_1^*}\| = \|L_{\bar{\theta}_{2c}\theta_{2c}^T \theta_1 \theta_1^*}\| = \|\bar{\theta}_{2c}\theta_{2c}^T \theta_1 \theta_1^*\|_\infty = \|\theta_{2c}^T \theta_1\|_\infty,$$

the last equation because θ_1 and $\bar{\theta}_{2c}$ are isometries. Now

$$\theta_{2c}^T \theta_1 = [\,-N_2 \quad M_2\,] \begin{bmatrix} M_1 \\ N_1 \end{bmatrix} = M_2 N_1 - N_2 M_1 = M_2(P_1 - P_2)M_1,$$

so that

$$\vec{\delta}(\theta_1 L^2, \theta_2 L^2) = \big\| M_2(P_1 - P_2)M_1 \big\|_\infty.$$

On the imaginary axis we have

$$1 = M^* M + N^* N = M^*(1 + P^* P)M$$

and hence

$$|M| = (1 + P^* P)^{-1/2}.$$

Thus

$$\vec{\delta}(\theta_1 L^2, \theta_2 L^2) = \big\| (1 + P_2^* P_2)^{-1/2}(P_1 - P_2)(1 + P_1^* P_1)^{-1/2} \big\|_\infty.$$

The gap between $\theta_1 L^2$ and $\theta_2 L^2$ is the maximum of the two directed gaps

$$\vec{\delta}(\theta_1 L^2, \theta_2 L^2) \quad \text{and} \quad \vec{\delta}(\theta_2 L^2, \theta_1 L^2),$$

and the formula (2–1) for $\delta_{L^2}(P_1, P_2)$ follows. A modification of this proof works for matrix-valued functions.

However, δ_{L^2} does *not* induce the graph topology—stabilisation is not a robust property with respect to this metric. Indeed, if P is a stable plant then it is stabilised by the controller $C = 0$, and yet every δ_{L^2}-neighbourhood of P contains unstable plants. It would appear that δ_{L^2} is no use for the study of feedback systems, which is a pity given that it is so manageable. Vinnicombe's bright idea was to rescue this metric by introducing a winding number constraint.

An important notion for the study of robustness of stabilisation is the *stability margin*. If a controller C stabilises a plant P it may do so with more or less to spare, and various quantitative measures of this notion are in use. A natural one is the stability margin $b_{P,C}$:

$$b_{P,C} = \begin{cases} \|H(P,C)\|_{H^\infty}^{-1} & \text{if } (P, C) \text{ is stable,} \\ 0 & \text{otherwise.} \end{cases}$$

That this quantity truly deserves the name of stability margin is shown by the following fact [Georgiou and Smith 1990].

THEOREM 2.1. *Let C be a controller that stabilises a plant P and let $\beta > 0$. Then C stabilises the closed ball of radius β about C (with respect to the gap metric) if and only if $b_{P,C} > \beta$.*

Thus robust stabilisation of P in the gap metric is equivalent to finding a stabilising controller C for which $b_{P,C}$ is large, or in other words, $\|H(P,C)\|_{H^\infty}$ is small. This turns out to be a Nehari problem. There are numerous further striking results about this range of questions, of which we mention only two.

- For sufficiently small radii, the gap-metric ball of radius β centred at P coincides with the set of functions NM^{-1} with $\left\|\theta_P - \left[\begin{smallmatrix} N \\ M \end{smallmatrix}\right]\right\|_{H^\infty} < \beta$. Thus closeness in the gap metric is equivalent to smallness of perturbations of the "numerator" and "denominator" in θ_P [Georgiou and Smith 1990, Theorem 4].
- The maximum stability margin $\sup_C b_{P,C}$ is equal to $\left(1 - \|H_{\theta_P^*}\|^2\right)^{1/2}$, where $H_{\theta_P^*}$ is the Hankel operator with symbol θ_P^* (see [McFarlane and Glover 1990, Theorem 4.3], or [Georgiou and Smith 1990, Theorem 2]).

It is worth mentioning that $b_{P,C}$ has an interpretation in terms of the geometry of $H^2 \oplus H^2$. For a stable system (P,C) let $T_{P,C} \in \mathcal{L}(H^2 \oplus H^2)$ be the operation of multiplication by

$$\begin{bmatrix} 0 & 1 \\ 1 & 0 \end{bmatrix} H(P,C) \begin{bmatrix} 1 & 0 \\ 0 & -1 \end{bmatrix} = \begin{bmatrix} 1 \\ P \end{bmatrix} (1 - CP)^{-1} \begin{bmatrix} 1 & -C \end{bmatrix}.$$

It is immediate that $T_{P,C}$ is idempotent. Its range is $\mathcal{G}_P$ and its kernel is

$$\tilde{\mathcal{G}}_C \overset{\text{def}}{=} \left\{ \begin{bmatrix} Cv \\ v \end{bmatrix} : v,\, Cv \in H^2 \right\}.$$

That is, $T_{P,C}$ is the projection on $\mathcal{G}_P$ along $\tilde{\mathcal{G}}_C$. Clearly $b_{P,C} = \|T_{P,C}\|^{-1}$.

3. Duality and Vinnicombe's Metric

Theorem 2.1 suggests a notion of duality between stability margins and metrics on plants. Let us say that a metric δ is *dual* to the stability margin $b_{P,C}$ when the following holds:

C stabilises the closed δ–ball of radius $\beta > 0$ about P if and only if $b_{P,C} > \beta$.

Another way of expressing this notion is to say that δ is dual to $b_{P,C}$ if, for any stable pair (P,C), the supremum of the radii of the δ-balls about P that are stabilised by C is $b_{P,C}$.

One might incline to think that Theorem 2.1 would be the last word on metrics dual to $b_{P,C}$ and that one needs to look no further than the gap metric. However, Vinnicombe argues that we can do better in two ways by the use of his metric δ_V, which is also dual to $b_{P,C}$. Firstly δ_V is simpler to calculate *and* to analyse than the gap metric. Secondly, δ_V is *smaller* than the gap metric, so that it enables us to establish a larger "uncertainty ball" of plants about a nominal plant P_0 that are all stabilised by a single controller. Indeed, δ_V is the best possible metric in this sense: it is the smallest metric that is dual to $b_{P,C}$. It enjoys a property converse to that of duality to $b_{P,C}$. For small enough $\beta > 0$ the closed δ_V ball of radius β about P is the largest set of linear time–invariant plants that can be guaranteed to be stabilised by a controller C if all we know about C is that $b_{P,C} > \beta$.

For a rational function f without zeros on the imaginary axis $i\mathbb{R}$ we define the *winding number of f on the imaginary axis* (denoted by $\mathrm{wno}(f)$) to be the winding number of f about 0 along the anticlockwise-oriented contour

$$\{Re^{i\theta} : -\pi/2 \le \theta \le \pi/2\} \cup \{iy : R \ge y \ge -R\}$$

indented round any pole of f on $i\mathbb{R}$, with R chosen so that all zeros and poles of f in the closed right half-plane lie inside the contour.

DEFINITION 3.1. For rational plants P_1 and P_2, we have

$$\delta_V(P_1, P_2) = \begin{cases} \delta_{L^2}(P_1, P_2) & \text{if } \theta_{P_2}^* \theta_{P_1} \text{ has no zero on } i\mathbb{R} \text{ and } \mathrm{wno}(\theta_{P_2}^* \theta_{P_1}) = 0, \\ 1 & \text{otherwise.} \end{cases}$$

Here θ_P^* denotes the unique rational function satisfying $\theta_P^*(iy) = \theta_P(iy)^*$ for all $y \in \mathbb{R}$ (so that $\theta_P^*(s) = \theta_P(-\bar{s})^*$).

Thus, in view of (2–1), we have

$$\delta_V(P_1, P_2) = \begin{cases} \left\| (1+P_2 P_2^*)^{-1/2}(P_2 - P_1)(1 + P_1^* P_1)^{-1/2} \right\|_\infty & \text{if } \mathrm{wno}(\theta_{P_2}^* \theta_{P_1}) = 0, \\ 1 & \text{otherwise.} \end{cases}$$

Vinnicombe shows that δ_V is a metric—indeed, he proves the stronger assertion that $\sin^{-1}\delta_V$ is a metric. The proof of the triangle inequality requires some calculation. He also shows that δ_V induces the graph topology on $\mathbb{C}(s)$. (An exercise for the reader: show that the metric $\sin^{-1}\delta_V$ is dual to the stability margin $\sin^{-1}b_{P,C}$.) The chief virtue of δ_V is its tight duality relationship with the stability margin $b_{P,C}$, which is in fact even further–reaching than is stated above. The metric δ_V gives precise information as to how adversely a perturbation of P can affect the stability margin $b_{P,C}$, as witness the following result [Vinnicombe 1993, Theorems 4.2 and 4.5].

THEOREM 3.2. *Let (P_0, C_0) be a rational stable pair and let $0 < \beta \le \alpha < \sup_C b_{P_0,C}$. Then these two conditions are equivalent:*

(i) $b_{P_0,C_0} > \alpha$.
(ii) $b_{P,C_0} > \sin(\sin^{-1}\alpha - \sin^{-1}\beta)$ *for all P in the δ_V-ball of radius β about P_0.*

The next two conditions are also equivalent:

(iii) $b_{P,C} > \sin(\sin^{-1}\alpha - \sin^{-1}\beta)$ *for all C satisfying $b_{P_0,C} > \alpha$.*
(iv) $\delta_V(P_0, P) \le \beta$.

Hence a perturbation of P_0 of $(\sin^{-1}\delta_V)$-magnitude ε reduces the stability margin $\sin^{-1}b_{.,C_0}$ by at most ε, and moreover this estimate is sharp.

The foregoing results about rational functions are so elegant that one can hardly resist the temptation to try to generalise them to non-rational functions.

But to what class of functions? Much attention has been devoted in the engineering literature to the identification of a class that is wide enough to encompass all functions of physical interest and yet enjoys the structural properties that allow analysis of the robust stabilisation problem. Not every measurable function on the imaginary axis can be stabilised: there exist P such that $H(P, C) \notin H^\infty$ for every function C. A necessary and sufficient condition that P be stabilisable is that its graph $\mathcal{G}_P$ be closed and have a left-invertible symbol [Georgiou and Smith 1993, Proposition 3]. However there are also conditions of an algebraic character that are desirable. In the rational case one often uses the fact that the ring S of stable rational functions has the *Bézout property*; that is, if f and g are coprime elements of S, there exist a and $b \in S$ such that $af + bg = 1$. This property fails in many natural classes of functions.

Since stabilisation is defined by the condition $H(P, C) \in H^\infty$ a plausible class of functions to analyse is the field of fractions of the integral domain H^∞. However this class does not have all the properties needed for the generalisation of the strong results obtained for rational functions. In particular, it does not have the Bézout property. It is in any case unnecessarily large. It includes functions that can hardly be held to represent any physical system, such as $e^{1/s}$. Georgiou and Smith [1992] obtained good results for plants belonging to the field of fractions of the algebra of functions analytic in the open right half-plane and continuous on its closure (in the Riemann sphere). However, they observe that an example of S. Treil shows that plants in this class do not necessarily have normalised coprime factor representations in the class. The same authors, in collaboration with C. Foiaş, have extended their geometric approach to time-varying systems [Foiaş et al. 1993]—but that ceases to be function theory.

A class that has been much studied in the present context is the *Callier–Desoer class* $\hat{\mathcal{B}}(\beta)$ for suitable $\beta \in \mathbb{R}$. A concise and accessible account of the properties of this class and the reasons for introducing it is given in [Curtain and Zwart 1995, Sec 7.1]. It is defined as follows. Denote by $\mathcal{A}(\beta)$ the space of measures μ on $[0, \infty)$ having trivial singular part with respect to Lebesgue measure and satisfying

$$\int_{[0,\infty)} e^{-\beta t} |\mu| \, dt < \infty.$$

Let $\hat{\mathcal{A}}(\beta)$ be the space of Laplace-Stieltjes transforms of elements of $\mathcal{A}(\beta)$. Let

$$\hat{\mathcal{A}}_-(\beta) = \bigcup_{\alpha < \beta} \hat{\mathcal{A}}(\alpha);$$

thus $\hat{\mathcal{A}}_-(\beta)$ is an algebra of functions analytic on the closed half-plane $\{\operatorname{Re} s \geq \beta\}$ under pointwise operations. Let $\hat{\mathcal{A}}_\infty(\beta)$ be the subset of $\hat{\mathcal{A}}_-(\beta)$ consisting of those members that are bounded away from 0 at infinity. Then $\hat{\mathcal{B}}(\beta)$ is defined to be the field of fractions of $\hat{\mathcal{A}}_-(\beta)$ by the multiplicative subset $\hat{\mathcal{A}}_\infty(\beta)$.

The class $\hat{\mathcal{B}}(\beta)$ has numerous good properties. It is a commutative algebra with identity. If $P \in \hat{\mathcal{B}}(\beta)$ then P is meromorphic in $\{\operatorname{Re} s \geq \alpha\}$ for some $\alpha < \beta$. It is not true that $\hat{\mathcal{B}}(\beta)$ has the Bézout property, because one can easily find coprime elements that both tend to zero along the same sequence of points of $\{\operatorname{Re} s \geq \beta\}$, whence the ideal they generate cannot contain 1. (Caveat: in the engineering literature two elements of an algebra are often *defined* to be coprime if and only if the ideal they generate is the whole algebra; in this article I have stuck to the usual mathematical definition, according to which two elements are coprime if and only if their only common factors are units of the algebra.)

There is, however, a good replacement for the Bézout property. Any $P \in \hat{\mathcal{B}}(\beta)$ has a factorisation $P = NM^{-1}$ where $N \in \hat{\mathcal{A}}_-(\beta)$, $M \in \hat{\mathcal{A}}_\infty(\beta)$ and the ideal of $\hat{\mathcal{B}}(\beta)$ generated by N and M is the whole of $\hat{\mathcal{B}}(\beta)$. This fact permits the extension to $\hat{\mathcal{B}}(\beta)$ of numerous techniques from the rational case.

Another approach to identifying a suitable class of functions is to start from a state space description of a system, or evolution equation, and see what functions arise as the corresponding transfer functions. Some very subtle questions arise in this way. G. Weiss [1994] has characterised the transfer functions of "regular systems", a class that probably includes all linear time-invariant state-space systems of practical interest; they are analytic in a half-plane $\{\operatorname{Re} s > \beta\}$ for some $\beta > 0$ and have a limit as $s \to \infty$ along the real axis. It will be interesting if this line of investigation leads eventually to the same conclusions as the function-theoretic viewpoint of Georgiou and Smith. So far, though, the description of the ideal holomorphic space for the analysis of robust stabilisation awaits the final word.

References

[Curtain and Zwart 1995] R. F. Curtain and H. Zwart, *An introduction to infinite-dimensional linear systems theory*, Texts in Applied Mathematics, Springer, New York, 1995.

[Doyle et al. 1992] J. C. Doyle, B. A. Francis, and A. R. Tannenbaum, *Feedback control theory*, Macmillan Publishing Company, New York, 1992.

[Foiaş et al. 1993] C. Foiaş, T. T. Georgiou, and M. C. Smith, "Robust stability of feedback systems: a geometric approach using the gap metric", *SIAM J. Control Optim.* **31**:6 (1993), 1518–1537.

[Francis 1987] B. A. Francis, *A course in H_∞ control theory*, Lecture Notes in Control and Information Sciences **88**, Springer, Berlin, 1987.

[Georgiou 1988] T. T. Georgiou, "On the computation of the gap metric", *Systems Control Lett.* **11**:4 (1988), 253–257.

[Georgiou and Smith 1990] T. T. Georgiou and M. C. Smith, "Optimal robustness in the gap metric", *IEEE Trans. Automat. Control* **35**:6 (1990), 673–686.

[Georgiou and Smith 1992] T. T. Georgiou and M. C. Smith, "Robust stabilization in the gap metric: controller design for distributed plants", *IEEE Trans. Automat. Control* **37**:8 (1992), 1133–1143.

[Georgiou and Smith 1993] T. T. Georgiou and M. C. Smith, "Graphs, causality, and stabilizability: linear, shift-invariant systems on $\mathcal{L}_2[0, \infty)$", *Math. Control Signals Systems* **6**:3 (1993), 195–223.

[McFarlane and Glover 1990] D. C. McFarlane and K. Glover, *Robust controller design using normalized coprime factor plant descriptions*, Lecture Notes in Control and Information Sciences **138**, Springer, Berlin, 1990.

[Rohrs et al. 1993] C. E. Rohrs, J. L. Melsa, and D. G. Schultz, *Linear control systems*, McGraw-Hill series in electrical and computer engineering, McGraw-Hill, New York, 1993.

[Vidyasagar 1984] M. Vidyasagar, "The graph metric for unstable plants and robustness estimates for feedback stability", *IEEE Trans. Automat. Control* **29**:5 (1984), 403–418.

[Vinnicombe 1993] G. Vinnicombe, "Frequency domain uncertainty and the graph topology", *IEEE Trans. Automat. Control* **38**:9 (1993), 1371–1383.

[Weiss 1994] G. Weiss, "Transfer functions of regular linear systems. I. Characterizations of regularity", *Trans. Amer. Math. Soc.* **342**:2 (1994), 827–854.

[Young 1988] N. Young, *An introduction to Hilbert space*, Cambridge Mathematical Textbooks, Cambridge University Press, Cambridge, 1988.

[Zames and El-Sakkary 1980] G. Zames and A. El-Sakkary, "Unstable systems and feeback: the gap metric", pp. 380–385 in *Eighteenth Annual Allerton Conference on Communication, Control, and Computing* (Monticello, IL, 1980), University of Illinois, Department of Electrical Engineering, Urbana-Champaign, IL, 1980.

NICHOLAS YOUNG
DEPARTMENT OF MATHEMATICS AND STATISTICS
LANCASTER UNIVERSITY
LANCASTER LA1 4YF
ENGLAND

Present address:
DEPARTMENT OF MATHEMATICS AND STATISTICS
UNIVERSITY OF NEWCASTLE
NEWCASTLE UPON TYNE NE1 7RU
ENGLAND
N.J.Young@ncl.ac.uk

Holomorphic Spaces
MSRI Publications
Volume **33**, 1998

The Abstract Interpolation Problem and Applications

ALEXANDER KHEIFETS

ABSTRACT. A number of classical interpolation problems can be reduced to the following scheme. One is interested in the finding Schur class operator functions $w(\zeta) : E_1 \to E_2$, with $\zeta \in \mathbb{D}$, that satisfy certain interpolation conditions. The data of the problem are encoded in the Lyapunov identity

$$D(T_2 x, T_2 y) - D(T_1 x, T_1 y) = \langle M_1 x, M_1 y \rangle - \langle M_2 x, M_2 y \rangle,$$

where x, y are elements of a vector space X, D is a positive semidefinite quadratic form on X, T_1 and T_2 are linear operators on X, and M_1, M_2 are linear operators from X to the separable Hilbert spaces E_1, E_2. After introducing the de Branges–Rovnyak function space H^w associated with w, we can formulate the interpolation conditions thus: w is a solution to the interpolation problem if and only if there exists a linear mapping $F : X \to H^w$ such that

$$\|F x\|_{H^w}^2 \le D(x, x)$$

and

$$(FT_1 x)(t) = t(FT_2 x)(t) - \begin{bmatrix} \mathbf{1} & w(t) \\ w(t)^* & \mathbf{1} \end{bmatrix} \begin{bmatrix} -M_2 x \\ M_1 x \end{bmatrix}$$

for a.e. t with $|t| = 1$. The solutions w turn out to be the scattering matrices of the unitary colligations that extend the isometric colligation defined by the Lyapunov identity. These extensions and their scattering functions can be described using a "universal" extension and its scattering operator function. The description formula for solutions looks like

$$w = s_0 + s_2 (1 - \omega s)^{-1} \omega s_1,$$

where

$$S = \begin{bmatrix} s & s_1 \\ s_2 & s_0 \end{bmatrix} : N_2 \oplus E_1 \to N_1 \oplus E_2$$

is the scattering matrix of the "universal" extension and $\omega : N_1 \to N_2$ is an arbitrary parameter from the Schur class. The matrix S is defined essentially uniquely by the data of the problem and is called the scattering matrix of the problem. Using the functional model and the Fourier representation of the "universal" extension one can investigate analytic properties of the scattering matrices S for classes of interpolation problems.

Work partly supported by MSRI under NSF grant DMS-9022140.

Lecture 1. The Abstract Interpolation Problem

I will begin with the formal setup of the Abstract Interpolation Problem, or AIP, then consider several examples and discuss the role of the AIP in their investigation.

The data of the AIP are encoded in the Fundamental Identity:

$$D(T_2x, T_2y) - D(T_1x, T_1y) = \langle M_1x, M_1y \rangle_{E_1} - \langle M_2x, M_2y \rangle_{E_2}, \qquad (1\text{--}1)$$

where x, y are elements of a vector space X (which has no a priori topological structure), D is a positive-semidefinite quadratic form on X, T_1 and T_2 are linear operators on X, E_1 and E_2 are separable Hilbert spaces, and M_1, M_2 are linear mappings from X into E_1, E_2.

De Branges–Rovnyak function spaces. Let w be a Schur class operator function; that is, w associates to each ζ in the unit disk $\mathbb{D} = \{\zeta : |\zeta| \leq 1\}$ a contraction $w(\zeta) : E_1 \to E_2$, varying analytically with ζ. The de Branges–Rovnyak space L^w consists of the functions $f : \mathbb{T} \to E_2 \oplus E_1$ (where $\mathbb{T} = \{\zeta : |\zeta| = 1\}$ is the unit circle) having the form

$$f = \begin{bmatrix} \mathbf{1}_{E_2} & w \\ w^* & \mathbf{1}_{E_1} \end{bmatrix}^{1/2} g \qquad (1\text{--}2)$$

for some $g \in L^2(E_2 \oplus E_1)$. We set

$$\|f\|_{L^w} \overset{\text{def}}{=} \inf \|g\|_{L^2},$$

where the infimum is taken over all $g \in L^2(E_2 \oplus E_1)$ such that (1–2) is satisfied. All such preimages g differ by the addition of (arbitrary) elements of $\mathrm{Ker}\begin{bmatrix} 1 & w \\ w^* & 1 \end{bmatrix}$, and the infimum is attained when

$$g(t) \perp \mathrm{Ker} \begin{bmatrix} \mathbf{1} & w(t) \\ w(t)^* & \mathbf{1} \end{bmatrix} \quad \text{a.e. on } \mathbb{T}.$$

Let $\pi_w^{-1}f$ denote the particular g that achieves the infimum. Thus

$$\|f\|_{L^w}^2 = \|\pi_w^{-1}f\|_{L^2}^2 = \int_{\mathbb{T}} \left\|(\pi_w^{-1}f)(t)\right\|_{E_2 \oplus E_1}^2 m(dt),$$

where $m(dt)$ is Lebesgue measure. The inner product in L^w is defined by

$$\langle f, h \rangle_{L^w} = \langle \pi_w^{-1} f, \pi_w^{-1} h \rangle_{L^2}.$$

Now L^w is a Hilbert space. As a set it is contained in L^2. It might happen that a function $f \in L^w$ admits the representation

$$f(t) = \begin{bmatrix} 1 & w(t) \\ w(t)^* & 1 \end{bmatrix} \check{g}(t) \quad \text{a.e.,}$$

where $\check{g}(t)$ need not be in L^2; then, for any $h \in L^w$,

$$\langle f, h \rangle_{L^w} = \left\langle \begin{bmatrix} 1 & w \\ w^* & 1 \end{bmatrix}^{1/2} \check{g}(t), \pi_w^{-1} h \right\rangle_{L^2} = \int_{\mathbb{T}} \langle \check{g}(t), h(t) \rangle_{E_2 \oplus E_1} m(dt).$$

We also define the space H^w as the subspace of L^w consisting of

$$f = \begin{bmatrix} f_2 \\ f_1 \end{bmatrix} \quad \text{with } f_2 \in H_+^2(E_2) \text{ and } f_1 \in H_-^2(E_1),$$

where H_+^2 and H_-^2 are the standard Hardy classes.

Setup of the AIP. The Schur class function $w : E_1 \to E_2$ is said to be a *solution of the AIP* (with the data specified above) if there exists a linear mapping $F : X \to H^w$ such that, for all $x \in X$, the following conditions are satisfied:

(i) $\|Fx\|_{H^w}^2 \leq D(x, x)$.

(ii) $tFT_2 x - FT_1 x = \begin{bmatrix} 1 & w \\ w^* & 1 \end{bmatrix} \begin{bmatrix} -M_2 x \\ M_1 x \end{bmatrix}$ for a.e. $t \in \mathbb{T}$.

Property (ii) is, in fact, an implicit formula for the mapping F. Sometimes it defines F uniquely, sometimes not; but any mapping F that possesses (ii) is very special. We will describe all such maps in Lecture 4.

If we write

$$Fx = \begin{bmatrix} F_+ x \\ F_- x \end{bmatrix},$$

the conditions $Fx \in H^w$ and $\|Fx\|_{H^w}^2 \leq D(x, x)$ are equivalent to the conjunction of three conditions:

(a) $F_+ x \in H_+^2(E_2)$.

(b) $F_- x \in H_-^2(E_1)$.

(c) $\|Fx\|_{L^w}^2 \leq D(x, x)$.

By the definition of the inner product in L^w, property (c) is actually an upper bound for the average boundary values of Fx.

My goal now is to explain why this abstract problem is an "interpolation" problem. Before passing to examples I would like to consider a special case of the data (which was considered earlier than the general case).

A special case. Assume that the operators $(\zeta T_2 - T_1)$ and $(T_2 - \bar{\zeta}T_1)$ are invertible for all $\zeta \in \mathbb{D}$ except possibly for a discrete set. Because the first components of the sides of relation (ii) above are H_+^2 functions, one can consider the analytic continuation of the relation inside the unit disk $\mathbb{D}$:

$$\zeta(F_+T_2x)(\zeta) - (F_+T_1x)(\zeta) = -M_2x + w(\zeta)M_1x. \qquad (1\text{--}3)$$

We emphasize that the vectors $M_2x \in E_2$ and $M_1x \in E_1$ are independent of ζ. Fix $\zeta \in \mathbb{D}$. Because the mapping F is linear, we have

$$\zeta(F_+T_2x)(\zeta) = \big(F_+(\zeta T_2x)\big)(\zeta).$$

Actually, for any complex number μ,

$$\mu(F_+T_2x)(\zeta) = \big(F_+(\mu T_2x)\big)(\zeta);$$

in particular, this is true for $\mu = \zeta$. We can reexpress (1–3) now as

$$\big(F_+((\zeta T_2 - T_1)x)\big)(\zeta) = \big(-M_2 + w(\zeta)M_1\big)x.$$

Replacing x by $(\zeta T_2 - T_1)^{-1}x$, we end up with

$$(F_+^w x)(\zeta) = (w(\zeta)M_1 - M_2)(\zeta T_2 - T_1)^{-1}x. \qquad (1\text{--}4)$$

One can see better now the interpolation meaning of the property $F_+x \in H_+^2$: the "zeros" of the numerator cancel the "zeros" of the denominator, which means that w obeys the interpolation constraint

$$w(\zeta)M_1x = M_2x$$

at certain "points" ζ.

In a similar way, the second components of the two sides of equality (ii) can be reexpresed, under the assumptions at hand, as

$$(F_-^w x)(\zeta) = \bar{\zeta}(M_1 - w(\zeta)^*M_2)(T_2 - \bar{\zeta}T_1)^{-1}x. \qquad (1\text{--}5)$$

The interpolation meaning of the property $F_-^w x \in H_-^2$ is similar to the one considered above, but now for $w(\zeta)^*$.

The meaning of property (c) will be discussed in the examples.

REMARK. Under the assumptions of this section, condition (ii) defines F *uniquely* and *explicitly* for any solution w. This allows us to write F^w instead of F. Thus, under these assumptions, one can give a more explicit setup for the AIP:

Let

$$F^w x = \begin{bmatrix} F_+^w x \\ F_-^w x \end{bmatrix}$$

be defined by the formulas (1–4) and (1–5). The Schur class function w is said to be a solution of the AIP if F^w possesses properties (a), (b), and (c).

Generally, condition (ii) does not define F in a unique way, but we will see in Lecture 4 how to describe all such mappings F.

References for Lecture 1 are [Katsnelson et al. 1987; Kheifets 1988a; 1988b; 1990b; Kheifets and Yuditskii 1994].

Lecture 2. Examples

Example 1: The Nevanlinna–Pick Problem. Let $\zeta_1, \ldots, \zeta_n, \ldots$ be a finite or infinite sequence of points in the unit disk $\mathbb{D}$; let $w_1, \ldots, w_n, \ldots$ be a sequence of complex numbers. One is interested in describing all the Schur class functions w such that

$$w(\zeta_k) = w_k.$$

The well-known solvability criterion is

$$\left[\frac{1 - \bar{w}_k w_j}{1 - \bar{\zeta}_k \zeta_j} \right]_{k,j=1}^n \geq 0 \quad \text{for all } n.$$

We define the data of the AIP associated with this problem. Because the functions w are scalar, $E_1 = E_2 = \mathbb{C}^1$. Consider the space X that consists of all sequences

$$x = \begin{bmatrix} x_1 \\ \vdots \\ x_n \\ \vdots \end{bmatrix}$$

that have only a finite number of nonvanishing components. No topology is assumed on X.

Define the sesquilinear form D on X by

$$D(x, y) = \sum_{k,j} \bar{y}_k \frac{1 - \bar{w}_k w_j}{1 - \bar{\zeta}_k \zeta_j} x_j, \quad \text{for } x, y \in X.$$

The (diagonal) linear operators

$$T = T_1 = \begin{bmatrix} \zeta_1 & & & \\ & \ddots & & \\ & & \zeta_n & \\ & & & \ddots \end{bmatrix}$$

and $T_2 = \mathbf{1}_X$ are well defined on the space X. We can now check the Fundamental Identity (1–1):

$$D(x, y) - D(Tx, Ty) = \sum_{k,j} \bar{y}_k (1 - \bar{w}_k w_j) x_j$$

$$= \sum_{k,j} \bar{y}_k x_j - \sum_{k,j} \bar{y}_k \bar{w}_k w_j x_j$$

$$= \sum_k \bar{y}_k \cdot \sum_j x_j - \sum_k \bar{y}_k \bar{w}_k \cdot \sum_j w_j x_j.$$

By defining $M_1 x = \sum_j x_j$ and $M_2 x = \sum_j w_j x_j$, we end up with

$$D(x, y) - D(Tx, Ty) = \overline{M_1 y} \cdot M_1 x - \overline{M_2 y} \cdot M_2 x.$$

The products on the right-hand side are actually the standard inner product in $\mathbb{C}^1$.

Consider now the AIP associated with these data. Let w be a solution of this AIP. This means that there exists a mapping $F : X \to H^w$ such that conditions (i) and (ii) on page 353 hold. Because $(\zeta \mathbf{1} - T)^{-1}$ exists for all ζ, such that $|\zeta| < 1$ and $\zeta \neq \zeta_j$, and because $(\mathbf{1} - \bar{\zeta} T)^{-1}$ exists for all ζ with $|\zeta| < 1$, we know that F has the following form (see Lecture 1, special case on page 354):

$$(F_+^w x)(\zeta) = (w(\zeta) M_1 - M_2)(\zeta \mathbf{1} - T)^{-1} x$$
$$(F_-^w x)(\zeta) = (M_1 - \overline{w(\zeta)} M_2)(\mathbf{1} - \bar{\zeta} T)^{-1} x.$$

Thus, F^w is defined uniquely for any solution w. It is easy to compute these expressions more explicitly for this example. Because

$$(\zeta \mathbf{1} - T)^{-1} \begin{bmatrix} x_1 \\ \vdots \\ x_n \\ \vdots \end{bmatrix} = \begin{bmatrix} \frac{1}{\zeta - \zeta_1} x_1 \\ \vdots \\ \frac{1}{\zeta - \zeta_n} x_n \\ \vdots \end{bmatrix},$$

we obtain

$$(F_+^w x)(\zeta) = w(\zeta) \sum_j \frac{1}{\zeta - \zeta_j} x_j - \sum_j \frac{w_j x_j}{\zeta - \zeta_j} = \sum_j \frac{w(\zeta) - w_j}{\zeta - \zeta_j} x_j;$$

in the same way

$$(F_-^w x)(\zeta) = \bar{\zeta} \sum_j \frac{1 - \overline{w(\zeta)} w_j}{1 - \bar{\zeta} \zeta_j} x_j.$$

The function w is a solution of the AIP if and only if $F^w x \in H^w$ and $\|F^w x\|_{H^w}^2 \leq D(x, x)$ for all $x \in X$; that is, if and only if properties (a), (b), (c) of page 353 hold. One can see from the explicit formula that property (a) holds if and only if $w(\zeta_j) = w_j$. Property (b) holds automatically. To see what property (c) means, compute the left-hand side. On the boundary $(for |t| = 1)$, $F_+^w x$ and $F_-^w x$ look like

$$(F_+^w x)(t) = w(t) \sum_j \frac{x_j}{t - \zeta_j} - \sum_j \frac{w_j x_j}{t - \zeta_j},$$
$$(F_-^w x)(t) = \sum_j \frac{x_j}{t - \zeta_j} - \overline{w(t)} \sum_j \frac{w_j x_j}{t - \zeta_j}.$$

Or we can put these two formulas together:

$$(F^w x)(t) = \begin{bmatrix} \mathbf{1} & w(t) \\ w(t)^* & \mathbf{1} \end{bmatrix} \begin{bmatrix} -\sum_j \frac{w_j x_j}{t - \zeta_j} \\ \sum_j \frac{x_j}{t - \zeta_j} \end{bmatrix}.$$

Now,

$$\|F^w x\|_{H^w}^2 = \left\langle \begin{bmatrix} F_+^w x \\ F_-^w x \end{bmatrix}, \begin{bmatrix} -\sum_j \frac{w_j x_j}{t - \zeta_j} \\ \sum_j \frac{x_j}{t - \zeta_j} \end{bmatrix} \right\rangle_{L^2} = \left\langle F_-^w x, \sum_j \frac{x_j}{t - \zeta_j} \right\rangle_{L^2}$$

$$= \sum_j \bar{x}_j \frac{(F_-^w x)(\zeta_j)}{\bar{\zeta}_j} = \sum_j \bar{x}_j \sum_k \frac{1 - \bar{w}_j w_k}{1 - \bar{\zeta}_j \zeta_k} x_k = D(x, x).$$

The latter computation depends on property (a), $F_+^w x \in H_+^2$. Thus, we can see that w *is a solution of the AIP with these data if and only if w is a solution of the Nevanlinna–Pick problem.* For this particular problem, condition (a) actually carries all the interpolation information, (b) holds automatically, and (c) follows from (a). Moreover, we emphasize that for any solution w of this problem we have the equality $\|F^w x\|_{H^w}^2 = D(x, x)$ for all $x \in X$, instead of inequality.

To continue on to the second (more general) example, I will reformulate the first one. Consider the closed linear span of the functions

$$\left\{ \frac{1}{t - \zeta_j} \right\}$$

and denote by $K_{\bar{\theta}} \subseteq H^2$ the space $H_-^2 \ominus \bar{\theta} H_-^2$, where θ is the Blaschke product with zeros ζ_j if they satisfy the Blaschke condition and $\theta \equiv 0$ otherwise (in this case $K_{\bar{\theta}} = H_-^2$). We can associate the function

$$\tilde{x}(t) = \sum_j \frac{x_j}{t - \zeta_j} \in K_{\bar{\theta}}$$

with any $x \in X$ of the first example. $K_{\bar{\theta}}$ is invariant under $P_- t$, where t is the independent variable, and P_- is the orthogonal projection onto H_-^2. It is easy to see that the operator T from the first example acts as $P_- t$ under this correspondence; in fact,

$$P_- t \sum_j \frac{x_j}{t - \zeta_j} = \sum_j \frac{\zeta_j x_j}{t - \zeta_j}.$$

Let W be the operator on the space X defined by

$$W \begin{bmatrix} x_1 \\ \vdots \\ x_n \\ \vdots \end{bmatrix} = \begin{bmatrix} w_1 x_1 \\ \vdots \\ w_n x_n \\ \vdots \end{bmatrix}.$$

Obviously $WT = TW$. Under the correspondence,

$$\widetilde{W x}(t) = \sum_j \frac{w_k x_j}{t - \zeta_j}.$$

A Schur class function w is a solution of the Nevanlinna–Pick problem if and only if $\widetilde{W}x = P_-w\tilde{x}$. In fact,

$$P_-w\frac{1}{t-\zeta_j} = P_-\frac{w-w(\zeta_j)}{t-\zeta_j} + P_-\frac{w(\zeta_j)}{t-\zeta_j} = \frac{w(\zeta_j)}{t-\zeta_j}.$$

Observe that

$$M_1x = \sum_j x_j = (\tilde{x})_{-1},$$

where the last notation stands for the Fourier coefficient of index -1 of the H^2_- function $\tilde{x}$. And $M_2x = (\widetilde{W}x)_{-1}$. Since $\widetilde{W}x = P_-w\tilde{x}$ for any solution w of the Nevanlinna–Pick problem, then $\|\widetilde{W}x\|_{H^2_-} \leq \|\tilde{x}\|_{H^2_-}$. We can reexpress the quadratic form D as

$$D(x,y) = \langle \tilde{x}, \tilde{y}\rangle_{H^2_-} - \langle \widetilde{W}x, \widetilde{W}y\rangle_{H^2_-}.$$

In fact, let

$$\tilde{x} = \frac{1}{t-\zeta_j} \quad \text{and} \quad \tilde{y} = \frac{1}{t-\zeta_k};$$

then

$$\widetilde{W}x = \frac{w_j}{t-\zeta_j}, \qquad\qquad \widetilde{W}y = \frac{w_i}{t-\zeta_i},$$

$$\langle \tilde{x}, \tilde{y}\rangle = \frac{1}{1-\bar{\zeta}_k\zeta_j}, \quad \langle \widetilde{W}x, \widetilde{W}y\rangle = \frac{\bar{w}_k w_j}{1-\bar{\zeta}_k\zeta_j}.$$

If $\tilde{x} = \sum_j \dfrac{x_j}{t-\zeta_j}$, then

$$\|\tilde{x}\|^2 - \|\widetilde{W}x\|^2 = \sum_{k,j} \bar{x}_k \frac{1-\bar{w}_k w_j}{1-\bar{\zeta}_k\zeta_j} x_j = D(x,x).$$

Define the operator $\widetilde{W}$ on $K_{\bar\theta}$ by

$$\widetilde{W}(\tilde{x}) = \widetilde{W}x.$$

This is well defined and $D \geq 0$ if and only if $\widetilde{W}$ is a contraction.

Example 2. Sarason's Problem. Now let θ be an arbitrary inner function (not necessarily a Blaschke product). Set $K_{\bar\theta} = H^2_- \ominus \bar\theta H^2_-$; this space is invariant under P_-t. Set $T = P_-t|K_{\bar\theta}$. Let W be a contractive operator on $K_{\bar\theta}$ with $WT = TW$. One is interested in finding all the Schur class functions w such that

$$Wx = P_-wx \quad \text{for all } x \in K_{\bar\theta}.$$

Associate the following AIP data to the Sarason problem: $X = K_{\bar\theta}$, $T_1 = T$, $T_2 = \mathbf{1}$, $D(x,x) = \|x\|^2 - \|Wx\|^2$, $M_1x = (x)_{-1}$, $M_2x = (Wx)_{-1}$; here $E_1 = E_2 = \mathbb{C}^1$. One can check the Fundamental Identity (see the beginning of Lecture 1). For this data the implicit definition (ii) of the mapping F yields an explicit formula for F:

$$F^w x = \begin{bmatrix} 1 & w \\ \bar{w} & 1 \end{bmatrix} \begin{bmatrix} -Wx \\ x \end{bmatrix} \quad \text{a.e. } t \in T,$$

for all $x \in X$ and any solution w of the AIP. Thus, F^w is unique for any solution w. Then again w is a solution of the AIP if and only if $F^w x \in H^w$ for all $x \in X$ and $\|F^w x\|^2_{H^w} \leq D(x, x)$, that is, if and only if conditions (a), (b) and (c) of page 353 hold.

For condition (a) we have

$$F^w_+ x = wx - Wx \in H^2_+ \iff P_-(wx - Wx) = 0 \iff P_- wx = Wx$$

for all $x \in X$. In other words, condition (a) is satisfied if w solves the Sarason problem.

Condition (b), $F^w_- x = x - \bar{w} \cdot Wx \in H^2_-$, holds automatically for any Schur class function w.

For condition (c) we get

$$\begin{aligned}
\|F^w x\|^2_{L^w} &= \left\langle \begin{bmatrix} F^w_+ x \\ F^w_- x \end{bmatrix}, \begin{bmatrix} -Wx \\ x \end{bmatrix} \right\rangle_{L^2} = \langle F^w_- x, x \rangle_{L^2} \\
&= \langle x - \bar{w} \cdot Wx, x \rangle = \langle x, x \rangle - \langle Wx, wx \rangle \\
&= \langle x, x \rangle - \langle Wx, P_- wx \rangle = \langle x, x \rangle - \langle Wx, Wx \rangle = D(x, x).
\end{aligned}$$

Thus, for this data w *is a solution of the AIP if and only if w is a solution of the Sarason problem*. Property (a) carries all the interpolation information, (b) holds automatically, and (c) follows from (a). And we again have the equality $\|F^w x\|^2_{H^w} = D(x, x)$ for all $x \in X$ and for all solutions w, instead of inequality.

Example 2'. We now associate another AIP to the same Sarason problem. This AIP is best considered as being different (nonequivalent) from the one in Example 2, though they have a common set of solutions. The reason is that the coefficient matrices in the description formulas for the solution sets (see the last theorem of Lecture 4) are different and the associated universal colligations (see Section 2 of Lecture 4) are nonequivalent.

Let θ be an inner function, and set $K_\theta = H^2_+ \ominus \theta H^2_+$ and $T^*_\theta x_2 = P_+ \bar{t} x_2$ for $x_2 \in K_\theta$. Let W^*_2 be a contractive operator on K_θ that commutes with $T^*_\theta : W^*_2 T^*_\theta = T^*_\theta W^*_2$. Find all the Schur class functions w such that

$$W^*_2 x_2 = P_+ \bar{w} x_2.$$

One can check that the solutions of this problem and of the one in the previous example coincide (if the operator W of Example 2 and the operator W_2 are properly connected). Consider the AIP with the data $X = K_\theta$, $T_1 = \mathbf{1}$, $T_2 = T^*_\theta$, $E_1 = E_2 = \mathbb{C}^1$, $M_1 x_2 = (W^* x_2)_0$, and $M_2 x_2 = (x_2)_0$, where the notation $(\cdot)_0$ stands for the Fourier coefficient of index 0 of an H^2_+ function. As in Example 2,

F can be expressed explicitly and uniquely for any solution w of the AIP with this data:

$$F^w x_2 = \begin{bmatrix} 1 & w \\ \bar{w} & 1 \end{bmatrix} \begin{bmatrix} x_2 \\ -W_2^* x_2 \end{bmatrix}.$$

Now consider properties (a), (b) and (c). For (a) we have

$$F_+^w x_2 = x_2 - w \cdot W_2^* x_2 \in H_+^2;$$

this holds automatically for any Schur class function w. Property (b) becomes

$$F_-^w x_2 = \bar{w} x_2 - W_2^* x_2 \in H_-^2;$$

this holds if and only if $P_+ \bar{w} x_2 = W_2^* x_2$; that is, if and only if w solves the Sarason problem. Finally, for (c) we can write

$$\|F^w x_2\|_{L^w}^2 = D(x_2, x_2) \quad \text{for all } x_2 \in X$$

for any solution w. Thus, for these data property (b) carries all the interpolation information, (a) holds automatically for any Schur class w, and (c) follows from (b). The equality

$$\|F^w x_2\|_{H^w}^2 = D(x_2, x_2) \quad \text{for all } x_2 \in X$$

holds for any solution w.

Example 3. The boundary interpolation problem. Property (c) dominates in this example and (a) and (b) follow. The equality in condition (c) holds for some solutions but not for all of them. We will need some preliminaries.

A Schur class function w in the unit disk $\mathbb{D}$ is said to have *an angular derivative in the sense of Carathéodory* at the point $t_0 \in \mathbb{T}$ if $w(\zeta)$ has a nontangential unimodular limit w_0 as ζ goes to t_0, $|w_0| = 1$, and

$$\frac{w(\zeta) - w_0}{\zeta - \zeta_0}$$

has a nontangential limit w_0' at t_0.

THEOREM (CARATHÉODORY). *A Schur class function $w(\zeta)$ has an angular derivative at $t_0 \in \mathbb{T}$ if and only if*

$$D_{w,t_0} \stackrel{\text{def}}{=} \liminf_{\zeta \to t_0} \frac{1 - |w(\zeta)|^2}{1 - |\zeta|^2} < \infty$$

(here $|\zeta| \leq 1$ and $\zeta \to t_0$ in an arbitrary way). In this case $w_0' = D_{w,t_0} \cdot \dfrac{w_0}{t_0}$ and

$$\frac{1 - |w(\zeta)|^2}{1 - |\zeta|^2} \longrightarrow D_{w,t_0}$$

as ζ goes to t_0 nontangentially. D_{w,t_0} vanishes if and only if w is a constant of modulus 1.

Moreover, a Schur class function w has an angular derivative in the sense of Carathéodory at the point $t_0 \in \mathbb{T}$ if and only if there exists a unimodular constant w_0 such that

$$\left| \frac{w(t) - w_0}{t - t_0} \right|^2 + \frac{1 - |w(t)|^2}{|t - t_0|^2} \in L^1;$$

that is, if and only if this function is integrable over $\mathbb{T}$ against Lebesgue measure. In particular, this guarantees that

$$\frac{w - w_0}{t - t_0} \in H_+^2,$$

because the denominator is an outer function. In that case

$$\int_{\mathbb{T}} \left(\left| \frac{w(t) - w_0}{t - t_0} \right|^2 + \frac{1 - |w(t)|^2}{|t - t_0|^2} \right) m(dt) = D_{w,t_0}.$$

Now consider the following interpolation problem. Let t_0 be a point of the unit circle $\mathbb{T}$, let w_0 be a complex number with $|w_0| = 1$, and let $0 \le D < \infty$ be a given nonnegative number. One wants to describe all the Schur class functions w such that $w(\zeta) \to w_0$ as $\zeta \to t_0$ (nontangentially) and $D_{w,t_0} \le D$.

Associate to this problem the AIP data $X = \mathbb{C}^1$, $D(x,x) = \bar{x}Dx$, $T_1 x = t_0 x$, $T_2 x = x$, $M_1 x = x$, $M_2 x = w_0 x$, $E_1 = E_2 = \mathbb{C}^1$; then we can check the Fundamental Identity (1–1). The left-hand side of the identity is

$$\bar{x}Dx - \bar{x}\bar{t}_0 D t_0 x = 0,$$

and the right-hand side is

$$\overline{M_1 x} \cdot M_1 x - \overline{M_2 x} \cdot M_2 x = \bar{x}x - \bar{x}\bar{w}_0 w_0 x = 0.$$

The mapping F of Lecture 1 (pages 353–354) is unique for any solution w and can be computed explicitly for this data:

$$F^w x = \begin{bmatrix} 1 & w \\ \bar{w} & 1 \end{bmatrix} \begin{bmatrix} -\frac{w_0}{t - t_0} \\ \frac{1}{t - t_0} \end{bmatrix} x$$

We can analyse now what conditions (a), (b), and (c) tell us. Condition (a) becomes

$$F_+ x = \frac{w - w_0}{t - t_0} x \in H_+^2; \quad \text{that is,} \quad \frac{w - w_0}{t - t_0} \in H_+^2.$$

Condition (b) becomes

$$F_- x = \frac{1 - \bar{w}w_0}{t - t_0} x = \bar{t}\frac{\bar{w} - \bar{w}_0}{\bar{t} - \bar{t}_0} \cdot \frac{w_0}{t_0} x \in H_-^2.$$

Hence (a) and (b) coincide. Finally, condition (c) becomes

$$\|Fx\|_{L^w}^2 = \int_{\mathbb{T}} \left\langle (Fx)(t), \begin{bmatrix} -\frac{w_0}{t-t_0} \\ \frac{1}{t-t_0} \end{bmatrix} x \right\rangle_{\mathbb{C}^2} m(dt)$$

$$= \int_{\mathbb{T}} \bar{x} \frac{-\bar{w}_0(w - w_0) + (1 - \bar{w}w_0)}{|t - t_0|^2} x \, m(dt)$$

$$= \bar{x} \int_{\mathbb{T}} \frac{(\bar{w} - \bar{w}_0)(w - w_0) + 1 - \bar{w}w}{|t - t_0|^2} m(dt) \, x$$

$$= \bar{x} \int_{\mathbb{T}} \left(\left| \frac{w - w_0}{t - t_0} \right|^2 + \frac{1 - |w|^2}{|t - t_0|^2} \right) m(dt) \, x = \bar{x} D_{w,t_0} x.$$

Thus, $\|Fx\|_{L^w}^2 \leq D$ if and only if $D_{w,t_0} \leq D$; that is, w is a solution of the AIP with this data if and only if w is a solution of the boundary interpolation problem.

A reference for Examples 2 and $2'$ is [Kheifets 1990a]. References for Example 3 are [Kheifets 1996; Sarason 1994].

Lecture 3. Solutions of the Abstract Interpolation Problem

Role of the AIP. In Lecture 2 we showed how some specific analytic problems can be included in the AIP scheme. The AIP, as formulated in Lecture 1, is very well adapted to this inclusion (actually it arose from the experience of treating a number of problems of this type). To include a specific problem in the scheme one has to realize (if it is possible) what the associated data is and to prove the coincidence of two solution sets: that of the original analytic problem and that of the AIP with the associated data. The mapping F suggests an algorithm for what exactly is to be checked to prove this coincidence.

In this lecture we will slightly reformulate the AIP to adapt it better to solving the problem (now the origin of the data is unimportant to a certain extent).

Thus, the AIP can be viewed as an intermediate link between specific interpolation problems and their resolution. It has two sides: one of them looks at the specific interpolation problem and is devoted to proving the equivalence of the original problem and associated AIP, the second concerns solutions. As soon as a specific problem is included in the scheme, the analysis of its solutions goes in a standard and universal way.

Isometric colligation associated with the AIP data. Reformulation of the AIP. Let $[x]$ stand for the equivalence class of x with respect to the quadratic form D. (The equivalence relation is defined as follows: $x \sim 0$ if and only if $D(x, y) = 0$ for all $y \in X$, and $x_1 \sim x_2$ if and only if $x_1 - x_2 \sim 0$). Consider the linear space of equivalence classes $\{[x] : x \in X\}$ and define an inner product in it by

$$\langle [x], [y] \rangle \stackrel{\mathrm{def}}{=} D(x, y).$$

This product is well defined. One can complete it and obtain a Hilbert space, which we denote by H_0.

Rewrite the Fundamental Identity (1–1) as

$$D(T_1x, T_1y) + \langle M_1x, M_1y \rangle_{E_1} = D(T_2x, T_2y) + \langle M_2x, M_2y \rangle_{E_2}.$$

Using the notations introduced above one can reexpress this as

$$\langle [T_1x], [T_1y] \rangle_{H_0} + \langle M_1x, M_1y \rangle_{E_1} = \langle [T_2x], [T_2y] \rangle_{H_0} + \langle M_2x, M_2y \rangle_{E_2}. \quad (3\text{–}1)$$

Set

$$d_v \stackrel{\text{def}}{=} \text{Clos} \left\{ \begin{bmatrix} [T_1x] \\ M_1x \end{bmatrix} : x \in X \right\} \subseteq H_0 \oplus E_1$$

and

$$\Delta_v \stackrel{\text{def}}{=} \text{Clos} \left\{ \begin{bmatrix} [T_2x] \\ M_2x \end{bmatrix} : x \in X \right\} \subseteq H_0 \oplus E_2.$$

Define a mapping $V : d_v \to \Delta_v$ by the formula

$$V : \begin{bmatrix} [T_1x] \\ M_1x \end{bmatrix} \longrightarrow \begin{bmatrix} [T_2x] \\ M_2x \end{bmatrix}. \quad (3\text{–}2)$$

Because of (3–1), V is an isometry. This implies that V is well-defined. In fact, if

$$\begin{bmatrix} [T_1x'] \\ M_1x' \end{bmatrix} = \begin{bmatrix} [T_1x''] \\ M_2x'' \end{bmatrix}, \quad \text{that is,} \quad \begin{bmatrix} [T_1(x' - x'')] \\ M_1(x' - x'') \end{bmatrix} = 0,$$

then (3–1) implies that

$$\begin{bmatrix} [T_2(x' - x'')] \\ M_2(x' - x'') \end{bmatrix} = 0.$$

Now, if x' and x'' generate the same vector on the left-hand side of (3–2), they generate the same vector on the right-hand side of (3–2), which shows that V is well-defined.

Let w be a solution of the AIP; that is, suppose there exists a mapping $F : X \to H^w$ possessing properties (i) and (ii) of page 353. Property (i) says that

$$\|Fx\|_{H^w}^2 \leq D(x, x) \equiv \big\| [x] \big\|_{H_0}^2.$$

Hence, Fx depends only on the equivalence class $[x]$, not on the representative x. This means that F generates a mapping G of the equivalence classes

$$G[x] \stackrel{\text{def}}{=} Fx, \quad (3\text{–}3)$$

and

$$\|G[x]\|_{H^w}^2 \equiv \|Fx\|_{H^w}^2 \leq \big\| [x] \big\|_{H_0}^2.$$

Thus G is a contraction. Since $\{[x] : x \in X\}$ is dense in H_0 and G is a contraction, it can be extended to a contraction of H_0 into H^w. Thus, *any $F : X \to H^w$ with properties* (i) *and* (ii) (actually up to now we used only (i)) *generates a contraction $G : H_0 \to H^w$ such that $G[x] = Fx$ for all $x \in X$.*

We now try to interpret property (ii). To this end we reexpress (ii) as

$$FT_2x + \bar{t}\begin{bmatrix} \mathbf{1} \\ w^* \end{bmatrix} M_2x = \bar{t}\left(FT_1x + \begin{bmatrix} w \\ \mathbf{1} \end{bmatrix} M_1x \right),$$

or

$$G[T_2x] + \bar{t}\begin{bmatrix} \mathbf{1} \\ w^* \end{bmatrix} M_2x = \bar{t}\left(G[T_1x] + \begin{bmatrix} w \\ \mathbf{1} \end{bmatrix} M_1x \right) \qquad (3\text{--}4)$$

Our further goal is to realize the latter equality as one of the form

$$\begin{bmatrix} G[T_2x] \\ M_2x \end{bmatrix} = A^w \begin{bmatrix} G[T_1x] \\ M_1x \end{bmatrix}, \qquad (3\text{--}5)$$

where $A^w : H^w \oplus E_1 \to H^w \oplus E_2$ is a linear operator. As we know, G maps H_0 into H^w, M_1 maps H_0 into E_1, and M_2 maps H_0 into E_2. The relation (3–4) is of the type

$$f'' + \bar{t}\begin{bmatrix} \mathbf{1} \\ w^* \end{bmatrix} e_2 = \bar{t}\left(f' + \begin{bmatrix} w \\ \mathbf{1} \end{bmatrix} e_1 \right), \qquad (3\text{--}6)$$

where $f', f'' \in H^w$ and $e_1 \in E_1$, $e_2 \in E_2$. Observe that the three subspaces

$$\left\{ \begin{bmatrix} w \\ \mathbf{1} \end{bmatrix} e_1 : e_1 \in E_1 \right\}, \quad H^w, \quad \left\{ \bar{t}\begin{bmatrix} \mathbf{1} \\ w^* \end{bmatrix} e_2 : e_2 \in E_2 \right\}$$

are mutually orthogonal in L^w (see page 352 for the definition of L^w). Let P^w be the orthogonal projection from L^w onto H^w; then, for any

$$f = \begin{bmatrix} f_2 \\ f_1 \end{bmatrix} \in L^w,$$

we have

$$P^w \begin{bmatrix} f_2 \\ f_1 \end{bmatrix} = \begin{bmatrix} f_2 \\ f_1 \end{bmatrix} - \begin{bmatrix} \mathbf{1} & w \\ w^* & \mathbf{1} \end{bmatrix} \begin{bmatrix} P_- f_2 \\ P_+ f_1 \end{bmatrix} \qquad (3\text{--}7)$$

Obviously this difference is in L^w. It can be rewritten as

$$P^w \begin{bmatrix} f_2 \\ f_1 \end{bmatrix} = \begin{bmatrix} P_+ f_2 - w P_+ f_1 \\ P_- f_1 - w^* P_- f_2 \end{bmatrix}.$$

Hence, it is in H^w. Because

$$\begin{bmatrix} \mathbf{1} & w \\ w^* & \mathbf{1} \end{bmatrix} \begin{bmatrix} P_- f_2 \\ P_+ f_1 \end{bmatrix}$$

is orthogonal to H^w, P^w is really the orthogonal projection.

Because

$$f' = \begin{bmatrix} f'_+ \\ f'_- \end{bmatrix} \in H^w,$$

we can obtain from (3–6) and (3–7)

$$\bar{t}\begin{bmatrix} \mathbf{1} \\ w^* \end{bmatrix} e_2 = (\mathbf{1}_{L^w} - P^w) \left\{ \bar{t}\left(f' + \begin{bmatrix} w \\ \mathbf{1} \end{bmatrix} e_1 \right) \right\} = \begin{bmatrix} \mathbf{1} & w \\ w^* & \mathbf{1} \end{bmatrix} \begin{bmatrix} P_- \bar{t}(f'_+ + we_1) \\ P_+ \bar{t}(f'_- + e_1) \end{bmatrix}$$

$$= \begin{bmatrix} \mathbf{1} & w \\ w^* & \mathbf{1} \end{bmatrix} \begin{bmatrix} \bar{t}(f'_+(0) + w(0))e_1 \\ 0 \end{bmatrix} = \bar{t}\begin{bmatrix} \mathbf{1} \\ w^* \end{bmatrix} \left(f'_+(0) + w(0)e_1 \right).$$

Thus

$$e_2 = f'_+(0) + w(0)e_1. \tag{3-8}$$

Now, from (3–6) and (3–8),

$$f'' = \bar{t}\left(f' + \begin{bmatrix} w \\ 1 \end{bmatrix} e_1\right) - \bar{t}\begin{bmatrix} 1 \\ w^* \end{bmatrix} e_2 = \bar{t}\left(f' + \begin{bmatrix} w \\ 1 \end{bmatrix} e_1\right) - \begin{bmatrix} 1 \\ w^* \end{bmatrix}(f'_+(0) + w(0)e_1)).$$

That is,

$$f'' = \bar{t}\left(f' - \begin{bmatrix} 1 & w \\ w^* & 1 \end{bmatrix}\begin{bmatrix} f'_+(0) \\ 0 \end{bmatrix}\right) + \bar{t}\begin{bmatrix} 1 & w \\ w^* & 1 \end{bmatrix}\begin{bmatrix} -w(0) \\ 1 \end{bmatrix} e_1. \tag{3-9}$$

Putting (3–8) and (3–9) together we obtain

$$\begin{bmatrix} f'' \\ e_2 \end{bmatrix} = A^w \begin{bmatrix} f' \\ e_1 \end{bmatrix}, \quad A^w = \begin{bmatrix} A^w_{\mathrm{in}} & (A^w_1)^* \\ A^w_2 & A^w_{12} \end{bmatrix} : \begin{bmatrix} H^w \\ E_1 \end{bmatrix} \to \begin{bmatrix} H^w \\ E_2 \end{bmatrix}$$

where

$$A^w_{\mathrm{in}} f = P^w \bar{t} f = \bar{t}\left(f - \begin{bmatrix} 1 & w \\ w^* & 1 \end{bmatrix}\begin{bmatrix} f_+(0) \\ 0 \end{bmatrix}\right) : H^w \to H^w,$$

$$(A^w_1)^* e_1 = \bar{t}\begin{bmatrix} 1 & w \\ w^* & 1 \end{bmatrix}\begin{bmatrix} -w(0) \\ 1 \end{bmatrix} : E_1 \to H^w,$$

$$A^w_2 f = f_+(0) : H^w \to E_2,$$

$$A^w_{12} e_1 = w(0)e_1 : E_1 \to E_2.$$

Thus, condition (ii) of the AIP, equivalently (3–4) or (3–5), can be expressed as

$$\begin{bmatrix} G & 0 \\ 0 & \mathbf{1}_{E_2} \end{bmatrix}\begin{bmatrix} [T_2 x] \\ M_2 x \end{bmatrix} = A^w \begin{bmatrix} G & 0 \\ 0 & \mathbf{1}_{E_1} \end{bmatrix}\begin{bmatrix} [T_1 x] \\ M_1 x \end{bmatrix}. \tag{3-10}$$

According to the definition (3–2) of the isometry V,

$$\begin{bmatrix} [T_2 x] \\ M_2 x \end{bmatrix} = V \begin{bmatrix} [T_1 x] \\ M_1 x \end{bmatrix}.$$

Combining this with (3–10) we conclude that

$$\begin{bmatrix} G & 0 \\ 0 & \mathbf{1}_{E_2} \end{bmatrix} V \mid d_v = A^w \begin{bmatrix} G & 0 \\ 0 & \mathbf{1}_{E_1} \end{bmatrix} \mid d_v. \tag{3-11}$$

To give (3–11) a further interpretation we digress to recall some basic facts related to unitary colligations, their characteristic functions, and functional models.

Digression on unitary colligations, characteristic functions, and functional models. Let H, E_1, E_2 be separable Hilbert spaces. A unitary mapping A of $H \oplus E_1$ onto $H \oplus E_2$ is said to be a *unitary colligation*. The space H is called the *state space* of the colligation, E_1 is called the *input space*, and E_2 is called the *response space*. Both E_1 and E_2 are called *exterior spaces*.

The operator-valued function $w(\zeta) : E_1 \to E_2$ defined by the formula

$$w(\zeta) = P_{E_2} A (\mathbf{1}_{H \oplus E_1} - \zeta P_H A)^{-1} \mid E_1$$

is called the *characteristic function* of the colligation A. Because A is unitary, $w(\zeta)$ is well defined for $\zeta \in \mathbb{D}$; w is a contractive-valued analytic operator-function in $\mathbb{D}$.

Using the block decomposition

$$A = \begin{bmatrix} A_{\text{in}} & A_1 \\ A_2 & A_{12} \end{bmatrix} : \begin{bmatrix} H \\ E_1 \end{bmatrix} \to \begin{bmatrix} H \\ E_2 \end{bmatrix},$$

the characteristic function can be reexpressed as

$$w(\zeta) = A_{12} + \zeta A_2 (\mathbf{1}_H - \zeta A_{\text{in}})^{-1} A_1.$$

The unitary colligation A is said to be *simple* with respect to the exterior spaces E_1 and E_2 if there is no nonzero reducing subspace for A in H; that is, if there is no nonzero subspace $H_{\text{res}} \subseteq H$ that is invariant for A and A^*. We shall call the maximal reducing subspace for A in H the *residual subspace* of the colligation A. Thus a unitary colligation is simple if the residual subspace is trivial.

Let $H_{\text{res}} \subseteq H$ be the residual subspace of A. Let $H_{\text{simp}} = H \ominus H_{\text{res}}$. Then $A : H_{\text{simp}} \oplus E_1 \to H_{\text{simp}} \oplus E_2$ is a simple unitary colligation, and $A \mid H_{\text{res}}$ is a unitary operator on H_{res}.

Two unitary colligations

$$A : H \oplus E_1 \to H \oplus E_2 \quad \text{and} \quad A' : H' \oplus E_1 \to H' \oplus E_2$$

with the same exterior spaces are said to be *unitarily equivalent* if there exists a unitary mapping $\mathcal{G} : H \to H'$ such that

$$\begin{bmatrix} \mathcal{G} & 0 \\ 0 & \mathbf{1}_{E_2} \end{bmatrix} A = A' \begin{bmatrix} \mathcal{G} & 0 \\ 0 & \mathbf{1}_{E_1} \end{bmatrix}.$$

THEOREM. *Two simple unitary colligations A and A' are unitarily equivalent if and only if their characteristic functions coincide.*

Let w be a Schur class operator function, $w(\zeta) : E_1 \to E_2$. The unitary colligation A^w considered above is simple and w is the characteristic function of this colligation. Thus any Schur class operator function is the characteristic function of a unitary colligation.

Let $A : H \oplus E_1 \to H \oplus E_2$ be a simple unitary colligation, with characteristic function w. Then a unitary mapping $\mathcal{G} : H \to H^w$ that performs an equivalence between A and A^w is defined as follows:

$$(\mathcal{G}h)(\zeta) = \begin{bmatrix} (\mathcal{G}_+ h)(\zeta) \\ (\mathcal{G}_- h)(\zeta) \end{bmatrix} = \begin{bmatrix} P_{E_2} A (\mathbf{1}_{H \oplus E_1} - \zeta P_H A)^{-1} h \\ \bar{\zeta} P_{E_1} A^* (\mathbf{1}_{H \oplus E_2} - \bar{\zeta} P_H A^*)^{-1} h \end{bmatrix} \tag{3-12}$$

The colligation A^w is called a *functional model* of A and the mapping $\mathcal{G}$ is called the *Fourier representation* of H.

If A is not simple, the Fourier representation $\mathcal{G}$ defined by the same formula (3–12) is a unitary mapping from H_{simp} onto H^w and performs an equivalence between the simple parts of A and A^w. It vanishes on H_{res}:

$$\begin{bmatrix} \mathcal{G} & 0 \\ 0 & \mathbf{1}_{E_2} \end{bmatrix} A = A^w \begin{bmatrix} \mathcal{G} & 0 \\ 0 & \mathbf{1}_{E_1} \end{bmatrix}.$$

Solutions of the AIP as characteristic functions. We are ready to proceed with the analysis of the AIP solutions. We begin with (3–11). Assume for simplicity that we have the equality

$$\|Fx\|^2_{H^w} = D(x, x) \quad \text{for all } x \in X.$$

We noticed in Lecture 2 that for some problems this is the case for any solution w. In other words, our assumption means that the map G defined in (3–3) is an isometry:

$$\|Gh_0\|^2_{H^w} = \|h_0\|^2_{H_0}.$$

Set $H_1 = H^w \ominus G H_0$ and $H = H_0 \oplus H_1$. Define a mapping $\mathcal{G} : H \to H^w$ by setting

$$\mathcal{G} \,|\, H_0 = G, \qquad \mathcal{G} \,|\, H_1 = \mathbf{1}_{H_1},$$

and observe that $\mathcal{G}$ is a unitary mapping of H onto H^w. We can write

$$\begin{bmatrix} \mathcal{G} & 0 \\ 0 & \mathbf{1}_{E_2} \end{bmatrix} V \,|\, d_V = A^w \begin{bmatrix} \mathcal{G} & 0 \\ 0 & \mathbf{1}_{E_1} \end{bmatrix} \,|\, d_V, \tag{3-13}$$

instead of (3–11), because

$$d_V \subseteq H_0 \oplus E_1 \subseteq H \oplus E_1,$$
$$\Delta_V \subseteq H_0 \oplus E_2 \subseteq H \oplus E_2.$$

Finally, we obtain

$$V = \begin{bmatrix} \mathcal{G}^* & 0 \\ 0 & \mathbf{1}_{E_2} \end{bmatrix} A^w \begin{bmatrix} \mathcal{G} & 0 \\ 0 & \mathbf{1}_{E_1} \end{bmatrix} \,|\, d_V. \tag{3-14}$$

Define the operator A by

$$A \overset{\text{def}}{=} \begin{bmatrix} \mathcal{G}^* & 0 \\ 0 & \mathbf{1}_{E_2} \end{bmatrix} A^w \begin{bmatrix} \mathcal{G} & 0 \\ 0 & \mathbf{1}_{E_1} \end{bmatrix}. \tag{3-15}$$

A is a simple unitary colligation from $H \oplus E_1$ onto $H \oplus E_2$, and is unitarily equivalent to A^w. Hence, the characteristic function of A is w. By definition,

$H_0 \subseteq H$ and $A \mid d_V = V$ (see (3–14)); that is, A is a *unitary extension* of V. Thus *any solution w of the AIP is the characteristic function of a unitary extension A of the isometry V associated with the data:*

$$A : H \oplus E_1 \to H \oplus E_2 \quad \text{with} \quad H \supseteq H_0, \quad A \mid d_V = V.$$

In general, the equality $\|Fx\|_{H^w}^2 = D(x, x)$ for all $x \in X$ does not hold, only the inequality $\|Fx\|_{H^w}^2 \leq D(x, x)$ for all $x \in X$. Nevertheless, the conclusion is still valid, but it is more difficult to prove. (A short and simple proof was recently found by J. Ball and T. Trent [Ball and Trent $\geq$ 1997].)

REMARK. I would like to emphasize one basic difference between the general case (inequality) and the special case (equality) considered in this section. We need one more definition: a unitary extension $A : H \oplus E_1 \to H \oplus E_2$ of an isometric colligation $V : d_V \to \Delta_V$, with $d_V \subseteq H_0 \oplus E_1$, $\Delta_V \subseteq H_0 \oplus E_2$, $H_0 \subseteq H$ is said to be a *minimal extension* if A has no nonzero reducing subspace in $H \ominus H_0$. If an extension A is nonminimal we can discard the reducing subspace in $H \ominus H_0$ and end up with a unitary colligation that has the same characteristic function and that *still extends* V, so we can consider minimal extensions only. But a minimal extension need not be a simple colligation. The absence of a reducing subspace for A in $H \ominus H_0$ does not mean that A has no reducing subspace in H at all. By discarding such a reducing subspace, we end up with a unitary colligation that has the same characteristic function but that *no longer extends* V. In the case of equality, $\|Fx\|_{H^w}^2 = D(x, x)$ for all $x \in X$, the corresponding minimal extension is a simple colligation, as it is equivalent to A^w (see (3–14)); in the case of inequality, there exists $x \in X$ such that $\|Fx\|_{H^w}^2 < D(x, x)$, and it is not simple.

A natural question arises now: does an arbitrary unitary extension A of the isometry V produce a solution of the AIP? The answer is yes, and it is easy to prove.

Let $A : H \oplus E_1 \to H \oplus E_2$ be a unitary extension of V. Let $\mathcal{G}$ be the Fourier representation associated with the colligation A; see (3–12). Then $\mathcal{G}$ maps H into H^w, where w is the characteristic function of A. Thus $\mathcal{G}$ is a contractive operator and

$$\begin{bmatrix} \mathcal{G} & 0 \\ 0 & \mathbf{1}_{E_2} \end{bmatrix} A = A^w \begin{bmatrix} \mathcal{G} & 0 \\ 0 & \mathbf{1}_{E_1} \end{bmatrix}. \tag{3–16}$$

Define the mapping $F : X \to H^w$ by

$$Fx \overset{\text{def}}{\equiv} G[x] \overset{\text{def}}{\equiv} \mathcal{G}[x].$$

Recall that $[x] \in H_0 \subseteq H$. Since $\mathcal{G}$ is a contraction, we have

$$\|Fx\|_{H^w}^2 \leq \left\| [x] \right\|_{H^0}^2 = D(x, x);$$

that is, F satisfies (i). Since A extends V, (3–16) yields

$$\begin{bmatrix} \mathcal{G} & 0 \\ 0 & \mathbf{1}_{E_2} \end{bmatrix} V \mid d_V = A^w \begin{bmatrix} \mathcal{G} & 0 \\ 0 & \mathbf{1}_{E_1} \end{bmatrix} \mid d_V.$$

Since $d_V \subseteq H_0 \oplus E_1$ and $\Delta_V \subseteq H_0 \oplus E_2$, we can replace $\mathcal{G}$ with G:

$$\begin{bmatrix} G & 0 \\ 0 & \mathbf{1}_{E_2} \end{bmatrix} V \mid d_V = A^w \begin{bmatrix} G & 0 \\ 0 & \mathbf{1}_{E_1} \end{bmatrix} \mid d_V.$$

As we have seen, the latter is equivalent to (ii). Thus w, the characteristic function of A, is a solution of the AIP.

REMARK. Starting with an extension A of V we do not need the unitarity of $\mathcal{G}$ to check that the characteristic function of A is a solution. Neither the characteristic function nor the mapping $\mathcal{G}$, much less F, feels the residual part. Starting with the solution w in the general case we actually are given information on the simple part of the corresponding extension of V only. But we have to restore the residual part also in order to obtain the extension.

I will finish this lecture with the following summary of the discussion.

THEOREM. *Let V be the isometric colligation associated with the AIP data*

$$V : d_V \to \Delta_V, \quad d_V \subseteq H_0 \oplus E_1, \quad \Delta_V \subseteq H_0 \oplus E_2.$$

(no special assumptions are made on the data at present). Let $A : H \oplus E_1 \to H \oplus E_2$ be a minimal unitary extension of V, where $H \supseteq H_0$, so $A \mid d_V = V$. Let $w(\zeta)$ be the characteristic function of the colligation A:

$$w(\zeta) \stackrel{\text{def}}{=} P_{E_2} A(\mathbf{1}_{H \oplus E_1} - \zeta P_H A)^{-1} \mid E_1.$$

Then w is a solution of the AIP. The corresponding mapping $F : X \to H^w$ is defined by

$$Fx = \mathcal{G}[x],$$

where

$$(\mathcal{G}h)(\zeta) = \begin{bmatrix} (\mathcal{G}_+ h)(\zeta) \\ (\mathcal{G}_- h)(\zeta) \end{bmatrix} = \begin{bmatrix} P_{E_2} A(\mathbf{1}_{H \oplus E_1} - \zeta P_H A)^{-1} h \\ \bar{\zeta} P_{E_1} A^*(\mathbf{1}_{H \oplus E_2} - \bar{\zeta} P_H A^*)^{-1} h \end{bmatrix} \quad \text{for } h \in H.$$

All solutions of the AIP and the corresponding mappings F that satisfy (i) and (ii) are of this form.

References to Lecture 3 are [Katsnelson et al. 1987; Kheifets 1988a; 1988b; 1990b; Kheifets and Yuditskii 1994].

Lecture 4. Description of the Solutions of the AIP

In a sense, the description of all the solutions of the AIP and the corresponding mappings F was given in the last theorem of Lecture 3. The goal of this lecture is to give a description that separates explicitly the common part (related to the data) and the free parameters. The first step is the description of the unitary extensions of the given isometric colligation V in this fashion.

Unitary extensions of isometric colligations Let V be an isometric colligation,

$$V : d_V \to \Delta_V \quad \text{with} \quad d_V \subseteq H_0 \oplus E_1, \quad \Delta_V \subseteq H_0 \oplus E_2.$$

Let A be a minimal unitary extension of V,

$$A : H \oplus E_1 \to H \oplus E_2 \quad \text{with} \quad H \supseteq H_0, \quad A \,|\, d_V = V.$$

Let $d_V^\perp$ and $\Delta_V^\perp$ be the orthogonal complements of d_V in $H_0 \oplus E_1$ and Δ_V in $H_0 \oplus E_2$. Let $H_1 = H \ominus H_0$. Then the orthogonal complement of d_V in $H \oplus E_1$ is $H_1 \oplus d_V^\perp$ and the orthogonal complement of Δ_V in $H \oplus E_2$ is $H_1 \oplus \Delta_V^\perp$. Since A is a unitary operator mapping d_V onto Δ_V (since $A \,|\, d_V = V$), A has to map the orthogonal complement $H_1 \oplus d_V^\perp$ onto the orthogonal complement $H_1 \oplus \Delta_V^\perp$. Denote by A_1 the restriction of A to $H_1 \oplus d_V^\perp$. Thus, $A_1 : H_1 \oplus d_V^\perp \to H_1 \oplus \Delta_V^\perp$ is a unitary colligation. Since A is a minimal extension, A_1 is a simple colligation.

Conversely, take an arbitrary simple unitary colligation A_1 with the same exterior spaces $d_V^\perp$ and $\Delta_V^\perp$ and an arbitrary admissible state space H_1:

$$A_1 : H_1 \oplus d_V^\perp \to H_1 \oplus \Delta_V^\perp. \tag{4–1}$$

Let $H = H_0 \oplus H_1$ and define an extension of V by

$$A \,\big|\, d_V = V, \qquad A \,\big|\, H_1 \oplus d_V^\perp = A_1.$$

The result is a minimal unitary extension of V. Thus, the free parameter of a minimal unitary extension of V is an arbitrary simple unitary colligation A_1 with fixed exterior spaces $d_V^\perp$ and $\Delta_V^\perp$ and arbitrary admissible state space H_1. The word "admissible" means here that there exists a unitary colligation with this state space. For example: If $d_V^\perp$ and $\Delta_V^\perp$ are finite dimensional but their dimensions are different, then H_1 cannot be of finite dimension; if the dimensions of $d_V^\perp$ and $\Delta_V^\perp$ are equal then H_1 can be of arbitrary dimension. To give a full explanation it is enough to consider model colligations. Let $\omega : d_V^\perp \to \Delta_V^\perp$ be an arbitrary Schur class operator function (contractive-valued and analytic in the unit disc). One can take as A_1 the (model) unitary colligation

$$A^\omega = \begin{bmatrix} A_{\mathrm{in}}^\omega & (A_1^\omega)^* \\ A_2^\omega & A_{12}^\omega \end{bmatrix} : \begin{bmatrix} H^\omega \\ d_V^\perp \end{bmatrix} \to \begin{bmatrix} H^\omega \\ \Delta_V^\perp \end{bmatrix},$$

where H^ω is the de Branges–Rovnyak space corresponding to the operator function ω (see page 352) and

$$A_{\text{in}}^\omega f = P^\omega \bar{t} f = \bar{t}\left(f - \begin{bmatrix} \mathbf{1} & \omega \\ \omega^* & \mathbf{1} \end{bmatrix} \begin{bmatrix} f_+(0) \\ 0 \end{bmatrix} \right) : H^\omega \to H^\omega,$$

$$(A_1^\omega)^* = \bar{t}\begin{bmatrix} \mathbf{1} & \omega \\ \omega^* & \mathbf{1} \end{bmatrix} \begin{bmatrix} -\omega(0) \\ \mathbf{1} \end{bmatrix} : d_V^\perp \to H^\omega,$$

$$A_2^\omega f = f_+(0) : H^\omega \to \Delta_V^\perp,$$

$$A_{12}^\omega = \omega(0) : d_V^\perp \to \Delta_V^\perp.$$

Let $A_1 : H_1 \oplus d_V^\perp \to H_1 \oplus \Delta_V^\perp$ and $A_1' : H_1' \oplus d_V^\perp \to H_1' \oplus \Delta_V^\perp$ be two simple unitary colligations. If they are unitarily equivalent, that is, if there exists a unitary mapping $\mathcal{G}_1 : H_1 \to H_1'$ such that

$$\begin{bmatrix} \mathcal{G}_1 & 0 \\ 0 & \mathbf{1}_{\Delta_V^\perp} \end{bmatrix} A_1 = A_1' \begin{bmatrix} \mathcal{G}_1 & 0 \\ 0 & \mathbf{1}_{d_V^\perp} \end{bmatrix},$$

the corresponding minimal unitary extensions of V are unitarily equivalent colligations:

$$\begin{bmatrix} \mathcal{G} & 0 \\ 0 & \mathbf{1}_{E_2} \end{bmatrix} A = A' \begin{bmatrix} \mathcal{G} & 0 \\ 0 & \mathbf{1}_{E_1} \end{bmatrix},$$

with $\mathcal{G} \,|\, H_1 = \mathcal{G}_1$ and $\mathcal{G} \,|\, H_0 = \mathbf{1}_{H_0}$.

We emphasize here that this is more than just equivalence of unitary colligations, as $\mathcal{G}$ is the identity on H_0. We will call such extensions *equivalent extensions*. The following proposition follows from the previous discussion.

CLAIM 1. *Two minimal unitary extensions A and A' of V are equivalent extensions if and only if the corresponding simple unitary colligations A_1 and A_1' are unitarily equivalent colligations.*

The next claim is a straightforward consequence of the formulas of the last theorem of Lecture 3.

CLAIM 2. *Two equivalent minimal extensions of V generate the same solution w and the same mapping $F : X \to H^w$.*

Thus, a solution w of the AIP and a corresponding mapping $F : X \to H^w$ represent equivalence classes of simple unitary colligations $A_1 : H_1 \oplus d_V^\perp \to H_1 \oplus \Delta_V^\perp$. Hence, the latter may serve as free parameters. But we know from the digression on page 366 that the only invariant of this equivalence class is the characteristic function of A_1. In other words:

CLAIM 3. *The free parameter of pairs (w, F) (consisting of a solution w and a corresponding mapping F) is an arbitrary Schur class (contractive-valued and analytic in $\mathbb{D}$) operator function $\omega(\zeta) : d_V^\perp \to \Delta_V^\perp$.*

Universal unitary colligations associated with isometric colligations.
In the previous section we extracted the free parameter ω that the solution w
and the corresponding mapping F depend on. It is clear that the common part
of all the extensions of the isometry V is the isometry V itself. To obtain nice
and explicit formulas it is convenient to associate a *universal unitary colligation*
to the isometry V. Let V be an isometric colligation,

$$V : d_V \to \Delta_V \quad \text{with} \quad d_V \subseteq H_0 \oplus E_1, \quad \Delta_V \subseteq H_0 \oplus E_2,$$

and let $d_V^\perp$ and $\Delta_V^\perp$ be the orthogonal complements of d_V and Δ_V. Let N_1 be
an isomorphic copy of $d_V^\perp$ (that is, there exists a unitary and surjective mapping
$u_1 : d_V^\perp \to N_1$). Let N_2 be an isomorphic copy of $\Delta_V^\perp$ (that is, there exists a
unitary and surjective mapping $u_2 : \Delta_V^\perp \to N_2$). Define a unitary colligation
$A_0 : H_0 \oplus E_1 \oplus N_2 \to H_0 \oplus E_2 \oplus N_1$ by setting

$$A_0 \,|\, d_V = V \qquad (d_V \subseteq H_0 \oplus E_1),$$
$$A_0 \,|\, d_V^\perp = u_1, \qquad A_0 \,|\, N_2 = u_2^*.$$

REMARK. A_0 is not a unitary extension of V in the sense considered earlier,
because we now do not extend the state space H_0, only the exterior spaces:
$E_1 \oplus N_2$ instead of E_1 and $E_2 \oplus N_1$ instead of E_2. Of course, A_0 extends V but
in a different sense. We will call this extension a *universal extension*; the name
is motivated by category theory. We will see the role of this colligation below.

From now on we fix the unitary mappings u_1 and u_2, and also their images N_1 and
N_2. Also we fix the notation A_1 for a simple unitary colligation $A_1 : H_1 \oplus N_1 \to$
$H_1 \oplus N_2$ (this class of colligations is obviously related to the colligations of the
form (4–1) defined earlier, and is denoted by the same symbol). Thus, now the
equivalence classes of the simple unitary colligations $A_1 : H_1 \oplus N_1 \to H_1 \oplus N_2$
will be free parameters of the solutions w and the corresponding mappings F;
that is, the Schur class functions $\omega(\zeta) : N_1 \to N_2$ are free parameters now.

**Coupling of unitary colligations and unitary extensions of isometric
colligations.** Let V be an isometric colligation,

$$V : d_V \to \Delta_V \quad \text{with} \quad d_V \subseteq H_0 \oplus E_1, \quad \Delta_V \in H_0 \oplus E_2.$$

Let A_0 be the universal colligation associated with V:

$$A_0 : H_0 \oplus E_1 \oplus N_2 \to H_0 \oplus E_2 \oplus N_1,$$

with $A_0 \,|\, d_V = V$, $A_0(d_V^\perp) = N_1$, $A_0(N_2) = \Delta_V^\perp$. Let A be a minimal unitary
extension of V:

$$A : H \oplus E_1 \to H \oplus E_2 \quad \text{with} \quad H \supseteq H_0, \quad A \,|\, d_V = V.$$

We have seen that one can associate a simple unitary colligation $A_1 : H_1 \oplus N_1 \to H_1 \oplus N_2$ with A, where $H_1 = H \ominus H_0$:

$$A_1 = \begin{bmatrix} \mathbf{1}_{H_1} & 0 \\ 0 & (A_0^*) \,|\, \Delta_V^\perp \end{bmatrix} (A \,|\, H_1 \oplus d_V^\perp) \begin{bmatrix} \mathbf{1}_{H_1} & 0 \\ 0 & (A_0^*) \,|\, N_1 \end{bmatrix} \qquad (4\text{-}2)$$

An arbitrary simple unitary colligation A_1 with the exterior spaces N_1 and N_2 arises this way.

I will now give a procedure for recovering A from A_0 (fixed colligation) and A_1 (arbitrary parameter). Let $A_1 : H_1 \oplus N_1 \to H_1 \oplus N_2$ be an arbitrary simple unitary colligation. Let

$$\begin{bmatrix} h_0' \\ e_2 \\ n_1 \end{bmatrix} = A_0 \begin{bmatrix} h_0 \\ e_1 \\ n_2 \end{bmatrix} \quad \text{and} \quad \begin{bmatrix} h_1' \\ n_2' \end{bmatrix} = A_1 \begin{bmatrix} h_1 \\ n_1' \end{bmatrix}. \qquad (4\text{-}3)$$

We can choose vectors on the right-hand sides of these relations in an arbitrary way and compute the vectors on the left-hand sides, or we can choose vectors on the left-hand sides in an arbitrary way and compute the vectors on the right-hand sides (because A_0 and A_1 are unitary operators).

Let us consider, in addition to the relations above, two more relations

$$n_1 = n_1' \quad \text{and} \quad n_2 = n_2', \qquad (4\text{-}4)$$

and see what can be chosen now in an arbitrary way. Observe that the colligation A_0 possesses the property

$$P_{N_1} A_0 \,|\, N_2 = 0,$$

because A_0 sends N_2 onto $\Delta_V^\perp \subseteq H_0 \oplus E_2$, which is orthogonal to N_1. This property of A_0 guarantees (although it is not necessary) that we can choose h_0, h_1, e_1 in an arbitrary way and compute h_0', h_1', e_2 from the system of equations (4–3), (4–4) in a unique way, along with $n_1 = n_1'$ and $n_2 = n_2'$ as well. Take the result of this computation as the definition of the new linear operator A:

$$A \begin{bmatrix} h_1 \\ h_0 \\ e_1 \end{bmatrix} \overset{\text{def}}{=} \begin{bmatrix} h_1' \\ h_0' \\ e_2 \end{bmatrix}. \qquad (4\text{-}5)$$

It is easy to see that A is a unitary colligation, $A : H_1 \oplus H_0 \oplus E_1 \to H_0 \oplus H_0 \oplus E_2$. Moreover, A extends V. To see this, take an arbitrary vector

$$\begin{bmatrix} h_0 \\ e_1 \end{bmatrix} \in d_V;$$

then take

$$\begin{bmatrix} h_0' \\ e_2 \end{bmatrix} = V \begin{bmatrix} h_0 \\ e_1 \end{bmatrix} \in \Delta_V,$$

and take also

$$h_1 = h_1' = 0, \quad n_1 = n_1' = 0, \quad n_2 = n_2' = 0.$$

This collection of vectors satisfies the system of equations (4–3), (4–4): it obviously fits (4–4) and the second equality in (4–3); it satisfies the first equality in (4–3) because $A_0 \,|\, d_V = V$; i.e.,

$$\begin{bmatrix} V\begin{bmatrix} h_0 \\ e_1 \end{bmatrix} \\ 0 \end{bmatrix} = A_0 \begin{bmatrix} h_0 \\ e_1 \\ 0 \end{bmatrix} \quad \text{if} \quad \begin{bmatrix} h_0 \\ e_1 \end{bmatrix} \in d_V.$$

Hence, by the definition of the colligation A, these vectors (actually, a part of them) satisfy (4–5):

$$\begin{bmatrix} 0 \\ V\begin{bmatrix} h_0 \\ e_1 \end{bmatrix} \end{bmatrix} = A \begin{bmatrix} 0 \\ h_0 \\ e_1 \end{bmatrix} \quad \text{if} \quad \begin{bmatrix} h_0 \\ e_1 \end{bmatrix} \in d_V.$$

This means that A extends V.

We will say that the colligation A is the *feedback coupling* of A_0 with A_1 (or the *loading* of A_0 with A_1). It is also easy to check that these two procedures—the extraction of A_1 from the extension A and the feedback coupling of A_0 with A_1—are mutually inverse: performing the two successively we come back to the colligation that we started with.

Our next goal is to express the characteristic function w and the Fourier representation $\mathcal{G}$ of the colligation A in terms of the characteristic functions and Fourier representations of the colligations A_0 and A_1. But we need a digression first.

Digression: Unitary colligations, characteristic functions, Fourier representation, and discrete time dynamics. Let $A : H \oplus E_1 \to H \oplus E_2$ be a unitary colligation. One can associate with A the discrete-time dynamical system

$$\begin{bmatrix} h(k+1) \\ e_2(k) \end{bmatrix} = A \begin{bmatrix} h(k) \\ e_1(k) \end{bmatrix}, \tag{4–6}$$

where k is a nonnegative integer, $h(0)$ is an arbitrary vector from H, and $\{e_1(k)\}_{k=0}^{\infty}$ is an arbitrary input signal.

Let $\zeta \in \mathbb{D}$ be the corresponding spectral parameter (complex frequency). Let $\tilde{h}^+(\zeta)$, $\tilde{e}_1^+(\zeta)$, and $\tilde{e}_2^+(\zeta)$ be the discrete Laplace transforms of $h(k)$, $e_1(k)$, and $e_2(k)$, respectively; that is,

$$\tilde{h}^+(\zeta) = \sum_{k=0}^{\infty} \zeta^k h(k), \quad \tilde{e}_1^+(\zeta) = \sum_{k=0}^{\infty} \zeta^k e_1(k), \quad \tilde{e}_2^+(\zeta) = \sum_{K=0}^{\infty} \zeta^k e_2(k).$$

If the input signal $\{e_1(k)\}$ is square summable,

$$\sum_{k=0}^{\infty} \|e_1(k)\|^2 < \infty,$$

then the state evolution $\{h(k)\}_{k=0}^{\infty}$ and the output signal $\{e_2(k)\}_{k=0}^{\infty}$ possess the same property.

Multiplying (4–6) by ζ^k and taking the summation over k, one obtains the spectral form of the dynamics equation:

$$\tilde{h}^+(\zeta) = (\mathbf{1}_H - \zeta A_{in})^{-1} \cdot h(0) + \mathcal{G}_-(\zeta)^* \tilde{e}_1^+(\zeta)$$
$$\tilde{e}_2^+(\zeta) = \mathcal{G}_+(\zeta) \cdot h(0) + w(\zeta) \cdot \tilde{e}_1^+(\zeta), \tag{4-7}$$

where $w(\zeta)$ is the characteristic function of the colligation A, and $\mathcal{G}_+, \mathcal{G}_-$ are the components of the Fourier representation $\mathcal{G}$ (see (3–12)).

One can also rewrite (4–6) as

$$A^* \begin{bmatrix} h(k+1) \\ e_2(k) \end{bmatrix} = \begin{bmatrix} h(k) \\ e_1(k) \end{bmatrix}. \tag{4-8}$$

Now consider the negative integers $k \le -1$. (This actually means inverting time and exchanging the roles of the input and the output). Considering the past Laplace transform,

$$\tilde{h}^-(\zeta) = \sum_{k=-\infty}^{-1} \bar{\zeta}^{|k|} h(k), \quad \tilde{e}_2^-(\zeta) = \sum_{k=-\infty}^{-1} \bar{\zeta}^{|k|} e_2(k), \quad \tilde{e}_1^-(\zeta) = \sum_{k=-\infty}^{-1} \bar{\zeta}^{|k|} e_1(k),$$

one can rewrite (4–8) as

$$\tilde{h}^-(\zeta) = \bar{\zeta}(\mathbf{1} - \bar{\zeta} A_{in}^*)^{-1} h(0) + \mathcal{G}_+(\zeta)^* \tilde{e}_2^-(\zeta),$$
$$\tilde{e}_1^-(\zeta) = \mathcal{G}_-(\zeta) \cdot h(0) + w(\zeta)^* \tilde{e}_2^-(\zeta). \tag{4-9}$$

The second formulas in (4–7) and (4–9) are the most convenient for our purposes:

$$\tilde{e}_2^+(\zeta) = \mathcal{G}_+(\zeta) h(0) + w(\zeta) \tilde{e}_1^+(\zeta)$$
$$\tilde{e}_1^-(\zeta) = \mathcal{G}_-(\zeta) h(0) + w(\zeta)^* \tilde{e}_2^-(\zeta). \tag{4-10}$$

Formulas describing the solutions w of the AIP and the corresponding mappings F. Let A be the feedback coupling of A_0 with A_1. Denote the characteristic function of A_0 by $S(\zeta) : E_1 \oplus N_2 \to E_2 \oplus N_1$. Denote also the entries of S corresponding to this decomposition of the spaces as follows:

$$S = \begin{bmatrix} s_0 & s_2 \\ s_1 & s \end{bmatrix} : \begin{bmatrix} E_1 \\ N_2 \end{bmatrix} \longrightarrow \begin{bmatrix} E_2 \\ N_1 \end{bmatrix}.$$

Let the characteristic function of A_1 be $\omega(\zeta) : N_1 \to N_2$. Now write the dynamics related to A_0 and the dynamics related to A_1 in the form (4–10). I am going to consider the "+" parts only now; the treatment of the "−" parts is analogous. We have

$$\tilde{n}_2^+(\zeta) = \mathcal{G}_+^1(\zeta) h_1(0) + \omega(\zeta) \tilde{n}_1^+(\zeta),$$
$$\begin{bmatrix} \tilde{e}_2^+(\zeta) \\ \tilde{n}_1'^+(\zeta) \end{bmatrix} = \mathcal{G}_+^0(\zeta) h_0(0) + S(\zeta) \begin{bmatrix} \tilde{e}_1^+(\zeta) \\ \tilde{n}_2'^+(\zeta) \end{bmatrix}.$$

The coupling condition is

$$\tilde{n}_1^+ = \tilde{n}_1'^+, \qquad \tilde{n}_2^+ = \tilde{n}_2'^+.$$

Express $\tilde{e}_2^+(\zeta)$ in terms of

$$\begin{bmatrix} h_1(0) \\ h_0(0) \end{bmatrix}$$

and $\tilde{e}_1^+(\zeta)$, excluding $\tilde{n}_1^+ = \tilde{n'}_1^+$ and $\tilde{n}_2^+ = \tilde{n'}_2^+$. What we obtain now has to coincide (see (4–10)) with

$$\tilde{e}_2^+(\zeta) = \mathcal{G}_+(\zeta) \begin{bmatrix} h_1(0) \\ h_0(0) \end{bmatrix} + w(\zeta)\tilde{e}_1^+(\zeta).$$

This leads to the formulas

$$w(\zeta) = s_0(\zeta) + s_2(\zeta)w(\zeta)(\mathbf{1}_{N_1} - s(\zeta)w(\zeta))^{-1}s_1(\zeta)$$

$$\mathcal{G}_+(\zeta)\begin{bmatrix} h_1(0) \\ h_0(0) \end{bmatrix} = [\psi(\zeta)w(\zeta), \, \mathbf{1}_{E_2}]\mathcal{G}_+^0(\zeta)h_0(0) + \psi(\zeta)\mathcal{G}_+^1(\zeta)h_1(0),$$

$$(4\text{–}11)$$

where

$$\psi(\zeta) = s_2(\zeta)(\mathbf{1}_{N_2} - \omega(\zeta)s(\zeta))^{-1}. \qquad (4\text{–}12)$$

REMARK. By the definition of the characteristic function,

$$S(0) = P_{E_2 \oplus N_1} A_0 \,|\, (E_1 \oplus N_2).$$

In particular, $s(0) = P_{N_1} A \,|\, N_2 = 0$. This guarantees that the formulas (4–11) and (4–12) make sense, since $\|s(\zeta)\| \le |\zeta|$ when $|\zeta| < 1$ by Schwarz's lemma.

Considering the "−" parts of (4–10) for A_0, A_1, and A we obtain

$$\mathcal{G}_-(\zeta) \begin{bmatrix} h_1(0) \\ h_0(0) \end{bmatrix} = [\varphi(\zeta)^*\omega(\zeta)^*, \, \mathbf{1}_{E_1}]\mathcal{G}_-^0(\zeta)h_0(0) + \varphi(\zeta)^*\mathcal{G}_-^1(\zeta)h_1(0), \quad (4\text{–}13)$$

where

$$\varphi(\zeta) = (\mathbf{1}_{N_1} - s(\zeta)\omega(\zeta))^{-1}s_1(\zeta). \qquad (4\text{–}14)$$

Combining the expressions (4–11) and (4–12) for $\mathcal{G}_+$ and $\mathcal{G}_-$ we arrive at

$$\mathcal{G}\begin{bmatrix} h_1 \\ h_0 \end{bmatrix} = \begin{bmatrix} \psi\omega & \mathbf{1}_{E_2} & 0 & 0 \\ 0 & 0 & \varphi^*\omega^* & \mathbf{1}_{E_1} \end{bmatrix}\mathcal{G}^0 h_0 + \begin{bmatrix} \psi & 0 \\ 0 & \varphi^* \end{bmatrix}\mathcal{G}^1 h_1. \qquad (4\text{–}15)$$

As we saw in the last theorem of Lecture 3,

$$Fx = \mathcal{G}[x] \quad \text{for all } x \in X,$$

where $[x] \in H_0$ is the equivalence class generated by the quadratic form D. Hence

$$Fx = \begin{bmatrix} \psi\omega & \mathbf{1}_{E_2} & 0 & 0 \\ 0 & 0 & \varphi^*\omega^* & \mathbf{1}_{E_1} \end{bmatrix}\mathcal{G}^0[x].$$

The following theorem summarizes this lecture.

THEOREM. *Let V be the isometric colligation associated with the AIP data (page 362). Let A_0 be the unitary colligation associated with V (page 372). Let S be the characteristic function of A_0:*

$$S(\zeta) = P_{E_2 \oplus N_1} A_0 (\mathbf{1}_{H_0 \oplus E_1 \oplus N_2} - \zeta P_{H_0} A_0)^{-1} \mid (E_1 \oplus N_2),$$

$$S(\zeta) = \begin{bmatrix} s_0(\zeta) & s_2(\zeta) \\ s_1(\zeta) & s(\zeta) \end{bmatrix} : \begin{bmatrix} E_1 \\ N_2 \end{bmatrix} \longrightarrow \begin{bmatrix} E_2 \\ N_1 \end{bmatrix}.$$

Then the solutions w of the AIP and the corresponding mappings $F : X \to H^w$ are described by

$$w = s_0 + s_2 \omega (\mathbf{1}_{N_1} - s\omega)^{-1} s_1$$

$$Fx = \begin{bmatrix} \psi\omega & \mathbf{1}_{E_2} & 0 & 0 \\ 0 & 0 & \varphi^* \omega^* & \mathbf{1}_{E_1} \end{bmatrix} \mathcal{G}^0[x] \quad \text{for } x \in X$$

where ω is an arbitrary Schur class function $\omega(\zeta) : N_1 \to N_2$, for $|\zeta| < 1$,

$$\psi = s_2(\mathbf{1}_{N_2} - \omega s)^{-1}, \qquad \varphi = (\mathbf{1}_{N_1} - s\omega)^{-1} s_1,$$

$$\mathcal{G}^0(\zeta)h_0 = \begin{bmatrix} \mathcal{G}^0_+(\zeta)h_0 \\ \mathcal{G}^0_-(\zeta)h_0 \end{bmatrix} = \begin{bmatrix} P_{E_2 \oplus N_1} A_0 (\mathbf{1} - \zeta P_{H_0} A_0)^{-1} h_0 \\ \bar{\zeta} P_{E_1 \oplus N_2} A_0^* (\mathbf{1} - \bar{\zeta} P_{H_0} A_0^*)^{-1} h_0 \end{bmatrix}.$$

S and $\mathcal{G}^0$ depend on the data of the AIP only, whereas ω is arbitrary.

We can see that the parameter ω defines uniquely not only the solution w but also the corresponding mapping $F : X \to H^w$. This suggests denoting the mappings F by $F^\omega : X \to H^w$. In particular cases, when the mapping F is unique for a solution w, that is, when the F^ω coincide if the corresponding solutions w coincide, it can be denoted by F^w.

References to Lecture 4 are [Arov and Grossman 1983; Katsnelson et al. 1987; Kheifets 1988a; 1988b; 1990b; Kheifets and Yuditskii 1994].

Acknowledgements

I am very grateful to Harry Dym for discussions and personal conversations that we had during my stay at the Weizmann Institute in Rehovot, Israel, from fall 1993 through fall 1996, and to that institution for its encouraging scientific atmosphere. The Shabbat dinners with Harry and Irena Dym are particularly worth mentioning.

I thank also the Mathematics Sciences Research Institute in Berkeley, California, for its support. My participation in the Holomorphic Spaces program in fall 1995 gave me a great opportunity to meet many colleagues and friends. I am indebted to Don Sarason, Joe Ball, D. Z. Arov, Victor Katsnelson, Victor Vinnikov, Daniel Alpay, Misha Cotlar, Cora Sadosky, Stefania Marcantognini, Jim Rovnyak, Nikolas Young, Nikolai Nikolskii, Vasilii Vasyunin, Sergei

Treil, Vladimir Peller, Boris Freidin and many others for fruitful discussions and friendly conversations.

Special thanks go to Don Sarason and Silvio Levy for a lot of suggestions that improved the manuscript, and to Miriam Abraham for her perfect and careful typing of the manuscript.

References

[Arov and Grossman 1983] D. Z. Arov and L. Z. Grossman, "Scattering matrices in the theory of extensions of isometric operators", *Dokl. Akad. Nauk SSSR* **270**:1 (1983), 17–20. In Russian; translation in *Soviet Math. Dokl.* **27**:3 (1983), 518–521.

[Ball and Trent ≥ 1997]   J. Ball and T. Trent, in *Proceedings of IWOTA-96* (Bloomington, IN), Oper. Theory Adv. Appl., Birkhäuser, Basel.

[Katsnelson et al. 1987] V. È. Katsnel'son, A. Y. Kheĭfets, and P. M. Yuditskiĭ, "An abstract interpolation problem and the theory of extensions of isometric operators", pp. 83–96, 146 in *Operators in function spaces and problems in function theory*, edited by V. A. Marchenko, Naukova Dumka, Kiev, 1987. In Russian; translation in *Topics in interpolation theory*, edited by Harry Dym et al., Oper. Theory Adv. Appl. **95**, Birkhäuser, Basel, 1997, pp. 283–298.

[Kheifets 1988a] A. Y. Kheĭfets, "The Parseval equality in an abstract problem of interpolation, and the coupling of open systems, I", *Teor. Funktsiĭ Funktsional. Anal. i Prilozhen.* **49** (1988), 112–120, 126. In Russian; translation in *J. Sov. Math.* **49**:4 (1990) 1114–1120.

[Kheifets 1988b] A. Y. Kheĭfets, "The Parseval equality in an abstract problem of interpolation, and the coupling of open systems, II", *Teor. Funktsiĭ Funktsional. Anal. i Prilozhen.* **50** (1988), 98–103, iv. In Russian; translation in *J. Sov. Math.* **49**:6 (1990) 1307–1310.

[Kheifets 1990a] A. Y. Kheĭfets, "A generalized bitangential Schur–Nevanlinna–Pick problem and its related Parseval equality", *Teor. Funktsiĭ Funktsional. Anal. i Prilozhen.* **54** (1990), 89–96. In Russian; translation in *J. Sov. Math.* **58**:4 (1992), 358–364.

[Kheifets 1990b] A. Y. Kheĭfets, *Scattering matrices and the Parseval equality in the abstract interpolation problem*, Ph.D. thesis, 1990. In Russian.

[Kheifets 1996] A. Kheifets, "Hamburger moment problem: Parseval equality and A-singularity", *J. Funct. Anal.* **141**:2 (1996), 374–420.

[Kheifets and Yuditskii 1994] A. Y. Kheĭfets and P. M. Yuditskiĭ, "An analysis and extension of V. P. Potapov's approach to interpolation problems with applications to the generalized bi-tangential Schur–Nevanlinna–Pick problem and J-inner-outer factorization", pp. 133–161 in *Matrix and operator valued functions*, edited by I. Gohberg and L. A. Sakhnovich, Oper. Theory Adv. Appl. **72**, Birkhäuser, Basel, 1994.

[Sarason 1994] D. Sarason, *Sub-Hardy Hilbert spaces in the unit disk*, University of Arkansas Lecture Notes in the Mathematical Sciences, Wiley, New York, 1994.

ALEXANDER KHEIFETS
MATHEMATICS DIVISION
INSTITUTE FOR LOW-TEMPERATURE PHYSICS AND ENGINEERING
47 LENIN AVENUE
KHARKOV 310164
UKRAINE
  kheifets@ilt.kharkov.ua

Holomorphic Spaces
MSRI Publications
Volume **33**, 1998

A Basic Interpolation Problem

HARRY DYM

ABSTRACT. A basic interpolation problem, which includes bitangential matrix versions of a number of classical interpolation problems, is formulated and solved. Particular attention is placed on the development of the problem in a natural way and upon the fundamental role played by a special class of reproducing kernel Hilbert spaces of vector-valued meromorphic functions that originate in the work of L. de Branges. Necessary and sufficient conditions for the existence of a solution to this problem, and a parametrization of the set of all solutions to this problem when these conditions are met, are presented. Some comparisons with the methods of Katsnelson, Kheifets, and Yuditskii are made. The presentation is largely self-contained and expository.

1. Introduction

This paper presents a largely self-contained expository introduction to a number of problems in interpolation theory for matrix-valued functions, including the classical problems of Schur, Nevanlinna–Pick (NP), and Carathéodory–Fejér (CF) as special cases. The development will use little more than the elementary properties of vector-valued Hardy spaces of exponent 2.

To illustrate the scope of the paper we shall begin with sample problems, all of which are formulated in the class $\mathcal{S}^{p \times q}(\Omega_+)$ of $p \times q$ matrix-valued functions (mvf) $S(\lambda)$ that are analytic and contractive in a given region Ω_+ in the complex plane $\mathbb{C}$.

EXAMPLE 1.1 (THE LEFT TANGENTIAL NP PROBLEM). The data for this problem is a set of points $\omega_1, \ldots, \omega_n$ in Ω_+ and two sets of vectors: $\xi_1, \ldots \xi_n$ in $\mathbb{C}^p$ and $\eta_1, \ldots \eta_n$ in $\mathbb{C}^q$. An mvf $S(\lambda) \in \mathcal{S}^{p \times q}(\Omega_+)$ is said to be a solution of this problem if

$$\xi_j^* S(\omega_j) = \eta_j^* \quad \text{for } j = 1, \ldots, n. \tag{1.1}$$

The author thanks Renee and Jay Weiss for endowing the chair that supports his research.

EXAMPLE 1.2 (THE RIGHT TANGENTIAL NP PROBLEM). The data for this problem is exactly the same as for the preceding example, but now the interpolation conditions are imposed on the right: An mvf $S(\lambda) \in \mathcal{S}^{p \times q}(\Omega_+)$ is said to be a solution of this problem if

$$S(\omega_j)\eta_j = \xi_j \quad \text{for } j = 1, \ldots, n. \tag{1.2}$$

EXAMPLE 1.3 (THE BITANGENTIAL NP PROBLEM). The data for this problem is exactly the same as for the preceding two examples, but now the first μ interpolation constraints are imposed on the left and the last $\nu = n - \mu$ constraints are imposed on the right. An mvf $S(\lambda) \in \mathcal{S}^{p \times q}(\Omega_+)$ is said to be a solution of this problem if

$$\xi_j^* S(\omega_j) = \eta_j^* \quad \text{for } j = 1, \ldots, \mu, \tag{1.3}$$

and

$$S(\omega_j)\eta_j = \xi_j \quad \text{for } j = \mu+1, \ldots, n. \tag{1.4}$$

EXAMPLE 1.4 (THE LEFT TANGENTIAL CF PROBLEM). The data for this problem is a point $\omega \in \Omega_+$, a vector $\xi \in \mathbb{C}^p$ and a set of vectors $\eta_1, \ldots, \eta_n$ in $\mathbb{C}^q$. An mvf $S(\lambda) \in \mathcal{S}^{p \times q}(\Omega_+)$ is said to be a solution of this problem if

$$\frac{\xi^* S^{j-1}(\omega)}{(j-1)!} = \eta_j^* \quad \text{for } j = 1, \ldots, n. \tag{1.5}$$

EXAMPLE 1.5 (A MIXED PROBLEM). The data for this problem is a pair of points ω_1, ω_2 in Ω_+, a set of vectors ξ_1 and $\xi_{21}, \ldots, \xi_{2\nu}$ in $\mathbb{C}^p$ and a set of vectors $\eta_{11}, \ldots, \eta_{1\mu}$ and η_2 in $\mathbb{C}^q$. An mvf $S(\lambda) \in \mathcal{S}^{p \times q}(\Omega_+)$ is said to be a solution of this problem if

$$\frac{\xi_1^* S^{j-1}(\omega_1)}{(j-1)!} = \eta_{1j} \quad \text{for } j = 1, \ldots, \mu, \tag{1.6}$$

and

$$\frac{S^{j-1}(\omega_2)\eta_2}{(j-1)!} = \xi_{2j} \quad \text{for } j = 1, \ldots, \nu. \tag{1.7}$$

The basic objective for all these problems is twofold:

(1) To formulate necessary and sufficient conditions in terms of the given data for the existence of at least one mvf $S(\lambda) \in \mathcal{S}^{p \times q}(\Omega_+)$ that meets the interpolation constraints.
(2) To describe the set of all mvf $S(\lambda) \in \mathcal{S}^{p \times q}(\Omega_+)$ that meet the interpolation constraints when the conditions for existence are met.

It turns out that when the region Ω_+ is chosen to be either the open unit disc

$$\mathbb{D} = \{\lambda \in \mathbb{C} : |\lambda| < 1\},$$

or the open upper half-plane

$$\mathbb{C}_+ = \{\lambda \in \mathbb{C} : \text{Im } \lambda > 0\},$$

Ω_+	$\mathbb{D}$	$\mathbb{C}_+$	Π_+
$a(\lambda)$	1	$\sqrt{\pi}(1 - i\lambda)$	$\sqrt{\pi}(1 + \lambda)$
$b(\lambda)$	λ	$\sqrt{\pi}(1 + i\lambda)$	$\sqrt{\pi}(1 - \lambda)$
$\rho_\omega(\lambda)$	$1 - \lambda\omega^*$	$-2\pi i(\lambda - \omega^*)$	$2\pi(\lambda + \omega^*)$
Ω_0	$\mathbb{T}$	$\mathbb{R}$	$i\mathbb{R}$
$\langle f, g \rangle$	$\frac{1}{2\pi}\int_0^{2\pi} g(e^{i\theta})^* f(e^{i\theta})d\theta$	$\int_{-\infty}^{\infty} g(x)^* f(x)dx$	$\int_{-\infty}^{\infty} g(iy)^* f(iy)dy$
λ°	$1/\lambda^*$ if $\lambda \neq 0$	λ^*	$-\lambda^*$
$f^{\#}(\lambda)$	$f(\lambda^\circ)^*$	$f(\lambda^\circ)^*$	$f(\lambda^\circ)^*$
$\delta_\omega(\lambda)$	$\lambda - \omega$	$2\pi i(\lambda - \omega)$	$-2\pi(\lambda - \omega)$
$ab' - ba'$	1	$2\pi i$	-2π
$\varphi_{j,\omega}(\lambda)$	$\lambda^j/(1 - \lambda\omega^*)^{j+1}$	$-1/2\pi i(\lambda - \omega^*)^{j+1}$	$(-1)^j/2\pi(\lambda + \omega^*)^{j+1}$
$(R_\alpha \rho_\omega^{-1})(\lambda)$	$\omega^*/\rho_\omega(\alpha)\rho_\omega(\lambda)$	$2\pi i/\rho_\omega(\alpha)\rho_\omega(\lambda)$	$-2\pi/\rho_\omega(\alpha)\rho_\omega(\lambda)$

Table 1. Bringing the classical regions into a unified framework.

or the open right half-plane

$$\Pi_+ = \{\lambda \in \mathbb{C} : \operatorname{Re}\lambda > 0\},$$

all these problems (as well as more complicated problems in more complicated regions) can be incorporated into a more general problem, which we call the Basic Interpolation Problem (BIP). Moreover, by exercising a little care in the choice of notation, most of the analysis for all three of the classical choices of Ω_+ mentioned above can be carried out in one stroke. Table 1 serves as a dictionary for the meaning of the symbol that is appropriate for the region Ω_+ in use.

In order to describe the BIP we need to introduce some notation.

Let $H_2^k(\Omega_+)$ denote the set of $(k \times 1)$-vector-valued functions with entries in the Hardy space $H_2(\Omega_+)$. This space is identified as a closed subspace of the Hilbert space $L_2^k(\Omega_0)$ of $(k \times 1)$-vector-valued functions that are measurable and square integrable (that is, $\langle f, f \rangle$, as defined in Table 1, is finite) on the boundary Ω_0 of Ω_+ in the usual way. The symbol $H_2^k(\Omega_+)^{\perp}$ designates the orthogonal complement of $H_2^k(\Omega_+)$ in $L_2^k(\Omega_0)$ with respect to the inner product indicated in Table 1,

$\underline{\underline{p}}$ denotes the orthogonal projection of $L_2^k(\Omega_0)$ onto $H_2^k(\Omega_+)$,

and

$\underline{\underline{q}}' = I - \underline{\underline{p}}$ denotes the orthogonal projection of $L_2^k(\Omega_0)$ onto $H_2^k(\Omega_+)^{\perp}$.

The dependence of the projections on the height k of the column vectors is suppressed in order to keep the notation simple.

For each of the three listed choices of the kernel function $\rho_\omega(\lambda)$, we have

$$\Omega_+ = \{\omega \in \mathbb{C} : \rho_\omega(\omega) > 0\}$$

and

$$\Omega_0 = \{\omega \in \mathbb{C} : \rho_\omega(\omega) = 0\}.$$

We shall take

$$\Omega_- = \{\omega \in \mathbb{C} : \rho_\omega(\omega) < 0\}.$$

The use of such a flexible notation to cover problems in both $\mathbb{D}$ and $\mathbb{C}_+$ more or less simultaneously was promoted in [Alpay and Dym 1984] and [Dym 1989a]. The observation that the kernels $\rho_\omega(\lambda)$ that intervene in these problems can be expressed in terms of a pair of polynomials $a(\lambda)$ and $b(\lambda)$ as

$$\rho_\omega(\lambda) = a(\lambda)a(\omega)^* - b(\lambda)b(\omega)^* \tag{1.8}$$

was first made by Lev-Ari and Kailath [1986]. They noticed that certain fast algorithms in which the term $\rho_\omega(\lambda)$ intervenes will work if and only if $\rho_\omega(\lambda)$ can be expressed in the form (1.8). A general theory of reproducing kernels with denominators of this form and their applications was developed in [Alpay and Dym 1992; 1993a; 1993b; 1996]; for related developments see [Nudelman 1993].

The rest of the notation is fairly standard: The symbol A^* denotes the adjoint of an operator A on a Hilbert space, with respect to the inner product of the space. If A is a finite matrix, the adjoint will always be computed with respect to the standard inner product, so that in this case A^* will be the Hermitian transpose, or just the complex conjugate if A is a number. The symbol $\sigma(A)$ denotes the spectrum of a matrix A and J stands for the $m \times m$ signature matrix

$$J = \begin{bmatrix} I_p & 0 \\ 0 & -I_q \end{bmatrix}$$

with $p \geq 1$, $q \geq 1$ and $p + q = m$.

The following evaluations, which depend basically on Cauchy's formula for $H_2(\Omega_+)$, will prove useful. For details, see [Dym 1994b, Section 2.2].

LEMMA 1.6. *If $\omega \in \Omega_+$, $u \in \mathbb{C}^k$ and $S(\lambda) \in \mathcal{S}^{p \times q}(\Omega_+)$, then:*

(1) $\dfrac{u}{\rho_\omega} \in H_2^k(\Omega_+)$ *and* $\dfrac{u}{\delta_\omega} \in H_2^k(\Omega_+)^\perp$.

(2) $\underline{p}S^* \dfrac{u}{\rho_\omega} = S(\omega)^* \dfrac{u}{\rho_\omega}$ *(when $k = p$).*

(3) $\underline{q}'S \dfrac{u}{\delta_\omega} = S(\omega) \dfrac{u}{\delta_\omega}$ *(when $k = q$).*

(4) $\left(\underline{p}S^* \varphi_{j,\omega} u\right)(\lambda) = \displaystyle\sum_{i=0}^{j} S^{(j-i)}(\omega)^* \varphi_{i,\omega}(\lambda) u$ *(when $k = p$).*

(5) $\left(\underline{q}'S(\cdot - \omega)^{-j} u\right)(\lambda) = \displaystyle\sum_{i=0}^{j-1} (\lambda - \omega)^{i-j} \dfrac{S^{(i)}(\omega)}{i!} u$ *(when $k = q$).*

2. The Basic Interpolation Problem

First formulation. The data for the Basic Interpolation Problem consists of a set of $2n$ vector-valued functions $g_1, \ldots, g_n, h_1, \ldots, h_n$, where

$$g_1, \ldots, g_\mu \in H_2^p(\Omega_+), \qquad g_{\mu+1}, \ldots, g_n \in H_2^p(\Omega_+)^\perp,$$
$$h_1, \ldots, h_\mu \in H_2^q(\Omega_+), \qquad h_{\mu+1}, \ldots, h_n \in H_2^q(\Omega_+)^\perp.$$

An mvf $S(\lambda) \in \mathcal{S}^{p \times q}(\Omega_+)$ is said to be a solution of this problem if

$$\underline{p} S^* g_j = h_j \quad \text{for } j = 1, \ldots, \mu$$

and

$$\underline{q}' S h_j = g_j \quad \text{for } j = \mu+1, \ldots, n.$$

EXAMPLE 2.1. Let there be given points $\omega_1, \ldots, \omega_n$ in Ω_+, vectors $\xi_1, \ldots, \xi_n$ in $\mathbb{C}^p$ and a second set of vectors $\eta_1, \ldots, \eta_n$ in $\mathbb{C}^q$, just as in Examples 1.1–1.3. Let

$$g_j(\lambda) = \begin{cases} \dfrac{\xi_j}{\rho_{\omega_j}(\lambda)} & \text{for } j = 1, \ldots, \mu, \\[2ex] \dfrac{\xi_j}{\delta_{\omega_j}(\lambda)} & \text{for } j = \mu+1, \ldots, n, \end{cases}$$

and

$$h_j(\lambda) = \begin{cases} \dfrac{\eta_j}{\rho_{\omega_j}(\lambda)} & \text{for } j = 1, \ldots, \mu, \\[2ex] \dfrac{\eta_j}{\delta_{\omega_j}(\lambda)} & \text{for } j = \mu+1, \ldots, n. \end{cases}$$

Then

$$g_j \in H_2^p(\Omega_+) \quad \text{and} \quad h_j \in H_2^q(\Omega_+) \quad \text{for } j = 1, \ldots, \mu,$$

whereas

$$g_j \in H_2^p(\Omega_+)^\perp \quad \text{and} \quad h_j \in H_2^q(\Omega_+)^\perp \quad \text{for } j = \mu+1, \ldots, n.$$

Therefore, by evaluations (2) and (3) in Lemma 1.6, it is readily seen that $S(\lambda)$ is a solution of the BIP based on this choice of $g_1, \ldots, g_n$ and $h_1, \ldots, h_n$ if and only if $S(\lambda)$ is a solution of the bitangential NP set forth in Example 1.3. Examples 1.1 and 1.2 correspond to choosing $\mu = n$ and $\nu = n$, respectively.

More elaborate examples involving derivatives can be constructed in much the same way by taking advantage of the evaluations (4) and (5) in Lemma 1.6.

A reformulation of the Basic Interpolation Problem. It is readily checked that the mvf $S(\lambda) \in \mathcal{S}^{p \times q}(\Omega_+)$ is a solution of the BIP if and only if

$$\underline{p}(S^* g_j - h_j) = 0 \quad \text{for } j = 1, \ldots, \mu$$

and

$$\underline{q}'(S h_j - g_j) = 0 \quad \text{for } j = \mu+1, \ldots, n.$$

In fact, since the first condition is automatically satisfied for $j = \mu+1, \ldots, n$ and the second condition is automatically satisfied for $j = 1, \ldots, \mu$, the indices may be allowed to run from 1 to n in both cases. In other words, an mvf $S(\lambda) \in \mathcal{S}^{p \times q}(\Omega_+)$ is a solution of the BIP if and only if

$$g_j - S h_j \in H_2^p(\Omega_+) \quad \text{for } j = 1, \ldots, n$$

and

$$-S^* g_j + h_j \in H_2^q(\Omega_+)^{\perp} \quad \text{for } j = 1, \ldots, n.$$

This last pair of constraints can be stacked, yielding our final formulation:

Final formulation of the BIP. As before, the data for this problem is two sets of vector-valued functions $g_1, \ldots, g_n$ and $h_1, \ldots, h_n$, where

$$g_1, \ldots, g_\mu \in H_2^p(\Omega_+), \qquad g_{\mu+1}, \ldots, g_n \in H_2^p(\Omega_+)^{\perp},$$

$$h_1, \ldots, h_\mu \in H_2^q(\Omega_+), \qquad h_{\mu+1}, \ldots, h_n \in H_2^q(\Omega_+)^{\perp}.$$

Then $S(\lambda) \in \mathcal{S}^{p \times q}(\Omega_+)$ is said to be a solution of the BIP based on this given set of data if

$$\begin{bmatrix} I_p & -S \\ -S^* & I_q \end{bmatrix} \begin{bmatrix} g_j \\ h_j \end{bmatrix} \in H_2^p(\Omega_+) \oplus H_2^q(\Omega_+)^{\perp}$$

for $j = 1, \ldots, n$.

This formulation is meaningful even if the g_j and h_j are not themselves in $L_2^p(\Omega_0)$ and $L_2^q(\Omega_0)$, because it is the difference that comes into play. This is significant in the study of interpolation problems with constraints on the boundary Ω_0. Thus for example, if $\Omega_+ = \mathbb{D}$ and $\omega_1 \in \mathbb{T}$, then

$$g_1(\lambda) = \frac{\xi_1}{\rho_{\omega_1}(\lambda)} = \frac{\xi_1}{1 - \lambda \omega_1^*}$$

does not belong to $H_2^p(\mathbb{D})$ and

$$h_1(\lambda) = \frac{\eta_1}{\rho_{\omega_1}(\lambda)} = \frac{\eta_1}{1 - \lambda \omega_1^*}$$

does not belong to $H_2^q(\mathbb{D})$. Nevertheless it is meaningful to investigate the set of mvfs $S(\lambda) \in \mathcal{S}^{p \times q}(\mathbb{D})$ for which

$$g_1 - S h_1 \in H_2^p(\mathbb{D}).$$

We shall not pursue this question here.

3. Necessary Conditions for the Existence of a Solution to the Basic Interpolation Problem

THEOREM 3.1. *If $S(\lambda) \in \mathcal{S}^{p \times q}(\Omega_+)$ is a solution of the BIP based on the given data $g_1, \ldots, g_n$ and $h_1, \ldots, h_n$, then the $n \times n$ Hermitian matrix Q with entries*

$$q_{ij} = \begin{cases} \langle g_j, g_i \rangle - \langle h_j, h_i \rangle & \text{for } i, j = 1, \ldots, \mu, \\ -\langle Sh_j, g_i \rangle & \text{for } i = 1, \ldots, \mu \text{ and } j = \mu{+}1, \ldots, n, \\ -\langle g_j, g_i \rangle + \langle h_j, h_i \rangle & \text{for } i, j = \mu{+}1, \ldots, n, \end{cases} \quad (3.1)$$

is positive semidefinite.

PROOF. Define

$$q_{ij} = \left\langle \begin{bmatrix} I_p & -S \\ -S^* & I_q \end{bmatrix} \begin{bmatrix} g_j \\ h_j \end{bmatrix}, \begin{bmatrix} g_i \\ h_i \end{bmatrix} \right\rangle \quad (3.2)$$

for $i, j = 1, \ldots, n$. Then

$$Q = [q_{ij}] \geq 0,$$

since

$$\begin{bmatrix} I_p & -S(\lambda) \\ -S^*(\lambda) & I_q \end{bmatrix} \geq 0$$

for a.e. point $\lambda \in \Omega_0$. The rest of the proof amounts to evaluating (3.2). This is where the assumption that $S(\lambda)$ is a solution of the BIP comes into play. There are three basic cases, but it is convenient to begin the analysis of the first two together.

Suppose first that $1 \leq i \leq \mu$. Then

$$q_{ij} = \langle g_j - Sh_j, g_i \rangle + \langle -S^* g_j + h_j, h_i \rangle$$
$$= \langle g_j - Sh_j, g_i \rangle = \langle g_j, g_i \rangle - \langle Sh_j, g_i \rangle.$$

Now, if also $1 \leq j \leq \mu$, then

$$\langle Sh_j, g_i \rangle = \langle h_j, \underline{p}S^* g_i \rangle = \langle h_j, h_i \rangle,$$

whereas, if $\mu{+}1 \leq j \leq n$, then $\langle Sh_j, g_i \rangle$ cannot be reduced but $\langle g_j, g_i \rangle = 0$. These evaluations lead easily to the first two sets of formulas for q_{ij}. To verify the last set, assume $\mu{+}1 \leq i, j \leq n$. Then

$$q_{ij} = \langle g_j - Sh_j, g_i \rangle + \langle -S^* g_j + h_j, h_i \rangle = \langle -S^* g_j + h_j, h_i \rangle$$
$$= -\langle g_j, \underline{q}' Sh_i \rangle + \langle h_j, h_i \rangle = -\langle g_j, g_i \rangle + \langle h_j, h_i \rangle,$$

as claimed. The proof is complete. $\qquad \square$

If $g_1, \ldots, g_n$ and $h_1, \ldots, h_n$ are chosen as in Example 2.1, it is readily checked (with the aid of Cauchy's formula) that

$$q_{ij} = \left\langle \frac{\xi_j}{\rho_{\omega_j}}, \frac{\xi_i}{\rho_{\omega_i}} \right\rangle - \left\langle \frac{\eta_j}{\rho_{\omega_j}}, \frac{\eta_i}{\rho_{\omega_i}} \right\rangle = \frac{\xi_i^* \xi_j - \eta_i^* \eta_j}{\rho_{\omega_j}(\omega_i)} \quad \text{for } i, j = 1, \ldots, \mu, \quad (3.3)$$

whereas

$$q_{ij} = -\left\langle \frac{\xi_j}{\delta_{\omega_j}}, \frac{\xi_i}{\delta_{\omega_i}} \right\rangle + \left\langle \frac{\eta_j}{\delta_{\omega_j}}, \frac{\eta_i}{\delta_{\omega_i}} \right\rangle = \frac{-\xi_i^* \xi_j + \eta_i^* \eta_j}{\rho_{\omega_i}(\omega_j)} \quad \text{for } i, j = \mu+1, \ldots, n.$$

(3.4)

These formulas are expressed totally in terms of the data of the problem. Thus a necessary condition for the existence of a solution to the bitangential NP problem of Example 1.3 is that the matrices exhibited in formulas (3.3) and (3.4) are positive semidefinite.

Theorem 3.1 gives a necessary condition for the existence of a solution to the BIP. In general this condition will not be sufficient unless additional structure is imposed on the data $g_1, \ldots, g_n$ and $h_1, \ldots, h_n$ of the problem. In particular we shall assume that the space

$$\text{span}\left\{ \begin{bmatrix} g_j \\ h_j \end{bmatrix} : j = 1, \ldots, \mu \right\}$$

is a μ-dimensional R_α-invariant subspace of $H_2^m(\Omega_+)$ for some point $\alpha \in \Omega_+$ and that

$$\text{span}\left\{ \begin{bmatrix} g_j \\ h_j \end{bmatrix} : j = \mu+1, \ldots, n \right\}$$

is a ν-dimensional R_α-invariant subspace of $H_2^m(\Omega_+)^\perp$ for some point $\alpha \in \Omega_-$, where R_α denotes the operator defined by the rule

$$(R_\alpha f)(\lambda) = \frac{f(\lambda) - f(\alpha)}{\lambda - \alpha},$$

(3.5)

wherever it is meaningful.

4. R_α Invariance

In this section we study finite-dimensional spaces of vector-valued functions that are invariant under the action of R_α for at least one appropriately chosen point $\alpha \in \mathbb{C}$. The contents are taken largely from [Dym 1994b, Section 3].

THEOREM 4.1. *Let $\mathcal{M}$ be an n-dimensional vector space of $(m \times 1)$-vector-valued functions that are meromorphic in some open nonempty set $\Delta \subset \mathbb{C}$ and suppose further that $\mathcal{M}$ is R_α-invariant for some point $\alpha \in \Delta$ in the domain of analyticity of $\mathcal{M}$. Then $\mathcal{M}$ is spanned by the columns of a rational $m \times n$ matrix-valued function of the form*

$$F(\lambda) = V\{M - \lambda N)^{-1},$$

(4.1)

where $V \in \mathbb{C}^{m \times n}$, $M, N \in \mathbb{C}^{n \times n}$,

$$MN = NM \quad and \quad M - \alpha N = I_n.$$

(4.2)

Moreover, $\lambda \in \Delta$ is a point of analyticity of F if and only if the $n \times n$ matrix $M - \lambda N$ is invertible.

PROOF. Let $f_1, \ldots, f_n$ be a basis for $\mathfrak{M}$ and let

$$F(\lambda) = \begin{bmatrix} f_1(\lambda) & \cdots & f_n(\lambda) \end{bmatrix}$$

be the $m \times n$ matrix-valued function with columns $f_1(\lambda), \ldots, f_n(\lambda)$. Then, because of the presumed R_α-invariance of the columns of F,

$$R_\alpha F(\lambda) = \frac{F(\lambda) - F(\alpha)}{\lambda - \alpha} = F(\lambda) E_\alpha$$

for some $n \times n$ matrix E_α independent of λ. Thus

$$F(\lambda)\big(I_n - (\lambda - \alpha)E_\alpha\big) = F(\alpha),$$

and hence, since $\det\big(I_n - (\lambda - \alpha)E_\alpha\big) \not\equiv 0$,

$$F(\lambda) = F(\alpha)\big(I_n + \alpha E_\alpha - \lambda E_\alpha\big)^{-1},$$

which is of the form (4.1) with $V = F(\alpha)$, $M = I_n + \alpha E_\alpha$ and $N = E_\alpha$.

Suppose next that F is analytic at a point $\omega \in \Delta$ and that $u \in \ker(M - \omega N)$. Then

$$F(\lambda)(M - \lambda N)u = Vu = 0,$$

first for $\lambda = \omega$, and then for every $\lambda \in \Delta$ in the domain of analyticity of F. Thus, for all such λ,

$$(\omega - \lambda)F(\lambda)Nu = F(\lambda)\big(M - \lambda N - (M - \omega N)\big)u = 0.$$

Since the columns of $F(\lambda)$ are linearly independent functions of λ, we get $Nu = 0$. But this in conjunction with the prevailing assumption $(M - \omega N)u = 0$ implies that

$$u \in \ker M \cap \ker N \implies u = 0 \implies M - \omega N \text{ is invertible.}$$

Thus we have shown that if F is analytic at ω, then $M - \omega N$ is invertible. Since the opposite implication is easy, this serves to complete the proof. $\square$

COROLLARY 4.2. *If* $\det(M - \lambda N) \not\equiv 0$ *and* $F(\lambda) = V(M - \lambda N)^{-1}$ *is a rational* $m \times n$ *matrix-valued function with* n *linearly independent columns, then*

(1) M *is invertible if and only if* F *is analytic at zero, and*
(2) N *is invertible if and only if* F *is analytic at infinity and* $F(\infty) = 0$.

Moreover, if M *is invertible* F *can be expressed in the form*

$$F(\lambda) = C(I_n - \lambda A)^{-1}, \tag{4.3}$$

whereas if N *is invertible* F *can be expressed in the form*

$$F(\lambda) = C(A - \lambda I_n)^{-1}. \tag{4.4}$$

PROOF. The first assertion is contained in the theorem; the second is obtained in much the same way. More precisely, if $\lim_{\lambda \to \infty} F(\lambda) = 0$ and $u \in \ker N$, then

$$F(\lambda)Mu = F(\lambda)(M - \lambda N)u = Vu.$$

Upon letting λ approach ∞, it follows that

$$Vu = 0 \implies F(\lambda)Mu = 0 \implies Mu = 0 \implies u \in \ker M \cap \ker N \implies u = 0.$$

Thus N is invertible. The other direction is easy, as are formulas (4.3) and (4.4). Just take $C = VM^{-1}$ and $A = NM^{-1}$ in the first case, and $C = VN^{-1}$ and $A = MN^{-1}$ in the second. $\square$

COROLLARY 4.3. *Let f be an $(m \times 1)$-vector-valued function that is meromorphic in some open nonempty set $\Delta \subset \mathbb{C}$ and let $\alpha \in \Delta$ be a point of analyticity of f. Then f is an eigenfunction of R_α if and only if it can be expressed in the form*

$$f(\lambda) = \frac{v}{\rho_\omega(\lambda)}$$

for one or more choices of $\rho_\omega(\lambda)$ in Table 1 with $\rho_\omega(\alpha) \neq 0$ and some nonzero constant vector $v \in \mathbb{C}^m$.

Linear independence. It seems worthwhile to emphasize that herein the n columns of an $m \times n$ matrix-valued function $F(\lambda)$ are said to be linearly independent if they are linearly independent in the vector space of continuous $(m \times 1)$-vector-valued functions on the domain of analyticity of F. If

$$F(\lambda) = C(I_n - \lambda A)^{-1} \quad \text{or} \quad F(\lambda) = C(A - \lambda I_n)^{-1},$$

this is easily seen to be equivalent to

$$\bigcap_{j=0}^{n-1} \ker CA^j = 0;$$

that is, to the pair (C, A) being observable. Such a realization for F is minimal in the sense of Kalman because (in the usual terminology; see [Kailath 1980], for example) the pair (A, B) is automatically controllable:

$$\bigcap_{j=0}^{n-1} \ker B^* A^{*j} = \{0\}$$

(or, equivalently, rank $[B \ \ AB \ \ \cdots \ \ A^{n-1}B] = n$), since $B = I_n$.

R_α-invariant subspaces of $H_2^m(\Omega_+)$ and $H_2^m(\Omega_+)^\perp$. Let

$$F_1(\lambda) = \begin{bmatrix} g_1(\lambda) & \cdots & g_\mu(\lambda) \\ h_1(\lambda) & \cdots & h_\mu(\lambda) \end{bmatrix} \tag{4.5}$$

be an $m \times \mu$ mvf with columns

$$f_j(\lambda) = \begin{bmatrix} g_j(\lambda) \\ h_j(\lambda) \end{bmatrix} \in H_2^p(\Omega_+) \oplus H_2^q(\Omega_+),$$

and suppose that $\{F_1(\lambda)u : u \in \mathbb{C}^\mu\}$ is a μ-dimensional R_α-invariant subspace of $H_2^m(\Omega_+)$ for some point $\alpha \in \Omega_+$. This is meaningful because $F_1(\lambda)$ is analytic in Ω_+. Then it follows from Theorem 3.1 that $F_1(\lambda)$ admits a representation of the form

$$F_1(\lambda) = V(M - \lambda N)^{-1}.$$

Hence, since $F_1(\lambda)$ is analytic at the point $\lambda = 0$ if $\Omega_+ = \mathbb{D}$ and at the point $\lambda = \infty$ with $F_1(\infty) = 0$ if $\Omega_+ = \mathbb{C}_+$ or $\Omega_+ = \Pi_+$, it follows also that

$$F_1(\lambda) = \begin{cases} \begin{bmatrix} C_{11} \\ C_{21} \end{bmatrix} (I_\mu - \lambda A_1)^{-1} & \text{with } \sigma(A_1) \subset \mathbb{D} \quad \text{if } \Omega_+ = \mathbb{D}, \\[3mm] \begin{bmatrix} C_{11} \\ C_{21} \end{bmatrix} (\lambda I_\mu - A_1)^{-1} & \text{with } \sigma(A_1) \subset \Omega_- \quad \text{if } \Omega_+ = \mathbb{C}_+ \text{ or } \Omega_+ = \Pi_+, \end{cases} \tag{4.6}$$

where $C_{11} \in \mathbb{C}^{p\times\mu}$, $C_{21} \in \mathbb{C}^{q\times\mu}$ and of course $A_1 \in \mathbb{C}^{\mu\times\mu}$.

Next let

$$F_2(\lambda) = \begin{bmatrix} g_{\mu+1}(\lambda) & \cdots & g_n(\lambda) \\ h_{\mu+1}(\lambda) & \cdots & h_n(\lambda) \end{bmatrix} \tag{4.7}$$

be an $m \times \nu$ mvf with columns

$$f_j(\lambda) = \begin{bmatrix} g_j(\lambda) \\ h_j(\lambda) \end{bmatrix} \in H_2^p(\Omega_+)^\perp \oplus H_2^q(\Omega_+)^\perp$$

for $j = \mu+1, \ldots, n$ and suppose that $\{F_2(\lambda)u : u \in \mathbb{C}^\nu\}$ is a ν-dimensional R_α-invariant subspace of $H_2^p(\Omega_+)^\perp$ for some choice of $\alpha \in \Omega_-$. This too is meaningful because $F_2(\lambda)$ is analytic in Ω_-. Then, by another application of Theorem 3.1, $F_2(\lambda)$ admits a representation of the form

$$F_2(\lambda) = \begin{bmatrix} C_{12} \\ C_{22} \end{bmatrix} (\lambda I_\nu - A_2)^{-1} \quad \text{with } \sigma(A_2) \subset \Omega_+, \tag{4.8}$$

where $C_{12} \in \mathbb{C}^{p\times\nu}$, $C_{22} \in \mathbb{C}^{q\times\nu}$ and $A_2 \in \mathbb{C}^{\nu\times\nu}$.

More explicit formulas, based on the Jordan decomposition of A_1 and A_2, may be found in [Dym 1994b, Section 3.3; Dym 1989b].

5. Back to the Basic Interpolation Problem

From now on we assume that the data of the BIP is chosen so that $\text{span}\{f_j(\lambda) : j = 1, \ldots, \mu\}$ is a μ-dimensional R_α-invariant subspace of $H_2^m(\Omega_+)$ for some point $\alpha \in \Omega_+$ and $\text{span}\{f_j(\lambda) : j = \mu+1, \ldots, n\}$ is a ν-dimensional R_α-invariant subspace of $H_2^m(\Omega_+)^\perp$ for some choice of $\alpha \in \Omega_-$. In view of the analysis in the previous section, this means that the data of the BIP is of the form

$$F(\lambda) = C(M - \lambda N)^{-1}, \tag{5.1}$$

where

$$C = \begin{bmatrix} C_1 \\ C_2 \end{bmatrix} = \begin{bmatrix} C_{11} & C_{12} \\ C_{21} & C_{22} \end{bmatrix} \begin{matrix} \}p \\ \}q \end{matrix}, \tag{5.2}$$

$$\underbrace{\phantom{C_{11}}}_{\mu} \underbrace{\phantom{C_{12}}}_{\nu}$$

$$M - \lambda N = \begin{cases} \begin{bmatrix} I_\mu - \lambda A_1 & 0 \\ 0 & \lambda I_\nu - A_2 \end{bmatrix} & \text{if } \Omega_+ = \mathbb{D}, \\[1em] \begin{bmatrix} \lambda I_\mu - A_1 & 0 \\ 0 & \lambda I_\nu - A_2 \end{bmatrix} & \text{if } \Omega_+ = \mathbb{C}_+ \text{ or } \Pi_+, \end{cases} \tag{5.3}$$

$\mu + \nu = n$,

$$\sigma(A_1) \subset \mathbb{D} \text{ if } \Omega_+ = \mathbb{D}, \quad \sigma(A_1) \subset \Omega_- \text{ if } \Omega_+ = \mathbb{C}_+ \text{ or } \Pi_+, \tag{5.4}$$

and

$$\sigma(A_2) \subset \Omega_+ \text{ for all three of the classical choices of } \Omega_+. \tag{5.5}$$

This means that the BIP is now fully specified in terms of the matrices C, M and N, or equivalently, in terms of their block decompositions C_{11}, C_{12}, C_{21}, C_{22}, A_1 and A_2.

THEOREM 5.1. *If the BIP that is specified in terms of C, M and N admits a solution, then there exists an $n \times n$ matrix $P \geq 0$ which solves the Lyapunov–Stein equation*

$$M^*PM - N^*PN = C^*JC \qquad \text{if } \Omega_+ = \mathbb{D}, \tag{5.6a}$$

$$M^*PN - N^*PM = 2\pi i C^*JC \quad \text{if } \Omega_+ = \mathbb{C}_+, \tag{5.6b}$$

$$M^*PN + N^*PM = -2\pi C^*JC \quad \text{if } \Omega_+ = \Pi_+. \tag{5.6c}$$

PROOF. Let $S(\lambda) \in \mathcal{S}^{p \times q}(\Omega_+)$ be a solution of the BIP and define P by the rule

$$v^*Pu = \left\langle \begin{bmatrix} I_p & -S(\lambda) \\ -S(\lambda)^* & I_q \end{bmatrix} F(\lambda)u, F(\lambda)v \right\rangle, \tag{5.7}$$

where $F(\lambda)$ is given by (5.1) and u and v are any vectors in $\mathbb{C}^n$. Then clearly $P \geq 0$. The rest of the proof depends largely upon Theorem 3.1 and the special form of the data. We shall present details only for the case $\Omega_+ = \mathbb{D}$. Details for the other two classical choices of Ω_+ may be found in [Dym 1994b]. Because of the special block form of M and n, it is convenient to write P in the block form

$$P = \begin{bmatrix} P_{11} & P_{12} \\ P_{21} & P_{22} \end{bmatrix} \begin{matrix} \}\mu \\ \}\nu \end{matrix}.$$

$$\underbrace{\phantom{P_{11}}}_{\mu} \quad \underbrace{\phantom{P_{12}}}_{\nu}$$

Then P is a solution of (5.6a) if and only if the following three equations are satisfied:

$$P_{11} - A_1^* P_{11} A_1 = C_{11}^* C_{11} - C_{21}^* C_{21}, \tag{5.8}$$

$$A_1^* P_{12} - P_{12} A_2 = C_{11}^* C_{12} - C_{21}^* C_{22}, \tag{5.9}$$

$$A_2^* P_{22} A_2 - P_{22} = C_{12}^* C_{12} - C_{22}^* C_{22}. \tag{5.10}$$

By Theorem 3.1,

$$y^* P_{11} x = \langle J F_1(\lambda) x, F_1(\lambda) y \rangle$$

for every choice of x and y in $\mathbb{C}^\mu$. Thus

$$y^* (P_{11} - A_1^* P_{11} A_1) x$$

$$= \langle J F_1(\lambda) x, F_1(\lambda) y \rangle - \langle J F_1(\lambda) A_1 x, F_1(\lambda) A_1 y \rangle$$

$$= \langle J F_1(\lambda) x, F_1(\lambda)(I_\mu - \lambda A_1) y \rangle + \langle J F_1(\lambda)(I_\mu - \lambda A_1) x, F_1(\lambda) \lambda A_1 y \rangle$$

$$= \left\langle J F_1(\lambda) x, \begin{bmatrix} C_{11} \\ C_{21} \end{bmatrix} y \right\rangle + \left\langle J \begin{bmatrix} C_{11} \\ C_{21} \end{bmatrix}, F_1(\lambda) \lambda A_1 y \right\rangle$$

$$= y^* [C_{11}^* \ C_{21}^*] J F_1(0) x + O$$

$$= y^* (C_{11}^* C_{11} - C_{21}^* C_{21}) x,$$

which serves to prove (5.8).

Next, by the middle formula of (3.1),

$$y^* (A_1^* P_{12} - P_{12} A_2) x = -\langle S(\lambda) C_{22} (\lambda I_\nu - A_2)^{-1} x, C_{11}(I_\mu - \lambda A_1)^{-1} A_1 y \rangle$$

$$+ \langle S(\lambda) C_{22} (\lambda I_\nu - A_2)^{-1} A_2 x, C_{11}(I_\mu - \lambda A_1)^{-1} y \rangle$$

$$= -\langle S(\lambda) C_{22} (\lambda I_\nu - A_2)^{-1} \lambda x, C_{11}(I_\mu - \lambda A_1)^{-1} \lambda A_1 y \rangle$$

$$+ \langle S(\lambda) C_{22} (\lambda I_\nu - A_2)^{-1} A_2 x, C_{11}(I_\mu - \lambda A_1)^{-1} y \rangle$$

for every choice of $x \in \mathbb{C}^\nu$ and $y \in \mathbb{C}^\mu$. But, by adding the term

$$\langle S(\lambda) C_{22} (\lambda I_\nu - A_2)^{-1} \lambda x, C_{11}(I_\mu - \lambda A_1)^{-1} y \rangle$$

to the first inner product and subtracting it from the second, the last expression can be rewritten as ① − ②, where

$$① = \langle \lambda S(\lambda)C_{22}(\lambda I_\nu - A_2)^{-1}x,\ C_{11}(I_\mu - \lambda A_1)^{-1}(I_\mu - \lambda A_1)y \rangle$$
$$= \langle \lambda \underline{\underline{q'}}\big(S(\lambda)C_{22}(\lambda I_\nu - A_2)^{-1}x\big),\ C_{11}y \rangle$$
$$= \langle \lambda C_{12}(\lambda I_\nu - A_2)^{-1}x,\ C_{11}y \rangle = y^* C_{11}^* C_{12} x$$

and

$$② = \langle S(\lambda)C_{22}(\lambda I_\nu - A_2)^{-1}(\lambda I_\nu - A_2)x,\ C_{11}(I_\mu - \lambda A_1)^{-1}y \rangle$$
$$= \langle S(\lambda)C_{22}x,\ C_{11}(I_\mu - \lambda A_1)^{-1}y \rangle$$
$$= \langle C_{22}x,\ \underline{\underline{p}}\big(S(\lambda)^* C_{11}(I_\mu - \lambda A_1)^{-1}y\big) \rangle$$
$$= \langle C_{22}x,\ C_{21}(I_\mu - \lambda A_1)^{-1}y \rangle = y^* C_{21}^* C_{22} x.$$

Thus

$$① - ② = y^*(C_{11}^* C_{12} - C_{21}^* C_{22})x,$$

which serves to verify (5.9).

Finally, (5.10) is obtained in much the same way from the formula

$$y^* P_{22} x = -\langle J F_2(\lambda)x,\ F_2(\lambda)y \rangle,$$

which is valid for every choice of x and y in $\mathbb{C}^q$ and is itself obtained from Theorem 3.1.   □

Theorem 5.1 admits a converse: If the Lyapunov–Stein equation (5.6) admits a nonnegative solution P, then the BIP is solvable. However, this is only part of the story because if both $\mu \geq 1$ and $\nu = n - \mu \geq 1$, then (5.6) can have many solutions. To be more precise, under the spectral conditions that were imposed on A_1 and A_2 in (5.4) and (5.5), the P_{11} and P_{22} blocks of every solution P of (5.6) are the same, however, the P_{12} block is not unique unless $\sigma(A_1^*) \cap \sigma(A_2) = \varnothing$. This extra freedom can be used to impose more interpolation conditions and leads to a more refined problem which we shall refer to as the augmented BIP. For an instructive example, see [Dym 1989b, Section 10].

6. The Augmented Basic Interpolation Problem

The augmented BIP is formulated in terms of the data C,

$$M = \begin{bmatrix} M_{11} & 0 \\ 0 & M_{22} \end{bmatrix} \begin{matrix} \}\mu \\ \}\nu \end{matrix},\qquad N = \begin{bmatrix} N_{11} & 0 \\ 0 & N_{22} \end{bmatrix} \begin{matrix} \}\mu \\ \}\nu \end{matrix},$$
$$\underbrace{\quad}_{\mu}\ \underbrace{\quad}_{\nu}\qquad\qquad \underbrace{\quad}_{\mu}\ \underbrace{\quad}_{\nu}$$

and a solution $P \geq 0$ of the Lyapunov–Stein equation (5.6): more precisely, $S(\lambda) \in \mathcal{S}^{p \times q}(\Omega_+)$ is said to be a solution of the augmented BIP if

(1) $S(\lambda)$ is a solution of the BIP, and

(2) $-\langle S(\lambda)C_{22}(M_{22}-\lambda N_{22})^{-1}y,\ C_{11}(M_{11}-\lambda N_{11})^{-1}x\rangle = x^*P_{12}y$ for every choice of $x \in \mathbb{C}^\mu$ and $y \in \mathbb{C}^\nu$.

LEMMA 6.1. *If $S(\lambda)$ is a solution of the BIP based on C, M and N (subject to the spectral conditions (5.4) and (5.5)) and if P is a nonnegative solution of the Lyapunov–Stein equation (5.6), the following conditions are equivalent:*

(1) $-\langle S(\lambda)C_{22}(M_{22} - \lambda N_{22})^{-1}y,\ C_{11}(M_{11} - \lambda N_{11})^{-1}x\rangle = x^*P_{12}y$

for every choice of $x \in \mathbb{C}^\mu$ and $y \in \mathbb{C}^\nu$.

(2) $\left\langle \begin{bmatrix} I_p & -S(\lambda) \\ -S(\lambda)^* & I_q \end{bmatrix} F(\lambda)v,\ F(\lambda)u \right\rangle = u^*Pv$

for every choice of u and v in $\mathbb{C}^n$.

(3) $\left\langle \begin{bmatrix} I_p & -S(\lambda) \\ -S(\lambda)^* & I_q \end{bmatrix} F(\lambda)u,\ F(\lambda)u \right\rangle = u^*Pu$

for every choice of $u \in \mathbb{C}^n$.

(4) $\left\langle \begin{bmatrix} I_p & -S(\lambda) \\ -S(\lambda)^* & I_q \end{bmatrix} F(\lambda)u,\ F(\lambda)u \right\rangle \leq u^*Pu$

for every choice of $u \in \mathbb{C}^n$.

PROOF. Suppose first that (4) holds for some given solution $S(\lambda)$ of the BIP and some given solution $P \geq 0$ of the Lyapunov–Stein equation (5.6). Let u^*Qv denote the inner product on the left-hand side of the equality in (2) for this choice of $S(\lambda)$. By (4), the matrix

$$X = P - Q = \begin{bmatrix} X_{11} & X_{12} \\ X_{21} & X_{22} \end{bmatrix}$$

is nonnegative. Therefore, since X_{11} is the $\mu \times \mu$ zero matrix and X_{22} is the $\nu \times \nu$ zero matrix, the matrix

$$\begin{bmatrix} 0 & X_{12} \\ X_{12}^* & 0 \end{bmatrix}$$

is nonnegative. But this is only possible if $X_{12} = 0$. This serves to establish the nontrivial half of the equivalence of (4) and (3). The equivalence of (3) and (2) is a standard argument; the equivalence of (1) and (2) rests heavily on the proof of Theorem 3.1 and the fact that $S(\lambda)$ is assumed to be a solution of the BIP. The details are left to the reader.   □

7. Reproducing Kernel Hilbert Spaces

A Hilbert space $\mathcal{H}$ of $(m \times 1)$-vector-valued functions defined on some subset Δ of $\mathbb{C}$ is said to be a reproducing kernel Hilbert space (RKHS) if there exists an $m \times m$ mvf $K_\omega(\lambda)$ on $\Delta \times \Delta$ such that, for every choice of $\omega \in \Delta$, $u \in \mathbb{C}^m$, and $f \in \mathcal{H}$, we have $K_\omega u \in \mathcal{H}$ (as a function of λ), and

$$\langle f, K_\omega u \rangle_{\mathcal{H}} = u^* f(\omega). \tag{7.1}$$

The main facts are these:

- The RK (reproducing kernel) is unique; that is, if $K_\omega(\lambda)$ and $L_\omega(\lambda)$ are both RK's for the same RKHS, then $K_\omega(\lambda) = L_\omega(\lambda)$ for every choice of ω and λ in Δ.
- $K_\alpha(\beta)^* = K_\beta(\alpha)$. $\hfill (7.2)$
- For every choice of $\omega_1, \ldots, \omega_n$ in Δ and $u_1, \ldots, u_n$ in $\mathbb{C}^m$, we have

$$\sum_{i,j=1}^{n} u_j^* K_i(\omega_j) u_i \geq 0. \tag{7.3}$$

EXAMPLE 7.1. $H_2^m(\Omega_+)$ is an RKHS with RK

$$K_\omega(\lambda) = I_m / \rho_\omega(\lambda)$$

for each of the classical choices of Ω_+, where $\rho_\omega(\lambda)$ and the corresponding inner product are specified in Table 1. Basically this is just Cauchy's theorem for $H_2(\Omega_+)$.

EXAMPLE 7.2 ($\mathcal{H}(S)$ SPACES). For each choice of $S(\lambda) \in \mathcal{S}^{p \times q}(\Omega_+)$, the kernel

$$L_\omega(\lambda) = \frac{I_p - S(\lambda)S(\omega)^*}{\rho_\omega(\lambda)} \tag{7.4}$$

is positive in the sense exhibited in inequality (7.3). Perhaps the easiest way to see this is to observe that

$$\sum_{i,j=1}^{n} \xi_i^* \Lambda_{\alpha_j}(\alpha_i) \xi_j = \langle g, g \rangle - \langle \underline{p} S^* g, \underline{p} S^* g \rangle$$

with $g = \sum_{j=1}^{n} \xi_j / \rho_{\alpha_j}$; see Lemma 1.6 for help with the evaluation, if need be.

Because of this positivity, it follows on general grounds (see [Aronszajn 1950], for example) that $L_\omega(\lambda)$ is the reproducing kernel of exactly one RKHS, which we shall designate by $\mathcal{H}(S)$. The following beautiful characterization of $\mathcal{H}(S)$ is due to de Branges and Rovnyak [1966].

THEOREM 7.3. *Let $S \in \mathcal{S}^{p \times q}(\Omega_+)$, and for $f \in H_2^p(\Omega_+)$ let*

$$\kappa(f) = \sup\{\|f + Sg\|^2 - \|g\|^2 : g \in H_2^q \Omega_+\}. \tag{7.5}$$

Then

$$\mathcal{H}(S) = \{f \in H_2^p(\Omega_+) : \kappa(f) < \infty\}$$

and

$$\|f\|^2_{\mathcal{H}(S)} = \kappa(f).$$

PROOF. A proof for $\Omega_+ = \mathbb{D}$ can be found in [de Branges and Rovnyak 1966], but it goes through for the other two cases in just the same way. □

It is an instructive exercise to check that if $S(\lambda)$ is isometric a.e. on Ω_0, then $p \geq q$ and

$$\mathcal{H}(S) = H_2^p \ominus SH_2^q.$$

A number of useful properties of the space $\mathcal{H}(S)$ as well as references to more extensive lists are provided in [Dym 1994b, Section 6].

EXAMPLE 7.4. Let $\mathcal{M} = \{F(\lambda)u : u \in \mathbb{C}^n\}$, where $F(\lambda)$ is an $m \times n$ mvf with linearly independent columns $f_1(\lambda), \ldots, f_n(\lambda)$, and let P be any $n \times n$ positive definite matrix (that is, $P > 0$). Then the space $\mathcal{M}$, endowed with the inner product

$$\langle F(\lambda)u, F(\lambda)v \rangle_{\mathcal{M}} = v^*Pu \tag{7.6}$$

for every choice of u and v in $\mathbb{C}^n$, is an RKHS with RK

$$K_\omega(\lambda) = F(\lambda)P^{-1}F(\omega)^*. \tag{7.7}$$

The verification is by direct computation.

8. A Special Class of Reproducing Kernel Hilbert Spaces

We shall be particularly interested in RKHS's of $(m \times 1)$-vector-valued meromorphic functions in $\mathbb{C}$ with RK's of a special form that will be described below in Theorem 8.1. The theorem is an elaboration of a fundamental result from [de Branges 1963]. It is formulated in terms of the polynomials $a(\lambda)$ and $b(\lambda)$ given in Table 1 in order to obtain a statement that is applicable to each of the three classical choices of Ω_+.

A set Δ is said to be symmetric with respect to Ω_0 (or $\rho_\omega(\lambda)$) if for every $\lambda \in \Delta$ (except 0 for $\Omega_0 = \mathbb{T}$) the point λ^o belongs to Δ; note that $\rho_\omega(\omega^o) = 0$. Recall that $\rho_\omega(\lambda) = a(\lambda)a(\omega)^* - b(\lambda)b(\omega)^*$.

THEOREM 8.1. *Let $\mathcal{H}$ be an RKHS of $(m \times 1)$-vector-valued functions that are analytic in an open nonempty subset Δ of $\mathbb{C}$ symmetric with respect to Ω_0. Then the reproducing kernel $K_\omega(\lambda)$ can be expressed in the form*

$$K_\omega(\lambda) = \frac{J - \Theta(\lambda)J\Theta(\omega)^*}{\rho_\omega(\lambda)}, \tag{8.1}$$

for some choice of $m \times m$ matrix-valued function $\Theta(\lambda)$ analytic in Δ and some signature matrix J, if and only if the following two conditions hold:

(1) *$\mathcal{H}$ is R_α-invariant for every $\alpha \in \Delta$.*

(2) *The structural identity*

$$\langle R_\alpha(bf), R_\beta(bg)\rangle_{\mathcal{H}} - \langle R_\alpha(af), R_\beta(ag)\rangle_{\mathcal{H}} = |ab' - ba'|^2 g(\beta)^* Jf(\alpha) \qquad (8.2)$$

holds for every choice of α, β in Δ and f, g in $\mathcal{H}$.

Moreover, in this case, the function $\Theta(\lambda)$ that appears in (8.1) is unique up to a J unitary constant factor on the right. If there exists a point $\gamma \in \Delta \cap \Omega_0$, it can be taken equal to

$$\Theta(\lambda) = I_m - \rho_\gamma(\lambda)K_\gamma(\lambda)J. \qquad (8.3)$$

This formulation is adapted from [Alpay and Dym 1993b]; see especially Theorems 4.1, 4.3, and 4.4 of that reference. The restriction to the three choices of $a(\lambda)$ and $b(\lambda)$ specified earlier permits some simplification in the presentation, because the terms $r(a, b; \alpha)f$ and $r(b, a; \alpha)f$ that intervene there are constant multiples of $R_\alpha(af)$ and $R_\alpha(bf)$, respectively.

For the three cases of interest, the structural identity (8.2) can be reexpressed as

$$\langle (I + \alpha R_\alpha)f, (I + \beta R_\beta)g\rangle_{\mathcal{H}} - \langle R_\alpha f, R_\beta g\rangle_{\mathcal{H}} = g(\beta)^* Jf(\alpha). \qquad (8.4)$$

if $\Omega_+ = \mathbb{D}$,

$$\langle R_\alpha f, g\rangle_{\mathcal{H}} - \langle f, R_\beta g\rangle_{\mathcal{H}} - (\alpha - \beta^*)\langle R_\alpha f, R_\beta g\rangle_{\mathcal{H}} = 2\pi i g(\beta)^* Jf(\alpha) \qquad (8.5)$$

if $\Omega_+ = \mathbb{C}_+$, and

$$\langle R_\alpha f, g\rangle_{\mathcal{H}} + \langle f, R_\beta g\rangle_{\mathcal{H}} + (\alpha + \beta^*)\langle R_\alpha f, R_\beta g\rangle_{\mathcal{H}} = -2\pi g(\beta)^* Jf(\alpha) \qquad (8.6)$$

if $\Omega_+ = \Pi_+$.

Formula (8.5) appears in [de Branges 1963]; formula (8.4) is equivalent to a formula of Ball [1975], who adapted de Branges' work to the disc, including an important technical improvement from [Rovnyak 1968].

The role of the two conditions in Theorem 8.1 becomes particularly transparent when $\mathcal{H}$ is finite-dimensional. Indeed, if the n-dimensional space $\mathcal{M}$ considered in Example 7.4 is R_α invariant for some point α in the domain of analyticity of $F(\lambda)$, then, by Theorem 4.1, $F(\lambda)$ can be expressed in the form

$$F(\lambda) = V(M - \lambda N)^{-1} \qquad (8.7)$$

with M and N satisfying (4.2). Thus R_α-invariance forces the elements of $\mathcal{M}$ to be rational of the indicated form. Since

$$(R_\beta F)(\lambda) = F(\lambda)N(M - \beta N)^{-1}$$

for every point β at which the matrix $M - \beta N$ is invertible, that is, for every $\beta \in \mathcal{A}_F$, the domain of analyticity of F, it is readily checked that

$$\langle R_\alpha Fu, Fv\rangle_{\mathcal{M}} = \langle FN(M - \alpha N)^{-1}u, Fv\rangle_{\mathcal{M}}$$

$$= v^* PN(M - \alpha N)^{-1}u, \qquad (8.8)$$

and similarly that

$$\langle Fu, R_\beta Fv \rangle_\mathcal{M} = v^*(M^* - \beta^* N^*)^{-1} N^* Pu \tag{8.9}$$

and

$$\langle R_\alpha Fu, R_\beta v \rangle_\mathcal{M} = v^*(M^* - \beta^* N^*)^{-1} N^* PN(M - \alpha N)^{-1} u \tag{8.10}$$

for every choice of α, β in $\mathcal{A}_F$ and u, v in $\mathbb{C}^n$. For each of the three special choices of Ω_+ under consideration, it is now readily checked that the structural identity (8.2) reduces to a matrix equation for P by working out (8.4)–(8.6) with the aid of (8.8)–(8.10). In other words, in a finite-dimensional R_α-invariant space $\mathcal{M}$ with Gram matrix P, the de Branges structural identity is equivalent to a Lyapunov–Stein equation for P. This was first established explicitly in [Dym 1989b] by a considerably lengthier calculation. If F is analytic at zero, we may presume that $M = I_n$ in (8.7) and take $\alpha = \beta = 0$ in the structural identity (8.2).

THEOREM 8.2. *Let $F(\lambda) = V(M - \lambda N)^{-1}$ be an $m \times n$ matrix-valued function with $\det(M - \lambda N) \not\equiv 0$ and linearly independent columns, and let the vector space*

$$\mathcal{M} = \{F(\lambda)u : u \in \mathbb{C}^n\}$$

be endowed with the inner product

$$\langle Fu, Fv \rangle_\mathcal{M} = v^* Pu,$$

based on an $n \times n$ positive definite matrix P. Then $\mathcal{M}$ is a finite-dimensional RKHS with RK $K_\omega(\lambda)$ given by (7.7).

The RK can be expressed in the form

$$K_\omega(\lambda) = \frac{J - \Theta(\lambda)J\Theta(\omega)^*}{\rho_\omega(\lambda)}$$

with $\rho_\omega(\lambda)$ as in Table 1 if and only if P is a solution of the equation

$$M^* PM - N^* PN = V^* JV \qquad \text{for } \Omega_+ = \mathbb{D}, \tag{8.11}$$

$$M^* PN - N^* PM = 2\pi i V^* JV \quad \text{for } \Omega_+ = \mathbb{C}_+, \tag{8.12}$$

$$M^* PN + N^* PM = -2\pi V^* JV \quad \text{for } \Omega_+ = \Pi_+. \tag{8.13}$$

Moreover, in each of these cases $\Theta(\lambda)$ is uniquely specified up to a J unitary constant multiplier on the right by the formula

$$\Theta(\lambda) = I_m - \rho_\gamma(\lambda) F(\lambda) P^{-1} F(\gamma)^* J \tag{8.14}$$

for any choice of the point $\gamma \in \Omega_0 \cap \mathcal{A}_F$.

PROOF. This is an easy consequence of Theorem 8.1 and the discussion preceding the statement of this theorem. The basic point is that, because of the special form of F, (8.4) holds if and only if P is a solution of (8.11); similarly (8.5) holds if and only if P is a solution of (8.12), and (8.6) if and only if P is a solution of (8.13).   □

It is well to note that formula (8.14) is a realization formula for $\Theta(\lambda)$, and that in the usual notation of (4.3) and (4.4) it depends only upon A, C and P. It can be reexpressed in one of the standard A, B, C, D forms by elementary manipulations.

Formulas (8.3) and (8.14) for $\Theta(\lambda)$ are obtained by matching the right-hand sides of (8.1) and (8.9). This leads to the formula

$$\Theta(\lambda)J\Theta(\omega)^* = J - \rho_\omega(\lambda)F(\lambda)P^{-1}F(\omega)^*,$$

which is clearly a necessary constraint on $\Theta(\lambda)$ since $\mathcal{M}$ has only one reproducing kernel, and hence any two recipes for it must agree. The final formula emerges upon setting $\omega = \gamma \in \Omega_0 \cap \mathcal{A}_F$ and then discarding J unitary constant factors on the right such as $\Theta(\gamma)^{-1}$ and J. Thus the general theory of "structured" reproducing kernel spaces as formulated in Theorem 8.1 yields formula (8.14). However, once the formula is available, it can be used to check that

$$F(\lambda)P^{-1}F(\omega)^* = \frac{J - \Theta(\lambda)J\Theta(\omega)^*}{\rho_\omega(\lambda)} \tag{8.15}$$

for every pair of points λ, ω in $\mathcal{A}_F$ by straightforward calculation, using only the fact that P is a solution of one of the equations (8.11)–(8.13), according to the choice of Ω_+. More information and references may be found in [Dym 1994b, Section 5], which was used heavily in the preparation of this section.

9. Diversion on J-Inner Functions

An $m \times m$ mvf $\Theta(\lambda)$ that is meromorphic in Ω_+ is said to be J-inner if it meets the following two conditions, where $\mathcal{A}_\Theta$ denotes the domain of analyticity of the mvf $\Theta(\lambda)$:

$$\Theta(\lambda)^*J\Theta(\lambda) \leq J \quad \text{for } \lambda \in \Omega_+ \cap \mathcal{A}_\Theta. \tag{9.1}$$

$$\Theta(\lambda)^*J\Theta(\lambda) = J \quad \text{for a.e. point} \lambda \in \Omega_0. \tag{9.2}$$

The evaluations in (9.2) are taken as nontangential boundary limits. Such limits exist, because the inequality (9.1) insures that every entry in the mvf $\Theta(\lambda)$ can be expressed as the ratio of two functions in $\mathcal{S}^{1\times 1}(\Omega_+)$; see, for example, [Dym 1989a, Theorem 1.1].

The identity (8.15) implies that

$$J - \Theta(\omega)J\Theta(\omega)^* = \rho_\omega(\omega)F(\omega)P^{-1}F(\omega)^*$$

and hence that

$$\Theta(\omega)J\Theta(\omega)^* \le J \quad \text{for } \omega \in \Omega_+ \cap \mathcal{A}_\Theta, \tag{9.3}$$

$$\Theta(\omega)J\Theta(\omega)^* = J \quad \text{for } \omega \in \Omega_0 \cap \mathcal{A}_\Theta. \tag{9.4}$$

The constraints (9.2) and (9.3) are equivalent to the assertion that $\Theta(\lambda)$ is J-inner, even though the "stars" are on the wrong side. The equivalence of (9.2) and (9.4) is self-evident. That of (9.1) and (9.3) takes a little more doing; see [Dym 1989a, pp. 16, 21], for example, for a couple of approaches. In any event, we shall be able to derive most of what we need directly from (9.3) and (9.4) without making use of the equivalence.

Upon writing

$$\Theta(\lambda) = \begin{bmatrix} \Theta_{11}(\lambda) & \Theta_{12}(\lambda) \\ \Theta_{21}(\lambda) & \Theta_{22}(\lambda) \end{bmatrix} \begin{matrix} \}p \\ \}q \end{matrix}$$
$$\underbrace{\phantom{\Theta_{11}(\lambda)}}_{p} \quad \underbrace{\phantom{\Theta_{12}(\lambda)}}_{q}$$

in the indicated block form, it is readily seen from the $(2,2)$ block of the inequality (9.3) that

$$\Theta_{22}(\lambda)\Theta_{22}(\lambda)^* \ge I_q + \Theta_{21}(\lambda)\Theta_{21}(\lambda)^*$$

for every point $\lambda \in \Omega_+ \cap \mathcal{A}_\Theta$. This implies that $\Theta_{22}(\lambda)$ is invertible for all such points and hence that the $m \times m$ mvf

$$\Sigma(\lambda) = \begin{bmatrix} I_p & -\Theta_{12}(\lambda) \\ 0 & -\Theta_{22}(\lambda) \end{bmatrix}^{-1} \begin{bmatrix} \Theta_{11}(\lambda) & 0 \\ \Theta_{21}(\lambda) & -I_q \end{bmatrix} \tag{9.5}$$

is well defined for $\lambda \in \Omega_+ \cap \mathcal{A}_\Theta$. Let $[\ ]^{-*}$ stand for $([\ ]^*)^{-1}$. The identity

$$I_m - \Sigma(\lambda)\Sigma(\omega)^* = \begin{bmatrix} I_p & -\Theta_{12}(\lambda) \\ 0 & -\Theta_{22}(\lambda) \end{bmatrix}^{-1} \left(J - \Theta(\lambda)J\Theta(\omega)^*\right) \begin{bmatrix} I_p & -\Theta_{21}(\omega) \\ 0 & -\Theta_{22}(\omega) \end{bmatrix}^{-*}, \tag{9.6}$$

which is readily checked by direct calculation, implies that $\Sigma(\lambda)$ is contractive and analytic in $\Omega_+ \cap \mathcal{A}_\Theta$ and hence in fact analytic in all of Ω_+. Thus $\Sigma(\lambda) \in \mathcal{S}^{m \times m}(\Omega_+)$. The mvf $\Sigma(\lambda)$ is termed the Potapov–Ginzburg transform of $\Theta(\lambda)$. It is also unitary a.e. on Ω_0 (when, as in the present case, $\Theta(\lambda)$ is J-inner; that is, $\Theta(\lambda)$ being J-inner implies that $\Sigma(\lambda)$ is inner).

By (9.5), the entries in the block decomposition

$$\Sigma(\lambda) = \begin{bmatrix} \Sigma_{11}(\lambda) & \Sigma_{12}(\lambda) \\ \Sigma_{21}(\lambda) & \Sigma_{22}(\lambda) \end{bmatrix}$$

are given by the formulas

$$\Sigma_{11}(\lambda) = \Theta_{11}(\lambda) - \Theta_{12}(\lambda), \Theta_{22}(\lambda)^{-1}\Theta_{21}(\lambda), \quad \Sigma_{12}(\lambda) = \Theta_{12}(\lambda)\Theta_{22}(\lambda)^{-1}$$

$$\Sigma_{21}(\lambda) = -\Theta_{22}(\lambda)^{-1}\Theta_{21}(\lambda), \quad\quad\quad\quad \Sigma_{22}(\lambda) = \Theta_{22}(\lambda)^{-1}. \tag{9.7}$$

Moreover, since $\Sigma(\lambda) \in \mathcal{S}^{m \times m}(\Omega_+)$, it follows that $\Sigma_{11}(\lambda) \in \mathcal{S}^{p \times p}(\Omega_+)$, $\Sigma_{12}(\lambda) \in \mathcal{S}^{p \times q}(\Omega_+)$, $\Sigma_{21}(\lambda) \in \mathcal{S}^{q \times p}(\Omega_+)$, and $\Sigma_{22}(\lambda) \in \mathcal{S}^{q \times q}(\Omega_+)$. Consequently,

$$\Theta_{21}(\lambda)\mathcal{E}(\lambda) + \Theta_{22}(\lambda) = \Theta_{22}(\lambda)\big(I_q - \Sigma_{21}(\lambda)\mathcal{E}(\lambda)\big)$$

is invertible in $\Omega_+ \cap \mathcal{A}_\Theta$ for every choice of $\mathcal{E}(\lambda) \in \mathcal{S}^{p \times q}(\Omega_+)$. Thus the linear fractional transformation

$$T_\Theta[\mathcal{E}] = \big(\Theta_{11}(\lambda)\mathcal{E}(\lambda) + \Theta_{12}(\lambda)\big)\big(\Theta_{21}(\lambda)\mathcal{E}(\lambda) + \Theta_{22}(\lambda)\big)^{-1} \qquad (9.8)$$

is well defined for $\mathcal{E}(\lambda) \in \mathcal{S}^{p \times q}(\Omega_+)$. The following facts are readily checked:

$$T_\Theta[\mathcal{E}] \in \mathcal{S}^{p \times q}(\Omega_+);$$

$$T_\Theta[0] = \Sigma_{12}(\lambda); \qquad (9.9)$$

$$T_\Theta[\mathcal{E}] = \Sigma_{12}(\lambda) + \Sigma_{11}(\lambda)\mathcal{E}(\lambda)\big(I_q - \Sigma_{21}(\lambda)\mathcal{E}(\lambda)\big)^{-1}\Sigma_{22}(\lambda). \qquad (9.10)$$

It is also easy to check that, if $\Psi(\lambda)$ also is J-inner, then

$$T_{\Theta\Psi}[\mathcal{E}] = T_\Theta\big[T_\Psi[\mathcal{E}]\big]. \qquad (9.11)$$

Linear fractional transformations of the form (9.10) were extensively studied in [Redheffer 1960].

THEOREM 9.1. *If $\Theta(\lambda)$ is J-inner and $\Sigma(\lambda)$ denotes the Potapov–Ginzburg transform of $\Theta(\lambda)$ (that is, if $\Sigma(\lambda)$ is given by (9.5)), and if*

$$f(\lambda) = \begin{bmatrix} g(\lambda) \\ h(\lambda) \end{bmatrix}$$

belongs to the RKHS $\mathcal{H}(\Theta)$, we have:

(1) $\begin{bmatrix} I_p & -\Sigma_{12}(\lambda) \\ -\Sigma_{12}(\lambda)^* & I_q \end{bmatrix} \begin{bmatrix} g(\lambda) \\ h(\lambda) \end{bmatrix} \in H_2^p(\Omega_+) \oplus H_2^q(\Omega_+)^\perp.$

(2) $\Sigma_{11}^* g \in H_2^p(\Omega_+)^\perp.$

(3) $\Sigma_{22} h \in H_2^q(\Omega_+)$ *and*

$$\|f\|_{\mathcal{H}(\Theta)}^2 = \|g - \Sigma_{12} h\|^2 + \|\Sigma_{22} h\|^2. \qquad (9.12)$$

PROOF. This is Theorem 2.7 of [Dym 1989a] restated in the current notation. $\square$

COROLLARY 9.2. *In the setting of the theorem,*

$$\|f\|_{\mathcal{H}(\Theta)}^2 = \left\langle \begin{bmatrix} I_p & -\Sigma_{12}(\lambda) \\ -\Sigma_{12}(\lambda)^* & I_q \end{bmatrix} \begin{bmatrix} g(\lambda) \\ h(\lambda) \end{bmatrix}, \begin{bmatrix} g(\lambda) \\ h(\lambda) \end{bmatrix} \right\rangle. \qquad (9.13)$$

PROOF. The right-hand side of (9.12) is equal to

$$\langle g - \Sigma_{12} h, g - \Sigma_{12} h \rangle + \langle \Sigma_{22} h, \Sigma_{22} h \rangle$$

$$= \langle g - \Sigma_{12} h, g \rangle - \langle \Sigma_{12}^* g, h \rangle + \big\langle (\Sigma_{12}^* \Sigma_{12} + \Sigma_{22}^* \Sigma_{22}) h,\, h \big\rangle.$$

But this is equal to the right-hand side of (9.13) since

$$\Sigma_{12}(\lambda)\Sigma_{12}(\lambda) + \Sigma_{22}(\lambda)^*\Sigma_{22}(\lambda) = I_q$$

for $\lambda \in \Omega_0$. □

10. Sufficiency When $P > 0$

To this point we know that if the BIP based on the matrices C, M and N (subject to the spectral constraints imposed in (5.4) and (5.5)) admits a solution, then there exists a solution $P \geq 0$ of the Lyapunov–Stein equation (5.6) and of the augmented BIP based on C, M, N and P. Our next objective is to establish a converse when $P > 0$.

THEOREM 10.1. *If P is a positive definite solution of the Lyapunov–Stein equation (5.6), and if $\Theta(\lambda)$ is the J-inner mvf defined by (8.14), then $T_\Theta[\mathcal{E}]$ is a solution of the augmented BIP for every choice of $\mathcal{E}(\lambda) \in \mathcal{S}^{p \times q}(\Omega_+)$.*

PROOF. Let $S(\lambda) = T_\Theta[\mathcal{E}]$ for some choice of $\mathcal{E}(\lambda) \in \mathcal{S}^{p \times q}(\Omega_+)$ and let

$$f_j(\lambda) = \begin{bmatrix} g_j(\lambda) \\ h_j(\lambda) \end{bmatrix}$$

denote the j'th element in the associated RKHS $\mathcal{H}(\Theta)$, where $f_j(\lambda) \in H_2^m(\Omega_+)$ for $j = 1, \ldots, \mu$ and $f_j(\lambda) \in H_2^m(\Omega_+)^\perp$ for $j = \mu+1, \ldots, n$. Also, let

$$f(\lambda) = \begin{bmatrix} g(\lambda) \\ h(\lambda) \end{bmatrix} = F(\lambda)u$$

denote an arbitrary element of the same RKHS. We now proceed in steps.

Step 1. $\big(g(\lambda) - S(\lambda)h(\lambda)\big) \in H_2^p(\Omega_+)$.

Proof of Step 1. Theorem 9.1 guarantees that

$$\big(g(\lambda) - \Sigma_{12}(\lambda)h(\lambda)\big) \in H_2^p(\Omega_+).$$

Therefore, in view of formula (9.10), it remains only to show that

$$\Sigma_{11}(\lambda)\mathcal{E}(\lambda)\big(I_q - \Sigma_{21}\mathcal{E}(\lambda)\big)^{-1}\Sigma_{22}(\lambda)h(\lambda) \in H_2^p(\Omega_+) \tag{10.1}$$

and hence, since

$$\Sigma_{11}(\lambda)\mathcal{E}(\lambda)\big(I_q - \Sigma_{21}(\lambda)\mathcal{E}(\lambda)\big)^{-1} \in H_\infty^{p \times q}(\Omega_+) \tag{10.2}$$

in the present setting, this follows from part (3) of Theorem 9.1.

Step 2. $\big(-S(\lambda)^*g(\lambda) + h(\lambda)\big) \in H_2^q(\Omega_+)^\perp$.

Proof of Step 2. The proof is similar to that of Step 1: Theorem 9.1 guarantees that

$$\left(-\Sigma_{21}(\lambda)^* g(\lambda) + h(\lambda)\right) \in H_2^q(\Omega_+)^\perp$$

and therefore, by (9.10), it remains only to show that

$$\Sigma_{22}(\lambda)^* \left(I_q - \Sigma_{21}(\lambda)\mathcal{E}(\lambda)\right)^{-*} \mathcal{E}(\lambda)^* \Sigma_{11}(\lambda)^* g(\lambda) \in H_2^q(\Omega_+)^\perp.$$

But this goes through much as before except that now we invoke assertion (2) instead of (3) of Theorem 9.1.

Step 3. $-\langle \Sigma_{12}(\lambda)h_j(\lambda), g_i(\lambda)\rangle = p_{ij}$ for $i = 1, \ldots, \mu$ and $j = \mu+1, \ldots, n$, where p_{st} denotes the (s,t) entry of P for $s, t = 1, \ldots, n$.

Proof of Step 3. By the corollary to Theorem 9.1,

$$p_{ij} = \langle f_j(\lambda), f_i(\lambda)\rangle_{\mathcal{H}(\Theta)}$$

$$= \left\langle \begin{bmatrix} I_p & -\Sigma_{12}(\lambda) \\ -\Sigma_{12}(\lambda)^* & I_q \end{bmatrix} \begin{bmatrix} g_j(\lambda) \\ h_j(\lambda) \end{bmatrix}, \begin{bmatrix} g_i(\lambda) \\ h_i(\lambda) \end{bmatrix} \right\rangle$$

$$= \langle g_j(\lambda) - \Sigma_{12}(\lambda)h_j(\lambda), g_i(\lambda)\rangle + \langle -\Sigma_{12}(\lambda)^* g_j(\lambda) + h_j(\lambda), h_i(\lambda)\rangle$$

for every choice of $i, j = 1, \ldots, n$. But now, if $i = 1, \ldots, \mu$ and $j = \mu+1, \ldots, n$, this is easily seen to reduce to the asserted identity with the help of Step 2.

Step 4. $-\langle S(\lambda)h_j(\lambda), g_i(\lambda)\rangle = p_{ij}$ for $i = 1, \ldots, \mu$ and $j = \mu+1, \ldots, n$.

Proof of Step 4. In view of Step 3 and formula (9.10), it remains to show that

$$-\langle \Sigma_{11}(\lambda)\mathcal{E}(\lambda)\left(I_q - \Sigma_{21}(\lambda)\mathcal{E}(\lambda)\right)^{-1}\Sigma_{22}(\lambda)h_j(\lambda), g_i(\lambda)\rangle = 0$$

for $i = 1, \ldots, \mu$ and $j = \mu+1, \ldots, n$. But this is easily checked since Theorem 9.1 guarantees that

$$\Sigma_{11}(\lambda)^* g_i(\lambda) \in H_2^p(\Omega_+)^\perp \quad \text{(even for } i = 1, \ldots, n),$$
$$\Sigma_{22}(\lambda)h_j(\lambda) \in H_2^q(\Omega_+) \quad \text{(even for } j = 1, \ldots, n),$$

and

$$\left(I_q - \Sigma_{21}(\lambda)\mathcal{E}(\lambda)\right)^{-1} \in H_\infty^{q \times q}(\Omega_+).$$

This completes the proof of the step and the theorem. $\square$

We remark that most of this analysis goes through in one form or another for problems with infinitely many interpolation constraints. Indeed Theorem 9.1 is applicable to infinite-dimensional spaces and the only point in the proof of Theorem 10.1 that depends critically on the specified form of the BIP (including the assumptions on $\sigma(A_1)$ and $\sigma(A_2)$) is the assertion that

$$\left(I_q - \Sigma_{21}(\lambda)\mathcal{E}(\lambda)\right)^{-1} \in H_\infty^{q \times q}(\Omega_+), \tag{10.3}$$

which was used implicitly in Steps 1 and 2 and explicitly in Step 4. In the present setting, (10.3) can be justified by direct estimates based on the specific formula for $\Sigma_{21}(\lambda)$ provided in (11.25). In more general settings, the most that can be said is that $\left(I_q - \Sigma_{21}(\lambda)\mathcal{E}(\lambda)\right)^{-1}$ is an outer function in the Smirnov class $\mathcal{N}_+^{q \times q}(\Omega_+)$. The needed estimates to justify the proof of Theorem 10.1 are then obtained by applications of the maximum principle in the Smirnov class. (Thus for example, in the proof of Step 1, (10.1) is established by showing first that $\Sigma_{11}\mathcal{E}\left(I_q - \Sigma_{21}\mathcal{E}\right)^{-1} \in \mathcal{N}_+^{p \times q}(\Omega_+)$. Then, since $\Sigma_{22}h \in H_2^q(\Omega_+)$, the product of these two terms belongs to $\mathcal{N}_+^p(\Omega_+) \cap L_2^p(\Omega_+)$ and hence by the maximum principle to $H_2^p(\Omega_+)$.) The author first learned the power of estimates in matrix Smirnov classes from [Arov 1973]; much useful information may also be found in [Katsnelson and Kirstein 1997].

The next theorem states that all solutions to the augmented BIP are obtained by the parametrization furnished in Theorem 10.1. In order to keep the length of the discussion under control, we shall rely a little more on outside references than has been our practice to this point.

THEOREM 10.2. *If P is a positive definite solution of the Lyapunov–Stein equation (5.6), then every solution $S(\lambda)$ of the augmented BIP based on this choice of P can be expressed as a linear fractional transformation of the form $S(\lambda) = T_\Theta[\mathcal{E}]$ in terms of the associated $\Theta(\lambda)$ for some choice of $\mathcal{E}(\lambda) \in \mathcal{S}^{p \times q}(\Omega_+)$.*

PROOF. Let $S(\lambda) \in \mathcal{S}^{p \times q}(\Omega_+)$ be a solution of the augmented BIP based on P and let $X(\lambda) = [I_p \;\; -S(\lambda)]$. Then, as follows from [Dym 1994b, pp. 205–206, 210], $X(\lambda)F(\lambda) \in \mathcal{H}(S)$ and the corresponding Gram matrix Q defined by the rule

$$u^*Qv = \langle X(\lambda)F(\lambda)v, \, X(\lambda)F(\lambda)u \rangle_{\mathcal{H}(S)}$$

for every choice of u and v in $\mathbb{C}^n$ is majorized by P, that is,

$$Q \leq P. \tag{10.4}$$

Now, in terms of the notation introduced in Example 7.2, the kernel

$$\frac{X(\lambda)\Theta(\lambda)J\Theta(\omega)^*X(\omega)^*}{\rho_\omega(\lambda)} = \frac{X(\lambda)JX(\omega)^*}{\rho_\omega(\lambda)} - X(\lambda)\left(\frac{J - \Theta(\lambda)J\Theta(\omega)^*}{\rho_\omega(\lambda)}\right)X(\omega)^*$$

$$= \Lambda_\omega(\lambda) - X(\lambda)F(\lambda)P^{-1}F(\omega)^*X(\omega)^*.$$

Our next objective is to show that this kernel is positive, that is, that

$$\sum_{i,j=1}^k y_i^*\left(\Lambda_{\alpha_j}(\alpha_i) - (XF)(\alpha_i)P^{-1}(XF)(\alpha_j)^*\right)y_j \geq 0$$

for every choice of vectors $y_1, \ldots, y_k$ in $\mathbb{C}^p$ and points $\alpha_1, \ldots, \alpha_k$ in Ω_+. In view of (10.4), it is enough to show that this inequality holds with Q^{-1} in place of P^{-1}. But now as

$$F(\lambda) = \begin{bmatrix} f_1(\lambda) & \cdots & f_n(\lambda) \end{bmatrix},$$

the second term in the sum with P replaced by Q can be reexpressed as

$$y_i^*(XF)(\alpha_i)Q^{-1}(XF)(\alpha_j)^*y_j = \sum_{u,v=1}^{n} y_i^*(Xf_u)(\alpha_i)(Q^{-1})_{uv}(Xf_v)(\alpha_j)^*y_j$$

$$= \sum_{u,v=1}^{n} \langle Xf_u, \Lambda_{\alpha_i}y_i\rangle_{\mathcal{H}(S)}(Q^{-1})_{uv}\langle \Lambda_{\alpha_j}y_j, Xf_v\rangle_{\mathcal{H}(S)}.$$

Thus, upon setting

$$f_0 = \sum_{ju=1}^{k} \Lambda_{\alpha_j}y_j,$$

it now follows readily that

$$\sum_{i,j=1}^{k} y_i^*(XF)(\alpha_i)P^{-1}(XF)(\alpha_j)^*y_j \le \sum_{u,v=1}^{n} \langle Xf_u, f_0\rangle_{\mathcal{H}(S)}(Q^{-1})_{uv}\langle f_0, Xf_v\rangle_{\mathcal{H}(S)}$$

$$= \left\|\Pi f_0(\lambda)\right\|_{\mathcal{H}(S)}^2,$$

where Π denotes the orthogonal projection of $\mathcal{H}(S)$ onto the subspace spanned by the elements $X(\lambda)f_u(\lambda)$, for $u = 1, \ldots, n$. The asserted inequality is now clear since

$$\sum_{i,j=1}^{k} y_i^*\Lambda_{\alpha_j}(\alpha_i)y_j = \left\|f_0(\lambda)\right\|_{\mathcal{H}(S)}^2.$$

The desired conclusion now follows directly from [Dym 1989a, Theorem 3.8], since the "admissibility" needed to invoke that theorem amounts to the kernel dealt with above being positive. $\square$

11. Explicit Formulas

In this section we shall provide explicit formulas for the mvfs $\Theta(\lambda)$ and $\Sigma(\lambda)$ in terms of the data C, M and N of the BIP and the solution P of the Lyapunov–Stein equation when $P > 0$. We shall do this in a more general setting than is needed for the three classical choices of Ω_+ that were considered earlier because it enhances the usefulness of the formulas and does not involve any extra work, just a little extra notation.

To this end, let $a(\lambda)$ and $b(\lambda)$ denote a pair of functions that are defined and analytic in an open nonempty connected set $\Omega \subset \mathbb{C}$ and assume that the subsets

$$\Omega_+ = \left\{\lambda \in \Omega : |a(\lambda)|^2 - |b(\lambda)|^2 > 0\right\}$$

and

$$\Omega_- = \left\{\lambda \in \Omega : |a(\lambda)|^2 - |b(\lambda)|^2 < 0\right\}$$

are both nonempty. This implies that

$$\Omega_0 = \left\{\lambda \in \Omega : |a(\lambda)|^2 - |b(\lambda)|^2 = 0\right\}$$

is nonempty and that in fact, as is shown in [Alpay and Dym 1996], Ω_0 contains an open arc. These definitions of $\Omega_\pm$ and Ω_0 are consistent with the earlier ones and the choices $\Omega = \mathbb{C}$ with $a(\lambda)$ and $b(\lambda)$ as in Table 1.

Let $\mathfrak{A} \in \mathbb{C}^{n \times n}$, $\mathfrak{B} \in \mathbb{C}^{n \times n}$, $C_1 \in \mathbb{C}^{p \times n}$, $C_2 \in \mathbb{C}^{q \times n}$ and $P \in \mathbb{C}^{n \times n}$ be fixed matrices such that $P \geq 0$,

$$\mathfrak{A}^* P \mathfrak{A} - \mathfrak{B}^* P \mathfrak{B} = C_1^* C_1 - C_2^* C_2, \tag{11.1}$$

and the determinant of the mvf

$$G(\lambda) = a(\lambda)\mathfrak{A} - b(\lambda)\mathfrak{B} \tag{11.2}$$

does not vanish identically: $\det G(\lambda) \not\equiv 0$ in Ω.

We shall also make frequent use of the $n \times n$ mvf

$$H(\lambda) = b(\lambda)\mathfrak{A}^* - a(\lambda)\mathfrak{B}^* \tag{11.3}$$

and the $m \times n$ mvf

$$F(\lambda) = CG(\lambda)^{-1}, \tag{11.4}$$

where

$$C = \begin{bmatrix} C_1 \\ C_2 \end{bmatrix} \quad \text{and} \quad m = p + q.$$

The functions $\rho_\omega(\lambda)$ and $\delta_\omega(\lambda)$ are expressed in terms of $a(\lambda)$ and $b(\lambda)$ just as before:

$$\rho_\omega(\lambda) = a(\lambda)a(\omega)^* - b(\lambda)b(\omega)^* \tag{11.5}$$

and

$$\delta_\omega(\lambda) = a(\omega)b(\lambda) - b(\omega)a(\lambda) \tag{11.6}$$

for every choice of λ and ω in Ω.

We shall refer to (11.1) as the General Lyapunov–Stein (GLS) equation. This usage too is easily seen to be consistent with the Lyapunov–Stein equation (5.6), which intervenes for the classical choices of Ω_+ when one defines $\mathfrak{A}$ and $\mathfrak{B}$ in terms of M and N via the formula

$$M - \lambda N = a(\lambda)\mathfrak{A} - b(\lambda)\mathfrak{B} \tag{11.7}$$

for the three sets of choices of $a(\lambda)$ and $b(\lambda)$ appearing in Table 1.

Some algebraic identities.

LEMMA 11.1. *The identity*

$$\rho_\omega(\lambda)G(\lambda)^* PG(\omega) + \rho_\omega(\lambda)^* H(\lambda) PH(\omega)^*$$

$$= \rho_\lambda(\lambda)G(\omega)^* PG(\omega) + \rho_\omega(\omega)H(\lambda)PH(\lambda)^* \tag{11.8}$$

holds for every pair of points λ and ω in Ω. It is independent of (11.1).

PROOF. This is a tedious but straightforward calculation, which does not depend upon the basic identity (11.1). The proof amounts to identifying the coefficients of like terms. Thus, for example, the coefficient of $\mathfrak{A}^*P\mathfrak{A}$ on the left hand side of (11.8) is equal to

$$\rho_\omega(\lambda)a(\lambda)^*a(\omega) + \rho_\omega(\lambda)^*b(\lambda)b(\omega)^* = \rho_\lambda(\lambda)a(\omega)^*a(\omega) + \rho_\omega(\omega)b(\lambda)b(\lambda)^*,$$

which is equal to the coefficient of $\mathfrak{A}^*P\mathfrak{A}$ on the right-hand side of (11.8). The coefficients of $\mathfrak{A}^*P\mathfrak{B}$, $\mathfrak{B}^*P\mathfrak{A}$ and $\mathfrak{B}^*P\mathfrak{B}$ are identified in the same way. □

LEMMA 11.2. *The identity*

$$\rho_\lambda(\lambda)G(\omega)^*PG(\omega) + \rho_\omega(\omega)H(\lambda)PH(\lambda)^*$$

$$= \rho_\omega(\omega)G(\lambda)^*PG(\lambda) + \rho_\lambda(\lambda)H(\omega)PH(\omega)^* \quad (11.9)$$

holds for every pair of points λ and ω in Ω. It is independent of (11.1).

PROOF. Let $L_\omega(\lambda)$ denote the left-hand side of (11.8). Then since

$$L_\omega(\lambda) = L_\lambda(\omega)^*,$$

the same invariance must hold true for the right-hand side of (11.8). □

LEMMA 11.3. *The identities*

$$\rho_\omega(\lambda)G(\lambda)^*PG(\omega) + \rho_\omega(\lambda)^*G(\omega)^*PG(\lambda)$$

$$= \rho_\lambda(\lambda)G(\omega)^*PG(\omega) + \rho_\omega(\omega)H(\lambda)PH(\lambda)^* + |\rho_\omega(\lambda)|^2(\mathfrak{A}^*P\mathfrak{A} - \mathfrak{B}^*P\mathfrak{B}) \quad (11.10)$$

and

$$\rho_\omega(\lambda)H(\omega)PH(\lambda)^* + \rho_\omega(\lambda)^*H(\lambda)PH(\omega)^*$$

$$= \rho_\lambda(\lambda)G(\omega)^*PG(\omega) + \rho_\omega(\omega)H(\lambda)PH(\lambda)^* - |\rho_\omega(\lambda)|^2(\mathfrak{A}^*P\mathfrak{A} - \mathfrak{B}^*P\mathfrak{B}) \quad (11.11)$$

hold for every pair of points λ and ω in Ω. They are both independent of (11.1).

This, too, can be verified by a tedious but straightforward calculation.

The mvf $\Delta_\omega(\lambda)$. Let

$$\Delta_\omega(\lambda) = G(\omega)^*PG(\lambda) + \rho_\omega(\lambda)C_2^*C_2 \quad (11.12)$$

for every pair of points λ and ω in Ω. Clearly

$$\Delta_\omega(\lambda) = \Delta_\lambda(\omega)^*. \quad (11.13)$$

LEMMA 11.4. *If P is a solution of the GLS equation (11.1), then*

$$\Delta_\omega(\lambda) = H(\lambda)PH(\omega)^* + \rho_\omega(\lambda)C_1^*C_1 \quad (11.14)$$

for every pair of points λ and ω in Ω.

PROOF. In view of (11.1), it suffices to show that

$$G(\omega)^*PG(\lambda) - H(\lambda)PH(\omega)^* = \rho_\omega(\lambda)\big(\mathfrak{A}^*P\mathfrak{A} - \mathfrak{B}^*P\mathfrak{B}\big). \tag{11.15}$$

But this is a straightforward computation. $\qquad\square$

THEOREM 11.5. *If $P \geq 0$ is a solution of the GLS equation* (11.1) *and if $\omega \in \Omega_+ \cup \Omega_0$, and*

$$\mathrm{rank} \begin{bmatrix} P^{1/2}G(\omega) \\ P^{1/2}H(\omega)^* \\ C \end{bmatrix} = n, \tag{11.16}$$

then the following conclusions hold:

(1) $\Delta_\omega(\lambda)$ *is invertible for every point* $\lambda \in \Omega_+$.
(2) *If $\omega \in \Omega_+$, then $\Delta_\omega(\lambda)$ is invertible for every point $\lambda \in \Omega_+$ at which*

$$\mathrm{rank} \begin{bmatrix} P^{1/2}G(\lambda) \\ C \end{bmatrix} = n. \tag{11.17}$$

(3) *If $\omega \in \Omega_0$ and $G(\omega)$ is invertible, then $\Delta_\omega(\lambda)$ is invertible for every point $\lambda \in \Omega_0 \setminus \{\omega\}$ at which the rank condition* (11.17) *holds.*
(4) *If $\omega \in \Omega_0$ and $G(\omega)$ is invertible, then $\Delta_\omega(\omega)$ is invertible if and only if $P > 0$.*

PROOF. Suppose first that

$$\Delta_\omega(\lambda)y = 0$$

for some choice of λ and ω in $\Omega_+ \cup \Omega_0$ and $y \in \mathbb{C}^n$. Then, by (11.12) and (11.14),

$$0 = y^*\big(\rho_\omega(\lambda)^*\Delta_\omega(\lambda) + \rho_\omega(\lambda)\Delta_\omega(\lambda)^*\big)y$$

$$= y^*\big(\rho_\omega(\lambda)^*(H(\lambda)PH(\omega)^* + \rho_\omega(\lambda)C_1^*C_1) + \rho_\omega(\lambda)(G(\lambda)^*PG(\omega) + \rho_\omega(\lambda)C_2^*C_2)\big)y$$

$$= y^*\big(\rho_\lambda(\lambda)G(\omega)^*PG(\omega) + \rho_\omega(\omega)H(\lambda)PH(\lambda)^* + |\rho_\omega(\lambda)|^2(C_1^*C_1 + C_2^*C_2)\big)y.$$

Therefore, since each of the summands is nonnegative for λ and ω in $\Omega_+ \cup \Omega_0$, it follows from Lemma 11.2 and the last line that

$$\rho_\lambda(\lambda)G(\omega)^*PG(\omega)y = 0, \qquad \rho_\omega(\omega)G(\lambda)^*PG(\lambda)y = 0,$$
$$\rho_\lambda(\lambda)H(\omega)PH(\omega)^*y = 0, \qquad \rho_\omega(\omega)H(\lambda)PH(\lambda)^*y = 0,$$
$$\rho_\omega(\lambda)C_1 y = 0, \qquad \rho_\omega(\lambda)C_2 y = 0.$$

The rest of the argument proceeds in cases that amount to figuring out which of the preceding six identities really come into play.

Case 1. If $\lambda \in \Omega_+$, then $\rho_\lambda(\lambda) > 0$ and $|\rho_\omega(\lambda)| > 0$, and hence

$$\Delta_\omega(\lambda)y = 0 \implies \begin{bmatrix} P^{1/2}G(\omega) \\ P^{1/2}H(\omega)^* \\ C \end{bmatrix} y = 0 \implies y = 0,$$

in view of the rank assumption (11.16). This serves to establish assertion (1).

Case 2. If $\lambda \in \Omega_0$ but $\omega \in \Omega_+$, then $|\rho_\omega(\omega)| > 0$ and $|\rho_\omega(\lambda)| > 0$ and hence

$$\Delta_\omega(\lambda)y = 0 \implies \begin{bmatrix} P^{1/2}G(\omega) \\ C \end{bmatrix} y = 0 \implies y = 0,$$

in view of the rank assumption (11.17). This serves to establish assertion (2).

Case 3. If $\lambda \in \Omega_0$ and $\omega \in \Omega_0$ but $\lambda \neq \omega$, then $|\rho_\omega(\lambda)| > 0$. Hence $\Delta_\omega(\lambda)y = 0$ implies $Cy = 0$. This conclusion comes from the last two of the six identities established above. It now follows further from the definition of $\Delta_\omega(\lambda)$ that $G(\omega)^*PG(\lambda)y = 0$ and hence, since $G(\omega)$ is assumed to be invertible, that $PG(\lambda)y = 0$. But this in turn implies that $P^{1/2}G(\lambda)y = 0$, since $P \geq 0$. Thus we see that, in this case,

$$\Delta_\omega(\lambda)y = 0 \implies \begin{bmatrix} P^{1/2}G(\omega) \\ C \end{bmatrix} y = 0 \implies y = 0$$

in view of the rank condition (11.17). This serves to complete the third assertion and so too the proof, since the fourth assertion is obvious. $\qquad\square$

COROLLARY 11.6. *If $P > 0$ and $G(\omega)$ is invertible for some point $\omega \in \Omega_+ \cup \Omega_0$, then*:

(1) $\Delta_\omega(\lambda)$ *is invertible for every point $\lambda \in \Omega_+$.*
(2) $\Delta_\omega(\lambda)$ *is invertible for every point $\lambda \in \Omega_0$ at which*

$$\mathrm{rank} \begin{bmatrix} G(\lambda) \\ C \end{bmatrix} = n. \tag{11.18}$$

We now assume that $P > 0$ is a positive definite solution of the GLS equation (11.1) and, for a fixed point $\gamma \in \Omega_0$ at which $G(\gamma)$ is invertible, define

$$\Theta(\lambda) = I_m - \rho_\gamma(\lambda)F(\lambda)P^{-1}F(\gamma)^*J$$

for every point $\lambda \in \Omega$ at which $G(\lambda)$ is invertible, just as in (8.3). (Strictly speaking, it would be better to write $\Theta_\gamma(\lambda)$ instead of $\Theta(\lambda)$ in order to indicate the dependence upon the "normalization" point γ, but this makes the formulas involving subblocks awkward.)

Thus, upon setting

$$\varphi_\omega(\lambda) = \rho_\omega(\lambda)G(\lambda)^{-1}P^{-1}G(\omega)^{-*}, \tag{11.20}$$

for those points λ and ω in Ω at which the indicated inverses exist, it is readily seen that

$$\Theta_{11}(\lambda) = I_p - C_1\varphi_\gamma(\lambda)C_1^*, \tag{11.21}$$

$$\Theta_{12}(\lambda) = C_1\varphi_\gamma(\lambda)C_2^*, \tag{11.22}$$

$$\Theta_{21}(\lambda) = -C_2\varphi_\gamma(\lambda)C_1^*, \tag{11.23}$$

$$\Theta_{22}(\lambda) = I_q + C_2\varphi_\gamma(\lambda)C_2^*. \tag{11.24}$$

Since $\Theta(\lambda)$ is J-inner with respect to Ω_+, the Potapov–Ginzburg transform $\Sigma(\lambda)$ is well defined (by formulas (9.5) and (9.7)) and is inner.

THEOREM 11.7. *If $P > 0$ is a solution of the GLS equation (11.1) and if $\gamma \in \Omega_0$ and $G(\gamma)$ is invertible, then*

$$\Sigma(\lambda) = I_m - \rho_\gamma(\lambda) \begin{bmatrix} C_1 \\ -C_2 \end{bmatrix} \Delta_\gamma(\lambda)^{-1} \begin{bmatrix} C_1^* & -C_2^* \end{bmatrix} \tag{11.25}$$

and

$$\Sigma(\lambda)\Sigma(\lambda)^* = I_m - \rho_\lambda(\lambda) \begin{bmatrix} C_1 \\ -C_2 \end{bmatrix} \Delta_\gamma(\lambda)^{-1} G(\gamma)^* P G(\gamma) \Delta_\gamma(\lambda)^{-*} \begin{bmatrix} C_1^* & -C_2^* \end{bmatrix} \tag{11.26}$$

for every point $\lambda \in \Omega$ at which the indicated inverses exist.

The verification of these formulas is by direct calculation, with the help of the following well known result:

LEMMA 11.8. *If $X \in \mathbb{C}^{n \times n}$, $Y \in \mathbb{C}^{p \times n}$, $Z \in \mathbb{C}^{n \times p}$ and if X and $X + ZY$ are invertible, then $I_p + YX^{-1}Z$ is invertible and*

$$(I_p + YX^{-1}Z)^{-1} = I_p - Y(X + ZY)^{-1}Z. \tag{11.27}$$

PROOF. It suffices to check that

$$\big(I_p + YX^{-1}Z\big)\big(I_p - Y(X + ZY)^{-1}Z\big) = I_p.$$

But this is a straightforward calculation. □

PROOF OF THEOREM 11.7. Suppose first that $G(\lambda)$ is invertible. Then, since

$$\Theta_{22}(\lambda) = I_q + \rho_\gamma(\lambda) C_2 \big(G(\gamma)^* P G(\lambda)\big)^{-1} C_2^*$$

and

$$G(\gamma)^* P G(\lambda) + \rho_\gamma(\lambda) C_2^* C_2 = \Delta_\gamma(\lambda)$$

is invertible for $\lambda \in \Omega_+ \cup \Omega_0$ under the present assumptions by Theorem 11.5, Lemma 11.8 guarantees that $\Theta_{22}(\lambda)$ is invertible and that

$$\Sigma_{22}(\lambda) = \Theta_{22}(\lambda)^{-1} = I_q - \rho_\gamma(\lambda) C_2 \Delta_\gamma(\lambda)^{-1} C_2^*. \tag{11.28}$$

This serves to verify the $(2, 2)$ block entry of (11.25).

Next, by (11.28) and (11.23),

$$\Sigma_{21}(\lambda) = -\Theta_{22}(\lambda)^{-1}\Theta_{21}(\lambda)$$

$$= \big(I_q - \rho_\gamma(\lambda) C_2 \Delta_\gamma(\lambda)^{-1} C_2^*\big) C_2 \varphi_\gamma(\lambda) C_1^*$$

$$= C_2 \big(I_q - \rho_\gamma(\lambda) \Delta_\gamma(\lambda)^{-1} C_2^* C_2\big) \varphi_\gamma(\lambda) C_1^*,$$

which is readily seen to confirm the $(2, 1)$ block entry of (11.25).

The verification of the formulas for $\Sigma_{12}(\lambda)$ and $\Sigma_{11}(\lambda)$ is similar, though the latter requires a bit more work (but just a bit, if you take advantage of the fact that $-\Theta_{12}\Theta_{22}^{-1}\Theta_{21} = \Theta_{12}\Sigma_{21}$).

Finally, the verification of (11.26) is left to the reader.   □

We remark that although $\Sigma(\lambda)$ has been derived from $\Theta(\lambda)$, the given formulas are meaningful at those points λ at which $\Delta_\gamma(\lambda)$ is invertible. It is evident from Theorem 11.5 that this can happen even if P and $G(\lambda)$ are not invertible. Thus, for example, Theorem 11.5 guarantees that:

THEOREM 11.9. *If $P \geq 0$ is a solution of the GLS equation (11.1) and if $\gamma \in \Omega_0$ and*

$$G(\gamma)^*PG(\gamma) + C_1^*C_1 + C_2^*C_2 > 0,$$

then:

(1) $\Delta_\gamma(\lambda)$ *is invertible at every point $\lambda \in \Omega_+$.*

(2) $\Delta_\gamma(\lambda)$ *is invertible at every point $\lambda \in \Omega_0$ at which*

$$G(\lambda)^*PG(\lambda) + |\rho_\gamma(\lambda)| (C_1^*C_1 + C_2^*C_2) > 0. \tag{11.29}$$

Nevertheless, we shall not pursue this level of generality here. It is instructive, however, to see how to work directly from formula (11.25). To this end, it is convenient to first summarize the key properties of $\Delta_\gamma(\lambda)$ that come into play.

THEOREM 11.10. *If $P > 0$ is a solution of the GLS equation (11.1) and if $G(\lambda)$ is invertible at every point $\lambda \in \Omega_0$, then:*

(1) $\Delta_\gamma(\lambda)$ *is invertible at every point $\lambda \in \Omega_+ \cup \Omega_0$.*

(2) $\Sigma(\lambda)$ *is analytic on $\Omega_+ \cup \Omega_0$ and unitary at every point $\lambda \in \Omega_0$.*

(3) $\Sigma_{11}(\lambda)$ *is invertible at every point $\lambda \in \Omega_0$ and*

$$\det \Sigma_{11}(\lambda) = \frac{\det\big(H(\lambda)PH(\gamma)^*\big)}{\det\big(\Delta_\gamma(\lambda)\big)}. \tag{11.30}$$

(4) $\Sigma_{22}(\lambda)$ *is invertible at every point $\lambda \in \Omega_0$ and*

$$\det \Sigma_{22}(\lambda) = \frac{\det\big(G(\gamma)^*PG(\lambda)\big)}{\det\big(\Delta_\gamma(\lambda)\big)}. \tag{11.31}$$

Moreover, if Ω_+ is chosen to be one of the three classical settings, then:

(5) $\Sigma_{12}(\lambda)$ *and $\Sigma_{21}(\lambda)$ are strictly contractive on $\Omega_+ \cup \Omega_0$; that is, there exists a positive number $\delta < 1$ such that*

$$\|\Sigma_{12}(\lambda)\| \leq \delta \quad \text{and} \quad \|\Sigma_{21}\| \leq \delta$$

for every point $\lambda \in \Omega_+ \cup \Omega_0$.

(6) $\big(I_q - \Sigma_{21}(\lambda)\mathcal{E}(\lambda)\big)^{-1} \in H_\infty^{q \times q}(\Omega_+)$ *for every choice of $\mathcal{E}(\lambda) \in \mathcal{S}^{p \times q}(\Omega_+)$.*

(7) $\Delta_\gamma(\lambda)^{-1}u \in H_2^n(\Omega_+)$ *for every choice of $u \in \mathbb{C}^n$.*

(8) $\rho_\gamma(\lambda)\Delta_\gamma(\lambda)^{-1} \in H_\infty^{n \times n}(\Omega_+)$.

PROOF. The first four assertions are immediate from Theorems 11.9 and 11.7, except perhaps for the formulas for the determinant. But these too are easy if you take advantage of the fact that for $X \in \mathbb{C}^{p \times q}$ and $Y \in \mathbb{C}^{q \times p}$,

$$\det(I_p - XY) = \det(I_q - YX).$$

Next, in view of the identities

$$\Sigma_{21}(\lambda)\Sigma_{21}(\lambda)^* = I_q - \Sigma_{22}(\lambda)\Sigma_{22}(\lambda)^*,$$
$$\Sigma_{12}(\lambda)\Sigma_{12}(\lambda)^* = I_q - \Sigma_{22}(\lambda)\Sigma_{22}(\lambda)^*,$$

valid for $\lambda \in \Omega_0$, it suffices to show that

$$\|\Sigma_{22}\| \geq \delta_1 > 0$$

for every point $\lambda \in \Omega_0$. Since $\Sigma_{22}(\lambda)$ is contractive, it follows readily from the singular value decomposition of $\Sigma_{22}(\lambda)$ that

$$\|\Sigma_{22}(\lambda)\| \geq |\det \Sigma_{22}(\lambda)|.$$

Now if Ω_0 is compact, then this does the trick, since $|\det \Sigma_{22}(\lambda)| \neq 0$ for any point $\lambda \in \Omega_0$ by formula (11.31) the assumptions on $G(\lambda)$ and the first conclusion. The final four assertions are easily established when $\Omega_+ = \mathbb{D}$. The analysis for the other two classical choices of Ω_+ is more subtle, but may be completed by a finer investigation of $\Delta_\gamma(\lambda)$ as in [Dym 1996, pp. 206–208]. The details are left to the reader. $\qquad\square$

With the aid of Theorem 11.10, it is now not too difficult to show directly that (under the hypotheses of that theorem) if

$$S(\lambda) = \Sigma_{12}(\lambda) + \Sigma_{11}(\lambda)\mathcal{E}(\lambda)\big(I_q - \Sigma_{21}(\lambda)\mathcal{E}(\lambda)\big)^{-1}\Sigma_{22}(\lambda), \qquad (11.35)$$

then $\big(C_1 - S(\lambda)C_2\big)G(\lambda)^{-1}u \in H_2^p(\Omega_+)$,

$$\big(-S(\lambda)^*C_1 + C_2\big)G(\lambda)^{-1}u \in H_2^q(\Omega_+)^\perp,$$

and

$$-\left\langle S(\lambda)C_2 G(\lambda)^{-1}\begin{bmatrix} 0 \\ y \end{bmatrix}, \ C_1 G(\lambda)^{-1}\begin{bmatrix} x \\ 0 \end{bmatrix}\right\rangle = [x^* \ 0] P \begin{bmatrix} 0 \\ y \end{bmatrix}$$

for every choice of $\mathcal{E}(\lambda) \in \mathcal{S}^{p \times q}(\Omega_+)$, $u \in \mathbb{C}^n$, $x \in \mathbb{C}^\mu$ and $y \in \mathbb{C}^\nu$. This yields an independent check of Theorem 10.1, and exhibits a strategy that can be imitated even when $P \geq 0$ is singular.

12. $P \geq 0$ and Other Remarks

We begin by looking backwards.

About the first ten sections. The story begins with the observation that matrix versions of a number of classical interpolation problems are all special cases of a single general problem described in Section 2: the BIP. This problem is formulated in terms of the columns $f_j(\lambda)$ of an $m \times n$ mvf

$$F(\lambda) = [f_1(\lambda) \ \cdots \ f_n(\lambda)]$$

whose first μ columns span a μ-dimensional subspace $\mathcal{M}_1$ of $H_2^m(\Omega_+)$ and whose last $\nu = n - \mu$ columns span a ν-dimensional subspace $\mathcal{M}_2$ of $H_2^m(\Omega_+)^\perp$.

In Section 3 it is shown that if the BIP admits at least one solution $S(\lambda) \in \mathcal{S}^{p \times q}(\Omega_+)$, then the $n \times n$ matrix Q defined by (3.1) must be positive semidefinite; that is, the condition $Q \geq 0$ is a necessary condition for the existence of a solution to the BIP. The remarkable fact is that if $\mathcal{M}_1$ is an R_α-invariant subspace of $H_2^m(\Omega_+)$ for some $\alpha \in \Omega_+$ and $\mathcal{M}_2$ is an R_α-invariant subspace of $H_2^m(\Omega_+)^\perp$ for some $\alpha \in \Omega_-$, then this condition is also sufficient. The point is that the R_α-invariance assumptions force $F(\lambda)$ to be of the special form

$$F(\lambda) = C(M - \lambda N)^{-1}, \tag{12.1}$$

where the matrices M and N are subject to certain spectral assumptions, as is explained in (5.3)–(5.5). It then follows further that Q is a solution of the Lyapunov–Stein equation (5.6). For one sided problems (that is, when $\mu = n$ or $\nu = n$) this is the whole story because under the special constraints (5.4) and (5.5), the Lyapunov–Stein equation has only one solution. For two sided problems this is not true unless $\sigma(A_1^*) \cap \sigma(A_2) = \varnothing$. This extra freedom is used to define the augmented BIP in Section 6. Now it turns out that if $P > 0$ is any positive definite solution of the Lyapunov–Stein equation (5.6), then the space

$$\mathcal{M} = \{F(\lambda)u : u \in \mathbb{C}^n\} \tag{12.2}$$

(with $F(\lambda)$ as in (12.1)) endowed with the inner product

$$\langle F(\lambda)v, F(\lambda)u \rangle_{\mathcal{M}} = u^*Pv \tag{12.3}$$

is an RKHS of the special kind considered in Section 8. In particular its RK is given in terms of a rational J-inner mvf $\Theta(\lambda)$ (of McMillan degree n) that is uniquely determined by the space (that is, by the elements and the inner product) up to a J unitary constant factor on the right. Moreover, as is explained in Section 10,

$$\{T_\Theta[\mathcal{E}] : \mathcal{E}(\lambda) \in \mathcal{S}^{p \times q}(\Omega_+)\}$$

is a complete description of the set of all solutions to the augmented BIP based on $F(\lambda)$ and P (that is, on the elements of $\mathcal{M}$ and the inner product imposed on $\mathcal{M}$).

What if $P \geq 0$ is singular? If the rank of the $n \times n$ matrix P is equal to k, where $k < n$, then the space $\mathcal{M}$ endowed with the inner product P is no longer a Hilbert space. However, it turns out that there exists a k-dimensional subspace $\widetilde{\mathcal{M}}$ of $\mathcal{M}$ that is R_α-invariant for every point $\alpha \in \mathcal{A}_F$ which is an RKHS with an RK of the special form (8.1) based on a rational J-inner mvf $\widetilde{\Theta}(\lambda)$ (of McMillan degree k). Moreover, in this instance,

$$\left\{ T_{\widetilde{\Theta}}[\mathcal{E}] : \mathcal{E}(\lambda) \in \mathcal{S}^{p \times q}(\Omega_+) \right\}$$

fulfill k of the n interpolation conditions. The remaining $n - k$ conditions (which are not all independent) are then met by imposing extra constraints on $\mathcal{E}(\lambda)$. Some special cases illustrating this are given in [Dym 1989a, Chapter 7]. For other approaches see [Ball and Helton 1986; Bruinsma 1991; Dubovoj 1984].

Existence and representation formulas for the case of singular P may also be obtained by the methods introduced by Katsnelson, Kheifets, and Yuditskii [Katsnelson et al. 1987]; see also [Kheifets and Yuditskii 1994; Kheifets 1988a; 1988b; 1990; 1998], the last of which appears in this volume. Later in this section (after the next paragraph) we will discuss applications of this method to the augmented BIP problem.

About Section 11. In this section explicit formulas are presented for $\Theta(\lambda)$ and its Potapov–Ginzburg transform $\Sigma(\lambda)$ in terms of the data C, M, N and P when $P > 0$. It is then indicated how to verify that

$$\Sigma_{12}(\lambda) + \Sigma_{11}(\lambda)\mathcal{E}(\lambda)\left(I_q - \Sigma_{21}(\lambda)\mathcal{E}(\lambda)\right)^{-1}\Sigma_{22}(\lambda) \tag{12.4}$$

is a solution of the augmented BIP for every choice of $\mathcal{E}(\lambda) \in \mathcal{S}^{p \times q}(\Omega_+)$ directly from these formulas. This gives a second independent proof of this fact in this more general setting. In fact every solution of the augmented BIP can be expressed in this form. Moreover, although we do not pursue this here, large parts of this analysis in terms of $\Sigma(\lambda)$ is valid for matrices $P \geq 0$ that are singular.

The Abstract Interpolation Problem of Katsnelson, Kheifets, and Yuditskii. The problem of establishing the existence and parametrization of solutions to the augmented BIP in the disc fits naturally into the domain of problems that can be resolved within the framework of the Abstract Interpolation Problem of [Katsnelson et al. 1987] that was referred to just above. The starting point is the assumption that there exists an $n \times n$ positive semidefinite solution P of the GLS equation (5.6a). The idea is to rewrite (5.6a) as

$$M^*PM + C_2^*C_2 = N^*PN + C_1^*C_1 \tag{12.5}$$

and then to define an isometric colligation $V : \mathcal{D}_V \to \mathcal{R}_V$, where

$$\mathcal{D}_V = \left\{ \begin{bmatrix} P^{1/2}M \\ C_2 \end{bmatrix} x : x \in \mathbb{C}^n \right\} \subset \mathbb{C}^n \oplus \mathbb{C}^q,$$

$$\mathcal{R}_V = \left\{ \begin{bmatrix} P^{1/2}N \\ C_1 \end{bmatrix} x : x \in \mathbb{C}^n \right\} \subset \mathbb{C}^n \oplus \mathbb{C}^p.$$

The following facts then follow from the general analysis in [Katsnelson et al. 1987]:

(1) The set of all solutions to the augmented BIP is equal to the set of characteristic functions of those unitary colligations U that extend V (that is, $U|_{\mathcal{D}_V} = V$) and have the same "input" space $\mathbb{C}^q$ and the same "output" space $\mathbb{C}^p$ as V.

(2) The set of all such characteristic functions (and hence the set of all solutions to the augmented BIP in the disc) is equal to the set of all $p \times q$ mvf's of the general form (12.4) except that here $\Omega_+ = \mathbb{D}$ and $\mathcal{E}(\lambda)$ is an element in $\mathcal{S}^{p' \times q'}(\mathbb{D})$, where $q' = \dim(\mathbb{C}^{n+q} \ominus \mathcal{D}_V)$, $p' = \dim(\mathbb{C}^{n+p} \ominus \mathcal{R}_V)$ and $\Sigma(\lambda)$ is the characteristic function of a very special unitary colligation from $\mathbb{C}^n \oplus \mathbb{C}^{p'} \oplus \mathbb{C}^q$ onto $\mathbb{C}^n \oplus \mathbb{C}^p \oplus \mathbb{C}^{q'}$ (which is not of the type referred to in (1) because the input and output spaces have been enlarged to $\mathbb{C}^{p'} \oplus \mathbb{C}^q$ and $\mathbb{C}^p \oplus \mathbb{C}^{q'}$, respectively).

The description of the characteristic functions of unitary colligations that extend a given isometric colligation originates in the work of Arov and Grossman [1983; 1992]. See also [Kheifets 1988a; 1988b; 1990] for applications to the Abstract Interpolation Problem of [Katsnelson et al. 1987], and [Dym and Freydin 1997a; 1997b] for an application of these methods to the BIP problem in the setting of upper triangular operators. A useful discussion of applications of unitary colligations and of the Arov–Grossman formula may be found in [Arocena 1994].

Analogues of the problem of Katsnelson, Kheifets, and Yuditskii in the setting of Section 11. Here we exhibit a connection between the formulas that emerge from the analysis in Section 11 and the formulas that emerge by adapting the strategy of [Katsnelson et al. 1987], as outlined in the preceding subsection, to the setting of Section 11. A full analysis of these calculations will appear elsewhere.

Let P be a nonnegative solution of the GLS equation (11.1). Then Lemma 11.4 guarantees that

$$\Delta_\omega(\omega) = G(\omega)^* P G(\omega) + \rho_\omega(\omega) C_2^* C_2 = H(\omega) P H(\omega)^* + \rho_\omega(\omega) C_1^* C_1. \quad (12.6)$$

Clearly $\Delta_\omega(\omega)$ is positive semidefinite for every point $\omega \in \Omega_+$. Fix such a point and let

$$W_1 = \begin{bmatrix} P^{1/2}G(\omega) \\ \rho_\omega(\omega)^{1/2}C_2 \end{bmatrix} \quad \text{and} \quad W_2 = \begin{bmatrix} -P^{1/2}H(\omega)^* \\ \rho_\omega(\omega)^{1/2}C_1 \end{bmatrix}.$$

Then, in view of (12.6), $V : W_1 x \to W_2 x$ is an isometry from

$$\mathcal{D}_V = \{W_1 x : x \in \mathbb{C}^n\} \subset \mathbb{C}^n \oplus \mathbb{C}^q$$

onto

$$\mathcal{R}_V = \{W_2 x : x \in \mathbb{C}^n\} \subset \mathbb{C}^n \oplus \mathbb{C}^p.$$

Let $k = \dim \mathcal{D}_V = \dim \mathcal{R}_V$. Then $q' = \dim \mathcal{D}_V^\perp = n + q - k$, $p' = \dim \mathcal{R}_V^\perp = n + p - k$, and $k \le n$, with equality if and only if $\Delta_\omega(\omega) > 0$. For the time being we shall assume only that $k \ge 1$ and shall let

$$W_1^\perp \in \mathbb{C}^{(n+q) \times q'} \quad \text{and} \quad W_2^\perp \in \mathbb{C}^{(n+p) \times p'}$$

be isometric matrices whose columns span $\mathcal{D}_V^\perp$ and $\mathcal{R}_V^\perp$, respectively.

The next step is to define the matrix

$$P = \begin{bmatrix} U_{11} & U_{12} & U_{13} \\ U_{21} & U_{22} & U_{23} \\ U_{31} & U_{32} & U_{33} \end{bmatrix} \begin{matrix} \}n \\ \}p \\ \}q' \end{matrix} \ .$$
$$\underbrace{\phantom{U_{11}}}_{n} \ \underbrace{\phantom{U_{12}}}_{q} \ \underbrace{\phantom{U_{13}}}_{p'}$$

by the formulas

$$\begin{bmatrix} U_{11} & U_{12} \\ U_{21} & U_{22} \end{bmatrix} = W_2(W_1^* W_1)^{-1} W_1^*, \qquad [U_{31} \ U_{32}] = (W_1^\perp)^*, \qquad \begin{bmatrix} U_{13} \\ U_{23} \end{bmatrix} = W_2^\perp,$$

and $U_{33} = 0$.

It is readily checked that the formulas $W_1(W_1^* W_1)^{-1} W_1^*$, $W_2(W_2^* W_2)^{-1} W_2^*$ and $W_2(W_1^* W_1)^{-1} W_1^*$ are meaningful single-valued mappings when the inverses are interpreted as inverse images even if $W_1^* W_1 = W_2^* W_2$ is not invertible and furthermore, since

$$I_{n+q} - W_1(W_1^* W_1)^{-1} W_1^* = W_1^\perp (W_1^\perp)^*$$

and

$$I_{n+p} - W_2(W_2^* W_2)^{-1} W_2^* = W_2^\perp (W_2^\perp)^*,$$

that U is unitary. This matrix U corresponds to the special unitary colligation singled out in the last subsection.

The next step is to define the mvf

$$\Sigma(\lambda) = \begin{bmatrix} \Sigma_{11}(\lambda) & \Sigma_{12}(\lambda) \\ \Sigma_{21}(\lambda) & \Sigma_{22}(\lambda) \end{bmatrix} \begin{matrix} \}p \\ \}q' \end{matrix}$$
$$\underbrace{\phantom{\Sigma_{11}}}_{p'} \ \underbrace{\phantom{\Sigma_{12}}}_{q}$$

$$= \begin{bmatrix} U_{23} & U_{22} \\ 0 & U_{32} \end{bmatrix} + \psi_\omega(\lambda) \begin{bmatrix} U_{21} \\ U_{31} \end{bmatrix} (I - \psi_\omega(\lambda) U_{11})^{-1} [U_{13} \ U_{12}], \qquad (12.7)$$

where

$$\psi_\omega(\lambda) = \frac{\delta_\omega(\lambda)}{\rho_\omega(\lambda)}.$$

This mvf is the natural analogue of the characteristic function of U in this setting. It is well defined for $\lambda \in \Omega_+$ since

$$\sigma(\lambda) = \frac{b(\lambda)}{a(\lambda)} \quad \text{and} \quad \psi_\omega(\lambda) = \frac{a(\omega)}{a(\omega)^*} \left(\frac{\sigma(\lambda) - \sigma(\omega)}{1 - \sigma(\lambda)\sigma(\omega)^*} \right)$$

are contractive for $\lambda \in \Omega_+$.

From now on we shall assume that $\Delta_\omega(\omega) > 0$, even though many of the formulas are meaningful without this restriction. Then, with the help of the identity

$$\rho_\omega(\lambda)\Delta_\omega(\omega) + \delta_\omega(\lambda)G(\omega)^*PH(\omega)^* = \rho_\omega(\omega)\Delta_\omega(\lambda), \tag{12.8}$$

it is readily checked that

$$\left(I_n - \psi_\omega(\lambda)U_{11}\right)^{-1} U_{12} = -\rho_\omega(\omega)^{-1/2}\rho_\omega(\lambda)P^{1/2}H(\omega)^*\Delta_\omega(\lambda)^{-1}C_2^*, \tag{12.9}$$

$$U_{21}\left(I_n - \psi_\omega(\lambda)U_{11}\right)^{-1} = \rho_\omega(\omega)^{-1/2}\rho_\omega(\lambda)C_1\Delta_\omega(\lambda)^{-1}G(\omega)^*P^{1/2}, \tag{12.10}$$

$$\left(I_n - \psi_\omega(\lambda)U_{11}\right)^{-1} = I_n - \rho_\omega(\omega)^{-1}\delta_\omega(\lambda)P^{1/2}H(\omega)^*\Delta_\omega(\lambda)^{-1}G(\omega)^*P^{1/2}. \tag{12.11}$$

Therefore, by direct substitution of these last three formulas into the entries of formula (12.7), we obtain

$$\Sigma_{12}(\lambda) = U_{22} + \psi_\omega(\lambda)U_{21}\left(I - \psi_\omega(\lambda)U_{11}\right)^{-1}U_{12} = \rho_\omega(\lambda)C_1\Delta_\omega(\lambda)^{-1}C_2^*,$$

$$\Sigma_{11}(\lambda) = U_{23} + \psi_\omega(\lambda)U_{21}\left(I_n - \psi_\omega(\lambda)U_{11}\right)^{-1}U_{13}$$

$$= [\psi_\omega(\lambda)U_{21}\left(I_n - \psi_\omega(\lambda)U_{11}\right)^{-1} : I_p]\begin{bmatrix} U_{13} \\ U_{23} \end{bmatrix}$$

$$= [\rho_\omega(\omega)^{-1/2}\delta_\omega(\lambda)C_1\Delta_\omega(\lambda)^{-1}G(\omega)^*P^{1/2} : I_p]W_2^\perp,$$

$$\Sigma_{22}(\lambda) = U_{32} + \psi_\omega(\lambda)U_{31}\left(I_n - \psi_\omega(\lambda)U_{11}\right)^{-1}U_{12}$$

$$= [U_{31} \ U_{32}]\begin{bmatrix} \psi_\omega(\lambda)\left(I_n - \psi_\omega(\lambda)U_{11}\right)^{-1}U_{12} \\ I_{q'} \end{bmatrix}$$

$$= (W_1^\perp)^*\begin{bmatrix} -\rho_\omega(\omega)^{-1/2}\delta_\omega(\lambda)P^{1/2}H(\omega)^*\Delta_\omega(\lambda)^{-1}C_2^* \\ I_q \end{bmatrix},$$

$$\Sigma_{21}(\lambda) = \psi_\omega(\lambda)U_{31}\left(I_n - \psi_\omega(\lambda)U_{11}\right)^{-1}U_{13}$$

$$= \psi_\omega(\lambda)(W_1^\perp)^*$$

$$\times \begin{bmatrix} I_n \\ 0 \end{bmatrix}\left(I_n - \rho_\omega(\omega)^{-1}\delta_\omega(\lambda)P^{1/2}H(\omega)^*\Delta_\omega(\lambda)^{-1}G(\omega)^*P^{1/2}\right)[I_n \ 0]W_2^\perp.$$

If rank $P = r$ and $r < n$, we can write the isometric matrices $W_1^\perp$ and $W_2^\perp$ as follows (I wish to thank Vladimir Bolotnikov for suggesting this decomposition, which led to some improved formulas in this subsection):

$$W_1^\perp = \begin{bmatrix} X & Y_1 \\ 0 & Z_1 \end{bmatrix}\begin{matrix}\}n \\ \}q\end{matrix} \quad \text{and} \quad W_2^\perp = \begin{bmatrix} X & Y_2 \\ 0 & Z_2 \end{bmatrix}\begin{matrix}\}n \\ \}p\end{matrix}, \quad (12.12)$$
$$\underbrace{}_{n-r}\ \underbrace{}_{q+r-k} \qquad\qquad \underbrace{}_{n-r}\ \underbrace{}_{p+r-k}$$

where the columns of X are an orthonormal basis for $\ker P$ and hence are independent of ω. We shall keep this notation for $r = n$ also.

THEOREM 12.1. *If $\omega \in \Omega_+$ and rank $P = r$, the block entries in the "characteristic function" $\Sigma(\lambda)$ defined by the formulas (12.7) can be expressed by*

$$\Sigma_{11}(\lambda) = [\ \underbrace{0}_{n-r}\ \ \underbrace{\tilde{\Sigma}_{11}(\lambda)}_{p+r-k}\ ]\}p, \qquad\qquad \Sigma_{12}(\lambda) = \rho_\omega(\lambda)C_1\Delta_\omega(\lambda)^{-1}C_2^*,$$

$$\Sigma_{21}(\lambda) = \begin{bmatrix} \psi_\omega(\lambda)I_{n-r} & 0 \\ 0 & \tilde{\Sigma}_{21}(\lambda) \end{bmatrix}\begin{matrix}\}n-r\\ \}q+r-k\end{matrix}, \quad \Sigma_{22}(\lambda) = \begin{bmatrix} 0 \\ \tilde{\Sigma}_{22}(\lambda) \end{bmatrix}\begin{matrix}\}n-r\\ \}q+r-k\end{matrix},$$
$$\underbrace{}_{p+r-k} \qquad\qquad\qquad \underbrace{}_{q}$$

where

$$\tilde{\Sigma}_{11}(\lambda) = Z_2 + \rho_\omega(\omega)^{-1/2}\delta_\omega(\lambda)C_1\Delta_\omega(\lambda)^{-1}G(\omega)^*P^{1/2}Y_2,$$

$$\tilde{\Sigma}_{21}(\lambda) = \psi_\omega(\lambda)Y_1^*\big(I_n - \rho_\omega(\omega)^{-1}\delta_\omega(\lambda)P^{1/2}H(\omega)^*\Delta_\omega(\lambda)^{-1}G(\omega)^*P^{1/2}\big)Y_2,$$

$$\tilde{\Sigma}_{22}(\lambda) = Z_1^* - \rho_\omega(\omega)^{-1/2}\delta_\omega(\lambda)Y_1^*P^{1/2}H(\omega)^*\Delta_\omega(\lambda)^{-1}C_2^*,$$

and the entries Y_1, Z_1, Y_2, Z_2 from the second block columns of $W_1^\perp$ and $W_2^\perp$ depend upon ω.

Thus, upon writing $\varepsilon(\lambda) \in \mathcal{S}^{p' \times q'}$ in the block form

$$\varepsilon(\lambda) = \begin{bmatrix} \varepsilon_{11}(\lambda) & \varepsilon_{12}(\lambda) \\ \varepsilon_{21}(\lambda) & \varepsilon_{22}(\lambda) \end{bmatrix}\begin{matrix}\}n-r\\ \}p+r-k\end{matrix},$$
$$\underbrace{}_{n-r}\ \underbrace{}_{q+r-k}$$

it is readily checked that

$$\Sigma_{21}(\lambda) + \Sigma_{11}(\lambda)\varepsilon(\lambda)\big(I_{q'} - \Sigma_{21}(\lambda)\varepsilon(\lambda)\big)^{-1}\Sigma_{22}(\lambda)$$
$$= \Sigma_{21}(\lambda) + \tilde{\Sigma}_{11}(\lambda)\tilde{\varepsilon}(\lambda)\big(I_{q'-(n-r)} - \tilde{\Sigma}_{21}(\lambda)\tilde{\varepsilon}(\lambda)\big)^{-1}\tilde{\Sigma}_{22}(\lambda),$$

where

$$\tilde{\varepsilon}(\lambda) = \varepsilon_{22}(\lambda) + \varepsilon_{21}(\lambda)\psi_\omega(\lambda)\big(I_{n-r} - \psi_\omega(\lambda)\varepsilon_{11}(\lambda)\big)^{-1}\varepsilon_{12}(\lambda).$$

This leads easily to the following result:

THEOREM 12.2. *If* rank $P = r$, *then*

$$\left\{ \Sigma_{12} + \Sigma_{11}\varepsilon(I_{q'} - \Sigma_{21}\varepsilon)^{-1}\Sigma_{22} : \varepsilon \in \mathcal{S}^{p' \times q'} \right\}$$
$$= \left\{ \Sigma_{12} + \tilde{\Sigma}_{11}\tilde{\varepsilon}(I_{q''} - \tilde{\Sigma}_{21}\tilde{\varepsilon})^{-1}\tilde{\Sigma}_{22} : \tilde{\varepsilon} \in \mathcal{S}^{p'' \times q''} \right\},$$

where $p'' = p' - n + r$ *and* $q'' = q' - n + r$.

The formulas

$$\left[-H(\omega)P^{1/2} \ \ \rho_\omega(\omega)^{1/2}C_1^* \right] W_2^\perp = 0 \quad \text{and} \quad (W_1^\perp)^* \begin{bmatrix} P^{1/2}G(\omega) \\ \rho_\omega(\omega)^{1/2}C_2 \end{bmatrix} = 0$$

permit additional simplifications.

Upon making use of the identities

$$\delta_\omega(\lambda)G(\omega)^* + \rho_\omega(\lambda)H(\omega) = \rho_\omega(\omega)H(\lambda)$$

and

$$\delta_\omega(\lambda)H(\omega)^* + \rho_\omega(\lambda)G(\omega) = \rho_\omega(\omega)G(\lambda),$$

and then letting ω tend to Ω_0, we obtain:

$$\tilde{\Sigma}_{11}(\lambda) = \left(I_p - \rho_\omega(\lambda)C_1\Delta_\omega(\lambda)^{-1}C_1^* \right)Z_2,$$
$$\tilde{\Sigma}_{21}(\lambda) = \psi_\omega(\lambda)Y_1^*Y_2 + \rho_\omega(\lambda)Z_1^*C_2\Delta_\omega(\lambda)^{-1}C_1^*Z_2,$$
$$\tilde{\Sigma}_{22}(\lambda) = Z_1^*\left(I_q - \rho_\omega(\lambda)C_2\Delta_\omega(\lambda)^{-1}C_2^* \right).$$

If $\Delta_\omega(\omega) > 0$ for $\omega \in \Omega_0$, then $n = r$ and $Y_1 = Y_2 = 0$, and we may choose $Z_1 = I_q$ and $Z_2 = I_p$. Then the formulas in Theorem 12.1 agree with the formula for $\Sigma(\lambda)$ in Theorem 11.7, which was derived by an entirely different strategy.

Other methods, other problems. The analysis presented here is based largely on [Dym 1989a; 1989b; 1994b] and subsequent extensions. There are many other approaches. The books [Ball et al. 1990; Foiaş and Frazho 1990; Helton et al. 1987] reflect three other schools of thought, and each contains an extensive bibliography and notes to the literature. Additional sources may be found in [Dubovoj et al. 1992; Dym 1994a]; see also [Arov 1993; Ivanchenko and Sakhnovich 1994].

References

[Alpay and Dym 1984] D. Alpay and H. Dym, "Hilbert spaces of analytic functions, inverse scattering and operator models, I", *Integral Equations Operator Theory* **7**:5 (1984), 589–641.

[Alpay and Dym 1992] D. Alpay and H. Dym, "On reproducing kernel spaces, the Schur algorithm, and interpolation in a general class of domains", pp. 30–77 in *Operator theory and complex analysis* (Sapporo, 1991), edited by T. Ando and I. Gohberg, Oper. Theory Adv. Appl. **59**, Birkhäuser, Basel, 1992.

[Alpay and Dym 1993a] D. Alpay and H. Dym, "On a new class of reproducing kernel spaces and a new generalization of the Iohvidov laws", *Linear Algebra Appl.* **178** (1993), 109–183.

[Alpay and Dym 1993b] D. Alpay and H. Dym, "On a new class of structured reproducing kernel spaces", *J. Funct. Anal.* **111**:1 (1993), 1–28.

[Alpay and Dym 1996] D. Alpay and H. Dym, "On a new class of realization formulas and their application", *Linear Algebra Appl.* **241/243** (1996), 3–84.

[Arocena 1994] R. Arocena, "Unitary colligations and parametrization formulas", *Ukraïn. Mat. Zh.* **46**:3 (1994), 147–154.

[Aronszajn 1950] N. Aronszajn, "Theory of reproducing kernels", *Trans. Amer. Math. Soc.* **68** (1950), 337–404.

[Arov 1973] D. Z. Arov, "Realization of matrix-valued functions according to Darlington", *Izv. Akad. Nauk SSSR Ser. Mat.* **37** (1973), 1299–1331. In Russian.

[Arov 1993] D. Z. Arov, "The generalized bitangent Carathéodory–Nevanlinna–Pick problem and (j, J_0)-inner matrix functions", *Izv. Ross. Akad. Nauk Ser. Mat.* **57**:1 (1993), 3–32. In Russian; translation in *Russian Acad. Sci. Izv. Math.* **42**:1 (1994), 1–26.

[Arov and Grossman 1983] D. Z. Arov and L. Z. Grossman, "Scattering matrices in the theory of extensions of isometric operators", *Dokl. Akad. Nauk SSSR* **270**:1 (1983), 17–20. In Russian; translation in *Soviet Math. Dokl.* **27**:3 (1983), 518–521.

[Arov and Grossman 1992] D. Z. Arov and L. Z. Grossman, "Scattering matrices in the theory of unitary extension of isometric operators", *Math. Nachr.* **157** (1992), 105–123.

[Ball 1975] J. A. Ball, "Models for noncontractions", *J. Math. Anal. Appl.* **52**:2 (1975), 235–254.

[Ball and Helton 1986] J. A. Ball and J. W. Helton, "Interpolation problems of Pick–Nevanlinna and Loewner types for meromorphic matrix functions: parametrization of the set of all solutions", *Integral Equations Operator Theory* **9**:2 (1986), 155–203.

[Ball et al. 1990] J. A. Ball, I. Gohberg, and L. Rodman, *Interpolation of rational matrix functions*, Operator Theory: Advances and Applications **45**, Birkhäuser Verlag, Basel, 1990.

[Bruinsma 1991] P. Bruinsma, "Degenerate interpolation problems for Nevanlinna pairs", *Indag. Math. (N.S.)* **2**:2 (1991), 179–200.

[de Branges 1963] L. de Branges, "Some Hilbert spaces of analytic functions, I", *Trans. Amer. Math. Soc.* **106** (1963), 445–468.

[de Branges and Rovnyak 1966] L. de Branges and J. Rovnyak, "Canonical models in quantum scattering theory", pp. 295–392 in *Perturbation theory and its application in quantum mechanics* (Madison, 1965), edited by C. H. Wilcox, Wiley, New York, 1966.

[Dubovoj 1984] V. K. Dubovoĭ, "Indefinite metric in Schur's interpolation problem for analytic functions, IV", *Teor. Funktsiĭ Funktsional. Anal. i Prilozhen.* **42** (1984), 46–57. In Russian.

[Dubovoj et al. 1992] V. K. Dubovoj, B. Fritzsche, and B. Kirstein, *Matricial version of the classical Schur problem*, Teubner-Texte zur Mathematik **129**, B. G. Teubner, Stuttgart, 1992.

[Dym 1989a] H. Dym, *J contractive matrix functions, reproducing kernel Hilbert spaces and interpolation*, CBMS Regional Conference Series in Mathematics **71**, Amer. Math. Soc., Providence, 1989.

[Dym 1989b] H. Dym, "On reproducing kernel spaces, *J* unitary matrix functions, interpolation and displacement rank", pp. 173–239 in *The Gohberg anniversary collection*, vol. 2, edited by H. Dym, S. Goldberg, M. A. Kaashoek, P. Lancaster, Oper. Theory Adv. Appl. **41**, Birkhäuser, Basel, 1989.

[Dym 1994a] H. Dym, "Review of 'The commutant lifting approach to interpolation problems' by C. Foias and A. E. Frazho", *Bull. Amer. Math. Soc. (N.S.)* **31** (1994), 125–140.

[Dym 1994b] H. Dym, "Shifts, realizations and interpolation, redux", pp. 182–243 in *Nonselfadjoint operators and related topics* (Beer Sheva, 1992), edited by A. Feintuch and I. Gohberg, Oper. Theory Adv. Appl. **73**, Birkhäuser, Basel, 1994.

[Dym 1996] H. Dym, "More on maximum entropy interpolants and maximum determinant completions of associated Pick matrices", *Integral Equations Operator Theory* **24**:2 (1996), 188–229.

[Dym and Freydin 1997a] H. Dym and B. Freydin, "Bitangential interpolation for upper triangular operators", pp. 143–164 in *Topics in interpolation theory*, edited by H. Dym, B. Fritzsche, V. Katsnelson, and B. Kirstein, Oper. Theory Adv. Appl. **95**, Birkhäuser, Basel, 1997.

[Dym and Freydin 1997b] H. Dym and B. Freydin, "Bitangential interpolation for triangular operators when the Pick operator is strictly positive", pp. 104–142 in *Topics in interpolation theory*, edited by H. Dym, B. Fritzsche, V. Katsnelson, and B. Kirstein, Oper. Theory Adv. Appl. **95**, Birkhäuser, Basel, 1997.

[Foiaş and Frazho 1990] C. Foiaş and A. E. Frazho, *The commutant lifting approach to interpolation problems*, Oper. Theory Adv. Appl. **44**, Birkhäuser, Basel, 1990.

[Helton et al. 1987] J. W. Helton, J. A. Ball, C. R. Johnson, and J. N. Palmer, *Operator theory, analytic functions, matrices, and electrical engineering*, CBMS Regional Conference Series in Mathematics, Amer. Math. Soc., 1987.

[Ivanchenko and Sakhnovich 1994] T. S. Ivanchenko and L. A. Sakhnovich, "An operator approach to the Potapov scheme for the solution of interpolation problems", pp. 48–86 in *Matrix and operator valued functions*, edited by I. Gohberg and L. Sakhnovich, Oper. Theory Adv. Appl. **72**, Birkhäuser, Basel, 1994.

[Kailath 1980] T. Kailath, *Linear systems*, Prentice-Hall Information and System Sciences Series, Prentice-Hall, Englewood Cliffs, NJ, 1980.

[Katsnelson and Kirstein 1997] V. E. Katsnelson and B. Kirstein, "On the theory of matrix-valued functions belonging to the Smirnov class", pp. 283–350 in *Topics in interpolation theory*, edited by H. Dym, B. Fritzsche, V. Katsnelson, and B. Kirstein, Oper. Theory Adv. Appl. **95**, Birkhäuser, Basel, 1997.

[Katsnelson et al. 1987] V. È. Katsnel'son, A. Y. Kheĭfets, and P. M. Yuditskiĭ, "An abstract interpolation problem and the theory of extensions of isometric operators", pp. 83–96, 146 in *Operators in function spaces and problems in function theory*, edited by V. A. Marchenko, Naukova Dumka, Kiev, 1987. In Russian; translation in *Topics in interpolation theory*, edited by Harry Dym et al., Oper. Theory Adv. Appl. **95**, Birkhäuser, Basel, 1997, pp. 283–298.

[Kheifets 1988a] A. Y. Kheĭfets, "The Parseval equality in an abstract problem of interpolation, and the coupling of open systems, I", *Teor. Funktsiĭ Funktsional. Anal. i Prilozhen.* **49** (1988), 112–120, 126. In Russian; translation in *J. Sov. Math.* **49**:4 (1990) 1114–1120.

[Kheifets 1988b] A. Y. Kheĭfets, "The Parseval equality in an abstract problem of interpolation, and the coupling of open systems, II", *Teor. Funktsiĭ Funktsional. Anal. i Prilozhen.* **50** (1988), 98–103, iv. In Russian; translation in *J. Sov. Math.* **49**:6 (1990) 1307–1310.

[Kheifets 1990] A. Y. Kheĭfets, "A generalized bitangential Schur–Nevanlinna–Pick problem and its related Parseval equality", *Teor. Funktsiĭ Funktsional. Anal. i Prilozhen.* **54** (1990), 89–96. In Russian; translation in *J. Sov. Math.* **58**:4 (1992), 358–364.

[Kheifets 1998] A. Y. Kheifets, "Abstract interpolation problem and applications", pp. 351–379 in *Holomorphic Spaces*, edited by S. Axler et al., Math. Sci. Res. Inst. Publications **33**, Cambridge Univ. Press, New York, 1998.

[Kheifets and Yuditskii 1994] A. Y. Kheĭfets and P. M. Yuditskiĭ, "An analysis and extension of V. P. Potapov's approach to interpolation problems with applications to the generalized bi-tangential Schur–Nevanlinna–Pick problem and J-inner-outer factorization", pp. 133–161 in *Matrix and operator valued functions*, edited by I. Gohberg and L. A. Sakhnovich, Oper. Theory Adv. Appl. **72**, Birkhäuser, Basel, 1994.

[Lev-Ari and Kailath 1986] H. Lev-Ari and T. Kailath, "Triangular factorization of structured Hermitian matrices", pp. 301–324 in *I. Schur methods in operator theory and signal processing*, edited by I. Gohberg, Oper. Theory Adv. Appl. **18**, Birkhäuser, Basel, 1986.

[Nudelman 1993] A. A. Nudelman, "Some generalizations of classical interpolation problems", pp. 171–188 in *Operator extensions, interpolation of functions and related topics* (Timişoara, 1992), edited by D. T. A. Gheondea and F.-H. Vascilescu, Oper. Theory Adv. Appl. **61**, Birkhäuser, Basel, 1993.

[Redheffer 1960] R. M. Redheffer, "On a certain linear fractional transformation", *J. Math. and Phys.* **39** (1960), 269–286.

[Rovnyak 1968] J. Rovnyak, "Characterization of spaces $\mathcal{H}(M)$", 1968. Available at http://wsrv.clas.virginia.edu/~jlr5m/papers/papers.html.

HARRY DYM
DEPARTMENT OF THEORETICAL MATHEMATICS
THE WEIZMANN INSTITUTE OF SCIENCE
REHOVOT 76100
ISRAEL

Holomorphic Spaces
MSRI Publications
Volume **33**, 1998

Reproducing Kernel Pontryagin Spaces

DANIEL ALPAY, AAD DIJKSMA, JAMES ROVNYAK,
AND HENDRIK S. V. DE SNOO

ABSTRACT. The theory of reproducing kernel Pontryagin spaces is surveyed. A new proof is given of an abstract theorem that constructs contraction operators on Pontryagin spaces from densely defined relations. The theory is illustrated with examples from the theory of generalized Schur functions.

1. Introduction

We present here the main results of the theory of reproducing kernel Pontryagin spaces [Schwartz 1964; Sorjonen 1975] including some recent improvements [Alpay et al. 1997]. The paper is expository and is intended for nonspecialists in the indefinite theory. We presume knowledge of the Hilbert space case, that is, Aronszajn's theory [1950] of reproducing kernel Hilbert spaces. The main point is that much of the experience with the Hilbert space theory is transferable to Pontryagin spaces.

Section 2 presents background from operator theory. A key result here is a theorem to construct contraction operators by specifying their action on dense sets; we give a new proof that reduces the result to the isometric case. The main results on reproducing kernels are in Section 3. Examples in Section 4 illustrate the theory with kernels of the form $(1 - S(z)\overline{S(w)})/(1 - z\bar{w})$ where $S(z)$ is a generalized Schur function.

Scalar-valued functions are assumed throughout. See [Alpay et al. 1997] for the extension to vector-valued functions and a detailed account of the theory of generalized Schur functions and associated colligations and reproducing kernel Pontryagin spaces.

1991 *Mathematics Subject Classification.* Primary 46C20, 46E22. Secondary 47B50.

Key words and phrases. Pontryagin space, reproducing kernel, negative squares, Schur class, contraction operator.

J. Rovnyak was supported by the National Science Foundation under DMS–9501304 and during a stay at the Mathematical Sciences Research Institute (Berkeley) under DMS–9022140.

2. Contraction operators on Pontryagin spaces

Inner products are assumed to be linear and symmetric and defined on a complex vector space. Orthogonality and direct sum are defined for any inner product space as in the Hilbert space case. The *antispace* of an inner product space $(\mathfrak{H}, \langle \cdot, \cdot \rangle_{\mathfrak{H}})$ is $(\mathfrak{H}, -\langle \cdot, \cdot \rangle_{\mathfrak{H}})$. We write $\mathfrak{H}$ for $(\mathfrak{H}, \langle \cdot, \cdot \rangle_{\mathfrak{H}})$ when the inner product is understood.

Linear and symmetric inner product spaces are too general, and strong results can only be proved in special cases. A *Pontryagin space* is an inner product space $\mathfrak{H}$ which can be written as the orthogonal direct sum

$$\mathfrak{H} = \mathfrak{H}_+ \oplus \mathfrak{H}_- \tag{2-1}$$

of a Hilbert space $\mathfrak{H}_+$ and the antispace $\mathfrak{H}_-$ of a finite-dimensional Hilbert space. In a natural way, $\mathfrak{H}$ becomes a Hilbert space when $\mathfrak{H}_-$ is replaced by its antispace in (2–1). Such decompositions are not unique, but any two norms obtained in this way turn out to be equivalent. Every Pontryagin space thus has a unique *strong topology*. We write $\mathfrak{L}(\mathfrak{H}, \mathfrak{K})$ for the space of continuous operators from $\mathfrak{H}$ into $\mathfrak{K}$ and $\mathfrak{L}(\mathfrak{H}) = \mathfrak{L}(\mathfrak{H}, \mathfrak{H})$. The dimensions of $\mathfrak{H}_\pm$ in (2–1) are independent of the choice of decomposition and called the *positive and negative indices* of $\mathfrak{H}$. The negative index κ of $\mathfrak{H}$ is the maximum dimension of a subspace $\mathfrak{H}_-$ which is the antispace of a Hilbert space in the inner product of $\mathfrak{H}$, and every such κ-dimensional subspace occurs in a decomposition (2–1).

LEMMA 2.1 (GRAM MATRICES). *Let* $g_1, \ldots, g_n$ *be vectors in a complex vector space* $\mathfrak{H}$ *with a linear and symmetric inner product* $\langle \cdot, \cdot \rangle_{\mathfrak{H}}$. *Then the number of negative eigenvalues of the Gram matrix* $G = \left(\langle g_j, g_i \rangle_{\mathfrak{H}} \right)_{i,j=1}^n$ *is equal to the maximum dimension of a subspace* $\mathfrak{N}$ *of the span of* $g_1, \ldots, g_n$ *which is the antispace of a Hilbert space in the inner product of* $\mathfrak{H}$.

A proof is given in [Alpay et al. 1997, Lemma 1.1.1′].

The *adjoint* of an operator $A \in \mathfrak{L}(\mathfrak{H}, \mathfrak{K})$ is the unique $A^* \in \mathfrak{L}(\mathfrak{K}, \mathfrak{H})$ such that $\langle Ah, k \rangle_{\mathfrak{K}} = \langle h, A^*k \rangle_{\mathfrak{H}}$ for all $h \in \mathfrak{H}$ and $k \in \mathfrak{K}$. An operator in $\mathfrak{L}(\mathfrak{H})$ is *selfadjoint* if it is equal to its adjoint and a *projection* if it is selfadjoint and idempotent. As in the Hilbert space case, the set of selfadjoint operators is partially ordered by writing $A \leq B$ to mean that $\langle Af, f \rangle_{\mathfrak{H}} \leq \langle Bf, f \rangle_{\mathfrak{H}}$ for all $f \in \mathfrak{H}$ whenever $A, B \in \mathfrak{L}(\mathfrak{H})$ are selfadjoint operators. Closed subspaces $\mathfrak{M}$ of a Pontryagin space $\mathfrak{H}$ which are themselves Pontryagin spaces in the inner product of $\mathfrak{H}$ behave much like closed subspaces of Hilbert spaces. They coincide with ranges of projections, and the projection theorem holds: $\mathfrak{H} = \mathfrak{M} \oplus \mathfrak{M}^\perp$. Such subspaces are called *regular*.

Identity operators are written as 1. We call $A \in \mathfrak{L}(\mathfrak{H}, \mathfrak{K})$ an *isometry* if $A^*A = 1$ and *unitary* if both A and A^* are isometric. More generally, $A \in \mathfrak{L}(\mathfrak{H}, \mathfrak{K})$ is a *partial isometry* if $AA^*A = A$. The properties of partial isometries are given in [Dritschel and Rovnyak 1996, Theorems 1.7, 1.8]. If $A \in \mathfrak{L}(\mathfrak{H}, \mathfrak{K})$ is a partial isometry, there exist regular subspaces $\mathfrak{M}$ of $\mathfrak{H}$ and $\mathfrak{N}$ of $\mathfrak{K}$ such that A maps $\mathfrak{M}$ isometrically onto $\mathfrak{N}$ and $\mathfrak{M}^{\perp}$ to the zero subspace. Conversely, any such operator $A \in \mathfrak{L}(\mathfrak{H}, \mathfrak{K})$ is a partial isometry, and A^*A and AA^* are projections onto $\mathfrak{M}$ and $\mathfrak{N}$. Unitary operators are isomorphisms and make the domain and range spaces abstractly indistinguishable.

We say that $A \in \mathfrak{L}(\mathfrak{H}, \mathfrak{K})$ is a *contraction* if

$$\langle Ah, Ah \rangle_{\mathfrak{K}} \leq \langle h, h \rangle_{\mathfrak{H}}$$

for all $h \in \mathfrak{H}$. If both A and A^* are contractions, A is called a *bicontraction*. If $\mathfrak{H}$ and $\mathfrak{K}$ are Pontryagin spaces having the same negative index, every contraction $A \in \mathfrak{L}(\mathfrak{H}, \mathfrak{K})$ is a bicontraction [Dritschel and Rovnyak 1996, Corollary 2.5]. Without the index condition, this is not true (consider the identity operator acting on a Hilbert space to its antispace). See also [Azizov and Iokhvidov 1986; Bognár 1974; Iohvidov et al. 1982] for the basic properties of these notions.

In concrete situations, we define contraction operators by specifying their graphs. Typically, to start we have only partial information and have to deal with sets that are perhaps not graphs, that is, they may contain elements of the form $(0, k)$ with nontrivial $k \in \mathfrak{K}$. A *linear relation* from $\mathfrak{H}$ to $\mathfrak{K}$ is a subspace $\mathbf{R}$ of $\mathfrak{H} \times \mathfrak{K}$. Its *domain* is the set of first members of pairs in $\mathbf{R}$, and the *range* is the set of second members. We call $\mathbf{R}$ *contractive* if $\langle k, k \rangle_{\mathfrak{K}} \leq \langle h, h \rangle_{\mathfrak{H}}$ for all (h, k) in $\mathbf{R}$, and *isometric* if equality holds in this inequality for all pairs.

The following result appears in [Alpay et al. 1997] with a proof based on a method of T. Ya. Azizov. An alternative proof is given below.

THEOREM 2.2. *Let $\mathfrak{H}$ and $\mathfrak{K}$ be Pontryagin spaces having the same negative index. Let $\mathbf{R}$ be a densely defined and contractive linear relation in $\mathfrak{H} \times \mathfrak{K}$. Then the closure of $\mathbf{R}$ is the graph of a contraction $T \in \mathfrak{L}(\mathfrak{H}, \mathfrak{K})$.*

The result is often applied when $\mathbf{R}$ is isometric, and then T is an isometry.

PROOF. *Step* 1: *Reduction to the isometric case.* Assume the result is known for densely defined isometric linear relations. Let $\mathbf{R}$ be any contractive linear relation from $\mathfrak{H}$ to $\mathfrak{K}$ having dense domain. Define $\mathfrak{A}_0$ to be $\mathbf{R}$ as a vector space but considered in the inner product

$$\langle u_1, u_2 \rangle_{\mathfrak{A}_0} = \langle f_1, f_2 \rangle_{\mathfrak{H}} - \langle g_1, g_2 \rangle_{\mathfrak{K}}$$

for any $u_1 = (f_1, g_1)$ and $u_2 = (f_2, g_2)$ in $\mathbf{R}$. The inner product is nonnegative since $\mathbf{R}$ is contractive. Let $\mathfrak{N}$ be the set of elements u in $\mathfrak{A}_0$ such that $\langle u, v \rangle_{\mathfrak{A}_0} = 0$ for all v in $\mathfrak{A}_0$. A strictly positive inner product is induced in $\mathfrak{A}_0/\mathfrak{N}$, and this space can be completed to a Hilbert space $\mathfrak{A}$. If $u \in \mathfrak{A}_0$, let $[u] = u + \mathfrak{N}$ be the corresponding coset in $\mathfrak{A}$. Define an isometric linear relation from $\mathfrak{H}$ to $\mathfrak{K} \oplus \mathfrak{A}$ by

$$\mathbf{V} = \left\{ \left(f, \begin{pmatrix} g \\ [(f, g)] \end{pmatrix} \right) : (f, g) \in \mathbf{R} \right\}.$$

We verify the isometric property: for any $(f, g) \in \mathbf{R}$,

$$\left\langle \left(\begin{array}{c} g \\ [(f,g)] \end{array} \right), \left(\begin{array}{c} g \\ [(f,g)] \end{array} \right) \right\rangle_{\mathfrak{K} \oplus \mathfrak{A}} = \langle g, g \rangle_{\mathfrak{K}} + \langle f, f \rangle_{\mathfrak{H}} - \langle g, g \rangle_{\mathfrak{K}} = \langle f, f \rangle_{\mathfrak{H}}.$$

If the result is known in the isometric case, $\overline{\mathbf{V}}$ is the graph of an isometry

$$V = \left(\begin{array}{c} T \\ A \end{array} \right) \in \mathfrak{L}(\mathfrak{H}, \mathfrak{K} \oplus \mathfrak{A}).$$

We show that $\overline{\mathbf{R}}$ is the graph of T. If $(f, g) \in \overline{\mathbf{R}}$, then $(f_n, g_n) \to (f, g)$ for some sequence $\{(f_n, g_n)\}_1^\infty$ in $\mathbf{R}$. By the continuity of V,

$$\left(\begin{array}{c} g_n \\ [(f_n, g_n)] \end{array} \right) = V f_n \to V f = \left(\begin{array}{c} Tf \\ Af \end{array} \right);$$

hence $g_n \to Tf$ and so $(f, g) = (f, Tf)$ is in the graph of T. Conversely, if $(f, g) = (f, Tf)$ is in the graph of T, then (f, Vf) is in $\overline{\mathbf{V}}$. Hence there is a sequence $\{(f_n, g_n)\}_1^\infty$ in $\mathbf{R}$ such that

$$\left(f_n, \left(\begin{array}{c} g_n \\ [(f_n, g_n)] \end{array} \right) \right) \to \left(f, \left(\begin{array}{c} Tf \\ Af \end{array} \right) \right).$$

Then $f_n \to f$ and $g_n \to Tf$, so $(f, g) = (f, Tf)$ is in $\overline{\mathbf{R}}$. Thus the result holds in general if it is true in the isometric case.

Step 2: *Proof in the isometric case.* The argument uses Pontryagin's Theorem [Dritschel and Rovnyak 1996, Theorem 2.9]: For any dense subspace $\mathfrak{H}_0$ of a Pontryagin space $\mathfrak{H}$, there is a decomposition (2–1) such that $\mathfrak{H}_- \subseteq \mathfrak{H}_0$.

Assume that $\mathbf{R}$ is an isometric linear relation from $\mathfrak{H}$ to $\mathfrak{K}$ with dense domain. Let $\mathfrak{H}$ and $\mathfrak{K}$ have negative index κ. By the polarization identity,

$$\langle f_1, f_2 \rangle_{\mathfrak{H}} = \langle g_1, g_2 \rangle_{\mathfrak{K}} \quad \text{for } (f_1, g_1), (f_2, g_2) \in \mathbf{R}. \tag{2–2}$$

By Pontryagin's theorem, we can choose a fundamental decomposition (2–1) such that $\mathfrak{H}_-$ is contained in the domain of $\mathbf{R}$. Choose $f_1, \ldots, f_\kappa \in \mathfrak{H}_-$ such that $\langle f_j, f_i \rangle_{\mathfrak{H}} = -\delta_{ij}$ for $i, j = 1, \ldots, \kappa$, and let $(f_1, g_1), \ldots, (f_\kappa, g_\kappa)$ be corresponding elements of $\mathbf{R}$. By (2–2), $\langle g_j, g_i \rangle_{\mathfrak{K}} = -\delta_{ij}$ for $i, j = 1, \ldots, \kappa$. Hence $g_1, \ldots, g_\kappa$ span a κ-dimensional subspace $\mathfrak{K}_-$ which is the antispace of a Hilbert space and part of a fundamental decomposition $\mathfrak{K} = \mathfrak{K}_+ \oplus \mathfrak{K}_-$.

We show that $\mathbf{R}$ is a graph. If $(0, g)$ belongs to $\mathbf{R}$, then by (2–2), $\langle g, v \rangle_{\mathfrak{K}} = 0$ for all v in the range of $\mathbf{R}$ and hence for all v in $\mathfrak{K}_-$. Thus $g \in \mathfrak{K}_+$. Since g itself is in the range of $\mathbf{R}$, we have $\langle g, g \rangle_{\mathfrak{K}} = 0$ and so $g = 0$. It follows that $\mathbf{R}$ is the graph of a densely defined operator T_0 from $\mathfrak{H}$ into $\mathfrak{K}$.

The restriction of T_0 to $\mathfrak{H}_-$ maps $\mathfrak{H}_-$ isometrically onto $\mathfrak{K}_-$. The restriction of T_0 to $\mathfrak{H}_+$ is a densely defined isometry from $\mathfrak{H}_+$ to $\mathfrak{K}_+$. Since these are Hilbert spaces, T_0 has an extension by continuity to an isometry $T \in \mathfrak{L}(\mathfrak{H}, \mathfrak{K})$. Then $\overline{\mathbf{R}}$ is the graph of T. $\square$

The hypothesis in Theorem 2.2 that $\mathfrak{H}$ and $\mathfrak{K}$ have the same negative index is essential. Examples show what can go wrong when this condition is not met.

EXAMPLE 2.3. Let $\mathfrak{H} = \mathbb{C}$ be the complex numbers in the Euclidean norm. Let $\mathfrak{K}$ be $\mathbb{C}^3$ in the inner product

$$\langle a, b \rangle_{\mathfrak{K}} = a_1 \bar{b}_1 + a_2 \bar{b}_2 - a_3 \bar{b}_3,$$

where $a = (a_1, a_2, a_3)$ and $b = (b_1, b_2, b_3)$. Let $\mathbf{R}$ be the set of pairs $(a, (a, c, c))$ with $a, c \in \mathbb{C}$. Then $\mathbf{R}$ is an isometric linear relation with domain $\mathfrak{H}$, but $\mathbf{R}$ contains elements $(0, g)$ with $g \neq 0$ and hence is not the graph of an operator.

EXAMPLE 2.4. Let $\mathfrak{K}$ be an infinite-dimensional Pontryagin space of negative index 1. Choose a nonclosed subspace $\mathfrak{H}$ of $\mathfrak{K}$ which is a Hilbert space in the inner product of $\mathfrak{K}$, that is, $\langle f, g \rangle_{\mathfrak{H}} = \langle f, g \rangle_{\mathfrak{K}}$ for $f, g \in \mathfrak{H}$. Such subspaces exist [Azizov and Iokhvidov 1986, Example 4.12, p. 27; Bognár 1974, Example 6.5, p. 111; Dritschel and Rovnyak 1996, Supplement]. The inclusion mapping V from $\mathfrak{H}$ into $\mathfrak{K}$ is thus an everywhere defined linear operator which preserves inner products, that is,

$$\langle Vf, Vg \rangle_{\mathfrak{K}} = \langle f, g \rangle_{\mathfrak{H}} \quad \text{for } f, g \in \mathfrak{H},$$

but V is not continuous relative to the strong topologies of $\mathfrak{H}$ and $\mathfrak{K}$ (if it were continuous, its range would be closed; see [Dritschel and Rovnyak 1996, p. 127]).

3. Reproducing Kernel Pontryagin Spaces

Let Ω be a nonempty set. By a *kernel* we mean a (complex-valued) function $K(s, t)$ on $\Omega \times \Omega$. Such a function is said to be *Hermitian* if $K(t, s) = \overline{K(s, t)}$ for $s, t \in \Omega$. In this case,

$$(K(s_k, s_j))_{j,k=1}^n = \begin{pmatrix} K(s_1, s_1) & K(s_2, s_1) & \ldots & K(s_n, s_1) \\ K(s_1, s_2) & K(s_2, s_2) & \ldots & K(s_n, s_2) \\ & & \ldots & \\ K(s_1, s_n) & K(s_2, s_n) & \ldots & K(s_n, s_n) \end{pmatrix}$$

is a selfadjoint matrix for any points $s_1, \ldots, s_n$ in Ω ($n = 1, 2, 3, \ldots$). We say that $K(s, t)$ has κ *negative squares*, κ a nonnegative integer, if every matrix of this form has at most κ negative eigenvalues and at least one such matrix has exactly κ negative eigenvalues (counted according to multiplicity). If this condition is satisfied with $\kappa = 0$, we call the kernel *nonnegative*. We remark that changing the matrix to $(K(s_k, s_j) c_k \bar{c}_j)_{j,k=1}^n$ for any complex numbers $c_1, \ldots, c_n$ cannot increase the number of negative eigenvalues, and so the definition of κ negative squares used here is equivalent to that of [Alpay et al. 1997] in the scalar case.

Let $\mathfrak{H}$ be a Pontryagin space whose elements are functions on Ω. Linear operations are understood to be defined pointwise, hence all evaluation mappings $E(s) : f \to f(s)$, for $s \in \Omega$, are linear functionals on $\mathfrak{H}$. By a *reproducing kernel* for $\mathfrak{H}$ we mean a function $K(s,t)$ on $\Omega \times \Omega$ such that for each $s \in \Omega$ the function $K(s, \cdot)$ belongs to $\mathfrak{H}$, and

$$\langle f(\cdot), K(s, \cdot) \rangle_{\mathfrak{H}} = f(s)$$

for every function f in $\mathfrak{H}$. As in the Hilbert space case, it is an easy fact that a reproducing kernel exists if and only if all evaluation mappings are continuous. In this case, $E(s) \in \mathfrak{L}(\mathfrak{H}, \mathbb{C})$ for every $s \in \Omega$. The reproducing kernel is unique and given by

$$K(s,t) = E(t)E(s)^* \quad \text{for } s,t \in \Omega,$$

where $E(s)^* \in \mathfrak{L}(\mathbb{C}, \mathfrak{H})$ is the adjoint of the evaluation mapping $E(s) \in \mathfrak{L}(\mathfrak{H}, \mathbb{C})$ for any fixed $s \in \Omega$. (In this equation, the right side is an operator on $\mathbb{C}$ to $\mathbb{C}$ and hence identified with a complex number. The equation can also be read to say that $K(s, \cdot)$ is the element of $\mathfrak{H}$ obtained from the element 1 of $\mathbb{C}$ under the action of $E(s)^*$.) Clearly, a reproducing kernel $K(s,t)$ is Hermitian. Applying Lemma 2.1 to the matrices

$$\left(\langle K(s_k, \cdot), K(s_j, \cdot) \rangle_{\mathfrak{H}} \right)_{j,k=1}^n = \left(K(s_k, s_j) \right)_{j,k=1}^n, \tag{3-1}$$

we see that $K(s,t)$ has at most κ negative squares, where κ is the negative index of $\mathfrak{H}$. A little further argument, using Pontryagin's Theorem on dense sets cited in the proof of Theorem 2.2, shows that $K(s,t)$ has exactly κ negative squares. Summarizing these properties, we have:

THEOREM 3.1. *A Pontryagin space $\mathfrak{H}$ of functions on Ω admits a reproducing kernel $K(s,t)$ if and only if all evaluation mappings are continuous. In this case, $K(s,t)$ is unique, and it is a Hermitian kernel having κ negative squares, where κ is the negative index of $\mathfrak{H}$.*

A converse result holds.

THEOREM 3.2. *If $K(s,t)$ is a Hermitian kernel on $\Omega \times \Omega$ having κ negative squares, there is a unique Pontryagin space $\mathfrak{H}$ of functions on Ω having $K(s,t)$ as reproducing kernel.*

PROOF. Let $\mathfrak{H}_0$ be the span of functions $K(s, \cdot)$ with $s \in \Omega$. By adding zero terms if necessary, any two functions f, g in $\mathfrak{H}_0$ can be represented in the form

$$f(\cdot) = \sum_{k=1}^n a_k K(s_k, \cdot), \qquad g(\cdot) = \sum_{j=1}^n b_j K(s_j, \cdot), \tag{3-2}$$

using the same points $s_1, \ldots, s_n$ in Ω. Define an inner product by setting

$$\langle f, g \rangle_{\mathfrak{H}_0} = \sum_{j,k=1}^n a_k \bar{b}_j K(s_k, s_j).$$

Since $\langle f, g \rangle_{\mathfrak{H}_0} = \sum_{j=1}^n \bar{b}_j f(s_j) = \sum_{k=1}^n a_k \overline{g(s_k)}$, the inner product is well defined, linear, and symmetric. The identity $\langle f(\cdot), K(s, \cdot) \rangle_{\mathfrak{H}_0} = f(s)$ holds for all $s \in \Omega$ and all f in $\mathfrak{H}_0$.

Lemma 2.1, applied to the inner product space $\mathfrak{H}_0$ and the Gram matrices (3–1), implies that $\mathfrak{H}_0$ contains a subspace $\mathfrak{H}_-$ which is the antispace of a Hilbert space, and no subspace of higher dimension has this property. A reproducing kernel for $\mathfrak{H}_-$ can be exhibited in terms of any functions $u_1, \ldots, u_\kappa$ in $\mathfrak{H}_-$ such that

$$\langle u_k, u_j \rangle_{\mathfrak{H}_0} = -\delta_{jk} \quad \text{for } j, k = 1, \ldots, \kappa,$$

namely, $K_-(s, t) = -\sum_{l=1}^\kappa u_l(t) \overline{u_l(s)}$. Define

$$K_+(s, t) = K(s, t) - K_-(s, t)$$

on $\Omega \times \Omega$, and let $\mathfrak{H}_{0+}$ be the span of all functions $K_+(s, \cdot)$ with $s \in \Omega$. For any $s \in \Omega$ and $j = 1, \ldots, \kappa$,

$$\langle K_+(s, \cdot), u_j(\cdot) \rangle_{\mathfrak{H}_0} = \langle K(s, \cdot), u_j(\cdot) \rangle_{\mathfrak{H}_0} + \sum_{l=1}^\kappa \langle u_l(\cdot) \overline{u_l(s)}, u_j(\cdot) \rangle_{\mathfrak{H}_0}$$
$$= \overline{u_j(s)} - \overline{u_j(s)} = 0,$$

and hence $\mathfrak{H}_{0+} \perp \mathfrak{H}_-$ in the inner product of $\mathfrak{H}_0$. Next consider two functions f, g in $\mathfrak{H}_{0+}$. Representing f, g in the form (3–2) with $K(s, t)$ replaced by $K_+(s, t)$, we obtain

$$\langle f, g \rangle_{\mathfrak{H}_0} = \left\langle \sum_{k=1}^n a_k K(s_k, \cdot) - \sum_{k=1}^n a_k K_-(s_k, \cdot), \sum_{j=1}^n b_j K(s_j, \cdot) - \sum_{j=1}^n b_j K_-(s_j, \cdot) \right\rangle_{\mathfrak{H}_0}$$
$$= \sum_{j,k=1}^n a_k \bar{b}_j \{ K(s_k, s_j) - K_-(s_k, s_j) - K_-(s_k, s_j) + K_-(s_k, s_j) \}$$
$$= \sum_{j,k=1}^n a_k \bar{b}_j K_+(s_k, s_j).$$

It follows that $K_+(s, t)$ is nonnegative, since otherwise we can find a $(\kappa + 1)$-dimensional subspace of $\mathfrak{H}_0$ which is the antispace of a Hilbert space in the inner product of $\mathfrak{H}_0$. For if $K_+(s, t)$ is not nonnegative, another application of Lemma 2.1 implies that $\mathfrak{H}_{0+}$ contains an element $u_{\kappa+1}$ such that $\langle u_{\kappa+1}, u_{\kappa+1} \rangle_{\mathfrak{H}_0} < 0$, and then the span of $u_1, \ldots, u_\kappa, u_{\kappa+1}$ has the stated properties.

The proof is completed using the known Hilbert space case [Aronszajn 1950]. Complete $\mathfrak{H}_{0+}$ to a functional Hilbert space $\mathfrak{H}_+$ with reproducing kernel $K_+(s, t)$. Define a Pontryagin space $\mathfrak{H}$ of functions on Ω by (2–1) with $\mathfrak{H}_\pm$ as above. We easily verify that $\mathfrak{H}$ has reproducing kernel $K(s, t)$. This establishes existence.

To prove uniqueness, consider a second Pontryagin space $\mathfrak{H}'$ of functions on Ω with reproducing kernel $K(s, t)$. Then $\mathfrak{H}'$ contains $\mathfrak{H}_0$ and $\mathfrak{H}_-$ isometrically, and so $\mathfrak{H}_{0+}$ is contained isometrically in $\mathfrak{H}' \ominus \mathfrak{H}_-$. Since $\mathfrak{H}' \ominus \mathfrak{H}_-$ and $\mathfrak{H}_+$ are two Hilbert space with reproducing kernel $K_+(s,t)$, they are equal isometrically by the Hilbert space version of the uniqueness result. Hence $\mathfrak{H}$ and $\mathfrak{H}'$ are equal isometrically. $\qquad\qquad\qquad\qquad\qquad\qquad\qquad\qquad\qquad\qquad\qquad\qquad\qquad\square$

THEOREM 3.3. *Let* $K(s,t) = K_+(s,t) + K_-(s,t)$ *on* $\Omega \times \Omega$, *where* $K_+(s,t)$ *is nonnegative and* $K_-(s,t) = -\sum_{l=1}^{m} u_l(t)\overline{u_l(s)}$ *for some linearly independent functions* $u_1, \ldots, u_m$ *on* Ω. *Then* $K(s,t)$ *has* κ *negative squares where* $\kappa \leq m$, *and* $\kappa = m$ *if no nonzero function in the span of* $u_1, \ldots, u_m$ *belongs to the Hilbert space with reproducing kernel* $K_+(s,t)$.

PROOF. Consider points $s_1, \ldots, s_n$ in Ω. Write $C = A + B$, where

$$C = (K(s_k, s_j))_{j,k=1}^n, \qquad A = (K_+(s_k, s_j))_{j,k=1}^n.$$

Then $B = -FF^*$ where $F = (u_k(s_j))_{n \times m}$ has rank at most m, and so B has at most m negative eigenvalues. Now B and C are Gram matrices for the standard basis of $\mathbb{C}^n$ relative to the inner products $\langle a, b \rangle_C = \langle Ca, b \rangle_{\mathbb{C}^n}$ and $\langle a, b \rangle_B = \langle Ba, b \rangle_{\mathbb{C}^n}$. Since $C = A + B \geq B$, $\langle u, u \rangle_C \geq \langle u, u \rangle_B$ for all $u \in \mathbb{C}^n$. Hence, by Lemma 2.1, C has at most m negative eigenvalues. Thus $K(s,t)$ has $\kappa \leq m$ negative squares.

Suppose that no nonzero function in the span $\mathfrak{H}_-$ of $u_1, \ldots, u_m$ belongs to the Hilbert space $\mathfrak{H}_+$ with reproducing kernel $K_+(s,t)$. Then a Pontryagin space having reproducing kernel $K(s,t) = K_+(s,t) + K_-(s,t)$ is defined by forming a direct sum (2–1) with the given spaces $\mathfrak{H}_-$ and $\mathfrak{H}_+$. Since $\mathfrak{H}$ has negative index m by construction, $K(s,t)$ has m negative squares by Theorem 3.1. $\qquad\square$

THEOREM 3.4. *Let* $\mathfrak{H}$ *be a Pontryagin space of functions on* Ω *with reproducing kernel* $K(s,t)$. *If* $\mathfrak{G}$ *is a closed subspace which is a Pontryagin space in the inner product of* $\mathfrak{H}$, *then* $\mathfrak{G}$ *has a reproducing kernel* $K_{\mathfrak{G}}(s,t)$ *such that, for each* $s \in \Omega$, $K_{\mathfrak{G}}(s, \cdot)$ *is the projection of* $K(s, \cdot)$ *into* $\mathfrak{G}$.

The proof is the same as in the Hilbert space case and omitted. We remark that a subspace $\mathfrak{G}$ as in Theorem 3.4 is the range of a projection [Dritschel and Rovnyak 1996, Theorem 1.3].

Recall the result on restrictions of reproducing kernels from the Hilbert space theory [Aronszajn 1950, p. 351]. Let $K(s,t)$ be the reproducing kernel for a Hilbert space $\mathfrak{H}$ of functions on Ω. Let $K_1(s,t)$ be the restriction of $K(s,t)$ to $\Omega_1 \times \Omega_1$ for some subset Ω_1 of Ω. Then $K_1(s,t)$ is the reproducing kernel for the Hilbert space $\mathfrak{H}_1$ whose elements consist of all restrictions h_1 of functions h in $\mathfrak{H}$ with

$$\|h_1\|_{\mathfrak{H}_1} = \inf \left\{ \|h\|_{\mathfrak{H}} : h \in \mathfrak{H}, \ h_1 = h|_{\Omega_1} \right\}.$$

In general, this fails for Pontryagin spaces.

EXAMPLE 3.5. The kernel $K(w,z) = 1 - z\bar{w}$ has one negative square on $\mathbb{C} \times \mathbb{C}$ by Theorem 3.3. By Theorem 3.2 it is the reproducing kernel for the two-dimensional Pontryagin space $\mathfrak{H}$ spanned by the functions $h_0(z) = 1$ and $h_1(z) =$

z, with

$$\langle h_0, h_0 \rangle_{\mathfrak{H}} = 1, \quad \langle h_1, h_1 \rangle_{\mathfrak{H}} = -1, \quad \langle h_0, h_1 \rangle_{\mathfrak{H}} = 0.$$

If $\Omega_1 = \{1\}$, the restriction of $K(w, z)$ to $\Omega_1 \times \Omega_1$ is $K_1(w, z) \equiv 0$. The set $\mathfrak{H}_1$ of restrictions of functions in $\mathfrak{H}$ to Ω_1 is a one-dimensional vector space, so there is no way to make $\mathfrak{H}_1$ a Pontryagin space with reproducing kernel $K_1(w, z)$. Note the decrease in number of negative squares: $K(w, z)$ has $\kappa = 1$ negative square, and $K_1(w, z)$ has $\kappa = 0$ negative squares.

THEOREM 3.6. *Let $\mathfrak{H}$ be a Pontryagin space of functions on Ω with reproducing kernel $K(s, t)$. If Ω_1 is a subset of Ω, the following conditions are equivalent:*

(a) *The set of functions in $\mathfrak{H}$ that vanish on Ω_1 forms a Hilbert space in the inner product of $\mathfrak{H}$.*

(b) *The restriction $K_1(s, t)$ of $K(s, t)$ to $\Omega_1 \times \Omega_1$ has the same number of negative squares as $K(s, t)$.*

In this case, $K_1(s, t)$ is the reproducing kernel for the Pontryagin space $\mathfrak{H}_1$ of functions on Ω_1 such that the restriction mapping $h \to h|_{\Omega_1}$ is a partial isometry from $\mathfrak{H}$ onto $\mathfrak{H}_1$.

Condition (a) is trivially satisfied if the only function in $\mathfrak{H}$ which vanishes on Ω_1 is the function identically zero.

PROOF. (b) $\Rightarrow$ (a) Let $\mathfrak{H}_1$ be the Pontryagin space with reproducing kernel $K_1(s, t)$. Our hypotheses imply (by Theorem 3.1) that $\mathfrak{H}_1$ and $\mathfrak{H}$ have the same negative index. Define a linear relation $\mathbf{R}$ in $\mathfrak{H}_1 \times \mathfrak{H}$ as the span of all pairs

$$(K_1(s, \cdot), K(s, \cdot)) \quad \text{for } s \in \Omega_1.$$

For any $s_1, \ldots, s_n \in \Omega_1$ and numbers $a_1, \ldots, a_n$ and $b_1, \ldots, b_n$,

$$\sum_{j,k=1}^{n} a_k \bar{b}_j K_1(s_k, s_j) = \sum_{j,k=1}^{n} a_k \bar{b}_j K(s_k, s_j).$$

Thus $\mathbf{R}$ is isometric. By Theorem 2.2, the closure of $\mathbf{R}$ is the graph of an isometry $V \in \mathfrak{L}(\mathfrak{H}_1, \mathfrak{H})$. Then $\mathfrak{H} = \mathfrak{K} \oplus \mathfrak{N}$, where $\mathfrak{K} = \operatorname{ran} V$ is a Pontryagin space having the same negative index as $\mathfrak{H}$. Hence $\mathfrak{N} = \ker V^*$ is a Hilbert space in the inner product of $\mathfrak{H}$. For any $s \in \Omega_1$ and $h \in \mathfrak{H}$,

$$(V^* h)(s) = \langle (V^* h)(\cdot), K_1(s, \cdot) \rangle_{\mathfrak{H}_1} = \langle h(\cdot), K(s, \cdot) \rangle_{\mathfrak{H}} = h(s),$$

so $\mathfrak{N} = \ker V^* = \{h : h \in \mathfrak{H}, h|_{\Omega_1} \equiv 0\}$, and (a) follows.

(a) $\Rightarrow$ (b) Assume that $\mathfrak{N} = \{h : h \in \mathfrak{H}, h|_{\Omega_1} \equiv 0\}$ is a Hilbert subspace of $\mathfrak{H}$. Then $\mathfrak{H} = \mathfrak{K} \oplus \mathfrak{N}$, where $\mathfrak{K}$ is a Pontryagin subspace of $\mathfrak{H}$ having the same negative index as $\mathfrak{H}$. Let $\mathfrak{H}_1$ be the Pontryagin space of all restrictions $h_1 = h|_{\Omega_1}$ of functions h in $\mathfrak{K}$ in the inner product such that the mapping $U : h \to h|_{\Omega_1}$ is an isometry from $\mathfrak{K}$ onto $\mathfrak{H}_1$. Then $\mathfrak{H}_1$ has the same negative index as $\mathfrak{H}$. If

$s \in \Omega_1$, then $K(s, \cdot)$ is orthogonal to $\mathfrak{N}$ and $U : K(s, \cdot) \to K_1(s, \cdot)$. Hence $K_1(s, \cdot)$ is in $\mathfrak{H}_1$, and for any h_1 of the form Uh, with $h \in \mathfrak{K}$, we have

$$\langle h_1(\cdot), K_1(s, \cdot) \rangle_{\mathfrak{H}_1} = \langle h(\cdot), K(s, \cdot) \rangle_{\mathfrak{H}} = h(s) = h_1(s).$$

Thus $\mathfrak{H}_1$ has reproducing kernel $K_1(s, t)$. Since $\mathfrak{H}_1$ and $\mathfrak{K}$ have the same negative index, $K_1(s, t)$ and $K(s, t)$ have the same number of negative squares by Theorem 3.1. This proves both (b) and the last statement in the theorem.   □

In the case of holomorphic kernels, the number of negative squares is propagated to arbitrarily large domains. Let Ω be a region in the complex plane. A kernel $K(w, z)$ on $\Omega \times \Omega$ is called *holomorphic* if it is holomorphic in z for each fixed w and holomorphic in $\bar{w}$ for each fixed z. As in the Hilbert space case, a reproducing kernel $K(w, z)$ for a Pontryagin space $\mathfrak{H}$ of functions on Ω is holomorphic if and only if the elements of $\mathfrak{H}$ are holomorphic functions on Ω.

THEOREM 3.7. *Let $K(w, z)$ be a holomorphic Hermitian kernel on $\Omega \times \Omega$ for some region Ω in the complex plane. Let Ω_1 be a subregion of Ω, and assume that the restriction $K_1(w, z)$ of $K(w, z)$ to $\Omega_1 \times \Omega_1$ has κ negative squares. Then $K(w, z)$ has κ negative squares on $\Omega \times \Omega$.*

This is proved in [Alpay et al. 1997, Theorem 1.1.4], in a version for operator-valued functions, and we omit the argument here.

The Hilbert space theorem on sums of reproducing kernels is given in [Aronszajn 1950, p. 353]. Let $K(s, t) = K_1(s, t) + K_2(s, t)$, where $K(s, t)$, $K_1(s, t)$, and $K_2(s, t)$ are reproducing kernels for Hilbert spaces $\mathfrak{H}$, $\mathfrak{H}_1$, and $\mathfrak{H}_2$. Then $\mathfrak{H}_1$ and $\mathfrak{H}_2$ are vector subspaces of $\mathfrak{H}$. Every element h of $\mathfrak{H}$ is of the form $h = h_1 + h_2$ with $h_1 \in \mathfrak{H}_1$ and $h_2 \in \mathfrak{H}_2$, and

$$\|h\|_{\mathfrak{H}}^2 = \min \left(\|h_1\|_{\mathfrak{H}_1}^2 + \|h_2\|_{\mathfrak{H}_2}^2 \right),$$

where the minimum is over all such representations. The minimum is uniquely attained. These assertions fail in general for Pontryagin spaces.

EXAMPLE 3.8. Let $L(s, t) \not\equiv 0$ be the reproducing kernel for a finite-dimensional Hilbert space of functions on Ω. Put

$$K_1(s, t) = L(s, t), \qquad K_2(s, t) = -L(s, t).$$

Then $K(s, t) = K_1(s, t) + K_2(s, t)$ vanishes identically and is the reproducing kernel for $\mathfrak{H} = \{0\}$. Thus the reproducing kernel Pontryagin spaces $\mathfrak{H}_1$ and $\mathfrak{H}_2$ corresponding to $K_1(s, t)$ and $K_2(s, t)$ are not contained in $\mathfrak{H}$.

Nevertheless, there are indefinite extensions of the Hilbert space results in this area. They are related to the complementation theory of de Branges [1988] for contractively contained spaces. A Pontryagin space $\mathfrak{H}_1$ is *contained contractively* in a Pontryagin space $\mathfrak{H}$ if $\mathfrak{H}_1$ is a vector subspace of $\mathfrak{H}$ and the inclusion mapping is a continuous and contractive operator from $\mathfrak{H}_1$ into $\mathfrak{H}$. If $\mathfrak{H}_1$ and $\mathfrak{H}_2$ are two Pontryagin spaces which are contained contractively in $\mathfrak{H}$, they are called *complementary* if

(a) $(h_1, h_2) \to h_1 + h_2$ is a partial isometry from $\mathfrak{H}_1 \times \mathfrak{H}_2$ onto $\mathfrak{H}$, and
(b) the kernel of the partial isometry is a Hilbert space.

In this case, every $h \in \mathfrak{H}$ is of the form $h = h_1 + h_2$ with $h_1 \in \mathfrak{H}_1$, $h_2 \in \mathfrak{H}_2$, and

$$\langle h, h \rangle_{\mathfrak{H}} = \min \left(\langle h_1, h_1 \rangle_{\mathfrak{H}_1} + \langle h_2, h_2 \rangle_{\mathfrak{H}_2} \right),$$

where the minimum is over all such representations and uniquely attained, and $\kappa = \kappa_1 + \kappa_2$ where $\kappa, \kappa_1, \kappa_2$ are the negative indices of $\mathfrak{H}, \mathfrak{H}_1, \mathfrak{H}_2$.

THEOREM 3.9. *Suppose that $K_1(s,t), K_2(s,t)$ are reproducing kernels for Pontryagin spaces $\mathfrak{H}_1, \mathfrak{H}_2$ of functions on Ω with negative indices κ_1, κ_2. Then*

$$K(s,t) = K_1(s,t) + K_2(s,t) \tag{3-3}$$

is the reproducing kernel for a Pontryagin space $\mathfrak{H}$ having negative index $\kappa \leq \kappa_1 + \kappa_2$. Equality holds if and only if $\mathfrak{R} = \mathfrak{H}_1 \cap \mathfrak{H}_2$ is a Hilbert space in the inner product

$$\langle h, k \rangle_{\mathfrak{R}} = \langle h, k \rangle_{\mathfrak{H}_1} + \langle h, k \rangle_{\mathfrak{H}_2} \quad \text{for } h, k \in \mathfrak{R}. \tag{3-4}$$

If $\kappa = \kappa_1 + \kappa_2$, then $\mathfrak{H}_1$ and $\mathfrak{H}_2$ are contained contractively in $\mathfrak{H}$ as complementary spaces.

A bit more is true. For example, the converse to the last statement holds: if $\mathfrak{H}_1$ and $\mathfrak{H}_2$ are contained contractively in $\mathfrak{H}$ as complementary spaces, then $\kappa = \kappa_1 + \kappa_2$. See [Alpay et al. 1997, Chapter 1].

PROOF. We obtain the first assertion from Lemma 2.1 as in the proof of Theorem 3.3. (Hint. First write $K_j(s,t) = K_{j+}(s,t) + K_{j-}(s,t)$ for $j = 1, 2$, as in Theorem 3.3.)

Assume $\kappa = \kappa_1 + \kappa_2$. Let $\mathbf{R}$ be the linear relation in $\mathfrak{H} \times (\mathfrak{H}_1 \times \mathfrak{H}_2)$ spanned by all pairs $\left(K(s,\cdot), (K_1(s,\cdot), K_2(s,\cdot)) \right)$ with $s \in \Omega$. Apply Theorem 2.2 to construct an isometry W^* from $\mathfrak{H}$ into $\mathfrak{H}_1 \times \mathfrak{H}_2$ such that

$$W^* : K(s,\cdot) \to (K_1(s,\cdot), K_2(s,\cdot)) \quad \text{for } s \in \Omega.$$

Then W is a partial isometry whose initial set is a Pontryagin subspace of $\mathfrak{H}_1 \times \mathfrak{H}_2$ having the negative index $\kappa = \kappa_1 + \kappa_2$, and whose kernel therefore is a Hilbert space. A short calculation shows that $W : (h_1, h_2) \to h_1 + h_2$. In fact, if $(h_1, h_2) \in \mathfrak{H}_1 \times \mathfrak{H}_2$ and $W : (h_1, h_2) \to h$, then

$$h(s) = \langle h(\cdot), K(s,\cdot) \rangle_{\mathfrak{H}} = \langle (h_1(\cdot), h_2(\cdot)), (K_1(s,\cdot), K_2(s,\cdot)) \rangle_{\mathfrak{H}_1 \times \mathfrak{H}_2}$$
$$= h_1(s) + h_2(s)$$

for all $s \in \Omega$. The kernel of W is naturally isomorphic with $\mathfrak{R} = \mathfrak{H}_1 \cap \mathfrak{H}_2$ in the inner product (3-4). Thus $\mathfrak{R}$ is a Hilbert space in the inner product (3-4). We have also proved the last assertion of the theorem.

Conversely, assume $\mathfrak{R} = \mathfrak{H}_1 \cap \mathfrak{H}_2$ is a Hilbert space in the inner product (3–4). Then the elements $(h, -h)$ in $\mathfrak{H}_1 \times \mathfrak{H}_2$ with $h \in \mathfrak{R}$ form a Hilbert space in the inner product of $\mathfrak{H}_1 \times \mathfrak{H}_2$. Hence there is a Pontryagin space $\mathfrak{H}'$ such that the mapping

$$W : (h_1, h_2) \to h_1 + h_2$$

is a partial isometry from $\mathfrak{H}_1 \times \mathfrak{H}_2$ onto $\mathfrak{H}'$. We easily check that (3–3) is a reproducing kernel for $\mathfrak{H}'$, hence $\mathfrak{H}' = \mathfrak{H}$ isometrically. Since $\ker W$ is a Hilbert space, $\mathfrak{H}_1 \times \mathfrak{H}_2$ and $\mathfrak{H}$ have the same negative index, namely, $\kappa = \kappa_1 + \kappa_2$.   □

THEOREM 3.10. *Suppose that $K(s,t), K_1(s,t)$ are reproducing kernels for Pontryagin spaces $\mathfrak{H}, \mathfrak{H}_1$ of functions on Ω having negative indices κ, κ_1. If $\mathfrak{H}_1$ is contained contractively in $\mathfrak{H}$, then $\kappa \geq \kappa_1$ and*

$$K_2(s,t) = K(s,t) - K_1(s,t)$$

has $\kappa_2 = \kappa - \kappa_1$ negative squares and is the reproducing kernel for a Pontryagin space $\mathfrak{H}_2$ which is contained contractively in $\mathfrak{H}$ such that $\mathfrak{H}_1$ and $\mathfrak{H}_2$ are complementary.

PROOF. By a theorem of de Branges [1988] (see also [Dritschel and Rovnyak 1991]), there is a unique Pontryagin space $\mathfrak{H}_2$ which is contained contractively in $\mathfrak{H}$ such that $\mathfrak{H}_1$ and $\mathfrak{H}_2$ are complementary. Evaluation mappings on $\mathfrak{H}_2$ are compositions of continuous mappings and thus continuous. Hence $\mathfrak{H}_2$ has a reproducing kernel $K_2(s,t)$. By the definition of complementary spaces, the mapping $W : (h_1, h_2) \to h_1 + h_2$ is a partial isometry from $\mathfrak{H}_1 \times \mathfrak{H}_2$ onto $\mathfrak{H}$, and $\ker W$ is a Hilbert space. Thus $\mathfrak{H}_2$ has negative index $\kappa - \kappa_1$. In particular, $\kappa \geq \kappa_1$.

For fixed $s \in \Omega$, $K_1(s, \cdot) + K_2(s, \cdot)$ belongs to $\mathfrak{H}$ and is the image under W of $(K_1(s, \cdot), K_2(s, \cdot))$, where the kernel function pair is in $\mathfrak{H}_1 \times \mathfrak{H}_2$ and orthogonal to $\ker W$. Every h in $\mathfrak{H}$ is the image under W of some pair (h_1, h_2) in $\mathfrak{H}_1 \times \mathfrak{H}_2$, and hence

$$\langle h(\cdot), K_1(s, \cdot) + K_2(s, \cdot) \rangle_{\mathfrak{H}} = \langle (h_1(\cdot), h_2(\cdot)), (K_1(s, \cdot), K_2(s, \cdot)) \rangle_{\mathfrak{H}_1 \times \mathfrak{H}_2}$$
$$= h_1(s) + h_2(s) = h(s).$$

Therefore $K_1(s,t) + K_2(s,t)$ is a reproducing kernel for $\mathfrak{H}$. Since a reproducing kernel is unique, $K_1(s,t) + K_2(s,t) = K(s,t)$, and so $K_2(s,t) = K(s,t) - K_1(s,t)$ is the reproducing kernel for $\mathfrak{H}_2$. The result follows.   □

One more result may be mentioned.

THEOREM 3.11. *Let $K_1(s,t)$ be the reproducing kernel for a Pontryagin space $\mathfrak{H}_1$ of functions on Ω having negative index κ_1. Define*

$$K_2(s,t) = A(t)K_1(s,t)\overline{A(s)} \quad \text{for } s, t \in \Omega,$$

where $A(t)$ is a fixed function on Ω. Then $K_2(s,t)$ is the reproducing kernel for a Pontryagin space $\mathfrak{H}_2$ of functions on Ω having negative index $\kappa_2 \leq \kappa_1$, and

$\kappa_1 = \kappa_2$ *if and only if the set of functions $h(t)$ in $\mathfrak{H}_1$ such that $A(t)h(t) \equiv 0$ is a Hilbert space in the inner product of $\mathfrak{H}_1$. In this case, multiplication by $A(t)$ is a partial isometry from $\mathfrak{H}_1$ onto $\mathfrak{H}_2$ whose kernel is a Hilbert space.*

The proof is another application of Theorem 2.2 and omitted (when $\kappa_1 = \kappa_2$, set up a linear relation **R** to define the adjoint of multiplication by $A(t)$).

We have presented the theory of reproducing kernel spaces in the context of Pontryagin spaces. The main definitions can be adapted to Kreĭn spaces, and some of the constructions carry over. However, new phenomena arise in this generality. A reproducing kernel is uniquely determined by the associated Kreĭn space, but, in contrast with Theorem 3.2, essentially different Kreĭn spaces can have the same reproducing kernel. For example, see [Alpay 1991; Schwartz 1964].

4. Examples Involving Generalized Schur Functions

The Schur class is the set of holomorphic functions $S(z)$ which are defined and satisfy $|S(z)| \leq 1$ on the unit disk $\mathbb{D}$ in the complex plane $\mathbb{C}$. For any such function, the associated kernel

$$K_S(w, z) = \frac{1 - S(z)\overline{S(w)}}{1 - z\bar{w}} \qquad (4\text{-}1)$$

is nonnegative and therefore the reproducing kernel for a Hilbert space $\mathfrak{H}(S)$ of holomorphic functions on $\mathbb{D}$. These spaces have been extensively studied, and many properties are known. For example, the transformation

$$T : h(z) \rightarrow \frac{h(z) - h(0)}{z} \qquad (4\text{-}2)$$

maps $\mathfrak{H}(S)$ into itself and satisfies the difference quotient inequality

$$\left\| (h(z) - h(0))/z \right\|^2_{\mathfrak{H}(S)} \leq \|h(z)\|^2_{\mathfrak{H}(S)} - |h(0)|^2 \qquad (4\text{-}3)$$

for all elements $h(z)$ of the space. There are operators

$$\begin{cases} F : c \rightarrow \dfrac{S(z) - S(0)}{z}\, c & \text{on } \mathbb{C} \text{ to } \mathfrak{H}(S), \\[2mm] G : h(z) \rightarrow h(0) & \text{on } \mathfrak{H}(S) \text{ to } \mathbb{C}, \\[2mm] H : c \rightarrow S(0)c & \text{on } \mathbb{C} \text{ to } \mathbb{C}, \end{cases} \qquad (4\text{-}4)$$

such that the colligation

$$V = \begin{pmatrix} T & F \\ G & H \end{pmatrix} \in \mathfrak{L}(\mathfrak{H}(S) \oplus \mathbb{C}) \qquad (4\text{-}5)$$

is coisometric, that is, $VV^* = 1$. The characteristic function of the colligation is the given Schur function $S(z)$:

$$S(z)c = Hc + zG(1 - zT)^{-1}Fc \quad \text{for } c \in \mathbb{C}. \qquad (4\text{-}6)$$

The colligation is unitary if and only if equality always holds in (4–3) and $S(z) \not\equiv 0$. An equivalent condition that V is unitary is that the function $S(z)$ itself does not belong to $\mathfrak{H}(S)$. If $S(z) = S_1(z)S_2(z)$ where $S_1(z)$ and $S_2(z)$ are Schur functions, then $\mathfrak{H}(S_1)$ is contained contractively in $\mathfrak{H}(S)$. If $\mathfrak{H}(S_1) \cap S_1\mathfrak{H}(S_2) = \{0\}$, the inclusion is isometric, and $\mathfrak{H}(S) = \mathfrak{H}(S_1) \oplus S_1\mathfrak{H}(S_2)$. The scalar case is treated in [de Branges and Rovnyak 1966; Sarason 1994].

We derive the preceding scalar results in a Pontryagin space setting by the methods of [Alpay et al. 1997].

DEFINITION 4.1. Let Ω be a region satisfying $0 \in \Omega \subseteq \mathbb{D}$. The *generalized Schur class* $\mathbf{S}_\kappa$ is the set of holomorphic functions $S(z)$ defined on Ω and such that the kernel (4–1) has κ negative squares on $\Omega \times \Omega$. For such a function, let $\mathfrak{H}(S)$ be the Pontryagin space of holomorphic functions with reproducing kernel (4–1).

The domain of a function $S(z) \in \mathbf{S}_\kappa$ will be denoted $\Omega = \Omega(S)$. By Theorem 3.7, this can be any region in $\mathbb{D}$ containing the origin on which $S(z)$ is holomorphic. Technically the spaces $\mathfrak{H}(S)$ are different for different regions, but any two such spaces can be identified in a natural way.

THEOREM 4.2. *For every $S(z)$ in $\mathbf{S}_\kappa$, the transformation (4–2) is everywhere defined on $\mathfrak{H}(S)$ and a bicontraction. The difference-quotient inequality*

$$\left\langle \frac{h(z) - h(0)}{z}, \frac{h(z) - h(0)}{z} \right\rangle_{\mathfrak{H}(S)} \leq \langle h(z), h(z) \rangle_{\mathfrak{H}(S)} - h(0)\overline{h(0)} \tag{4–7}$$

holds for every $h(z)$ in $\mathfrak{H}(S)$. There exist operators (4–4) such that (4–5) is a coisometry. The identity (4–6) holds in a neighborhood of the origin.

It is interesting to ask for a converse to Theorem 4.2, that is, to say when a Pontryagin space of holomorphic functions is of the form $\mathfrak{H}(S)$ for some function $S(z)$ in $\mathbf{S}_\kappa$. When $\kappa = 0$, such characterizations are known [Guyker 1991; Leech 1969], but it is an open problem to obtain a similar result in the general case. When more general spaces involving operator-valued functions are allowed, a complete characterization is possible and given in [Alpay et al. 1997, Theorem 3.1.2].

PROOF. Let the domain of $S(z)$ be Ω. Theorem 2.2 will be used to construct an isometry which turns out to be the adjoint of the operator V that we seek. The construction is based on knowing how V^* must act on special elements. This motivates the definition of a linear relation $\mathbf{R}$ in $(\mathfrak{H}(S) \oplus \mathbb{C}) \times (\mathfrak{H}(S) \oplus \mathbb{C})$ as the span of all pairs

$$\left(\begin{pmatrix} K_S(\alpha, \cdot)a_1 \\ \\ a_2 \end{pmatrix}, \begin{pmatrix} \dfrac{K_S(\alpha, \cdot) - K_S(0, \cdot)}{\bar\alpha} a_1 + K_S(0, \cdot)a_2 \\ \\ \dfrac{\overline{S(\alpha)} - \overline{S(0)}}{\bar\alpha} a_1 + \overline{S(0)}a_2 \end{pmatrix} \right)$$

with $0 \neq \alpha \in \Omega$ and $a_1, a_2 \in \mathbb{C}$. The inner product of the right members of any two such pairs with nonzero $\alpha, \beta \in \Omega$ and $a_1, a_2, b_1, b_2 \in \mathbb{C}$ is

$$
\frac{K_S(\alpha,\beta) - K_S(0,\beta) - K_S(\alpha,0) + K_S(0,0)}{\bar{\alpha}\beta} a_1\bar{b}_1 + \frac{K_S(0,\beta) - K_S(0,0)}{\beta} a_2\bar{b}_1
$$

$$
+ \frac{K_S(\alpha,0) - K_S(0,0)}{\bar{\alpha}\beta} a_1\bar{b}_2 + K_S(0,0)a_2\bar{b}_2
$$

$$
+ \frac{S(\beta) - S(0)}{\beta} \frac{\overline{S(\alpha)} - \overline{S(0)}}{\bar{\alpha}\beta} a_1\bar{b}_1 + \frac{S(\beta) - S(0)}{\beta} \overline{S(0)}\, a_2\bar{b}_1
$$

$$
+ S(0)\frac{\overline{S(\alpha)} - \overline{S(0)}}{\bar{\alpha}\beta} a_1\bar{b}_2 + S(0)\overline{S(0)}\, a_2\bar{b}_2
$$

$$
= K_S(\alpha,\beta)a_1\bar{b}_1 + a_2\bar{b}_2
$$

$$
= \left\langle \begin{pmatrix} K_S(\alpha,\cdot)a_1 \\ a_2 \end{pmatrix}, \begin{pmatrix} K_S(\beta,\cdot)b_1 \\ b_2 \end{pmatrix} \right\rangle_{\mathfrak{H}(S)\oplus\mathbb{C}} .
$$

Therefore $\mathbf{R}$ is isometric, hence $\bar{\mathbf{R}}$ is the graph of an isometry

$$
V^* = \begin{pmatrix} T^* & G^* \\ F^* & H^* \end{pmatrix} \in \mathfrak{L}(\mathfrak{H}(S) \oplus \mathbb{C}).
$$

The definition of $\mathbf{R}$ is made so that T, F, G, H are given by (4–2) and (4–4).

On solving the equation $(1 - wT)^{-1}h = g$, we see that

$$
(1 - wT)^{-1} : h(z) \to \frac{zh(z) - wh(w)}{z - w}
$$

for any $h(z)$ in $\mathfrak{H}(S)$ and w in a neighborhood of the origin. Therefore

$$
\left(H + wG(1 - wT)^{-1}F\right)c = S(0)c + w \left\{ \frac{z\frac{S(z)-S(0)}{z}c - w\frac{S(w)-S(0)}{w}c}{z - w} \right\}_{z=0}
$$

$$
= S(0)c + w\frac{S(w)-S(0)}{w}c = S(w)c,
$$

for w in a neighborhood of the origin and $c \in \mathbb{C}$, yielding (4–6).

Since V^* is an isometry and hence a contraction acting on a Pontryagin space into itself, it is a bicontraction. Hence $V^*V \leq 1$ and $T^*T + G^*G \leq 1$, which implies (4–7). By (4–7), T is a contraction. Since T maps a Pontryagin space into itself, it is a bicontraction. $\qquad\square$

THEOREM 4.3. *In Theorem 4.2, the following conditions are equivalent:*

(a) *Equality always holds in (4–7) and $S(z) \not\equiv 0$.*

(b) *$S(z)$ does not belong to $\mathfrak{H}(S)$.*

(c) *V is unitary.*

PROOF. (a) $\Rightarrow$ (b) Assume (a). Then $T^*T + G^*G = 1$ and so

$$1 - V^*V = \begin{pmatrix} 0 & -T^*F - G^*H \\ -F^*T - H^*G & 1 - F^*F - H^*H \end{pmatrix}.$$

Since V^* is an isometry and hence a bicontraction, V is a contraction. Hence $1 - V^*V \geq 0$. In view of the zero in the upper left entry of the operator matrix, this implies $T^*F + G^*H = 0$ by properties of nonnegative quadratic forms. The action of G^* is computed from the uniqueness of a reproducing kernel: $G^*c = K_S(0, \cdot)c$ for every $c \in \mathbb{C}$. Therefore the identity $T^*F + G^*H = 0$ yields

$$T^* : \frac{S(z) - S(0)}{z} \to -K_S(0, z)S(0).$$

If $S(z)$ belongs to $\mathfrak{H}(S)$, the last relation yields

$$(1 - G^*G)S(\cdot) = T^*TS(\cdot) = -K_S(0, \cdot)S(0).$$

Then $S(z) - K_S(0, z)S(0) \equiv -K_S(0, z)S(0)$ and $S(z) \equiv 0$, which is excluded in (a). Therefore $S(z)$ does not belong to $\mathfrak{H}(S)$.

(b) $\Rightarrow$ (c) Assume (b). Since V^* is an isometry, to prove that V is unitary we need only show that $\ker V = \{0\}$. If

$$V \begin{pmatrix} f \\ c \end{pmatrix} = 0,$$

then

$$\frac{f(z) - f(0)}{z} + \frac{S(z) - S(0)}{z} c = 0,$$

$$f(0) + S(0)c = 0,$$

identically. Therefore $f(z) = -S(z)c$ belongs to $\mathfrak{H}(S)$. By (b), $c = 0$. Thus $\ker V = \{0\}$ and so V is unitary.

(c) $\Rightarrow$ (a) The unitarity of V implies $V^*V = 1$. Hence $T^*T + G^*G = 1$, and therefore equality always holds in (4–7). If $S(z) \equiv 0$, then $F = 0$ and $H = 0$ by (4–4), contradicting the assumption that V is unitary. Therefore $S(z) \not\equiv 0$. $\square$

In the case $\kappa = 0$, condition (b) in Theorem 4.3 has been much studied; see [de Branges and Rovnyak 1966; Sarason 1994]. The condition can also be pursued using Leech's theorem on the factorization of operator-valued functions. The result obtained in this way states that $S(z)$ belongs to $\mathfrak{H}(S)$ and $\|S(z)\|_{\mathfrak{H}(S)} \leq m$ if and only if

$$S(z) = \frac{mC_1(z)}{\sqrt{m^2 + 1} - zC_2(z)}$$

where $C_1(z)$ and $C_2(z)$ are holomorphic and satisfy $|C_1(z)|^2 + |C_2(z)|^2 \leq 1$ on $\mathbb{D}$. See [Alpay et al. 1997] for details and some of the results that are known in the indefinite case.

THEOREM 4.4. *Suppose*

$$S = S_1 S_2,$$

where $S_1 \in \mathbf{S}_{\kappa_1}$, $S_2 \in \mathbf{S}_{\kappa_2}$. Then $S \in \mathbf{S}_\kappa$ for some $\kappa \leq \kappa_1 + \kappa_2$. Equality holds if and only if the set $\mathfrak{N}$ of elements $(h_1, h_2) \in \mathfrak{H}(S_1) \times \mathfrak{H}(S_2)$ such that $h_1 + S_1 h_2 \equiv 0$ form a Hilbert space in the inner product of $\mathfrak{H}(S_1) \times \mathfrak{H}(S_2)$.

The proof of the theorem and general results on reproducing kernels in §3 yield additional conclusions. Let $S_1\mathfrak{H}(S_2)$ be the Pontryagin space with reproducing kernel $S_1(z)K_{S_2}(w,z)\overline{S_1(w)}$ (notice that when $S_1 \not\equiv 0$, multiplication by S_1 maps $\mathfrak{H}(S_2)$ isometrically onto $S_1\mathfrak{H}(S_2)$ by Theorem 3.11). Then, for example, when $\kappa = \kappa_1 + \kappa_2$, $\mathfrak{H}(S_1)$ and $S_1\mathfrak{H}(S_2)$ are contractively contained in $\mathfrak{H}(S)$ as complementary spaces.

PROOF. We may suppose that the functions $S_1 \in \mathbf{S}_{\kappa_1}$, $S_2 \in \mathbf{S}_{\kappa_2}$, and $S = S_1 S_2$ are defined on a common region Ω with $0 \in \Omega \subseteq \mathbb{D}$. The result is easily checked when $S_1 \equiv 0$, so let us exclude this degenerate case. Then

$$K_S(w,z) = K_{S_1}(w,z) + S_1(z)K_{S_2}(w,z)\overline{S_1(w)},$$

where $K_{S_1}(w,z)$ has κ_1 negative squares and $S_1(z)K_{S_2}(w,z)\overline{S_1(w)}$ has κ_2 negative squares. By Theorem 3.9, $K_S(w,z)$ has $\kappa \leq \kappa_1 + \kappa_2$ negative squares. Define $S_1\mathfrak{H}(S_2)$ as in the remarks preceding the proof. Theorem 3.9 also says that $\kappa = \kappa_1 + \kappa_2$ if and only if $\mathfrak{R} = \mathfrak{H}(S_1) \cap S_1\mathfrak{H}(S_2)$ is a Hilbert space in the inner product defined by

$$\langle h, k \rangle_{\mathfrak{R}} = \langle h, k \rangle_{\mathfrak{H}(S_1)} + \langle h, k \rangle_{S_1\mathfrak{H}(S_2)} \quad \text{for } h, k \in \mathfrak{R}.$$

We show that $\mathfrak{R}$ is in one-to-one isometric correspondence with $\mathfrak{N}$ in the inner product of $\mathfrak{H}(S_1) \times \mathfrak{H}(S_2)$. In fact, the mapping $(h_1, h_2) \to h_1$ from $\mathfrak{N}$ onto $\mathfrak{R}$ is such a correspondence. For if $(h_1, h_2) \in \mathfrak{N}$, then $h_1 + S_1 h_2 \equiv 0$ and so $h_1 \in \mathfrak{H}(S_1) \cap S_1\mathfrak{H}(S_2) = \mathfrak{R}$. Clearly every element of $\mathfrak{R}$ arises in this way from a unique element of $\mathfrak{N}$, because as noted above multiplication by S_1 maps $\mathfrak{H}(S_2)$ isometrically onto $S_1\mathfrak{H}(S_2)$. Suppose $(k_1, k_2) \to k_1$ for a second pair in $\mathfrak{N}$. Then

$$\langle (h_1, h_2), (k_1, k_2) \rangle_{\mathfrak{H}(S_1) \times \mathfrak{H}(S_2)} = \langle h_1, k_1 \rangle_{\mathfrak{H}(S_1)} + \langle h_2, k_2 \rangle_{\mathfrak{H}(S_2)}$$
$$= \langle h_1, k_1 \rangle_{\mathfrak{H}(S_1)} + \langle S_1 h_2, S_1 k_2 \rangle_{S_1\mathfrak{H}(S_2)}$$
$$= \langle h_1, k_1 \rangle_{\mathfrak{H}(S_1)} + \langle h_1, k_1 \rangle_{S_1\mathfrak{H}(S_2)} = \langle h_1, k_1 \rangle_{\mathfrak{R}}.$$

Thus $\kappa = \kappa_1 + \kappa_2$ if and only if $\mathfrak{N}$ is a Hilbert space in the inner product of $\mathfrak{H}(S_1) \times \mathfrak{H}(S_2)$. $\square$

We conclude with the answer to a question that will no doubt have occurred to the reader: what are the analytic continuation properties of functions in $\mathbf{S}_\kappa$? For the case $\kappa = 0$, it is well known that positivity of the kernel (4–1) for w, z in a neighborhood of the origin implies that $S(z)$ has an extension to a holomorphic function which is bounded by one on the unit disk, and so $\mathbf{S}_0$ is naturally identified with the usual Schur class of complex analysis. The general result is a well-known theorem of Kreĭn and Langer [1972].

THEOREM 4.5. *Every function* $S(z)$ *in* $\mathbf{S}_\kappa$ *has a factorization*

$$S(z) = B(z)^{-1} S_0(z),$$

where $B(z)$ *is a Blaschke product having* κ *factors and* $S_0(z)$ *belongs to the classical Schur class* $\mathbf{S}_0$ *and is nonvanishing at the zeros of* $B(z)$. *Conversely, every function of this form belongs to* $\mathbf{S}_\kappa$.

To say that $B(z)$ is a Blaschke product of κ factors means that it has the form

$$B(z) = C \prod_{j=1}^{\kappa} \frac{z - \alpha_j}{1 - \bar{\alpha}_j z},$$

where $\alpha_1, \ldots, \alpha_\kappa$ are (not necessarily distinct) points of the unit disk and C is a constant of unit modulus. The case $\kappa = 0$ is included by interpreting an empty product as one. See [Alpay et al. 1997, § 4.2], for a proof of Theorem 4.5 in a version for operator-valued functions.

The generalized Schur class $\mathbf{S}_\kappa$ has deep connections with interpolation and operator theory. It has been studied by Kreĭn and Langer in a series of papers including [Kreĭn and Langer 1981]. However, a more complete account is beyond the scope of the present introductory survey. We refer the reader to [Alpay et al. 1997] for additional results and literature notes.

References

[Alpay 1991]   D. Alpay, "Some remarks on reproducing kernel Kreĭn spaces", *Rocky Mountain J. Math.* **21**:4 (1991), 1189–1205.

[Alpay et al. 1997]   D. Alpay, A. Dijksma, J. Rovnyak, and H. S. V. de Snoo, *Schur functions, operator colligations, and reproducing kernel Pontryagin spaces*, Operator Theory: Advances and Applications **96**, Birkhäuser Verlag, Basel, 1997.

[Aronszajn 1950]   N. Aronszajn, "Theory of reproducing kernels", *Trans. Amer. Math. Soc.* **68** (1950), 337–404.

[Azizov and Iokhvidov 1986]   T. Y. Azizov and I. S. Iokhvidov, Основы теории линейных операторов в пространствах с индефинитной метрикой, Nauka, Moscow, 1986. Translated as *Linear operators in spaces with an indefinite metric*, Wiley, Chichester, 1989.

[Bognár 1974]   J. Bognár, *Indefinite inner product spaces*, Ergebnisse der Mathematik und ihrer Grenzgebiete **78**, Springer, New York, 1974.

[de Branges 1988]   L. de Branges, "Complementation in Kreĭn spaces", *Trans. Amer. Math. Soc.* **305**:1 (1988), 277–291.

[de Branges and Rovnyak 1966]   L. de Branges and J. Rovnyak, *Square summable power series*, Holt, Rinehart and Winston, New York, 1966.

[Dritschel and Rovnyak 1991]   M. A. Dritschel and J. Rovnyak, "Julia operators and complementation in Kreĭn spaces", *Indiana Univ. Math. J.* **40**:3 (1991), 885–901.

[Dritschel and Rovnyak 1996]   M. A. Dritschel and J. Rovnyak, "Operators on indefinite inner product spaces", pp. 141–232 in *Lectures on operator theory and its applications* (Waterloo, 1994), edited by P. Lancaster, Fields Institute Monographs **3**, American Mathematical Society, Providence, RI, 1996. Available at http://wsrv.clas.virginia.edu/~jlr5m/papers/papers.html. Supplement and errata, June 1996, same URL.

[Guyker 1991]   J. Guyker, "The de Branges-Rovnyak model", *Proc. Amer. Math. Soc.* **111**:1 (1991), 95–99.

[Iohvidov et al. 1982]   I. S. Iohvidov, M. G. Kreĭn, and H. Langer, *Introduction to the spectral theory of operators in spaces with an indefinite metric*, Mathematical Research **9**, Akademie-Verlag, Berlin, 1982.

[Kreĭn and Langer 1972]   M. G. Kreĭn and H. Langer, "Über die verallgemeinerten Resolventen und die charakteristische Funktion eines isometrischen Operators im Raume Π_κ", pp. 353–399 in *Hilbert space operators and operator algebras* (Tihany, 1970), Colloq. Math. Soc. János Bolyai **5**, North-Holland, Amsterdam, 1972.

[Kreĭn and Langer 1981]   M. G. Kreĭn and H. Langer, "Some propositions on analytic matrix functions related to the theory of operators in the space Π_κ", *Acta Sci. Math. (Szeged)* **43** (1981), 181–205.

[Leech 1969]   R. B. Leech, "On the characterization of $\mathcal{H}(B)$ spaces", *Proc. Amer. Math. Soc.* **23** (1969), 518–520.

[Sarason 1994]   D. Sarason, *Sub-Hardy Hilbert spaces in the unit disk*, University of Arkansas Lecture Notes in the Mathematical Sciences, Wiley, New York, 1994.

[Schwartz 1964]   L. Schwartz, "Sous-espaces hilbertiens d'espaces vectoriels topologiques et noyaux associés (noyaux reproduisants)", *J. Analyse Math.* **13** (1964), 115–256.

[Sorjonen 1975]   P. Sorjonen, "Pontrjaginräume mit einem reproduzierenden Kern", *Ann. Acad. Sci. Fenn. Ser. A I Math.* **594** (1975), 30 pp.

Daniel Alpay
Department of Mathematics
Ben-Gurion University of the Negev
P. O. Box 653
84105 Beer-Sheva
Israel
  dany@ivory.bgu.ac.il

Aad Dijksma
Department of Mathematics
University of Groningen
P. O. Box 800
9700 AV Groningen
The Netherlands
  dijksma@math.rug.nl

James Rovnyak
Department of Mathematics
University of Virginia
Charlottesville, Virginia 22903-3199
United States
  rovnyak@virginia.edu

Hendrik S. V. de Snoo
Department of Mathematics
University of Groningen
P. O. Box 800
9700 AV Groningen
The Netherlands
  desnoo@math.rug.nl

Holomorphic Spaces
MSRI Publications
Volume **33**, 1998

Commuting Operators and Function Theory on a Riemann Surface

VICTOR VINNIKOV

ABSTRACT. In the late 70's M. S. Livšic has discovered that a pair of commuting nonselfadjoint operators in a Hilbert space, with finite nonhermitian ranks, satisfy a polynomial equation with constant (real) coefficients; in particular the joint spectrum of such a pair of operators lies on a certain algebraic curve in the complex plane, the so called discriminant curve of the pair of operators. More generally, it turns out that much in the same way as the study of a single nonselfadjoint operator is intimately related to the function theory on the complex plane, more specifically on the upper half-plane, the study of a system of commuting nonselfadjoint operators, at least with finite nonhermitian ranks, is related to the function theory on a compact Riemann surface of a higher genus, more specifically on a compact real Riemann surface. From a different perspective, while the study of a single nonselfadjoint operator leads to one-variable continuous time linear systems, the study of a pair of commuting nonselfadjoint operators leads to two-variable continuous time systems, which are necessarily overdetermined, hence must be considered together with an additional structure of compatibility conditions at the input and at the output. In this survey we give an introduction to the spectral theory of commuting nonselfadjoint operators and its interplay with system theory and the theory of Riemann surfaces and algebraic curves, including some recent results and open problems.

Introduction

It is fair to say that until the 1940's operator theory was mostly concerned with selfadjoint or unitary operators; several commuting selfadjoint or unitary operators do not present any essential new problems because such operators possess commuting resolutions of the identity. Starting with the work of Livšic and his associates in the 1940's and 1950's [Brodskiĭ and Livšic 1958; Brodskiĭ 1969], and later that of Sz.-Nagy and Foiaş [1967] and of de Branges and Rovnyak [1966a; 1966b], a comprehensive study of nonselfadjoint and nonunitary operators began, especially for operators that are not "too far" from being selfadjoint

The author is an incumbent of Graham and Rhona Beck career development chair, Weizmann Institute, and was supported by Mathematical Sciences Research Institute.

or unitary (i.e., the nonhermitian part $A - A^*$ or the defect operators $I - AA^*$, $I - A^*A$ are of finite rank, or trace class). This study has revealed deep connections with the theory of bounded analytic functions on the upper half-plane or on the unit disk, and more generally with the theory of matrix-valued and operator-valued functions on these domains possessing metric properties such as contractivity, and with system theory. The main point is that there is a relation between invariant subspaces of an operator and factorizations of its so-called characteristic function. The characteristic function turns out to be the transfer function of a certain associated linear time-invariant conservative dynamical system.

Initial attempts to generalize these results to several commuting nonselfadjoint or nonunitary operators ran into serious difficulties, since it was natural now to define the characteristic function as a function of several complex variables, which therefore does not admit a good factorization theory. However, it was discovered by Livšic in the late 1970's that a pair of commuting nonselfadjoint operators with finite nonhermitian ranks satisfy a polynomial equation with constant (real) coefficients. Therefore the joint spectrum of such a pair of operators lies on a (real) algebraic curve in $\mathbb{C}^2$, called the discriminant curve; and it seems natural that their spectral study would lead to function theory on the corresponding compact real Riemann surface (i.e., compact Riemann surface endowed with an antiholomorphic involution coming from the complex conjugation on the curve) rather than to function theory of two independent complex variables. This is indeed the case, and the proper analogue of the notion of the characteristic function of a single nonselfadjoint operator is the so-called joint characteristic function of a pair of operators, which is a mapping of certain vector bundles (or more generally, of certain sheaves) on the discriminant curve. It turns out again to have a system-theoretic interpretation as the transfer function of the associated linear time-invariant conservative dynamical system, which is now two-dimensional: the input, the state, and the output depend on two (continuous) parameters rather than one.

The objective of this survey is to give an introduction to the spectral theory of commuting nonselfadjoint operators and its interplay with system theory and the theory of Riemann surfaces and algebraic curves, including some recent results and open problems. In Section 1 we review the basic constructions as established by Livšic [1979; 1980; 1983; 1986a]; see also [Livšic 1987; Waksman 1987; Kravitsky 1983; Vinnikov 1992]. Section 2 discusses the joint characteristic function introduced in [Livšic 1986b] and further investigated in [Vinnikov 1992; 1994]; see also [Ball and Vinnikov 1996]. Much of the material discussed in Sections 1 and 2 appears in [Livšic et al. 1995]. In Section 3 we discuss semicontractive and semiexpansive functions on a compact real Riemann surface—the analogues of contractive and expansive functions on the upper half-plane or unit disk—and their canonical factorizations and we present functional models for commuting nonselfadjoint operators constructed by Alpay and Vinnikov [1994; a].

All the results obtained up to now deal with commuting operators close to self-adjoint. Of course, commuting operators close to unitary have to be studied as well. System-theoretically this means studying conservative discrete-time, rather than continuous-time, systems in two or more dimensions. It turns out to be quite nontrivial how to transfer various notions from the nonselfadjoint to the nonunitary case. As an indication (see Section 1 for a motivation), it is obvious what to require of the matrices σ_1, σ_2, γ if we want the curve $\det(\lambda_1\sigma_2 - \lambda_2\sigma_1 + \gamma) = 0$ to be invariant under the anti-holomorphic involution $(\lambda_1, \lambda_2) \mapsto (\bar{\lambda}_1, \bar{\lambda}_2)$ (namely, the matrices should be self-adjoint); but what if we consider instead the anti-holomorphic involution

$$(\lambda_1, \lambda_2) \mapsto \left(\frac{1}{\bar{\lambda}_1}, \frac{1}{\bar{\lambda}_2}\right) ?$$

Some progress on the corresponding proper framework for the study of commuting nonunitary operators has been achieved recently in joint work with J. Ball. Once the basic notions are fixed, given two commuting contractions A_1, A_2 with finite defects, the corresponding compact real Riemann surface X should be necessarily dividing (see Section 3 for the definition), and the functional model would also yield an $H^\infty(X_+)$ functional calculus for A_1, A_2. In particular if X is the double of a finitely connected planar domain S, and we denote by Z the global planar coordinate on $S = X_+$, then $T = Z(A_1, A_2)$ is an operator with spectral set S. This should provide a link to the work of Abrahamse and Douglas [1976], and may be also a useful approach to the well-known question whether an operator with a multiply connected spectral set admits a normal boundary dilation [Agler 1985].

1. Commuting Nonselfadjoint Operators, Two-Dimensional Systems, and Algebraic Curves

It is well-known (see, for example, [Brodskiĭ 1969; Livšic and Yantsevich 1971; Ball and Cohen 1991]) that the most natural object to consider in the study of a single (bounded) nonselfadjoint operator A in a Hilbert space H is not the operator A itself, but rather an *operator colligation* (or *node*) $\mathcal{C} = (A, H, \Phi, E, \sigma)$. Here E is an auxiliary Hilbert space called the external space of the colligation (H is called the inner space), $\Phi : H \to E$ and $\sigma : E \to E$ are bounded linear mappings with $\sigma^* = \sigma$, and

$$\frac{1}{i}(A - A^*) = \Phi^*\sigma\Phi. \tag{1–1}$$

We shall be considering only operators with a finite nonhermitian rank ($\dim(A - A^*)H < \infty$), so we assume $\dim E = M < \infty$. Note that a given operator A in H (with a finite nonhermitian rank) can be always embedded in a colligation by setting

$$E = (A - A^*)H, \qquad \Phi = P_E, \qquad \sigma = \frac{1}{i}(A - A^*)\Big|_E, \tag{1–2}$$

where P_E is the orthogonal projection of H onto E. (Another possible embedding is obtained by setting

$$\Phi = \left| \frac{A - A^*}{i} \right|_E \Big|^{1/2}, \qquad \sigma = \text{sign} \left. \frac{A - A^*}{i} \right|_E,$$

where the absolute value and the sign functions are understood in the sense of the usual functional calculus for self-adjoint operators; this is used more often in single-operator theory because of the added convenience of $\sigma^2 = I$, but it does not admit a good generalization to the two-operator case.)

The advantage of the notion of colligation is that it allows us to "isolate" the nonhermitian part of the operator. In particular, given two colligations

$$\mathcal{C}' = (A', H', \Phi', E, \sigma) \quad \text{and} \quad \mathcal{C}'' = (A'', H'', \Phi'', E, \sigma),$$

with the same external part (E, σ), we define their *coupling*

$$\mathcal{C} = \mathcal{C}' \vee \mathcal{C}'' = (A, H, \Phi, E, \sigma),$$

where $H = H' \oplus H''$ and

$$A = \begin{pmatrix} A' & 0 \\ i\,\Phi''^* \sigma \Phi' & A'' \end{pmatrix}, \qquad \Phi = (\Phi' \quad \Phi''), \tag{1-3}$$

the operators being written in the block form with respect to the direct sum decomposition $H = H' \oplus H''$. The coupling procedure allows us to construct operators with a more complicated spectral data out of operators with a simpler one, while preserving the nonhermitian part. Note that H'' is an invariant subspace of A. Conversely, if $H'' \subset H$ is an invariant subspace of the operator A in a colligation $\mathcal{C} = (A, H, \Phi, E, \sigma)$ and $H' = H \ominus H''$, it is easy to see that we can write $\mathcal{C} = \mathcal{C}' \vee \mathcal{C}''$, where $\mathcal{C}'$, $\mathcal{C}''$ are the *projections* of $\mathcal{C}$ onto the subspaces H', H'' respectively, given by

$$\mathcal{C}' = (P'A|_{H'}, H', \Phi|_{H'}, E, \sigma), \tag{1-4}$$

$$\mathcal{C}'' = (A|_{H''}, H'', \Phi|_{H''}, E, \sigma). \tag{1-5}$$

Here P' is the orthogonal projection of H onto H'.

The notion of a colligation has also a system-theoretic significance: a colligation $\mathcal{C} = (A, H, \Phi, E, \sigma)$ defines a (linear time-invariant) conservative system

$$i\frac{df}{dt} + Af = \Phi^* \sigma u, \tag{1-6}$$

$$v = u - i\,\Phi f. \tag{1-7}$$

Here $f = f(t)$ is the state, with values in the inner space H, and $u = u(t)$, $v = v(t)$ are respectively the input and the output, with values in the external space E. Conservativeness means that the difference in energy between the input and the output equals the change in the energy of the state; here we use the inner product as the energy form on the state (inner) space H and the hermitian form induced by σ as the (possibly indefinite) energy form on the input/output (external) space E. Thus the conservation law is

$$\frac{d}{dt}(f, f) = (\sigma u, u) - (\sigma v, v). \tag{1-8}$$

The coupling of colligations corresponds to the cascade connection of systems, i.e., forming a new system by feeding the output of the first one as the input into the second one: if u', f' and v' and u'', f'' and v'' are the input, the state and the output of the first and the second system respectively, then the input, the state and the output u, f and v of the new system are given by

$$u = u', \qquad f = \begin{pmatrix} f' \\ f'' \end{pmatrix}, \qquad v = v'',$$

while setting $u'' = v'$. A substitution into the system equations shows that we get exactly the formula (1–3) for the coupling.

We pass now to the study of a pair A_1, A_2 of (bounded) commuting non-selfadjoint operators in a Hilbert space H, with finite nonhermitian ranks. As a first try we may consider a *commutative (two-operator) colligation*

$$\mathcal{C} = (A_1, A_2, H, \Phi, E, \sigma_1, \sigma_2);$$

here again E is another Hilbert space (the external space, whose dimension we assume is $M < \infty$), $\Phi : H \to E$ and $\sigma_1, \sigma_2 : E \to E$ are bounded linear mappings with $\sigma_1^* = \sigma_1$, $\sigma_2^* = \sigma_2$, and

$$\frac{1}{i}(A_k - A_k^*) = \Phi^* \sigma_k \Phi \quad \text{for } k = 1, 2. \tag{1-9}$$

However, the notion of a commutative colligation does not possess enough structure: there is nothing in it to reflect the interplay between the two operators A_1, A_2. More concretely, the coupling of two commutative colligations with the same external part (E, σ_1, σ_2) (defined as in (1–3) except that one uses σ_1 and σ_2 instead of σ in the formulas for A_1 and A_2 respectively) is in general *not* commutative. In fact, even in a finite dimensional Hilbert space H it is not at all clear how to construct commuting nonselfadjoint operators with given nonhermitian parts.

It turns out that the correct object to consider in the study of a pair of commuting nonselfadjoint operators is a (*commutative two-operator*) *vessel* $\mathcal{V} = (A_1, A_2, H, \Phi, E, \sigma_1, \sigma_2, \gamma, \tilde{\gamma})$. Here $(A_1, A_2, H, \Phi, E, \sigma_1, \sigma_2)$ is a commutative two-operator colligation as in (1–9), and $\gamma, \tilde{\gamma} : E \to E$ are (bounded) self-adjoint operators such that

$$\sigma_1 \Phi A_2^* - \sigma_2 \Phi A_1^* = \gamma \Phi, \tag{1-10}$$

$$\sigma_1 \Phi A_2 - \sigma_2 \Phi A_1 = \tilde{\gamma} \Phi, \tag{1-11}$$

$$\tilde{\gamma} - \gamma = i(\sigma_1 \Phi \Phi^* \sigma_2 - \sigma_2 \Phi \Phi^* \sigma_1). \tag{1-12}$$

(The term "vessel" was coined in [Livšic et al. 1995]; earlier papers use instead
the term "regular colligation".) Upon multiplying (1–10) and (1–11) by Φ^* on
the left and using (1–9) and the obvious identities

$$(A_1 - A_1^*)A_2^* - (A_2 - A_2^*)A_1^* = A_1 A_2^* - A_2 A_1^*, \tag{1–13}$$

$$(A_1 - A_1^*)A_2 - (A_2 - A_2^*)A_1 = A_2^* A_1 - A_1^* A_2, \tag{1–14}$$

which follow from the commutativity of A_1 and A_2, we obtain

$$\tfrac{1}{i}\,(A_1 A_2^* - A_2 A_1^*) = \Phi^* \gamma \Phi, \tag{1–15}$$

$$\tfrac{1}{i}\,(A_2^* A_1 - A_1^* A_2) = \Phi^* \tilde{\gamma} \Phi. \tag{1–16}$$

Therefore the self-adjoint operators γ and $\tilde{\gamma}$ are related to the nonhermitian parts
of $A_1 A_2^*$ and $A_2^* A_1$ respectively, and thus carry information about the interaction
of A_1 and A_2. In the case when $\Phi : H \to E$ is onto, equations (1–15)–(1–16) are
equivalent to (1–10)–(1–12), but in general the stronger relations (1–10)–(1–12)
are needed for subsequent development. Note that analogously to (1–2), a given
pair A_1, A_2 of commuting operators in H (with finite nonhermitian ranks) can
be always embedded in a commutative vessel by setting

$$E = (A_1 - A_1^*)H + (A_2 - A_2^*)H, \qquad \Phi = P_E, \tag{1–17}$$

$$\sigma_1 = \tfrac{1}{i}\,(A_1 - A_1^*)|_E, \qquad \sigma_2 = \tfrac{1}{i}\,(A_2 - A_2^*)|_E,$$

$$\gamma = \tfrac{1}{i}\,(A_1 A_2^* - A_2 A_1^*)|_E, \qquad \tilde{\gamma} = \tfrac{1}{i}\,(A_2^* A_1 - A_1^* A_2)|_E;$$

the subspace E is invariant under $A_1 A_2^* - A_2 A_1^*$ and $A_2^* A_1 - A_1^* A_2$ because of
(1–13)–(1–14).

Given a commutative vessel $\mathcal{V} = (A_1, A_2, H, \Phi, E, \sigma_1, \sigma_2, \gamma, \tilde{\gamma})$, we define a
polynomial in two complex variables λ_1, λ_2 by setting

$$p(\lambda_1, \lambda_2) = \det(\lambda_1 \sigma_2 - \lambda_2 \sigma_1 + \gamma). \tag{1–18}$$

We assume that $p(\lambda_1, \lambda_2) \not\equiv 0$, so that $p(\lambda_1, \lambda_2)$ is a polynomial with real coefficients of degree $M = \dim E$ at most. We call $p(\lambda_1, \lambda_2)$ the *discriminant
polynomial* of the vessel $\mathcal{V}$, and the real (affine) plane algebraic curve C_0 with an
equation $p(\lambda_1, \lambda_2) = 0$—the *(affine) discriminant curve*. To state the first fundamental result discovered by Livšic we have to introduce the principal subspace
$\hat{H} \subseteq H$ of the vessel $\mathcal{V}$,

$$\hat{H} = \bigvee_{k_1, k_2 = 0}^{\infty} A_1^{k_1} A_2^{k_2} \Phi^*(E) = \bigvee_{k_1, k_2 = 0}^{\infty} A_1^{*k_1} A_2^{*k_2} \Phi^*(E). \tag{1–19}$$

Then $\hat{H}$ is reducing for A_1 and A_2, and the restrictions of A_1 and A_2 to $H \ominus \hat{H}$
are self-adjoint operators (the restriction of Φ to $H \ominus \hat{H}$ is 0); hence it is enough,
at least in principle, to consider the restriction of our operators to the principal
subspace.

THEOREM 1.1 (GENERALIZED CAYLEY–HAMILTON THEOREM [Livšic 1979; 1983]). *The operator $p(A_1, A_2)$ vanishes on the principal subspace $\hat{H}$.*

This theorem contains as special cases the classical Cayley–Hamilton Theorem and the theorem of Burchnall and Chaundy [1928] stating that a pair of commuting linear differential operators satisfy a polynomial equation with constant coefficients—a result that plays an important role in the study of finite-zone solutions of the KdV equation and other completely integrable nonlinear PDEs [Dubrovin 1981]. (To be precise, Theorem 1.1 implies the theorem of Burchnall and Chaundy only for formally self-adjoint differential operators; the general case follows from a more general version of Theorem 1.1, due to Kravitsky [1983].)

Theorem 1.1 implies that *the joint spectrum of the operators A_1 and A_2, restricted to the principal subspace $\hat{H}$, lies on the affine discriminant curve C_0.* (The joint spectrum is the set of all points $\lambda = (\lambda_1, \lambda_2) \in \mathbb{C}^2$ such that there exists a sequence $h_1, h_2, \dots$ of vectors of unit length in H satisfying

$$\lim_{n \to \infty} (A_k - \lambda_k I) h_n = 0 \quad \text{for } k = 1, 2;$$

it was proved in [Livšic and Markus 1994] that for a pair of commuting operators with finite-dimensional (or more generally, compact) nonhermitian parts this is equivalent to any other reasonable definition of the joint spectrum; see [Harte 1972; Taylor 1970].) This is a first indication that the spectral analysis of a pair of commuting nonselfadjoint operators with finite nonhermitian ranks should be developed on a compact real Riemann surface (the normalization of the projective closure of C_0) rather than on a domain in $\mathbb{C}^2$.

In the definition (1–18) of the discriminant polynomial we have discriminated in favour of γ and against $\tilde{\gamma}$. However, we have the following remarkable equality.

THEOREM 1.2 [Livšic 1979; 1983]. $\det(\lambda_1 \sigma_2 - \lambda_2 \sigma_1 + \gamma) = \det(\lambda_1 \sigma_2 - \lambda_2 \sigma_1 + \tilde{\gamma})$.

The proof is based on the theory of characteristic functions; we will give a system-theoretic explanation of why Theorem 1.2 is true in Section 2 below. We see that associated to the vessel $\mathcal{V}$ we have the discriminant polynomial $p(\lambda_1, \lambda_2)$ and two *self-adjoint determinantal representations* of it, $\lambda_1 \sigma_2 - \lambda_2 \sigma_1 + \gamma$ and $\lambda_1 \sigma_2 - \lambda_2 \sigma_1 + \tilde{\gamma}$ (called, for system-theoretic reasons, the input and the output determinantal representation respectively); more geometrically, we have the affine discriminant curve C_0 (or rather its projective closure) and a pair of sheaves on it, given by the kernels of the matrices $\lambda_1 \sigma_2 - \lambda_2 \sigma_1 + \gamma$ and $\lambda_1 \sigma_2 - \lambda_2 \sigma_1 + \tilde{\gamma}$. This will turn out to provide a proper algebro-geometrical framework for the study of the pair of operators A_1, A_2.

We now proceed to a system-theoretic interpretation. We start with a commutative two-operator colligation $\mathcal{C} = (A_1, A_2, H, \Phi, E, \sigma_1, \sigma_2)$ as in (1–9), and we write the corresponding (linear time-invariant) commutative conservative two-dimensional system

$$i\frac{\partial f}{\partial t_1} + A_1 f = \Phi^* \sigma_1 u, \tag{1-20}$$

$$i\frac{\partial f}{\partial t_2} + A_2 f = \Phi^* \sigma_2 u, \tag{1-21}$$

$$v = u - i\,\Phi f. \tag{1-22}$$

Here, as in (1–6)–(1–7), $f = f(t_1, t_2)$, $u = u(t_1, t_2)$ and $v = v(t_1, t_2)$ are the state, the input and the output respectively; the difference is that now we have a two-dimensional parameter $t = (t_1, t_2)$. One may think of t_1 as a time variable and of t_2 as a spatial variable, so that (1–20)–(1–22) describes a continuum of interacting temporal systems distributed in space; see [Livšic 1986a]. The energy conservation law is

$$\left(\xi_1 \frac{\partial}{\partial t_1} + \xi_2 \frac{\partial}{\partial t_2}\right)(f, f) = ((\xi_1 \sigma_1 + \xi_2 \sigma_2)u, u) - ((\xi_1 \sigma_1 + \xi_2 \sigma_2)v, v); \tag{1-23}$$

that is, the system conserves energy in any direction (ξ_1, ξ_2) in the (t_1, t_2) plane, where the (possibly indefinite) energy form in the input/output space in the direction (ξ_1, ξ_2) is induced by $\xi_1 \sigma_1 + \xi_2 \sigma_2$.

Unlike the usual one-dimensional system (1–6)–(1–7), the system (1–20)–(1–22) is overdetermined, the compatibility conditions arising from the equality of the mixed partials $\dfrac{\partial^2 f}{\partial t_1 \partial t_2} = \dfrac{\partial^2 f}{\partial t_2 \partial t_1}$. The commutativity $A_1 A_2 = A_2 A_1$ means precisely that the system is consistent for an arbitrary initial state $f(0,0) = f_0$ and the identically zero input. For an arbitrary input u the system (1–20)–(1–22) will not in general be consistent; using the system equations twice in the equality of the mixed partials we obtain

$$\Phi^*\left(\sigma_2 \frac{\partial u}{\partial t_2} - \sigma_1 \frac{\partial u}{\partial t_1}\right) + iA_2 \Phi^* \sigma_1 u - iA_1 \Phi^* \sigma_2 u = 0. \tag{1-24}$$

We see thus that if we assume the vessel condition (1–10), then a necessary and sufficient condition for the input to be compatible is given by

$$\Phi^*\left(\sigma_2 \frac{\partial}{\partial t_1} - \sigma_1 \frac{\partial}{\partial t_2} + i\gamma\right)u = 0. \tag{1-25}$$

In particular we get a sufficient condition for the compatibility of the input entirely in terms of the external data of the system

$$\left(\sigma_2 \frac{\partial}{\partial t_1} - \sigma_1 \frac{\partial}{\partial t_2} + i\gamma\right)u = 0. \tag{1-26}$$

If we use (1–22) to express the input in terms of the output and the state, and substitute into (1–26), we obtain

$$\left(\sigma_2 \frac{\partial}{\partial t_1} - \sigma_1 \frac{\partial}{\partial t_2} + i\left(\gamma + i\,\sigma_1 \Phi\Phi^* \sigma_2 - i\,\sigma_2 \Phi\Phi^* \sigma_1\right)\right)v$$

$$+i\left(\sigma_1 \Phi A_2 - \sigma_2 \Phi A_1 - (\gamma + i\,\sigma_1 \Phi\Phi^* \sigma_2 - i\,\sigma_2 \Phi\Phi^* \sigma_1)\Phi\right)f = 0. \tag{1-27}$$

So we see that if we assume in addition the vessel conditions (1–11) and (1–12), then corresponding to (1–26) there is a sufficient condition for the compatibility of the output

$$\left(\sigma_2 \frac{\partial}{\partial t_1} - \sigma_1 \frac{\partial}{\partial t_2} + i\tilde{\gamma}\right)v = 0. \tag{1–28}$$

Therefore *a commutative two-operator vessel is a commutative two-dimensional system* (1–20)–(1–22) *together with the compatibility PDEs* (1–26) *and* (1–28) *at the input and at the output respectively.*

To illustrate how well the notion of vessel suits the needs of the theory, let us consider the problem of coupling of two commutative two-operator colligations, i.e., cascade connection of two commutative two-dimensional systems. As we have noticed, the result will not in general be commutative. Now assume that we have two commutative vessels

$$\mathcal{V}' = (A_1', A_2', H', \Phi', E, \sigma_1, \sigma_2, \gamma', \tilde{\gamma}'),$$
$$\mathcal{V}'' = (A_1'', A_2'', H'', \Phi'', E, \sigma_1, \sigma_2, \gamma'', \tilde{\gamma}'')$$

with the same (E, σ_1, σ_2), and we want their coupling

$$\mathcal{V} = \mathcal{V}' \vee \mathcal{V}'' = (A_1, A_2, H, \Phi, E, \sigma_1, \sigma_2, \gamma, \tilde{\gamma})$$

to be a commutative vessel, where, as in (1–3), $H = H' \oplus H''$ and

$$A_k = \begin{pmatrix} A_k' & 0 \\ i\,\Phi''^*\sigma_k\Phi' & A_k'' \end{pmatrix} \quad \text{for } k = 1, 2, \qquad \Phi = (\Phi' \quad \Phi''). \tag{1–29}$$

Since, when forming the cascade connection, the output of the first system is fed into the second, the procedure makes sense only when the output compatibility PDE of the first system coincides with the input compatibility PDE of the second (and in this case the input compatibility PDE of the new system coincides with the input compatibility PDE of the first system, and the output compatibility PDE of the new system coincides with the output compatibility PDE of the second system). This explains the following result.

THEOREM 1.3 (MATCHING THEOREM [Livšic 1979; 1983]). $\mathcal{V}$ *(with $\gamma = \gamma'$ and $\tilde{\gamma} = \tilde{\gamma}''$) is a commutative vessel if and only if $\tilde{\gamma}' = \gamma''$.*

We can consider now the following inverse problem. Suppose we are given a real polynomial $p(\lambda_1, \lambda_2)$ defining a real (affine) plane curve C_0, a self-adjoint determinantal representation $\lambda_1 \sigma_2 - \lambda_2 \sigma_1 + \gamma$ of $p(\lambda_1, \lambda_2)$, and a subset S of C_0 which is closed and bounded in $\mathbb{C}^2$ and all of whose accumulation points are real points of C_0. We want *to construct, up to the unitary equivalence on the principal subspace, all commutative two-operator vessels with discriminant polynomial* $p(\lambda_1, \lambda_2)$, *input determinantal representation* $\lambda_1 \sigma_2 - \lambda_2 \sigma_1 + \gamma$, *and the operators* A_1, A_2 *in the vessel having, on the principal subspace, joint spectrum* S. Here two commutative two-operator vessels $\mathcal{V}^{(\alpha)} = (A_1^{(\alpha)}, A_2^{(\alpha)}, H^{(\alpha)}, \Phi^{(\alpha)}, E, \sigma_1, \sigma_2, \gamma, \tilde{\gamma})$ ($\alpha = 1, 2$) are said to be unitary equivalent on their principal subspaces $\hat{H}^{(1)}$

and $\hat{H}^{(2)}$ respectively if there is an isometric mapping U of $\hat{H}^{(1)}$ onto $\hat{H}^{(2)}$ such that

$$A_k^{(2)}|_{\hat{H}^{(2)}} = UA_k^{(1)}|_{\hat{H}^{(1)}}U^{-1} \quad (k = 1, 2), \qquad \Phi^{(2)}|_{\hat{H}^{(2)}} = \Phi^{(1)}|_{\hat{H}^{(1)}}U. \quad (1\text{--}30)$$

Suppose $p(\lambda_1, \lambda_2)$ is an irreducible polynomial (and is of degree $M = \dim E$, so that there are no factors "hidden" at infinity: this amounts to the condition $\det(\xi_1\sigma_1 + \xi_2\sigma_2) \not\equiv 0$). Suppose also that C_0—more precisely, its projective closure C—is a smooth (irreducible) curve (of degree M). Then a complete and explicit solution of the inverse problem stated above was obtained in [Vinnikov 1992]; see [Livšic et al. 1995, Chapter 12] for a detailed elementary exposition in the simplest nontrivial case $M = 3$. (The assumption that C is a smooth curve may be replaced by the assumptions that $\lambda_1\sigma_2 - \lambda_2\sigma_1 + \gamma$ is a maximal determinantal representation of $p(\lambda_1, \lambda_2)$ (i.e., its kernel has a maximal possible dimension at the singular points of C; see Section 2 below), and that the prescribed set $\mathcal{S}$ does not contain any singular points.) This solution leads to triangular models for the corresponding pairs of operators A_1, A_2 with finite nonhermitian ranks, generalizing the well-known triangular models (see [Brodskiĭ and Livšic 1958; Brodskiĭ 1969], for example) for a single nonselfadjoint operator. The solution is based on first constructing elementary objects—vessels with one-dimensional inner space corresponding to the points of the joint spectrum, and then coupling them using Theorem 1.3. It follows from the vessel condition (1–12) that in a vessel with one-dimensional inner space the output determinantal representation is determined by the input determinantal representation and the spectral data; the successive matching of output and input determinantal representations in Theorem 1.3 then gives a system of nonlinear difference (for the discrete part of the spectrum) and differential (for the continuous part of the spectrum) equations for self-adjoint determinantal representations of the polynomial $p(\lambda_1, \lambda_2)$. The geometric assumptions on the curve C imply that self-adjoint determinantal representations can be parametrized by certain points in the Jacobian variety of C [Vinnikov 1993]; and it turns out that passing from a self-adjoint determinantal representation to the corresponding point in the Jacobian variety linearises the systems of nonlinear difference and differential equations alluded to above. Actually, the system can be even solved explicitly using the theta functions, yielding explicit formulas for the operators A_1, A_2 in a triangular model. These formulas contain as a special case the "algebro-geometrical" formulas for finite-zone solutions of completely integrable nonlinear PDEs [Dubrovin 1981].

The fact that triangular models give us all the solutions to the inverse problem, i.e., that every commutative two-operator vessel with a smooth irreducible discriminant curve is unitarily equivalent (on its principal subspace) to a triangular model vessel (on its principal subspace) is related to the fact that the operators A_1, A_2 in the given vessel possess a "sufficiently nice" maximal chain of joint invariant subspaces; compare [Brodskiĭ 1969], for example, for the single-operator case. More important, this is related to the canonical factorization of the (normalized) joint characteristic function of the vessel — we will return to this point in Section 3.

We end this section by discussing several generalizations. First, even though we restrict our attention in this paper to pairs of commuting nonselfadjoint operators, the same framework can be applied to l-tuples for any l; much work here remains to be done, and we shall just review briefly the basic constructions, following [Livšic et al. 1995, Chapters 2,3,4,7]. We start with l commuting operators $A_1, \ldots, A_l$ in a Hilbert space H (with finite nonhermitian ranks), and consider a *commutative l-operator vessel*

$$\mathcal{V} = \big(A_k \ (k = 1, \ldots, l), \ H, \ \Phi,$$
$$E, \ \sigma_k \ (k = 1, \ldots, l), \ \gamma_{kj} \ (k, j = 1, \ldots, l), \ \tilde{\gamma}_{kj} \ (k, j = 1, \ldots, l) \big),$$

where again E is the external space of the vessel (of dimension $M < \infty$), and $\Phi : H \to E$, $\sigma_k, \gamma_{kj}, \tilde{\gamma}_{kj} : E \to E$ are bounded linear mappings with $\sigma_k^* = \sigma_k$, $\gamma_{kj}^* = \gamma_{kj}$, $\tilde{\gamma}_{kj}^* = \tilde{\gamma}_{kj}$, $\gamma_{kj} = -\gamma_{jk}$, $\tilde{\gamma}_{kj} = -\tilde{\gamma}_{jk}$, and

$$\tfrac{1}{i} \left(A_k - A_k^* \right) = \Phi^* \sigma_k \Phi, \tag{1--31}$$

$$\sigma_k \Phi A_j^* - \sigma_j \Phi A_k^* = \gamma_{kj} \Phi, \tag{1--32}$$

$$\sigma_k \Phi A_j - \sigma_j \Phi A_k = \tilde{\gamma}_{kj} \Phi, \tag{1--33}$$

$$\tilde{\gamma}_{kj} - \gamma_{kj} = i \left(\sigma_k \Phi \Phi^* \sigma_j - \sigma_j \Phi \Phi^* \sigma_k \right), \tag{1--34}$$

for $k, j = 1, \ldots, l$. Analogously to the two-operator case, $\mathcal{V}$ defines a (linear time-invariant) commutative conservative lD system together with appropriate compatibility PDEs at the input and at the output. We define the *input discriminant ideal* $\mathfrak{I}$ of the vessel $\mathcal{V}$ to be the ideal in the polynomial ring $\mathbb{C}[\lambda_1, \ldots, \lambda_l]$ generated by all polynomials of the form

$$p(\lambda_1, \ldots, \lambda_l) = \det \left(\sum_{k,j=1}^{l} M^{kj} (\lambda_k \sigma_j - \lambda_j \sigma_k + \gamma_{kj}) \right), \tag{1--35}$$

where $M^{kj} = -M^{jk}$ are arbitrary operators on E; the *(affine) input discriminant variety* $\mathcal{D}$ is the zero variety of the ideal $\mathfrak{I}$, or what turns out to be the same, the set of all points $(\lambda_1, \ldots, \lambda_l) \in \mathbb{C}^l$ such that

$$\bigcap_{k,j=1}^{l} \ker(\lambda_k \sigma_j - \lambda_j \sigma_k + \gamma_{kj}) \neq \{0\}. \tag{1--36}$$

The *output discriminant ideal* $\tilde{\mathfrak{I}}$ of $\mathcal{V}$ and the *(affine) output discriminant variety* $\tilde{\mathcal{D}}$ are defined similarly replacing γ_{kj} by $\tilde{\gamma}_{kj}$. The generalized Cayley–Hamilton theorem states that

$$p(A_1^*, \ldots, A_l^*) = 0 \tag{1--37}$$

for all $p \in \mathfrak{I}$ and

$$p(A_1, \ldots, A_l) = 0 \tag{1--38}$$

for all $p \in \tilde{\mathfrak{I}}$, on the appropriately defined principal subspace $\hat{H}$. The analogue of Theorem 1.2 in general fails, i.e., we may have $\mathfrak{I} \neq \tilde{\mathfrak{I}}$; however—see Livšic and Markus [1994]—the discriminant varieties $\mathcal{D}$ and $\tilde{\mathcal{D}}$ may differ only by a finite number of isolated points, which have to be nonreal joint eigenvalues of finite multiplicity of either $A_1^*, \ldots, A_l^*$ or $A_1, \ldots, A_l$.

At least if we assume a nondegeneracy condition $\det\left(\sum_{k=1}^{l} \xi_k \sigma_k\right) \not\equiv 0$, it follows from (1–36) that the discriminant varieties $\mathcal{D}$ and $\tilde{\mathcal{D}}$ cannot contain components of (complex) dimension greater than 1 (e.g., if $\det \sigma_1 \neq 0$, then it follows that the projection onto the first coordinate λ_1 is (at most) finite to one). Hence $\mathcal{D}$ and $\tilde{\mathcal{D}}$ consist of one and the same (affine) algebraic curve in $\mathbb{C}^l$, and two possibly distinct finite collections of isolated points. It seems that these isolated points are related to various well-known pathologies for l commuting nonselfadjoint or nonunitary operators with $l > 2$, such as the failure of von Neumann's inequality and the nonexistence of commuting unitary dilations for three or more commuting contractions.

Linear time-invariant one-dimensional systems without energy balance condition (1–8) are the basic object of study in system theory, starting with the work of Kalman; see, for example, [Kailath 1980; Bart et al. 1979]. There is a similar nonconservative analogue of the notion of vessel; it has been worked out by Kravitsky [1983]; see also [Livšic et al. 1995, Chapter 8; Vinnikov 1994; Ball and Vinnikov 1996]. In particular, we may use these nonconservative vessels to study meromorphic matrix functions on a compact Riemann surface via their realizations as the (normalized) joint transfer function of a vessel; see below in Section 2.

Another interesting generalization is a time-varying analogue of the notion of vessel that was considered by Gauchman [1983b; 1983a] (in a very general setting of Hilbert bundles on differentiable manifolds) and recently by Livšic [1996]; for simplicity we restrict ourselves again to the conservative two-dimensional case. We consider a linear time-varying conservative two-dimensional system exactly as in (1–20)–(1–22), except that $A_1 = A_1(t_1, t_2)$, $A_2 = A_2(t_1, t_2)$, $\Phi = \Phi(t_1, t_2)$, $\sigma_1 = \sigma_1(t_1, t_2)$ and $\sigma_2 = \sigma_2(t_1, t_2)$ are functions of $t = (t_1, t_2)$; we still assume that $\sigma_1(t)$ and $\sigma_2(t)$ are self-adjoint and the colligation conditions (1–9) hold (for all t). The condition for the system to be compatible for identically zero input and arbitrary initial state becomes the so-called zero-curvature condition:

$$\frac{\partial A_1}{\partial t_2} - \frac{\partial A_2}{\partial t_1} + i[A_1, A_2] = 0. \qquad (1\text{–}39)$$

Repeating the derivation of (1–24) and (1–27) (taking into account various partial derivatives of system operators coming in) we see that we obtain again linear compatibility PDEs (but with variable coefficients) (1–26) and (1–28) at the input and at the output respectively if we assume that we have a *time-varying zero-curvature two-operator vessel*

$$\mathcal{V} = (A_1(t), A_2(t), H, \Phi(t), E, \sigma_1(t), \sigma_2(t), \gamma(t), \tilde{\gamma}(t)),$$

where $\gamma(t), \tilde{\gamma}(t) : E \to E$ are (bounded) operators such that

$$\sigma_1 \Phi A_2^* - \sigma_2 \Phi A_1^* - i(\sigma_1 \frac{\partial \Phi}{\partial t_2} - \sigma_2 \frac{\partial \Phi}{\partial t_1}) = \gamma \Phi, \qquad (1\text{--}40)$$

$$\sigma_1 \Phi A_2 - \sigma_2 \Phi A_1 - i(\sigma_1 \frac{\partial \Phi}{\partial t_2} - \sigma_2 \frac{\partial \Phi}{\partial t_1}) = \tilde{\gamma} \Phi, \qquad (1\text{--}41)$$

$$\tilde{\gamma} - \gamma = i(\sigma_1 \Phi \Phi^* \sigma_2 - \sigma_2 \Phi \Phi^* \sigma_1), \qquad (1\text{--}42)$$

$$\frac{1}{i}(\gamma - \gamma^*) = \frac{\partial \sigma_2}{\partial t_1} - \frac{\partial \sigma_1}{\partial t_2}. \qquad (1\text{--}43)$$

Note that $\gamma(t)$ and $\tilde{\gamma}(t)$ are generally not self-adjoint. An analogue of Theorem 1.3 holds for time-varying zero-curvature vessels, so that we may construct zero-curvature vessels by coupling elementary objects—zero-curvature vessels with one-dimensional state space. A more detailed study of the resulting "coupling chains", both discrete and continuous, remains to be done. Another basic problem is to study the input-output map of the vessel, which goes from the solution space of the input compatibility PDE to the solution space of the output compatibility PDE, and to describe the class of all input-output maps.

An especially important situation is when all the operators depend on only one of the two variables, let us say on t_1. In this case it seems reasonable to consider a perturbation problem, i.e., our time-varying vessel is either compactly supported or rapidly decaying perturbation of a usual time-invariant vessel.

2. The Joint Characteristic Function

The fundamental interplay between the spectral theory of a single nonselfadjoint operator in a Hilbert space and function theory is based on the notion of the *characteristic function* of an operator, more precisely of an operator colligation $\mathcal{C} = (A, H, \Phi, E, \sigma)$, defined by

$$S(\lambda) = I - i\Phi(A - \lambda I)^{-1}\Phi^*\sigma. \qquad (2\text{--}1)$$

See, for example, [Brodskiĭ and Livšic 1958; Brodskiĭ 1969; Sz.-Nagy and Foiaş 1967; de Branges and Rovnyak 1966b; 1966a]. It is an analytic function of $\lambda \in \mathbb{C}$ for λ outside the spectrum of A, whose values are operators on E—or, since we are assuming $\dim E = M < \infty$, matrices. Equation (2–1) has the following consequences:

(1) $S(\lambda)$ is analytic in a neigbourhood of $\lambda = \infty$, and $S(\infty) = I$.

(2) $S(\lambda)$ is meromorphic on $\mathbb{C} \setminus \mathbb{R}$, and satisfies in its domain of analyticity the following metric properties with respect to the self-adjoint operator σ:

$$S^*(\lambda)\sigma S(\lambda)\geq \sigma \quad \text{when } \operatorname{Im} \lambda > 0, \qquad (2\text{--}2)$$

$$S^*(\lambda)\sigma S(\lambda)\leq \sigma \quad \text{when } \operatorname{Im} \lambda < 0, \qquad (2\text{--}3)$$

$$S^*(\lambda)\sigma S(\lambda)= \sigma \quad \text{when } \operatorname{Im} \lambda = 0. \qquad (2\text{--}4)$$

Conversely, given E and σ with $\det \sigma \neq 0$, any function satisfying (1) and (2) is the characteristic function of some colligation with external part (E, σ). Furtermore, the characteristic function determines the corresponding colligation uniquely, up to the unitary equivalence on the principal subspace.

We mention two basic facts relating multiplicative properties of the characteristic function to the spectral properties of the operator A in the colligation.

- The set of singularities of $S(\lambda)$ (i.e., the set of points in the complex plane to a neighborhood of which $S(\lambda)$ cannot be continued analytically) coincides with the spectrum of A restricted to the principal subspace of the colligation.
- If $\mathcal{C} = \mathcal{C}' \vee \mathcal{C}''$ and $S'(\lambda)$, $S''(\lambda)$ are the characteristic functions of the colligations $\mathcal{C}'$, $\mathcal{C}''$ respectively, then $S(\lambda) = S''(\lambda)S'(\lambda)$.

It follows from the second fact that the canonical factorization of $S(\lambda)$ (the Riesz–Nevanlinna–Smirnov factorization for $\dim E = 1$, when $S(\lambda)$ is just a bounded analytic function in the lower or the upper half-plane, and the Potapov [1955] factorization for $\dim E > 1$) is related to the reduction of the operator A to a triangular form; more generally, factorizations of $S(\lambda)$ are related to invariant subspaces of A.

System-theoretically, the characteristic function of the colligation is the *transfer function* of the corresponding system (1–6)–(1–7). There are many equivalent ways to define the transfer function of a linear time-invariant system. The simplest one for our purposes is to assume that the input, the state and the output of the system are waves with the same frequency λ: $u(t) = u_0 e^{it\lambda}$, $f(t) = f_0 e^{it\lambda}$, $v(t) = v_0 e^{it\lambda}$, where $u_0, v_0 \in E$, $f_0 \in H$. Substitution into the system equations (1–6)–(1–7) shows that

$$v_0 = S(\lambda)u_0; \qquad (2\text{--}5)$$

that is, the transfer function (the characteristic function) maps the input amplitude to the output amplitude.

We now consider a commutative vessel $\mathcal{V} = (A_1, A_2, H, \Phi, E, \sigma_1, \sigma_2, \gamma, \tilde{\gamma})$ with the discriminant polynomial

$$p(\lambda_1, \lambda_2) = \det(\lambda_1\sigma_2 - \lambda_2\sigma_1 + \gamma) = \det(\lambda_1\sigma_2 - \lambda_2\sigma_1 + \tilde{\gamma}),$$

and the (affine) discriminant curve C_0 with the equation $p(\lambda_1, \lambda_2) = 0$. We first define the *complete characteristic function* of the vessel by

$$W(\xi_1, \xi_2, z) = I - i\,\Phi(\xi_1 A_1 + \xi_2 A_2 - zI)^{-1}\Phi^*(\xi_1\sigma_1 + \xi_2\sigma_2). \qquad (2\text{--}6)$$

It is an analytic function of $\xi_1, \xi_2, z \in \mathbb{C}$ for z outside the spectrum of $\xi_1 A_1 + \xi_2 A_2$. Since it consists essentially of usual characteristic functions of single-operator colligations obtained from $\mathcal{V}$ by averaging in all possible directions, it is not hard to show that the complete characteristic function determines the corresponding vessel uniquely, up to the unitary equivalence on the principal subspace. We also have appropriate metric properties (where $\xi_1, \xi_2 \in \mathbb{R}$):

$$W^*(\xi_1,\xi_2,z)(\xi_1\sigma_1 + \xi_2\sigma_2)W(\xi_1,\xi_2,z)\geq \xi_1\sigma_1 + \xi_2\sigma_2 \quad \text{when } \operatorname{Im} z > 0, \quad (2\text{--}7)$$

$$W^*(\xi_1,\xi_2,z)(\xi_1\sigma_1 + \xi_2\sigma_2)W(\xi_1,\xi_2,z)\leq \xi_1\sigma_1 + \xi_2\sigma_2 \quad \text{when } \operatorname{Im} z < 0, \quad (2\text{--}8)$$

$$W^*(\xi_1,\xi_2,z)(\xi_1\sigma_1 + \xi_2\sigma_2)W(\xi_1,\xi_2,z)= \xi_1\sigma_1 + \xi_2\sigma_2 \quad \text{when } \operatorname{Im} z = 0. \quad (2\text{--}9)$$

But since $W(\xi_1,\xi_2,z)$ is a function of two independent complex variables (two, because of the homogeneity), it does not admit a good factorization theory to relate to the spectral theory of the pair of operators A_1, A_2.

However it turns out that the complete characteristic function fits perfectly into the algebro-geometrical framework associated to the vessel, given by the discriminant polynomial $p(\lambda_1, \lambda_2)$ and its two determinantal representations. For each point $\lambda = (\lambda_1, \lambda_2) \in C_0$ we may define two nontrivial subspaces of the external space E:

$$\mathcal{E}(\lambda)= \ker(\lambda_1\sigma_2 - \lambda_2\sigma_1 + \gamma), \qquad (2\text{--}10)$$

$$\tilde{\mathcal{E}}(\lambda)= \ker(\lambda_1\sigma_2 - \lambda_2\sigma_1 + \tilde{\gamma}). \qquad (2\text{--}11)$$

THEOREM 2.1 [Livšic 1986b]. *For any point $\lambda = (\lambda_1, \lambda_2)$ on C_0 and for arbitrary complex numbers ξ_1, ξ_2 (such that $\xi_1\lambda_1 + \xi_2\lambda_2$ is outside the spectrum of $\xi_1 A_1 + \xi_2 A_2$), $W(\xi_1, \xi_2, \xi_1\lambda_1 + \xi_2\lambda_2)$ maps $\mathcal{E}(\lambda)$ into $\tilde{\mathcal{E}}(\lambda)$, and the restriction $W(\xi_1, \xi_2, \xi_1\lambda_1 + \xi_2\lambda_2)|\mathcal{E}(\lambda)$ is independent of ξ_1, ξ_2.*

This theorem allows us to define the *joint characteristic function* of the vessel by restricting the complete characteristic function to the discriminant curve and to the fibres of the "input family of subspaces" (2–10):

$$S(\lambda) = W(\xi_1,\xi_2,\xi_1\lambda_1 + \xi_2\lambda_2)|\mathcal{E}(\lambda) : \mathcal{E}(\lambda) \longrightarrow \tilde{\mathcal{E}}(\lambda), \qquad (2\text{--}12)$$

where $\lambda = (\lambda_1, \lambda_2) \in C_0$ and ξ_1, ξ_2 are free complex parameters such that $\xi_1\lambda_1 + \xi_2\lambda_2$ is outside the spectrum of $\xi_1 A_1 + \xi_2 A_2$.

To clarify the definition of the joint characteristic function we shall interpret it as the *joint transfer function* of the corresponding commutative two-dimensional system (1–20)–(1–22) together with the compatibility PDEs (1–26) and (1–28) at the input and at the output. We assume as before that the input, the state and the output of the system are (planar) waves with the same (double) frequency $\lambda = (\lambda_1, \lambda_2) \in \mathbb{C}^2$:

$$u(t_1,t_2) = u_0 e^{it_1\lambda_1 + it_2\lambda_2},$$

$$f(t_1,t_2) = f_0 e^{it_1\lambda_1 + it_2\lambda_2},$$

$$v(t_1,t_2) = v_0 e^{it_1\lambda_1 + it_2\lambda_2},$$

where $u_0, v_0 \in E$, $f_0 \in H$. The input compatibility PDE (1–26) yields

$$(\lambda_1\sigma_2 - \lambda_2\sigma_1 + \gamma)u_0 = 0; \qquad (2\text{–}13)$$

hence $\lambda \in C_0$ and $u_0 \in \mathcal{E}(\lambda)$. The output compatibility PDE (1–28) yields

$$(\lambda_1\sigma_2 - \lambda_2\sigma_1 + \tilde{\gamma})v_0 = 0, \qquad (2\text{–}14)$$

hence again $\lambda \in C_0$ (*this explains the equality of determinants in Theorem* 1.2) and $v_0 \in \tilde{\mathcal{E}}(\lambda)$. Substituting into the system equations, multiplying (1–20) and (1–21) by free complex parameters ξ_1 and ξ_2 respectively and adding, we obtain

$$v_0 = S(\lambda)u_0. \qquad (2\text{–}15)$$

Hence as before the joint transfer function (the joint characteristic function) maps the input amplitude at a given (double) frequency to the output amplitude, except that because of the compatibility PDEs the double frequency is restricted to lie on the (affine) discriminant curve and the input and output amplitudes must lie in the fibres of the input and the output families of subspaces (2–10) and (2–11) respectively at this double frequency.

Unlike the complete characteristic function, which depends on two independent complex variables, the joint characteristic function depends on a point on a one-dimensional complex variety, namely the discriminant curve. We would like to claim that no information is lost by passing to the joint characteristic function. We make two geometric assumptions. The first one is mainly for the simplicity of exposition. We assume that the discriminant polynomial $p(\lambda_1, \lambda_2)$ has only one, possibly multiple, irreducible factor (and is of degree $M = \dim E$, so that there are no factors hidden at infinity). Thus $p(\lambda_1, \lambda_2) = \big(f(\lambda_1, \lambda_2)\big)^r$ for some $r \geq 1$, where $f(\lambda_1, \lambda_2) = 0$ is the irreducible affine equation of a real irreducible projective plane curve C—the projective closure of the affine discriminant curve C_0—of degree m, where $M = mr$.

The second assumption is deeper. We assume that for all smooth points μ on C, we have $\dim \mathcal{E}(\mu) = \dim \tilde{\mathcal{E}}(\mu) = r$; if $\mu \in C$ is a singular point of multiplicity s, we assume $\dim \mathcal{E}(\mu) = \dim \tilde{\mathcal{E}}(\mu) = rs$. In general we have only inequalities:

$$1 \leq \dim \mathcal{E}(\mu) \leq r, \quad 1 \leq \dim \tilde{\mathcal{E}}(\mu) \leq r$$

at smooth points, and

$$1 \leq \dim \mathcal{E}(\mu) \leq rs, \quad 1 \leq \dim \tilde{\mathcal{E}}(\mu) \leq rs$$

at a singular point of multiplicity s. We refer to this second assumption as the *maximality* of the input and the output determinantal representations of $p(\lambda_1, \lambda_2)$. Note that it holds automatically if $r = 1$, i.e., if $p(\lambda_1, \lambda_2)$ is irreducible, and C is smooth.

It follows from the maximality that the subspaces $\mathcal{E}(\mu)$, $\tilde{\mathcal{E}}(\mu)$ for different points μ on C (including, of course, the points at infinity) fit together to form two complex holomorphic rank r vector bundles $\mathcal{E}$, $\tilde{\mathcal{E}}$ on a compact Riemann surface X which is the desingularization (normalization) of C; here $X = C$ when C is smooth, and when C is singular X is obtained from C by resolving the singularities (see [Fulton 1969] or [Griffiths 1989], for example). Note that, since C is a real curve, X is a *real* Riemann surface, that is, a Riemann surface equipped with an anti-holomorphic involution (the complex conjugation on C). The joint characteristic function $S : \mathcal{E} \to \tilde{\mathcal{E}}$ is (after the natural extension to the points of C at infinity) simply a bundle mapping, holomorphic outside the joint spectrum of A_1, A_2. It is meromorphic on $X \setminus X_{\mathbb{R}}$, where $X_{\mathbb{R}}$ is the set of real points of X (fixed points of the anti-holomorphic involution). The following basic fact was established by Livšic in the dissipative case (i.e., when $\xi_1\sigma_1 + \xi_2\sigma_2 > 0$ for some ξ_1, ξ_2) and by Vinnikov in general.

THEOREM 2.2 [Vinnikov 1992; Ball and Vinnikov 1996]. *The joint characteristic function of a vessel (having maximal input and output determinantal representations) determines uniquely the complete characteristic function.*

PROOF. Since C is a plane curve of degree m, for $(\xi_1, \xi_2, z) \in \mathbb{C}^3$ generic the straight line $\xi_1\lambda_1 + \xi_2\lambda_2 = z$ intersects the curve C in m distinct (affine) points $\lambda^1, \ldots, \lambda^m$ (that are all smooth points of C). These points correspond to the m distinct eigenvalues of the (one variable) matrix pencil obtained by restricting $\lambda_1\sigma_2 - \lambda_2\sigma_1 + \gamma$ to the given line. The corresponding eigenspaces are just $\mathcal{E}(\lambda^1), \ldots, \mathcal{E}(\lambda^m)$; by the maximality assumption each one of them has dimension r, so that the sum of their dimensions equals mr and coincides with the dimension M of the ambient space E. Thus we have a (nonorthogonal) direct sum decomposition

$$\mathcal{E}(\lambda^1) \dotplus \cdots \dotplus \mathcal{E}(\lambda^m) = E. \tag{2-16}$$

Let $P(\xi_1, \xi_2, \lambda^i)$ be the corresponding projections of E onto $\mathcal{E}(\lambda^i)$, so that

$$P(\xi_1, \xi_2, \lambda^1) + \cdots + P(\xi_1, \xi_2, \lambda^m) = I. \tag{2-17}$$

Since $W(\xi_1, \xi_2, z)|_{\mathcal{E}(\lambda^i)} = S(\lambda^i)$ for all i by the definition of the joint characteristic function, we obtain from (2–17) an explicit formula, called the *restoration formula*

$$W(\xi_1, \xi_2, z) = \sum_{i=1}^{m} S(\lambda^i) P(\xi_1, \xi_2, \lambda^i) \tag{2-18}$$

for the complete characteristic function in terms of the joint (on an open dense subset of the domain of analyticity of $W(\xi_1, \xi_2, z)$). $\quad\square$

The next question is how to express the metric properties (2–7)–(2–9) in terms of the joint characteristic function. To this end we introduce an (indefinite) scalar product on the fibres of the input bundle $\mathcal{E}$ over nonreal (affine) points, by setting

$$[u, v]_\lambda^\mathcal{E} = i \, \frac{v^*(\xi_1\sigma_1 + \xi_2\sigma_2)u}{\xi_1(\lambda_1 - \bar{\lambda}_1) + \xi_2(\lambda_2 - \bar{\lambda}_2)} \quad \text{for } u, v \in \mathcal{E}(\lambda), \tag{2-19}$$

and similarly for the output bundle $\tilde{\mathcal{E}}$; here $\lambda = (\lambda_1, \lambda_2) \in C_0$, $\bar{\lambda} \neq \lambda$, and ξ_1, ξ_2 are free parameters—the value of (2–19) turns out to be independent of ξ_1, ξ_2. Note that this metric on the bundle generalizes the Poincaré metric on the upper half-plane. There is also a version of (2–19) at the real points, taking the limit and renormalizing it to be finite, namely

$$[u, v]_\lambda^{\mathcal{E}} = \frac{v^*(\xi_1 \sigma_1 + \xi_2 \sigma_2)u}{\xi_1 \, d\lambda_1(\lambda) + \xi_2 \, d\lambda_2(\lambda)} \quad \text{for } u, v \in \mathcal{E}(\lambda); \tag{2–20}$$

here $\lambda = (\lambda_1, \lambda_2) \in C_0$, $\bar{\lambda} = \lambda$, and ξ_1, ξ_2 are free parameters—the value of (2–20) turns out again to be independent of ξ_1, ξ_2. More generally, we introduce a hermitian pairing between the fibres of the bundle over nonconjugate (affine) points

$$[u, v]_{\lambda^1, \lambda^2}^{\mathcal{E}} = i \frac{v^*(\xi_1 \sigma_1 + \xi_2 \sigma_2)u}{\xi_1(\lambda_1^1 - \bar{\lambda}_1^2) + \xi_2(\lambda_2^1 - \bar{\lambda}_2^2)} \quad \text{for } u \in \mathcal{E}(\lambda^1), \ v \in \mathcal{E}(\lambda^2) \tag{2–21}$$

(with $\lambda^1 = (\lambda_1^1, \lambda_2^1), \lambda^2 = (\lambda_1^2, \lambda_2^2) \in C_0$, $\bar{\lambda}^2 \neq \lambda^1$), and over conjugate (affine) points

$$[u, v]_{\lambda, \bar{\lambda}}^{\mathcal{E}} = \frac{v^*(\xi_1 \sigma_1 + \xi_2 \sigma_2)u}{\xi_1 \, d\lambda_1(\lambda) + \xi_2 \, d\lambda_2(\lambda)} \quad \text{for } u \in \mathcal{E}(\lambda), \ v \in \mathcal{E}(\bar{\lambda}) \tag{2–22}$$

(with $\lambda = (\lambda_1, \lambda_2) \in C_0$). Then it can be shown, using the restoration formula (2–18), that *the properties* (2–7)–(2–9) *are equivalent to the following metric properties of the joint characteristic function in its domain of analyticity:*

$$[S(\lambda)u, S(\mu)v]_{\lambda, \mu}^{\tilde{\mathcal{E}}} \geq [u, v]_{\lambda, \mu}^{\mathcal{E}} \quad \text{for } u \in \mathcal{E}(\lambda), \ v \in \mathcal{E}(\mu), \tag{2–23}$$

$$[S(\lambda)u, S(\bar{\lambda})v]_{\lambda, \bar{\lambda}}^{\tilde{\mathcal{E}}} = [u, v]_{\lambda, \bar{\lambda}}^{\mathcal{E}} \quad \text{for } u \in \mathcal{E}(\lambda), \ v \in \mathcal{E}(\bar{\lambda}). \tag{2–24}$$

See [Vinnikov 1992] and, for details, [Livšic et al. 1995, Chapter 10]. The inequality in (2–23) means as usual that the expression appearing on the left-hand side is a positive definite kernel, i.e., for any N points $\lambda^1, \ldots, \lambda^N$ on C_0 in the domain of analyticity of $S(\lambda)$ $(\lambda^i \neq \bar{\lambda}^j)$ and any $u_i \in \mathcal{E}(\lambda^i)$ we have

$$\left([S(\lambda^i)u_i, S(\lambda^j)u_j]_{\lambda^i, \lambda^j}^{\tilde{\mathcal{E}}} - [u_i, u_j]_{\lambda^i, \lambda^j}^{\mathcal{E}} \right)_{i,j=1,\ldots,N} \geq 0.$$

In particular, the bundle map S is expansive at nonreal points and isometric at real points with respect to the scalar product (2–19)–(2–20) on the input and the output bundles.

The joint characteristic function is not quite the end of the quest for a proper generalization of the usual characteristic function, since the kernel bundles $\mathcal{E}$ and $\tilde{\mathcal{E}}$ and the scalar product (2–19)–(2–20) are hard to deal with analytically. However, it follows from the theory of determinantal representations of plane algebraic curves [Vinnikov 1989; 1993; Ball and Vinnikov 1996] that these bundles are isomorphic (up to an inessential twist) to certain vector bundles of the form $V_\chi \otimes \Delta$, where V_χ is the flat vector bundle corresponding to a representation $\chi : \pi_1(X) \to \mathrm{GL}(r, \mathbb{C})$ of the fundamental group of the Riemann surface X, and

Δ is a line bundle of half-order differentials on X (a square root of the canonical bundle). Sections of $V_\chi \otimes \Delta$ are thus multiplicative $\mathbb{C}^r$-valued half-order differentials on X; here a half-order differential is an expression locally of the form $f(t)\sqrt{dt}$ where t is a local parameter on X, and "multiplicative" means that our vector-valued half-order differential picks up a multiplier (factor of automorphy) $\chi(R)$ when we go around a closed loop R on X. We see then that we obtain from the joint characteristic function of the vessel a so-called *normalized joint characteristic function*, which is a mapping of flat vector bundles on X, i.e., a multiplicative $r \times r$ matrix function on X, with appropriate multipliers on the left and on the right. We proceed, following [Vinnikov 1992], to describe the scalar case $r = 1$ (line bundles); in this case the results are the most complete since we may use the classical theory of Jacobian varieties and theta functions (see also [Livšic et al. 1995, Chapters 10–11] for a detailed exposition, using elliptic functions only, of the simplest nontrivial case when the discriminant curve C is a smooth cubic, $m = 3$).

We let g be the genus (the "number of handles") of the compact Riemann surface X; when C is a smooth curve, the genus is given in terms of the degree of C by the formula $g = (m-1)(m-2)/2$. We choose a canonical homology basis on X consisting of the A-cycles $A_1, \ldots, A_g$ and the B-cycles $B_1, \ldots, B_g$ (and satisfying certain symmetry requirements with respect to the anti-holomorphic involution on X: see [Vinnikov 1993]). We can then construct a basis $\omega_1, \ldots, \omega_g$ for the space of holomorphic differentials on X which is normalized with repect to our homology basis: $\int_{A_j} \omega_i = \delta_{ij}$; the so-called period matrix $\Omega = \left(\int_{B_j} \omega_i \right) \in \mathbb{C}^{g \times g}$; the period lattice $\Lambda = \mathbb{Z}^g + \Omega\mathbb{Z}^g \subset \mathbb{C}^g$ (this is the lattice in $\mathbb{C}^g$ formed by integrals of the column with entries $\omega_1, \ldots, \omega_g$ over all possible closed loops on X); the Jacobian variety $J(X) = \mathbb{C}^g/\Lambda$, and the associated Riemann's theta function $\theta(w) = \theta(w; \Omega)$, $w \in \mathbb{C}^g$. See [Mumford 1983], for example, for all these classical notions. For $\zeta \in J(X)$ we let L_ζ be the flat line bundle with multipliers of absolute value 1 corresponding to ζ: we write $\zeta = b + \Omega a$ where $a, b \in \mathbb{R}^g$ have coordinates a_j, b_j, and the multipliers over the cycles A_j and B_j are given by $e^{-2\pi i a_j}$ and $e^{2\pi i b_j}$ respectively, for $j = 1, \ldots, g$. We let Δ be the line bundle of half-order differentials corresponding to $-\kappa$, where $\kappa \in J(X)$ is the so-called Riemann's constant. Then $\mathcal{E}$ is isomorphic to the kernel bundle associated to a maximal self-adjoint determinantal representation of the irreducible defining polynomial $f(\lambda_1, \lambda_2)$ of C if and only if

$$\mathcal{E} \otimes \mathcal{O}(m-2)(-D) \cong L_\zeta \otimes \Delta, \qquad (2\text{-}25)$$

where D is the divisor of singularities of C on X ($D = 0$ when C is smooth) and $\zeta \in J(X)$ satisfies

$$\theta(\zeta) \neq 0 \qquad (2\text{-}26)$$

(this is equivalent, by Riemann's Theorem, to the fact that the line bundle $L_\zeta \otimes \Delta$ has no global holomorphic sections, and is a necessary and sufficient condition for $\mathcal{E}$ to be isomorphic to the kernel bundle associated to some—not necessarily self-adjoint—determinantal representation of $f(\lambda_1, \lambda_2)$), and

$$\zeta + \bar{\zeta} = \bar{\kappa} - \kappa \tag{2-27}$$

(this condition ensures the self-adjointness). The solution set of (2–26)–(2–27) in the g-dimensional complex torus $J(X)$ is a finite disjoint union of punctured g-dimensional real tori, the punctures coming from the zeroes of the theta function. In concrete terms, the isomorphism (2–25) means that there exists a nowhere zero section $u^{\times}(p)$ of $\mathcal{E}$ on X, whose entries are multiplicative meromorphic half-order differentials with multipliers corresponding to $-\zeta$ and with simple poles (at most) at the points of C at infinity; the isomorphism is then given by

$$y(p) \mapsto \frac{1}{\omega(p)} y(p) u^{\times}(p), \tag{2-28}$$

where $y(p)$ is a holomorphic multiplicative half-order differential on an open subset U of X with multipliers corresponding to ζ (a holomorphic section of $L_{\zeta} \otimes \Delta$ on U), and

$$\omega = \frac{d\lambda_1}{\partial f / \partial \lambda_2} = -\frac{d\lambda_2}{\partial f / \partial \lambda_1}$$

is a fixed meromorphic differential on X with zeroes of order $m-3$ at infinity and poles on the divisor of singularities D; note that the right-hand side of (2–28) is a section of $\mathcal{E}$ on U whose entries are meromorphic functions with poles of order $m-2$ (at most) at the points of C at infinity and vanishing on D (a holomorphic section of $\mathcal{E} \otimes \mathcal{O}(m-2)(-D)$ on U), as required. We call $u^{\times}(p)$ a normalized section of $\mathcal{E}$; it is determined uniquely up to a nonzero constant factor.

It follows that, if the input and the output line bundles $\mathcal{E}$ and $\tilde{\mathcal{E}}$ of the vessel $\mathcal{V}$ correspond as in (2–25) to the points ζ and $\tilde{\zeta}$ in $J(X)$ respectively, then the joint characteristic function $S : \mathcal{E} \to \tilde{\mathcal{E}}$ yields, under the corresponding isomorphisms, a scalar multiplicative function T on X with multipliers corresponding to $\tilde{\zeta} - \zeta$, called the *normalized joint characteristic function*. $T(p)$ is holomorphic and nonzero in a neighborhood of the points of C at infinity, and is meromorphic on $X \setminus X_{\mathbb{R}}$. In terms of normalized sections $u^{\times}(p)$ and $\tilde{u}^{\times}(p)$ of $\mathcal{E}$ and $\tilde{\mathcal{E}}$ respectively we have

$$S(p) u^{\times}(p) = T(p) \tilde{u}^{\times}(p). \tag{2-29}$$

Now—and this is the main point—the scalar product (2–19)–(2–20) on the line bundles $\mathcal{E}$ and $\tilde{\mathcal{E}}$, and more generally the pairing (2–21)–(2–22), can be expressed analytically in terms of theta functions. Explicitly, we have (after adjusting $u^{\times}(p)$ by an appropriate constant factor)

$$[u^{\times}(p), u^{\times}(q)]_{p,q}^{\mathcal{E}} = \varepsilon \, \frac{\theta \begin{bmatrix} a \\ b \end{bmatrix}(p - \bar{q})}{i\,\theta \begin{bmatrix} a \\ b \end{bmatrix}(0)\, E(p, \bar{q})} \quad \text{when } p \neq \bar{q}, \tag{2-30}$$

$$[u^\times(p), u^\times(p)]^{\tilde{\mathcal{E}}}_{p,\bar{p}} = \varepsilon, \tag{2-31}$$

and similarly for the output line bundle $\tilde{\mathcal{E}}$, the notation being this: $\varepsilon = \pm 1$ is the so-called sign of the input determinantal representation, which distinguishes between two self-adjoint determinantal representations that differ by a factor of -1 and hence have the same kernel bundle (it turns out that the output determinantal representation has the same sign ε);

$$\theta\begin{bmatrix} a \\ b \end{bmatrix}(w), \quad \text{for } w \in \mathbb{C}^g,$$

is the theta function with characteristics $a, b \in \mathbb{R}^g$ corresponding to ζ in $J(X)$, i.e., $\zeta = b + \Omega a$ (whenever we write a point on the Riemann surface X in the argument of a theta function, we mean the image of the point in $J(X)$ under the Abel–Jacobi map—more precisely, some lifting of the image to $\mathbb{C}^g$); and finally, $E(\cdot, \cdot)$ is the *prime form* on X, whose main property is that $E(p, s) = 0$ if and only if $p = s$ (this is a generalization to a compact Riemann surface of higher genus of the difference between two numbers in the complex plane). For more on the prime form, see [Mumford 1984; Fay 1973].

It follows from (2–30)–(2–31) that the metric properties (2–23)–(2–24) become, in terms of T

$$\varepsilon\, T(p)\overline{T(q)}\, \frac{\theta\begin{bmatrix} \tilde{a} \\ \tilde{b} \end{bmatrix}(p - \bar{q})}{i\,\theta\begin{bmatrix} \tilde{a} \\ \tilde{b} \end{bmatrix}(0)\, E(p, \bar{q})} - \varepsilon\, \frac{\theta\begin{bmatrix} a \\ b \end{bmatrix}(p - \bar{q})}{i\,\theta\begin{bmatrix} a \\ b \end{bmatrix}(0)\, E(p, \bar{q})} \geq 0, \tag{2-32}$$

$$T(p)\overline{T(\bar{p})} = 1 \tag{2-33}$$

in the domain of analyticity of $T(p)$ on X. Here $a, b \in \mathbb{R}^g$ and $\tilde{a}, \tilde{b} \in \mathbb{R}^g$ are the characteristics corresponding to ζ and $\tilde{\zeta}$ in $J(X)$ respectively, i.e., $\zeta = b + \Omega a$ and $\tilde{\zeta} = \tilde{b} + \Omega \tilde{a}$. The inequality in (2–32) means as in (2–23) that the expression appearing on the left-hand side is a positive definite kernel.

We can give now a complete analytic description of the class of normalized joint characteristic functions of vessels with an irreducible discriminant polynomial (and maximal input and output determinantal representations).

THEOREM 2.3. *A multiplicative function $T(p)$ on X with multipliers corresponding to $\tilde{\zeta} - \zeta$ is the normalized joint characteristic function of a vessel with discriminant polynomial $f(\lambda_1, \lambda_2)$ and having maximal input and output determinantal representations of sign ε and corresponding to points ζ and $\tilde{\zeta}$ in $J(X)$ respectively if and only if $T(p)$ is holomorphic and nonzero in a neighborhood of the points of C at infinity, is meromorphic on $X \setminus X_\mathbb{R}$, and satisfies (2–33) and (2–32).*

It is worthwhile to mention that the corresponding point in $J(X)$ and the sign determine a maximal self-adjoint determinantal representation of $f(\lambda_1, \lambda_2)$ up to equivalence, where the equivalence relation is defined by multiplying a self-adjoint determinantal representation on the right and on the left by a (constant) invertible operator on the external space E and by its adjoint respectively. In Theorem 2.3 we may choose arbitrarily the input determinantal representation of the vessel within the equivalence class corresponding to $\zeta \in J(X)$ and ε, and then the output determinantal representation is uniquely (and explicitly) determined (within the equivalence class corresponding to $\tilde{\zeta} \in J(X)$ and ε) by the given normalized joint characteristic function $T(p)$ (actually, by the values of $T(p)$ at the points of C at infinity). This fact is of fundamental importance in the construction of triangular models; see [Livšic et al. 1995, Chapter 12].

Finally, it can be also shown that *the set of singularities of the normalized joint characteristic function coincides with the joint spectrum of the operators A_1 and A_2 in the vessel restricted to the principal subspace.*

It is natural to try to generalize all these results to the case when the discriminant polynomial has a multiple irreducible factor and the normalized joint characteristic function is a multiplicative matrix function on a compact real Riemann surface. An appropriate tool for such a generalization seems to be the notion of the *Cauchy kernel* $K(V_\chi; p, s)$ for a flat vector bundle V_χ of rank r on a compact Riemann surface X, where it is assumed that $V_\chi \otimes \Delta$ has no global holomorphic sections. The Cauchy kernel is defined as the unique meromorphic section of $\pi_1^* V_\chi \otimes \pi_2^* V_\chi^\vee \otimes \pi_1^* \Delta \otimes \pi_2^* \Delta$ on $X \times X$, where π_1 and π_2 denote the projections of $X \times X$ onto the first and the second factor respectively and $F^\vee$ denotes the dual of a vector bundle F, holomorphic except for a simple pole with residue I_r along the diagonal $p = s$. This notion was introduced in [Ball and Vinnikov 1996]; similar kernels were also considered in [Fay 1992]. In the scalar case $r = 1$, when $V_\chi = L_\zeta$ is the unitary flat line bundle corresponding to a point ζ in $J(X)$ with characteristics $a, b \in \mathbb{R}^g$ as before, we have

$$K(V_\chi; p, s) = \frac{\theta \begin{bmatrix} a \\ b \end{bmatrix} (s - p)}{\theta \begin{bmatrix} a \\ b \end{bmatrix} (0) \, E(s, p)}. \tag{2–34}$$

There are so far no similar explicit formulas for $r > 1$, but the Cauchy kernels themselves seem to provide the basic building blocks for the theory. For instance, Fay's trisecant identity—the fundamental identity satisfied by theta functions on a compact Riemann surface—was generalized in [Ball and Vinnikov] to vector bundles of higher rank in terms of the Cauchy kernels. Using the Cauchy kernels it should be possible to generalize Theorem 2.3 and the functional models of Section 3 below to the case of a matrix valued normalized joint characteristic function.

In the usual one-dimensional case, the notion of the transfer function is very important also for nonconservative systems, especially when the state space is finite dimensional, so that the transfer function is a rational matrix function;

in particular, realization of a rational matrix function as the transfer function is a basic tool in the study of various factorization and interpolation problems; see, for example, [Bart et al. 1979; Ball et al. 1990]. It is possible to introduce the joint transfer function and (assuming the maximality of determinantal representations) the normalized joint transfer function also for nonconservative vessels; see [Vinnikov 1994; Ball and Vinnikov 1996]. In the latter reference a nonconservative analog of Theorem 2.3 for multiplicative meromorphic $r \times r$ matrix functions on a compact Riemann surface was obtained, i.e., realization as the normalized joint transfer function of a (nonconservative) vessel with a finite-dimensional state space. The proof, at least for the case of simple poles only, is linear-algebraic in the spirit of [Bart et al. 1979], constructing the vessel explicitly from the poles and the residues of the meromorphic matrix function. Realizations are then used in [Ball and Vinnikov 1996] to solve completely, for the case of simple zeroes and poles, the problem of reconstructing a multiplicative meromorphic matrix function from its zero-pole data (including directional information), i.e., to solve the "homogenous interpolation problem" of [Ball et al. 1990] on a compact Riemann surface of a higher genus.

3. Semiexpansive Functions and Functional Models

Let X be a compact real Riemann surface, let $\zeta, \tilde{\zeta} \in J(X)$ satisfy (2–26) and (2–27), and let $T(p)$ be a multiplicative function on X with multipliers corresponding to $\tilde{\zeta} - \zeta$ that is meromorphic on $X \setminus X_{\mathbb{R}}$; we call $T(p)$ a *semiexpansive*, or more specifically $(\zeta, \tilde{\zeta})$-*expansive*, function if it satisfies (2–33) and (2–32) with $\varepsilon = +1$ (respectively semicontractive or $(\zeta, \tilde{\zeta})$-contractive for $\varepsilon = -1$). Theorem 2.3 suggests that the class of semiexpansive (or semicontractive) functions should be a proper generalization to the case of a compact real Riemann surface of a higher genus of the class of expansive (or contractive) functions on the upper half-plane, or on the unit disk. We first list some basic properties.

We assume the set $X_{\mathbb{R}}$ of real points of X is nonempty; it follows that $X_{\mathbb{R}}$ is a disjoint union of $k > 0$ topological circles $X_0, \ldots, X_{k-1}$. It turns out— see [Vinnikov 1993]—that the real tori comprising the solution set of (2–27) can be naturally indexed as T_ν, where $\nu = (\nu_1, \ldots, \nu_{k-1}) \in \{0, 1\}^{k-1}$. Now, there can be two different situations: either $X_{\mathbb{R}}$ disconnects X, necessarily into two connected components interchanged by the anti-holomorphic involution, X_+ (the "interior" with $\partial X_+ = X_{\mathbb{R}}$ relative to a chosen orientation of $X_{\mathbb{R}}$) and X_- (the "exterior" with $\partial X_- = -X_{\mathbb{R}}$)—the *dividing case*, or $X \setminus X_{\mathbb{R}}$ remains connected—the *nondividing case*. Note that in the dividing case X is simply the double of a finite bordered Riemann surface X_+.

It is easy to see that ζ and $\tilde{\zeta}$ belong to the same real torus T_ν. Assume that X is dividing and that $\zeta, \tilde{\zeta} \in T_0$. Then it turns out that $T(p)$ is $(\zeta, \tilde{\zeta})$-expansive if and only if $|T(p)| \geq 1$ for $p \in X_+$ (equivalently, by (2–33), $|T(p)| \leq 1$ for $p \in X_-$), i.e., $T(p)$ is simply an expansive multiplicative function on X_+. This follows in a standard way from the fact that

$$-\frac{1}{2\pi} \frac{\theta\begin{bmatrix} a \\ b \end{bmatrix}(\bar{q} - p)}{i\,\theta\begin{bmatrix} a \\ b \end{bmatrix}(0)\,E(\bar{q}, p)}$$

is (in the variable p) the reproducing kernel for the Hardy space $H^2(L_\zeta \otimes \Delta, X_+)$ of holomorphic sections of $L_\zeta \otimes \Delta$ on X_+ with the norm $\|y\| = \int_{X_\mathbb{R}} y\bar{y}$. Note that since y is a section of $L_\zeta \otimes \Delta$, $y\bar{y}$ is locally of the form $f(t)|dt|$ where t is a local parameter on X, so the above integral makes sense; in fact, Hardy spaces of half-order differentials on a finite bordered Riemann surface, being invariantly defined without any additional choices, turn out to be more convenient to handle than more traditional Hardy spaces of functions, which require a choice of some measure on the boundary; see [Alpay and Vinnikov b].

It is possible to give an operator-theoretic criterion for the above situation, i.e., when for a vessel $\mathcal{V}$ with irreducible discriminant polynomial and maximal input and output determinantal representations, X (the compact real Riemann surface which is the desingularization of the discriminant curve) is dividing, $\zeta, \tilde{\zeta} \in T_0$ and so the normalized joint characteristic function is simply expansive (or contractive if $\varepsilon = -1$) on X_+. This happens (assuming the mapping $\Phi : H \to E$ in the vessel is surjective) *if and only if there exists a real rational function* $r(\lambda_1, \lambda_2)$ *of two variables such that the operator* $r(A_1, A_2)$ *is defined and dissipative (i.e., has nonnegative imaginary part).* The proof uses the functional model and the description (see [Vinnikov 1993]) of definite self-adjoint determinantal representations of a real plane curve; the real rational function r is defined by $z(p) = r(\lambda_1(p), \lambda_2(p))$, where $\lambda_1(p)$ and $\lambda_2(p)$ are the affine coordinate functions on the discriminant curve and $z(p)$ is a meromorphic function on X mapping X_+ onto the upper half-plane. The existence of such functions was established in [Ahlfors 1950].

In general, if X is dividing and $\zeta, \tilde{\zeta} \in T_\nu$, it turns out that the nontangential boundary value (from the left) on $X_\mathbb{R}$ of a $(\zeta, \tilde{\zeta})$-expansive function $T(p)$ (which exists almost everywhere) satisfies $|T(p)| \geq 1$ if $\nu_j = 0$ and $|T(p)| \leq 1$ if $\nu_j = 1$ for $p \in X_j$ ($j = 0, \ldots, k-1$, we set $\nu_0 = 1$). Assuming that $T(p)$ has no zeroes in X_+ and $|T(p)|$ is bounded away from zero on $X_\mathbb{R}$, it follows that multiplication by $T(p)$ defines a contraction from $H^2_\nu(L_\zeta \otimes \Delta, X_-)$ to $H^2_\nu(L_{\tilde{\zeta}} \otimes \Delta, X_-)$, where $H^2_\nu(\cdot, X_-)$ is the Hardy space of holomorphic sections of an appropriate bundle on X_- with an indefinite inner product $[y, y]_\nu = \sum_{j=0}^{k-1} (-1)^{\nu_j} \int_{X_j} y\bar{y}$. This indefinite inner product space is actually a Kreĭn space, that is, an orthogonal direct sum of a Hilbert space and an anti-Hilbert space [Alpay and Vinnikov b]; this fact is entirely nonobvious even in the simplest case when X_+ is an annulus (and X is a torus). The reproducing kernel for this space is given by the same formula in terms of theta functions as in the Hilbert space case ($\nu = 0$) above.

We discuss briefly the canonical factorization theorem for semiexpansive functions, generalizing the Riesz–Nevanlinna–Smirnov factorization for expansive functions on the upper half-plane; see [Vinnikov 1992], and for an exposition in the case of genus 1 also [Livšic et al. 1995, Chapter 12]. Every semiexpansive function on X can be factored into the Blaschke product constructed from its poles on $X \setminus X_{\mathbb{R}}$, and the singular inner function and outer function constructed from the singular and absolutely continuous parts respectively (with respect to the local Lebesgue measure on $X_{\mathbb{R}}$) of a uniquely determined finite positive Borel measure on $X_{\mathbb{R}}$; furthermore, $\tilde{\zeta}$ can be computed from ζ, the poles and the measure. When X is dividing and $\zeta \in T_0$, so that we have an expansive function on X_+, the measure on $X_{\mathbb{R}}$ is arbitrary and the poles p_i are arbitrary points in X_+ satisfying an appropriate Blaschke condition ($\sum_{i=1}^{\infty}(\bar{p}_i - p_i)$ converges in $J(X)$); this is a fairly straightforward generalization of the usual canonical factorization to multiply connected domains and finite bordered Riemann surfaces; see [Voichick and Zalcman 1965; Hasumi 1966]. In general, the poles and the measure must satisfy certain explicit admissibility conditions with respect to the given ζ having to do with possible zeroes of the theta function; for example, we must have $\theta(\zeta + \sum_{i=1}^{\infty}(\bar{p}_i - p_i)) \neq 0$. The canonical factorization of the normalized joint characteristic function corresponds to the construction of the triangular model for the pair of operators A_1, A_2 in the vessel; and the admissibility conditions for the poles and the measure are precisely the solvability conditions for the systems of nonlinear difference and differential equations for self-adjoint determinantal representations that arise in the construction of the triangular model and were mentionned in Section 1.

We now turn to the description of functional models for commutative two-operator vessels with irreducible discriminant polynomial (and maximal input and output determinantal representations); see [Alpay and Vinnikov 1994; a]. For the single-operator case see [Sz.-Nagy and Foias 1967; de Branges and Rovnyak 1966a; 1966b], as well as the more recent surveys [Nikolskii and Vasyunin 1986; Ball and Cohen 1991]. We shall describe an analog of the de Branges–Rovnyak functional model; at least in the case when X is dividing and $\zeta, \tilde{\zeta} \in T_0$, an analog of Sz.-Nagy–Foias functional model can be constructed as well. It would be very interesting to further investigate, in this case, the geometry of the minimal joint unitary dilation of the corresponding commuting continuous semigroups of contractions, and also to find an analogue of the "coordinate free" functional model of Nikolskii and Vasyunin [1986; 1998].

For a given $(\zeta, \tilde{\zeta})$-expansive function $T(p)$ on X we let the corresponding model space $H(T)$ be the reproducing kernel Hilbert space with reproducing kernel

$$K_T(p, q) = \frac{\theta \begin{bmatrix} \tilde{a} \\ \tilde{b} \end{bmatrix}(\bar{q} - p)}{i\,\theta \begin{bmatrix} \tilde{a} \\ \tilde{b} \end{bmatrix}(0)\, E(\bar{q}, p)} - T(p)\overline{T(q)}\,\frac{\theta \begin{bmatrix} a \\ b \end{bmatrix}(\bar{q} - p)}{i\,\theta \begin{bmatrix} a \\ b \end{bmatrix}(0)\, E(\bar{q}, p)}. \tag{3-1}$$

By the definition of semiexpansive functions $K_T(p, q)$ is a positive definite kernel (here, as in (2–32), the characteristics $a, b \in \mathbb{R}^g$ and $\tilde{a}, \tilde{b} \in \mathbb{R}^g$ correspond to ζ and $\tilde{\zeta}$ in $J(X)$ respectively); therefore $H(T)$ exists and its elements are sections of $L_{\tilde{\zeta}} \otimes \Delta$ holomorphic on the domain of analyticity of $T(p)$. Assume that X is dividing and $\zeta, \tilde{\zeta} \in T_0$. If in addition we assume that $T(p)$ is $(\zeta, \tilde{\zeta})$-inner, i.e., the nontangential boundary values satisfy $|T(p)| = 1$ on $X_{\mathbb{R}}$ almost everywhere, then it follows in a standard way that

$$H(T) = H^2(L_{\tilde{\zeta}} \otimes \Delta, X_-) \ominus TH^2(L_{\zeta} \otimes \Delta, X_-). \tag{3–2}$$

(The equality sign here is a bit sloppy since the elements of the right-hand side space are defined only on X_-, while the elements of $H(T)$ are defined on X_+ as well, except for the poles of $T(p)$; we mean a natural isomorphism given by the restriction of an element of $H(T)$ to X_-.) If we don't assume that $T(p)$ is $(\zeta, \tilde{\zeta})$-inner, then $H(T)$ is the generalized orthogonal complement, in the sense of de Branges, of $TH^2(L_{\zeta} \otimes \Delta, X_-)$ in $H^2(L_{\tilde{\zeta}} \otimes \Delta, X_-)$. If X is dividing and $\zeta, \tilde{\zeta} \in T_\nu$ we have similar formulas using the Kreĭn spaces $H_\nu^2(L_{\zeta} \otimes \Delta, X_-)$ and $H_\nu^2(L_{\tilde{\zeta}} \otimes \Delta, X_-)$ instead (assuming that $T(p)$ has no zeroes in X_+ and $|T(p)|$ is bounded away from zero on $X_{\mathbb{R}}$).

We proceed to define the model operators. Let $z(p)$ be a meromorphic function on X whose poles are contained in the domain of analyticity of $T(p)$. For any section y of $L_{\tilde{\zeta}} \otimes \Delta$ which is holomorphic in a neighborhood of the poles of $z(p)$ we define

$$(M^z y)(p) = z(p)y(p) - \sum_{i=1}^n c_i y(p_i) \frac{\theta\begin{bmatrix} \tilde{a} \\ \tilde{b} \end{bmatrix}(p_i - p)}{\theta\begin{bmatrix} \tilde{a} \\ \tilde{b} \end{bmatrix}(0) \, E(p_i, p)}. \tag{3–3}$$

Here n is the degree of the meromorphic function $z(p)$, $p_1, \ldots, p_n$ are the poles of $z(p)$—assumed to be all simple for the ease of notation, and c_i is the residue of $z(p)$ at p_i (in terms of some local parameter—since the other two factors in each term in the sum on the right-hand side of (3–3) are half-order differentials, the product is well-defined independently of the choice of local parameter). It follows that $M^z y$ is again holomorphic in a neighborhood of $p_1, \ldots, p_n$. Actually, M^z is a bounded linear operator on $H(T)$, and for two meromorphic functions $z(p)$ and $w(p)$ the operators M^z and M^w commute. It is worthwhile to write down the resolvent $R_\alpha^z = (M^z - \alpha I)^{-1}$ of the operator M^z:

$$(R_\alpha^z y)(p) = \frac{y(p)}{z(p) - \alpha} - \sum_{i=1}^n \frac{1}{dz(p_i(\alpha))} \, y(p_i(\alpha)) \frac{\theta\begin{bmatrix} \tilde{a} \\ \tilde{b} \end{bmatrix}(p_i(\alpha) - p)}{\theta\begin{bmatrix} \tilde{a} \\ \tilde{b} \end{bmatrix}(0) \, E(p_i(\alpha), p)}, \tag{3–4}$$

where n is as before the degree of $z(p)$ and $p_1(\alpha), \ldots, p_n(\alpha)$ are the points on X with $z(p) = \alpha$ (assumed to be all distinct, and to lie in the domain of analyticity of $T(p)$). Note that this is a natural generalization of the usual difference quotients transformation to a compact Riemann surface represented as a (ramified) covering of the Riemann sphere by means of the meromorphic function $z(p)$.

We now pick a pair of real meromorphic functions $\lambda_1(p)$, $\lambda_2(p)$ on X (i.e., meromorphic functions satisfying $\overline{\lambda_k(\bar{p})} = \lambda_k(p)$ for $k = 1, 2$) that generate the whole field of meromorphic functions on X. By standard results in the theory of compact Riemann surfaces, $\lambda_1(p)$ and $\lambda_2(p)$ satisfy an irreducible polynomial equation $f(\lambda_1(p), \lambda_2(p)) = 0$ of some degree m (with real coefficients) and X is the desingularization of the real irreducible projective plane curve C with the irreducible affine equation $f(\lambda_1, \lambda_2) = 0$. We assume that $T(p)$ is holomorphic and invertible at the poles of λ_1 and λ_2 on X, which are the points of C at infinity. Then M^{λ_1} and M^{λ_2} are commuting bounded linear operators on $H(T)$. Furthermore:

THEOREM 3.1. $\mathcal{V}_T = (M^{\lambda_1}, M^{\lambda_2}, H(T), \Phi, \mathbb{C}^m, \sigma_1, \sigma_2, \gamma, \tilde{\gamma})$ is a commutative two-operator vessel with discriminant polynomial $f(\lambda_1, \lambda_2)$ and normalized joint characteristic function $T(p)$. Here $\Phi : H(T) \to \mathbb{C}^m$ is the evaluation at the poles of λ_1 and λ_2 (assuming all the poles to be simple—for a pole of order h we have to evaluate the derivatives up to order $h - 1$, with respect to some local parameter, as well), and $\sigma_1, \sigma_2, \gamma, \tilde{\gamma}$ are given by certain explicit formulas in terms of theta functions with characteristics a, b and $\tilde{a}, \tilde{b}$ corresponding to ζ and $\tilde{\zeta}$ in $J(X)$ respectively, so that $\lambda_1\sigma_2 - \lambda_2\sigma_1 + \gamma$ and $\lambda_1\sigma_2 - \lambda_2\sigma_1 + \tilde{\gamma}$ are maximal determinantal representations of $f(\lambda_1, \lambda_2)$ corresponding to ζ and $\tilde{\zeta}$ (and having sign $+1$).

It can be shown that the vessel $\mathcal{V}_T$ is irreducible (or minimal), i.e., the principal subspace coincides with all of the inner space $H(T)$.

We call $\mathcal{V}_T$ the *model vessel* corresponding to the semiexpansive function $T(p)$. To justify this name we have to show that any commutative two-operator vessel $\mathcal{V} = (A_1, A_2, H, \Phi, E, \sigma_1, \sigma_2, \gamma, \tilde{\gamma})$ with discriminant polynomial $f(\lambda_1, \lambda_2)$ and maximal input and output determinantal representations corresponding to ζ and $\tilde{\zeta}$ in $J(X)$ (and having sign $+1$, say) is unitarily equivalent, on its principal subspace, to the model vessel corresponding to its normalized joint characteristic function (up to an automorphism of the external space E). The mapping from the inner space H of the given vessel $\mathcal{V}$ to the model space $H(T)$ is given explicitly by

$$h \mapsto \frac{\xi_1 d\lambda_1(p) + \xi_2 d\lambda_2(p)}{\omega(p)} P(\xi_1, \xi_2, p) \Phi(\xi_1 A_1 + \xi_2 A_2 - \xi_1\lambda_1(p) - \xi_2\lambda_2(p))^{-1}h.$$

$$(3\text{--}5)$$

Here $h \in H$, $p \in X$, and ξ_1, ξ_2 are free parameters — the right-hand side of (3–5) turns out to be independent of ξ_1, ξ_2. $P(\xi_1, \xi_2, p)$ is the projection of E onto the fibre $\tilde{\mathcal{E}}(p)$ of the output bundle at p "in the direction" ξ_1, ξ_2, appearing in the restoration formula (2–18), and $\omega(p)$ is a meromorphic differential with zeroes of order $m - 3$ at the points of C at infinity and poles on the divisor of singularities D as in (2–28). The right-hand side of (3–5) is a section of $\tilde{\mathcal{E}}$ with poles of order $m - 2$ at the points of C at infinity and vanishing on D, and the isomorphism (2–28) gives us a section of $L_{\tilde{\zeta}} \otimes \Delta$ holomorphic outside the joint spectrum of A_1, A_2. We may arrive at the functional model by starting with the mapping (3–5) (restricted to the principal subspace), imposing the range norm on the image and then verifying that we actually obtain the reproducing kernel Hilbert space with reproducing kernel $K_T(p, q)$ of (3–1). The mapping (3–5) has a system-theoretic significance, since it is (at least in the stable dissipative case) the output of the two-dimensional system (1–20)–(1–22) with identically zero input and initial state $f(0, 0) = h$ after taking a suitably defined "Laplace transform along the discriminant curve".

We make two final remarks on functional models. First, it may be checked that *the mapping $z \mapsto M^z$ defines a homomorphism from the algebra of meromorphic functions $z(p)$ on X, whose poles are contained in the domain of analyticity of $T(p)$, to the algebra of bounded linear operators on $H(T)$.* Thus when we construct the functional model for a given vessel $\mathcal{V}$, we obtain model operators not only for A_1 and A_2, but for all the operators in the algebra of rational functions in A_1, A_2.

Second, it is possible to characterize the spaces of the form $H(T)$ for a $(\zeta, \tilde{\zeta})$-expansive T as *reproducing kernel Hilbert spaces whose elements are meromorphic sections of $L_{\tilde{\zeta}} \otimes \Delta$ on $X \setminus X_{\mathbb{R}}$, which are invariant under a pair of operators of the form $R_{\alpha_1}^{\lambda_1}$ and $R_{\alpha_2}^{\lambda_2}$ as in (3–4) and such that a certain identity, generalizing de Branges identity for difference quotients, holds.* (Here $\tilde{\zeta}$ is fixed, while ζ may be arbitrary.) In particular, this yields a generalization of Beurling's Theorem on invariant subspaces of H^2 on the unit disk to multiply connected domains and finite bordered Riemann surfaces, proved by Sarason [1965], Voichick [1964] and Hasumi [1966].

Acknowledgements

It is a great pleasure to thank M. S. Livšic for sharing his insight during many discussions, and D. Alpay and J. Ball for a fruitful collaboration on many of the topics discussed here. This paper is based on a minicourse delivered in the framework of the "Holomorphic Spaces" program at Mathematical Sciences Research Institute at Berkeley in Fall 1995; it is a pleasant duty to thank both the organizers and the participants of the program for a lively and enthusiastic mathematical environment, and the management of MSRI for their hospitality.

References

[Abrahamse and Douglas 1976]   M. B. Abrahamse and R. G. Douglas, "A class of subnormal operators related to multiply-connected domains", *Advances in Math.* **19**:1 (1976), 106–148.

[Agler 1985]   J. Agler, "Rational dilation on an annulus", *Ann. of Math.* (2) **121**:3 (1985), 537–563.

[Ahlfors 1950]   L. L. Ahlfors, "Open Riemann surfaces and extremal problems on compact subregions", *Comment. Math. Helv.* **24** (1950), 100–134.

[Alpay and Vinnikov 1994]   D. Alpay and V. Vinnikov, "Analogues d'espaces de de Branges sur des surfaces de Riemann", *C. R. Acad. Sci. Paris Sér. I Math.* **318**:12 (1994), 1077–1082.

[Alpay and Vinnikov a]   D. Alpay and V. Vinnikov, "Finite-dimensional de Branges spaces on Riemann surfaces", preprint. Contact authors at dany@math.bgu.ac.il or vinnikov@wisdom.weizmann.ac.il.

[Alpay and Vinnikov b]   D. Alpay and V. Vinnikov, "Indefinite Hardy spaces on finite bordered Riemann surfaces", preprint. Contact authors at dany@math.bgu.ac.il or vinnikov@wisdom.weizmann.ac.il.

[Ball and Cohen 1991]   J. A. Ball and N. Cohen, "de Branges-Rovnyak operator models and systems theory: a survey", pp. 93–136 in *Topics in matrix and operator theory* (Rotterdam, 1989), edited by H. Bart et al., Oper. Theory Adv. Appl. **50**, Birkhäuser, Basel, 1991.

[Ball and Vinnikov 1996]   J. A. Ball and V. Vinnikov, "Zero-pole interpolation for meromorphic matrix functions on an algebraic curve and transfer functions of 2D systems", *Acta Appl. Math.* **45**:3 (1996), 239–316.

[Ball and Vinnikov]   J. A. Ball and V. Vinnikov, "Zero-pole interpolation for meromorphic matrix functions on a compact Riemann surface and a matrix Fay trisecant identity", preprint. Available at http://eprints.math.duke.edu/alg-geom/9712 as #9712028.

[Ball et al. 1990]   J. A. Ball, I. Gohberg, and L. Rodman, *Interpolation of rational matrix functions*, Operator Theory: Advances and Applications **45**, Birkhäuser, Basel, 1990.

[Bart et al. 1979]   H. Bart, I. Gohberg, and M. A. Kaashoek, *Minimal factorization of matrix and operator functions*, Operator Theory: Adv. Appl. **1**, Birkhäuser, Basel, 1979.

[Brodskiĭ 1969]   M. S. Brodskiĭ, Треугольные и жордановы представления линейных операторов, Nauka, Moscow, 1969. Translated as *Triangular and Jordan representations of linear operators*, Transl. Math. Monographs **32**, Amer. Math. Soc., Providence, 1971.

[Brodskiĭ and Livšic 1958]   M. S. Brodskiĭ and M. S. Livšic, "Spectral analysis of non-self-adjoint operators and intermediate systems", *Uspehi Mat. Nauk (N.S.)* **13**:1(79) (1958), 3–85. In Russian; translation in *Amer. Math. Soc. Transl. Ser. 2* **13** (1960), 265–346.

[Burchnall and Chaundy 1928]   J. L. Burchnall and T. W. Chaundy, *Commutative ordinary differential operators*, 1928.

[de Branges and Rovnyak 1966a]   L. de Branges and J. Rovnyak, "Canonical models in quantum scattering theory", pp. 295–392 in *Perturbation theory and its application in quantum mechanics* (Madison, 1965), edited by C. H. Wilcox, Wiley, New York, 1966.

[de Branges and Rovnyak 1966b]   L. de Branges and J. Rovnyak, *Square summable power series*, Holt, Rinehart and Winston, New York, 1966.

[Dubrovin 1981] B. A. Dubrovin, "Theta-functions and nonlinear equations", *Uspekhi Mat. Nauk* **36**:2(218) (1981), 11–80. In Russian; translation in *Russian Math. Surveys* **36**:2 (1981), 11–92.

[Fay 1973] J. D. Fay, *Theta functions on Riemann surfaces*, Lecture Notes in Mathematics **352**, Springer, Berlin, 1973.

[Fay 1992] J. Fay, "Kernel functions, analytic torsion, and moduli spaces", *Mem. Amer. Math. Soc.* **96**:464 (1992), vi+123.

[Fulton 1969] W. Fulton, *Algebraic curves: An introduction to algebraic geometry*, Mathematics Lecture Notes Series, W. A. Benjamin, New York and Amsterdam, 1969.

[Gauchman 1983a] H. Gauchman, "Connection colligations of the second order", *Integral Equations Operator Theory* **6**:2 (1983), 184–205.

[Gauchman 1983b] H. Gauchman, "Connection colligations on Hilbert bundles", *Integral Equations Operator Theory* **6**:1 (1983), 31–58.

[Griffiths 1989] P. A. Griffiths, *Introduction to algebraic curves*, Translations of Mathematical Monographs **76**, American Mathematical Society, Providence, RI, 1989. Translated from the Chinese by Kuniko Weltin.

[Harte 1972] R. E. Harte, "Spectral mapping theorems", *Proc. Roy. Irish Acad. Sect. A* **72** (1972), 89–107.

[Hasumi 1966] M. Hasumi, "Invariant subspace theorems for finite Riemann surfaces", *Canad. J. Math.* **18** (1966), 240–255.

[Kailath 1980] T. Kailath, *Linear systems*, Prentice-Hall Information and System Sciences Series, Prentice-Hall, Englewood Cliffs, NJ, 1980.

[Kravitsky 1983] N. Kravitsky, "Regular colligations for several commuting operators in Banach space", *Integral Equations Operator Theory* **6**:2 (1983), 224–249.

[Livšic 1979] M. S. Livšic, "Operator waves in Hilbert space and related partial differential equations", *Integral Equations Operator Theory* **2**:1 (1979), 25–47.

[Livšic 1980] M. S. Livšic, "A method for constructing triangular canonical models of commuting operators based on connections with algebraic curves", *Integral Equations Operator Theory* **3**:4 (1980), 489–507.

[Livšic 1983] Livšic, "Cayley-Hamilton theorem, vector bundles and divisors of commuting operators", *Integral Equations Operator Theory* **6**:2 (1983), 250–373.

[Livšic 1986a] M. S. Livšic, "Collective motions of spatio-temporal systems", *J. Math. Anal. Appl.* **116**:1 (1986), 22–41.

[Livšic 1986b] M. S. Livšic, "Commuting nonselfadjoint operators and mappings of vector bundles on algebraic curves", pp. 255–277 in *Operator theory and systems* (Amsterdam, 1985), edited by H. Bart et al., Oper. Theory: Adv. Appl. **19**, Birkhäuser, Basel, 1986.

[Livšic 1987] M. S. Livšic, "Commuting nonselfadjoint operators and collective motions of systems", pp. 1–38 in *Commuting nonselfadjoint operators in Hilbert space: two independent studies*, Lecture Notes in Math. **1272**, Springer, New York, 1987.

[Livšic 1996] M. S. Livšic, "System theory and geometry", preprint, Ben-Gurion University of the Negev, POBox 653, 84105 Beer-Sheva, Israel, 1996.

[Livšic and Markus 1994]   M. S. Livšic and A. S. Markus, "Joint spectrum and discriminant varieties of commuting nonselfadjoint operators", pp. 1–29 in *Nonselfadjoint operators and related topics* (Beer Sheva, 1992), edited by A. Feintuch and I. Gohberg, Oper. Theory Adv. Appl. **73**, Birkhäuser, Basel, 1994.

[Livšic and Yantsevich 1971]   M. S. Livšic and A. A. Yantsevich, Теория операторных узлов в гильбертовых пространствах, Izdat. Har'kov. Univ., Kharkov, 1971. Translated as *Operator colligations in Hilbert spaces*, Wiley, New York, 1979.

[Livšic et al. 1995]   M. S. Livšic, N. Kravitsky, A. S. Markus, and V. Vinnikov, *Theory of commuting nonselfadjoint operators*, Mathematics and its Applications **332**, Kluwer, Dordrecht, 1995.

[Mumford 1983]   D. Mumford, *Tata lectures on theta, I*, Progress in Mathematics **28**, Birkhäuser, Boston, Mass., 1983.

[Mumford 1984]   D. Mumford, *Tata lectures on theta, II: Jacobian theta functions and differential equations*, Progress in Mathematics **43**, Birkhäuser, Boston, MA, 1984.

[Nikolskii and Vasyunin 1986]   N. K. Nikol'skiĭ and V. I. Vasyunin, "Notes on two function models", pp. 113–141 in *The Bieberbach conjecture* (West Lafayette, IN, 1985), edited by A. Baernstein et al., Math. Surveys Monographs **21**, Amer. Math. Soc., Providence, 1986.

[Nikolskii and Vasyunin 1998]   N. Nikolski and V. Vasyunin, "Elements of spectral theory in terms of the free function model, I: Basic constructions", in *Holomorphic Spaces*, edited by S. Axler et al., Math. Sci. Res. Inst. Publications **33**, Cambridge Univ. Press, 1998.

[Potapov 1955]   V. P. Potapov, "The multiplicative structure of J-contractive matrix functions", *Trudy Moskov. Mat. Obšč.* **4** (1955), 125–236. In Russian; translation in *Amer. Math. Soc. Transl.* (2) **15** (1960), 131–243.

[Sarason 1965]   D. Sarason, "The H^p spaces of an annulus", *Mem. Amer. Math. Soc.* **56** (1965), 78.

[Sz.-Nagy and Foiaş 1967]   B. Sz.-Nagy and C. Foiaş, *Analyse harmonique des opérateurs de l'espace de Hilbert*, Masson, Paris, and Akadémiai Kiadó, Budapest, 1967. Translated as *Harmonic analysis of operators on Hilbert space*, North-Holland, Amsterdam, and Akadémiai Kiadó, Budapest, 1970.

[Taylor 1970]   J. L. Taylor, "A joint spectrum for several commuting operators", *J. Functional Analysis* **6** (1970), 172–191.

[Vinnikov 1989]   V. Vinnikov, "Complete description of determinantal representations of smooth irreducible curves", *Linear Algebra Appl.* **125** (1989), 103–140.

[Vinnikov 1992]   V. Vinnikov, "Commuting nonselfadjoint operators and algebraic curves", pp. 348–371 in *Operator theory and complex analysis* (Sapporo, 1991), edited by T. Ando and I. Gohberg, Oper. Theory Adv. Appl. **59**, Birkhäuser, Basel, 1992.

[Vinnikov 1993]   V. Vinnikov, "Selfadjoint determinantal representations of real plane curves", *Math. Ann.* **296**:3 (1993), 453–479.

[Vinnikov 1994]   V. Vinnikov, "2D systems and realization of bundle mappings on compact Riemann surfaces", pp. 909–912 in *Systems and Networks: Mathematical Theory and Applications*, vol. 2, edited by U. Helmke et al., Math. Res. **79**, Akademie-Verlag, Berlin, 1994.

[Voichick 1964] M. Voichick, "Ideals and invariant subspaces of analytic functions", *Trans. Amer. Math. Soc.* **111** (1964), 493–512.

[Voichick and Zalcman 1965] M. Voichick and L. Zalcman, "Inner and outer functions on Riemann surfaces", *Proc. Amer. Math. Soc.* **16** (1965), 1200–1204.

[Waksman 1987] L. L. Waksman, "Harmonic analysis of multi-parameter semigroups of contractions", pp. 39–115 in *Commuting nonselfadjoint operators in Hilbert space: two independent studies*, Lecture Notes in Math. **1272**, Springer, New York, 1987.

VICTOR VINNIKOV
DEPARTMENT OF THEORETICAL MATHEMATICS
WEIZMANN INSTITUTE OF SCIENCE
REHOVOT 76100
ISRAEL
vinnikov@wisdom.weizmann.ac.il